Lecture Notes in Computer Scien

5434

Commenced Publication in 1973
Founding and Former Series Editors:
Gerhard Goos, Juris Hartmanis, and Jan van Leeuwen

Svetozar Margenov Lubin G. Vulkov
Jerzy Waśniewski (Eds.)

Numerical Analysis and Its Applications

4th International Conference, NAA 2008
Lozenetz, Bulgaria, June 16-20, 2008
Revised Selected Papers

Springer

Volume Editors

Svetozar Margenov
Institute for Parallel Processing
Bulgarian Academy of Sciences
25A Acad. G. Bonchev St., 1113 Sofia, Bulgaria
E-mail: margenov@parallel.bas.bg

Lubin G. Vulkov
University of Rousse
FNSE, Department of Numerical Methods and Statistics
8 Studentska St., 7017 Rousse, Bulgaria
E-mail: lvalkov@ru.acad.bg
E-mail: vulkov@ami.ru.acad.bg

Jerzy Waśniewski
Technical University of Denmark
Department of Informatics and Mathematical Modelling
2800 Kongens Lyngby, Denmark
E-mail: jw@imm.dtu.dk

Library of Congress Control Number: 2009921163

CR Subject Classification (1998): G.1, F.2.1, G.4, I.6, G.2, J.2

LNCS Sublibrary: SL 1 – Theoretical Computer Science and General Issues

ISSN 0302-9743
ISBN-10 3-642-00463-6 Springer Berlin Heidelberg New York
ISBN-13 978-3-642-00463-6 Springer Berlin Heidelberg New York

springer.com

Printed in Germany

Typesetting: Camera-ready by author, data conversion by Scientific Publishing Services, Chennai, India
Printed on acid-free paper SPIN: 12623679 06/3180 5 4 3 2 1 0

Preface

This volume of the *Lecture Notes in Computer Science* series comprises the proceedings of the 4th International Conference on Numerical Analysis and Applications, which was held at the hotel Sunset Beach, Lozenetz, Bulgaria, June 15–20, 2008. The conference was organized by the Department of Numerical Analysis and Statistics at the University of Rousse.

This conference continued the tradition of three previous meetings (1996, 2000, 2004 in Rousse) as a forum, where scientists from leading research groups from the "East" and "West" are provided with the opportunity to meet and exchange ideas and establish research cooperation. More than 100 scientists from all over the world participated in the conference.

The key lectures reviewed some of the advanced achievements in the field of numerical methods and their efficient applications. The conference lectures were presented by university researchers and industry engineers including applied mathematicians, numerical analysts and computer experts. Two special sessions were organized:

- Robust Numerical Methods for Multiscale Singular Perturbation Problems - G.I. Shishkin and I. Tselishcheva
- Reliable Numerical Modelling in Science and Engineering - I. Farago and S. Korotov

A wide range of problems concerning recent achievements in numerical analysis and its applications in physics, chemistry, engineering, and economics were discussed. An extensive exchange of ideas between scientists who develop and study numerical methods, and researchers who use them for solving real-life problems, took place during the conference.

We recognize the effort required to prepare these key lectures and to organize the minisymposia. We appreciate the contribution of the authors who shared their knowledge of modern high-performance computing numerical methods with the conference participants. We also thank M. Koleva for the help in putting together the book.

The 5th International Conference on Numerical Analysis and Its Applications will be held in June 2012.

November 2008

Svetozar Margenov
Lubin Vulkov
Jerzy Waśniewski

Table of Contents

Invited Papers

Contributed Talks

The Transmission Problem for Elliptic Second Order Equations in a Domain with Conical Boundary Points

Mikhail V. Borsuk

Department of Mathematics and Informatics
University of Warmia and Mazury in Olsztyn
10-957 Olsztyn-Kortowo, Poland
borsuk@uwm.edu.pl

Abstract. We investigate the behavior of weak solutions to the transmission problem for linear and weak quasi-linear elliptic divergence second order equations in a neighborhood of the boundary conical point. We obtain **best possible** estimates of the weak solutions to the transmission problem near conical boundary point.

The transmission problems often appear in different fields of physics and technics. For instance, one of the importance problem of the electrodynamics of solid media is the electromagnetic processes research in ferromagnetic media with different dielectric constants. These problems appear as well as in solid mechanics if a body consists of composite materials. Let us quote also vibrating folded membranes, composite plates, folded plates, junctions in elastic multi-structures etc.

The present article is a survey of our last results. We consider **best possible** estimates of the weak solutions to the transmission problem near conical boundary point. Analogous results were established in [2] for the elliptic boundary value problems in the non-smooth domains without interfaces.

A principal new feature of our work is the consideration of estimates of weak solutions for *linear* elliptic second-order equations with *minimal smooth coefficients* in $n-$ *dimensional conic* domains. Our examples demonstrate this fact.

Let $G \subset \mathbb{R}^n$, $n \geq 2$ be a bounded domain with boundary ∂G that is a smooth surface everywhere except at the origin $\mathcal{O} \in \partial G$ and near the point $\mathcal{O}$ it is a conical surface with vertex at $\mathcal{O}$ and the opening ω_0. We assume that $G = \bigcup_{i=1}^{N} G_i$ is divided into $N \geq 2$ subdomains G_i, $i = 1, \ldots, N$ by $(N-1)$ hyperplanes Σ_k, $k = 1, \ldots, N-1$ (by hyperplane Σ_0 in the case $N = 2$), where $\mathcal{O}$ belongs to every $\overline{\Sigma_k}$ and $G_i \cap G_j = \emptyset$, $i \neq j$. We shall study following elliptic transmission problems:

S. Margenov, L.G. Vulkov, and J. Waśniewski (Eds.): NAA 2008, LNCS 5434, pp. 1–12, 2009.

- • problem (LN) for the Laplace operator with N different media and mixed boundary condition

$$\begin{cases} \mathcal{L}_i[u] \equiv a_i \triangle u_i - p_i u_i(x) = f_i(x), & x \in G_i,\ i = 1, \dots, N; \\ [u]_{\Sigma_k} = 0, \quad \mathcal{S}_k[u] \equiv \left[a \frac{\partial u}{\partial n_k}\right]_{\Sigma_k} + \frac{1}{|x|}\beta_k(\omega)u(x) = h_k(x), & x \in \Sigma_k, \\ & k = 1, \dots, N-1; \\ \mathcal{B}[u] \equiv \alpha(x) a \frac{\partial u}{\partial \boldsymbol{n}} + \frac{1}{|x|}\gamma(\omega)u(x) = g(x), & x \in \partial G \setminus \mathcal{O}, \end{cases}$$

where $\omega = \frac{x}{|x|}$, $a_i > 0$, $p_i \geq 0$, $\quad (i = 1, \dots, N)$ are constants;

$\alpha(x) = \begin{cases} 0, & if\ x \in \mathcal{D}, \\ 1, & if\ x \notin \mathcal{D}, \end{cases}$ and $\mathcal{D} \subseteq \partial G$ is the part of the boundary ∂G where we consider the Dirichlet boundary condition;

- • problem (WL) for weak nonlinear equations

$$\begin{cases} -\frac{d}{dx_i}\left(|u|^q a^{ij}(x) u_{x_j}\right) + b(x, u, \nabla u) = 0, & q \geq 0, x \in G \setminus \Sigma_0; \\ [u]_{\Sigma_0} = 0, \quad \mathcal{S}[u] \equiv \left[\frac{\partial u}{\partial \nu}\right]_{\Sigma_0} + \frac{\beta(\omega)}{|x|} u|u|^q = h(x, u), & x \in \Sigma_0; \\ \mathcal{B}[u] \equiv \frac{\partial u}{\partial \nu} + \frac{\gamma(\omega)}{|x|} u|u|^q = g(x, u), & x \in \partial G \setminus \mathcal{O} \end{cases}$$

(summation over repeated indices from 1 to n is understood; $\frac{\partial u}{\partial \nu}$ is the conormal derivative of $u(x)$). If $q = 0$ and $b(x, u, \nabla u) = a^i(x) u_{x_i} + a(x)u - f(x)$ then we have *linear transmission problem.*

Problem (LN)**.** Let ϕ_i be openings at the vertex $\mathcal{O}$ in domains G_i. Let us define the value $\theta_k = \phi_1 + \phi_2 + \cdots + \phi_k$, thus $\omega_0 = \theta_N$. We introduce following notations:

- • Ω_i : a domain on the unit sphere S^{n-1} with boundary $\partial \Omega_i$ obtained by the intersection of the domain G_i with the sphere S^{n-1}, $(i = 1, \dots, N)$; thus $\Omega = \bigcup_{i=1}^{N} \Omega_i$;
- • $\Sigma = \sum_{k=1}^{N-1} \Sigma_k$, $\Sigma_k = G \cap \{\omega_1 = \frac{\omega_0}{2} - \theta_k\}$, $k = 1, \dots, N-1$; $\sigma = \sum_{k=1}^{N-1} \sigma_k$, $\sigma_k = \Sigma_k \cap \Omega$;
- $(G_i)_a^b = \{(r, \omega) \mid 0 \leq a < r < b; \omega \in \Omega\} \cap G_i$ $i = 1, \dots, N$;
- • $(\Sigma_k)_a^b = G_a^b \cap \Sigma_k$, $k = 1, \dots, N-1$;
- • $u(x) = u_i(x)$, $\quad f(x) = f_i(x)$, $x \in G_i$; $a\Big|_{\overline{G_i}} = a_i$, $\quad$ etc.;
- • $[u]_{\Sigma_k}$ denotes the saltus of the function $u(x)$ on crossing Σ_k, i.e.

$$[u]_{\Sigma_k} = u_k(\overline{x})\Big|_{\Sigma_k} - u_{k+1}(\overline{x})\Big|_{\Sigma_k}, \quad u_k(\overline{x})\Big|_{\Sigma_k} = \lim_{G_k \ni x \to \overline{x} \in \Sigma_k} u(x),$$

$$u_{k+1}(\overline{x})\Big|_{\Sigma_k} = \lim_{G_{k+1} \ni x \to \overline{x} \in \Sigma_k} u(x);$$

– • $\left[a\frac{\partial u}{\partial \boldsymbol{n}_k}\right]_{\Sigma_k}$ denotes the saltus of the co-normal derivative of the function $u(x)$ on crossing Σ_k, i.e. $\left[a\frac{\partial u}{\partial \boldsymbol{n}_k}\right]_{\Sigma_k} = a_k\frac{\partial u_k}{\partial \boldsymbol{n}_k}\Big|_{\Sigma_k} - a_{k+1}\frac{\partial u_{k+1}}{\partial \boldsymbol{n}_k}\Big|_{\Sigma_k}$, where $\boldsymbol{n}_k$ denotes the unite outward with respect G_k normal to Σ_k.

We assume without loss of generality that there exists $d > 0$ such that G_0^d is *a convex rotational cone* with the vertex at $\mathcal{O}$ and the aperture $\omega_0 \in (0,\pi)$ (see Figure 1.) We use the standard function spaces: $C^k(\overline{G_i})$ with the norm $|u_i|_{k,G_i}$, Lebesgue space $L_p(G_i), q \geq 1$ with the norm $\|u_i\|_{p,G_i}$, the Sobolev space $W^{k,p}(G_i)$ with the norm $\|u_i\|_{k,p;G_i}$, and introduce their direct sums $\mathbf{C}^k(\overline{G}) = C^k(\overline{G_1}) \dotplus \cdots \dotplus C^k(\overline{G_N})$ with the norm $|u|_{k,G} = \sum\limits_{i=1}^{N} |u_i|_{k,G_i}$; $\mathbf{L}_p(G) = L_p(G_1) \dotplus \cdots \dotplus L_p(G_N)$ with the norm $\|u\|_{\mathbf{L}_p(G)} = \sum\limits_{i=1}^{N}\left(\int\limits_{G_i}|u_i|^q dx\right)^{\frac{1}{p}}$; $\mathbf{W}^{k,p}(G) = W^{k,p}(G_1) \dotplus \cdots \dotplus W^{k,p}(G_N)$ with the norm $\|u\|_{k,p;G} = \sum\limits_{i=1}^{N}\left(\int\limits_{G_i}\sum\limits_{|\beta|=0}^{k}|D^\beta u_i|^p dx\right)^{\frac{1}{p}}$. We define the weighted Sobolev spaces: $\mathbf{V}_{p,\alpha}^k(G) = V_{p,\alpha}^k(G_1) \dotplus \cdots \dotplus V_{p,\alpha}^k(G_N)$ for integer $k \geq 0$ and real α, where $V_{p,\alpha}^k(G_i)$ denotes the space of all distribution $u \in \mathcal{D}'(G_i)$ satisfying $r^{\frac{\alpha}{p}+|\beta|-k}|D^\beta u_i| \in L_p(G_i)$, $i = 1,\ldots,N$. $\mathbf{V}_{p,\alpha}^k(G)$ is a Banach space

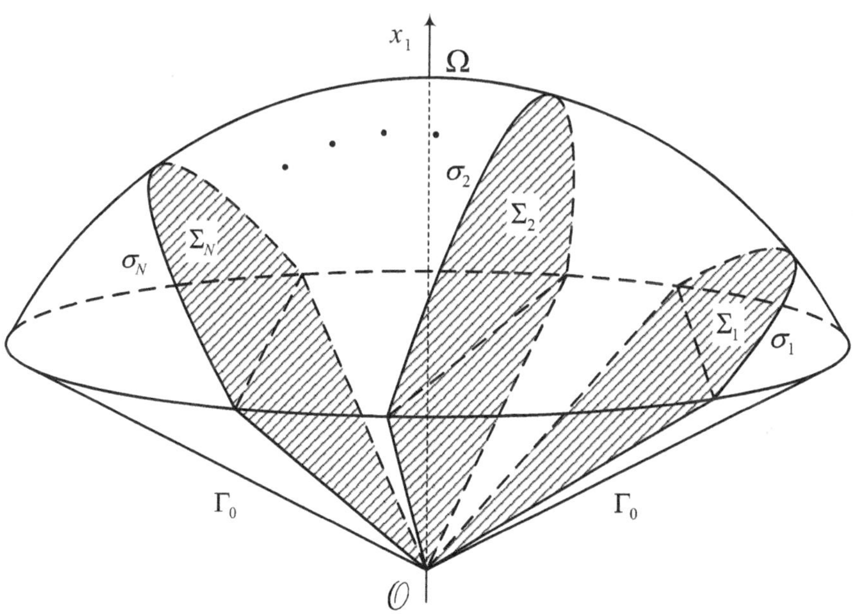

Fig. 1. Domain Ω, G_0^d, $d > 0$ is a convex rotational cone with the vertex at $\mathcal{O}$

for the norm $\|u\|_{\mathbf{V}^k_{p,\alpha}(G)} = \sum\limits_{i=1}^{N}\left(\int\limits_{G_i}\sum\limits_{|\beta|=0}^{k} r^{\alpha+p(|\beta|-k)}|D^\beta u_i|^p\,dx\right)^{\frac{1}{p}}$. $\mathbf{V}^{k-\frac{1}{p}}_{p,\alpha}(\partial G)$ is the space of functions φ, given on ∂G, with the norm $\|\varphi\|_{\mathbf{V}^{k-\frac{1}{p}}_{p,\alpha}(\partial G)} = \inf\|\Phi\|_{\mathbf{V}^k_{p,\alpha}(G)}$, where the infimum is taken over all functions Φ such that $\Phi\Big|_{\partial G} = \varphi$ in the sense of traces. We denote $\mathbf{W}^k(G) \equiv \mathbf{W}^{k,2}(G)$, $\overset{\circ}{\mathbf{W}}{}^k_\alpha(G) \equiv \mathbf{V}^k_{2,\alpha}(G)$.

We assume that $M_0 = \max\limits_{x\in\overline{G}}|u(x)|$ is known. Let us define numbers

$$\begin{cases} a_* = \min\{a_1,\ldots,a_N\} > 0,\ a^* = \max\{a_1,\ldots,a_N\} > 0; \\ p^* = \max\{p_1,\ldots,p_N\} \geq 0;\ [a]_{\Sigma_k} = a_k - a_{k+1},\ k = 1,\ldots,N-1; \\ a_0 = \max\limits_{1\leq k\leq N-1}\left|[a]_{\Sigma_k}\right|;\quad \widetilde{a} = \max(a^*, a_0). \end{cases}$$

We assume that:

(a) $f(x) \in \mathbf{L}_{q/2}(G) \cap \mathbf{L}_2(G);\quad q > n;$
(b) $\gamma(\phi) \geq \gamma_0 > \widetilde{a}\tan\frac{\omega_0}{2}$ *on* ∂G; $\beta_k(\phi) \geq \beta_0 > \widetilde{a}\tan\frac{\omega_0}{2}$ *on* Σ_k, $k = 1,\ldots,N-1$;
(c) *there exist numbers* $f_0 \geq 0,\ g_0 \geq 0,\ h_0 \geq 0,\ s > 1, \beta \geq s-2$ *such that*

$$|f(x)| \leq f_0|x|^\beta,\ |g(x)| \leq g_0|x|^{s-1}, |h_k(x)| \leq h_0|x|^{s-1},\ k = 1,\ldots,N-1.$$

Our main result is the following theorem. Let $\lambda = \frac{2-n+\sqrt{(n-2)^2+4\vartheta}}{2}$, where ϑ is the smallest positive eigenvalue of **the eigenvalue problem** (EVP)**:**

Let $\Omega \subset S^{n-1}$ with smooth boundary $\partial\Omega$ be the intersection of the cone $\mathcal{C}$ with the unit sphere S^{n-1}. Let $\vec{\nu}$ be the exterior normal to $\partial\mathcal{C}$ at points of $\partial\Omega$ and $\vec{\tau}_k$ be the exterior with respect Ω_k normal to Σ_k (lying in the tangent to Ω_k plane), $k = 1,\ldots,N-1$. Let $\gamma(\phi)$, $\phi \in \partial\Omega$ be a positive bounded piecewise smooth function, $\beta_k(\phi)$ be a positive continuous function on Σ_k, $k = 1,\ldots,$ $N-1$. We consider the eigenvalue problem for the Laplace-Beltrami operator $\triangle_\phi$ on the unit sphere:

$$\begin{cases} a_i\left(\triangle_\phi\psi_i + \vartheta\psi_i\right) = 0, \quad \phi \in \Omega_i,\ a_i \text{ are positive constants;} \\ \qquad\qquad\qquad\qquad\qquad\qquad i = 1,\ldots,N, \\ [\psi]_{\sigma_k} = 0, \quad \left[a\frac{\partial\psi}{\partial\vec{\tau}_k}\right]_{\sigma_k} + \beta_k(\phi)\psi\Big|_{\sigma_k} = 0, \quad k = 1,\ldots,N-1, \\ \alpha(\phi)a\frac{\partial\psi}{\partial\vec{\nu}} + \gamma(\phi)\psi\Big|_{\partial\Omega} = 0, \end{cases} \qquad (EVP)$$

which consists of the determination of all values ϑ (eigenvalues) for which (EVP) has a non-zero weak solutions (eigenfunctions).

Theorem 1. *Let* u *be a weak solution of the problem* (L) *and assumptions* $(a)-(c)$ *are satisfied. Let* λ *be as above. Let domain* G *and parameters in* $(a)-(c)$ *be such that* $\lambda > 1$. *Then there are* $d \in (0,1)$ *and constants* $C_0 > 0$,

$c>0$ depending only on $n, a_*, a^*, p^*, \lambda, q, \omega_0, f_0, h_0, g_0, \beta_0, \gamma_0, s, M_0$, meas G, diam G such that $\forall x \in G_0^d$

$$|u(x)| \le C_0 \begin{cases} |x|^\lambda, & \text{if } s > \lambda, \\ |x|^\lambda \ln^c\left(\frac{1}{|x|}\right), & \text{if } s = \lambda, \\ |x|^s, & \text{if } s < \lambda. \end{cases}$$

Suppose, in addition, that $\gamma(\omega) \in \mathbf{C}^1(\partial G)$, $f(x) \in \mathbf{V}^0_{q,2q-n}(G)$, $h(x) \in V^{1-1/q}_{q,2q-n}(\Sigma), g(x) \in \mathbf{V}^{1-1/q}_{q,2q-n}(\partial G)$; $q > n$ and there is a number $\tau_s =: \sup_{\varrho>0} \varrho^{-s}\left(\|h\|_{\mathbf{V}^{1-\frac{1}{q}}_{q,2q-n}(\Sigma^\varrho_{\varrho/2})} + \|g\|_{V^{1-\frac{1}{q}}_{q,2q-n}(\Gamma^\varrho_{\varrho/2})}\right)$. Then for $\forall x \in G_0^d$

$$|\nabla u(x)| \le C_1 \begin{cases} |x|^{\lambda-1}, & \text{if } s > \lambda, \\ |x|^{\lambda-1} \ln^c\left(\frac{1}{|x|}\right), & \text{if } s = \lambda, \\ |x|^{s-1}, & \text{if } s < \lambda. \end{cases}$$

Furthermore, the following is true

– • $u \in \mathbf{V}^2_{q,2q-n}(G)$, *$q > n$ and* $\|u\|_{\mathbf{V}^2_{q,2q-n}(G_0^\varrho)} \le C_2 \begin{cases} \varrho^\lambda, & \text{if } s > \lambda, \\ \varrho^\lambda \ln^c\left(\frac{1}{\varrho}\right), & \text{if } s = \lambda, \\ \varrho^s, & \text{if } s < \lambda; \end{cases}$

– • *if* $f(x) \in \overset{\circ}{\mathbf{W}}{}^0_\alpha(G)$, $\int_\Sigma r^{\alpha-1} h^2(x) ds + \int_{\partial G} r^{\alpha-1} g^2(x) ds < \infty$,

where $4 - n - 2\lambda < \alpha \le 2$, *then* $u(x) \in \overset{\circ}{\mathbf{W}}{}^1_{\alpha-2}(G)$ *and*

$$\int_G a\left(r^{\alpha-2}|\nabla u|^2 + r^{\alpha-4}u^2\right) dx + \int_\Sigma r^{\alpha-3}\beta(\phi)u^2(x)ds +$$
$$+ \int_{\partial G} \alpha(x) r^{\alpha-3}\gamma(\phi)u^2(x)ds \le C\Big\{\int_G \left(u^2 + (1+r^\alpha)f^2(x)\right) dx +$$
$$+ \int_\Sigma r^{\alpha-1}h^2(x)ds + \int_{\partial G} \alpha(x) r^{\alpha-1} g^2(x) ds\Big\},$$

where the constant $C > 0$ depends only on $q, n, a_*, a^*, \alpha, \lambda$ and the domain G.

Eigenvalue transmission problem in a composite plane domain with an angular point. Let $G \subset \mathbf{R}^2$ be bounded domain with the boundary ∂G that is a smooth curve everywhere except at the origin $\mathcal{O} \in \partial G$. Near the point $\mathcal{O}$ it is the fan that consists N corners with vertexes at $\mathcal{O}$. Thus $G = \bigcup_{i=1}^N G_i$; $\partial G = \bigcup_{j=0}^{N+1} \Gamma_j$; $\Sigma = \bigcup_{k=1}^{N-1} \Sigma_k$. Here Σ_k, $k = 1, \ldots, N-1$ are rays

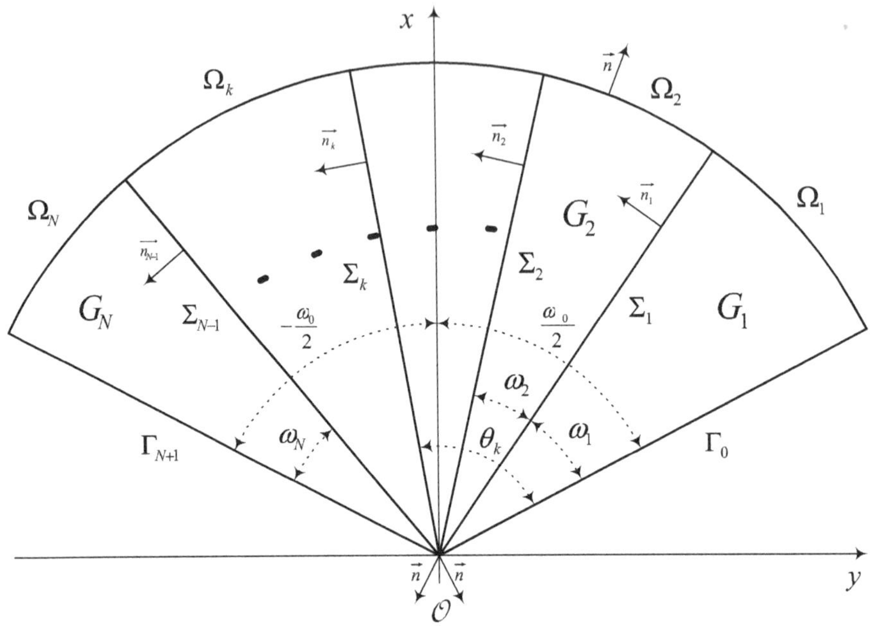

Fig. 2. Angular subdomains G_i, $i = 1, \dots, N$ of G

that G divided into angular domains G_i, $i = 1, \dots, N$ (see Figure 2). Let ω_i be apertures at the vertex $\mathcal{O}$ in domains G_i, $i = 1, \dots, N$. We define the value $\theta_k = \omega_1 + \omega_2 + \dots + \omega_k$. Let $\varGamma = \bigcup_{j=1}^{N} \varGamma_j$ be the curvilinear portion of the boundary ∂G. In this case we have $\vartheta = \lambda^2$. We assume also that $\varGamma_0 = \{(r,\omega) \big| r > 0,\ \omega = 0\}$; $\varGamma_{N+1} = \{(r,\omega) \big| r > 0,\ \omega = \theta_N\}$; $\beta_k \Big|_{\sigma_k} = \beta_k(\theta_k) = \beta_k = const$; $\gamma(0) = \gamma_1 = const$, $\gamma(\omega_0) = \gamma_N = const$. The eigenvalue problem in this case has the form

$$\begin{cases} \psi_i'' + \lambda^2 \psi_i(\omega) = 0, \quad \omega \in \varOmega_i = \{\omega_{i-1} < \omega < \omega_i\}, & i = 1, \dots, N; \\ \psi_i(\theta_i) = \psi_{i+1}(\theta_i), & i = 1, \dots, N-1; \\ a_i \psi_i'(\theta_i) - a_{i+1} \psi_{i+1}'(\theta_i) + \beta_i \psi_i(\theta_i) = 0, & i = 1, \dots, N-1; \\ \alpha_1 a_1 \psi_1'(0) + \gamma_1 \psi_1(0) = 0, & \\ \alpha_N a_N \psi_N'(\omega_0) + \gamma_N \psi_N(\omega_0) = 0, & \end{cases}$$

where $a_1, \alpha_N \in \{0, 1\}$. By direct calculation, we get $\psi_i(\omega) = A_i \cos(\lambda\omega) + B_i \sin(\lambda\omega)$, $i = 1, \dots, N$ and constants $A_1, \dots A_N$; $B_1, \dots B_N$ are defined from the algebraic homogeneous system

$$\begin{cases} \lambda\alpha_1 a_1 B_1 + \gamma_1 = 0, \\ A_{i+1} \quad = \Big(\cos^2(\lambda\theta_i) + \frac{a_i}{a_{i+1}}\sin^2(\lambda\theta_i) - \frac{\beta_i}{\lambda a_{i+1}}\sin(\lambda\theta_i)\cos(\lambda\theta_i)\Big)\cdot A_i + \\ \quad + \Big(\sin(\lambda\theta_i)\cos(\lambda\theta_i)\Big(1-\frac{a_i}{a_{i+1}}\Big) - \frac{\beta_i}{\lambda a_{i+1}}\sin^2(\lambda\theta_i)\Big)\cdot B_i, \\ \qquad\qquad i = 1,\dots,N-1; \\ B_{i+1} \quad = \Big(\sin(\lambda\theta_i)\cos(\lambda\theta_i)\Big(1-\frac{a_i}{a_{i+1}}\Big) + \frac{\beta_i}{\lambda a_{i+1}}\cos^2(\lambda\theta_i)\Big)\cdot A_i + \\ \quad + \Big(\sin^2(\lambda\theta_i) + \frac{a_i}{a_{i+1}}\cos^2(\lambda\theta_i) + \frac{\beta_i}{\lambda a_{i+1}}\sin(\lambda\theta_i)\cos(\lambda\theta_i)\Big)\cdot B_i, \\ \qquad\qquad i = 1,\dots,N-1; \\ (\gamma_N\cos(\lambda\omega_0) - \lambda\alpha_N a_N\sin(\lambda\omega_0))\cdot A_N + \\ \qquad + (\gamma_N\sin(\lambda\omega_0) + \lambda\alpha_N a_N\cos(\lambda\omega_0))\cdot B_N = 0. \end{cases}$$

The least positive eigenvalue λ is defined from the vanishing of the determinant of this system.

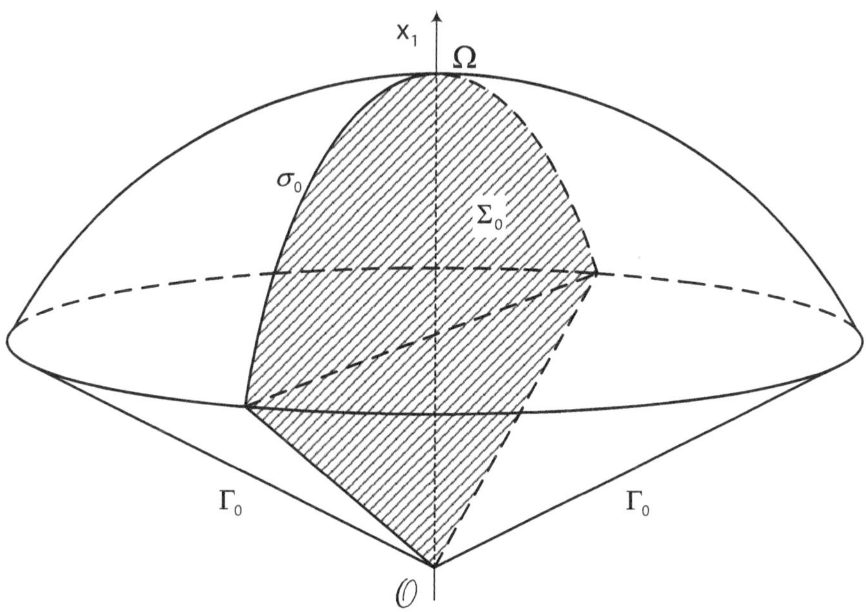

Fig. 3. Domain Ω with $G = G_+ \cup G_- \cup \Sigma_0$, $\Sigma_0 = G \cap \{x_n = 0\}$, $\mathcal{O} \in \overline{\Sigma_0}$

Problem (WL)**.** We consider problem (WL) that is the transmission problem for a quasi-linear equation with **semi-linear principal part**. We assume (see Figure 3) that $G = G_+ \cup G_- \cup \Sigma_0$ is divided into two subdomains G_+ and G_- by a $\Sigma_0 = G \cap \{x_n = 0\}$, where $\mathcal{O} \in \overline{\Sigma_0}$. We assume also without loss of generality that there exists $d > 0$ such that G_0^d is *a rotational cone* with the vertex at $\mathcal{O}$ and the aperture $\omega_0 \in (0, 2\pi)$. Regarding the equation we assume that the following **conditions** are satisfied:

Let $q \geq 0,\ 0 \leq \mu < q+1,\ s > 1,\ f_1 \geq 0,\ g_1 \geq 0, h_1 \geq 0,\ \beta \geq s-2$ *be given numbers;*

(a) *the condition of the uniform ellipticity:*

$$a_\pm \xi^2 \leq a_\pm^{ij}(x)\xi_i\xi_j \leq A_\pm \xi^2, \quad \forall x \in \overline{G_\pm}, \quad \forall \xi \in \mathbb{R}^n;\ a_\pm, A_\pm = const > 0,$$

$$a^{ij}(0) = a\delta_i^j, \text{ where } \delta_i^j \text{ is the Kronecker symbol; } a = \begin{cases} a_+, & x \in G_+, \\ a_-, & x \in G_-; \end{cases}$$

$$\text{we denote} \begin{cases} a_* = \min\{a_+, a_-\} > 0, \\ a^* = \max\{a_+, a_-\} > 0, \\ A^* = \max(A_-, A_+); \end{cases}$$

(b) $a^{ij}(x) \in \mathbf{C}^0(\overline{G})$ *and the inequality*

$$\left(\sum_{i,j=1}^{n} |a_\pm^{ij}(x) - a_\pm^{ij}(y)|^2\right)^{\frac{1}{2}} \leq \mathcal{A}(|x-y|)$$

holds for $x, y \in \overline{G}$, *where* $\mathcal{A}(r)$ *is a monotonically increasing, nonnegative function,* **continuous at 0**, $\mathcal{A}(0) = 0$*;*

(c) $|b(x, u, u_x)| \leq a\mu|u|^{q-1}|\nabla u|^2 + b_0(x);\ 0 \leq \mu < 1+q,\ b_0(x) \in L_{p/2}(G),$ $n < p < 2n$*;*

(d) $\beta(\omega) \geq \nu_0 > 0$ *on* σ_0*;* $\gamma(\omega) \geq \nu_0 > 0$ *on* ∂G*;*

(e) $\frac{\partial h(x,u)}{\partial u} \leq 0,\ \frac{\partial g(x,u)}{\partial u} \leq 0$*;*

(f) $|b_0(x)| \leq f_1|x|^\beta,\ |g(x,0)| \leq g_1|x|^{s-1},\ |h(x,0)| \leq h_1|x|^{s-1}$.

Our main result is the following statement.

Theorem 2. *Let* u *be a weak solution of the problem* (WL), *the assumptions* $(a)-(f)$ *are satisfied with* $\mathcal{A}(r)$ *Dini-continuous at zero. Let us assume that* $M_0 = \max_{x \in \overline{G}} |u(x)|$ *is known. Let* λ *be as above. Then there are* $d \in (0,1)$ *and constants* $C_0 > 0,\ c > 0$ *depending only on* $n, a_*, A^*, p, q, \lambda, \mu, f_1, h_1, g_1, \nu_0, s, M_0,$ *meas* G, *diam* G *and on the quantity* $\int_0^1 \frac{\mathcal{A}(r)}{r} dr$ *such that* $\forall x \in G_0^d$

$$|u(x)| \leq C_0 \begin{cases} |x|^{\frac{\lambda(1+q-\mu)}{(q+1)^2}}, & \text{if } s > \lambda\frac{1+q-\mu}{1+q}, \\ |x|^{\frac{\lambda(1+q-\mu)}{(q+1)^2}} \ln^c\left(\frac{1}{|x|}\right), & \text{if } s = \lambda\frac{1+q-\mu}{1+q}, \\ |x|^{\frac{s}{q+1}}, & \text{if } s < \lambda\frac{1+q-\mu}{1+q}. \end{cases}$$

Suppose, in addition, that coefficients of the problem (WL) are satisfied such conditions, which guarantee the local a-priori estimate $|\nabla u|_{0,G'} \le M_1$ for any smooth $G' \subset\subset \overline{G} \setminus \{\mathcal{O}\}$ (see for example §4 [1]). Then for $\forall x \in G_0^d$

$$|\nabla u(x)| \le C_1 \begin{cases} |x|^{\frac{\lambda(1+q-\mu)}{(q+1)^2}-1}, & \text{if } s > \lambda\frac{1+q-\mu}{1+q}, \\ |x|^{\frac{\lambda(1+q-\mu)}{(q+1)^2}-1} \ln^c\left(\frac{1}{|x|}\right), & \text{if } s = \lambda\frac{1+q-\mu}{1+q}, \\ |x|^{\frac{s}{q+1}-1}, & \text{if } s < \lambda\frac{1+q-\mu}{1+q} \end{cases}$$

with $C_1 = c_1\Big(\|u\|_{2(q+1),G} + f_1 + g_1 + h_1\Big)$, where c_1 depends on M_0, M_1 and C_0 from above.

There the idea of the Theorems 1 - 2 proof is based on the deduction a new inequality of the Friedrichs-Wirtinger type with the **exact** constant as well as other integro-differential inequalities adapted to the transmission problem. The precise exponent of the solution decrease rate depends on this exact constant. We obtain *the Friedrichs - Wirtinger type inequality* by the variational principle:

Lemma 1. *Let ϑ be the smallest positive eigenvalue of problem (EVP). Let $\Omega \subset S^{n-1}$ be a bounded domain. Let $\psi \in \mathbf{W}^1(\Omega)$ and satisfies the boundary and conjunction conditions from (EVP) in the weak sense. Let $\gamma(\omega)$ be a positive bounded piecewise smooth function on $\partial\Omega$, $\beta(\omega)$ be a positive continuous function on σ_0. Then*

$$\vartheta \int\limits_{\Omega} a\psi^2(\omega)d\Omega \le \int\limits_{\Omega} a|\nabla_\omega \psi(\omega)|^2 d\Omega + \int\limits_{\sigma_0} \beta(\omega)\psi^2(\omega)d\sigma + \int\limits_{\partial\Omega} \alpha(x)\gamma(\omega)\psi^2(\omega)d\sigma.$$

Example. Here we consider two dimensional transmission problem for the weak nonlinear equation with absorbtion term in an angular domain and investigate the corresponding eigenvalue problem. Suppose $n = 2$, the domain G lies inside the corner $G_0 = \{(r,\omega)\,|r > 0;\ -\frac{\omega_0}{2} < \omega < \frac{\omega_0}{2}\}$, $\omega_0 \in]0, 2\pi[$; $\mathcal{O} \in \partial G$ and in some neighborhood of $\mathcal{O}$ the boundary ∂G coincides with the sides of the corner $\omega = -\frac{\omega_0}{2}$ and $\omega = \frac{\omega_0}{2}$. We denote $\Gamma_\pm = \{(r,\omega) \mid r > 0;\ \omega = \pm\frac{\omega_0}{2}\}$, $\Sigma_0 = \{(r,\omega) \mid r > 0;\ \omega = 0\}$ and we put $\beta(\omega)\Big|_{\Sigma_0} = \beta = const \ge 0$, $\gamma(\omega)\Big|_{\omega=\pm\frac{\omega_0}{2}} = \gamma_\pm = const > 0$. We consider the following problem:

$$\begin{cases} \dfrac{d}{dx_i}\left(|u|^q u_{x_i}\right) = a_0 r^{-2} u|u|^q - \mu u|u|^{q-2}|\nabla u|^2, & x \in G_0 \setminus \Sigma_0; \\ [u]_{\Sigma_0} = 0, \quad \left[a|u|^q \frac{\partial u}{\partial n}\right]_{\Sigma_0} + \frac{\beta}{|x|} u|u|^q = 0, & x \in \Sigma_0; \\ \alpha_\pm a_\pm |u_\pm|^q \frac{\partial u_\pm}{\partial n} + \frac{\gamma_\pm}{|x|} u_\pm |u_\pm|^q = 0, & x \in \Gamma_\pm \setminus \mathcal{O}, \end{cases}$$

where $a = \begin{cases} a_+, & x \in G_+, \\ a_-, & x \in G_-, \end{cases}$ $a_\pm$ are positive constants; $a_0 \ge 0,\ 0 \le \mu < 1 + q$, $q \ge 0$; $\alpha_\pm \in \{0; 1\}$. We make the function change $u = v|v|^{\varsigma - 1}$ with $\varsigma = \frac{1}{q+1}$ and consider our problem for the function $v(x)$:

$$\begin{cases} \triangle v + \mu\varsigma v^{-1}|\nabla v|^2 = a_0(1+q)r^{-2}v; \quad \varsigma = \frac{1}{1+q}, & x \in G_0 \setminus \Sigma_0; \\ [v]_{\Sigma_0} = 0, \quad \left[a\frac{\partial v}{\partial n}\right]_{\Sigma_0} + (1+q)\beta\frac{v(x)}{|x|} = 0, & x \in \Sigma_0; \\ \alpha_\pm a_\pm \frac{\partial v_\pm}{\partial n} + (1+q)\gamma_\pm \frac{v_\pm(x)}{|x|} = 0, & x \in \Gamma_\pm \setminus \mathcal{O}. \end{cases}$$

We want find the exact solution of this problem in the form $v(r,\omega) = r^{\varkappa}\psi(\omega)$. For $\psi(\omega)$ we obtain the problem

$$\begin{cases} \psi''(\omega) + \frac{\mu\varsigma}{\psi(\omega)}\psi'^2(\omega) + \left\{(1+\mu\varsigma)\varkappa^2 - a_0(1+q)\right\} \cdot \psi(\omega) = 0, \\ \qquad\qquad \omega \in \left(-\frac{\omega_0}{2}, 0\right) \cup \left(0, \frac{\omega_0}{2}\right); \\ [\psi]_{\omega=0} = 0, \quad [a\psi'(0)] = (1+q)\beta\psi(0); \\ \pm\alpha_\pm a_\pm \psi'_\pm\left(\pm\frac{\omega_0}{2}\right) + (1+q)\gamma_\pm\psi_\pm\left(\pm\frac{\omega_0}{2}\right) = 0. \end{cases}$$

We assume that $\varkappa^2 > a_0\frac{(1+q)^2}{1+q+\mu}$ and define the value $\Upsilon = \sqrt{\varkappa^2 - a_0\frac{(1+q)^2}{1+q+\mu}}$. We consider separately two cases: $\mu = 0$ and $\mu \neq 0$.

$$\boxed{\mu = 0.}$$

In this case we get $\psi_\pm(\omega) = A\cos(\Upsilon\omega) + B_\pm \sin(\Upsilon\omega)$, where constants $A, B_\pm$ it should be determined from conjunction and boundary conditions.

The Dirichlet problem: $\alpha_\pm = 0,\ \gamma_\pm \neq 0$.

Direct calculations will give

$$\psi_\pm(\omega) = \cos(\Upsilon\omega) \mp \cot\left(\Upsilon\frac{\omega_0}{2}\right) \cdot \sin(\Upsilon\omega), \quad \Upsilon = \begin{cases} \frac{\pi}{\omega_0}, & \text{if } \beta = 0; \\ \Upsilon^*, & \text{if } \beta \neq 0, \end{cases}$$

where Υ^* is the least positive root of the transcendental equation

$$\Upsilon \cdot \cot\left(\Upsilon\frac{\omega_0}{2}\right) = -\frac{1+q}{a_+ + a_-}\beta$$

and from the graphic solution we obtain $\frac{\pi}{\omega_0} < \Upsilon^* < \frac{2\pi}{\omega_0}$. The corresponding eigenfunctions are

$$\psi_\pm(\omega) = \begin{cases} \cos\left(\frac{\pi\omega}{\omega_0}\right), & \text{if } \beta = 0; \\ \cos(\Upsilon^*\omega) \mp \cot\left(\Upsilon^*\frac{\omega_0}{2}\right) \cdot \sin(\Upsilon^*\omega), & \text{if } \beta \neq 0. \end{cases}$$

The Neumann problem: $\alpha_\pm = 1,\ \gamma_\pm = 0.$

Direct calculations will give $\Upsilon = \begin{cases} \frac{\pi}{\omega_0}, & \text{if } \beta = 0; \\ \Upsilon^*, & \text{if } \beta \neq 0, \end{cases}$ where Υ^* is the least positive root of the transcendental equation $\Upsilon \cdot \tan\left(\Upsilon \frac{\omega_0}{2}\right) = \frac{1+q}{a_+ + a_-}\beta$ and from the graphic solution we obtain $0 < \Upsilon^* < \frac{\pi}{\omega_0}$. The corresponding eigenfunctions are

$$\psi_\pm(\omega) = \begin{cases} a_\mp \sin\left(\frac{\pi\omega}{\omega_0}\right), & \text{if } \beta = 0; \\ \cos(\Upsilon^*\omega) \pm \tan\left(\Upsilon^* \frac{\omega_0}{2}\right) \cdot \sin(\Upsilon^*\omega), & \text{if } \beta \neq 0. \end{cases}$$

Mixed problem: $\alpha_+ = 1,\ \alpha_- = 0;\ \gamma_+ = 0,\ \gamma_- = 1.$

Direct calculations will give: $\Upsilon = \Upsilon^*$, where Υ^* is the least positive root of the transcendental equation $a_+ \tan\left(\Upsilon \frac{\omega_0}{2}\right) - a_- \cot\left(\Upsilon \frac{\omega_0}{2}\right) = \frac{1+q}{\Upsilon}\beta.$ The corresponding eigenfunctions are

$$\psi_+(\omega) = \cos(\Upsilon^*\omega) + \tan\left(\Upsilon^* \frac{\omega_0}{2}\right) \cdot \sin(\Upsilon^*\omega), \quad \omega \in \left[0, \frac{\omega_0}{2}\right];$$
$$\psi_-(\omega) = \cos(\Upsilon^*\omega) + \cot\left(\Upsilon^* \frac{\omega_0}{2}\right) \cdot \sin(\Upsilon^*\omega), \quad \omega \in \left[-\frac{\omega_0}{2}, 0\right].$$

The Robin problem: $\alpha_\pm = 1,\ \gamma_\pm \neq 0.$

Direct calculations of the above system will give:

1) $\frac{\gamma_+}{\gamma_-} = \frac{a_+}{a_-}.$ $\Longrightarrow$ $\psi_\pm(\omega) = a_\mp \sin(\Upsilon^*\omega)$, where Υ^* is the least positive root of the transcendental equation $\Upsilon \cdot \cot\left(\Upsilon \frac{\omega_0}{2}\right) = -(1+q)\frac{\gamma_+}{a_+}$ and from the graphic solution we obtain $\frac{\pi}{\omega_0} < \Upsilon^* < \frac{2\pi}{\omega_0}$.

2) $\frac{\gamma_+}{\gamma_-} \neq \frac{a_+}{a_-}.$ $\Longrightarrow$ $A \neq 0$ and $\psi_\pm(0) \neq 0$; further see below the general case $\mu \neq 0$.

$\boxed{\mu \neq 0.}$

It is obvious that in this case $\psi(0) \neq 0$. By setting $y(\omega) = \frac{\psi'(\omega)}{\psi(\omega)}$, we arrive at the problem for $y(\omega)$

$$\begin{cases} y' + (1+\mu\varsigma)y^2(\omega) + (1+\mu\varsigma)\varkappa^2 - a_0(1+q) = 0, & \omega \in \left(-\frac{\omega_0}{2}, 0\right) \cup \left(0, \frac{\omega_0}{2}\right); \\ a_+y_+(0) - a_-y_-(0) = (1+q)\beta; \\ \pm\alpha_\pm a_\pm y_\pm\left(\pm\frac{\omega_0}{2}\right) + (1+q)\gamma_\pm = 0. \end{cases}$$

Integrating the equation of our problem we find

$$y_\pm(\omega) = \Upsilon \tan\left\{\Upsilon\left(C_\pm - (1+\mu\varsigma)\omega\right)\right\}, \ \forall C_\pm.$$

From boundary conditions we have $C_\pm = \pm(1+\mu\varsigma)\frac{\omega_0}{2} \mp \frac{1}{\Upsilon}\arctan\frac{(1+q)\gamma_\pm}{\alpha_\pm a_\pm \Upsilon}$. Finally, in virtue of the conjunction condition, we get the equation for required $\varkappa$:

$$a_+ \cdot \frac{\alpha_+ a_+ \Upsilon \tan\left\{(1+\mu\varsigma)\Upsilon\frac{\omega_0}{2}\right\} - (1+q)\gamma_+}{\alpha_+ a_+ \Upsilon + (1+q)\gamma_+ \tan\left\{(1+\mu\varsigma)\Upsilon\frac{\omega_0}{2}\right\}} +$$
$$+\, a_- \cdot \frac{\alpha_- a_- \Upsilon \tan\left\{(1+\mu\varsigma)\Upsilon\frac{\omega_0}{2}\right\} - (1+q)\gamma_-}{\alpha_- a_- \Upsilon + (1+q)\gamma_- \tan\left\{(1+\mu\varsigma)\Upsilon\frac{\omega_0}{2}\right\}} = \frac{1+q}{\Upsilon}\beta,$$

where $1+\mu\varsigma = \frac{1+q+\mu}{1+q}$. Thus we obtain

$$y_\pm(\omega) = \Upsilon \tan\left\{\Upsilon\frac{1+q+\mu}{1+q}\left(\pm\frac{\omega_0}{2} - \omega\right) \mp \arctan\frac{(1+q)\gamma_\pm}{\alpha_\pm a_\pm \Upsilon}\right\}$$

and, because of $(\ln\psi(\omega))' = y(\omega)$, hence it follows

$$\psi_\pm(\omega) = \cos^{\frac{1+q}{1+q+\mu}}\left\{\Upsilon\frac{1+q+\mu}{1+q}\left(\pm\frac{\omega_0}{2} - \omega\right) \mp \arctan\frac{(1+q)\gamma_\pm}{\alpha_\pm a_\pm \Upsilon}\right\}.$$

At last, returning to the function u we establish a solution our problem

$$\boxed{u_\pm(r,\omega) = r^{\frac{\varkappa}{1+q}} \cos^{\frac{1}{1+q+\mu}}\left\{\Upsilon\frac{1+q+\mu}{1+q}\left(\pm\frac{\omega_0}{2} - \omega\right) \mp \arctan\frac{(1+q)\gamma_\pm}{\alpha_\pm a_\pm \Upsilon}\right\}.}$$

Acknowledgment

This work was supported by the Polish Ministry of Science and Higher Education through the grant Nr N201 381834.

References

1. Borsuk, M.V.: A priori estimates and solvability of second order quasilinear elliptic equations in a composite domain with nonlinear boundary conditions and conjunction condition. Proc. Steklov Inst. of Math. 103, 13–51 (1970)
2. Borsuk, M., Kondratiev, V.: Elliptic Boundary Value Problems of Second Order in Piecewise Smooth Domains. North-Holland Mathematical Library 69, 531 (2006)

Some Contributions of Homotopic Deviation to the Theory of Matrix Pencils

Françoise Chatelin[1] and Morad Ahmadnasab[2]

[1] Université Toulouse 1 and CERFACS, 42, av. G. Coriolis, 31057 Toulouse Cedex 1
chatelin@cerfacs.fr
[2] Department of Mathematics, University of Kurdistan, Pasdaran boulevard, Sanandaj, Iran, Postal Code 66177–15175, and CERFACS, 42 Avenue G. Coriolis, 31057 Toulouse Cedex 1, France
Morad.Ahmadnasab@cerfacs.fr

Abstract. Let $A, E \in \mathbb{C}^{n\times n}$ be two given matrices, where rank$E = r \leq n$. The matrix E is written in the form (derived from SVD) $E = UV^H$ where $U, V \in \mathbb{C}^{n\times r}$ have rank $r \leq n$. For $0 < r < n$, 0 is an eigenvalue of E with algebraic (resp. geometric) multiplicity m $(g = n - r \leq m)$.

We consider the pencil $P_z(t) = (A - zI) + tE$, defined for $t \in \hat{\mathbb{C}} = \mathbb{C} \cup \{\infty\}$ which depends on the complex parameter $z \in \mathbb{C}$. We analyze how its structure evolves as the parameter z varies, by means of conceptual tools borrowed from Homotopic Deviation theory [1,8]. The new feature is that, because t varies in $\hat{\mathbb{C}}$, we can look at what happens in the limit when $|t| \to \infty$. This enables us to propose a remarkable connection between the algebraic theory of Weierstrass and the Cauchy analytic theory in $\mathbb{C}$ as $|t| \to \infty$.

Keywords: Homotopic Deviation, observation point, frontier point, communication matrix, induction matrix, matrix pencil, Weierstrass, Cauchy.

1 A Brief Survey of Homotopic Deviation

Given the matrices A and E in $\mathbb{C}^{n\times n}$, the family $A(t) = A + tE$ represents the coupling between A and E by the complex parameter t. We denote the *spectrum* of A by $\sigma(A)$ and the *resolvent* set of A by $re(A) = \mathbb{C} \backslash \sigma(A)$. Homotopic Deviation (HD) theory [1,8] studies the singularities in $\mathbb{C}$ of the linear coupling $A(t) = A + tE$, which depend on the parameter t which varies in the completed complex plane $\hat{\mathbb{C}} = \mathbb{C} \cup \{\infty\}$. The parameter t defines the intensity of the coupling (A, E). When $|t| \to \infty$, $\sigma(A(t)) \to \infty$ when E is full rank. But when rank $E = r < n$, it is possible that some eigenvalues of $A(t)$ stay at finite distance, rather than escaping to ∞. A physical example in Acoustics is described in [5].

1.1 The Communication Matrix M_z [5,8]

Let $E = UV^H$ be derived from the SVD of E, where $U, V \in \mathbb{C}^{n\times r}$ have rank $r \leq n$. The singularities of $R(t, z) = (A + tE - zI)^{-1}$ are the eigenvalues $\lambda(t)$ of

S. Margenov, L.G. Vulkov, and J. Waśniewski (Eds.): NAA 2008, LNCS 5434, pp. 13–19, 2009.

$A+tE$. The point z is the observation point for HD. When $z \in re(A)$, $\sigma(A+tE)$ is easily related to A, U, V by the communication matrix $M_z = V^H(zI-A)^{-1}U \in \mathbb{C}^{r\times r}$, since we have the following fundamental relation for $z \in re(A)$

$$\boxed{z \in \sigma(A+tE) \iff t = 1/\mu_z \in \hat{\mathbb{C}}, \text{ where } \mu_z \in \sigma(M_z).}$$

1.2 The Frontier Set $F(A,E) \subset re(A)$ [8]

When M_z is invertible, any z given in $re(A)$ is an eigenvalues of r matrices $A(t_i)$, for $t_i = 1/\mu_{iz}$, $i = 1, \cdots, r$. When z in $re(A)$ is such that rank $M_z < r$, there are *less* than r such matrices. This is possible when $r < n$.

Definition 1. *The frontier set is the subset of* $re(A)$ *defined by* $F(A,E) = \{z \in re(A);\ \text{rank} M_z < r\}$, *for* $r < n$.

When $z \in re(A)\backslash F(A,E)$, the resolvent matrix $R(t,z)$ is analytic in t around 0 ($|t| < 1/\rho(M_z)$) *and* around ∞ ($|t| > \rho(M_z^{-1})$). When $z \in F(A,E)$, the analyticity around ∞ *disappears.*

It is possible that $\rho(M_z) = 0$, that is M_z is nilpotent for $z \in F(A,E)$. Such a particular frontier point is called *critical* and $R(t,z)$ is a polynomial in t of degree $\leq r$. The critical points form the critical set $F_c(A,E) \subseteq F(A,E) \subseteq re(A)$. For a critical z, then $z \in re(A(t))$ for all $t \in \mathbb{C}$: z repels all eigenvalues $\lambda(t) \in \sigma(A(t))$ for all t. However, it is a limit point: $z = \lim_{|t|\to\infty} \lambda(t)$ for at least one $\lambda(t)$. The nature of the resolvent $R(t,z)$, which is a polynomial in t, *changes* as $|t| \to \infty$ to become singular.

We mention that when $r = n$, $F(A,E) = \emptyset$ and $\lim_{|t|\to\infty} R(t,z) = 0$: there is *no* effect at ∞.

1.3 A Characterization of $F(A,E)$ [1,8]

Let $\hat{A}(z)$ be the augmented matrix $\hat{A}(z) = \begin{bmatrix} zI-A & -U \\ V^H & 0 \end{bmatrix}$ of order $n+r$ defined for $z \in \mathbb{C}$. $\hat{A}(z)$ is the homotopic pencil $\begin{bmatrix} -A & -U \\ V^H & 0 \end{bmatrix} + z\begin{bmatrix} I & 0 \\ 0 & 0 \end{bmatrix}$. We set $\hat{\pi}(z) = \det \hat{A}(z)$. This *homotopic* polynomial has degree $\leq g = n-r$; it is such that for $z \in re(A)$, $\hat{\pi}(z) = \pi(z)\det M_z$, where $\pi(z) = \det(zI-A)$ is the characteristic polynomial of A. Let $\hat{Z}$ be the set of roots of $\hat{\pi}(z)$ in $\mathbb{C}$: $\hat{Z} = \{z \in \mathbb{C};\ \hat{\pi}(z) = 0\}$. Then $F(A,E) = \hat{Z} \cap re(A)$. When $\hat{\pi}(z) \equiv 0$, then $\hat{Z} = \mathbb{C}$ and $F(A,E) = re(A)$ is continuous. The critical set $F_c(A,E)$ can be either continuous or discrete. When it is continuous, then $F_c(A,E) = F(A,E) = re(A)$ and $\sigma(A(t)) = \sigma(A)$ for any $t \in \mathbb{C}$: the spectrum is invariant under t. Below, we assume that $\hat{Z} \neq \mathbb{C}$.

It is possible to extend $F(A,E)$ defined for $z \notin \sigma(A)$ to certain eigenvalues of A in $\sigma^f \subset \sigma(A)$ [8]. For $\lambda \in \sigma^f$, at least one $\mu_z \in \sigma(M_z)$ is such that $\lim_{z\to\lambda} \mu_z = 0$. The frontier set in $\mathbb{C}$ is the closure $\bar{F}(A,E) = F(A,E) \cup \sigma^f$.

Similarly, $F_c(A,E)$ may be expanded into $\sigma(A)$ at certain eigenvalues which constitute $\sigma_c \subset \sigma^f \subset \sigma(A)$: $\sigma_c = \{\lambda \in \sigma^f;\ \lim_{z\to\lambda} \sigma(M_z) = \{0\}\} \subset \sigma^f$. Then the critical set in $\mathbb{C}$ is $\bar{F}_c(A,E) = F_c(A,E) \cup \sigma_c$.

1.4 The Limit Set Lim and the Kernel Set $\tilde{Z} \subseteq$ Lim

The set Lim consists of the limits $\xi \in \mathbb{C}$ for $\lim_{|t|\to\infty} \lambda(t)$. When $r = n$, the set is empty: all eigenvalues of $A(t)$ escape to ∞. But when $r < n$, Lim may be non empty. When $\lambda(t) \to \xi$, then ξ is an eigenvalue of the synthesis $A(\infty)$: this represents the completed coupling with infinite intensity. $A(\infty)$ denotes a concept (*not* a matrix) which can be explained as follows. Consider $st = 1$, $s = 1/t \in \hat{\mathbb{C}}$, then $A(t) = A + tE = t(E + sA) = \frac{E(s)}{s} \iff E(s) = \frac{A(t)}{t}$ for any $s, t \in \mathbb{C}\backslash\{0\}$ related by $st = 1$. The spectral properties of $A(\infty)$ can be analyzed by means of those of $E(s) = E + sA$ when $s \to 0$.

One has the identity $\text{Lim} \cap re(A) = F(A, E)$ in $re(A)$. A subset of Lim may be defined by means of the kernel pencil as follows [8]. We consider the geometric structure of $0 \in \sigma(E)$. It has $g = n - r$ eigenvectors, where g_1 of them define a trivial Jordan block of dimension 1, and g_2 start the non trivial Jordan blocks of dimension ≥ 2: $g = g_1 + g_2$. We assume that $g_1 \geq 1$, so that $g_2 < g$. The map A is restricted to the eigenspace Ker$E$ of dimension g by means of the eigenvectors for E and E^H properly sorted by nondecreasing size of the Jordan blocks (Lidskii). This produces the matrix $\tilde{\Pi}$ of order g in 2×2 block form and the *kernel* pencil

$$\tilde{\Pi}(z) = \tilde{\Pi} - z \begin{bmatrix} I_{g_1} & \\ & 0_{g_2} \end{bmatrix}.$$

Set $\tilde{Z} = \{z \in \mathbb{C}, \det\tilde{\Pi}(z) = 0\}$. The kernel set $\tilde{Z}$ (when $\neq \mathbb{C}$) satisfies the inclusion $\tilde{Z} \subseteq$ Lim.

When $0 \in \sigma(E)$ is semi-simple, then $g_1 = g$ and $\tilde{\Pi} = \Pi = PAP_{|\text{Ker}E}$ where P is the eigenprojection on Ker$E$. Then $\tilde{Z} = \sigma(\Pi) =$ Lim [8]. When 0 is defective in $\sigma(E)$, the generic case is $\tilde{Z} =$ Lim.

1.5 About the Convergence of $\lambda(t)$ to $\xi \in F(A, E)$

The spectral field for $A(t)$ is the complex vector function defined by: $t \in \hat{\mathbb{C}} \mapsto \sigma(A(t)) \in \mathbb{C}^n$.

For $t \neq 0$, let $t = |t|e^{i\theta}$ with $|t| = h > 0$ and $\theta \in [0, 2\pi[$. To analyse the isophasic evolution of the spectral field, we fix θ in $[0, 2\pi[$ and consider the map: $h \mapsto \Lambda(\theta) = \{\lambda_i(he^{i\theta}), i = 1, \cdots, n\}$ consisting of n *spectral rays.*

Let be given the frontier point $\xi \in F(A, E)$ and let $1 \leq \hat{m}_\xi \leq g$ be its multiplicity as a zero of $\hat{\pi}$. There are $c_\xi(\theta)$ spectral rays in $\Lambda(\theta)$ which end at ξ when $|t| \to \infty$. When $\hat{Z} \neq \mathbb{C}$, $c_\xi = \hat{m}_\xi$ is *independent* of θ: $c_\xi(\theta) = c_\xi$. This shows that the homotopic polynomial $\hat{\pi}$ rules analytically the convergence of a subset of $\sigma(A(t))$ to its zeros in $re(A)$, which are *not* eigenvalues of A: $\hat{\pi}$ is a *global* ruler.

1.6 The Induction Matrix B_ξ at $\xi \in F(A, E) \neq re(A)$

Let $\xi \in F(A, E)$ be isolated by a Jordan curve Γ traced in $re(A)\backslash F(A, E)$. The map

$$z \in re(A) \mapsto R(\infty, z) = R(0, z)[I + UM_z^{-1}V^H R(0, z)] = \lim_{|t|\to\infty} R(t, z)$$

does not exist at any zero eigenvalue of M_z, that is, at each frontier point in $F(A, E)$.

In analogy with the spectral projection [7], we define the *induction matrix* associated with (A, E) at $\xi \in F(A, E)$ by

$$B_\xi = R(0,\xi) U P_{0\xi} V^H R(0,\xi)$$

where $P_{0\xi} = P_0(M_\xi)$ denotes the spectral projection for M_ξ associated with $0 \in \sigma(M_\xi)$ of algebraic multiplicity $a_\xi = \text{rank} P_{0\xi} \leq r$. Then $\text{rank} B_\xi = a_\xi$. How is a_ξ related to $\hat{m}_\xi = c_\xi$? Examples in [8] show that can occur any of the 3 possibilities:

$$1 \leq a_\xi < \hat{m}_\xi, \quad a_\xi = \hat{m}_\xi, \quad a_\xi > \hat{m}_\xi \geq 1.$$

There can exist a shortage of algebra ($a_\xi < \hat{m}_\xi$), a balance ($a_\xi = \hat{m}_\xi$), or an *excess* of algebra ($a_\xi > \hat{m}_\xi$). The optimal situation is a balance between algebra and analysis. In this case, induction from level r to level n and deduction from level $r+n$ to level n are equivalent. This is not true when $\hat{m}_\xi \neq a_\xi$. The possible shortage of algebra is well-known (Gödel, Turing, Chaitin). But the possibility of an *excess* of algebra over analysis has been overlooked, despite its computational significance. It shows that analytic computation can be algebraically creative, a property which is not shared by Turing machine computation. A numerical illustration is provided in Section 3: the two frontier points exhibit an excess and a balance of algebra respectively.

Remark: Another computational example of the limitations of Turing machines is presented in [6].

2 The Family of Pencils $P_z(t) = (A - zI) + tE$ Where the Parameter z Varies in $\mathbb{C}$

We consider the family of matrix pencils $z \mapsto P_z(t) = (A - zI) + tE$, where the parameter z varies in $\mathbb{C}$.

2.1 $z \in re(A)$

We consider the polynomial $(-1)^n \pi(t, z) = \det(A - zI + tE) = \det P_z(t)$ which has degree at most r in t and constant coefficient $\det(A - zI)$.

We recall the identity $\pi(t, z) = \pi(z) \det(I - tM_z)$ valid for $z \in re(A)$. The pencil $P_z(t)$ is regular for $\pi(t, z) \not\equiv 0$ in t. This is the case when $z \in re(A)$. Then the Weierstrass canonical form [9] of $P_z(t)$ depends on whether z is frontier or not. When $z \in re(A) \backslash F(A, E)$, then $P_z(t)$ has exactly r finite eigenvalues, and the infinite eigenvalue is semi-simple of multiplicity $g = n - r$. Such regular pencils are said to have *index* 1 (referring to the infinite eigenvalue). This is the generic situation.

We now assume that z is not critical in $F(A, E)$. The canonical form consists of the *two* diagonal blocks given below, with $\varepsilon \in \{0, 1\}$, of respective size $g + a_z$ and $e_z = r - a_z$, where $1 \leq a_z < r$ is the algebraic multiplicity of $0 \in \sigma(M_z)$:

$$\begin{pmatrix} 1 & t\varepsilon & & 0 \\ & 1 & \ddots & \\ & & \ddots & t\varepsilon \\ 0 & & & 1 \end{pmatrix} \quad \text{and} \quad \begin{pmatrix} \frac{1}{\mu_{1z}} & \varepsilon & & 0 \\ & \ddots & \ddots & \\ & & \ddots & \varepsilon \\ 0 & & & \frac{1}{\mu_{e_z z}} \end{pmatrix}$$

Observe that a_z depends on the location of z in $F(A,E)$. The nonzero values μ_{iz}, $i = 1, \cdots, e_z = r - a_z \geq 1$ are the nonzero eigenvalues of M_z, for z frontier and non critical. The first block of order $g + a_z \leq n - 1$ corresponds to the infinite eigenvalue of $P_z(t)$ which is defective. The pencil has an index > 1. The second block of order $e_z = r - a_z$ corresponds to the finite eigenvalues of $P_z(t)$ given by $1/\mu_z$, $\mu_z \neq 0$.

When $z \in re(A)\backslash F(A,E)$, $a_z = 0$ and $e_z = r$, $\varepsilon = 0$ in the first block reduced to I_g: the infinite eigenvalue is semi-simple. The generic structure of $P_z(t)$ does *not* depend on the observation point z.

When z is critical, $a_z = r$ and there is a unique block corresponding to the infinite eigenvalue of $P_z(t)$: $\pi(t,z) \neq 0$ for all $t \in \mathbb{C}$. Moreover z critical is a finite eigenvalue of $A(\infty)$. This result goes beyond the theory of Weierstrass. The change in the nature of $R(t,z)$ as $|t| \to \infty$ is expressed dynamically by the fact that the status of z changes from being repelling to becoming attractive in the limit for the spectral field.

We emphasize that the new HD theory complements the Weierstrass approach not only at the critical points, but also at all the frontier points in $F(A,E)$. At $\xi \in F(A,E)$, the structure of $P_\xi(t)$ is nongeneric. And an important parameter is $\delta_\xi = \hat{m}_\xi - a_\xi$.

When $\delta_\xi \geq 0$, the computational situation is ruled deductively by the homotopic polynomial $\hat{\pi}(z)$ at its zero ξ, when $|t| \to \infty$. When $\delta_\xi < 0$ on the other hand, the situation is inductively creative by means of M_ξ and B_ξ.

2.2 $\lambda \in \sigma(A)$

The situation when the observation point is an eigenvalue $\lambda \in \sigma(A)$ is more complex. Its exposition is beyond the limited scope of this paper. For a complete account, the reader is referred to [8].

2.3 The Generalized Eigenvalue Problem for $P_0(t) = A + tE$

One of us (M. Ahmadnasab) has used HD to propose an algorithm for finding the eigenvalues of $P_0(t) = A + tE$ based on the communication matrix $M_0 = V^H A^{-1} U$ of order r when $0 \in re(A)$. The interested reader is referred to [2].

3 A Numerical Illustration

Let

$$A = \begin{bmatrix} 0 & 0 & 0 & 0 & 1 & 0 \\ -1 & 0 & 0 & 0 & 1 & 0 \\ 0 & 1 & 0 & 0 & 0 & 0 \\ 0 & 0 & 0 & 1 & 0 & 0 \\ 0 & 0 & 1 & 0 & 0 & 1 \\ 0 & 1 & 1 & 0 & 0 & 0 \end{bmatrix}, \quad E = \begin{bmatrix} 0 & 0 & 0 & 1 & 0 & -1 \\ 1 & 0 & 0 & 0 & 0 & 0 \\ 0 & 0 & 0 & 0 & 0 & 0 \\ -1 & 0 & 0 & 1 & 0 & 0 \\ 0 & 0 & 0 & 0 & 0 & 0 \\ 0 & 0 & 0 & 0 & 0 & 0 \end{bmatrix}.$$

Then rank$A = 6 = n$, rank$E = r = 3$, and $\det(A + tE) = -1$, for any $t \in \mathbb{C}$.

$\pi(z) = (z-1)(z^5 - 2z^2 + z + 1)$, $0 \notin \sigma(A)$,
$\hat{\pi}(z) = z(1+z)$, $\hat{Z} = \{0, -1\} = F(A, E)$,
$\sigma(M_0) = \{0^3\}$, $\{0\} = F_c(A, E) \subset F(A, E)$.

$R(t, 0)$ is the following matrix polynomial in t of degree 3:

$$R(t,0) = \begin{bmatrix} 1+t & -t-1 & t^2+t & -t & t^2+t & -t^2-t \\ 0 & 0 & 1 & 0 & 0 & 0 \\ 0 & 0 & -1 & 0 & 0 & 1 \\ t & -t & t^2 & 1-t & t^2 & -t^2 \\ -t^2+1 & t^2 & -t^3+t & t^2-t & -t^3+t & t^3-t \\ 0 & 0 & 1 & 0 & 1 & -1 \end{bmatrix}.$$

Because 0 is critical, HD shows that the pencil $A + tE$ has no finite eigenvalue. This is confirmed when we apply the QZ method in Matlab to this pencil.

$z = -1$ is not critical and $\sigma(M_{-1}) = \{0,\ 0.08 \pm 0.4i\}$. We find that $\hat{m}_0 = 1 < a_0 = 3 = r$, and $\hat{m}_{-1} = a_{-1} = 1$. There is an *excess* of algebra at the critical point 0.

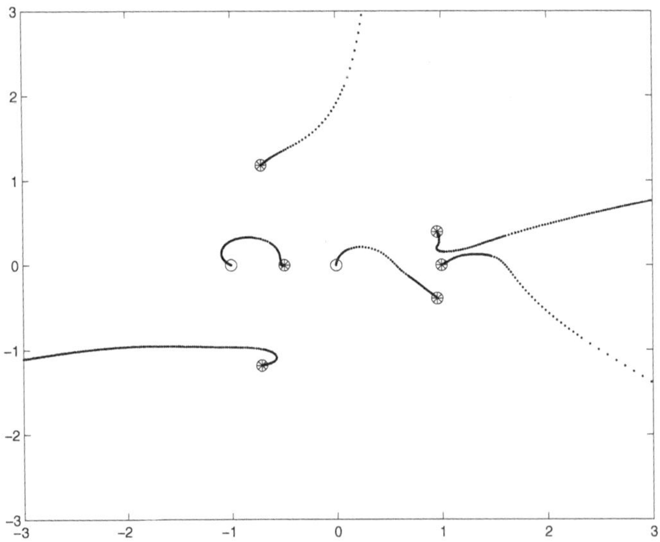

Fig. 1. The plot of $t = |t|e^{i\theta} \mapsto \sigma(A(t))$ for $0 \le |t| \le 10^{18}$ and $\theta = \pi/6$

The computation of $t \mapsto \sigma(A(t))$ illustrates the analytic convergence of $\sigma(A(t))$. The 6 eigenvalues in $\sigma(A)$ are marked $\circledast$. Figure 1 shows that there are two rays which originate in two eigenvalues of the matrix A and converge to $F(A, E) = \{-1,\ 0\}$ (marked $\odot$) as $|t|$ varies between $0 \le |t| \le 10^{18}$. The other 4 spectral rays escape to ∞ extremely fast.

Even though only *one* spectral ray converges nicely to 0 the matrix M_0 is nilpotent, making its unique defective eigenvalue 0 extremely ill-conditioned ($a_0 = 3$ with index 3). In finite arithmetic with machine precision $\sim 10^{-15}$, any u such that $|u| \sim 10^{-5}$ is seen *locally* as frontier (the spectral backward error for 10^{-5} as an eigenvalue of $M_{10^{-5}}$ is of order 10^{-15}). On the other hand, $\hat{\pi}(10^{-5}) \sim 10^{-5} \gg 10^{-15}$: u is *not* seen *globally* as frontier. The local and global conclusions at u disagree.

4 Conclusion

The contributions of HD to the theory of matrix pencils that we have presented are promising both from a theoretical and an algorithmic point of view. When the observation point $z = 0 \notin \sigma(A)$ happens to be frontier for $A + tE$, the deductive and inductive computational viewpoints may clash whenever there is an excess or a shortage of algebra over analysis. This remarkable connection between the Weierstrass and Cauchy viewpoints opens new avenues for the theory of computability. Mechanical computations cannot match the mathematical ones.

References

1. Ahmadnasab, M.: Homotopic Deviation theory: A qualitative study. PhD thesis, Université Toulouse 1 and CERFACS, Toulouse, France, October 24 (2007)
2. Ahmadnasab, M.: An order reduction method for computing the finite eigenvalues of regular matrix pencils. Technical Report TR/PA/08/23, CERFACS, Toulouse, France (2008)
3. Ahmadnasab, M., Chaitin-Chatelin, F.: Parameter analysis of the structure of matrix pencils by Homotopic Deviation theory. In: Proceedings ICIAM 2007. Wiley, Chichester (2007) (to appear)
4. Ahmadnasab, M., Chaitin-Chatelin, F.: Matrix pencils under Homotopic Deviation theory. Technical Report TR/PA/07/108, CERFACS, Toulouse, France (2007)
5. Chaitin-Chatelin, F., van Gijzen, M.B.: Analysis of parameterized quadratic eigenvalue problems in computational acoustics with homotopic deviation theory. Numerical Linear Algebra with Applications 13, 487–512 (2006)
6. Chaitin-Chatelin, F.: Computing beyond classical logic: SVD computation in nonassociative Dickson algebras. In: Calude, C. (ed.) Randomness and Complexity, pp. 13–23. World Scientific, Singapore
7. Chatelin., F.: Eigenvalues of matrices. Wiley, Chichester (1993)
8. Chatelin, F.: Homotopic Deviation in Linear algebra. In: Qualitative Computing: a computational journey into nonlinearity, vol. 7. World Scientific, Singapore (to appear, 2009)
9. Gantmacher, F.: The theory of matrices. Chelsea Publishing Company, New York (1960)

Numerical Integration with Complex Jacobi Weight Function

Gradimir V. Milovanović[1] and Aleksandar S. Cvetković[2]

[1] Faculty of Computer Sciences, Megatrend University Belgrade, Bulevar umetnosti 29, 11070 Novi Beograd, Serbia

[2] Faculty of Sciences and Mathematics, University of Niš, P.O. Box 224, 18000 Niš, Serbia

Abstract. In this paper we study the numerical integration on $(-1, 1)$ with respect to the Jacobi weight function $(1-x)^\alpha(1+x)^\beta$, where α and β are complex parameters. The problem arises in some applications of computational models in quantum mechanics. We discuss two methods for integration. One is suitable for integration of analytic functions and the other is applicable to the general Riemann integrable functions.

1 Introduction

In this paper we are interested in the numerical integration of the following integrals

$$\int_{-1}^{1} (1-x)^\alpha (1+x)^\beta f(x)dx, \tag{1}$$

where $\mathrm{Re}(\alpha) > -1$ and $\mathrm{Re}(\beta) > -1$.

Similar types of integrals appear in the theoretical physics (see [10]). In this paper we give two algorithms which can be used for constructing the quadrature rules for the numerical computation of (1). First algorithm uses the Gaussian quadrature rules and the second one uses an interpolatory quadrature rule. Applicability of those algorithms depends on the properties of the function f. If the function f can be extended to a function holomorphic in in some neighborhood of the interval $[-1, 1]$ we can use either method. To the contrary, assuming that f is merely integrable, it would imply the application of the interpolatory quadrature rules.

It is easy to understand that an application of the Gauss-Legendre quadrature rule would result in the poor performance, since we have an oscillatory part in the weight function, i.e.,

$$(1-x)^\alpha(1+x)^\beta = (1-x)^{\mathrm{Re}(\alpha)}(1+x)^{\mathrm{Re}(\beta)}\exp\big[\mathrm{i}\big(\mathrm{Im}(\alpha)\log(1-x)+\mathrm{Im}(\beta)\log(1+x)\big)\big].$$

2 Gaussian Quadrature Rule

In this section we want to construct the Gaussian quadrature for the complex Jacobi measure

$$d\mu_J^{(\alpha,\beta)}(x) = d\mu_J(x) = \chi_{[-1,1]}(x)(1-x)^\alpha(1+x)^\beta dx = w_J^{\alpha,\beta}(x)dx = w_J(x)dx, \tag{2}$$

with parameters $\mathrm{Re}(\alpha), \mathrm{Re}(\beta) > -1$.

S. Margenov, L.G. Vulkov, and J. Waśniewski (Eds.): NAA 2008, LNCS 5434, pp. 20–31, 2009.

In order to be able to construct the Gaussian quadrature rule for the integral given in (1), we need the orthogonal polynomials with respect to μ_J.

Definition 1. *By the monic orthogonal polynomials with respect to the (complex) measure* μ*, we assume the monic sequence of polynomials* p_n*,* $\deg(p_n)$*,* $n \in \mathbb{N}_0$*, which satisfies the following orthogonality conditions*

$$\int p_n(x)p_k(x)d\mu(x) = ||p_n||^2\delta_{n,k}, \quad n, k \in \mathbb{N}_0, \tag{3}$$

where $||p_n|| \neq 0$*,* $n \in \mathbb{N}_0$*.*

Such a sequence of orthogonal polynomials is usually termed as *formal*, and the original term the *orthogonal polynomial sequence* is reserved for the sequences orthogonal with respect to the positive measures. It can be proved that the sequence of monic (formal) orthogonal polynomials with respect to a general measure μ, satisfies the three-term recurrence relation given by

$$p_{n+1}(x) = (x - \alpha_n)p_n(x) - \beta_n p_{n-1}(x), \quad n \in \mathbb{N}_0, \tag{4}$$

with $p_{-1} = 0$, $p_0 = 1$, provided that the sequence p_n, $n \in \mathbb{N}_0$, exists. Usually for the convenience we choose $\beta_0 = \mu(\mathbb{R})$.

Unfortunately, the measure μ_J, given in (2), is complex, so that the sequence of (formal) orthogonal polynomials, with respect to μ_J need not exist. In contrast to the case of positive measures μ, when the existence of the sequence of orthogonal polynomials is granted (see [3]). A complex measure μ, for which the sequence of (formal) orthogonal polynomials exists, we call the *regular measure.*

Define a sequence of moments of the measure μ as $\mu_k = \int x^k d\mu(x)$, $k \in \mathbb{N}_0$. We have the following simple lemma.

Lemma 1. *The measure* μ *is regular if for every* $n \in \mathbb{N}_0$ *the so-called Hankel determinants are different from zero, i.e.,* $H_n = |\mu_{i+j}|_{i,j=0}^n \neq 0$*.*

It can be proved (see [3]) that in the case μ is regular sequence of (formal) orthogonal polynomials with respect to μ satisfies $\beta_n \neq 0$, $n \in \mathbb{N}_0$.

Lemma 2. *The measure* μ_J *is regular.*

Proof. Let $\gamma_k = \alpha + \beta + k + 1$. The moments of the measure μ_J can be expressed in the following form

$$\mu_k = \int x^k d\mu_J(x) = 2^{\gamma_k} \sum_{\nu=0}^{k} \binom{k}{\nu} \left(-\frac{1}{2}\right)^{k-\nu} B(\beta + k + 1, \alpha + 1).$$

Since the beta function B is an analytic function of its arguments, for $\mathrm{Re}(\alpha) > -1$ and $\mathrm{Re}(\beta) > -1$, we see that the moments are analytic functions of α and β for each $\mathrm{Re}(\alpha) > -1$ and $\mathrm{Re}(\beta) > -1$.

The three-term recurrence coefficients are rational functions of the moments (see [5], [11]). For real α and β, where $\alpha, \beta > -1$, the recursion coefficients are known explicitly (see [11], [14])

$$\alpha_k = \frac{\beta^2 - \alpha^2}{(2k+\alpha+\beta)(2k+\alpha+\beta+2)}, \beta_k = \frac{4k(k+\alpha)(k+\beta)(k+\alpha+\beta)}{(2k+\alpha+\beta)^2((2k+\alpha+\beta)^2-1)}.$$

As we can see the three-term recurrence coefficients depends analytically on α and β. In total, using the principle of analytical continuation (cf. [8]), we must have the same expressions for complex α and β, with $\mathrm{Re}(\alpha) > -1$ and $\mathrm{Re}(\beta) > -1$. Then, obviously $\beta_n \neq 0$, $n \in \mathbb{N}_0$. □

Lemma 3. *Suppose a regular (complex) measure μ is given and the monic (formal) orthogonal polynomials p_n, $n \in \mathbb{N}_0$, with respect to μ satisfy the following three-term recurrence relation $p_{n+1}(x) = (x - \alpha_n)p_n(x) - \beta_n p_{n-1}(x)$, $n \in \mathbb{N}_0$, with $p_{-1} = 0$ and $p_0 = 1$. Denoting by r_n, $n \in \mathbb{N}_0$, the second linearly independent solution of this three-term linear recurrence relation determined by the initial conditions $r_{-1} = -1/\beta_0$ and $r_1 = 0$, then the linear functional $G_n : \mathcal{P}_{2n-1} \mapsto \mathbb{C}$, defined by*

$$G_n(p) = \frac{1}{2\pi \mathrm{i}} \oint_C \frac{\beta_0 r_n(z)}{p_n(z)} p(z)dz, \quad p \in \mathcal{P}_{2n-1},$$

where all zeros of p_n lie in the interior of the curve C, has the following property $\int p(x)d\mu(z) = G_n(p)$, $p \in \mathcal{P}_{2n-1}$.

In the case of the positive measure the all zeros of p_n, $n \in \mathbb{N}$, are simple (see [3], [11]). Then a simple application of Cauchy's residue theorem leads to the well known form of the Gaussian quadrature rule, i.e.,

$$G_n(p) = \sum_{k=1}^{n} w_k p(x_k), \tag{5}$$

where x_k, $k = 1, \ldots, n$, are zeros of p_n, known also as nodes of the Gaussian quadrature rule, and where $w_k > 0$, $k = 1, \ldots, n$, are weights of the Gaussian quadrature rule. In the case when the regular measure is complex, we do not have the statement that all zeros of p_n are simple or that weights w_k, $k = 1, \ldots, n$ are positive. An assumption of the non-simplicity of the zeros, after an application of Cauchy's residue theorem, will produce terms with the derivatives of p in (5).

For our measure μ_J we can prove that zeros of orthogonal polynomials are simple.

Theorem 1. *Let p_n, $n \in \mathbb{N}$, be the sequence of orthogonal polynomials with respect to μ_J. Then for every $n \in \mathbb{N}$, the zeros of p_n are simple.*

Proof. This proof is due to the fact that the weight $w_J(x) = (1-x)^\alpha(1+x)^\beta$, is classical. The weight function w is classical provided there exist two polynomials ϕ and ψ, of degrees at most two and one, respectively, such that $(\phi w_J)'(x) =$

$(\psi w_J)(x)$, for x in the support of the measure μ_J. In this case we can check directly that $\phi(x) = 1 - x^2$, $\psi(x) = -(1+\alpha)(1+x) + (1+\beta)(1-x)$.

Now, consider the expansion of the following polynomial $((\phi p_n w_J)'/w_J)(x) = (\phi p_n' + \psi p_n)(x)$, of degree $n+1$, over the basis $\{1, p_1, \ldots, p_{n+1}\}$. Using an integration by parts, for $k < n-1$, we have

$$\int_{-1}^{1} (\phi p_n w_J)'(x) p_k(x) dx = (\phi p_n p_k w_J)(x) \Big|_{-1}^{1} - \int (p_n \phi p_k')(x) d\mu_J(x) = 0,$$

due to the orthogonality relation (3). Also for $k > n+1$, we have

$$\int_{-1}^{1} (\phi p_n w_J)'(x) p_k(x) dx = \int (\phi p_n' + \psi p_n)(x) p_k(x) d\mu_J(x) = 0,$$

due to the orthogonality relation (3). Using this we have

$$(\phi p_n' + \psi p_n)(x) = a_n^{(1)} p_{n+1}(x) + a_n^{(2)} p_n(x) + a_n^{(3)} p_{n-1}(x), \quad n \in \mathbb{N}_0,$$

which can be rewritten, using the three-term recurrence relation (4), as

$$\phi(x) p_n'(x) = [a_n^{(1)}(x - \alpha_n) - \psi(x) + a_n^{(2)}] p_n(x) + (a_n^{(3)} - a_n^{(1)} \beta_n) p_{n-1}(x). \quad (6)$$

We need to prove that $a_n^{(3)} - a_n^{(1)}\beta_n \neq 0$. Suppose it is, then we have

$$(1 - x^2) p_n'(x) = [a_n^{(1)}(x - \alpha_n) - \psi(x) + a_n^{(2)}] p_n(x),$$

for every $x \in \mathbb{C}$, from which we conclude that $p_n(1) = 0$ or $p_n(-1) = 0$. We elaborated in the proof of Lemma 2, that the three-term recurrence coefficients for the sequence of polynomials p_n, $n \in \mathbb{N}_0$, have the same analytic expressions as ones for the Jacobi polynomials for the real parameters. Furthermore, the polynomials p_n, $n \in \mathbb{N}_0$, have the same analytic expressions as the Jacobi polynomials with real parameters. Such expressions can be used to calculate $p_n(1) = (1+\alpha)_n/n!$ and $p_n(-1) = (-1)^n (1+\beta)_n/n!$, (see [11, p. 135]). Obviously, we cannot have $p_n(1) = 0$ or $p_n(-1) = 0$, so that $a_n^{(3)} - a_n^{(1)}\beta_n \neq 0$.

Now, we prove that the polynomials p_n and p_{n-1}, $n \in \mathbb{N}$, cannot have zeros in common. Namely, supposing that for some t, $p_n(t) = p_{n-1}(t) = 0$, using the three-term recurrence we have

$$0 = p_n(t) = (t - \alpha_{n-1}) p_{n-1}(t) - \beta_{n-1} p_{n-2}(t) = -\beta_{n-1} p_{n-2}(t),$$

i.e., $p_{n-2}(t) = 0$, because the measure μ_J is regular and $\beta_{n-1} \neq 0$. Using the same arguments, now applied to p_{n-1} and p_{n-2}, we get $p_{n-3}(t) = 0$. Iterating the same arguments we obtain $p_0(t) = 0$, but that is impossible since $p_0(t) = 1$.

Now assume that p_n has a multiple zero at t, then using (6), we find

$$\begin{aligned} 0 = (1 - t^2) p_n'(t) &= [a_n^{(1)}(t - \alpha_n) - \psi(t) + a_n^{(2)}] p_n(t) + (a_n^{(3)} - a_n^{(1)} \beta_n) p_{n-1}(t) \\ &= (a_n^{(3)} - a_n^{(1)} \beta_n) p_{n-1}(t), \end{aligned}$$

i.e., $p_{n-1}(t) = 0$, since $a_n^{(3)} - a_n^{(1)}\beta_n \neq 0$, which is a contradiction.

Hence, if t is zero of the polynomial p_n, it must be simple. □

This theorem combined with Lemma 3, provides the Gaussian quadrature rule for the measure (2) in the form (5) where, as usual, x_k, $k = 1, \dots, n$, are the zeros of p_n.

Using results about the convergence of the Pade approximation (see [2]), we have the following result (see [12]).

Theorem 2. *Let $N \in \mathbb{N}$ be given and $\mathcal{D}$ be the open set in the complex plane with the compact closure. Let all zeros of polynomials p_n, $n > N$, orthogonal with respect to regular measure μ, be contained in the domain $\mathcal{D}$, and let* $\mathrm{supp}(\mu) \subset \mathcal{D}$. *If function f is analytic in $\mathcal{D}$ then $\lim_{n \to +\infty} G_n(f) = \int f d\mu$, where G_n is the Gaussian quadrature rule for the measure μ.*

It is essential to know the asymptotic distribution of the zeros of polynomials orthogonal with respect to μ_J. Actually for $\mathrm{Re}(\alpha) = \mathrm{Re}(\beta) = 0$, an application of theorem given in [9] guaranties that for every open set $\mathcal{D} \supset [-1, 1]$ there exists an $N \in \mathbb{N}$ such that all zeros of polynomials p_n, $n > N$, orthogonal with respect to μ_J, are contained in $\mathcal{D}$. However, for $\mathrm{Re}(\alpha) \neq 0$ or $\mathrm{Re}(\beta) \neq 0$ the mentioned theorem cannot be applied. In that case, we can only give a weaker result, that the absolute value of all zeros of orthogonal polynomials is lower than the norm of the Jacobi operator constructed for the measure μ_J (see [2]).

Theorem 3. *Let $x \in \mathbb{C}$ be the zero of the polynomial p_n, $n \in \mathbb{N}$, orthogonal with respect to $\mu_J^{\alpha,\beta}$. Then*

$$|x| \leq \sqrt{3}\sqrt{\frac{1}{2} + \frac{\delta}{|\gamma+2|^2}\left(\frac{1}{2} + \frac{\delta}{|\gamma+4|^2}\right) + \frac{|((\alpha-\beta)^2-1)(\gamma-1)|}{4}\left(1 + \frac{|\gamma+1|}{|\gamma+3|}\right)},$$

where $\delta = |\beta^2 - \alpha^2|$ and $\gamma = \alpha + \beta$.

Proof. We only give a sketch of the proof. It uses the fact that the norm of the Jacobi operator is bounded by $\sup_{k \in \mathbb{N}} \left(\sqrt{|\beta_{k-1}|} + |\alpha_k| + \sqrt{|\beta_k|}\right)$, which can be found in [2]. An estimate by the Cauchy-Schwartz inequality gives the desired result. □

3 Interpolatory Quadrature Rule

In this section we describe a procedure which can be used for the construction of the interpolatory quadrature rule

$$\int p(x) d\mu_J(x) = \sum_{k=1}^{n} w_k p(x_k), \quad p \in \mathcal{P}_{n-1}. \tag{7}$$

We call such a quadrature rule interpolatory, since the nodes x_k, $k = 1, \dots, n$, are fixed and weights w_k, $k = 1, \dots, n$, are constructed such that the quadrature rule is exact for all polynomials from $\mathcal{P}_{n-1}$. Define $\omega(x) = \prod_{k=1}^{n}(x - x_\nu)$, and $\omega_\nu(x) = \omega(x)/(x - x_\nu)$, $\nu = 1, \dots, n$, then the fundamental Lagrange interpolating polynomials are defined by $\ell_\nu(x) = \omega_\nu(x)/\omega_\nu(x_\nu)$, $\nu = 1, \dots, n$. All ℓ_ν are of degree $n - 1$. Applying the quadrature rule (7), we get

$$w_\nu = \int \ell_\nu(x) d\mu_J(x), \quad \nu = 1, \ldots, n, \tag{8}$$

Since ℓ_ν, $\nu = 1, \ldots, n$, make a basis of $\mathcal{P}_{n-1}$, the quadrature rule (7), with the weights w_ν given by (8), is exact on $\mathcal{P}_{n-1}$.

Before going deeper into the construction of weights, we need to give some directions about the way how to chose the nodes x_ν, $\nu = 1, \ldots, n$, of the quadrature rule (7). For that purpose we need an extension on a convergence theorem given in [15].

Theorem 4. *Let Q_n, $n \in \mathbb{N}_0$, be a sequence of polynomials orthogonal with respect to the nonnegative weight function W on $[-1, 1]$, which vanishes only on the set of the Lebesgue measure zero and which is integrable in the Lebesgue sense on $[-1, 1]$. Let a complex weight w on $[-1, 1]$, be a Lebesgue integrable on $[-1, 1]$. If*

$$I_W(w) = \int_{-1}^{1} \frac{|w(x)|^2}{W(x)} dx < +\infty, \tag{9}$$

and x_ν, $\nu = 1, \ldots, n$, are chosen to be zeros of the polynomial Q_n and the weights w_ν, $\nu = 1, \ldots, n$, are calculated according to (8), for every bounded Riemann integrable function f, we have $\lim_{n\to+\infty} \sum_{\nu=1}^{n} w_\nu f(x_\nu) = \int_{-1}^{1} w(x) f(x)\, dx$.

Proof. Let w_r and w_i denote real and imaginary parts of the function w. Since, w is a Lebesgue integrable, then w_r and w_i are too. Denote by $w_{\nu,r}$ and $w_{\nu,i}$ real and imaginary parts of the weights w_ν, $\nu = 1, \ldots, n$. We can think about the formula (7) as two real formulas: one for real and the other for imaginary part of the weight w. Since in addition $I_W(w) = I_W(w_r) + I_W(w_i)$, the functions w_r and w_i also satisfy condition (9). According to Theorem 1 from [15], we have that quadrature rules are convergent for the weights w_r and w_i. Since quadrature rule for the weight w is linear combination of the rules for weights w_r and w_i, we get the statement. □

This theorem completely solves the problem of choosing nodes in our interpolatory quadrature rule (7). Denote by $P_n^{\alpha,\beta}$, $n \in \mathbb{N}_0$, the monic Jocobi polynomials orthogonal with respect to the weight function $w_J^{(\alpha,\beta)}$, $\mathrm{Re}(\alpha) > -1$, $\mathrm{Re}(\beta) > -1$. In the literature the term Jacobi polynomial usually means polynomial $A_n^{(\alpha,\beta)} P_n^{(\alpha,\beta)}$, where

$$A_n^{(\alpha,\beta)} = \frac{1}{2^n} \sum_{k=0}^{n} \binom{n+\alpha}{k} \binom{n+\beta}{n-k} = \frac{1}{2^n} \frac{\Gamma(2n+\alpha+\beta+1)}{\Gamma(n+1)\Gamma(n+\alpha+\beta+1)}.$$

Lemma 4. *If we choose for the nodes x_ν, $\nu = 1, \ldots, n$, of the interpolatory quadrature rule (7), with complex Jacobi weight $w_J^{(\alpha,\beta)}$, $\mathrm{Re}(\alpha) > -1$, $\mathrm{Re}(\beta) > -1$, to be zeros of the Jacobi polynomial $P_n^{\alpha_1,\beta_1}$,*

$$-1 < \alpha_1 < 1 + 2\mathrm{Re}(\alpha), \quad -1 < \beta_1 < 1 + 2\mathrm{Re}(\beta), \tag{10}$$

then this quadrature rule converges for every bounded Riemann integrable function f on $[-1, 1]$.

Proof. First, if $\alpha > -1$, i.e., $2\alpha + 1 > -1$, it is easy to see that the Jacobi polynomials $P_n^{(\alpha_1,\beta_1)}$, mentioned in the statement, exist with real parameters bigger than -1 and which zeros belong to the interval $[-1,1]$. We need only to check (9), which gives $\int_{-1}^{1} |w_J^{(\alpha,\beta)}(x)|^2/w_J^{(\alpha_1,\beta_1)}(x)dx = \int_{-1}^{1}(1-x)^{2\mathrm{Re}(\alpha)-\alpha_1}(1+x)^{2\mathrm{Re}(\beta)-\beta_1}dx < +\infty$, according to $2\mathrm{Re}(\alpha) - \alpha_1 > -1$, $2\mathrm{Re}(\beta) - \beta_1 > -1$. □

We are interested in the convergent quadrature rules, henceforth, we choose nodes of the quadrature rule to be zeros of the polynomial $P_n^{\alpha_1,\beta_1}$ for some $-1 < \alpha_1 < 1 + 2\mathrm{Re}(\alpha)$ and $-1 < \beta_1 < 1 + 2\mathrm{Re}(\beta)$. Note that if $\alpha > -1$ then $-1 < \alpha < 1 + 2\alpha$, so that we are free to choose $\alpha_1 = \mathrm{Re}\,\alpha$ and $\beta_1 = \mathrm{Re}\,\beta$. An effective construction of such nodes can be achieved using QR-algorithm (see [7], [5], [6]).

Theorem 5. *Let the nodes x_ν, $\nu = 1,\ldots,n$, in the interpolatory quadrature rule (7) be the zeros of the Jacobi polynomial $P_n^{(\alpha_1,\beta_1)}$, $-1 < \alpha_1 < 1 + 2\mathrm{Re}\alpha$, $-1 < \beta_1 < 1 + 2\mathrm{Re}\beta$. The weights w_ν, $\nu = 1,\ldots,n$, can be computed as $w_\nu = a_\nu^n/(nP_{n-1}^{(\alpha_1+1,\beta_1+1)}(x_\nu))$, $\nu = 1,\ldots,n$, where*

$$a_\nu^{k+1} = \left(x_\nu - \alpha_k^{(\alpha_1,\beta_1)}\right) a_\nu^k - \beta_k^{(\alpha_1,\beta_1)} a_\nu^{k-1} + b^k,$$

$$a_\nu^0 = \text{v.p.} \int_{-1}^{1} \frac{w_J^{(\alpha,\beta)}(x)}{x - x_\nu} dx, \quad a_\nu^{-1} = 0,$$

$$b^k = \frac{(\alpha_1+1)_k \Gamma(\alpha+\beta+2)}{A_k^{(\alpha_1,\beta_1)} k!}\, {}_3F_2\left(\begin{matrix} -k, k+\alpha_1+\beta_1+1, \alpha+1 \\ \alpha_1+1, \alpha+\beta+2 \end{matrix}; 1\right).$$

Proof. First, we note that the fundamental Lagrange polynomials, in the case x_ν, $\nu = 1,\ldots,n$, are zeros of $P_n^{(\alpha_1,\beta_1)}$, are $\ell_\nu(x) = P_n^{(\alpha_1,\beta_1)}(x)/((x - x_\nu)\left(P_n^{(\alpha_1,\beta_1)}\right)'(x_\nu))$, $\nu = 1,\ldots,n$, If we devide the three-term recurrence relation for the monic Jacobi polynomials by $x - x_\nu$, we get

$$\frac{P_k^{(\alpha_1,\beta_1)}(x)}{x - x_\nu} = P_{k-1}^{(\alpha_1,\beta_1)}(x) + \left(x_\nu - \alpha_{k-1}^{(\alpha_1,\beta_1)}\right) \frac{P_{k-1}^{(\alpha_1,\beta_1)}(x)}{x - x_\nu} - \beta_{k-1}^{(\alpha_1,\beta_1)} \frac{P_{k-2}^{(\alpha_1,\beta_1)}(x)}{x - x_\nu}.$$

If we integrate it with respect to $\mu_J^{(\alpha,\beta)}$, we get $a_\nu^k = \left(x_\nu - \alpha_k^{(\alpha_1,\beta_1)}\right) a_\nu^{k-1} - \beta_{k-1}^{(\alpha_1,\beta_1)} a_\nu^{k-2} + b^{k-1}$, where $a_\nu^k = \text{v.p.} \int P_k^{(\alpha_1,\beta_1)}(x)/(x - x_\nu) d\mu_J(x)$ and $b^k = \int P_k^{(\alpha_1,\beta_1)}(x) d\mu_J(x)$. Using an expansion of $P_k^{(\alpha_1,\beta_1)}$ with respect to $P_\nu^{(\alpha,\beta)}$, $\nu = 0,1,\ldots,k$, from [1, p. 357], we get

$$\begin{aligned} b^k &= \int P_k^{(\alpha_1,\beta_1)}(x) d\mu_J^{(\alpha,\beta)}(x) = \int_{-1}^{1} \sum_{\nu=0}^{k} c_{k,\nu} P_\nu^{(\alpha,\beta)} w_J^{(\alpha,\beta)}(x) dx \\ &= \frac{(\alpha_1+1)_k \Gamma(\alpha+\beta+2)}{A_k^{(\alpha_1,\beta_1)} k!}\, {}_3F_2\left(\begin{matrix} -k, k+\alpha_1+\beta_1+1, \alpha+1 \\ \alpha_1+1, \alpha+\beta+2 \end{matrix}; 1\right). \end{aligned}$$

Since $P_{-1}^{(\alpha_1,\beta_1)}(x) = 0$ and $P_0^{(\alpha_1,\beta_1)}(x) = 1$, we get what is stated. □

This theorem establishes a way to compute the weights in the quadrature rule (7). However, it requires computation of the integrals which define a_ν^0, $\nu = 1,\ldots,n$, which are equally hard to compute. Therefore, some other way to compute the weights is needed.

Suppose the fundamental Legendre polynomials ℓ_ν, $\nu = 1,\ldots,n$, have the expansions

$$\ell_\nu(x) = \sum_{k=0}^{n-1} \ell_\nu^k x^k, \quad \nu = 1,\ldots,n, \tag{11}$$

and define the following sequence of polynomials $q_k(x) = -(k+\alpha+\beta+1)x^k + (\beta-\alpha)x^{k-1} + (k-1)x^{k-2}$, $k \geq 1$, and $q_0(x) = 1$.

Theorem 6. *For each $\nu = 1,\ldots,n$, there exist the complex numbers γ_ν^k, $k = 0,1,\ldots,n-1$, such that*

$$\ell_\nu(x) = \sum_{k=0}^{n-1} \gamma_\nu^k q_k(x), \tag{12}$$

which are the solutions of the following linear system of equations

$$\begin{aligned} \ell_\nu^0 &= \gamma_\nu^0 + (\beta-\alpha)\gamma_\nu^1 + \gamma_\nu^2, \\ \ell_\nu^k &= -(k+\alpha+\beta+1)\gamma_\nu^k + (\beta-\alpha)\gamma_\nu^{k+1} + (k+1)\gamma_\nu^{k+2}, \quad k = 1,\ldots,n-1, \end{aligned} \tag{13}$$

where $\gamma_\nu^{n+1} = \gamma_\nu^n = 0$. The weights in (7) can be calculated as $w_\nu = \gamma_\nu^0 \mu_0$, $\nu = 1,\ldots,n$, where $\mu_0 = \mu_J(\mathbb{R})$.

Proof. It is clear that q_k, $k = 0,1,\ldots,n$, constitute a basis for $\mathcal{P}_n$, since $\deg(q_k) = k$, so that the expansions in (12) do exist. To identify a linear system just compare the coefficients with x^k, $k = 0,1,\ldots,n-1$, at both sides of the equation (12), and note that all ℓ_ν have degree $n-1$.

Next, we have $\int q_k(x)d\mu_J(x) = (x^{k-1}(1-x)^{\alpha+1}(1+x)^{\beta+1})\big|_{-1}^{1} = 0$, for every $k \geq 1$. An application of (8), for $\nu = 1,\ldots,n$, gives

$$w_\nu = \int \ell_\nu(x)d\mu_J(x) = \int d\mu_J(x) \sum_{k=0}^{n-1} \gamma_\nu^k q_k(x) = \gamma_\nu^0 \int q_0(x)d\mu_J(x) = \gamma_\nu^0 \mu_0. \quad \square$$

This theorem means that in order to construct the weights in (7), we need only to determine expansions of the fundamental Legendre polynomials $\ell_\nu(x)$, $\nu = 1\ldots,n$, over the basis q_k, $k = 0,1,\ldots,n-1$, even more we need only coefficients γ_ν^0, $\nu = 1,\ldots,n$, standing with the polynomial $q_0(x)$.

Next question we want to address is the way we can compute the weights. A construction of weights can be performed using Theorem 6. If the nodes of the quadrature rule (7) are zeros of the Jacobi polynomial $P_n^{(\alpha_1,\beta_1)}$, obviously node polynomial ω is exactly monic version of $P_n^{(\alpha_1,\beta_1)}$. The Jacobi polynomial can be expressed explicitly as (see [1], [11])

$$P_n^{(\alpha,\beta)}(x) = \frac{1}{2^n A_n^{(\alpha,\beta)}} \sum_{i=0}^{n} \binom{n+\alpha}{i}\binom{n+\beta}{n-i}(x-1)^{n-i}(x+1)^i.$$

Let us denote by a_k, $k = 0,1,\ldots,n$, the coefficients of the monic Jacobi polynomial, i.e., coefficients in the expansion $P_n^{(\alpha,\beta)} = \sum_{k=0}^{n} a_k x^k$.

Lemma 5. *The coefficients of the fundamental Lagrange polynomial* ℓ_ν *satisfy the following two systems of linear equations*

$$\ell_\nu^{k-1} - x_\nu \ell_\nu^k = \frac{a_k}{nP_{n-1}^{(\alpha_1+1,\beta_1+1)}(x_\nu)}, \quad k = 1, \ldots, n, \quad \ell_\nu^n = 0, \tag{14}$$

and

$$\ell_\nu^{k-1} - x_\nu \ell_\nu^k = \frac{a_k}{nP_{n-1}^{(\alpha_1+1,\beta_1+1)}(x_\nu)}, \quad k = 0, 1, \ldots, n-1, \quad \ell_\nu^{-1} = 0, \tag{15}$$

where $P_{n-1}^{(\alpha_1,\beta_1+1)}$, $n \in \mathbb{N}_0$, *are monic versions of Jacobi polynomials with parameters* $\alpha_1 + 1$ *and* $\beta_1 + 1$.

Proof. We obtain the both systems comparing the coefficients with x^k in the following equation $(x - x_\nu)\sum_{k=0}^{n-1} \ell_\nu^k x^k = P_n^{(\alpha_1,\beta_1)}(x)/(nP_{n-1}^{(\alpha_1+1,\beta_1+1)}(x_\nu))$, where we are using the fact that the derivative of the Jacobi polynomials are again Jacobi polynomials with parameters increased by one (see [1, p. 304], [11]), so that for monic polynomials we have $(P_n^{(\alpha_1,\beta_1)})'(x)P_{n-1}^{(\alpha_1+1,\beta_1+1)}(x)$. The first system is obtained comparing coefficients with x^k, $k = 0, 1, \ldots, n-1$, and the second one comparing coefficients with x^k, $k = 1, \ldots, n$. □

As we can see the matrices of the systems are transposes one of another, so that the following lemma gives the condition numbers for the both systems.

Lemma 6. *The condition numbers of the matrices of the linear systems* (*14*) *and* (*15*) *are lower than* $n\sqrt{n+1}$ *and* $x_\nu^{-n} n\sqrt{n+1}$, *respectively.*

Proof. The condition number of a matrix A is defined by the product of norms of A and A^{-1}. Denote by A the matrix of the linear system (14). If we choose the Schmidt norms, for the norm of A we get $||A|| = \sum_{\nu=1}^{n} 1^2 + \sum_{\nu=1}^{n-1}(-x_\nu)^2 + (n-1)x_\nu^2 \le 2n$. It can be proved directly that the inverse matrix $A^{-1} = ||a_{i,j}||$, with elements $a_{i,j} = (-x_\nu)^{j-i}$ for, $i \le j$, and $a_{ij} = 0$ for $i > j$, and has the Schmidt norm $||A^{-1}|| = \sum_{k=1}^{n}\sum_{\nu=k}^{n}(-x_\nu)^{2(k-1)} = \sum_{k=1}^{n}(n-k+1)(-x_\nu)^{2(k-1)} \le \frac{n(n+1)}{2}$, for $x_\nu^2 \le 1$. If we multiply both quantities we get what is stated. Using similar arguments we get the condition number for the other matrix. □

4 Numerical Examples

In this section we present a few numerical examples. First, we consider the construction of the Gaussian quadrature rule for the weight function $\alpha = 1 - 10^6\mathrm{i}$ and $\beta = 10^6\mathrm{i}$. We constructed the Gaussian quadrature rule, using our software 'OrthogonalPolynomials' implemented in *Mathematica* (see [4]).

As we can see from Table 1, the weights are quite small quantities, which is due to the fact that values $\mathrm{Im}(\alpha)$ and $\mathrm{Im}(\beta)$ are considerable. Numbers in the parentheses represent decimal exponents.

Table 1. Nodes x_k and weights w_k, $k = 1, \ldots, 20$, of the Gaussian quadrature rule for the weight function $(1-x)^{1-10^6 \mathrm{i}}(1+x)^{10^6 \mathrm{i}}$

nodes	weights
$\mp 5.240680553 + 2.379145793(4)\mathrm{i}$	$(\pm 2.924973463 + 3.389547049)(-1364362)\mathrm{i}$
$\mp 5.174773926 + 1.300457828(4)\mathrm{i}$	$(\mp 2.499951602 + 0.7365135981)(-1364364)\mathrm{i}$
$\mp 5.079519102 + 3.343701818(4)\mathrm{i}$	$(\pm 0.08735046372 - 1.828918372)(-1364362)\mathrm{i}$
$\mp 4.744269224 + 4.217282884(4)\mathrm{i}$	$(\mp 1.233803697 + 2.955943718)(-1364359)\mathrm{i}$
$\mp 4.263313041 + 4.995187569(4)\mathrm{i}$	$(\pm 1.983287952 - 2.299173063)(-1364358)\mathrm{i}$
$\mp 3.659269891 + 5.666671359(4)\mathrm{i}$	$(\mp 1.434226827 + 1.030113182)(-1364357)\mathrm{i}$
$\mp 2.953528913 + 6.220169658(4)\mathrm{i}$	$(\pm 6.135836502 - 2.886644226)(-1364357)\mathrm{i}$
$\mp 2.167686969 + 6.645241073(4)\mathrm{i}$	$(\mp 0.1724425368 + 5.182220448)(-1364357)\mathrm{i}$
$\mp 1.323969764 + 6.933497616(4)\mathrm{i}$	$(\pm 0.3356774578 - 5.649886801)(-1364357)\mathrm{i}$
$\mp 0.4452337957 + 7.079114682(4)\mathrm{i}$	$(\mp 0.4654519522 + 2.526324603)(-1364357)\mathrm{i}$

Next, we address the question of the construction of the interpolatory quadrature rule (7). An immediate possibility is to construct the weights directly using the constructed Gaussian quadrature rule and formula (8). For the interpolatory rule with n points we need a Gaussian rule with $[n/2]+1$ nodes to have precise results. In order to apply this method we need to compute the derivative of the polynomial $P_n^{\alpha_1,\beta_1}$, where α_1 and β_1 satisfy the condition (10). This can be done quite easily since $(P_n^{\alpha_1,\beta_1})' P_{n-1}^{\alpha_1+1,\beta_1+1}$ and we can simply use the three-term recurrence relation for $P_k^{\alpha_1+1,\beta_1+1}$ to perform the computation and to obtain accurate results. This computation also requires the constructed Gaussian quadrature rules. In the case latter cannot be obtained, we need to find some more suitable choice. There are several other algorithms which can be employed.

There is a possibility of using recurrence relations given in Theorem 5. Although this could be the most preferred computation there is a quite enormous problem in the computation of starting values, i.e., the integrals representing quantities a_ν^0, $\nu = 1, \ldots, n$, given in Theorem 5.

According to the previous argumentation we prefer computation using the system of linear equations given in (13). In order to be able to start a computation, using this system of equations, we need first to compute the quantities ℓ_ν^k, where $k = 1, \ldots, n-1$ and $\nu = 1, \ldots, n$. This can be accomplished using linear systems (14) and (15).

As we can see the condition number of the system (14) is of order $\approx n^{3/2}$, which is a good condition number. Despite of this fact the computation can be ill-conditioned for the following reason. For the computation of ℓ_ν^0, we have

$$P_{n-1}^{(\alpha_1+1,\beta_1+1)}(x_\nu)\ell_\nu^0 = \sum_{k=0}^{n-1} a_{k+1} x_\nu^k. \tag{16}$$

If we look closely to this quantity it is easy to conclude that $P_{n-1}^{(\alpha_1+1,\beta_1+1)}(x_\nu)\ell_\nu^0 = (P_n^{(\alpha_1,\beta_1)}(x_\nu) - a_0)/x_\nu = -a_0/x_\nu$. Now suppose $x_\nu \approx 1$, we see that ℓ_ν^0, i.e., $-a_0/x_\nu \approx -a_0$, is a very small compared to the other coefficients a_k, $k \geq 1$. It turns that the computation using formula $\ell_\nu^0 = a_1/(nP_{n-1}^{\alpha,\beta_1}(x_\nu)) + x_\nu \ell_\nu^1$, must necessarily reduce the significant digits in ℓ_ν^0.

Table 2. Nodes and weights for the interpolatory quadrature rule (7) for $n = 10$ and $\alpha = 1 - 10^6\mathrm{i}$, $\beta = 10\mathrm{i}$, $\alpha_1 =$ and $\beta_1 = 0$

	nodes	weights
1	−0.9761647731351688	−8.456146645609039(−20) + 1.134836409397270(−18)i
2	−0.8765358562457037	5.290057427151006(−20) − 7.105879673763432(−19)i
3	−0.7057771007138595	−3.783390876366845(−20) + 5.082692901149730(−19)i
4	−0.4776806479830875	2.740479695075248(−20) − 3.681770216584216(−19)i
5	−0.2107203062284263	−1.927308941792804(−20) + 2.589339728803175(−19)i
6	0.7347753143132127(−1)	1.277752790667559(−20) − 1.716676559852106(−19)i
7	0.3518889233533302	−7.719281217662997(−21) + 1.037100289844088(−19)i
8	0.6019578420737977	4.025614545506861(−21) − 5.408507521849520(−20)i
9	0.8034219755802935	−1.621156695872623(−21) + 2.178066137468430(−20)i
10	0.9399419356770270	3.645692657081417(−22) − 3.645692657081417(−21)i

Table 3. Relative errors in computtaion of the integral (17)

rule	$n = 4$	$n = 8$	$n = 16$	$n = 32$
(5)	1.7(−45)	m.p.	m.p.	m.p.
(7)	1.7(−2)	5(−8)	2.7(−19)	2.4(−46)

Consider now the system (15). Using this system of linear equations formula for ℓ_ν^0 simply reads $-a_0/x_\nu$, i.e., the computation of ℓ_ν^0 is accurate. Nevertheless, using this system of linear equations the computation of the quantity ℓ_ν^{n-1}, can be ill-conditioned for the same reasons as a computation of ℓ_ν^0, was with the linear system (14). Note that using the linear system (14) we simply have that $P_{n-1}^{(\alpha_1+1,\beta_1+1)}(x_\nu)\ell_\nu^{n-1} = 1$.

We note that the system (15) is singular for $x_\nu = 0$, and that the condition number is very bad when x_ν close to zero. However, we do not need to solve this system for x_ν close to zero, since, in that case the first system will suffice cause it is ill-conditioned only for x_ν close to ± 1.

According to the previous argumentation, the best strategy is to solve both linear systems, and then calculate relative error of one with respect to the other solution. Choose the index of the solution, suppose it is j, where we have the biggest relative error and then take as the solution the following values: for $k = 1, \ldots, j$ take ℓ_ν^k, calculated using the linear system (15), and for $k = j + 1, n - 1$, take ℓ_ν^k, to be the values calculated using the linear system (14). If the minimum of relative errors is too big, then desirable value we have to use extended arithmetics.

This procedure is highly successful. For example, if computations are performed in double precision arithmetics (mach. eps. $\approx 2.2 \times 10^{-16}$), for $\alpha_1 = 3/2$ and $\beta_1 = -1/2$, one can obtain all ℓ_ν^k, $k = 0, 1, \ldots, 99$, $\nu = 1, \ldots, 99$, with at least 14 digits precision.

Having the quantities ℓ_ν^k, $k = 0, 1, \ldots, n-1$, $\nu = 1, \ldots, n$, we are ready to apply Theorem 6 to calculate quantities γ_ν^k, $k = 0, 1, \ldots, n-1$, $\nu = 1, \ldots, n$, and finally the weights w_ν, $\nu = 1, \ldots, n$. Table 2 holds values of the weights computed using this method for $n = 10$, $\alpha = 1 - 10^6\mathrm{i}$, $\beta = 10\mathrm{i}$, $\alpha_1 = 1$ and $\beta_1 = 0$.

Finally, Table 3 gives results of the computation of the integral in a multi-precision arithmetic (with 50 decimal digits)

$$\begin{aligned} I &= \int_{-1}^{1} (1-x)^{1-10^6\mathrm{i}}(1+x)^{\mathrm{i}} \sin x \, dx \qquad (17) \\ &\approx 4.5048641365260291681943894027349126989037900726366(-20) \\ &\quad -6.0428037920457445678627280329812687845810247448800(-19)\mathrm{i}, \end{aligned}$$

using the Gaussian rule (5) and the interpolatory rule (7). As we may inspect the Gaussian rule is far more efficient when it is possible to use it. Here, m.p. ($\approx 10^{-50}$) stands for machine precision.

Acknowledgments. The authors were supported in parts by the Serbian Ministry of Science and Technological Development (No. #144004).

References

1. Andrews, G.E., Askey, R., Roy, R.: Special Functions. In: Encyclopedia of mathematics and its applications, vol. 71. Cambridge University Press, Cambridge (1999)
2. Beckermann, B.: Complex Jacobi matrices. J. Comput. Appl. Math. 127, 17–65 (2001)
3. Chihara, T.S.: An Introduction to Orthogonal Polynomials. Gordon and Breach, New York (1978)
4. Cvetković, A.S., Milovanović, G.V.: The Mathematica Package Orthogonal-Polynomials. Facta Univ. Ser. Math. Inform. 19, 17–36 (2004)
5. Gautschi, W.: Algorithm 726: ORTHPOL – A package of routines for generating orthogonal polynomials and Gauss-type quadrature rules. ACM Trans. Math. Software 10, 21–62 (1994)
6. Gautschi, W.: Orthogonal Polynomials: Computation and Approximation. Clarendon Press, Oxford (2004)
7. Golub, G.H., Welsch, J.H.: Calculation of Gauss quadrature rule. Math. Comput. 23, 221–230 (1986)
8. Heins, M.: Complex Function Theory. Academic Press, London (1968)
9. Magnus, A.P.: Toeplitz matrix techniques and convergence of complex weight Pade approximation. J. Comput. Appl. Math. 19, 23–38 (1987)
10. Mancev, I.: Continuum distorted wave - Born initial state (CDW - BIS) model for single charge exchange. J. Comput. Meth. Sci. & Engineer. 5, 73–89 (2005)
11. Milovanović, G.V.: Numerical Analysis, Part I. Naučna Knjiga, Belgrade (Serbian) (1991)
12. Milovanović, G.V., Cvetković, A.S.: Complex Jacobi matrices and quadrature rules. Filomat. 17, 117–134 (2003)
13. Mukherjee, S.C., Roy, K., Sil, N.C.: Evaluation of the Coulomb integrals for scatering problems. Phys. Review 12(4) (1975)
14. Koekoek, R., Swarttouw, R.P.: The Askey-scheme of hypergeometric orthogonal polynomials and its q-analogue. Report 98–17, TU Delft (1998)
15. Sloan, I.H., Smith, W.E.: Properties of interpolatory product integration rules. SIAM J. Numer. Anal. 19, 427–442 (1982)

Surface Reconstruction via L^1-Minimization

Veselin Dobrev, Jean-Luc Guermond, and Bojan Popov

Department of Mathematics, Texas A&M University,
College Station, TX-77843, USA
{dobrev,guermond,popov}@math.tamu.edu

Abstract. A surface reconstruction technique based on the L^1-minimization of the variation of the gradient is introduced. This leads to a non-smooth convex programming problem. Well-posedness and convergence of the method is established and an interior point based algorithm is introduced. The L^1-surface reconstruction algorithm is illustrated on various test cases including natural and urban terrain data.

1 Introduction

In geometric modeling and image reconstruction, one often tries to *extract* a shape or recover a piece-wise smooth surface from a set of measurements. That is, one wants to find a surface that satisfies constraints or measurements and is visually good looking. The objectives could vary with the applications but the intuitive goal is to preserve the *shape* of the object. For example, one may want to reconstruct a convex body if the underlying data comes from a convex object, a flat surface if the data is locally flat, or preserve a particular structure of the level sets. Sometimes, this type of problems are solved by minimizing a L^p-norm of the curvature or the total variation of the gradient, see for example [1,2,3,4]. In this paper we take a different approach which we think is well suited for man made surfaces and Digital Elevation Maps (DEM). Namely, we minimize the total variation of the gradient of a function constructed on a finite element space satisfying interpolatory constraints. Similar minimization problems have been introduced by Lavery [5,1] and are hereafter referred to as the L^1-spline techniques. Minimizing the total variation of the gradient of a smooth function amounts to minimizing the L^1-norm of its second derivatives. The key observation from Lavery's work is that using the L^1-norm in the minimization process produces oscillation free surfaces.

In recent years, the idea of using the L^1-metric instead of the usual L^2 metric was exploited in many different areas with great success. For example, in compressed sensing [6,7] l^1-metric is used in the decoding step and in Partial Differential Equations the L^1-norm is used to measure the residual of the equation [8,9,10,11,12,13]. In all of the above applications, using L^1 is critical to obtain good numerical results and prove theoretical estimates.

One key ingredient in Lavery's work is the use of C^1-splines. The novelty of the approach in the present paper is to *relax* the C^1-smoothness on the finite element space which is used in the data reconstruction process. The discrete space is

S. Margenov, L.G. Vulkov, and J. Waśniewski (Eds.): NAA 2008, LNCS 5434, pp. 32–43, 2009.

composed of continuous finite elements with possibly discontinuous gradients. This is the natural discretization setting for functions that are in $W^{1,1}$ and whose gradient has bounded total variation.

The paper is organized as follows. In Section 2 we describe our scheme and in Section 3 we present different numerical tests for various types of data.

2 L^1-Minimization Problem

2.1 The Semi-discretized Functional

Let Ω be a bounded polygonal domain in $\mathbb{R}^2$ and let $\mathcal{T}_h$ be a partition of Ω composed of open triangles and quadrilaterals

$$\overline{\Omega} = \bigcup_{T \in \mathcal{T}_h} \overline{T}.$$

The mesh $\mathcal{T}_h$ is conforming in the sense that for any pair of distinct elements T, T' in $\mathcal{T}_h$, the intersection $T \cap T'$ is empty and $\overline{T} \cap \overline{T}'$ is either a common vertex or a common edge. For any element T in $\mathcal{T}_h$, we denote by h_T the diameter of T.

We introduce the discrete space X_h composed of continuous functions that are piecewise cubic on the mesh $\mathcal{T}_h$:

$$\begin{aligned} X_h = \{u \in \mathcal{C}(\overline{\Omega}) : u|_T \in \mathbb{P}_3 \text{ if } T \text{ is a triangle or,} \\ u|_T \in \mathbb{F}_T(\mathbb{Q}_3) \text{ if } T \text{ is a quadrilateral}, \forall T \in \mathcal{T}_h\} \end{aligned} \quad (1)$$

where

$$\mathbb{P}_p = \left\{ \sum_{i=0}^{p} \sum_{j=0}^{p-i} c_{ij} x^i y^j : c_{ij} \in \mathbb{R} \right\}, \qquad \mathbb{Q}_{pq} = \left\{ \sum_{i=0}^{p} \sum_{j=0}^{q} c_{ij} x^i y^j : c_{ij} \in \mathbb{R} \right\}$$

and the mapping $\mathbb{F}_T$ is defined by

$$(\mathbb{F}_T \hat{q})(x) = \hat{q}(F_T^{-1}(x)), \quad \forall x \in T, \hat{q} \in \mathcal{C}([0,1]^2),$$

where F_T is the transformation that maps the reference unit square $(0,1)^2$ to the quadrilateral T. We henceforth denote $\mathbb{Q}_p := \mathbb{Q}_{pp}$.

The set of all the interior edges of the partition $\mathcal{T}_h$ is denoted by $\mathcal{F}_h^i$. Let $F \in \mathcal{F}_h^i$ be one of the interior edges and let $T, T' \in \mathcal{T}_h$ be the two elements whose intersection is $F = \overline{T} \cap \overline{T}'$. Also, let $\mathbf{n}_{TF}$ denote the normal vector to F pointing from T to T'. We define the jump of the normal derivative of a function u to be

$$[\![u_{\mathbf{n}}]\!]|_F = (\nabla u|_T) \cdot \mathbf{n}_{TF} + (\nabla u|_{T'}) \cdot \mathbf{n}_{T'F}.$$

The set of all the vertices of the triangulation $\mathcal{T}_h$ is be denoted by $\mathcal{V}_h$.

We now assume that we are given a real-valued function (data) taking values over the vertices of the mesh, $d_h : \mathcal{V}_h \to \mathbb{R}$. We denote by Y_h the affine set of functions in X_h interpolating the data

$$Y_h = \{u \in X_h : u(x) = d_h(x),\ \forall x \in \mathcal{V}_h\}.$$

Our goal is now to find a function in Y_h that oscillates as little as possible. We think of such a function as one that best fits the data map d_h. For this purpose, we introduce the following functional

$$\widetilde{J}_h(u) = \sum_{T\in\mathcal{T}_h} \int_T (|u_{xx}| + 2|u_{xy}| + |u_{yy}|) + \alpha \sum_{F\in\mathcal{F}_h^i} \int_F |[\![u_{\mathbf{n}}]\!]|, \quad u \in X_h$$

representing the total variation of the gradient of u with a weight, $\alpha > 0$. Note that $\widetilde{J}_h$ defines a semi-norm which vanishes if and only if its argument is a linear function on Ω.

The data reconstruction problem is formulated as follows: Find $u_h \in Y_h$ such that

$$\widetilde{J}_h(u_h) = \min_{v_h\in Y_h} \widetilde{J}_h(v_h). \tag{2}$$

Whenever we have at hand a family of meshes $(\mathcal{T}_h)_{h>0}$ and a corresponding family of data functions $(d_h)_{h>0}$, we say that a sequence $(v_h)_{h>0}$, with $v_h \in Y_h$, is a sequence of almost minimizers if there is a constant C_a, uniform with respect to h, so that

$$\widetilde{J}_h(u_h) \le C_a \min_{v\in Y_h} \widetilde{J}_h(v).$$

The following result clarifies the approximation properties of (2):

Proposition 1. *Assume that the mesh family $(\mathcal{T}_h)_{h>0}$ is shape regular. Assume that there is $u \in W^{2,1}(\Omega)$ so that $d_h(x) := u(x)$, $\forall x \in \mathcal{V}_h$. Let $(u_h)_{h>0}$ be a sequence of almost minimizers, then the following error estimates hold:*

$$\sum_{T\in\mathcal{T}_h} h_T^{j-2} |u - u_h|_{j,1,T} \le C|u|_{2,1,\Omega}, \qquad j = 0, 1.$$

2.2 Quadratures

The computation of the functional $\widetilde{J}_h$ is not practical due to the integration of absolute values. Therefore we discretize $\widetilde{J}_h$ by replacing the integrals with quadrature rules $I = \{(p, \omega)\}$ which we view as sets of pairs (p, ω) of points $p \in \mathbb{R}^2$ and weights $\omega > 0$. The terms of the functional $\widetilde{J}_h$ are be approximated using quadrature rules $I(S, \mathcal{L})$:

$$\int_S |\mathcal{L}u| \approx \sum_{(p,\omega)\in I(S,\mathcal{L})} \omega |(\mathcal{L}u)(p)|$$

where either $S \in \mathcal{T}_h$ and $\mathcal{L}$ is one of the linear operators $\{\partial_{xx}, 2\partial_{xy}, \partial_{yy}\}$, or $S \in \mathcal{F}_h^i$ and $\mathcal{L} = \alpha\,[\![\partial_{\mathbf{n}}]\!]$. We require that the integration rules $I(S, \mathcal{L})$ satisfy the following two conditions:

1. Be exact when the sign of the integrant $\mathcal{L}u$ does not change;
2. Give an approximation that is equivalent to the exact integral, i.e., there are constants c_1, c_2 independent of S, $\mathcal{L}$, and h so that:

$$c_1 \int_S |\mathcal{L}u| \leq \sum_{(p,\omega)\in I(S,\mathcal{L})} \omega|(\mathcal{L}u)(p)| \leq c_2 \int_S |\mathcal{L}u|, \qquad \forall u \in X_h.$$

In general the second condition requires the use of integration rules with more points than required by the first one. For example, if T is a triangle and $\mathcal{L} = \partial_{xx}$ then u_{xx} is linear and the midpoint rule satisfies the first condition but not the second. The following proposition gives a natural construction of quadrature rules satisfying both the above conditions under an easily verifiable assumptions:

Proposition 2. *Let $\widehat{S}$ be a (closed) reference element (e.g. triangle, square, segment), and T be an invertible* affine *transformation mapping $\widehat{S}$ to S. Also, let $\widehat{\mathcal{P}}$ be a finite-dimensional subspace of $\mathcal{C}(\widehat{S})$ (e.g. polynomials) and $\mathcal{P} = \mathbb{T}\widehat{\mathcal{P}}$ be its image under the transformation $\mathbb{T} : \mathcal{C}(\widehat{S}) \longrightarrow \mathcal{C}(S)$ defined by*

$$u(x) := \mathbb{T}(\hat{u})(x) = \hat{u}(T^{-1}(x)), \qquad \forall x \in S.$$

Let $\hat{I} = \{(\hat{p}_i, \hat{\omega}_i)\}_{i=1}^n$ be an integration rule with positive weights on $\widehat{S}$. If $\hat{I}$ is exact for every function in $\widehat{\mathcal{P}}$ and the quadrature points are such that

$$\left[\hat{u} \in \widehat{\mathcal{P}} \quad \text{and} \quad \hat{u}(\hat{p}_i) = 0, \quad i = 1, \ldots, n\right] \qquad \text{implies} \qquad \left[\hat{u}(\hat{x}) = 0, \quad \forall \hat{x} \in \widehat{S}\right],$$

then the integration rule $I = \{(p_i, \omega_i)\}_{i=1}^n$ with $p_i = T(\hat{p}_i)$ and $\omega_i = \frac{|S|}{|\widehat{S}|}\hat{\omega}_i$ (where $|\cdot|$ denotes the measure of the corresponding set) is exact for every function in $\mathcal{P}$ and

$$c_1 \int_S |u| \leq \sum_{i=1}^n \omega_i |u(p_i)| \leq c_2 \int_S |u|, \qquad \forall u \in \mathcal{P}$$

with constants $c_2 > c_1 > 0$ that depend on $\widehat{S}$ and $\widehat{P}$ but do not depend on the transformation T.

Based on the above proposition we use the following quadrature rules:

- When $S \in \mathcal{T}_h$ is a triangle and $\mathcal{L} \in \{\partial_{xx}, 2\partial_{xy}, \partial_{yy}\}$ then $\mathcal{L}(X_h|_S) = \mathbb{P}_1 = \widehat{\mathcal{P}} = \mathcal{P}$ and therefore the 3-point quadrature rule using the midpoints of the sides of the triangle satisfies the conditions of the proposition (this rule is exact for $\mathbb{P}_2$).
- When $S \in \mathcal{T}_h$ is a rectangle with sides parallel to the coordinate axes we use three different quadrature rules for the three different second derivatives. For $\mathcal{L} = \partial_{xx}$ we have $\mathcal{L}(X_h|_S) = \mathbb{Q}_{1,3} = \widehat{\mathcal{P}} = \mathcal{P}$ and therefore we could use the 2×4 tensor product Gaussian rule; however, numerical experiments show some undesired oscillations which can be avoided by using the 3×4 tensor product Gaussian rule. For $\mathcal{L} = 2\partial_{xy}$ we have $\mathcal{L}(X_h|_S) = \mathbb{Q}_{2,2} = \widehat{\mathcal{P}} = \mathcal{P}$ and we use the 3×3 tensor product Gaussian rule. For $\mathcal{L} = \partial_{yy}$, $\mathcal{L}(X_h|_S) = \mathbb{Q}_{3,1} = \widehat{\mathcal{P}} = \mathcal{P}$ and we use the 4×3 tensor product Gaussian rule.

- When $S \in \mathcal{T}_h$ is not a rectangle with sides parallel to the coordinate axes we have $\widehat{\mathcal{P}} \neq \mathcal{P}$ and it is more convenient to replace the second derivatives in $\widetilde{J}_h$ by second derivatives in directions parallel to the sides of S. This case is not considered in the numerical experiments reported in this paper.
- When $S \in \mathcal{F}_h^i$ and $\mathcal{L} = \alpha \llbracket \partial_{\mathbf{n}} \rrbracket$ we have two cases: 1) S is the edge of two triangles and 2) S is a side in a quadrilateral. $\mathcal{P}$ and $\widehat{\mathcal{P}}$ are composed of one-dimensional quadratic polynomials in the first case and cubic polynomials in the second case. Therefore, we use the 3 point Gaussian rule in the first case and the 4 point Gaussian rule in the second.

Using the above quadrature rules we obtain the approximate functional

$$J_h(u) = \sum_{\substack{T\in\mathcal{T}_h \\ \mathcal{L}\in\{\partial_{xx},2\partial_{xy},\partial_{yy}\}}} \sum_{(p,\omega)\in I(T,\mathcal{L})} \omega|(\mathcal{L}u)(p)| + \alpha \sum_{F\in\mathcal{F}_h^i} \sum_{(p,\omega)\in I(F,\llbracket\partial_{\mathbf{n}}\rrbracket)} \omega|\llbracket u_{\mathbf{n}}\rrbracket(p)|.$$

Note that J_h defines a semi-norm on X_h which is equivalent to that induced by $\widetilde{J}_h$ with constants independent of h.

The fully discretized version of problem (2) is the following: Find $u_h \in Y_h$ such that

$$J_h(u_h) = \min_{v_h\in Y_h} J_h(v_h). \tag{3}$$

2.3 Matrix Formulation

Let $\{\phi_i\}_{i=1}^{\hat{n}}$ be a basis for X_h. The functional J_h can be re-written as follows

$$J_h(u) = |\widehat{A}x|_1 \qquad \text{where} \qquad x \in \mathbb{R}^{\hat{n}} : u = \sum_{i=1}^{\hat{n}} x_i\phi_i$$

and the entries of the matrix $\widehat{A}$ are given by

$$\widehat{A}_{ij} = \omega_i(\mathcal{L}_i\phi_j)(p_i) \qquad i = 1,\ldots,m \quad j = 1,\ldots,\hat{n}.$$

Here $\{(p_i,\omega_i)\}_{i=1}^m$ is an enumeration of all the quadrature points (and weights) in all the quadrature rules used in the discretization of $\widetilde{J}_h$ and $\mathcal{L}_i$ is the linear operator corresponding to the quadrature rule. Thus, the total number of quadrature points is given by

$$m = \sum_{\substack{T\in\mathcal{T}_h \\ \mathcal{L}\in\{\partial_{xx},2\partial_{xy},\partial_{yy}\}}} \#\left(I(T,\mathcal{L})\right) + \sum_{F\in\mathcal{F}_h^i} \#\left(I(F,\llbracket\partial_{\mathbf{n}}\rrbracket)\right)$$

where $\#(I)$ denotes the cardinal number of I.

Let us further assume that $\{\phi_i\}$ is the standard nodal basis for X_h and the basis functions corresponding to the vertices in $\mathcal{V}_h$ are the first $\hat{n}_1$ functions $\phi_1,\ldots,\phi_{\hat{n}_1}$. The matrix $\widehat{A}$ can be written in 1×2 block form $\widehat{A} = (\widehat{A}_1 A)$ where $\widehat{A}_1$ is $m\times\hat{n}_1$ and A is $m\times n$, $(n = \hat{n}-\hat{n}_1)$. Let $d \in \mathbb{R}^{\hat{n}_1}$ be the vector representing the data d_h at the vertices and set $b = -\widehat{A}_1 d$. Then the discrete problem (3) can be re-written as follows: Find $x \in \mathbb{R}^n$ such that

$$|Ax - b|_1 = \min_{y \in \mathbb{R}^n} |Ay - b|_1. \tag{4}$$

It can be shown that A is full rank.

2.4 Discrete Problem

In this section we study properties of ℓ_1-minimization problems of generic form (4). Let A be an $m \times n$ real matrix ($m > n$) and $b \in \mathbb{R}^m$. We define the Lagrangian

$$L(x, \lambda) = (b - Ax)^t \lambda, \qquad x \in \mathbb{R}^n, \lambda \in \mathbb{R}^m$$

and the primal and dual functions, f and g, respectively

$$f(x) = \max_{\substack{\lambda \in \mathbb{R}^m \\ |\lambda|_\infty \leq 1}} L(x, \lambda) = |b - Ax|_1$$

$$g(\lambda) = \min_{x \in \mathbb{R}^n} L(x, \lambda) = \begin{cases} b^t \lambda & A^t \lambda = 0 \\ -\infty & A^t \lambda \neq 0. \end{cases}$$

It is clear that for all $x \in \mathbb{R}^n$ and all $\lambda \in \mathbb{R}^m$, $|\lambda|_\infty \leq 1$ we have

$$f(x) \geq L(x, \lambda) \geq g(\lambda).$$

The primal problem is defined to be

$$\text{minimize} \quad f(x) = |b - Ax|_1 \tag{5}$$

and the dual problem is defined to be

$$\begin{aligned} &\text{maximize} \quad g(\lambda) = b^t \lambda \\ &\text{subject to} \quad A^t \lambda = 0, |\lambda|_\infty \leq 1. \end{aligned} \tag{6}$$

Proposition 3. *(Strong duality) For any pair of solutions x^* and λ^* to (5) and (6), respectively, we have $f(x^*) = g(\lambda^*)$.*

Corollary 1. *If x^* is a solution of (5) and $(b - Ax^*)_i \neq 0$ for some index i then every solution λ^* of (6) satisfies $\lambda_i^* = \operatorname{sign}(b - Ax^*)_i$. In particular, if λ^* is a solution of (6) and $|\lambda_i^*| < 1$ then for every solution x^* of (5) we have $(b - Ax^*)_i = 0$.*

We now assume that A and b have the following block structure

$$A = \begin{pmatrix} A_1 \\ \alpha A_2 \end{pmatrix} \qquad b = \begin{pmatrix} b_1 \\ \alpha b_2 \end{pmatrix}$$

which is exactly the structure they have in problem (4) where A_2 and b_2 correspond to the rows generated by the terms $\int_F |[\![u_{\mathbf{n}}]\!]|$, $F \in \mathcal{F}_h^i$. The primal function has the form

$$f(x) = |b - Ax|_1 = |b_1 - A_1 x|_1 + \alpha |b_2 - A_2 x|_1.$$

Proposition 4. *Assume the rows of A_2 are linearly independent. Then there exists a number $\bar{\alpha}$ such that when $\alpha > \bar{\alpha}$ every solution x^* of (5) satisfies*

$$b_2 - A_2 x^* = 0.$$

Proof. We will show that when α is large enough the feasible set of the dual problem (6) (and therefore any solution) satisfies $|\lambda_2|_\infty < 1$ which, in view of Corollary 1, implies the proposition. Indeed, if λ is dual feasible we have

$$0 = A^t \lambda = A_1^t \lambda_1 + \alpha A_2^t \lambda_2.$$

The assumption on A_2 implies the existence of right inverse R of A_2:

$$A_2 R = I \qquad \text{or} \qquad R^t A_2^t = I$$

and thus we have $\lambda_2 = -\frac{1}{\alpha} R^t A_1^t \lambda_1$. Now, if we define $\bar{\alpha} = |R^t A_1^t|_\infty$ and take $\alpha > \bar{\alpha}$ we get

$$|\lambda_2|_\infty = \frac{1}{\alpha} |R^t A_1^t \lambda_1|_\infty \leq \frac{1}{\alpha} |R^t A_1^t|_\infty |\lambda_1|_\infty < 1.$$

Proposition 5. *There exists a number $\bar{\alpha}$ such that when $\alpha > \bar{\alpha}$ and $b_2 \in \mathrm{Im} A_2$ every solution x^* of (5) satisfies $b_2 - A_2 x^* = 0$.*

Proof. Let $\widetilde{A}_2$ denote the matrix whose rows are a maximal linearly independent set of rows of A_2. Without loss of generality we can write

$$A_2 = \begin{pmatrix} \widetilde{A}_2 \\ A_3 \end{pmatrix} \qquad b_2 = \begin{pmatrix} \tilde{b}_2 \\ b_3 \end{pmatrix}.$$

We have the following property: if $b_2 \in \mathrm{Im} A_2$ and $\widetilde{A}_2 x = \tilde{b}_2$ then $A_3 x = b_3$. Let us now define

$$\widetilde{A} = \begin{pmatrix} A_1 \\ \alpha \widetilde{A}_2 \end{pmatrix} \qquad \tilde{b} = \begin{pmatrix} b_1 \\ \alpha \tilde{b}_2 \end{pmatrix}$$

and consider the reduced minimization problem

$$\text{minimize } \tilde{f}(x) = |\tilde{b} - \widetilde{A} x|_1 = |b_1 - A_1 x|_1 + \alpha |\tilde{b}_2 - \widetilde{A}_2 x|_1 \tag{7}$$

obtained from (5) by replacing A and b with $\widetilde{A}$ and $\tilde{b}$, respectively. Since the rows of $\widetilde{A}_2$ are linearly independent we can apply the previous proposition to this problem and conclude that for $\alpha > \bar{\alpha}$ every solution $\tilde{x}$ of (7) satisfies $\widetilde{A}_2 \tilde{x} = \tilde{b}_2$. We now assume that $\alpha > \bar{\alpha}$ and $b_2 \in \mathrm{Im} A_2$, and we want to show that problems (5) and (7) are equivalent. First we note that $\forall x \in \mathbb{R}^n$

$$\begin{aligned} \tilde{f}(x) &= |b_1 - A_1 x|_1 + \alpha |\tilde{b}_2 - \widetilde{A}_2 x|_1 \\ &\leq |b_1 - A_1 x|_1 + \alpha |\tilde{b}_2 - \widetilde{A}_2 x|_1 + \alpha |b_3 - A_3 x|_1 = f(x) \end{aligned}$$

and therefore for any two solutions x^* and $\tilde{x}$ of (5) and (7), respectively, we have $\tilde{f}(\tilde{x}) \leq f(x^*)$. Since $\widetilde{A}_2\tilde{x} = \tilde{b}_2$ and we assumed that $b_2 \in \mathrm{Im}A_2$ we conclude that $A_3\tilde{x} = b_3$ and therefore

$$f(\tilde{x}) = \tilde{f}(\tilde{x}) \leq f(x^*) \leq f(\tilde{x})$$

which shows that $\tilde{f}(\tilde{x}) = f(x^*)$ and $\tilde{x}$ is a solution to (5). For $\tilde{f}(x^*)$ we have

$$\tilde{f}(x^*) \leq f(x^*) = \tilde{f}(\tilde{x}) \leq \tilde{f}(x^*)$$

which shows that $\tilde{f}(x^*) = \tilde{f}(\tilde{x})$ and therefore x^* is a solution to (7). Since we already saw that $\widetilde{A}_2\tilde{x} = \tilde{b}_2$ and $A_3\tilde{x} = b_3$, that is $A_2\tilde{x} = b_2$, and since $\tilde{x}$ was an arbitrary solution to (7) (or as we just proved, to (5)) this completes the proof.

Corollary 2. *Assume that all elements of the mesh $\mathcal{T}_h$ are quadrilaterals. Then there exists $\bar{\alpha}$ such that when $\alpha > \bar{\alpha}$ every solution u_h to (3) is in $\mathcal{C}^1(\overline{\Omega})$.*

Proof. Note that $u_h \in \mathcal{C}^1(\overline{\Omega})$ is equivalent to $A_2x = b_2$ where (d, x) is the coefficient vector of u_h. Thus, $b_2 \in \mathrm{Im}A_2$ is equivalent to the existence of $v_h \in Y_h \cap \mathcal{C}^1(\overline{\Omega})$. In the case of quadrilateral elements such v_h can be constructed using Bogner-Fox-Schmit type interpolation in which one can prescribe not only the values of the function at the vertices but also its gradient and mixed second derivative. For triangular meshes, all solutions u_h are $\mathcal{C}^1$ if and only if the data and the mesh allow it, that is $Y_h \cap \mathcal{C}^1(\overline{\Omega}) \neq \emptyset$.

Remark 1. *In the above proof the value of $\bar{\alpha}$ is not a priori uniform with respect to the typical mesh-size h. However, numerical tests indicate that using $\alpha = 5$ guarantees $\mathcal{C}^1$-smoothness independently of h.*

2.5 Primal-Dual Interior-Point Method

We now describe an approach for solving the minimization problem (4). First, we reformulate (4) as a linear programming problem: Find $y \in \mathbb{R}^m$ and $x \in \mathbb{R}^n$ so that

$$\text{minimize } y^t 1 = \sum_{i=1}^{m} y_i, \qquad \text{subject to } \begin{cases} y \geq b - Ax \\ y \geq Ax - b. \end{cases}$$

Then we apply the primal-dual interior-point method described in [14,15].

After some simplifications, which we omit here, the above problem is solved using the following algorithm:

input: A, b, x, λ; μ, ϵ
$r = b - Ax$;
$a = (|r|_1 - r^t\lambda)/m$; $\quad y_i = |r_i| + a$, $i = 1, \ldots, m$;
while $(|r|_1/(r^t\lambda) - 1 \geq \epsilon)$
  $t^{-1} = (y^t 1 - r^t\lambda)/(2m\mu)$;
  $s_1 = y + r$; $\quad s_2 = y - r$;
  $d_1 = (1 - \lambda)/(2s_1)$; $\quad d_2 = (1 + \lambda)/(2s_2)$;

$d = 4d_1d_2/(d_1 + d_2)$;
$v = t^{-1}(s_2^{-1} - s_1^{-1}) + (d_2 - d_1)/(d_1 + d_2)[1 - t^{-1}(s_1^{-1} + s_2^{-1})]$;
$w = A^t v$;
$\Delta x = (A^t \mathrm{diag}(d) A)^{-1} w$;
$v = A\Delta x$;
$\Delta y = [-1 + t^{-1}(s_1^{-1} + s_2^{-1}) + (d_1 - d_2)v]/(d_1 + d_2)$;
$\Delta\lambda = -\lambda + t^{-1}(s_2^{-1} - s_1^{-1}) - (d_1 + d_2)v + (d_1 - d_2)\Delta y$;
$s = \max\{\sigma \in (0, 2] : \lambda + \sigma\Delta\lambda \geq -1, \lambda + \sigma\Delta\lambda \leq 1,$
$y + \sigma\Delta y \geq r - \sigma v, y + \sigma\Delta y \geq -r + \sigma v\}$;
$s = \min\{1,\ 0.99s\}$;
$x = x + s\Delta x; \quad y = y + s\Delta y; \quad r = r - sv; \quad \lambda = \lambda + s\Delta\lambda$;
end while
output: x, λ;

The input parameter μ is a positive real number (we use $\mu = 10$) and ϵ is a given tolerance. The initial input value of the dual variable λ is assumed to be strictly dual feasible, that is $A^t\lambda = 0$ and $|\lambda|_\infty < 1$ (we use $\lambda = 0$). In the algorithm, a, t, and s are scalar variables; $r, y, d, v, \Delta y, \Delta\lambda \in \mathbb{R}^m$; $w, \Delta x \in \mathbb{R}^n$; the vectors s_1, s_2, d_1, d_2 do not need to be stored since their components can be evaluated one by one when needed (one time when computing d and v, and another time when computing Δy and $\Delta\lambda$). All operations in the definitions of $d_1, d_2, d, v, \Delta y$, and $\Delta\lambda$ are component-wise. We use $\mathrm{diag}(d)$ to denote the diagonal matrix with main diagonal given by the vector d.

It can be shown that all vectors λ generated by this algorithm are strictly dual feasible provided that the input λ is strictly dual feasible. Thus, the stopping criterion we use guarantees that

$$f(x) - f(x^*) \leq f(x) - g(\lambda) = |r|_1 - r^t\lambda < \epsilon\, r^t\lambda \leq \epsilon f(x^*)$$

which means that x is an almost minimizer for (4) with a tolerance $1 + \epsilon$.

The most expensive step at each iteration of the **while** loop is the solution of the equation for Δx. Since direct solution methods are not practical for large n, we use an iterative method to solve the linear system approximately. In the resulting algorithm the vectors λ do not satisfy $A^t\lambda = 0$. However, numerically we observe that solving iteratively with relative tolerance $\epsilon/10$ produces results that are very similar the results obtained by solving almost exactly. The iterative method we use is the preconditioned conjugate gradient (PCG) method with a simple symmetric Gauss-Seidel preconditioner.

3 Numerical Examples

We illustrate our data reconstruction technique in this section. In all numerical experiments Ω is the unit square and we use a uniform rectangular mesh with equal step size in both x and y directions. The tolerance in the interior-point (IP) method is $\epsilon = 10^{-2}$ and the linear systems for Δx are solved with relative tolerance 10^{-3}. The initial approximation for the vector x in the IP method is obtained from the $\mathbb{Q}_1$ interpolant of the data.

3.1 Test 1: Piece-Wise Smooth Data

The data for this set of experiments is obtained from a single function

$$u(x,y) = f(\max\{|x-1/2|, |y-1/2|\}),$$

where

$$f(r) = \begin{cases} 5/3 & r \in [0, 1/8] \\ 1 & r \in (1/8, 5/16] \\ 16(1/2-r)/3 & r \in (5/16, 1/2] \end{cases}$$

and we use meshes with step size, h, varying from 1/16 to 1/256. Note that $u(x,y)$ is discontinuous at $\Gamma_1 = \{r = 1/8\}$ and its gradient also has jumps at $\Gamma_2 = \{r = 5/16\}$ and at $\Gamma_3 = (\{x = y\} \cup \{x + y = 1\}) \cap \{5/16 \leq r \leq 1/2\}$. Away from those discontinuities the function is linear. Figure 1 shows the two reconstructed surfaces obtained with $\alpha = 3$ and $\alpha = 5$ on 16×16 mesh. The solution obtained with $\alpha = 5$ is $\mathcal{C}^1$ everywhere and that obtained with $\alpha = 3$ is $\mathcal{C}^1$ almost everywhere but around the edges defined by Γ_3. In Table 1 (left side), we present results for the convergence of the IP method as we refine the mesh. We see a very small increase in the number of IP iterations of order $\ln(1/h)$. The total number of PCG iterations is given along with the increase in those numbers from one level to the next and we multiply that ratio by 4 which roughly gives the increase in the computational cost per level. If we compare these ratios with the actual increase in computing time, we see that both are fairly close. These numbers indicate an order $\mathcal{O}(n^\beta)$, $\beta = \ln(6)/\ln(4) \approx 1.29$, for the computational complexity and time.

Fig. 1. Test 1: reconstructed surfaces with $\alpha = 3$ and $\alpha = 5$, $h = 1/16$

3.2 Test Cases: Real Terrain Data

Next, we present results for two data sets taken from real terrain data. Test 2 is defined on a 20×20 mesh and it is one of the reference tests in [16]; Test 3

Table 1. Results for test 1 (left), 2, and 3 (right) with $\alpha = 3$

$1/h$	16	32	64	128	256	20×20	100×100
IP iter.	15	16	17	18	19	21	28
$\hat{n}$	2 401	9 409	37 249	148 225	591 361	3 721	90 601
m	10 368	41 728	167 424	670 720	2 684 928	16 240	409 200
PCG iter.	1 512	1 599	2 437	3 628	4 751	4 656	3 923
Ratio×4	—	4.23	6.10	5.95	5.24	—	—
Time, sec.	4.62	21.22	126.68	754.37	3 908.27	21.78	498.55
Ratio	—	4.59	5.97	5.95	5.18	—	—

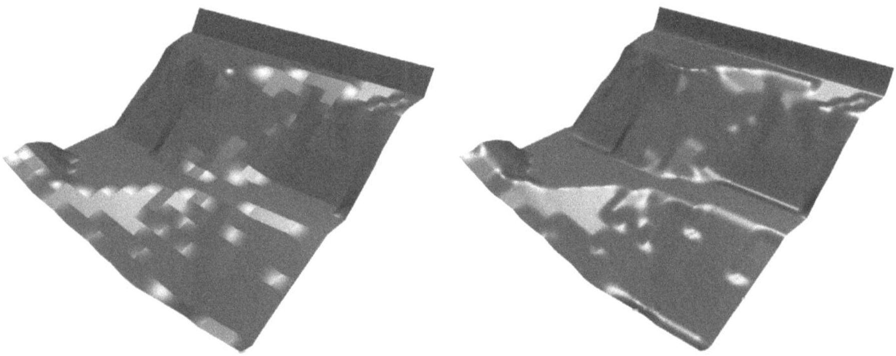

Fig. 2. Test 2: $\mathbb{Q}_1$ interpolant and reconstructed surface with $\alpha = 3$

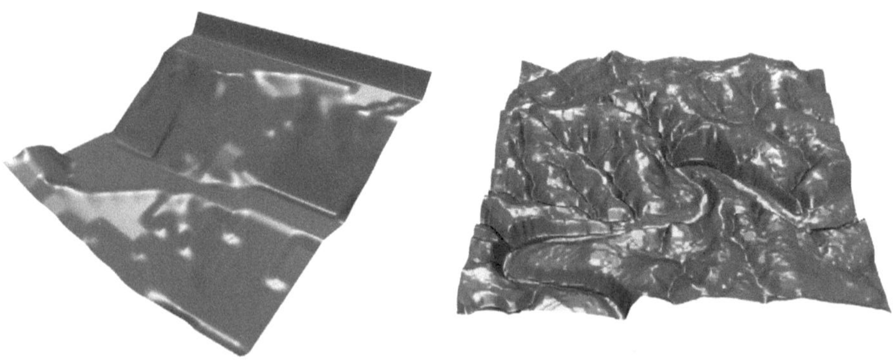

Fig. 3. Test 2, reconstructed with $\alpha = 5$, and Test 3, reconstructed with $\alpha = 3$

is defined on a 100×100 mesh and represents a 3×3 km terrain near Barton Creek in Austin, Texas. The results for test 2 are shown in Figures 2 and 3, and Figure 3 also shows the reconstructed terrain for test 3. In Table 1 (right), we present the computational results for the IP method applied to tests 2 and 3.

4 Conclusion

As claimed by Lavery [5,1], we have observed that the L^1-metric is suitable for reconstructing piecewise smooth data in the sense that it is non-oscillatory. We have proposed a finite element technique which is more flexible than cubic splines. We have proposed a preconditioned interior-point technique whose complexity scales like $n^{5/4}$.

Acknowledgement. This material is based upon work supported by the National Science Foundation grant DMS-0510650.

References

1. Lavery, J.E.: Shape-preserving interpolation of irregular data by bivariate curvature-based cubic L_1 splines in spherical coordinates. Comput. Aided Geom. Design 22(9), 818–837 (2005)
2. Darbon, J., Sigelle, M.: Image restoration with discrete constrained total variation. I. Fast and exact optimization. J. Math. Imaging Vision 26(3), 261–276 (2006)
3. Chambolle, A., Lions, P.-L.: Image recovery via total variation minimization and related problems. Numer. Math. 76(2), 167–188 (1997)
4. Rudin, L., Osher, S., Fatemi, E.: Nonlinear total variation based noise removal algorithms. Physica D. 60, 259–268 (1992)
5. Lavery, J.E.: Univariate cubic L_p splines and shape-preserving, multiscale interpolation by univariate cubic L_1 splines. Comput. Aided Geom. Design 17(4), 319–336 (2000)
6. Candes, E.J., Tao, T.: Decoding by linear programming. IEEE Trans. Inform. Theory 51(12), 4203–4215 (2005)
7. Candès, E.J., Romberg, J.K., Tao, T.: Stable signal recovery from incomplete and inaccurate measurements. Comm. Pure Appl. Math. 59(8), 1207–1223 (2006)
8. Lavery, J.E.: Solution of steady-state one-dimensional conservation laws by mathematical programming. SIAM J. Numer. Anal. 26(5), 1081–1089 (1989)
9. Lavery, J.E.: Solution of steady-state, two-dimensional conservation laws by mathematical programming. SIAM J. Numer. Anal. 28(1), 141–155 (1991)
10. Guermond, J.L.: A finite element technique for solving first-order PDEs in L^P. SIAM J. Numer. Anal. 42(2), 714–737 (2004) (electronic)
11. Guermond, J.L., Popov, B.: Linear advection with ill-posed boundary conditions via L^1 minimization. Int. J. Numer. Anal. Model. 4(1), 39–47 (2007)
12. Guermond, J.L., Popov, B.: L^1-minimization methods for Hamilton-Jacobi equations: the one-dimensional case. Numer. Math. 109(2), 269–284 (2008)
13. Guermond, J.L., Popov, B.: l^1-minimization methods for Hamilton-Jacobi equations. SIAM J. Numer. Anal. (accepted)
14. Boyd, S., Vandenberghe, L.: Convex optimization. Cambridge University Press, Cambridge (2004)
15. Wright, S.J.: Primal-dual interior-point methods. Society for Industrial and Applied Mathematics (SIAM), Philadelphia (1997)
16. Wang, Y., Fang, S.C., Lavery, J.E.: A compressed primal-dual method for generating bivariate cubic L_1 splines. J. Comput. Appl. Math. 201(1), 69–87 (2007)

Qualitative Analysis of the Crank-Nicolson Method for the Heat Conduction Equation

István Faragó

Eötvös Loránd University,
Pázmány P. s. 1/c, 1117 Budapest, Hungary

Abstract. The preservation of the basic qualitative properties – besides the convergence – is a basic requirement in the numerical solution process. For solving the heat conduction equation, the finite difference/linear finite element Crank-Nicolson type full discretization process is a widely used approach. In this paper we formulate the discrete qualitative properties and we also analyze the condition w.r.t. the discretization step sizes under which the different qualitative properties are preserved. We give exact conditions for the discretization of the one-dimensional heat conduction problem under which the basic qualitative properties are preserved.

1 Introduction

Complex (coupled) problems are usually led to a sequence of sub-problems by using the operator splitting technique. As a rule, each of these sub-problems means an easier task (cf. [11,7]). For example, in air pollution models, which are indispensable for the protection of the environment, a complex model can be split into different sub-problems, one of which is the diffusion problem, also called the heat equation. Clearly, the un-coupled model has several qualitative properties which should be preserved by the split problem, too. (E.g., in an air pollution model the solution methods should not result in negative concentration values as such values would have no physical meaning.) When each sub-problem possesses some given qualitative property, then the combined split method also has this property.

The heat conduction operator in $\mathbb{R}^d$, which we will discuss, is defined as

$$L \equiv \frac{\partial}{\partial t} - \Delta \equiv \frac{\partial}{\partial t} - \sum_{m=1}^{d} \frac{\partial^2}{\partial x_m^2}, \tag{1}$$

where Δ denotes the d-dimensional Laplace operator. For this operator the basic qualitative properties are valid. Namely, this operator is non-negativity preserving, satisfies the maximum/minimum principle and it is contractive in the maximum norm. (For the details, we refer to the recent papers [2,3,4].) We emphasize that these qualitative properties are physically based, and their violation contradicts to the real-life modelling.

S. Margenov, L.G. Vulkov, and J. Waśniewski (Eds.): NAA 2008, LNCS 5434, pp. 44–55, 2009.

Therefore, the preservation of the continuous qualitative properties in the corresponding discretized models is very important, and it is a quite natural requirement for really reliable and meaningful numerical models of various real-life phenomena, like the heat conduction process.

In this work we analyze the qualitative properties of the discrete model corresponding to the discrete heat conduction operator, where the discretization is done by finite difference and linear finite element methods w.r.t. the space variable and by the middle trapezoidal rule w.r.t. the time variable. This type of discretization is usually referred to as Crank-Nicolson method.

The paper is organized as follows.

In Section 2 we formulate the basic discrete qualitative properties and we give the connection between them. In the Section 3 we introduce the notion of the two-level discrete mesh operator and formulate the condition for its qualitative properties. In Section 4 we introduce the discrete mesh operator discretized by the Crank-Nicolson finite difference and linear finite element methods, and we give the conditions under which the discrete qualitative properties are equivalent. We show that the discrete non-negativity preservation property guarantees all the other required properties. Therefore, in Section 5 we give the exact condition of the non-negativity preservation in the one-dimensional case. In the last section we characterize the behavior of the Crank-Nicolson finite difference discrete mesh operator in the maximum norm, when it is not qualitatively adequate.

2 Discrete Mesh Operators and Their Qualitative Properties in the Discretized Time-Dependent Problems

In this part, first we formulate the discrete equivalent definitions of the qualitative properties. For more details and the proofs we refer to [4].

We will use the following notations.

The sets $\mathcal{P} = \{\mathbf{x}_1, \mathbf{x}_2, \ldots, \mathbf{x}_N\}$ and $\mathcal{P}_\partial = \{\mathbf{x}_{N+1}, \mathbf{x}_{N+2}, \ldots, \mathbf{x}_{N+N_\partial}\}$ consist of different vertices in Ω and on $\partial\Omega$, respectively. We set $\bar{N} = N + N_\partial$ and $\bar{\mathcal{P}} = \mathcal{P} \cup \mathcal{P}_\partial$. Let T as before, and we define $\Delta t < T$ an arbitrary positive number. Moreover, let us suppose that the natural number M satisfies the condition $M\Delta t \leq T < (M+1)\Delta t$, and introduce the set $\mathcal{R} = \{t_n = n\Delta t \,|\, n = 0, 1, \ldots, M\}$. For any values τ from the set $\mathcal{R}$ we introduce the notations

$$\mathcal{R}_\tau = \{t \in \mathcal{R} \,|\, 0 < t < \tau\}, \;\; \mathcal{R}_{\bar{\tau}} = \{t \in \mathcal{R} \,|\, 0 < t \leq \tau\}, \;\; \mathcal{R}_{\bar{\tau}}^0 = \{t \in \mathcal{R} \,|\, 0 \leq t \leq \tau\},$$

and the sets

$$\mathcal{Q}_\tau = \mathcal{P} \times \mathcal{R}_\tau, \qquad \bar{\mathcal{Q}}_\tau = \bar{\mathcal{P}} \times \mathcal{R}_{\bar{\tau}}^0, \qquad \mathcal{Q}_{\bar{\tau}} = \mathcal{P} \times \mathcal{R}_{\bar{\tau}}, \qquad \mathcal{G}_\tau = (\mathcal{P}_\partial \times \mathcal{R}_{\bar{\tau}}^0) \cup (\mathcal{P} \times \{0\}).$$

Definition 1. *Linear mappings that map from the space of real-valued functions defined on $\bar{\mathcal{Q}}_{t_M}$ to the space of real-valued functions defined on $\mathcal{Q}_{t_M}$ are called discrete (linear) mesh operators.*

The discrete maximum-minimum principles can be formulated as follows.

Definition 2. *We say that the discrete mesh operator $\mathcal{L}$ satisfies the discrete weak boundary maximum-minimum principle (DWBMP) if for any function $\nu \in$* $\operatorname{dom}\mathcal{L}$ *and $t^\star \in \mathcal{R}_{t_M}$ such that $\mathcal{L}\nu|_{\mathcal{Q}_{\bar{t}^\star}} \geq 0$ the inequality*

$$\min\{0, \min_{\mathcal{G}_{t^\star}} \nu\} \leq \min_{\bar{\mathcal{Q}}_{t^\star}} \nu \tag{2}$$

is valid, while for mesh-functions $\nu \in \operatorname{dom}\mathcal{L}$ with the property $\mathcal{L}\nu|_{\mathcal{Q}_{\bar{t}^\star}} \leq 0$ the relation

$$\max_{\bar{\mathcal{Q}}_{t^\star}} \nu \leq \max\{0, \max_{\mathcal{G}_{t^\star}} \nu\} \tag{3}$$

holds.

Definition 3. *We say that the discrete mesh operator $\mathcal{L}$ satisfies the discrete strong boundary maximum-minimum principle (DSBMP) if for any function $\nu \in$* $\operatorname{dom}\mathcal{L}$ *and $t^\star \in \mathcal{R}_{t_M}$ such that $\mathcal{L}\nu|_{\mathcal{Q}_{\bar{t}^\star}} = 0$ the relations*

$$\min_{\mathcal{G}_{t^\star}} \nu = \min_{\bar{\mathcal{Q}}_{t^\star}} \nu, \quad \max_{\bar{\mathcal{Q}}_{t^\star}} \nu = \max_{\mathcal{G}_{t^\star}} \nu \tag{4}$$

hold.

Definition 4. *We say that a discrete mesh operator $\mathcal{L}$ satisfies the discrete weak maximum-minimum principle (DWMP) if for any function $\nu \in \operatorname{dom}\mathcal{L}$ and $t^\star \in \mathcal{R}_{t_M}$ the inequalities*

$$\begin{aligned} \min\{0, \min\nolimits_{\mathcal{G}_{t^\star}} \nu\} + t^\star \cdot \min\{0, \min\nolimits_{\mathcal{Q}_{\bar{t}^\star}} \mathcal{L}\nu\} &\leq \min\nolimits_{\bar{\mathcal{Q}}_{t_M}} \nu \leq \\ \leq \max\nolimits_{\bar{\mathcal{Q}}_{t_M}} \nu \leq \max\{0, \max\nolimits_{\mathcal{G}_{t^\star}} \nu\} &+ t^\star \cdot \max\{0, \max\nolimits_{\mathcal{Q}_{\bar{t}^\star}} \mathcal{L}\nu\} \end{aligned} \tag{5}$$

hold.

Definition 5. *We say that a discrete mesh operator $\mathcal{L}$ satisfies the discrete strong maximum-minimum principle (DSMP) if for any function $\nu \in \operatorname{dom}\mathcal{L}$ and $t^\star \in \mathcal{R}_{t_M}$ the inequalities*

$$\begin{aligned} \min\nolimits_{\mathcal{G}_{t^\star}} \nu + t^\star \cdot \min\{0, \min\nolimits_{\mathcal{Q}_{\bar{t}^\star}} \mathcal{L}\nu\} &\leq \min\nolimits_{\bar{\mathcal{Q}}_{t_M}} \nu \leq \\ \leq \max\nolimits_{\bar{\mathcal{Q}}_{t_M}} \nu \leq \max\nolimits_{\mathcal{G}_{t^\star}} \nu &+ t^\star \cdot \max\{0, \max\nolimits_{\mathcal{Q}_{\bar{t}^\star}} \mathcal{L}\nu\} \end{aligned} \tag{6}$$

hold.

The further two qualitative properties are the following.

Definition 6. *The discrete mesh operator $\mathcal{L}$ is called non-negativity preserving (DNP) if for any $\nu \in \operatorname{dom}\mathcal{L}$ and any $t^\star \in \mathcal{R}_{t_M}$ such that $\min_{\mathcal{G}_{t^\star}} \nu \geq 0$ and $\mathcal{L}\nu|_{\mathcal{Q}_{\bar{t}^\star}} \geq 0$, the relation $\nu|_{\mathcal{Q}_{\bar{t}^\star}} \geq 0$ holds.*

Definition 7. *The discrete mesh operator $\mathcal{L}$ is called contractive in the maximum norm (DMNC) when for any arbitrary two functions $\hat{\nu}$, $\tilde{\nu} \in \operatorname{dom}\mathcal{L}$ and $t^\star \in \mathcal{R}_{t_M}$ such that $\mathcal{L}\hat{\nu}|_{\mathcal{Q}_{\bar{t}^\star}} = \mathcal{L}\tilde{\nu}|_{\mathcal{Q}_{\bar{t}^\star}}$ and $\hat{\nu}|_{\mathcal{P}_\partial \times \mathcal{R}^0_{\bar{t}^\star}} = \tilde{\nu}|_{\mathcal{P}_\partial \times \mathcal{R}^0_{\bar{t}^\star}}$, the relation*

$$\max_{\mathbf{x}\in\bar{\mathcal{P}}} |\hat{\nu}(\mathbf{x}, t^\star) - \tilde{\nu}(\mathbf{x}, t^\star)| \leq \max_{\mathbf{x}\in\bar{\mathcal{P}}} |\hat{\nu}(\mathbf{x}, 0) - \tilde{\nu}(\mathbf{x}, 0)|$$

is valid.

The implications between the discrete qualitative properties are shown in Figure 1. The solid arrows mean the implications without any condition, while the dashed ones are true only under the indicated assumptions. Here we have used the notations $\mathbb{1}$ and $\mathbb{t}$ for the mesh functions defined on $\bar{\mathcal{Q}}_{t_M}$ with the equalities $\mathbb{1}(\mathbf{x}_i, t_n) = 1$, $\mathbb{t}(\mathbf{x}_i, t_n) = n\Delta t$ for all $(\mathbf{x}_i, t_n) \in \bar{\mathcal{Q}}_{t_M}$.

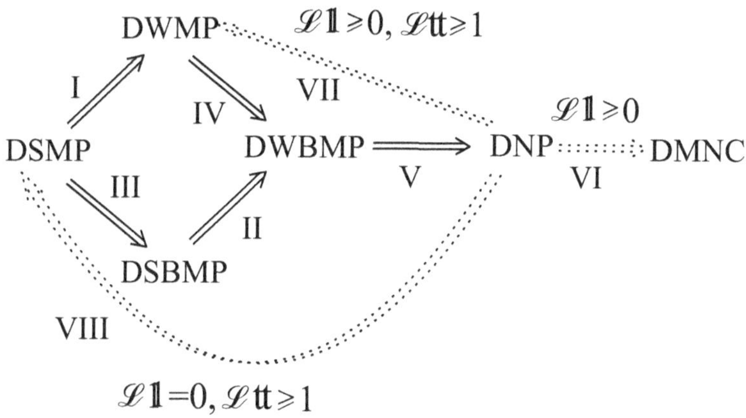

Fig. 1. Implications between the various discrete qualitative properties

3 Two-Level Discrete Mesh Operators

In the sequel, the values $\nu(\mathbf{x}_i, n\Delta t)$ of the function ν defined in $\bar{\mathcal{Q}}_{t_M}$ will be denoted by ν_i^n. Similar notation is applied to the function $\mathcal{L}\nu$. We introduce the vectors

$$\boldsymbol{\nu}^n = [\nu_1^n, \dots, \nu_{\bar{N}}^n] \in \mathbb{R}^{\bar{N}}, \quad \mathbf{e} = [1, \dots, 1] \in \mathbb{R}^{\bar{N}}; \quad \mathbf{e}_0 = [1, \dots, 1] \in \mathbb{R}^N.$$

In many numerical methods, the discrete mesh operator, called *two-level discrete mesh operator*, has a special form, namely, it is defined as

$$(\mathcal{L}\nu)_i^n = (\mathbf{X}_1 \boldsymbol{\nu}^n - \mathbf{X}_2 \boldsymbol{\nu}^{n-1})_i, \quad i = 1, \dots, N, \quad n = 1, \dots, M, \tag{7}$$

where $\mathbf{X}_1, \mathbf{X}_2 \in \mathbb{R}^{N\times\bar{N}}$ are given matrices.

Remark 1. The term "two-level discretization" refers to the fact that two discrete time levels are involved into the definition of the mesh operator. Sometimes such a method is also called "one-step method".

Our aim is to guarantee all the discrete qualitative properties. Therefore, we require all the possible implications in Figure 1. The condition $\mathcal{L}\mathbb{1} = 0$ reads as $\mathbf{X}_1\mathbf{e} - \mathbf{X}_2\mathbf{e} = (\mathbf{X}_1 - \mathbf{X}_2)\mathbf{e} = \mathbf{0}$, while the condition $\mathcal{L}t \geq 1$ means that

$$\mathbf{X}_1(\Delta tn\mathbf{e}) - \mathbf{X}_2(\Delta t(n-1)\mathbf{e}) = \Delta t(n(\mathbf{X}_1 - \mathbf{X}_2)\mathbf{e} + \mathbf{X}_2\mathbf{e}) \geq \mathbf{e}_0.$$

Since $(\mathbf{X}_1 - \mathbf{X}_2)\mathbf{e} = \mathbf{0}$, the above condition reduces to $\Delta t\mathbf{X}_2\mathbf{e} \geq \mathbf{e}_0$. Hence, we have

Theorem 1. *Let us assume that the discrete mesh operator of type* (7) *is non-negativity preserving. Then under the conditions*

$$(\mathbf{X}_1 - \mathbf{X}_2)\,\mathbf{e} = \mathbf{0} \quad \textit{and} \quad \Delta t\mathbf{X}_2\,\mathbf{e} \geq \mathbf{e}_0$$

the operator possesses all the discrete qualitative properties introduced in Section 2.

In the typical numerical applications the matrices $\mathbf{X}_1$ and $\mathbf{X}_2$ are derived from the approximation of the given continuous operator L. When we use a one-parameter family of the approximation (which is called the θ-method), these matrices are defined by the matrices $\mathbf{M}$, $\mathbf{K}$ (called mass and stiffness matrices, respectively), and a real parameter θ, as follows

$$\mathbf{X}_1 = \frac{1}{\Delta t}\mathbf{M} + \theta\mathbf{K}, \quad \mathbf{X}_2 = \frac{1}{\Delta t}\mathbf{M} - (1-\theta)\mathbf{K}. \tag{8}$$

The matrices $\mathbf{M}$ and $\mathbf{K}$ have the size $N \times \bar{N}$. Hence, the discrete mesh operator $\mathcal{L}$ in (7) can be written in the following (so-called canonical) form:

$$(\mathcal{L}\nu)_i^n = (\mathbf{M}\frac{\boldsymbol{\nu}^n - \boldsymbol{\nu}^{n-1}}{\Delta t} + \theta\mathbf{K}\boldsymbol{\nu}^n + (1-\theta)\mathbf{K}\boldsymbol{\nu}^{n-1})_i. \tag{9}$$

Then Theorem 1 can be re-formulated as follows:

Theorem 2. *Let us assume that the discrete mesh operator of type* (9) *is non-negativity preserving. Then, under the conditions*

$$\mathbf{Ke} = \mathbf{0}, \qquad \mathbf{Me} \geq \mathbf{e}_0 \tag{10}$$

the operator possesses all the discrete qualitative properties introduced in Section 2., i.e., all the implications, indicated in Figure 1, are valid.

4 Crank-Nicolson Discretized Heat Equation in 1D and Relation between Its Qualitative Properties

For the discretization of the heat conduction operator L, defined in (1), all qualitative properties are valid. Therefore, some discrete mesh operator $\mathcal{L}$ can be viewed as its adequate approximations, if the implications of Figure 1 are valid for it. When $\mathcal{L}$ is assumed to be a two-level discrete mesh operator of the form (9), then, according to Theorem 2, the required conditions are (10).

In the sequel, we analyze the discrete mesh operator in 1D when the space discretization is done by finite difference / linear finite element method and when for the time discretization we choose the θ-method with the fixed value $\theta = 0.5$. This combined method is usually called *Crank-Nicolson method* and the defined discrete mesh operator will be denoted by $\mathcal{L}_{CN}$. This means that the discrete mesh operator $\mathcal{L}_{CN}$ reads as

$$(\mathcal{L}_{CN}\nu)_i^n = (\mathbf{M}\frac{\boldsymbol{\nu}^{n+1} - \boldsymbol{\nu}^n}{\Delta t} + 0.5\mathbf{K}\boldsymbol{\nu}^{n+1} + 0.5\mathbf{K}\boldsymbol{\nu}^n)_i. \tag{11}$$

Remark 2. John Crank (1915 - 2006) originally worked in industry on the modelling and numerical solution of diffusion in polymers. In 1943, working with Phyllis Nicolson (1917-1968) on finite difference methods for the time-dependent heat equation, he proposed the Crank-Nicolson method, which has been incorporated universally in the solution of time-dependent problems since then. Their first result on this method [1] was published in 1947.

In the sequel we check the validity of the conditions, given in (10) for the different cases.

For the finite difference method on uniform mesh with step-size h the mass and stiffness matrices have the form:

$$\mathbf{M} = \left[\begin{array}{ccccc|cc} 1 & 0 & & \dots & 0 & 0 & 0 \\ 0 & 1 & 0 & & \vdots & 0 & 0 \\ & \ddots & \ddots & \ddots & & \vdots & \vdots \\ \vdots & & 0 & 1 & 0 & 0 & 0 \\ 0 & \dots & & 0 & 1 & 0 & 0 \end{array}\right], \quad \mathbf{K} = \frac{1}{h^2}\left[\begin{array}{ccccc|cc} 2 & -1 & & \dots & 0 & -1 & 0 \\ -1 & 2 & -1 & & \vdots & 0 & 0 \\ & \ddots & \ddots & \ddots & & \vdots & \vdots \\ \vdots & & -1 & 2 & -1 & 0 & 0 \\ 0 & \dots & & -1 & 2 & 0 & -1 \end{array}\right]. \tag{12}$$

Obviously, $\mathbf{Ke} = \mathbf{0}$ and $\mathbf{Me} = \mathbf{e}_0$, hence the conditions (10) are satisfied.

For the finite element method on uniform mesh with step-size h the mass and stiffness matrices are usually defined as

$$\begin{aligned} \widetilde{\mathbf{M}} &= [\widetilde{M}_{ij}]_{N\times\bar{N}}, \quad \widetilde{M}_{ij} = \int_{\Omega} \phi_j\phi_i \, dx =< \phi_j, \phi_i >, \\ \widetilde{\mathbf{K}} &= [\widetilde{K}_{ij}]_{N\times\bar{N}}, \quad \widetilde{K}_{ij} =< grad\phi_j, grad\phi_i > \end{aligned} \tag{13}$$

where $\phi_i(x)$ denote the finite element basis functions with the property

$$\sum_{i=1}^{\bar{N}} \phi_i(x) = 1. \tag{14}$$

Introducing the diagonal matrix $\mathbf{D} = diag[\text{mes}(\phi_i)] \in \mathbb{R}^{\bar{N}\times\bar{N}}$, we define the matrices $\mathbf{M}$ and $\mathbf{K}$ in (11) as follows:

$$\mathbf{M} = \mathbf{D}^{-1}\widetilde{\mathbf{M}}; \quad \mathbf{K} = \mathbf{D}^{-1}\widetilde{\mathbf{K}}. \tag{15}$$

Then, by using the relation (14), we have the relations:

$$(\mathbf{Me})_i = (\mathbf{D}^{-1}\widetilde{\mathbf{M}}\mathbf{e})_i = \frac{1}{\mathrm{mes}(\phi_i)} \sum_{j=1}^{\bar{N}} < \phi_j, \phi_i >= \\ = \frac{1}{\mathrm{mes}(\phi_i)} < \sum_{j=1}^{\bar{N}} \phi_j, \phi_i >= \frac{1}{\mathrm{mes}(\phi_i)} < 1, \phi_i >= 1; \quad (16)$$

and

$$(\mathbf{Ke})_i = (\mathbf{D}^{-1}\widetilde{\mathbf{K}}\mathbf{e})_i = \frac{1}{\mathrm{mes}(\phi_i)} \sum_{j=1}^{\bar{N}} < \mathrm{grad}\phi_j, \mathrm{grad}\phi_i >= \\ = \frac{1}{\mathrm{mes}(\phi_i)} < \mathrm{grad} \left(\sum_{j=1}^{\bar{N}} \phi_j \right), \phi_i >= \frac{1}{\mathrm{mes}(\phi_i)} < \mathrm{grad}\, 1, \phi_i >= 0. \quad (17)$$

This yields the validity of the conditions, given in (10) and hence we obtained

Theorem 3. *For the Crank-Nicolson discrete mesh operator, obtained by finite difference and by arbitrary finite element discretization for the heat conduction operator L, the implications of Figure 1 are valid.*

Remark 3. In the finite element method, in order to get the suitable mass and stiffness matrices, we multiplied the usual matrices by $\mathbf{D}^{-1}$, which, in fact, can be considered as a scaling process. It is motivated by the following. For the typical applications we a priori know $(Lu)(x_i, t_n)$, denoted by $f(x_i, t_n)$ (forcing term). Therefore, $(\mathcal{L}_{CN}\nu)_i^n$ serves as its approximation, i.e., $(\mathcal{L}_{CN}\nu)_i^n \simeq f(x_i, t_n)$. (This property is also related to the property called *consistency*.) Clearly, in the weak formulation of the problem, the forcing term is transformed into $\widetilde{\boldsymbol{\nu}}_i^n := < f(\cdot, t_n), \phi_i >$. With the choice $\widetilde{\mathbf{M}}$ and $\widetilde{\mathbf{K}}$ the operator $\mathcal{L}_{CN}$ does not have this approximation property, even for the constant forcing functions f. Therefore, we modify the matrices which define $\mathcal{L}_{CN}$, and we put

$$\boldsymbol{\nu}_i^n := \frac{1}{mes(\phi_i)} < f(\cdot, t_n), \phi_i >,$$

which yields the above scaling process.

5 Qualitative Properties of the Discretized Heat Equation

In Figure 2 we briefly summarize the implication results, obtained in Section 4, for the Crank-Nicolson discretization. This means, that, proving the DNP property,

$$\begin{array}{ccc} \text{DSMP} & \Leftrightarrow & \text{DNP} \\ \Downarrow & & \\ \text{DMNC} & & \end{array}$$

Fig. 2. Implications between the qualitative properties for the Crank-Nicolson discretization

we obtain all the other qualitative properties, too. Therefore, in the sequel we investigate the DNP property of the operator $\mathcal{L}_{CN}$.

In the FEM we will restrict to the linear basis functions, which means that the property

$$\phi_i(x) \geq 0, \quad i = 1, 2, \dots \bar{N} \tag{18}$$

holds. Then, for the mass and stiffness matrices we have the form

$$\mathbf{M} = \frac{1}{6}\left[\begin{array}{cccc|cc} 4 & 1 & & \dots 0 & 1 & 0 \\ 1 & 4 & 1 & \vdots & 0 & 0 \\ & \ddots & \ddots & \ddots & \vdots & \vdots \\ \vdots & & 1 & 4\ 1 & 0 & 0 \\ 0 \dots & & & 1\ 4 & 0 & 1 \end{array}\right], \quad \mathbf{K} = \frac{1}{h^2}\left[\begin{array}{cccc|cc} 2 & -1 & \dots & 0 & -1 & 0 \\ -1 & 2 & -1 & \vdots & 0 & 0 \\ & \ddots & \ddots & \ddots & \vdots & \vdots \\ \vdots & & -1 & 2\ -1 & 0 & 0 \\ 0 \dots & & -1 & 2 & 0 & -1 \end{array}\right]. \tag{19}$$

In order to give the DNP condition in a linear algebraic form, we introduce the following convenient partitions of the matrices $\mathbf{M}$ and $\mathbf{K}$:

$$\mathbf{M} = [\mathbf{M}_0 | \mathbf{M}_\partial], \quad \mathbf{K} = [\mathbf{K}_0 | \mathbf{K}_\partial]. \tag{20}$$

Then $\mathbf{X}_1$ and $\mathbf{X}_2$ are also partitioned as

$$\mathbf{X}_1 = [\mathbf{X}_{10} | \mathbf{X}_{1\partial}], \quad \mathbf{X}_2 = [\mathbf{X}_{20} | \mathbf{X}_{2\partial}]. \tag{21}$$

In the above formulas $\mathbf{M}_0, \mathbf{K}_0, \mathbf{X}_{10}$ and $\mathbf{X}_{20}$ are square matrices from $\mathbb{R}^{N\times N}$, and $\mathbf{M}_\partial, \mathbf{K}_\partial \mathbf{X}_{1\partial}, \mathbf{X}_{2\partial} \in \mathbb{R}^{N\times 2}$.

For the DNP of the two-level discrete mesh operator in the form (7), we have the following

Theorem 4. *The finite difference and the linear finite element CN discrete mesh operators are DNP if and only if*

$$\mathbf{X}_{10}^{-1} \geq \mathbf{0} \text{ and } \mathbf{X}_{10}^{-1}\mathbf{X}_{20} \geq \mathbf{0}. \tag{22}$$

The proof follows directly from the non-negativity of the matrices $-\mathbf{X}_{1\partial}$ and $\mathbf{X}_{2\partial}$.

Remark 4. Theorem 4 yields that $\mathcal{L}_{CN}$ is non-negativity preserving if and only if the matrices $\mathbf{X}_{10}$ and $\mathbf{X}_{20}$ form a weak regular splitting for the matrix $\mathbf{X}_{10} - \mathbf{X}_{20} = \mathbf{K}_0$.

Let us notice that the matrices $\mathbf{X}_{10}$ and $\mathbf{X}_{20}$ in (21) can be written in the form of uniformly continuant tridiagonal matrices

$$\mathbf{X}_{10} = z \cdot \text{tridiag}[-1, 2w, -1], \ \mathbf{X}_{20} = s \cdot \text{tridiag}[1, p, 1] \in \mathbb{R}^{N \times N}, \tag{23}$$

where

$$z = \frac{q}{2\Delta t}, \quad s = \frac{q}{2\Delta t}, \quad w = \frac{1+q}{q}, \quad p = 2\frac{1-q}{q} \tag{24}$$

for the finite difference Crank-Nicolson discretization, and

$$z = \frac{1/6 - q/2}{\Delta t}, \quad s = \frac{1/6 + q/2}{\Delta t}, \quad w = \frac{1/3 + q/2}{q/2 - 1/6}, \quad p = \frac{2/3 - q}{q/2 + 1/6} \tag{25}$$

for the linear finite element Crank-Nicolson discretization. (Here the notation $q = \Delta t/h^2$ was used.)

First we formulate the conditions in (22) for the matrices (23) in general form. We can give the exact conditions, which depend on the number of the division N, too.

The next theorem gives the condition to (22) by fixed dimension.

Theorem 5. *Assume that $z > 0$, $s > 0$ and $w > 1$. Then, $\mathbf{X}_{10}$ is an M-matrix. Moreover, $\mathbf{X}_{10}^{-1}\mathbf{X}_{20} \in \mathbb{R}^{N \times N}$ is non-negative for arbitrary fixed N if and only if the conditions*

$$2w + p > 0 \tag{26}$$

and

$$a(N) := \frac{\text{sh}(N\vartheta)}{\text{sh}((N+1)\vartheta)} \geq \frac{1}{2w + p} \tag{27}$$

are satisfied, where $\vartheta = \text{arch}(w)$.

For varying dimension the following statement holds.

Theorem 6. *Assume that $z > 0$, $s > 0$ and $w > 1$. If, for some number $N_0 \in \mathbb{N}$, the conditions (26) and (27) are satisfied, then all matrices $\mathbf{X}_{10}^{-1}\mathbf{X}_{20} \in \mathbb{R}^{N \times N}$ with $N \geq N_0$, are non-negative. Moreover, there exists such a number N_0, if and only if the conditions (26) and*

$$p > -w + \sqrt{w^2 - 1} \tag{28}$$

hold.

According to Theorem 6, the condition obtained for $N = 1$ guarantees the DNP for any values $N = 2, 3, \ldots$. Since

$$a(1) = \frac{\text{sh}\vartheta}{\text{sh}(2\vartheta)} = \frac{1}{2\text{ch}\vartheta} = \frac{1}{2w},$$

therefore, (27) results in the condition

$$p \geq 0. \tag{29}$$

Hence, the matrix $\mathbf{X}_{10}^{-1}\mathbf{X}_{20} \in \mathbb{R}^{N\times N}$ is non-negative for all $N = 1, 2, \dots$ if and only if $\mathbf{X}_{10}$ is an M-matrix and $\mathbf{X}_{10} - \mathbf{X}_{20}$ is a regular splitting of the matrix $\mathbf{K}_0$.

Similarly, the condition obtained for $N = 2$ is a sufficient condition for all values $N = 3, 4, \dots$, too. Due to the relation

$$a(2) = \frac{\mathrm{sh}(2\vartheta)}{\mathrm{sh}(3\vartheta)} = \frac{2\mathrm{ch}(\vartheta)}{4\mathrm{ch}^2(\vartheta) - 1} = \frac{2w}{4w^2 - 1},$$

condition (27) results in the assumption

$$p \geq -\frac{1}{2w}. \tag{30}$$

That is, $\mathbf{X}_{10}^{-1}\mathbf{X}_{20} \in \mathbb{R}^{N\times N}$ is non-negative for all $N = 2, 3, \dots$ if and only if $\mathbf{X}_{10}$ is an M-matrix and (30) is valid.

Now, we define the conditions for $\mathcal{L}_{CN}$ under which it has the DNP property. To this aim, we use Theorem 5. First we notice that, under the choice of the parameters (24) and (25), the condition (26) is always satisfied. Hence, we should check the validity of (27), only. Taking into account Theorem 6, and the formulas (29), (30), (28), and the requirement $z > 0$, we arrive at our main result.

Theorem 7. *The finite difference CN heat discrete mesh operator $\mathcal{L}_{CN}$ is DNP (and hence, it possesses all the other qualitative properties), under the following conditions:*

- *for any number of space partition (i.e., $N \geq 1$) if and only $q \leq 1$;*
- *for the number of space partition $N \geq 2$ if and only if $q \leq 2/\sqrt{3} \simeq 1.1547$;*
- *there exists a number of space partition N_0 such that $\mathcal{L}_{CN}$ is non-negativity preserving for all $N \geq N_0$ if an only if the condition $q < 2(2-\sqrt{2}) \simeq 1.17157$ holds.*

Theorem 8. *The linear finite element CN heat discrete mesh operator $\mathcal{L}_{CN}$ is DNP (and hence, it possesses all the other qualitative properties), under the following conditions:*

- *for any number of space partition (i.e., $N \geq 1$) if and only $1/3 \leq q \leq 2/3 \simeq 0.667$;*
- *for the number of space partition $N \geq 2$ if and only if $1/3 \leq q \leq \sqrt{5}/3 \simeq 0.745$;*
- *there exists a number of space partition N_0 such that $\mathcal{L}_{CN}$ is non-negativity preserving for all $N \geq N_0$ if an only if the condition $1/3 \leq q \leq\simeq 0.748$ holds.*

We note that the exact bound of the DMNC for the finite difference CN heat discrete mesh operator is defined in [8] and [6], and the condition reads as $q \leq 1.5$. This means that the implication *DNP ⇒ DMNC* in Figures 1 and 2 cannot be reversed, i.e., the discrete maximum norm contractivity does not imply the other qualitative properties, in general.

Finally, we summarize our results for the finite difference discretization.

Theorem 9. *The finite difference Crank-Nicolson mesh operator, which corresponds to the one-dimensional continuous differential operator L defined in* (1), *has the following qualitative properties:*

- *It satisfies all listed qualitative properties for any number of the uniform space partition, if and only if the condition $q \in (0, 1]$ holds.*
- *It satisfies all listed qualitative properties for a sufficiently large number of the uniform space partition if and only if the condition $q \in (0, 2(2 - \sqrt{2})]$ holds.*
- *For the values $q \in (0, 1.5]$ it is contractive in the maximum norm.*

In the sequel, we analyze the qualitative behavior of the finite difference mesh operator "after the death", i.e., in the case $q > 1.5$.

6 Stability Constant of the CN-Discretized Heat Equation

The stability function for the CN-method reads as

$$r_{CN}(q\mathbf{K}_0) = (\mathbf{I} - 0.5q\mathbf{K}_0)^{-1}(\mathbf{I} + 0.5q\mathbf{K}_0). \tag{31}$$

For $1 \leq p \leq +\infty$, let l_p be the Banach space of all the p-summable sequences of complex numbers $\mathbf{z} = \{z_j\}_{j=-\infty}^{+\infty}$, endowed with the standard norm

$$\| \mathbf{z} \|_p = h^{1/p} \left(\sum_{j=-\infty}^{+\infty} |z_j|^p \right)^{1/p}, \qquad 1 \leq p < +\infty,$$

and

$$\| \mathbf{z} \|_\infty = \sup_{-\infty < j < +\infty} |z_j|.$$

It is also well known (see, e.g., [10]) that for all $1 \leq p \leq +\infty$ the numerical method is stable independently of the step-sizes, i.e., independently of q. This means that there exists $C_p > 0$ such that for all $q > 0$ the estimation

$$\| r_{CN}^n(q\mathbf{K}_0) \|_p \leq C_p \tag{32}$$

holds. The number C_p denotes the smallest possible constant fulfilling (32), which is called *the stability constant* in the corresponding norm. Since the operator $\mathcal{L}_{CN}$ with (11)-(12) is absolute contractive in the l_2-norm, we have $C_2 = 1$. Due to Theorem 9, the operator $\mathcal{L}_{CN}$ with (11)-(12) is contractive in the maximum norm only for the values $0 < q \leq 1.5$. Therefore, clearly $C_\infty > 1$. In [9] it is proved that $C_\infty < 23$. The following statement gives a sharp characterization of the stability constant [5].

Theorem 10. *The finite difference Crank-Nicolson discrete mesh operator $\mathcal{L}_{CN}$ with (11)-(12) is stable in l_∞-norm (the maximum norm) for any step sizes. It is contractive only for the values $q \in (0, 1.5]$. However, for any choice of the step sizes it is not necessarily contractive, and the maximum norm of the initial function can increase by a factor of C_∞ at most, where $C_\infty \in [3, 4.324)$.*

Finally, we note that the lower bound in the estimation for C_∞ is sharp.

Acknowledgement

Part of the work was done during the visit at the Otago University (New Zealand) and it was supported by the Hungarian Research Grant OTKA K 67819 and by Project HS-MI-106/2005 of NSF of Bulgaria.

References

1. Crank, J., Nicolson, P.: A practical method for numerical evaluation of solutions of partial differential equations of the heat conduction type. Proc. Cambridge Philosophical Society 43, 50–64 (1947)
2. Faragó, I., Horváth, R.: Discrete maximum principle and adequate discretizations of linear parabolic problems. SIAM Sci. Comput. 28, 2313–2336 (2006)
3. Faragó, I., Horváth, R.: Qualitative properties of monotone linear operators. Electronic Journal of Qualitative Theory of Differential Equations 8, 1–15 (2008)
4. Faragó, I., Horváth, R.: Continuous and discrete parabolic operators and their qualitative properties, IMA Numerical Analysis (to appear)
5. Faragó, I., Palencia, C.: Sharpening the estimate of the stability bound in the maximum-norm of the Crank–Nicolson scheme for the one-dimensional heat equation. Appl. Numer. Math. 42, 133–140 (2002)
6. Horváth, R.: Maximum norm contractivity in the numerical solution of the one-dimensional heat equation. Appl. Numer. Math. 31, 451–462 (1999)
7. Hunsdorfer, W., Verwer, J.G.: Numerical solution of time-dependent advection-diffusion-reaction equations. Springer, Berlin (2003)
8. Kraaijevanger, J.: Maximum norm contractivity of discretization schemes for the heat equation. Appl. Numer. Math. 9, 475–492 (1992)
9. Serdyukova, S.J.: The uniform stability with respect to initial data of a sixpoint symmetrical scheme for the heat conduction equation. In: Numerical Methods for the Solution of Differential and Integral Equations and Quadrature Formulae, Nauka, Moscow, pp. 212–216 (1964) (in Russian)
10. Thomée, V.: Finite difference methods for linear parabolic equations. Elsevier, North-Holland (1990)
11. Zlatev, Z., Dimov, I.: Computational and numerical challenges in environmental modelling. Studies in Computational Mathematics 13 (2006)

Finite Element Approximation of an Elliptic Boundary Value Problem with Interface

Boško S. Jovanović[1] and Lubin G. Vulkov[2]

[1] University of Belgrade, Faculty of Mathematics
Studentski trg 16, 11000 Belgrade, Serbia
bosko@matf.bg.ac.yu

[2] University of Rousse, Center of Applied Mathematics and Informatics
Studentska str. 8, Rousse 7017, Bulgaria
vulkov@ami.ru.acad.bg

Abstract. For elliptic boundary value problem in domain with smooth curvilinear boundary and interface a finite element approximation is constructed. Convergence is proved in Sobolev like spaces $\widetilde{W}_2^1$ and $\widetilde{L}_2$.

1 Introduction

Interface problems occur in many applications in science and engineering [17,20]. Let us mention, for example, heat transfer in presence of concentrated capacity, oscillations with concentrated mass, Pupin's induction coils etc. Such kind of problems can be modelled by partial differential equations with discontinuous or singular input data. First partial derivatives of their solutions have discontinuities across one or several interfaces, which have lower dimension than the domain where the problem is defined. Numerical methods designed for the solution of problems with smooth solutions do not work efficiently for the interface problems.

In the present paper we consider an elliptic boundary value problem (BVP) in curvilinear domain with closed curvilinear interface. The Sobolev like spaces containing the solutions of this problem are introduced. The basic a priori estimates are obtained for the solutions from these spaces. The BVP is approximated by finite element method and the corresponding convergence rate estimates are obtained.

Conforming finite element method (FEM) for elliptic problems with discontinuous coefficients and homogeneous interface conditions are addressed in [1]. Finite element methods for nonhomogeneous elliptic interface problems are analyzed in [2], and it is shown that the discretization error is of optimal order for linear finite elements in quasi-uniform triangulations. Least-squares FEM for elliptic interface problems with Dirichlet and Neumann boundary data is proposed and analyzed in [3]. Elliptic and parabolic interface problems with non-zero jump in the flux across the smooth interface are considered in [4,10,14]. In [10], nearly optimal estimates in the energy-norm and in the L_2-norm are established under

S. Margenov, L.G. Vulkov, and J. Waśniewski (Eds.): NAA 2008, LNCS 5434, pp. 56–67, 2009.

reasonable assumptions on the original solutions, whereas some new estimates are presented in [14].

The immersed interface method (IIM) is based on using the jumps in the solution and its derivatives in finite difference schemes modifying the standard schemes in the neighborhood of the interface, see [11]. The IIM with FEM formulation is considered in [12,15].

Analogous problem in rectangular domain with straight line interface is considered in [9], where the finite difference method were used for the numerical solution. An abstract operator method for investigation of interface problems is proposed in [6]. Non-stationary interface problems are investigated in [7,8]. A extensive review of the numerical methods for solving such problems can be found in [13].

2 Formulation of Elliptic Interface Problem

Let Ω be a bounded convex domain in $\mathbb{R}^2$ with the boundary $\Gamma = \partial\Omega$. Let $S \subset \Omega$ be a convex closed curve dividing Ω into two disjoint parts Ω_1 and Ω_2 (Fig. 1).

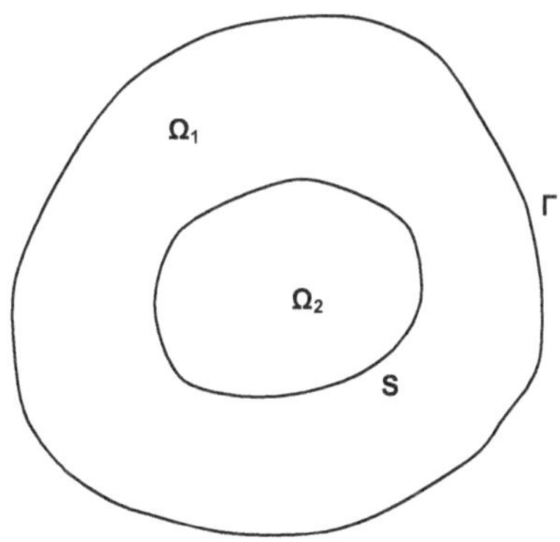

Fig. 1. Domain Ω with interface S

As a model problem we consider the following elliptic boundary value problem

$$-\sum_{i,j=1}^{2} \frac{\partial}{\partial x_i}\left(a_{ij}(x)\,\frac{\partial u}{\partial x_j}\right) + \big[c(x) + k(x)\delta_S(x)\big]u = f(x), \quad x \in \Omega,$$
$$u(x) = 0, \quad x \in \Gamma, \tag{1}$$

where

$$a_{ij} = a_{ji}, \quad \sum_{i,j=1}^{2} a_{ij}y_iy_j \geq a_0 \sum_{i=1}^{2} y_i^2, \quad a_0 > 0,$$

$$c(x) \geq c_0 > 0, \quad k(x) \geq k_0 > 0,$$

and $\delta_S(x)$ is the Dirac distribution [21] concentrated on S.

The considered interface problem can be formulated in an alternative manner, without explicit use of the Dirac distribution. Because of $\delta_S(x) = 0$ for $x \notin S$ from (1) immediately follows

$$-\sum_{i,j=1}^{2} \frac{\partial}{\partial x_i}\left(a_{ij}(x)\,\frac{\partial u}{\partial x_j}\right) + c(x)u = f(x), \quad x \in \Omega_1 \cup \Omega_2, \qquad u(x) = 0, \quad x \in \Gamma, \tag{2}$$

It is well known that for a piecewise smooth function $\varphi \in C^1[a,\xi] \cap C^1[\xi,b]$ the derivative in distributional sense can be expressed in the following way [21]:

$$\varphi'(x) = \{\varphi'(x)\} + [\varphi]_\xi\, \delta(x-\xi)$$

where $\{\varphi'(x)\}$ is the derivative in the classic sense and $[\varphi]_\xi = \varphi(\xi+0) - \varphi(\xi-0)$ is the jump of the function φ in the point ξ. Taking into account (1), we conclude that on the interface S the following conjugation conditions are satisfied

$$[u]_S = 0, \quad \left[\frac{\partial u}{\partial \nu}\right]_S = ku, \quad x \in S, \tag{3}$$

where $\frac{\partial u}{\partial \nu}$ denotes the co-normal derivative:

$$\frac{\partial u}{\partial \nu} = \sum_{i,j=1}^{2} a_{ij}(x)\,\frac{\partial u}{\partial x_j}\,\cos(\nu, x_i).$$

Dirac distribution δ_S belongs to Sobolev space $W_2^{-\lambda}(\Omega)$, with $\lambda > 1/2$ [23]. In such a way, equation (1) must be treated as an equation in this space. For $\lambda = 1$ this means that

$$\left\langle -\sum_{i,j=1}^{2} \frac{\partial}{\partial x_i}\left(a_{ij}(x)\,\frac{\partial u}{\partial x_j}\right) + \big[c(x) + k(x)\delta_S(x)\big]u,\ v\right\rangle = \langle f, v\rangle, \qquad \forall v \in \overset{\circ}{W}{}_2^1(\Omega), \tag{4}$$

where $\langle f, v\rangle$ denotes duality pairing between spaces $W_2^{-1}(\Omega)$ and $\overset{\circ}{W}{}_2^1(\Omega)$. Using standard rules for differentiation of distributions (see [21]) from (4) we obtain the following weak form of BVP (1): *find* $u \in \overset{\circ}{W}{}_2^1(\Omega)$ *such that*

$$a(u, v) = \langle f, v\rangle\,, \qquad \forall\, v \in \overset{\circ}{W}{}_2^1(\Omega)\,, \tag{5}$$

where

$$a(u, v) = \int_\Omega \left(\sum_{i,j=1}^{2} a_{ij}\,\frac{\partial u}{\partial x_j}\,\frac{\partial v}{\partial x_i} + cuv\right)dx + \int_S kuv\,dS\,. \tag{6}$$

One easily checks that (5), (6) is also the weak form of the BVP with conjugation conditions on the interface (2), (3). In this sense, problems (1) and (2), (3) are equivalent.

Through the paper by C, C_i we will denote positive generic constants which not depend on the solution of the problem or the other input data.

3 Abstract Model

Let H be a real separable Hilbert space endowed with the inner product $(\cdot,\cdot)$ and norm $\|\cdot\|$. Let A be an unbounded, selfadjoint, positive definite linear operator acting in H, with domain $D(A)$ dense in H. The product $(u,\, v)_A = (Au,\, v)$ $(u,\, v \in D(A))$ satisfies the inner product axioms. Reinforcing $D(A)$ in the norm $\|u\|_A = (u,\, u)_A^{1/2}$ we obtain so called energy space $H_A \subset H$. The inner product $(u,\, v)$ continuously extends to $H_A^* \times H_A$, where $H_A^* = H_{A^{-1}}$ is the adjoint space for H_A. The spaces H_A, H and $H_{A^{-1}}$ form Gel'fand triple $H_A \subset H \subset H_{A^{-1}}$, with continuous and dense embeddings. Operator A extends to a mapping $A: H_A \to H_{A^{-1}}$ (see [16,19]).

Let B be another unbounded, selfadjoint, positive definite linear operator acting in H, such that $D(A) \subset D(B) \subset H$. In general, A and B are noncommutative. We assume that the quotient $\|u\|_A/\|u\|_B$ is unbounded on $D(A)$. Under these assumptions there exists a countable set of eigenvalues $\{\lambda_i\}$ of the spectral problem [22]

$$Au = \lambda Bu. \tag{7}$$

All eigenvalues are positive and $\lambda_i \to \infty$ when $i \to \infty$. Further,

$$\|u\|_A \geq \sqrt{\lambda_1}\, \|u\|_B\,, \quad \|u\|_{AB^{-1}A} \geq \sqrt{\lambda_1}\, \|u\|_A\,,$$

where λ_1 is the first (minimal) eigenvalue of the spectral problem (7).

We consider the following linear equation of the first kind in H

$$(A+B)u = f, \tag{8}$$

which can be treated as an abstract model for the boundary value problem (1). Its solution satisfies the following a priori estimates:

$$\|u\|_B \leq \|f\|_{A^{-1}BA^{-1}}, \tag{9}$$

$$\|u\|_A \leq \|f\|_{A^{-1}}, \tag{10}$$

$$\|u\|_{AB^{-1}A} \leq \|f\|_{B^{-1}}. \tag{11}$$

The corresponding weak form of equation (8) is

$$a(u,v) \equiv (u,v)_A + (u,v)_B = (f,v)\,, \qquad \forall\, v \in H_A\,. \tag{12}$$

The bilinear form $a(u,v)$ is H_A-symmetric:

$$a(u,v) = a(v,u)\,, \qquad \forall\, u,\, v \in H_A\,,$$

H_A-elliptic:

$$a(u,u) \geq \|u\|_A^2\,, \qquad \forall\, u \in H_A\,,$$

and H_A-bounded:

$$a(u,v) \leq \Big(1 + \frac{1}{\lambda_1^2}\Big)\|u\|_A\|v\|_A\,, \qquad \forall\, u,\, v \in H_A\,.$$

4 Identification of Function Spaces and Norms

Let us choose $H = L_2(\Omega)$. Then the boundary value problem (1) reduces to the abstract form (8), where

$$Au = -\sum_{i,j=1}^{2} \frac{\partial}{\partial x_i}\Big(a_{ij}(x)\,\frac{\partial u}{\partial x_j}\Big), \qquad Bu = \big[c(x) + k(x)\delta_S(x)\big]u.$$

For $u \in D(A) = W_2^2(\Omega) \cap \overset{\circ}{W}{}_2^1(\Omega)$ using partial integration we get

$$\|u\|_A^2 = \int_\Omega \sum_{i,j=1}^{2} a_{ij}(x)\,\frac{\partial u}{\partial x_i}\,\frac{\partial u}{\partial x_j}\,dx.$$

For $a_{ij} \in L_\infty(\Omega)$ under the assumption of strong ellipticity one immediately obtains

$$\|u\|_A \asymp \|u\|_{W_2^1(\Omega)} \quad \text{i.e.} \quad C_1\,\|u\|_{W_2^1(\Omega)} \leq \|u\|_A \leq C_2\,\|u\|_{W_2^1(\Omega)}.$$

In such a way, we conclude that $H_A = \overset{\circ}{W}{}_2^1(\Omega)$. Consequently, $H_{A^{-1}} = H_A^* = \big(\overset{\circ}{W}{}_2^1(\Omega)\big)^* = W_2^{-1}(\Omega)$ and

$$\|u\|_{A^{-1}} \asymp \|u\|_{W_2^{-1}(\Omega)} = \sup_{v \in \overset{\circ}{W}{}_2^1(\Omega)} \frac{|\langle u,v\rangle|}{\|v\|_{W_2^1(\Omega)}}\,.$$

If $c \in L_\infty(\Omega)$, $k \in L_\infty(S)$, $c(x) \geq c_0 > 0$ and $k(x) \geq k_0 > 0$ then

$$\|u\|_B^2 = \int_\Omega c(x)\,u^2(x)\,dx + \int_S k(x)\,u^2(x)\,dS \asymp \|u\|_{L_2(\Omega)}^2 + \|u\|_{L_2(S)}^2 \equiv \|u\|_{\widetilde{L}_2(\Omega)}^2.$$

In such a way $H_B = \widetilde{L}_2(\Omega) = L_2(\Omega) \cap L_2(S)$, $H_{B^{-1}} = \big(\widetilde{L}_2(\Omega)\big)^*$ and

$$\|u\|_{B^{-1}} \asymp \sup_{v \in \widetilde{L}_2(\Omega)} \frac{\big|{}_{\widetilde{L}_2^*}\langle u,v\rangle_{\widetilde{L}_2}\big|}{\|v\|_{\widetilde{L}_2(\Omega)}}\,.$$

Here ${}_{\widetilde{L}_2^*}\langle u,v\rangle_{\widetilde{L}_2}$ denotes the duality pairing in $\big(\widetilde{L}_2(\Omega)\big)^* \times \widetilde{L}_2(\Omega)$.

In addition to the previous assumptions, let $a_{ij} \in W^1_\infty(\Omega)$. Then

$$\|u\|^2_{AB^{-1}A} \asymp |u|^2_{W^2_2(\Omega_1)} + |u|^2_{W^2_2(\Omega_2)} + \|u\|^2_{W^1_2(\Omega)} \equiv \|u\|^2_{\widetilde{W}^2_2(\Omega)}.$$

In such a way $H_{AB^{-1}A} = \widetilde{W}^2_2(\Omega) = W^2_2(\Omega_1) \cap W^2_2(\Omega_2) \cap \overset{\circ}{W}{}^1_2(\Omega)$, $H_{A^{-1}BA^{-1}} = \left(\widetilde{W}^2_2(\Omega)\right)^*$ and

$$\|u\|_{A^{-1}BA^{-1}} \asymp \sup_{v \in \widetilde{W}^2_2(\Omega)} \frac{\left|{}_{\widetilde{W}^{2*}_2}\langle u, v\rangle_{\widetilde{W}^2_2}\right|}{\|v\|_{\widetilde{W}^2_2(\Omega)}}.$$

Let us denote also $\widetilde{W}^1_2(\Omega) = \overset{\circ}{W}{}^1_2(\Omega)$. Then a priori estimates (9)-(11) can be rewritten as

$$\|u\|_{\widetilde{L}_2(\Omega)} \leq C\,\|f\|_{\left(\widetilde{W}^2_2(\Omega)\right)^*},$$

$$\|u\|_{\widetilde{W}^1_2(\Omega)} \leq C\,\|f\|_{\left(\widetilde{W}^1_2(\Omega)\right)^*},$$

$$\|u\|_{\widetilde{W}^2_2(\Omega)} \leq C\,\|f\|_{\left(\widetilde{L}_2(\Omega)\right)^*}.$$

5 Finite Element Approximation

Let coefficients of equation (1) satisfy the previous assumptions, in particular: $a_{ij} \in W^1_\infty(\Omega)$, $c \in L_\infty(\Omega)$ and $k \in L_\infty(S)$.

In a standard manner (see [5]) we define quasi-uniform triangulation $\mathcal{T}_h$ of domain Ω and the corresponding finite element space V_h. We suppose that the assumptions $(\mathcal{T}_h 2)$-$(\mathcal{T}_h 4)$ and $(H1)$-$(H4)$ from [5] are satisfied. Because the domain Ω is curvilinear, instead of $(\mathcal{T}_h 1)$ and $(\mathcal{T}_h 5)$ we demand the following conditions to be satisfied:

$(\widetilde{\mathcal{T}}_h 1)$: $\bar{\Omega}_h = \cup_{K \in \mathcal{T}_h} \bar{K} \subset \bar{\Omega}$ *and vertices of* $\bar{\Omega}_h$ *belong to* Γ; $\bar{\Omega}_{2,h} = \cup_{K \in \mathcal{T}_h,\, K \subset \Omega_2} \bar{K} \subset \bar{\Omega}_2$ *and vertices of* $\bar{\Omega}_{2,h}$ *belong to* S.

$(\widetilde{\mathcal{T}}_h 5)$: *Any edge of every triangle* $K_1 \in \mathcal{T}_h$ *is either edge of another triangle* $K_2 \in \mathcal{T}_h$ *or it join two vertices from* Γ.

An example of such triangulation is presented on Fig. 2.

The space V_h consists of piecewise linear functions vanishing on $\partial\Omega_h$. For $v_h \in V_h$ we set $v_h(x) = 0$ for $x \in \Omega \setminus \Omega_h$. In such a way, $V_h \subset \overset{\circ}{W}{}^1_2(\Omega)$.

Finite element approximation of BVP (1), i.e. (5)- (6), we define in a standard way using Ritz-Galerkin method: *find* $u_h \in V_h$ *such that*

$$a(u_h, v_h) = \langle f, v_h \rangle\,, \qquad \forall\, v_h \in V_h\,. \tag{13}$$

Approximate solution u_h satisfies the following optimality condition:

$$a(u - u_h, u - u_h) \leq a(u - v_h, u - v_h)\,, \qquad \forall\, v_h \in V_h\,. \tag{14}$$

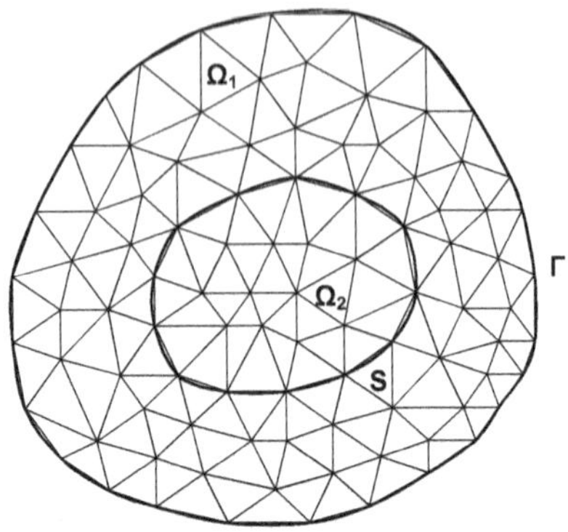

Fig. 2. Triangulation of Ω

6 Convergence in $\widetilde{W}_2^1(\Omega)$

Let Π_h be interpolation operator from $\overset{\circ}{W}{}_2^1(\Omega) \cap C(\bar{\Omega})$ into V_h, i.e. for $v \in \overset{\circ}{W}{}_2^1(\Omega) \cap C(\bar{\Omega})$, $\Pi_h v \in V_h$ and $\Pi_h v = v$ in vertices of every finite element $K \in \mathcal{T}_h$.

The following assertion holds true.

Lemma 1. *If the solution u of BVP (1) belongs to $\overset{\circ}{W}{}_2^1(\Omega) \cap C(\bar{\Omega})$ then*

$$\|u - u_h\|_{W_2^1(\Omega)} \leq C \inf_{v_h \in V_h} \|u - v_h\|_{W_2^1(\Omega)} \leq C \|u - \Pi_h u\|_{W_2^1(\Omega)},$$

Proof. Follows immediately from (14), using ellipticity and boundedness of bilinear form $a(u, v)$ in $\widetilde{W}_2^1(\Omega) = \overset{\circ}{W}{}_2^1(\Omega)$.

In such a way, in order to prove the convergence of finite element method (13) it is enough to estimate the interpolation error $u - \Pi_h u$. ◇

Lemma 2. *If $u \in \widetilde{W}_2^2(\Omega)$ then*

$$\|u - \Pi_h u\|_{W_2^1(\Omega)} \leq Ch \|u\|_{\widetilde{W}_2^2(\Omega)}.$$

Proof. Because $\Pi_h u = 0$ in the domain $\Omega \setminus \Omega_h$ we have

$$\|u - \Pi_h u\|_{W_2^1(\Omega)}^2 = \|u\|_{W_2^1(\Omega \setminus \Omega_h)}^2 + \sum_{K \in \mathcal{T}_h} \|u - \Pi_h u\|_{W_2^1(K)}^2. \tag{15}$$

Using the known inequality [18]

$$\|g\|_{L_2(0,\varepsilon)} \leq C\sqrt{\varepsilon} \|g\|_{W_2^1(0,1)}, \quad 0 < \varepsilon < 1$$

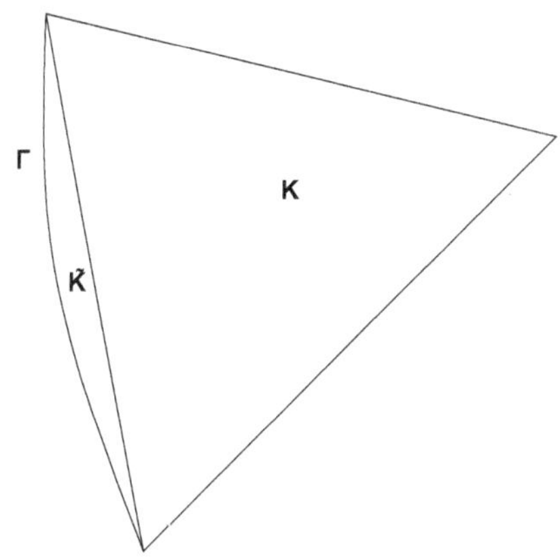

Fig. 3. Near border finite element

and taking in mind that the thickness of $\Omega \setminus \Omega_h$ is of order $O(h^2)$ (see Fig. 3), we immediately obtain

$$\|u\|_{W_2^1(\Omega\setminus\Omega_h)} \leq Ch\|u\|_{W_2^2(\Omega_1)} \, . \tag{16}$$

The sum $\sum_{K\in\mathcal{T}_h} \|u - \Pi_h u\|^2_{W_2^1(K)}$ can be represented in the following manner:

$$\sum_{K\in\mathcal{T}_h} \|u - \Pi_h u\|^2_{W_2^1(K)} = \sum_{K\in\mathcal{T}_h,\, K\subset\Omega_1} \|u - \Pi_h u\|^2_{W_2^1(K)}$$
$$+ \sum_{K\in\mathcal{T}_h,\, K\subset\Omega_2} \|u - \Pi_h u\|^2_{W_2^1(K)} + \sum_{K\in\mathcal{T}_h,\, K\cap S\neq\emptyset} \|u - \Pi_h u\|^2_{W_2^1(K)} \, .$$

The first and the second sum in the right-hand side can be estimated in the standard way:

$$\begin{aligned} \sum_{K\in\mathcal{T}_h,\, K\subset\Omega_i} \|u - \Pi_h u\|^2_{W_2^1(K)} &\leq \sum_{K\in\mathcal{T}_h,\, K\subset\Omega_i} Ch^2 \|u\|^2_{W_2^2(K)} \\ &\leq Ch^2 \|u\|^2_{W_2^2(\Omega_i)} \, , \qquad i = 1, 2. \end{aligned} \tag{17}$$

The last sum is taken on finite elements intersected by interface S (see Fig. 4). Let us choose the constant $\varepsilon_0 \geq C\,h^2$ and define the domain $\Omega_{1,\varepsilon_0} = \Omega_1 \cup (\cup_{x\in S} B(x,\varepsilon_0))$, where $B(x,\varepsilon_0)$ is the circle with center x and radius ε_0. Let Pu be an extension of function $u \in W_2^2(\Omega_1)$ in domain Ω_{1,ε_0}, such that $\|Pu\|_{W_2^2(\Omega_{1,\varepsilon_0})} \leq C\,\|u\|_{W_2^2(\Omega_1)}$. For a finite element $K \in \mathcal{T}_h$, $K \cap S \neq \emptyset$ we have:

$$\|u - \Pi_h u\|_{W_2^1(K)} \leq \|Pu - \Pi_h u\|_{W_2^1(K)} + \|u\|_{W_2^1(K\setminus\Omega_1)} + \|Pu\|_{W_2^1(K\setminus\Omega_1)} \, .$$

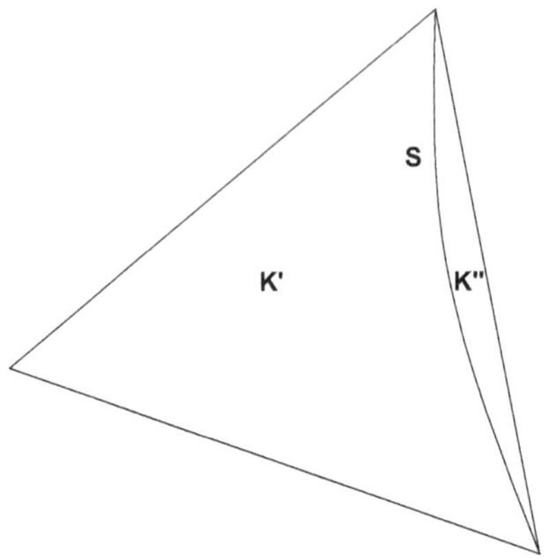

Fig. 4. Finite element intersecting interface S

Because of $Pu = u$ in the vertices of K it follows that $\Pi_h u = \Pi_h P u$. In such a way, the first summand on the right-hand side can be estimated analogously as (17):

$$\begin{aligned}\sum_{K\in\mathcal{T}_h,\, K\cap S\neq\emptyset} \|Pu - \Pi_h u\|^2_{W_2^1(K)} &\leq \sum_{K\in\mathcal{T}_h,\, K\cap S\neq\emptyset} Ch^2 \|Pu\|^2_{W_2^2(K)} \\ &\leq Ch^2 \|Pu\|^2_{W_2^2(\Omega_{1,\varepsilon_0})} \leq Ch^2 \|u\|^2_{W_2^2(\Omega_1)}. \end{aligned} \tag{18}$$

The other two summands can be estimated analogously as (16):

$$\sum_{K\in\mathcal{T}_h,\, K\cap S\neq\emptyset} \|u\|^2_{W_2^1(K\backslash\Omega_1)} = \|u\|^2_{W_2^1(\Omega_2\backslash\Omega_{2,h})} \leq Ch^2 \|u\|^2_{W_2^2(\Omega_2)}, \tag{19}$$

$$\begin{aligned}\sum_{K\in\mathcal{T}_h,\, K\cap S\neq\emptyset} \|Pu\|^2_{W_2^1(K\backslash\Omega_1)} &= \|Pu\|^2_{W_2^1(\Omega_2\backslash\Omega_{2,h})} \\ &\leq Ch^2 \|Pu\|^2_{W_2^2(\Omega_{1,\varepsilon_0})} = Ch^2 \|u\|^2_{W_2^2(\Omega_1)}. \end{aligned} \tag{20}$$

The assertion follows from (15)-(20). ◊

From Lemma 1 and 2, taking in mind that $\widetilde{W}_2^1(\Omega) = \overset{\circ}{W}{}_2^1(\Omega)$, we immediately obtain the following result.

Theorem 1. *If the solution u of BVP (1) belong sto $\widetilde{W}_2^2(\Omega)$ then the finite element scheme (13) converges and the following convergence rate estimate holds:*

$$\|u - u_h\|_{\widetilde{W}_2^1(\Omega)} \leq Ch \|u\|_{\widetilde{W}_2^2(\Omega)}.$$

7 Convergence in $\widetilde{L}_2(\Omega)$

Let V_h be a subspace of H_A. We consider the following abstract approximation of the problem (12): *find* $u_h \in V_h$ *such that*

$$a(u_h, v_h) \equiv (u_h, v_h)_A + (u_h, v_h)_B = (f, v_h)\,, \qquad \forall\, v_h \in V_h\,. \tag{21}$$

From (12) and (21) follows

$$a(u - u_h, v_h) = 0\,, \qquad \forall\, v_h \in V_h\,. \tag{22}$$

Error norm $\|u - u_h\|_B$ can be represented as

$$\|u - u_h\|_B = \sup_{g \in H_B} \frac{|(g, u - u_h)_B|}{\|g\|_B} = \sup_{g \in H_B} \frac{|(Bg, u - u_h)|}{\|g\|_B}\,. \tag{23}$$

Let w be the solution of equation

$$(A + B)w = Bg\,.$$

Because of $g \in H_B$ it follows that $Bg \in H_{B^{-1}}$ and $w \in H_{AB^{-1}A}$. Taking inner product of the last equation with $v \in H_A$ we obtain:

$$a(v, w) = a(w, v) = (Bg, v)\,, \qquad \forall\, v \in H_A\,,$$

wherefrom, setting $v = u - u_h$ follows

$$a(u - u_h, w) = (Bg, u - u_h)\,. \tag{24}$$

From (22) and (24) follows

$$a(u - u_h, w - v_h) = (Bg, u - u_h)\,. \tag{25}$$

From (25), using H_A-boundedness of bilinear form $a(u, v)$, we obtain

$$|(Bg, u - u_h)| = |a(u - u_h, w - v_h)| \le C\, \|u - u_h\|_A \|w - v_h\|_A\,,$$

and because the left-hand side does not depend on v_h:

$$|(Bg, u - u_h)| \le C\, \|u - u_h\|_A \inf_{v_h \in V_h} \|w - v_h\|_A\,. \tag{26}$$

Let the Hilbert space H and linear operators A and B are defined as in Section 4 and let V_h be the finite element space defined in Section 5. Then the problem (21) coincides with (13). From Theorem 1 follows

$$\|u - u_h\|_A \le C\, \|u - u_h\|_{\widetilde{W}_2^1(\Omega)} \le C\, h\, \|u\|_{\widetilde{W}_2^2(\Omega)}\,, \tag{27}$$

while from the Lemma 2 one obtains

$$\begin{aligned} \inf_{v_h \in V_h} \|w - v_h\|_A &\le C \inf_{v_h \in V_h} \|w - v_h\|_{\widetilde{W}_2^1(\Omega)} \\ &\le C\, \|w - \Pi_h w\|_{\widetilde{W}_2^1(\Omega)} \le C\, h\, \|w\|_{\widetilde{W}_2^2(\Omega)}\,. \end{aligned} \tag{28}$$

Moreover, from (11) follows

$$\|w\|_{\widetilde{W}_2^2(\Omega)} \leq C\,\|w\|_{AB^{-1}A} \leq C\,\|Bg\|_{B^{-1}} = C\,\|g\|_B\,. \tag{29}$$

From (23) and (26)-(29) we finally obtain the desired convergence rate estimate in the norm $\widetilde{L}_2(\Omega)$:

$$\|u - u_h\|_{\widetilde{L}_2(\Omega)} \leq Ch^2\,\|u\|_{\widetilde{W}_2^2(\Omega)}\,. \tag{30}$$

In such a way, we proved the next assertion.

Theorem 2. *If the solution u of BVP (1) belongs to $\widetilde{W}_2^2(\Omega)$ then the finite element scheme (13) converges in the norm $\widetilde{L}_2(\Omega)$ and the convergence rate estimate (30) is satisfied.*

Remark 1. Let us denote that the obtained convergence rate estimates for the finite element scheme (13) approximating the BVP with interface (1) are analogous to the standard estimates where the spaces $W_2^k(\Omega)$ are replaced by $\widetilde{W}_2^k(\Omega)$.

Acknowledgement

The research of the first author was supported by Ministry of Science of Republic of Serbia under project 144005A, while the research of the second author was supported by Bulgarian National Fund of Sciences under project VU-MI-106/2005.

References

1. Babuska, I.: The finite element method for elliptic equations with discontinuous coefficients. Computing 5, 207–213 (1970)
2. Bramble, J., King, J.: A finite element method for interface problems in domains with smooth boundaries and interfaces. Advances in Comput. Math. 6, 109–138 (1996)
3. Cao, Y., Gunzburger, M.: Least-squares finite element approximations to solutions of interface problems. SIAM J. Numer. Anal. 35, 393–405 (1998)
4. Chen, Z., Zou, J.: Finite element methods and their convergence for elliptic and parabolic problems. Numer. Math. 79, 175–202 (1998)
5. Ciarlet, P.G.: The finite element method for elliptic problems. North–Holland, Amsterdam (1978)
6. Jovanović, B.S., Vulkov, L.G.: Operator's approach to the problems with concentrated factors. In: Vulkov, L.G., Waśniewski, J., Yalamov, P. (eds.) NAA 2000. LNCS, vol. 1988, pp. 439–450. Springer, Heidelberg (2001)
7. Jovanović, B.S., Vulkov, L.G.: On the convergence of finite difference schemes for the heat equation with concentrated capacity. Numer. Math. 89(4), 715–734 (2001)
8. Jovanović, B.S., Vulkov, L.G.: On the convergence of difference schemes for hyperbolic problems with concentrated data. SIAM J. Numer. Anal. 41(2), 516–538 (2003)

9. Jovanović, B.S., Vulkov, L.G.: Finite difference approximation of an elliptic interface problem with variable coefficients. In: Li, Z., Vulkov, L.G., Waśniewski, J. (eds.) NAA 2004. LNCS, vol. 3401, pp. 46–55. Springer, Heidelberg (2005)
10. Huang, J., Zou, J.: Some new a priori estimates for second-order elliptic and parabolic interface problems. J. Differ. Equations 184, 570–586 (2002)
11. LeVeque, R., Li, Z.: The immersed interface method for elliptic equations with discontinuous coefficients and singular sources. SIAM. J. Numer. Anal. 31, 1019–1044 (1994)
12. Li, Z.: The immersed interface method using a finite elemet formulation. Appl. Numer. Math. 27, 253–267 (1998)
13. Li, Z.: An overview of the immersed interface method and its applications. Taiwanese Journal of Mathematics 7(1), 1–49 (2003)
14. Li, Z., Ito, K.: The Immersed Interface Method – Numerical Solutions of PDEs Involving Interfaces and Irregular Domains. SIAM Frontier Series in Applied mathematics, FR33 (2006)
15. Li, Z., Lin, T., Wu, X.: New Cartesian grid methods for interface problem using finite element formulation. Numer. Math. 96, 61–98 (2003)
16. Lions, J.L., Magenes, E.: Non homogeneous boundary value problems and applications. Springer, Berlin (1972)
17. Lykov, A.V.: Heat and mass transfer. Energiya, Moscow (1978) (Russian)
18. Oganesyan, L. A., Rukhovets, L. A.: Variational-difference methods for solution of elliptic equations. Armenian AS, Erevan (1979) (Russian)
19. Riesz, F., Sz.-Nagy, B.: Leçons d'analyse fonctionelle. Akadémiai Kiadó, Budapest (1972)
20. Tikhonov, A. N., Samarskiĭ, A. A.: Equations of Mathematical Physics. GITTL, Moscow (1953) (Russian)
21. Vladimirov, V. S.: Equations of mathematical physics. Nauka, Moscow (1988) (Russian)
22. Vulkov, L.: Application of Steklov–type eigenvalues problems to convergence of difference schemes for parabolic and hyperbolic equation with dynamical boundary conditions. In: Vulkov, L.G., Yalamov, P., Waśniewski, J. (eds.) WNAA 1996. LNCS, vol. 1196, pp. 557–564. Springer, Heidelberg (1997)
23. Wloka, J.: Partial differential equations. Cambridge Univ. Press, Cambridge (1987)

Perturbation Bounds for Certain Matrix Expressions and Numerical Solution of Matrix Equations

M.M. Konstantinov[1], P.Hr. Petkov[2], and N.D. Christov[3]

[1] University of Architecture, Civil Engineering and Geodesy, 1046 Sofia, Bulgaria
mmk_fte@uacg.bg
[2] Technical University of Sofia, 1000 Sofia, Bulgaria
php@tu-sofia.bg
[3] Université des Sciences et Technologies de Lille, 59655 Villeneuve d'Ascq, France
ndchr@tu-sofia.bg

Abstract. The paper deals with the derivation of improved perturbation bounds for the matrix expression $A_1A_2^{-1}A_3$ and their application to the sensitivity analysis and the solution of fractional–affine matrix equations. An estimate of the overall error in the solution of matrix equations is also given.

1 Introduction

When solving matrix equations in a computing environment there are several sources of errors in the computed solution. Among them the following three are worth mentioning: (i) the machine arithmetic and in particular its *rounding unit*; (ii) the given equation and in particular its *sensitivity* to perturbations in the data, and (iii) the computational algorithm and in particular its *numerical stability*. The proper accounting of these three factors leads to derivation of an error bound on the computed solution. A numerical solution of a matrix equation may not be recognized as reliable without taking into account this error bound.

The sensitivity of matrix equations may be revealed and taken into account by the methods and techniques of perturbation analysis, see e.g. [9,3]. We stress also that the perturbation analysis of problems and processes is also of independent theoretical interest.

In what follows we present a technique to derive *improved perturbation bounds* for matrix expressions of the form $F(A) = A_1A_2^{-1}A_3$, $A = (A_1, A_2, A_3)$, under perturbations $A_k \to A_k + E_k$, where A_k are matrices of compatible sizes and A_2 is invertible.

Below we use the following notation: $\mathbb{R}^{m\times n}$ and $\mathbb{C}^{m\times n}$ – the spaces of $m \times n$ matrices over the field of real $\mathbb{R}$ and complex $\mathbb{C}$ numbers; $\mathbb{R}_+$ – the set of non–negative real numbers; $\mathbb{R}^n = \mathbb{R}^{n\times 1}$; I – an identity matrix of corresponding size; $A^\top$ and A^{H} – the transpose and the complex conjugate transpose of the matrix A; $\mathrm{vec}(A)$ – the column–wise vectorization of the matrix A; $A \otimes B$ – the Kronecker product of the matrices A and B; $\|\cdot\|$ – a vector or a matrix norm; $\|\cdot\|_{\mathrm{F}}$ and $\|\cdot\|_2$ – the Frobenius and the 2–norm of a matrix or a vector,

S. Margenov, L.G. Vulkov, and J. Waśniewski (Eds.): NAA 2008, LNCS 5434, pp. 68–79, 2009.

respectively; $\mathbf{u}$ – the rounding unit of the finite arithmetic. The notation ":=" means "equal by definition".

2 Problem Statement

Consider the matrix expression

$$F(A) := A_1 A_2^{-1} A_3, \; A := (A_1, A_2, A_3), \tag{1}$$

with A_2 being invertible, which arises in fractional–affine matrix equations such as the discrete–time algebraic Riccati equation $X = \Gamma + \Phi^{\mathrm{H}} X (I_n + \Sigma X)^{-1} \Phi$ in $\mathbb{C}^{n\times n}$.

Let $\varepsilon_k > 0$ be given quantities and suppose that the matrices A_k are subject to perturbations $A_k \to A_k + E_k$ such that $\|E_k\|_{\mathrm{F}} \leq \varepsilon_k$. We suppose also that

$$\varepsilon_2 < \sigma, \; \sigma := \|A_2^{-1}\|_2^{-1}$$

which guarantees that the matrix $A_2 + E_2$ is invertible. The last inequality should not seem too restrictive since for matrices E_2 with $\|E_2\|_{\mathrm{F}} = \sigma$ the matrix $A_2 + E_2$ may be singular.

Denoting $E = (E_1, E_2, E_3)$ and

$$\varphi(E) := \|F(A+E) - F(A)\|_{\mathrm{F}}, \; \varepsilon = [\varepsilon_1, \varepsilon_2, \varepsilon_3]^{\top} \in \mathbb{R}_+^3,$$

we may formulate our main problem as follows.

Main problem. *Determine the quantity*

$$\alpha(\varepsilon) := \sup\{\varphi(E) : \|E_k\|_{\mathrm{F}} \leq \varepsilon_k, \; k = 1, 2, 3\}$$

which yields the perturbation estimate $\varphi(E) \leq \alpha(\varepsilon)$.

Unfortunately, to find $\alpha(\varepsilon)$ explicitly is practically impossible except in some special cases, e.g. $A_k \in \mathbb{R}$.

Example 1. Let $A_k \neq 0$ and E_k be real scalars. Then

$$\alpha(\varepsilon) = \frac{|A_2 A_3|\varepsilon_1 + |A_1 A_3|\varepsilon_2 + |A_1 A_2|\varepsilon_3 + |A_2|\varepsilon_1\varepsilon_3}{|A_2|(|A_2| - \varepsilon_2)}, \; \varepsilon_2 < |A_2|.$$

We also have $\alpha(\varepsilon) \leq \alpha_1\varepsilon_1 + \alpha_2\varepsilon_2 + \alpha_3\varepsilon_3 + \beta\varepsilon_1\varepsilon_3$, where

$$\alpha_1 := \frac{|A_3|}{|A_2|(1-q)}, \; \alpha_2 := \frac{|A_1 A_3|}{A_2^2(1-q)}, \; \alpha_3 := \frac{|A_1|}{|A_2|(1-q)}, \; \beta := \frac{1}{|A_2|(1-q)},$$

provided $|E_2/A_2| \leq q$ for some positive constant $q < 1$.

So we come to the next problem.

Realistic problem. *Find a rigorous computable perturbation estimate*

$$\varphi(E) \leq b(\varepsilon) := b_1(\varepsilon) + b_2(\varepsilon)$$

such that the asymptotic relations $b_i(\varepsilon) = \mathrm{O}(\|\varepsilon\|^i)$, $\varepsilon \to 0$, $i = 1, 2$, *are valid.*

Moreover, our aim is to determine the bound $b_1(\varepsilon) + b_2(\varepsilon)$ as tight as possible.

3 Main Results

Denoting $B_2 := (A_2 + E_2)^{-1}$ we have $\|B_2\|_2 \le (\sigma - \varepsilon_2)^{-1}$. Next we represent the perturbed quantity $F(A+E) = (A_1 + E_1)B_2(A_3 + E_3)$ as

$$F(A+E) = A_1B_2A_3 + A_1B_2E_3 + E_1B_2A_3 + E_1B_2E_3.$$

We shall express B_2 in four different ways in the above four summands as follows:

in $A_1B_2A_3$ we write $B_2 = A_2^{-1} - A_2^{-1}E_2A_2^{-1} + A_2^{-1}E_2B_2E_2A_2^{-1}$;
in $A_1B_2E_3$ we write $B_2 = A_2^{-1} - A_2^{-1}E_2B_2$;
in $E_1B_2A_3$ we write $B_2 = A_2^{-1} - B_2E_2A_2^{-1}$;
in $E_1B_2E_3$ we leave B_2 as it is.

As a result we get

$$\begin{aligned} F(A+E) - F(A) = {} & \mathcal{L}_1(E_1) + \mathcal{L}_2(E_2) + \mathcal{L}_3(E_3) + E_1B_2E_3 \\ & - \mathcal{L}_1(E_1B_2E_2) - \mathcal{L}_2(E_2B_2E_2) - \mathcal{L}_3(E_2B_2E_3), \end{aligned} \tag{2}$$

where the linear matrix operators $\mathcal{L}_k$ are defined from

$$\mathcal{L}_1(E_1) := E_1A_2^{-1}A_3,\ \mathcal{L}_2(E_2) := -A_1A_2^{-1}E_2A_2^{-1}A_3,\ \mathcal{L}_3(E_3) := A_1A_2^{-1}E_3.$$

Note that the expression (2) is exact since there are no neglected terms. It is also symmetric and nicely looking.

Next we shall derive improved first order perturbation bounds. Consider the general expression

$$y = N_1z_1 + N_2z_2 + \cdots + N_mz_m,$$

where y and $z_1, z_2, \ldots, z_m$ are real or complex vectors and $N_1, N_2, \ldots, N_m$ are constant matrices of compatible size. Suppose that $\|z_i\|_2 \le \delta_i$, where $\delta := [\delta_1, \delta_2, \ldots, \delta_m]^\top$ is a given non–negative vector, and define the quantity

$$\gamma(N, \delta) := \sup\{\|y\|_2 : \|z_i\|_2 \le \delta_i,\ i = 1, 2, \ldots, m\},$$

where $N := \begin{bmatrix} N_1\ N_2 \ldots\ N_m \end{bmatrix}$. Unfortunately it is impossible to find $\gamma(N;\delta)$ explicitly in the general case. That is why we shall give an improved upper bound for this quantity.

We have the immediate estimate

$$\gamma(N;\delta) \le \min\{\mathrm{est}_1(N;\delta), \mathrm{est}_2(N;\delta)\},$$

where

$$\mathrm{est}_1(N;\delta) := \sum_{i=1}^{m} \|N_i\|_2\, \delta_i,\ \ \mathrm{est}_2(N;\delta) := \|N\|_2\|\delta\|_2.$$

We also have the less known bound

$$\gamma(N;\delta) \le \mathrm{est}_3(N;\delta) := \sqrt{\delta^\top M\delta},$$

where $M = [M_{ij}] \in \mathbb{R}_+^{m\times m}$ is a non–negative matrix with elements

$$M_{ij} := \left\|N_i^{\mathrm{H}}N_j\right\|_2,\ i, j = 1, 2, \ldots, m.$$

Since $\mathrm{est}_3(N;\delta) \le \mathrm{est}_1(N;\delta)$ we obtain the following result [5].

Theorem 1. *The improved estimate*

$$\gamma(N;\delta) \le \mathrm{est}(N;\delta) := \min\{\mathrm{est}_2(N;\delta), \mathrm{est}_3(N;\delta)\}$$

is valid.

Note that $\mathrm{est}(N;\delta)$ is not a linear but a first order homogeneous expression in δ. It is also interesting that the bound $\mathrm{est}_1(N;\delta)$ based on the absolute condition numbers $\|N_i\|_2$ and widely used in practice "disappeared" from the final bound $\mathrm{est}(N;\delta)$ because it may not be smaller than $\mathrm{est}_3(N;\delta)$.

Now setting $m = 3$, $e_i := \mathrm{vec}(E_i)$,

$$y := \mathrm{vec}(\mathcal{L}_1(E_1) + \mathcal{L}_2(E_2) + \mathcal{L}_3(E_3)) =: L_1 e_1 + L_2 e_2 + L_3 e_3$$

and

$$L_1 := \left(A_2^{-1}A_3\right)^\top \otimes I,\ L_2 := -\left(A_2^{-1}A_3\right)^\top \otimes \left(A_1 A_2^{-1}\right),\ L_3 := I \otimes \left(A_1 A_2^{-1}\right),$$

we obtain our main result.

Theorem 2. *The improved perturbation bound for the matrix expression* $F(A) = A_1 A_2^{-1} A_3$ *is*

$$\|F(A+E) - F(A)\|_{\mathrm{F}} \le b(\varepsilon) := b_1(\varepsilon) + b_2(\varepsilon),\ \varepsilon_2 < \sigma,$$

where

$$b(\varepsilon) := \frac{\sigma\,\mathrm{est}(L;\varepsilon) + \varepsilon_1\varepsilon_3}{\sigma - \varepsilon_2} = \mathrm{est}(L;\varepsilon) + \frac{\varepsilon_1\varepsilon_3 + \varepsilon_2 \mathrm{est}(L;\varepsilon)}{\sigma - \varepsilon_2}.$$

An interesting observation is that in the scalar case Theorem 2 gives the exact perturbation bound.

A special case arises in complex matrix equations with complex conjugate coefficients such as the Lyapunov equation

$$XA_1 + A_1^{\mathrm{H}} X + A_2 = 0.$$

Here $e_1 = \mathrm{vec}(E_1)$ and $e_2 = \overline{e}_1$. It is worth mentioning that until the year 1999 most published perturbation results for such equations [10] were either not tight or contained a systematic error, see [4], where the correct result was given. Below we show how to obtain correct perturbation bounds in such cases.

Consider the complex expression $z = Gy + H\overline{y}$, where y, z are vectors and $G = G_0 + \imath G_1$, $H = H_0 + \imath H_1$ are matrices of compatible sizes with G_i and H_i real. Suppose that we want to estimate $\|z\|_2$ under the restriction $\|y\|_2 \le \eta$. We have $\|z\|_2 \le \|C\|_2 \eta$, where the matrix C is defined from [4]

$$C := \begin{bmatrix} G_0 + H_0 & H_1 - G_1 \\ G_1 + H_1 & G_0 - H_0 \end{bmatrix}.$$

Consider now the more general matrix expression

$$F(A) := A_1 F_2(A_2) A_3,$$

where F_2 is a differentiable matrix function in a neighborhood of a given matrix argument A_2 (in the previous case $F_2(A_2) = A_2^{-1}$). Then we have

$$F_2(A_2 + E_2) = F_2(A_2) + \Phi_1(A_2, E_2) = F_2(A_2) + F_2'(A_2)(E_2) + \Phi_2(A_2, E_2),$$

where

$$\Phi_1(A_2, E_2) := F_2(A_2 + E_2) - F_2(A_2),\ \Phi_2(A_2, E_2) := \Phi_1(A_2, E_2) - F_2'(A_2)(E_2)$$

and $F_2'(A_2)$ is the Fréchet derivative of the function F_2 computed at the point A_2.

Suppose that for some $\sigma > 0$ and $\|E_2\|_{\mathrm{F}} < \sigma$ it is fulfilled

$$\|F_2(A_2 + E_2)\|_{\mathrm{F}} = \mathrm{O}(1),\ \|\Phi_i(A_2, E_2)\|_{\mathrm{F}} = \mathrm{O}\left(\|E_2\|^i\right),\ E_2 \to 0.$$

Then we may apply the above technique in order to obtain

$$F(A + E) - F(A) = \Psi_1(A, E) + \Psi_2(A, E),$$

where

$$\begin{aligned}
\Psi_1(A, E) &:= E_1 F_2(A_2 + E_2) A_3 + A_1 F_2'(A_2)(E_2) A_3 + A_1 F_2(A_2 + E_2) E_3, \\
\Psi_2(A, E) &:= E_1 \Phi_1(A_2, E_2) A_3 + A_1 \Phi_1(A_2, E_2) E_3 + A_1 \Phi_2(A_2, E_2) A_3 \\
&\quad + E_1 F_2(A_2 + E_2) E_3.
\end{aligned}$$

Like in the previous case $\|\Psi_i(A, E)\| = \mathrm{O}(\|\varepsilon\|^i)$, $\varepsilon \to 0$.

4 Application to Matrix Equations

4.1 Equivalent Operator Equation

The above technique is applicable to the perturbation analysis of many algebraic matrix equations. Consider the algebraic matrix equation

$$F(A, X) = 0, \tag{3}$$

where $A := (A_1, A_2, \ldots, A_m)$ and A_i are matrix coefficients subject to perturbations $A_i \to A_i + E_i$. Let $X + Y$ be the solution of the perturbed equation

$$F(A + E, X + Y) = 0, \tag{4}$$

where $E := (E_1, E_2, \ldots, E_m)$.

The aim of norm–wise perturbation analysis here is to estimate the norm $\|Y\|_{\mathrm{F}}$ of Y as a function of the perturbation vector

$$\delta = [\delta_1, \delta_2, \ldots, \delta_m]^\top,\ \delta_i := \|E_i\|_{\mathrm{F}}.$$

If the function F is differentiable in X and differentiable or pseudo–differentiable in A, the perturbed equation may be written as an equivalent operator equation

$$Y = \Pi(E, Y), \tag{5}$$

where

$$\Pi(E,Y) := -F_X^{-1}(A,X)(F_A(A,X)(E) + G(A,X,E,Y)),$$
$$G(A,X,E,Y) := F(A+E,X+Y) - F(A,X) - F_A(A,X)(E) - F_X(A,X)(Y).$$

In turn, the matrix equation (5) may be transformed into a vector equation

$$y = P(e,y), \tag{6}$$

where $y := \mathrm{vec}(Y)$, $e := \mathrm{vec}(E)$ and $P(e,y) := \mathrm{vec}(\Pi(\mathrm{vec}^{-1}(e), \mathrm{vec}^{-1}(y)))$.

It may be shown that the operator $P(e,\cdot)$ transforms into itself a "small" set $\mathcal{B}_\rho$ of radius ρ vanishing together with $\|E\|$. Thus according to the Schauder fixed point principle there is a small solution $Y \in \mathcal{B}_\rho$ with $\|Y\|_{\mathrm{F}} \leq \rho$. The last inequality is the desired non–local perturbation estimate.

4.2 Lyapunov Majorants

The technique of Lyapunov majorants goes back to the monographs [7,1] and, of course, to the original works of A. Lyapunov. Further developments on this subject may be found in [3].

Consider the operator equation (5) in Y, where

$$A = (A_1, A_2, \ldots, A_m), \; E = (E_1, E_2, \ldots, E_m),$$

and let $\delta := [\delta_1, \delta_2, \ldots, \delta_m]^\top$ be a non–negative vector.

Definition 1. *The function $l : \mathbb{R}_+^m \times \mathbb{R}_+ \to \mathbb{R}_+$, defined by*

$$l(\delta,\rho) := \max\{\|\Pi(E,Y)\|_{\mathrm{F}} : \|E_i\|_{\mathrm{F}} \leq \delta_i, \; \|Y\|_{\mathrm{F}} \leq \rho\}$$

is said to be the exact Lyapunov majorant *for the operator Π.*

Usually the exact Lyapunov majorant l is difficult to be constructed and we use an easily computable function h such that $l(\delta,\rho) \leq h(\delta,\rho)$.

The function h is non–decreasing and convex (strictly convex for non–linear equations) in all its arguments, differentiable in ρ and satisfies the conditions $h(0,0) = 0$ and $h'_\rho(0,0) < 1$.

Definition 2. *The function h, satisfying the above conditions, is said to be an* approximate Lyapunov majorant *for the operator Π.*

The technique of Lyapunov majorants is based on the *majorant equation*

$$\rho = h(\delta,\rho). \tag{7}$$

Under the above conditions we have the following result.

Theorem 3. *For non–linear equations there exists a domain $\Omega \subset \mathbb{R}^m_+$ such that:*

(i) for $\delta \in \Omega$ the majorant equation (7) in ρ has two roots $\rho_1(\delta) < \rho_2(\delta)$ which are continuous in δ and $\rho_1(0) = 0$;

(ii) for some points δ on the boundary $\partial\Omega$ of Ω it is fulfilled $\rho_1(\delta) = \rho_2(\rho)$.

Denote by $\overline{\Omega}$ the closure of Ω and set $f(\delta) := \rho_1(\delta)$ and $\mathcal{B}_r := \{y : \|y\|_2 \leq r\}$.

For $\delta \in \overline{\Omega}$ and $\|y\|_2 \leq f(\delta)$ we have $\|P(e, y)\|_2 \leq f(\delta)$. Thus the operator $P(e, \cdot)$ transforms the set $\mathcal{B}_{f(\delta)}$ into itself. Hence, according to the Schauder fixed point principle, there is a solution $y \in \mathcal{B}_{f(\delta)}$ of the operator equation $y = P(e, y)$. As a corollary we have the following important result.

Theorem 4. *For $\delta \in \overline{\Omega}$ the perturbed equation (4) has a solution for which the non–local non–linear perturbation estimate*

$$\|Y\|_{\mathrm{F}} \leq f(\delta) \tag{8}$$

holds.

We stress two important issues about Theorem 4. Firstly, the inclusion $\delta \in \overline{\Omega}$ guarantees that the perturbed equation (4) has a solution. Secondly, the estimate (8) is rigorous. This estimate, however, may be pessimistic in certain cases when e.g. the domain Ω is small. Nevertheless, this domain is finite and the estimate (8) is non–local.

In practice the domain Ω is not constructed explicitly. Rather, the inclusion $\delta \in \overline{\Omega}$ is checked directly by a single inequality.

4.3 Fractional–Affine Equations

Fractional–affine matrix equations involve inversions of affine expressions in X. Typical example here is the discrete–time matrix Riccati equation $Q - X + A^{\mathrm{H}}X(I + MX)^{-1}A = 0$, where the matrices $Q = Q^{\mathrm{H}}$ and $M = M^{\mathrm{H}}$ are non–negative definite, the pair $[A, M)$ is controllable and the pair $(Q, A]$ is detectable. The Lyapunov majorant $h(\rho, \delta)$ for such equations may be constructed as a fractional–affine function in ρ.

Consider the Lyapunov majorant function

$$h(\delta, \rho) := b_0(\delta) + b_1(\delta)\rho + \frac{b_2(\delta) + b_3(\delta)\rho + b_4(\delta)\rho^2}{b_5(\delta) - b_6(\delta)\rho}, \tag{9}$$

where $\delta \in \mathbb{R}^m_+$ and $b_k(\delta) \geq 0$. Suppose that (i) all functions b_k are continuous; (ii) the functions b_k are non–decreasing in δ for $k \neq 5$; (iii) the function b_5 is positive and non–increasing; (iv) the relations

$$b_0(0) = b_2(0) = 0,\ b_1(0) < 1,\ b_5(0) > 0,\ b_6(0) > 0$$

and

$$h'_\rho(0, 0) = b_1(0) + \frac{b_3(0)}{b_5(0)} < 1$$

are fulfilled.

Denote

$$\begin{aligned} c_0(\delta) &:= b_2(\delta) + b_0(\delta)b_5(\delta), \\ c_1(\delta) &:= b_5(\delta)(1 - b_1(\delta)) + b_0(\delta)b_6(\delta) - b_3(\delta), \\ c_2(\delta) &:= b_4(\delta) + b_6(\delta)(1 - b_1(\delta)). \end{aligned}$$

Then we have $c_0(0) = 0$ and $c_1(0) = b_5(0)(1 - h'_\rho(0,0)) > 0$. Moreover, for small δ we have $c_1(\delta) > 0$ and $c_1^2(\delta) > 4c_0(\delta)c_2(\delta)$.

The majorant equation $\rho = h(\delta, \rho)$ may be written as

$$c_2(\delta)\rho^2 - c_1(\delta)\rho + c_0(\delta) = 0$$

and hence we have the following result.

Theorem 5. *The set*

$$\Omega := \left\{\delta \in \mathbb{R}^m_+ : c_1(\delta) > 0,\ c_1^2(\delta) \geq 4c_0(\delta)c_2(\delta)\right\}$$

has interior points δ, i.e. all elements of δ are positive, and the perturbation bound is

$$f(\delta) := \frac{2c_0(\delta)}{c_1(\delta) + \sqrt{c_1^2(\delta) - 4c_0(\delta)c_2(\delta)}},\quad \delta \in \Omega.$$

Consider for example the matrix fractional–affine equation

$$F(A, X) := A_1 + A_2X + XA_3 + A_4X^{-1}A_5 = 0, \tag{10}$$

where $A := (A_1, A_2, A_3, A_4, A_5)$, $A_i, X \in \mathbb{C}^{n\times n}$. This a generalization of the well studied matrix equation $X = Q \pm R^\top X^{-1}R$, arising in system theory.

As before, let the matrix coefficients A_i be perturbed to $A_i + E_i$ and let $X + Y$ be the solution of the perturbed equation

$$F(A + E, X + Y) = 0,\ E := (E_1, E_2, E_3, E_4, E_5).$$

Suppose that $\|Y\|_\mathrm{F} \leq \rho$ and $\rho < \sigma := \|X^{-1}\|_2^{-1}$. Then the matrix $Z := X + Y$ is invertible, and

$$\begin{aligned} Z^{-1} &= X^{-1} - X^{-1}YZ^{-1} = X^{-1} - Z^{-1}YX^{-1} \\ &= X^{-1} - X^{-1}YX^{-1} + X^{-1}YZ^{-1}YX^{-1}. \end{aligned}$$

Moreover, we have $\|Z^{-1}\|_2 \leq (\sigma - \rho)^{-1}$.

The perturbation analysis presented below is based on the identity

$$F(A + E, X + Y) = F(A, X) + \mathcal{K}(Y) + F_0(E) + F_1(E, Y) + F_2(Y),$$

where

$$\begin{aligned} \mathcal{K}(Y) &:= F_X(A, X)(Y) = A_2Y + YA_3 - A_4X^{-1}YX^{-1}A_5, \\ F_0(E) &:= E_1 + E_2X + XE_3 + A_4X^{-1}E_5 + E_4X^{-1}A_5 + E_4Z^{-1}E_5, \\ F_1(E, Y) &:= E_2Y + YE_3 - A_4X^{-1}YZ^{-1}E_5 - E_4Z^{-1}YX^{-1}A_5, \\ F_2(Y) &:= A_4X^{-1}YZ^{-1}YX^{-1}A_5. \end{aligned}$$

Suppose that the linear matrix operator $\mathcal{K}$ is invertible and denote

$$B := \left(X^{-1}A_5\right)^{\top} \otimes \left(A_4X^{-1}\right).$$

Then the matrix $K := I_n \otimes A_2 + A_3^{\top} \otimes I_n - B$ of $\mathcal{K}$ is also invertible. Hence the perturbed equation may be written as

$$y = P(e,y) := P_0(e) + P_1(e,y) + P_2(y),$$

where $y := \mathrm{vec}(Y)$, $e := \mathrm{vec}(E)$, $e_k := \mathrm{vec}(E_k)$ and

$$\begin{aligned}
P_0(e) &:= N_1e_1 + N_2e_2 + N_3e_3 + N_4e_4 + N_5e_5 + N_1\mathrm{vec}(E_4Z^{-1}E_5),\\
P_1(e,y) &:= N_1\mathrm{vec}(E_2Y + YE_3) - \; N_4\mathrm{vec}(E_4Z^{-1}Y) - N_5\mathrm{vec}(YZ^{-1}E_5)\\
&\quad + N_1B\mathrm{vec}(YZ^{-1}Z),\\
P_2(y) &:= A_4X^{-1}YZ^{-1}YX^{-1}A_5.
\end{aligned}$$

Here the matrices N_k are defined from

$$\begin{aligned}
&N_1 := -K^{-1},\; N_2 := N_1(I_n \otimes X),\; N_3 := N_1(I_n \otimes X),\\
&N_4 := N_1\left(\left(X^{-1}A_5\right)^{\top} \otimes I_n\right),\; N_5 := N_1\left(I_n \otimes \left(A_4X^{-1}\right)\right).
\end{aligned}$$

For $\|y\|_2 \le \rho < \sigma$ and after standard calculations we get

$$\begin{aligned}
\|P_0(e)\|_2 &\le \mathrm{est}(N_1,N_2,N_3,N_4;\delta_1,\delta_2,\delta_3,\delta_4) + \frac{\nu\delta_4\delta_5}{\sigma-\rho},\\
\|P_1(e,y)\| &\le \rho\nu(\delta_2+\delta_3) + \frac{\rho\,\mathrm{est}(L_4,L_5;\delta_4,\delta_5)}{\sigma-\rho},\\
\|P_2(y)\|_2 &\le \frac{\beta\rho^2}{\sigma-\rho},
\end{aligned}$$

where $\nu := \|N_1\|_2$ and $\beta := \|N_1B\|_2$.

Using these inequalities we obtain a Lyapunov majorant of type (9) with

$$\begin{aligned}
&b_0(\delta) := \mathrm{est}(N_1,N_2,N_3,N_4;\delta_1,\delta_2,\delta_3,\delta_4),\; b_1(\delta) := \nu(\delta_2+\delta_3),\\
&b_2(\delta) := \nu\delta_4\delta_5,\; b_3(\delta) := \mathrm{est}(N_4,N_5;\delta_4,\delta_5),\; b_4 := \beta,\; b_5 := \sigma,\; b_6 := 1.
\end{aligned}$$

In this case

$$c_0(0) = 0,\; c_1(0) = \sigma > 0,\; c_2(0) = 1 + \beta$$

and the domain Ω is correctly defined.

An important particular case arises when $\delta_k \le \mathbf{u}\alpha_k$, where $\alpha_k := \|A_k\|_{\mathrm{F}}$. Here it is fulfilled

$$\begin{aligned}
&b_0(\delta) := \mathbf{u}\,\mathrm{est}(N_1,N_2,N_3,N_4;\alpha_1,\alpha_2,\alpha_3,\alpha_4),\; b_1(\delta) := \mathbf{u}\nu(\alpha_2+\alpha_3),\\
&b_2(\delta) := \mathbf{u}^2\nu\alpha_4\alpha_5,\; b_3(\delta) := \mathbf{u}\,\mathrm{est}(N_4,N_5;\alpha_4,\alpha_5),\; b_4 := \beta,\; b_5 := \sigma,\; b_6 := 1.
\end{aligned}$$

4.4 An Accuracy Estimate

Consider again the matrix equation (3) and its perturbed version (4), where A is considered as a single matrix for simplicity. Suppose that the partial Frécher derivatives F_X and F_A of F in X and A, respectively, exist. Suppose also that the linear operator F_X is invertible. Then, within terms of first order in E, Y we have

$$Y \approx -F_X^{-1}(A,X) \circ F_A(A,X)(E)$$

and hence the absolute condition number of the problem is

$$\mathrm{Cond} = \left\| F_X^{-1}(A,X) \circ F_A(A,X) \right\|.$$

If Mat is the matrix of the linear operator $-F_X^{-1}(A,X) \circ F_A(A,X)$ and the Frobenius norm is used for the perturbations then $\mathrm{Cond} = \|\mathrm{Mat}\|_2$.

When $A \neq 0$ and $X \neq 0$ we define the *relative condition number* from

$$\mathrm{cond} := \mathrm{Cond}\frac{\|A\|}{\|X\|}.$$

For small E we have

$$\delta_X \leq \mathrm{cond}\, \delta_A + \mathrm{o}(\delta_A),\ \delta_A \to 0,$$

where

$$\delta_X := \frac{\|Y\|}{\|X\|},\ \delta_A := \frac{\|E\|}{\|A\|}$$

are the relative perturbations in the solution and the data, respectively.

For complex equations, however, the dependence of X on A may not be Frécher differentiable but only Fréchet pseudo–differentiable, see [6]. In this case

$$Y = \mathcal{L}(E) + \mathcal{M}(E^{\mathrm{H}}) + \mathrm{o}(\|E\|),\ E \to 0,$$

where $\mathcal{L}$ and $\mathcal{M}$ are linear matrix operators with matrices L and M, respectively. Vectorizing both sides of this equality we get

$$y = Le + MV\overline{e} + \mathrm{o}(\|e\|),\ e \to 0.$$

Here V is the vec–permutation matrix of corresponding size such that $\mathrm{vec}(Z^\top) = V\mathrm{vec}(Z)$.

Denote by L_0 and L_1 the real and imaginary parts of L, and by M_0 and M_1 the real and imaginary parts of M. Then the absolute condition number of the problem in the Frobenius norm is

$$\mathrm{Cond} := \left\| \begin{bmatrix} L_0 + M_0V & M_1V - L_0 \\ L_1 + M_1V & L_0 - M_0V \end{bmatrix} \right\|_2,$$

see [4]. It is interesting to note that perturbation bounds for such equations (see [10] for example) published before the paper [4] were either not very tight, or incorrect.

Set $X = \Phi(A)$ and let a numerically stable algorithm be applied to solve the equation in a computing environment with rounding unit $\mathbf{u}$. Then, within first order terms in $\mathbf{u}$, the computed solution X^* shall be close to the exact solution of a near problem in the sense that

$$\|X^* - \Phi(A^*)\| \le a\mathbf{u}\|X\|, \ \|A^* - A\| \le b\mathbf{u}\|A\|,$$

where the constants a and b depend on the algorithm.

Note that for $a = 0$ the computed solution X^* is the exact solution $\Phi(A^*)$ of a near problem since A^* and A are $\mathbf{u}$–close. Thus the algorithm is backwardly stable in the sense of Wilkinson. When $b = 0$ we have $A^* = A$ and the algorithm is forwardly stable.

Now we may estimate the actual error in the computed solution as [8,2]

$$\begin{aligned}\|X^* - X\| &= \|X^* - \Phi(A)\| = \|X^* - \Phi(A^*) + \Phi(A^*) - \Phi(A)\| \\ &\le \|X^* - \Phi(A^*)\| + \|\Phi(A^*) - \Phi(A)\| \\ &\le a\mathbf{u}\|X\| + \mathrm{Cond}\|A^* - A\| \le a\mathbf{u}\|X\| + \mathrm{Cond}\, b\mathbf{u}\|A\|.\end{aligned}$$

Dividing both sides of the last inequality by $\|X\|$ (when $X \neq 0$) we get

$$\frac{\|X^* - X\|}{\|X\|} \le \mathbf{u}(a + b\mathrm{cond}). \tag{11}$$

The inequality (11) clearly reveals the three main factors determining the relative error in the computed solution:

- the parameters of the computing environment via the rounding unit $\mathbf{u}$;
- the sensitivity of the equation via the relative condition number cond;
- the properties of the computational algorithm via the constants a and b.

In practice it is difficult to estimate a and b. If we set $a = 0$ and $b = 1$ the heuristic inequality

$$\frac{\|X^* - X\|}{\|X\|} \le \mathbf{u}\,\mathrm{cond} \tag{12}$$

is obtained. Thus we have the following well known heuristic rule.

When $\mathbf{u}\,\mathrm{cond} < 1$ *we may expect about*

$$-\log_{10}(\mathbf{u}\,\mathrm{cond})$$

true decimal digits in the computed solution.

References

1. Grebenikov, E., Ryabov, Y.: Constructive Methods for Analysis of Nonlinear Systems. Nauka, Moscow (1979) (in Russian)
2. Higham, N., Konstantinov, M., Mehrmann, V., Petkov, P.: The Sensitivity of Computational Control Problems. IEEE Control Syst. Magazine 24, 28–43 (2004)

3. Konstantinov, M., Gu, D., Mehrmann, V., Petkov, P.: Perturbation Theory for Matrix Equations. Elsevier, Amsterdam (2003)
4. Konstantinov, M., Petkov, P.: Note on Perturbation theory for algebaic Riccati equations (SIAM J. Matrix Anal. Appl. 19, 39–65, by J.G. Sun (1998)). SIAM J. Matrix Anal. Appl. 21, 327 (1999)
5. Konstantinov, M., Petkov, P., Gu, D.: Improved Perturbation Bounds for General Quadratic Matrix Equations. Numer. Funct. Anal. Opim. 20, 717–736 (1999)
6. Konstantinov, M., Stanislavova, M., Petkov, P.: Perturbation Bounds and Characterisation of the Solution of the Associated Algebraic Riccati Equation. Linear Algebra Appl. 285, 7–31 (1998)
7. Lika, D., Ryabov, Y.: Iterative Methods and Lyapunov Majorant Equations in Non–Linear Osillation Theory. Shtiinca, Kishinev (in Russian) (1974)
8. Petkov, P., Christov, N., Konstantinov, M.: Computational Methods for Linear Control Problems. Prentice Hall, Hemel Hempstead (1991)
9. Stewart, S., Sun, J.: Matrix Perturbation Theory. Academic Press, New York (1990)
10. Sun, J.: Perturbation Theory for Algebraic Riccati Equations. SIAM J. Matrix Anal. Appl. 19, 39–65 (1998)

Numerical Analysis of a 2d Singularly Perturbed Semilinear Reaction-Diffusion Problem

Natalia Kopteva

Department of Mathematics and Statistics, University of Limerick, Limerick, Ireland
natalia.kopteva@ul.ie
www.staff.ul.ie/natalia/

Abstract. A semilinear reaction-diffusion equation with multiple solutions is considered in a smooth two-dimensional domain. Its diffusion parameter ε^2 is arbitrarily small, which induces boundary layers. We extend the numerical method and its maximum norm error analysis of the paper [N. Kopteva: Math. Comp. 76 (2007) 631–646], in which a parametrization of the boundary $\partial\Omega$ is assumed to be known, to a more practical case when the domain is defined by an ordered set of boundary points. It is shown that, using layer-adapted meshes, one gets second-order convergence in the discrete maximum norm, uniformly in ε for $\varepsilon \le Ch$. Here $h > 0$ is the maximum side length of mesh elements, while the number of mesh nodes does not exceed Ch^{-2}. Numerical results are presented that support our theoretical error estimates.

1 Introduction

Consider the singularly perturbed semilinear reaction-diffusion problem

$$Fu \equiv -\varepsilon^2 \triangle u + b(x,u) = 0, \qquad x = (x_1, x_2) \in \Omega \subset \mathbb{R}^2, \tag{1a}$$

$$u(x) = g(x), \qquad x \in \partial\Omega, \tag{1b}$$

where ε is a small positive parameter, $\triangle = \partial^2/\partial x_1^2 + \partial^2/\partial x_2^2$ is the Laplace operator, and Ω is a bounded two-dimensional domain whose boundary $\partial\Omega$ is sufficiently smooth. Assume also that the functions b and g are sufficiently smooth. We shall examine solutions of (1) that exhibit boundary layer behaviour.

The aim of the present paper is to extend the numerical method and its maximum norm error analysis of the recent paper [4], in which a parametrization of the boundary $\partial\Omega$ is assumed to be known, to a more practical case when the domain is defined by an ordered set of boundary points $\{(\varphi_j, \psi_j)\}_{j=0}^{M}$, where $(\varphi_0, \psi_0) = (\varphi_M, \psi_M)$ and the distance between any two consecutive points $(\varphi_{j-1}, \psi_{j-1})$ and (φ_j, ψ_j) does not exceed Ch for some constant C, while $C^{-1}h \le M \le Ch$. A preliminary presentation of our results was given in [5]; now we introduce a more intricate analysis that allows a less accurate, but simpler approximation of the domain boundary curvature.

S. Margenov, L.G. Vulkov, and J. Waśniewski (Eds.): NAA 2008, LNCS 5434, pp. 80–91, 2009.

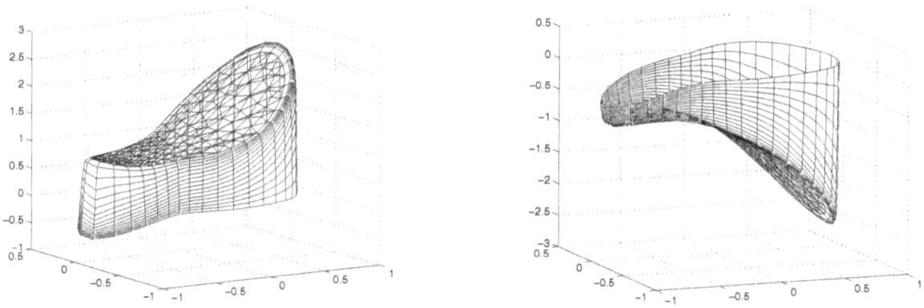

Fig. 1. Multiple boundary-layer solutions of model problem (32); in the interior subdomain $u(x) \approx \bar{u}_0(x)$ (left) or $u(x) \approx -\bar{u}_0(x)$ (right), where $\pm\bar{u}_0(x)$ are stable solutions of the reduced problem (2)

The reduced problem of (1) is defined by formally setting $\varepsilon = 0$ in (1a), i.e.

$$b(x, u_0(x)) = 0 \quad \text{for } x \in \Omega. \tag{2}$$

Any solution u_0 of (2) does not in general satisfy the boundary condition (1b).

In the numerical analysis literature it is often assumed that $b_u(x,u) > \gamma^2 > 0$ for all $(x,u) \in \Omega \times \mathbb{R}^1$, for some positive constant γ. Under this condition the reduced problem has a unique solution u_0, which is sufficiently smooth in $\bar{\Omega}$. This global condition is nevertheless rather restrictive. E.g., mathematical models of biological and chemical processes frequently involve problems related to (1) with $b(x,u)$ that is *non-monotone* with respect to u [3, §2.3], [7, §14.7]. Hence, following [4], we consider problem (1) under the following weaker *assumptions* from [2,8]:

– it has a *stable reduced solution*, i.e., there exists a sufficiently smooth solution u_0 of (2) such that

$$b_u(x, u_0) > \gamma^2 > 0 \quad \text{for all } x \in \Omega; \tag{A1}$$

– the boundary condition satisfies

$$\int_{u_0(x)}^{v} b(x,s)\, ds > 0 \qquad \text{for all } v \in \big(u_0(x), g(x)\big]', \qquad x \in \partial\Omega. \tag{A2}$$

Here the notation $(a,b]'$ is defined to be $(a,b]$ when $a < b$ and $[b,a)$ when $a > b$, while $(a,b]' = \emptyset$ when $a = b$.

If $g(x) \approx u_0(x)$, then (A2) follows from (A1) combined with (2); if $g(x) = u_0(x)$ for some $x \in \partial\Omega$, then (A2) does not impose any restriction on g at this point.

Conditions (A1), (A2) intrinsically arise from the asymptotic analysis of problem (1) and guarantee that there exists a boundary-layer solution u of (1) such that $u \approx u_0$ in the interior subdomain of Ω away from the boundary, while the boundary layer is of width $O(\varepsilon|\ln\varepsilon|)$ [2,8,4]. Note that assumption (A1) is local. Furthermore, if multiple stable solutions of the reduced problem satisfy (A2), problem (1) has *multiple* boundary-layer solutions; see Figure 1.

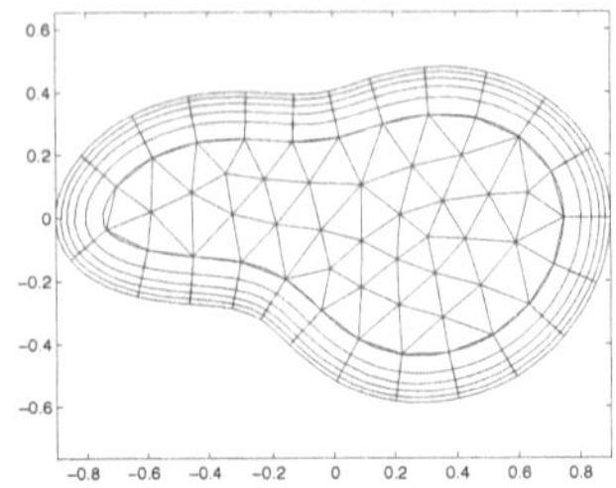

Fig. 2. Layer-adapted mesh of Bakhvalov/Shishkin type

We discretize the domain as in Figure 2—see §§3.1, 4.1 for details—using layer-adapted meshes of Bakhvalov and Shishkin types whose number of mesh nodes does not exceed Ch^{-2}. Here $h > 0$ is the maximum side length of mesh elements of the layer-adapted meshes that we consider.

Then we discretize equation (1a) combining finite differences on the curvilinear tensor-product part of the mesh and lumped mass linear finite elements on a quasiuniform Delaunay triangulation in the interior region. Imitating the analysis of [4], we construct discrete sub- and super-solutions and then invoke the theory of Z-fields to prove existence and estimate the accuracy of multiple discrete solutions of problem (1). Our main result, Theorem 3, states that our numerical method is second-order convergent (with, in the case of the Shishkin mesh, a logarithmic factor) in the discrete maximum norm, uniformly in ε.

Throughout our analysis we assume that

$$\varepsilon \leq Ch. \tag{$A3$}$$

This is not a practical restriction, and from a theoretical viewpoint the analysis of a nonlinear problem such as (1) would be very different if ε were not small.

Note that similar two-dimensional problems were considered by Schatz and Wahlbin (1983), Blatov (1992), Melenk (2002), Clavero *et al.* (2005); see [4] for these references and a further discussion.

The paper is organized as follows. In §2 we discuss asymptotic properties of solutions of (1) and describe sub- and super-solutions. In §3 we recall the layer-adapted meshes and the numerical method from [4], which explicitly uses a parametrization of the boundary $\partial\Omega$. In §4 the above method is extended to a more practical case when the domain is defined by an ordered set of boundary points. Precise convergence results for the numerical method are then derived on Bakhvalov and Shishkin meshes. Finally, in §5, we present numerical results that support our error estimates.

Notation. Throughout this paper we let C denote a generic positive constant that may take different values in different formulas, but is always independent of h and ε. A subscripted C (e.g., C_1) denotes a positive constant that is independent of h and ε and takes a fixed value. For any two quantities w_1 and w_2, the notation $w_1 = O(w_2)$ means $|w_1| \leq Cw_2$.

2 Local Curvilinear Coordinates. Asymptotic Expansion. Sub- and Super-Solutions

Given a sufficiently smooth boundary $\partial\Omega$, let its unknown arc-length parametrization with the counterclockwise orientation be defined by

$$x_1 = \varphi(l), \qquad x_2 = \psi(l), \qquad 0 \le l \le L, \tag{3}$$

where L is the arc-length of $\partial\Omega$. Hence the tangent vector (φ', ψ') has magnitude $\tau = \sqrt{\varphi'^2 + \psi'^2} = 1$ for all l. Furthermore, $(\varphi(0), \psi(0)) = (\varphi(L), \psi(L))$ and all functions that are defined for l beyond $[0, L]$ are understood as extended L-periodically. We also use the curvature κ of the boundary at $(\varphi(l), \psi(l))$ given by

$$\kappa = \kappa(l) = \varphi'\psi'' - \psi'\varphi''. \tag{4}$$

In a narrow neighbourhood of $\partial\Omega$ that will be specified later, introduce the curvilinear local coordinates (r, l) by

$$x_1 = \varphi(l) - r\psi'(l), \qquad x_2 = \psi(l) + r\varphi'(l), \tag{5}$$

where $(-\psi', \varphi')$ is the inward unit normal to $\partial\Omega$ at $(\varphi(l), \psi(l))$, which is orthogonal to the tangent vector (φ', ψ'). Since $\partial\Omega$ is smooth, there exists a sufficiently small constant C_1 such that in the subdomain $\{0 < r < C_1\}$ the new coordinates are well-defined. Below we shall use a smooth positive cut-off function $\omega(x)$ that equals 1 for $r \le C_1/2$ and vanishes in the interior of the closed curve $\{r = C_1\}$.

Lemma 1 ([4, Lemma 2.1]). *For the Laplace operator we have*

$$\triangle u = \eta^{-1}\Big[\frac{\partial}{\partial r}\Big(\eta\frac{\partial u}{\partial r}\Big) + \frac{\partial}{\partial l}\Big(\eta^{-1}\frac{\partial u}{\partial l}\Big)\Big], \quad \textit{where } \eta := 1 - \kappa r. \tag{6}$$

To obtain an asymptotic expansion, introduce the stretched variable $\xi := r/\varepsilon$ and the function $v_0(\xi, l)$ defined by $-\partial^2 v_0/\partial\xi^2 + b(\bar{x}, u_0(\bar{x}) + v_0) = 0$ for $\xi > 0$, with the boundary conditions $v_0(0, l) = g(\bar{x}) - u_0(\bar{x})$ and $v_0(\infty, l) = 0$. Here $\bar{x} = \bar{x}(l) := (\varphi(l), \psi(l))$. Our conditions (A1),(A2) are precisely what is needed to ensure existence and asymptotic properties of v_0 [2,8,6].

Theorem 1 ([8, Theorem 3]). *Under hypotheses (A1), (A2), for sufficiently small ε there exists a solution $u(x)$ of (1) in an $O(\varepsilon)$-neighbourhood of the zero-order asymptotic expansion $u_0(x) + v_0(\xi, l)\,\omega(x)$.*

Our sub- and super-solutions will invoke the function $\beta(x; p)$ constructed in [4, §2.3], which is a modified first-order asymptotic expansion such that

$$\beta(x; p) = u(x) + O(\varepsilon^2 + p). \tag{7}$$

The value p in the definition of β is a small real number that will be chosen later and is typically $o(h)$. To be more precise,

$$\beta(x; p) = u_0(x) + \bar{v}(\xi, l; p)\,\omega(x) + C_0 p, \qquad \Big|\frac{\partial^{k+m}}{\partial\xi^k\,\partial^m l}\,\bar{v}(\xi, l; p)\Big| \le Ce^{-\gamma_0\xi}, \tag{8}$$

where C_0 and γ_0 are positive constants and $\gamma_0^2 < \min_{x\in\partial\Omega} b_u(x, u_0(x))$.

Now we describe sub- and super-solutions $\beta(x; -p)$ and $\beta(x; p)$ of problem (1).

Lemma 2 ([4, Corollaries 2.7, 2.9]). *There exists $p_0 \in (0, \gamma_0^2)$ such that for all $|p| \le p_0$ the function $\beta(x;p)$ is well-defined. Furthermore, there exists $C_0 > 0$ and $C_2 > 0$ such that $C_2\varepsilon^2 \le p \le p_0$ implies $\beta(x;-p) \le \beta(x;p)$ and*

$$F\beta(x;-p) \le -C_0|p|\,\gamma^2/2, \qquad F\beta(x;p) \ge C_0 p\,\gamma^2/2.$$

3 Numerical Method from [4]

3.1 Layer-Adapted Meshes

Throughout this section we assume that the unknown parametrization (3) is available. Introduce a small positive parameter σ that will be specified later. Let $\sigma \le C_1$ so that the closed curve $\partial\Omega_\sigma$ that is defined by the equation $r = \sigma$ is well-defined and does not intersect itself. Furthermore, let Ω_σ be the interior of $\partial\Omega_\sigma$. Our problem will be discretized separately in Ω_σ and $\Omega \setminus \Omega_\sigma$, to which we shall refer as the interior region and the layer region respectively; see Figure 2.

The boundary-layer region $\Omega \setminus \Omega_\sigma$ is the rectangle $(0,\sigma) \times [0,L]$ in the coordinates (r,l). Hence in this subdomain introduce the tensor-product mesh $\{(r_i,l_j), i = 0,\dots,N, j = -1,\dots M\}$, where, as usual, $r_0 = 0$, $r_N = \sigma$, $l_0 = 0$, and $l_M = L$, while $l_{-1} = l_{M-1} - L$. Furthermore, let $\{l_j\}$ be a quasiuniform mesh on $[0,L]$, i.e., $C^{-1}h \le l_j - l_{j-1} \le Ch$. The choice of the *layer-adapted* mesh $\{r_i\}$ on $[0,\sigma]$ is crucial and will be discussed later; see (a),(b). Now assume only that $r_i - r_{i-1} \le h$ and $C^{-1}h^{-1} \le N \le Ch^{-1}$.

In the interior region Ω_σ introduce a quasiuniform Delaunay triangulation, i.e. the maximum side length of any triangle is at most h, the area of any triangle is bounded below by Ch^2, and the sum of the angles opposite to any edge is less than or equal to π (while any angle opposite to $\partial\Omega_\sigma$ does not exceed $\pi/2$).

Furthermore, let the union of all the triangles define a polygonal domain Ω_σ^h whose boundary vertices lie on $\partial\Omega_\sigma$. Note that we do not replace our original domain Ω by a similar polygonal domain Ω^h, since a significant part of the boundary layer would be lost in $\Omega \setminus \Omega^h$. We also require that both the interior and layer meshes have the same sets of nodes on $\partial\Omega_\sigma$.

We focus on two particular choices of $\{r_i\}$:

3.1(a) Bakhvalov mesh. [1] Set $\sigma := 2\gamma_0^{-1}\varepsilon|\ln\varepsilon|$ and define the mesh $\{r_i\}$ by $r_i := r\big([1-\varepsilon]\,i/N\big)$, $i = 0\dots,N$, $\quad r(t) := -2\gamma_0{}^{-1}\varepsilon\ln(1-t)$ for $t \in [0, 1-\varepsilon]$.

3.1(b) Shishkin mesh. [10] Set $\sigma = 2\gamma_0^{-1}\varepsilon\ln N$ and introduce a uniform mesh $\{r_i\}_{i=0}^N$ on $[0,\sigma]$, i.e. $r_i - r_{i-1} = \sigma/N = 2\gamma_0^{-1}\varepsilon N^{-1}\ln N$.

Note that if ε is sufficiently small—recall (A3)—the condition $\sigma \le C_1$ is satisfied and the meshes (a) and (b) are well-defined. If (A3) is not satisfied, but (a) $\sigma \le C_1$ and $\varepsilon \le 1/e$, or (b) $\sigma \le C_1$, the meshes 3.1(a) and 3.1(b) remain well-defined. Otherwise we have $\varepsilon > C$, i.e. our problem is not singularly perturbed.

3.2 Discretization in the Boundary-Layer Region

Recall that $\Omega \setminus \Omega_\sigma$ is the rectangle $(0,\sigma) \times [0,L]$ in the coordinates (r,l). Hence rewrite (1a) in (r,l) coordinates, by (6), and then discretize it using the standard finite differences on the tensor-product mesh $\{(r_i, l_j)\}$ [9]. In the interior of $\Omega \setminus \Omega_\sigma$, i.e. for $i = 1, \ldots, N-1$, $j = 0, \ldots M-1$, set

$$\begin{aligned} F^h U_{ij} &:= -\varepsilon^2 \eta_{ij}^{-1} \Big(D_r[\zeta_{ij} D_r^- U_{ij}] + D_l[\vartheta_{ij}^{-1} D_l^- U_{ij}] \Big) + b(x_{ij}, U_{ij}) = 0, \\ U_{i,M} &= U_{i,0}, \qquad U_{i,-1} = U_{i,M-1}, \qquad U_{0,j} = g(x_{0,j}). \end{aligned} \tag{9}$$

Here U_{ij} is the computed solution at the mesh node $x_{ij} = (\varphi_j, \psi_j) + r_i(-\psi'_j, \varphi'_j)$,

$$D_r^- v_{ij} := \frac{v_{ij} - v_{i-1,j}}{r_i - r_{i-1}}, \qquad D_r v_{ij} := \frac{v_{i+1,j} - v_{ij}}{(r_{i+1} - r_{i-1})/2}, \tag{10}$$

$$D_l^- v_{ij} := \frac{v_{ij} - v_{i,j-1}}{H_j}, \qquad D_l v_{ij} := \frac{v_{i,j+1} - v_{ij}}{(H_j + H_{j+1})/2}, \qquad H_j := l_j - l_{j-1},$$

$$\eta_{ij} := 1 - \kappa_j r_i, \qquad \zeta_{ij} := 1 - \kappa_j r_{i-1/2}, \qquad \vartheta_{ij} := 1 - \tfrac{1}{2}(\kappa_{j-1} + \kappa_j) r_i,$$

while $\kappa_j = \kappa(l_j)$, $\varphi_j = \varphi(l_j)$, $\psi_j = \psi(l_j)$, $\varphi'_j = \varphi'(l_j)$, and $\psi'_j = \psi'(l_j)$.

On the *interface boundary* $\partial\Omega_\sigma$ introduce the fictitious Neumann condition

$$\frac{\partial u}{\partial r} = \phi(x) \qquad \text{for } x \in \partial\Omega_\sigma. \tag{11}$$

For $i = N$, $j = 0, \ldots M-1$, following [9], we discretize (1a), (6), (11) as follows:

$$\begin{aligned} F_-^h U_{Nj} &:= -\varepsilon^2 \eta_{Nj}^{-1} \Big(\delta_r^2 U_{Nj} + D_l[\vartheta_{Nj}^{-1} D_l^- U_{Nj}] \Big) + b(x_{Nj}, U_{Nj}) = 0 \quad \forall\, x_{Nj} \in \partial\Omega_\sigma, \\ U_{N,M} &= U_{N,0}, \qquad U_{N,-1} = U_{N,M-1}, \end{aligned} \tag{12a}$$

where we use $h_N := r_N - r_{N-1}$, $\phi_j := \phi(x_{Nj})$ and

$$\delta_r^2 U_{Nj} := \frac{\eta_{Nj}\,\phi_j - \zeta_{Nj}\, D_r^- U_{Nj}}{h_N/2} = \eta_{Nj}\,\frac{2}{h_N}\phi_j - \frac{2}{h_N}\zeta_{Nj}\, D_r^- U_{Nj}\,. \tag{12b}$$

Note that F_-^h involves an unknown function ϕ. The actual discretization on the interface boundary $\partial\Omega_\sigma^h$ is obtained by combining (12a) with (15) eliminating ϕ.

3.3 Discretization in the Interior Region. Existence and Accuracy

Let $S^h \subset W_2^1(\Omega_\sigma^h)$ be the standard finite element space of continuous functions that are linear on each of the triangles of our mesh in Ω_σ^h. In Ω_σ^h define the approximate solution $U \in S^h$ by

$$\varepsilon^2 (\nabla U, \nabla \chi_i) + \varepsilon^2 \phi_i \oint_{\partial\Omega_\sigma^h} \chi_i\, ds + b(X_i, U_i)\,(1, \chi_i) = 0 \qquad \forall\, \chi_i \in S^h, \tag{13}$$

where X_i is a mesh node in $\bar\Omega_\sigma^h$, while $U_i = U(X_i)$, $\phi_i = \phi(X_i)$, and $\chi_i \in S^h$ are the nodal basis functions, i.e. $\chi_i(X_j)$ equals 1 if $i = j$ and 0 otherwise. Here we used the *lumped mass* discretization of both the boundary integral and the integral involving b.

At *interior meshnodes* X_i of Ω_σ, our discretization (13) implies

$$F^h U_i := \frac{\varepsilon^2}{(1, \chi_i)} (\nabla U, \nabla \chi_i) + b(X_i, U_i) = 0 \qquad \forall\, X_i \in \Omega_\sigma. \tag{14}$$

Similarly, at the mesh nodes X_j on the *interface boundary* $\partial\Omega_\sigma$, we get

$$F_+^h U_j := \frac{\varepsilon^2}{(1,\chi_j)}(\nabla U, \nabla\chi_j) + \frac{\varepsilon^2 a_j}{h}\phi_j + b(X_j, U_j) = 0 \qquad \forall\, X_j \in \partial\Omega_\sigma, \quad (15)$$

where

$$a_j := \frac{h}{(1,\chi_j)}\oint_{\partial\Omega_\sigma^h} \chi_j\, ds, \qquad 0 < C^{-1} < a_j < C. \quad (16)$$

Finally the discretizations F_-^h (12a) and F_+^h (15) are compiled, eliminating ϕ, as in [4, (3.14)]; see also a similar formula (28) below.

Theorem 2 ([4, Theorem 3.20]). *Let the mesh $\{r_i\}_{i=0}^N$ be the Bakhvalov mesh of §3.1(a), or the Shishkin mesh of §3.1(b). There exists a discrete solution U of (9), (12), (14), (15) such that for h sufficiently small,*

$$\left|U(X_i) - u(X_i)\right| \le Ch^2|\ln h|^m \qquad \forall \text{ mesh nodes } X_i \in \bar\Omega,$$

where $m = 0$ for the Bakhvalov mesh (a) and $m = 2$ for the Shishkin mesh (b).

4 Numerical Method Using Approximate Curvature

We cannot implement the numerical method of §3 since no explicit parametrization (3) is available. Instead we are given a set of boundary points $\{(\varphi_j, \psi_j)\}_{j=0}^M$ ordered in the counterclockwise direction. Using these data, we modify our method as follows.

4.1 Modified Layer-Adapted Meshes. Approximate Curvature

We imitate the layer-adapted meshes of §3.1 that use the arc-length parametrization (3), in which $l = 0$ is associated with (φ_0, ψ_0) and $l = L$ is associated with (φ_M, ψ_M), where $(\varphi_0, \psi_0) = (\varphi_M, \psi_M)$. The mesh $\{l_j\}$ is chosen on $[0, L]$ so that $(\varphi_j, \psi_j) = (\varphi(l_j), \psi(l_j))$. Clearly, this mesh $\{l_j\}$ exists and is unique, but the exact values of l_j and L will remain unknown. Therefore we define

$$\begin{aligned} &\tilde H_j := \sqrt{(\varphi_j - \varphi_{j-1})^2 + (\psi_j - \psi_{j-1})^2}, \\ &\tilde D_l^- v_{ij} := \frac{v_{ij} - v_{i,j-1}}{\tilde H_j}, \qquad \tilde D_l v_{ij} := \frac{v_{i,j+1} - v_{ij}}{(\tilde H_j + \tilde H_{j+1})/2}, \end{aligned} \quad (17)$$

to replace H_j, D_l^- and D_l, respectively, in (9) and (12). Note that (17) implies that both $(\tilde D_l^- \varphi_j, \tilde D_l^- \psi_j)$ and its orthogonal vector $n_{j-1/2} := (-\tilde D_l^- \psi_j, \tilde D_l^- \varphi_j)$ are unit vectors. Imitating (5), we associate (r_i, l_j) with the point $\tilde x_{ij} \approx x_{ij} = x(r_i, l_j)$ defined by

$$\tilde x_{ij} := (\varphi_j, \psi_j) + r_i \tilde n_j, \quad \tilde n_j := \frac{\hat n_j}{|\hat n_j|}, \quad \hat n_j := \frac{\tilde H_{j+1}\, n_{j-1/2} + \tilde H_j\, n_{j+1/2}}{\tilde H_j + \tilde H_{j+1}}. \quad (18)$$

Here we normalize $\hat n_j$, whose length $|\hat n_j| = 1 + O(h^2)$, to get the unit vector $\tilde n_j$ that approximates the unit vector $n_j = (-\psi'(l_j), \varphi'(l_j))$.

Next, let the ordered set of vertices $\{\tilde{x}_{Nj}\}_{j=0}^{M}$ define the polygonal domain Ω_σ^h, in which we introduce a quasiuniform Delaunay triangulation, whose set of the boundary nodes is precisely the set $\{\tilde{x}_{Nj}\}_{j=1}^{M}$. Note that $\tilde{x}_{Nj} \approx x_{Nj} \in \partial\Omega_\sigma$ implies that $\partial\Omega_\sigma^h$ is an $O(h^2)$-perturbation of $\partial\Omega_\sigma$ (see Lemma 3 below).

Furthermore, our method will invoke the approximate curvature $\tilde{\kappa}_j \approx \kappa(l_j)$:

$$\tilde{\kappa}_j := \tfrac{1}{2}\big(\tilde{D}_l^-\varphi_j + \tilde{D}_l^-\varphi_{j+1}\big)\tilde{D}_l\tilde{D}_l^-\psi_j - \tfrac{1}{2}\big(\tilde{D}_l^-\psi_j + \tilde{D}_l^-\psi_{j+1}\big)\tilde{D}_l\tilde{D}_l^-\varphi_j \qquad (19)$$

—compare with (4)—for which a calculation shows that

$$\tilde{\kappa}_j = \frac{(\varphi_j - \varphi_{j-1})(\psi_{j+1} - \psi_j) - (\varphi_{j+1} - \varphi_j)(\psi_j - \psi_{j-1})}{\tilde{H}_j\tilde{H}_{j+1}(\tilde{H}_j + \tilde{H}_{j+1})/2}. \qquad (20)$$

Lemma 3. *Let the arc-length H_j of $\partial\Omega$ between any two consecutive points $(\varphi_{j-1}, \psi_{j-1})$ and (φ_j, ψ_j) satisfy $C^{-1}h \le H_j \le Ch$. Then for $\tilde{H}_j$, $\tilde{x}_{ij}$ and $\tilde{\kappa}_j$ defined by (17), (18) and (19), we have*

$$H_j = \tilde{H}_j[1 + O(h^2)], \quad \tilde{x}_{ij} - x_{ij} = O(r_i h^2) = O(\sigma h^2), \quad \tilde{\kappa}_j - \kappa(l_j) = O(h).$$

Proof. Recall that (3) is an arc-length parametrization, i.e. $\sqrt{\varphi'^2(l) + \psi'^2(l)} = 1$ for all l. Combining this with

$$(\varphi_j - \varphi_{j-1})/H_j = \varphi'(l_{i-1/2}) + O(h^2), \quad (\psi_j - \psi_{j-1})/H_j = \psi'(l_{i-1/2}) + O(h^2),$$

we get $\tilde{H}_j = H_j\sqrt{\varphi'^2(l_{j-1/2}) + \psi'^2(l_{j-1/2}) + O(h^2)} = H_j[1 + O(h^2)]$, which yields the desired estimate for $\tilde{H}_j$.

Since $H_j = \tilde{H}_j[1 + O(h^2)]$ implies $\tilde{D}_l^- = [1 + O(h^2)]D_l^-$, $\tilde{D}_l = [1 + O(h^2)]D_l$, it suffices to prove the desired estimates for $\tilde{x}_{ij}$ and $\tilde{\kappa}_j$ with $\tilde{D}_l^-$ and $\tilde{D}_l$ replaced by D_l^- and D_l in the definitions of $\tilde{n}_j$ and $\tilde{\kappa}_j$. Such estimates follow immediately from Taylor series expansions. In particular, $\tilde{x}_{ij} - x_{ij} = r_i[\tilde{n}_j - n_j]$, where $\tilde{n}_j$ is an $O(h^2)$ approximation of the unit vector $n_j = (-\psi'(l_j), \varphi'(l_j))$. □

Remark 1. If $\tilde{H}_{j+1} - \tilde{H}_j = O(h^2)$, then (19) implies that $\tilde{\kappa}_j - \kappa(l_j) = O(h^2)$. If $\tilde{H}_{j+1} - \tilde{H}_j = O(h)$ and, furthermore, condition (A3) is violated, i.e. $\varepsilon > Ch$, our method would remain second-order accurate provided that (19) is modified to some second-order approximation of $\kappa(l_j)$, which involves $\{(\varphi_i, \psi_i)\}_{i=j-2}^{j+2}$.

4.2 Modified Discretization in the Boundary-Layer Region

In $\Omega \setminus \Omega_\sigma^h$, i.e. for $i = 1, \ldots, N-1$, $j = 0, \ldots M-1$, we modify (9) as follows:

$$\begin{aligned} \tilde{F}^h\tilde{U}_{ij} &:= -\varepsilon^2\tilde{\eta}_{ij}^{-1}\Big(D_r[\tilde{\zeta}_{ij}D_r^-\tilde{U}_{ij}] + \tilde{D}_l[\tilde{\vartheta}_{ij}^{-1}\tilde{D}_l^-\tilde{U}_{ij}]\Big) + b(\tilde{x}_{ij}, \tilde{U}_{ij}) = 0, \\ \tilde{U}_{i,M} &= \tilde{U}_{i,0}, \qquad \tilde{U}_{i,-1} = \tilde{U}_{i,M-1}, \qquad \tilde{U}_{0,j} = g(\tilde{x}_{0,j}). \end{aligned} \qquad (21)$$

Here $\tilde{U}_{ij}$ is the discrete computed solution at the mesh node $\tilde{x}_{ij}$, the finite difference operators D_r^-, D_r, $\tilde{D}_l^-$, $\tilde{D}_l$ and the quantities $\tilde{x}_{ij}$, $\tilde{\kappa}_j$ are defined by (10), (17), (18) and (19) (see also Remark 1), while

$$\tilde{\eta}_{ij} := 1 - \tilde{\kappa}_j r_i\,, \qquad \tilde{\zeta}_{ij} := 1 - \tilde{\kappa}_j r_{i-1/2}\,, \qquad \tilde{\vartheta}_{ij} := 1 - \tfrac{1}{2}(\kappa_{j-1} + \tilde{\kappa}_j) r_i\,.$$

For $i = N$, $j = 0, \dots M-1$, imitating (12), we discretize (1a), (6), (11) using

$$\tilde{F}^h_- \tilde{U}_{Nj} := -\varepsilon^2 \tilde{\eta}_{Nj}^{-1} \Big(\tilde{\delta}_r^2 \tilde{U}_{Nj} + \tilde{D}_l [\tilde{\vartheta}_{Nj}^{-1} \tilde{D}_l^- \tilde{U}_{Nj}] \Big) + b(\tilde{x}_{Nj}, \tilde{U}_{Nj}) = 0 \quad \forall\, \tilde{x}_{Nj} \in \partial\Omega_\sigma^h,$$
$$\tilde{U}_{N,M} = \tilde{U}_{N,0}, \qquad \tilde{U}_{N,-1} = \tilde{U}_{N,M-1}, \tag{22a}$$

where $h_N := r_N - r_{N-1}$, $\phi_j = \phi(\tilde{x}_{Nj})$ and

$$\tilde{\delta}_r^2 \tilde{U}_{Nj} := \frac{\tilde{\eta}_{Nj}\,\phi_j - \tilde{\zeta}_{Nj}\, D_r^- \tilde{U}_{Nj}}{h_N/2} = \tilde{\eta}_{Nj} \frac{2}{h_N} \phi_j - \frac{2}{h_N} \tilde{\zeta}_{Nj}\, D_r^- \tilde{U}_{Nj}. \tag{22b}$$

Lemma 4. *Let $\beta(x;p)$ be described by (8), and the mesh $\{r_i\}_{i=0}^N$ be either the Bakhvalov mesh of §3.1(a), or the Shishkin mesh of §3.1(b). Then for all $|p| \le p_0$ at all interior mesh nodes $x_{ij} \approx \tilde{x}_{ij}$, $i = 1, \dots, N-1$, $j = 0, \dots, M-1$, we have*

$$\big|\tilde{F}^h \beta(x_{ij}) - F\beta(x_{ij})\big| \le Ch^2 |\ln h|^m, \tag{23a}$$

while at all interface-boundary mesh nodes $x_{Nj} \approx \tilde{x}_{Nj} \in \partial\Omega_\sigma^h$ we have

$$\tilde{F}^h_- \beta(x_{Nj}) - F\beta(x_{Nj}) = \frac{2\varepsilon^2}{h_N} \Big(\frac{\partial \beta}{\partial r}\Big|_{x_{Nj}} - \phi_j \Big) + O\big(h^2\big), \tag{23b}$$

where $m = 0$ for the Bakhvalov mesh (a) and $m = 2$ for the Shishkin mesh (b).

Proof. [4, Lemma 3.11, Lemma 3.13] state (23) with $\tilde{F}^h$ replaced by F^h. Hence it remains to estimate $\tilde{F}^h \beta(x_{ij}) - F^h \beta(x_{ij})$. Throughout this proof, we use $(1 + r/\varepsilon)|\partial^k \beta/\partial r^k| \le C\varepsilon^{-k}$ and $|\partial^k \beta/\partial l^k| \le C$, $k = 1, 2$, which follow from (8).

First, invoking the estimate for $\tilde{x}_{ij}$ of Lemma 3, we get

$$b(\tilde{x}_{ij}, \beta(x_{ij})) - b(x_{ij}, \beta(x_{ij})) = O(\sigma h^2) = O(h^2). \tag{24}$$

Next, the estimates for $\tilde{H}_j$ and $\tilde{\kappa}_j$ of Lemma 3 imply that $\tilde{D}_l^- = [1 + O(h^2)] D_l^-$, $\tilde{D}_l = [1 + O(h^2)] D_l$, and $\tilde{\vartheta}_{ij} - \vartheta_{ij} = O(h)$, $\tilde{\eta}_{ij} - \eta_{ij} = O(h)$. Hence we have

$$\tilde{D}_l [\tilde{\vartheta}_{ij}^{-1} \tilde{D}_l^- \beta(x_{ij})] = \tilde{D}_l [\vartheta_{ij}^{-1} D_l^- \beta(x_{ij}) + O(h)] = D_l [\vartheta_{ij}^{-1} D_l^- \beta(x_{ij})] + O(1).$$

Here we used $D_l[O(h)] = O(1)$, which follows from $H_j \ge Ch$. Therefore we get

$$\tilde{\eta}_{ij}^{-1} \tilde{D}_l [\tilde{\vartheta}_{ij}^{-1} \tilde{D}_l^- \beta(x_{ij})] - \eta_{ij}^{-1} D_l [\vartheta_{ij}^{-1} D_l^- \beta(x_{ij})] = O(1). \tag{25}$$

Furthermore, a calculation using $\zeta_{ij} = 1 - \kappa_j\, r_{i-1/2}$ and $\tilde{\zeta}_{ij} = 1 - \tilde{\kappa}_j\, r_{i-1/2}$ yields

$$\begin{aligned} &\tilde{\eta}_{ij}^{-1} D_r [\tilde{\zeta}_{ij} D_r^- \beta(x_{ij})] - \eta_{ij}^{-1} D_r [\zeta_{ij} D_r^- \beta(x_{ij})] \\ &= (\tilde{\eta}_{ij}^{-1} - \eta_{ij}^{-1}) D_r D_r^- \beta(x_{ij}) - (\tilde{\eta}_{ij}^{-1} \tilde{\kappa}_j - \eta_{ij}^{-1} \kappa_j) D_r [r_{i-1/2} D_r^- \beta(x_{ij})] \\ &= O(\varepsilon^{-1} h). \end{aligned} \tag{26}$$

Here we invoked $\tilde{\eta}_{ij} - \eta_{ij} = O(r_i h)$ and $\tilde{\eta}_{ij}^{-1} \tilde{\kappa}_j - \eta_{ij}^{-1} \kappa_j = O(h)$ (which follow from Lemma 3) combined with $r_i D_r D_r^- \beta(x_{ij}) = O(\varepsilon^{-1})$ and $D_r[r_{i-1/2} D_r^- \beta(x_{ij})] = D_r^- \beta(x_{i+1,j}) + r_{i-1/2} D_r D_r^- \beta(x_{ij}) = O(\varepsilon^{-1})$.

Combining estimates (24), (25), (26), we arrive at $\tilde{F}^h \beta(x_{ij}) - F^h \beta(x_{ij}) = O(\varepsilon h + \varepsilon^2 + h^2) = O(h^2)$, where we also used (A3). Thus (23a) is established.

Estimate (23b) is obtained similarly, observing that

$$\left[\tilde{\eta}_{Nj}^{-1}\tilde{\delta}_r^2 - \eta_{Nj}^{-1}\delta_r^2\right]\beta(x_{Nj}) = -\left(\tilde{\eta}_{Nj}^{-1}\tilde{\zeta}_{Nj} - \eta_{Nj}^{-1}\zeta_{Nj}\right)\frac{2}{h_N}D_r^-\beta(x_{Nj}) = O(1+h^2/\varepsilon^2),$$

where we combined $\tilde{\eta}_{Nj}^{-1}\tilde{\zeta}_{Nj} - \eta_{Nj}^{-1}\zeta_{Nj} = O(r_N h) = O(\sigma/N)$ with $\sigma/N \le Ch_N$. We also invoked $D_r^-\beta(x_{Nj}) = O(1+h^2/\varepsilon^2)$, which follows from (8) by [4, Lemma 3.12]. □

4.3 Discretization in the Interior Region

In the interior part of the domain Ω_σ^h we use the lumped-mass finite elements (14), (15), (16); see also [4]:

$$\tilde{F}^h\tilde{U}_i := F^h\tilde{U}_i = 0 \quad \forall\, X_i \in \Omega_\sigma^h; \qquad \tilde{F}_+^h\tilde{U}_j := F_+^h\tilde{U}_j = 0 \quad \forall\, X_j \in \partial\Omega_\sigma^h. \quad (27)$$

Finally, the discretization $\tilde{F}_-^h$ (22) and the above discretization $\tilde{F}_+^h$ are compiled as in [4] by eliminating the auxiliary unknown function ϕ:

$$\tilde{F}^h\tilde{U}_j := \frac{(h_N/2)\,\tilde{F}_-^h\tilde{U}_j + (h/a_j)\,\tilde{F}_+^h\tilde{U}_j}{h_N/2 + h/a_j} \qquad \forall\, X_j \in \partial\Omega_\sigma^h. \quad (28)$$

Lemma 5. *Let $\beta^I \in S^h$ be a non-standard piecewise linear interpolant of $\beta(x;p)$ such that $\beta^I(X_i;p) := \beta(X_i;p)$ at all mesh nodes $X_i \in \Omega_\sigma^h$, while at all mesh nodes $X_j = \tilde{x}_{Nj} \in \partial\Omega_\sigma^h$ we have $\beta^I(X_j;p) := \beta(x_{Nj};p)$. Furthermore, let σ be chosen as in either §3.1(a) or §3.1(b). Then for all $|p| \le p_0$ we have*

$$\left|\tilde{F}^h\beta_i^I - F\beta(X_i)\right| \le Ch^2 \qquad \forall\, X_i \in \Omega_\sigma^h; \quad (29a)$$

at all mesh nodes $X_j = \tilde{x}_{Nj}$ on $\partial\Omega_\sigma^h$ we have

$$\tilde{F}_+^h\beta_j^I - F\beta(x_{Nj}) = -a_j\frac{\varepsilon^2}{h}\Big(\frac{\partial\beta}{\partial r}\Big|_{x_{Nj}} - \phi_j\Big) + O(h^2) \qquad \forall\, X_j \in \partial\Omega_\sigma^h; \quad (29b)$$

and for $\tilde{F}^h$ of (28) at all mesh nodes $X_j = \tilde{x}_{Nj}$ on $\partial\Omega_\sigma^h$ we have

$$\left|\tilde{F}^h\beta(x_{Nj}) - F\beta(x_{Nj})\right| \le Ch^2 \qquad \forall\, X_j \in \partial\Omega_\sigma^h. \quad (30)$$

Proof. [4, Lemmas 3.15, 3.16] give the desired estimates (29) in the case of β^I being the standard interpolant of β in $\bar{\Omega}_\sigma^h$, and x_{Nj} replaced by $X_j = \tilde{x}_{Nj}$. Note that the proof of [4, Lemma 3.16] is applicable to the domain Ω_σ^h, since $\partial\beta/\partial n = -\partial\beta/\partial r\big|_{X_j} + O(h)$ on $\partial\Omega_\sigma^h$ within $O(h)$-distance from X_j, which follows from $\partial\Omega_\sigma^h$ being an $O(h^2)$-perturbation of $\partial\Omega_\sigma$. Hence to prove (29), it suffices to show that (i) the values of $\tilde{F}_+^h\beta_j^I$ for the standard interpolant and the non-standard interpolant of this lemma differ by $O(h^2)$; (ii) the values of $\tilde{F}^h\beta_j^I$ enjoy a similar property; (iii) $F\beta(x_{Nj}) - F\beta(\tilde{x}_{Nj}) = O(h^2)$; (iv) similarly the values of $\partial\beta/\partial r$ at $\tilde{x}_{Nj}$ and x_{Nj} differ by $O(h^2)$. These assertions (i)-(iv) follow from $\|\beta\|_{C^2(\bar{\Omega}_\sigma \cup \bar{\Omega}_\sigma^h)} \le C$ combined with $|\tilde{x}_{Nj} - x_{Nj}| \le Ch^2$ and (A3).

In view of (28), to get estimate (30), we add (23b) multiplied by $(h_N/2)$ to (29b) multiplied by (h/a_j) and divide the result by $(h_N/2 + h/a_j)$. □

4.4 Existence and Accuracy. Discrete Sub- and Super-Solutions

Theorem 3. *Let the mesh* $\{r_i\}_{i=0}^N$ *be either the Bakhvalov mesh of §3.1(a), or the Shishkin mesh of §3.1(b). Then there exists a discrete solution* $\tilde{U}$ *of (21), (22), (27), (28) such that for* h *sufficiently small,*

$$\big|\tilde{U}(X_i) - u(X_i)\big| \le Ch^2 |\ln h|^m \qquad \forall \text{ mesh nodes } X_i \in \bar{\Omega}, \tag{31}$$

where $m = 0$ *for the Bakhvalov mesh (a) and* $m = 2$ *for the Shishkin mesh (b).*

Proof. We invoke the theory of Z-fields, imitating the proofs of [4, Lemma 3.19, Theorem 3.20]. Set $\bar{p} := C_3 h^2 |\ln h|^m$, where $C_3 > 0$ is a sufficiently large constant. Now combining Lemma 2 with (23a), (29a) and (30), we conclude that the functions equal to $\beta(x_{ij}; \pm p)$ at $\tilde{x}_{ij}$ and $\beta(X_i; \pm p)$ at $X_i \in \Omega_\sigma^h$ are discrete sub- and supers-solutions, where '−' is used for the sub-solution and '+' is used for the super-solution. Since, by [4, Lemma 3.6], our discrete operator $\tilde{F}^h$ is a Z-field, there exists a discrete solution $\tilde{U}$ between our sub- and super-solutions. In particular, $\beta(x_{ij}; -p) \le \tilde{U}(\tilde{x}_{ij}) \le \beta(x_{ij}; p)$. Using (8) and Lemma 3, we observe that $|x_{ij} - \tilde{x}_{ij}| = O(r_i h^2)$ combined with $|\nabla \beta_{ij}| \le C[1 + \varepsilon^{-1} e^{-\gamma_0 r_i/\varepsilon}]$ implies $\beta(x_{ij}) = \beta(\tilde{x}_{ij}) + O(h^2)$. Hence our discrete solution $\tilde{U}(X_i)$ is between $\beta(X_i; -p) - O(h^2)$ and $\beta(X_i; p) + O(h^2)$ for *all* mesh nodes $\{X_i\} \supset \{\tilde{x}_{ij}\}$, which, combined with (7) and (A3), yields the desired error estimate. □

5 Numerical Results

Our model problem is (1) in the domain Ω—see Figure 2 and [4, §7]—in which

$$b(x, u) = \big(u - \bar{u}_0(x)\big) u \big(u + \bar{u}_0(x)\big), \qquad \bar{u}_0(x) = x_1^2 + x_1 + 1. \tag{32}$$

Here $\pm\bar{u}_0(x)$ are two stable solutions and 0 is an unstable solution of the corresponding reduced problem. The boundary condition $g(x) = (x_1 - x_1^2)/3$ satisfies

Table 1. Maximum nodal errors $|U - u|$ in the numerical method [4] outlined in §3 (upper part), and additional errors $|\tilde{U} - U|$ induced by using an ordered set of boundary points instead of an explicit parametrization of the domain (lower part)

	Bakhvalov mesh			Shishkin mesh		
N	$\varepsilon = 10^{-2}$	$\varepsilon = 10^{-4}$	$\varepsilon = 10^{-8}$	$\varepsilon = 10^{-2}$	$\varepsilon = 10^{-4}$	$\varepsilon = 10^{-8}$
32	3.741e-3	3.842e-3	3.843e-3	3.914e-2	3.947e-2	3.948e-2
64	9.335e-4	9.534e-4	9.536e-4	1.317e-2	1.325e-2	1.325e-2
128	2.333e-4	2.388e-4	2.388e-4	4.004e-3	4.400e-3	4.401e-3
256	5.854e-5	5.967e-5	5.968e-5	1.008e-3	1.430e-3	1.430e-3
32	4.021e-3	7.597e-6	1.536e-9	2.605e-3	5.343e-6	5.284e-10
64	1.041e-3	2.022e-6	4.023e-10	9.235e-4	1.423e-6	1.406e-10
128	2.835e-4	5.068e-7	1.023e-10	2.905e-4	3.575e-7	3.527e-11
256	7.099e-5	1.463e-7	2.566e-11	7.278e-5	1.458e-7	1.443e-11

(A2) for both $\pm\bar{u}_0$; see Figure 1. We present numerical results for the solution u near $\bar{u}_0$; see Figure 1 (left); the results for the solution near $-\bar{u}_0$ are similar.

Table 1 gives numerical results for the Bakhvalov and Shishkin meshes with the parameter $\gamma_0 := 3\sqrt{2}/5$. The upper part of the table shows maximum nodal errors $\max_i |U_i - u(X_i)|$—which are computed as described in [6, §4]—for the numerical method [4] outlined in §3, which requires an explicit parametrization of the domain. The lower part of the table shows the additional errors $\max_i |\tilde{U}_i - U_i|$ induced by switching to the method of §4, which instead uses an ordered set of boundary points. The errors in the lower part of the table are comparable with the errors in the upper part and decay very fast as ε tends to 0.

In summary, the numerical results support our error estimates of Theorems 2 and 3. Thus, we observe that even if no explicit parametrization of the domain is available, the modification of the numerical method [4], which we presented in this paper, produces reliable computed solutions.

Acknowledgement. This publication has emanated from research conducted with the financial support of Science Foundation Ireland under the Basic Research Grant Programme 2004.

References

1. Bakhvalov, N.S.: On the optimization of methods for solving boundary value problems with boundary layers. Zh. Vychisl. Mat. Mat. Fis. 9, 841–859 (1969) (in Russian)
2. Fife, P.C.: Semilinear elliptic boundary value problems with small parameters. Arch. Ration. Mech. Anal. 52, 205–232 (1973)
3. Grindrod, P.: Patterns and Waves: the Theory and Applications of Reaction-Diffusion Equations. Clarendon Press (1991)
4. Kopteva, N.: Maximum norm error analysis of a 2d singularly perturbed semilinear reaction-diffusion problem. Math. Comp. 76, 631–646 (2007)
5. Kopteva, N.: Pointwise error estimates for 2nd singularly perturbed semilinear reaction-diffusion problems. In: Farago, I., Vabishchevich, P., Vulkov, L. (eds.) Finite Difference Methods: Theory and Applications. Proceedings of the 4th International Conference, Lozenetz, Bulgaria, pp. 105–114 (2006)
6. Kopteva, N., Stynes, M.: Numerical analysis of a singularly perturbed nonlinear reaction-diffusion problem with multiple solutions. Appl. Numer. Math. 51, 273–288 (2004)
7. Murray, J.D.: Mathematical Biology. Springer, Heidelberg (1993)
8. Nefedov, N.N.: The method of differential inequalities for some classes of nonlinear singularly perturbed problems with internal layers. Differ. Uravn. 31, 1142–1149 (1995) (in Russian) (Translation in Differ. Equ. 31, 1077–1085 (1995)
9. Samarski, A.A.: Theory of Difference Schemes. Nauka (1989) (in Russian)
10. Shishkin, G.I.: Grid Approximation of Singularly Perturbed Elliptic and Parabolic Equations. Ur. O. Ran, Ekaterinburg (1992) (in Russian)
11. Vasil'eva, A.B., Butuzov, V.F., Kalachev, L.V.: The Boundary Function Method for Singular Perturbation Problems. SIAM, Philadelphia (1995)

Weight Uniform Accuracy Estimates of Finite Difference Method for Poisson Equation, Taking into Account Boundary Effect

V.L. Makarov[1] and L.I. Demkiv[2]

[1] Institute of Mathematics of Ukrainian National Academy of Sciences, 3, Tereshchenkivska str., Kyiv, Ukraine
[2] National University, Lvivska Polytechnica, 12, St. Bandera str. Lviv, Ukraine

Abstract. Poisson equation in polyhedral domain $\Omega \subset \mathrm{R}^n, n = 2,3$ with boundary Γ, when Dirichlet conditions are given on all faces or on all but one where Neimann conditions are given, is considered.

Traditional difference schemes with semi-constant steps along axes precisely approximate Dirichlet conditions hence it is expected that their accuracy order increases approaching to corresponding part of boundary γ. This paper is dedicated to quantitative investigation of this boundary effect. It is also shown that analogous boundary effect in the mesh knots takes place also for finite-element method (super convergence).

1 Introduction

Let $\Omega \subset \mathrm{R}^2$ be bounded polygonal convex domain with boundary Γ. Consider Dirichlet problem for Poisson equation

$$-\Delta u(x) = f(x),\ x \in \Omega,\ u(x) = u_0(x),\ x \in \Gamma, \tag{1}$$

where $x = (x_1, x_2),\ f(x), u_0(x)$ are given functions. Suppose, that the domain Ω may be covered by the mesh $\hat{\omega}$ with semi-constant step by each direction along axes in such a way that if a polygon side crosses a mesh cell then it necessarily goes through it's two knots. Then domain γ of the mesh polygon $\hat{\omega}$ totally coincides with the boundary Γ (boundary compliance).

Approximate problem (1) with the following difference scheme

$$-\Delta_h y(x) = -y(x)_{\bar{x}_1\hat{x}_1} - y(x)_{\bar{x}_2\hat{x}_2} = f(x),\ x \in \hat{\omega},\ y(x) = u_0(x),\ x \in \gamma, \tag{2}$$

where

$$y(x)_{\bar{x}_1\hat{x}_1} = \frac{2}{h_1^+ + h_1^-}\left(\frac{y(x_1 + h_1^+, x_2) - y(x_1, x_2)}{h_1^+} - \frac{y(x_1, x_2) - y(x_1 - h_1^-, x_2)}{h_1^-}\right)$$

is an ordinary three point approximation of the second derivative on x_1 on nonuniform mesh, $y(x)_{\bar{x}_2\hat{x}_2}$ is determined analogously. Let us write difference scheme for the error function $z(x) = y(x) - u(x)$.

S. Margenov, L.G. Vulkov, and J. Waśniewski (Eds.): NAA 2008, LNCS 5434, pp. 92–103, 2009.

We obtain

$$-\Delta_h z(x) = \psi(x),\ x \in \hat{\omega},\quad z(x) = 0,\ x \in \gamma,\ \psi(x) = \Delta_h u(x) - \Delta u(x) \qquad (3)$$

where $\psi(x)$ is approximation error, which has $O(|h|^2)$-th order, if the point $x \in \hat{\omega}$ is center of regular template. (i.e. $h_1^+ = h_1^-, h_2^+ = h_2^-$) $u(x) \in C^4(\bar{\Omega})$ and it has $O(|h|)$-th order , if template is nonregular or $u(x) \in C^3(\bar{\Omega})$. Here $|h| = \max\{h_1^+, h_1^-, h_2^+, h_2^-\}$.

Difference operator of the scheme (2) fulfills maximum principle and hence well known estimate (see for. ex. [1]) is valid here

$$\|z\|_\infty \leq C\,|h|^{m-2},\ u(x) \in C^m\left(\bar{\Omega}\right),\ m = 3, 4.$$

where constant C depends only on corresponding norm of $u(x)$ and does not depend on $|h|$. But adduced estimate has one defect: it does not take into account the fact that the error $z(x)$ on the mesh boundary γ equals zero. Hence, one expects, that when point $x \in \hat{\omega}$ approaches to domain γ accuracy order of the difference scheme increases. This fact was noticed in [2], where Poisson equation in parallelepiped was considered. On the one side Neumann condition was given and on the other Dirichlet one. One proved that when point $x \in \hat{\omega}$ approaches to the boundary opposite to the one with Neumann conditions, accuracy order of the corresponding difference scheme in uniform metrics increases by one order. More precisely this result will be given in Section 4.

In the paper [3] for more general elliptic type equation of divergent form the following accuracy estimate

$$\max_{x \in \bar{\omega}} \left|\rho^{-1/2}(x)\, z(x)\right| \leq M h^2 \|u\|_{W_4^2(\Omega)},\quad x \in \bar{\omega}$$

was obtained, where

$$\rho(x) = \rho(x_1, x_2) = min\left[x_1 x_2, x_1(1 - x_2), (1 - x_1) x_2, (1 - x_1)(1 - x_2)\right]$$

Similar estimate was obtained in [4] for quasilinear elliptic type equations with main non-linear part of non-divergent form.

Current paper is devoted to strengthening of the results mentioned above. In the Section 2 for the case when Ω is unit square and $u(x) \in C^4\bar{\Omega}$ it is shown that approximation order 2 of the difference scheme (3) approaching to the square side increases by one and approaching to square vertexes by two with respect to $ln1$. In Section 3 the case when domain Ω is rectangular trapezoid is considered. It is proved that accuracy order of corresponding difference scheme with semi-constant step along abscissa when $u(x) \in C^4\left(\bar{\Omega}\right)$ increases by one approaching to vertical and horizontal sides of trapezoid and increases by two (with respect to $ln\frac{1}{|h|}$) approaching to vertexes of the angles less then $\frac{\Pi}{2}$.

It is shown that results of the Section 2 and 3 are valid both for finite element method. Hereby estimates from [8] are improved. Section 4 is dedicated to strengthening of the results from [2] for the case when Ω is unit cube.

2 Unit Square Domain Case

Consider the case when the domain Ω is unit square, $f(x_1, x_2) \equiv 1$ and the mesh ω is chosen to be uniform with equal steps along the axes Ox_1, Ox_2, $h_1 = h = 1/N$. Then difference scheme (2) accuracy is characterized by the estimate

$$|z(x)| \leq Ch^2 \, |u|_{W^4_\infty(\Omega)} \, v(x)$$

where function $v(x)$ is solution of the problem

$$-\Delta_h v(x) = 1, \; x \in \omega, \quad v(x) = 0, \; x \in \gamma$$

which can be obtained in the analytical form:

$$v(x_{1,s}, x_{2,t}) = \\ = \sum_{i=1}^{N-1} \sum_{j=1}^{N-1} \frac{(1-(-1)^i)(1-(-1)^j)}{2N^4(1-\cos(\frac{i\pi}{N}))(1-\cos(\frac{j\pi}{N}))} \frac{\sin(\frac{i\pi}{N})\sin(\frac{j\pi}{N})\sin(\frac{si\pi}{N})\sin(\frac{tj\pi}{N})}{(2-\cos(\frac{i\pi}{N})-\cos(\frac{j\pi}{N}))} \quad (4)$$

Note that function $2v(x_{1,s}, x_{2,t})$ is finite difference approximation of the stress function during torsion of the square profile shank.

The following theorem is valid

Theorem 1. *Let solution of the problem* (1) *belong to the space* $C^4(\bar{\Omega})$, *then accuracy of the difference scheme* (2) *is characterized by the estimate*

$$|z(x)| \leq Ch^2 v(x) \, |u|_{C^{(4)}(\bar{\Omega})} \quad (5)$$

where constant C does not depend on $h, u(x)$, and function $v(x)$ approaching ti the sides of the mesh square ω behaves itself as $\bar{O}(h)$ and approaching to its vertex - as $O(h^2 \ln(\frac{1}{h}))$.

Remark 1. Obtained results allows to derive new estimates of the finite elements method which adduce to traditional difference schemes. Let us illustrate this by the example of the problem (4), when domain Ω is a square and $u_0(x) \equiv 0$. Let for each $h \in (0,1)$, T_h implies triangulation Ω as it is shown on the Figure 1. Let $S_1^h(\Omega) = S_1^h$ imply finite-measurable space of continuous functions on $\bar{\Omega}$, confinement of which on each triangular $\tau \in T_h$ is polynomial of the first degree and $\overset{0}{S}_{1^h}(\Omega)$ implies subspace $S_1^h(\Omega)$, which contains functions which vanish on Γ. It is obvious that $S_1^h(\Omega) \subset W^1_\infty(\Omega)$.

System of functions (see. [7])

$$\omega_{k,l}(x_1, x_2) = \begin{cases} 1 - \frac{1}{h}(x_{1,k} - x_1) - \frac{1}{h}(x_{2,l} - x_2), & (x_1, x_2) \in \Omega^h_{k,l,1} \\ 1 - \frac{1}{h}(x_{1,k} - x_1), & (x_1, x_2) \in \Omega^h_{k,l,2} \\ 1 + \frac{1}{h}(x_{2,k} - x_2), & (x_1, x_2) \in \Omega^h_{k,l,3} \\ 1 + \frac{1}{h}(x_{1,k} - x_1) + \frac{1}{h}(x_{2,l} - x_2), & (x_1, x_2) \in \Omega^h_{k,l,4} \\ 1 + \frac{1}{h}(x_{1,k} - x_1), & (x_1, x_2) \in \Omega^h_{k,l,5} \\ 1 - \frac{1}{h}(x_{2,k} - x_2), & (x_1, x_2) \in \Omega^h_{k,l,6} \end{cases} \quad (6)$$

$$k, l = 1, 2, ..., N-1$$

is a basis in $\overset{0}{S}{}^h_1(\Omega)$.

Finite-element approximation of $u(x)$ is

$$y(x) = \sum_{k,l=1}^{N-1} y(x_{k,l})\omega_{k,l}(x) \tag{7}$$

where $y(x_{k,l})$ is defined as solution of difference scheme

$$\begin{aligned} &-\Delta_h y(x) = -y(x)_{\bar{x}_1 x_1} - y(x)_{\bar{x}_2 x_2} = \varphi(x),\ x \in \omega, y(x) = 0,\ x \in \gamma, \\ &\varphi(x_{1,k}, x_{2,l}) = \frac{1}{h^2} \iint_{\Omega_{k,l}} f(x)\omega_{k,l}(x)dx_1 dx_2,\ P_{k,l} = (x_{1,k}, x_{2,l}) \in \omega \end{aligned} \tag{8}$$

We use a technique from [7] for investigation of accuracy of difference scheme (8) and write down the integral consequence of Poisson equation. We obtain

$$-\frac{1}{h^2} \iint\limits_{\Omega_{k,l}} \Delta u(x)\omega_{k,l}(x)dx_1 dx_2 = \varphi(x_{1,k}, x_{2,l}) = \frac{1}{h^2} \iint\limits_{\Omega_{k,l}} f(x)\omega_{k,l}(x)dx_1 dx_2 \tag{9}$$

Now write down the difference scheme for the error

$$-\Delta_h z(x) = -z(x)_{\bar{x}_1 x_1} - z(x)_{\bar{x}_2 x_2} = \psi(x),\ x \in \omega, z(x) = 0,\ x \in \gamma, \tag{10}$$

where local truncation error $\psi(x)$ has the following form

$$\psi(x_{1,k}, x_{2,l}) = l_{k,l}(u(\cdot)) = \Delta_h u(x) - \frac{1}{h^2} \iint_{\Omega_{k,l}} \Delta u(x)\omega_{k,l}(x)dx_1 dx_2 \tag{11}$$

For the solution of the problem (10) the following estimate holds true

$$|z(x)| \leq \|\psi\|_\infty v(x) \tag{12}$$

where function $v(x)$ has the same meaning as in Theorem 1.

Linear functional $l_{k,l}(u(\cdot))$ which is defined and bounded on the space of functions $W_\infty^2(\Omega_{k,l}^h)$ and vanishes on polynomials of the third degree. Using the Bremble-Hilbert lemma (see. [7]), we obtain

$$|\psi(x_{1,k}, x_{2,l})| \leq Ch^2 |u|_{W_\infty^4(\Omega)},$$

which together with (12) proves the following statement

Theorem 2. *Let solution of the problem* (4) *belong to the space* $W_\infty^4(\Omega)$, *then accuracy of the finite-element method in the knots of the mesh* ω *will be characterized by the estimate*

$$|z(x)| \leq Ch^2 v(x) |u|_{W_\infty^4(\Omega)},\ x \in \omega \tag{13}$$

where constant C *does not depend on* h *and* $u(x)$, *and function* $v(x)$ *approaching to the sides of the mesh square* ω *behaves itself as* $O(h)$, *and approaching to its vertexes - as* $O(h^2 \ln(\frac{1}{h}))$.

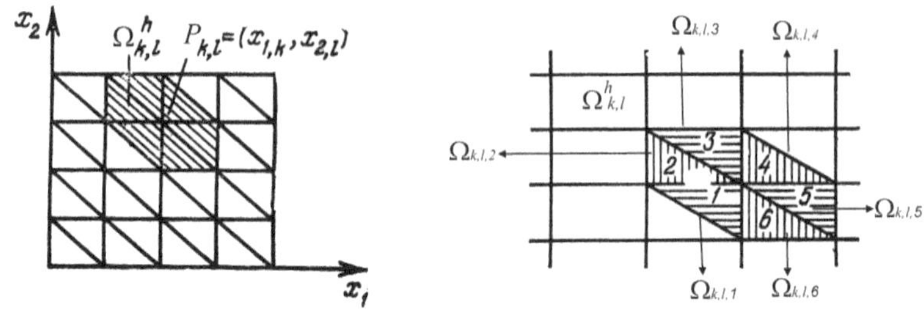

Fig. 1. Triangulation Ω

3 Rectangular Trapezoid Domain Case

Consider the domain $ABCD$, which is represented in the Figure 2a). Its left part is a unit square $ABCE$, which we designate as Ω_1. Let its mesh step along the axis Ox_1 be h_1. And the right side of the domain is a triangle ECD which we designate as Ω_2. Let its mesh step along the axis Ox_1 be $h_2 = (a-1)/N_2$. and let the mesh step along the axis Ox_2 be h_1 . Introduce the following notations $\omega_1 = \Omega_1 \cap \hat{\omega}$, $\omega_2 = \Omega_2 \cap \hat{\omega}$, $\gamma_1 = (\bar{\Omega}_1 \cap \bar{\Omega}_2) \cap \omega$. Then for the local truncation error $\psi(x)$ (3) we obtain

$$\psi(x) = \begin{cases} O(h_1^2),\ x \in \omega_1, \\ O(h_1^2 + h_2^2),\ x \in \omega_2, \\ O(|h| + h_1^2),\ x \in \gamma_1, \quad |h| = \max(h_1, h_2) \end{cases} \tag{14}$$

if $u(x) \in C^4(\bar{\Omega})$ and $\psi(x) = O(|h|),\ x \in \omega$, if $u(x) \in C^3(\bar{\Omega})$.

Technique of investigation of the difference scheme accuracy in Tchebyshev norm for Dirichlet problem for Poisson equation in the complex form domain which is based on the maximum principle is adduced in monograph [1]. Thus the unevenness of mesh is concentrated only near boundary of domain. Some

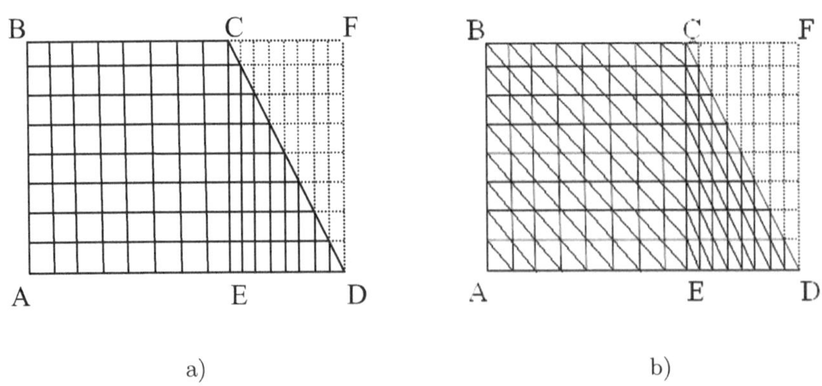

Fig. 2. Trapezoid domain

other situation is examined here. An unevenness of mesh is in the middle of area (around line CE) and although the technique of receipt of a priori estimates of exactness is also based on principle of maximum, obtained a priori estimates take into account boundary effect.

For the case $u(x) \in C^4(\bar{\Omega})$ present solution of the problem (3) in the form

$$z(x) = z_1(x) + z_2(x)$$

where $z_i(x),\ i = 1, 2$ solutions of the following problems

$$\hat{\Delta}_h z_i(x) = -z(x)_{\bar{x}_1 \hat{x}_1} - z(x)_{\bar{x}_2 x_2} = \psi_i(x),\ x \in \hat{\omega}, z_i(x) = 0, x \in \gamma \tag{15}$$

and

$$\psi_1(x) = \begin{cases} \psi(x),\ x \in \omega \backslash \gamma_1 \\ 0,\ x \in \gamma_1 \end{cases} \qquad \psi_2(x) = \begin{cases} 0,\ x \in \omega \backslash \gamma_1 \\ \psi(x),\ x \in \gamma_1 \end{cases}$$

For the estimation of $z_2(x)$ we use the following majorant function

$$v_2(x) = \hbar \begin{cases} x_1,\ 0 \le x_1 \le 1 \\ 2 - x_1,\ 1 \le x_1 \le a \end{cases}, \qquad \hbar = \frac{h_1 + h_2}{2}$$

It fulfills the conditions

$$\Delta_h v_2(x) = \begin{cases} -2,\ x \in \gamma_1, \\ 0,\ \ x \in \omega \backslash \gamma_1 \end{cases}$$
$$v_2(x) > 0, x \in \gamma,$$

Then the following estimate

$$z_2(x) \le \max_{x \in \gamma_1}(|\psi(x)|) v_2(x) = O(|h| + h^2)\hbar \le C\,|h|^2 \tag{16}$$

is valid. Further we obtain

$$\begin{aligned} &|z_1(x)| \le \max_{x \in \hat{\omega} \backslash \gamma_1} (|\psi(x)|\,)\times \\ &\times \max(x_1(a - x_1), x_2(1 - x_2), x_1(a - x_1 - (a-1)x_2)\,) \le \\ &\le C\,|h|^2 \max(x_1(a - x_1), x_2(1 - x_2), x_1(a - x_1 - (a-1)x_2)\,) \end{aligned} \tag{17}$$

Thus, taking into account definitions and inequalities (16),(17) we obtain

$$\max_{x \in \gamma_1}(|z(x)|) \le C\,|h|^2 \tag{18}$$

To estimate the behavior of $z(x)$ more precise, when x approaches to vertexes A, B, D, present $z(x)$ through the solutions of the two following problems

$$\Delta_h w_1(x) = -\psi(x),\ x \in \hat{\omega}_1,\ \ w_1(x) = 0, x \in (\bar{\omega}_1 \backslash \omega_1)\backslash \gamma_1,\ \ w_1(x) = z(x), x \in\ \gamma_1,$$
$$\Delta_h w_2(x) = -\psi(x),\ x \in \hat{\omega}_2,\ \ w_2(x) = 0, x \in (\bar{\omega}_2 \backslash \omega_2)\backslash \gamma_1,\ \ w_2(x) = z(x), x \in\ \gamma_1$$

namely

$$z(x) = \begin{cases} w_1(x), x \in \omega_1, \\ w_2(x), x \in \omega_2 \end{cases}$$

For estimation of functions $w_1(x), w_2(x)$ consider the following auxiliary problems

$$\begin{array}{ll} \Delta_h w_{1,1}(x) = -2,\ x \in \omega_1, & w_1(x) = 0, x \in \bar{\omega}_1 \backslash \omega_1, \\ \Delta_h w_{1,2}(x) = 0,\ x \in \hat{\omega}_1, & w_{1,2}(x) = 0, x \in (\bar{\omega}_1 \backslash \omega_1) \backslash \gamma_1,\ w_{1,2}(x) = 1, x \in \gamma_1, \\ \Delta_h w_{2,1}(x) = -2,\ x \in \omega_2, & w_{2,1}(x) = 0, x \in \bar{\omega}_2 \backslash \omega_2, \\ \Delta_h w_{2,2}(x) = 0,\ x \in \omega_2, & w_{2,2}(x) = 0, x \in (\bar{\omega}_2 \backslash \omega_2) \backslash \gamma_1,\ \ w_{2,2}(x) = 1, x \in \ \gamma_1 \end{array}$$

It is not difficult to see that taking into account (18) the following estimate

$$|z(x)| \le C\,|h|^2 \begin{cases} w_{1,1}(x) + w_{1,2}(x), x \in \omega_1 \\ w_{2,1}(x) + w_{2,2}(x), x \in \omega_2 \end{cases} \tag{19}$$

is valid. It is obvious that $w_{1,1}(x) \equiv v(x)$, and function of discrete arguments $w_{1,2}(x)$ has the following form

$$w_{1,2}(x_{1,i}, x_{2,k}) = \frac{1}{N} \sum_{n=1}^{N-1} \frac{1-(-1)^n}{1-\cos(\frac{n\pi}{N})} \sin(\frac{n\pi}{N}) \sin(\frac{nk\pi}{N}) \frac{U_{i-1}(2-\cos(\frac{n\pi}{N}))}{U_{N-1}(2-\cos(\frac{n\pi}{N}))} \tag{20}$$

Approaching to the knots of the mesh boundary γ which are situated near middle of the intervals AB, BC, AE accuracy order of difference scheme (2) increases to three and approaching to vertexes A, B it increases almost to four (with respect to logarithm).

At last one must specify the behavior of the error approaching to vertex D. According to (19) one must obtain estimates for $w_{2,1}(x), w_{2,2}(x)$ for this. We have

$$w_{2,1}(x) \le \min\{(a-x_1)(x_1-1),\ x_2(a-x_1-(a-1)x_2)\}$$

The function $w_{2,2}(x)$ we estimate through solution of the problem

$$\Delta_h w_{2,2}(x) = 0,\ x \in \omega_2^{\rightarrow},\quad w_{2,2}(x) = 0, x \in (\bar{\omega}_2^{\rightarrow} \backslash \omega_2^{\rightarrow}) \backslash \gamma_1,\quad w_{2,2}(x) = 1, x \in \ \gamma_1$$

where $\omega_2^{\rightarrow}$ is mesh domain, which covers the rectangle $CFDE$. The last one has following form

$$w_{2,2}(1+ih_2, x_{2,k}) = \frac{1}{N} \sum_{n=1}^{N-1} \frac{1-(-1)^n}{1-\cos(\frac{n\pi}{N})} \sin(\frac{n\pi}{N}) \sin(\frac{nk\pi}{N}) \frac{U_{N_1-i-1}(2-\cos(\frac{n\pi}{N}))}{U_{N_1-1}(2-\cos(\frac{n\pi}{N}))}$$

and is estimated similar to function $w_{1,2}(x)$

Thus we proved

Theorem 3. *Let solution of the problem* (1) *belong to the space* $C^4(\bar{\Omega})$, *then accuracy of the difference scheme* (2) *is characterized by the estimate*

$$|z(x)| \le C\,|h|^2 \rho(x)$$

where function $\rho(x)$ approaching to the side AB of the quadrangle $ABCD$ and to the middle of the intervals BC, AE, ED behaves itself as $O(|h|)$ and approaching to vertexes A, B, D - as $O(|h|^2 \ln(\frac{1}{|h|}))$.

A little easier is to prove

Theorem 4. *Let solution of the problem (1) belongs to space $C^3(\bar{\Omega})$, then accuracy of the difference scheme (2) is characterized by the estimate*

$$|z(x)| \leq C\,|h| \begin{cases} \min\{x_1(a-x_1), x_2(1-x_2), w_{1,2}(x)\}, x \in \omega_1 \\ \min\{x_2(1-x_2), x_1(a-x_1-(a-1)x_2)\}, x \in \omega_2 \end{cases}$$

i.e. accuracy of the difference scheme (2) when the point $x \in \hat{\omega}$ approaches to sides AB, BC, CD, AD behaves itself as $O(|h|^2)$ and approaching to the vertex D - as $O(|h|^3)$ and approaching to vertexes A, B - as $O(|h|^3 \ln(\frac{1}{|h|}))$.

Theorems 3, 4 in common way can be generalized on the case when polygon Ω is such that it can be covered with semi-even mesh and its sides goes through either sides or diagonals of the mesh cells.

Remark 2. We show that analogue of the Theorems 3, 4 are valid for finite elements method also. Let T_h be triangulation Ω as it is shown in the Figure 2b). Let, same as previously, $S_1^h(\Omega) = S_1^h$ be finite-dimensional space of continuous on $\bar{\Omega}$ functions which narrowing on every triangle $\tau \in T_h$ is polynomial of the first degree and $\overset{0}{S}{}_1^h(\Omega)$ defines subspace $S_1^h(\Omega)$ which contains the functions which vanishes on Γ. Base in $\overset{0}{S}{}_1^h(\Omega)$ are functions $\omega_{k,l}(x_1, x_2)$ which for $(x_1, x_2) \in \omega_1$ are defined by formula (6) and for $(x_1, x_2) \in \gamma_1 \cup \omega_2$ by formula

$$\omega_{k,l}(x_1,x_2) = \begin{cases} 1 - \frac{1}{h_\alpha}(x_{1,k} - x_1) - \frac{1}{h_1}(x_{2,l} - x_2), & (x_1,x_2) \in \Omega^h_{k,l,1} \\ 1 - \frac{1}{h_\alpha}(x_{1,k} - x_1), & (x_1,x_2) \in \Omega^h_{k,l,2} \\ 1 + \frac{1}{h_1}(x_{2,k} - x_2), & (x_1,x_2) \in \Omega^h_{k,l,3} \\ 1 + \frac{1}{h_2}(x_{1,k} - x_1) + \frac{1}{h_1}(x_{2,l} - x_2), & (x_1,x_2) \in \Omega^h_{k,l,4} \\ 1 + \frac{1}{h_2}(x_{1,k} - x_1), & (x_1,x_2) \in \Omega^h_{k,l,5} \\ 1 - \frac{1}{h_1}(x_{2,k} - x_2), & (x_1,x_2) \in \Omega^h_{k,l,6} \end{cases} \tag{21}$$

$$\alpha = \begin{cases} 1, & (x_1,x_2) \in \gamma_1 \\ 2, & (x_1,x_2) \in \omega_2 \end{cases}$$

$$k, l = 1, 2, ..., N-1, \ (x_1, x_2) \in \gamma_1 \cup \omega_2$$

Finite-element approximation of $u(x)$ is

$$y(x) = \sum_{(x_{1,k}, x_{2,l}) \in \hat{\omega}} y(x_{1,k}, x_{2,l})\omega_{k,l}(x) \tag{22}$$

where $y(x_{k,l}) = y(x_{1,k}, x_{2,l})$ is defined as solution of the difference scheme

$$\begin{gathered} -\hat{\Delta}_h y(x) = -y(x)_{\bar{x}_1\hat{x}_1} - y(x)_{\bar{x}_2 x_2} = \varphi(x), \ x \in \hat{\omega}, y(x) = 0, \ x \in \gamma, \\ \varphi(x_{1,k}, x_{2,l}) = \frac{1}{h_1 \hbar_1} \iint_{\Omega_{k,l}} f(x)\omega_{k,l}(x)dx_1 dx_2, \ (x_{1,k}, x_{2,l}) \in \hat{\omega} \end{gathered} \tag{23}$$

Here expression for $y(x)_{\bar{x}_1\hat{x}_1}$ and $\hbar_1$ are adduced in p. 1.

For investigation of accuracy of the difference scheme (23) we again use technique from [7] and write down integral conclusion of Poisson equation. We obtain

$$-\frac{1}{h_1\hbar_1}\iint\limits_{\Omega_{k,l}} \Delta u(x)\omega_{k,l}(x)dx_1dx_2 = \varphi(x_{1,k}, x_{2,l}) = h_1\hbar_1 \iint\limits_{\Omega_{k,l}} f(x)\omega_{k,l}(x)dx_1dx_2 \tag{24}$$

Write down difference scheme for the error

$$-\hat{\Delta}_h z(x) = -z(x)_{\bar{x}_1\hat{x}_1} - z(x)_{\bar{x}_2x_2} = \psi(x),\ x \in \hat{\omega}, z(x) = 0,\ x \in \gamma, \tag{25}$$

where $\psi(x)$ is local truncation error of the following form

$$\psi(x_{1,k}, x_{2,l}) = l_{k,l}(x_{k,l}, u(\cdot)) = \hat{\Delta}_h u(x) - \frac{1}{h_1\hbar_1}\iint_{\Omega_{k,l}} \Delta u(x)\omega_{k,l}(x)dx_1dx_2 \tag{26}$$

Linear functional $l_{k,l}(x_{k,l}, u(\cdot))$ for $x_{k,l} \in \omega\backslash\gamma_1$ is defined and bounded in the space of functions $W_\infty^4(\Omega_{k,l}^h)$ and vanishes on polynomials of the third degree, and for $x_{k,l} \in \gamma_1$ it is defined and bounded in the space of functions $W_\infty^3(\Omega_{k,l}^h)$ and vanishes on polynomials of the second degree. Using Bremble-Hilbert lemma (see. [7]) one obtains

$$|\psi(x_{1,k}, x_{2,l})| \le C \begin{cases} |h|^2\, |u|_{W_\infty^4(\Omega)}, (x_{1,k}, x_{2,l}) \in \hat{\omega}\backslash\gamma_1 \\ |h|\ \ |u|_{W_\infty^3(\Omega)}, (x_{1,k}, x_{2,l}) \in \gamma_1 \end{cases}. \tag{27}$$

Further, repeating reasonings which were used for proving of the Theorem 3 we convince in correctness of the statement

Theorem 5. *Let solution of the problem* (1) *belong to the space* $W_\infty^4(\Omega)$, *then accuracy of the finite-element method in the knots of the mesh* $\hat{\omega}$ *is characterized by the estimate*

$$|z(x)| \le C\,|h|^2\,\rho(x)\,\|u\|_{W_\infty^4(\Omega)},\ x \in \hat{\omega}$$

where constant C *does not depend on* h_1, h_2 *and* $u(x)$, *besides function* $\rho(x)$ *has the same meaning as in Theorem 3.*

Analogously to the Theorem 4 one proves

Theorem 6. *Let solution of the problem* (1) *belong to the space* $W_\infty^3(\Omega)$, *then accuracy of the finite-element method in the knots of the mesh* $\hat{\omega}$ *is characterized by the estimate*

$$|z(x)| \le C\,|h|\,\|u\|_{W_\infty^3(\Omega)} \begin{cases} \min\{x_1(a-x_1), x_2(1-x_2), w_{1,2}(x)\}, x \in \omega_1 \\ \min\{x_2(1-x_2), x_1(a-x_1-(a-1)x_2)\}, x \in \omega_2 \end{cases}$$

where constant C *does not depend on* $h_1, h_2, u(x)$, *i.e. when point* $x \in \hat{\omega}$ *approaches to the sides* AB, BC, CD, AD *accuracy of the finite-element method in the knots of the mesh behaves itself as* $O(|h|^2)$ *and when this point approaches to vertex* D *it behaves as* $O(|h|^3\ln(\frac{1}{|h|}))$.

4 Unit Cube Domain Case

Consider the case when the domain Ω is unique cube, $f(x_1, x_2, x_3) \equiv 1$ and mesh ω is chosen to be even with equal steps along the axes Ox_1, Ox_2, Ox_3, $h_1 = h_2 = h_3 = 1/N$. Then fir error of the difference scheme (2) $u(x) \in C^{(4)}(\bar{\Omega})$ the following estimate holds true

$$|z(x)| \leq C\,|u|_{C^{(4)}(\bar{\Omega})}\, v(x)$$

where $v(x)$ is solution of the problem

$$-\Delta_h v(x) = 1\;, x \in \omega,\;\; v(x) = 0, x \in \gamma,$$

which can be written in analytical form. It can be presented as follows

$$v(x_{1,s}, x_{2,t}, x_{3,l}) = \sum_{i=1}^{N-1}\sum_{j=1}^{N-1}\sum_{k=1}^{N=1} \frac{(1-(-1)^i)(1-(-1)^j)(1-(-1))^k}{2N^5(1-\cos(\frac{i\pi}{N}))(1-\cos(\frac{j\pi}{N}))(1-\cos(\frac{k\pi}{N}))}$$
$$\frac{\sin(\frac{i\pi}{N})\sin(\frac{j\pi}{N})\sin(\frac{k\pi}{N})\sin(\frac{si\pi}{N})\sin(\frac{tj\pi}{N})\sin(\frac{lk\pi}{N})}{(3-\cos(\frac{i\pi}{N})-\cos(\frac{j\pi}{N})-\cos(\frac{k\pi}{N}))} \tag{28}$$

The following theorem is valid

Theorem 7. *Let solution of the problem* (1) *belong to space* $C^{(4)}(\bar{\Omega})$*, then accuracy of the difference scheme* (2) *is characterized by the estimate*

$$|z(x)| \leq Ch^2 v(x)\,|u|_{C^{(4)}(\bar{\Omega})}$$

where constant C *does not depend on* $h, u(x)$ *and function* $v(x)$ *behaves as* $O(h)$ *approaching to faces of cube and approaching to sides as* $O(h^2\ln^2(1/h))$*, approaching to vertexes as* $O(h^2\ln(1/h))$*.*

Now consider mixed boundary problem

$$\begin{aligned}-\Delta u(x) &= -\frac{\partial^2 u(x)}{\partial x_1^2} - \frac{\partial^2 u(x)}{\partial x_2^2} - \frac{\partial^2 u(x)}{\partial x_3^2} = f(x),\; x \in \Omega,\\ u(x) &= u_0(x),\; x \in \Gamma_1,\; \frac{\partial u(x)}{\partial n} = g(x),\; x \in \Gamma_2\end{aligned} \tag{29}$$

where $\Gamma = \Gamma_1 \cup \Gamma_2$ is boundary of the cube Ω, $\Gamma_2 = \{x = (x_1, x_2, x_3) : x_3 = 0, 0 < x_\alpha < 1, \alpha = 1, 2\}$, n is external normal line to Γ. Introduce in $\bar{\Omega}$ the following cube mesh with the step h:

$$\begin{aligned}\bar{\omega} = \{&x = (x_{1,i_1}, x_{2,i_2}, x_{3,i_3}) \in \bar{\Omega} :\\ &x_{\alpha,i_\alpha} = i_\alpha h,\; i_\alpha = 0, 1, ..., N, \alpha = 1, 2, 3,\;\; h = 1/N\}\end{aligned}$$

Designate the set of internal knots of the mesh $\bar{\omega}$ as $\omega = \{x = (x_1, x_2, x_3) \in \Omega\}$ and the set of limiting points as $\gamma = \bar{\omega}\backslash\omega$. Let

$$\gamma_2 = \bar{\omega} \cap \Gamma_2,\;\; \gamma_1 = \gamma\backslash\gamma_2,\; \omega^* = \omega \cup \gamma_2$$

Approximate the problem (29) on the mesh $\bar{\omega}$ with a difference scheme

$$-\Lambda y = -\sum_{\alpha=1}^{3} \Lambda_\alpha y = \varphi(x),\ x \in \omega^*,\ y(x) = u_0(x), x \in \gamma_1, \tag{30}$$

where

$$\Lambda_\alpha y(x) = \begin{cases} y(x)_{\bar{x}_\alpha x_\alpha},\ x \in \omega, \\ \frac{2}{h} y(x)_{x_3},\ x \in \gamma_2, \end{cases} \quad \varphi(x) = \begin{cases} f(x),\ x \in \omega, \\ f(x) + \frac{2}{h} g(x)\,, x \in \gamma_2. \end{cases}$$

It has been shown in [2] that under condition $u(x) \in C^{(4)}(\bar{\Omega})$ for the error $z(x) = y(x) - u(x)$ of this difference scheme the following estimate is valid

$$|z(x)| \le C(1 - x_{3,i_3})h^2\ , x = (x_{1,i_1}, x_{2,i_2}, x_{3,i_3}) \in \omega^*. \tag{31}$$

The estimate testifies that approaching to the cube face which lies in the plane $x_3 = 1$ accuracy order increases by one order.

Further we present substantial strengthening of the obtained result.

Present $z(x)$ in the form $z(x) = z_1(x) + z_2(x)$, where functions $z_i(x),\ i = 1, 2$ are solutions of the following Dirichlet difference problems

$$-\Lambda z_1(x) = -\sum_{\alpha=1}^{3} \Lambda_\alpha z_1(x) = \psi(x),\ x \in \omega,\ y(x) = 0, x \in \gamma,$$

$$\psi(x) = \Lambda u(x) - \Delta u(x) = O(h^2),\ u(x) \in C^{(4)}(\bar{\Omega}),$$

$$-\Lambda z_2(x) = -\sum_{\alpha=1}^{3} \Lambda_\alpha z_2(x) = 0,\ x \in \omega,\ z_2(x) = 0, x \in \gamma_1,\ z_2(x) = z(x), x \in \gamma_2.$$

It is obvious that to estimate $z_1(x)$ one can use Theorem 7 while taking for the function $z_2(x)$ taking into account (31) the following estimate is valid

$$|z_2(x)| \le Ch^2 v_2(x),$$

where $v_2(x)$ is solution of the following problem

$$-\Lambda v_2(x) = -\sum_{\alpha=1}^{3} \Lambda_\alpha v_2(x) = 0,\ x \in \omega,\ v_2(x) = 0, x \in \gamma_1,\ v_2(x) = 1, x \in \gamma_2,$$

which has the following form

$$\begin{aligned} &v_2(x_{1,s}, x_{2,k}, x_{3,i}) = \\ &= \frac{1}{N^2} \sum_{n=1}^{N-1} \sum_{j=1}^{N-1} \frac{(1-(-1)^n)(1-(-1)^j)}{(1-\cos(\frac{n\pi}{N}))(1-\cos(\frac{j\pi}{N}))} \sin(\frac{n\pi}{N}) \sin(\frac{nk\pi}{N}) \\ &\sin(\frac{j\pi}{N}) \sin(\frac{sj\pi}{N}) \frac{U_{N-i-1}(3-\cos(\frac{n\pi}{N})-\cos(\frac{j\pi}{N}))}{U_{N-1}(3-\cos(\frac{n\pi}{N})-\cos(\frac{j\pi}{N}))} \end{aligned} \tag{32}$$

The following theorem holds true

Theorem 8. *Let solutions of the problem* (29) *belong to space* $C^{(4)}(\bar{\Omega})$ *then accuracy of the difference scheme* (30) *is characterized by the estimate*

$$|z(x)| \leq Ch^2(v(x) + v_2(x))\, |u|_{C^{(4)}(\bar{\Omega})}$$

where function $v(x) + v_2(x)$ *approaching to faces of cube behaves itself as* $O(h)$, *approaching to sides - as* $O(h^2 \ln^2(1/h))$ *and approaching to vertexes - as* $O(h^2 \ln(1/h))$. *Here constant* C *does not depend on* $h, u(x)$.

References

1. Samarskii, A.A.: Introduction in the difference scheme theory. Nauka, Moskva (1971) (in Russian)
2. Halba, E.F.: On accuracy order of difference scheme for Poisson equation with mixed boundary conditions. coll. Computation optimization, Ins. cybern. NASU, Kiev, 30–34 (1985) (in Russian)
3. Makarov, V.: On a priori estimates of difference schemes giving an account of boundary effect C.R. Acad. Bulgare Sci. 42(5), 41–44 (1989)
4. Makarov, V.L., Demkiv, L.I.: Accuracy estimates of difference schemes for quasi-linear elliptic equations with variable coefficients taking into account boundary effect. In: Li, Z., Vulkov, L.G., Waśniewski, J. (eds.) NAA 2004. LNCS, vol. 3401, pp. 80–90. Springer, Heidelberg (2005)
5. Espinosa, O., Moll, V.: A Generalized Polygamma Function. Integral Transforms and Special Functions, 101–115 (2004)
6. Marchuk, G.I.: Methods of computational mathematics, Nauka, Sibir. dep., Novosib (1980) (in Russian)
7. Samarskii A.A., Lazarov R.D., Makarov V.L.: Difference schemes for differential equations with generalized solutions. Moskau, Vyssh. shkola (1987) (in Russian)
8. Ciarlet, P.G., Raviart, P.A.: Maximum principle and uniform convergence for the finite element method. Computer methods in appl. mechenics and engineering 2, 17–31 (1973)

An Iterative Numerical Algorithm for a Strongly Coupled System of Singularly Perturbed Convection-Diffusion Problems

E. O'Riordan[1], J. Stynes[2], and M. Stynes[3]

[1] School of Mathematical Sciences, Dublin City University, Ireland
eugene.oriordan@dcu.ie
[2] Department of Computing, Cork Institute of Technology, Cork, Ireland
jeanne.stynes@cit.ie
[3] Department of Mathematics, National University of Ireland, Cork, Ireland
m.stynes@ucc.ie

Abstract. An iterative numerical method is constructed for a coupled system of singularly perturbed convection-diffusion-reaction two-point boundary value problems. It combines a standard finite difference operator with a piecewise-uniform Shishkin mesh, and uses a Jacobi-type iteration to compute a solution. Under certain assumptions on the coefficients in the differential equations, a bound on the maximum-norm error in the computed solution is established; this bound is independent of the values of the singular perturbation parameter. Numerical results are presented to illustrate the performance of the numerical method.

1 Introduction and Background

Consider the system

$$L_E \boldsymbol{u} := (-E\boldsymbol{u}'' - B\boldsymbol{u}' + A\boldsymbol{u})(x) = \boldsymbol{f}(x), \quad x \in (0,1) \tag{1}$$

of $m \geq 2$ singularly perturbed convection-diffusion-reaction two-point boundary value problems in the unknown vector function $\boldsymbol{u} = (u_1, u_2, \ldots, u_m)^T$, where the boundary values $\boldsymbol{u}(0)$ and $\boldsymbol{u}(1)$ are given. The $m \times m$ coefficient matrices $A = (a_{ij})$ and $B = (b_{ij})$ have entries in $C^3[0,1]$, $\boldsymbol{f} = (f_1, \ldots, f_m)^T \in (C^3[0,1])^m$ and $E = diag\{(\varepsilon_1, \varepsilon_2 \ldots \varepsilon_m)\}$ is a constant diagonal matrix with $0 < \varepsilon_i \leq 1$. In general this is a system of coupled singularly perturbed differential equations with m singular perturbation parameters $\{\varepsilon_i\}_{i=1}^m$ and its solution will exhibit boundary layers. In this paper we study the special case where $\varepsilon_i = \varepsilon$ for all i.

If $B \equiv 0$ in (1), then the system is said to be of reaction-diffusion type. Several recent publications [1,2,3,4,5,6,7] have studied reaction-diffusion systems and on layer-adapted meshes have constructed and analysed numerical methods for these problems that converge, in the maximum norm, uniformly with respect to the parameters ε_i. Bakhvalov [8] established second order convergence under the assumptions that $\varepsilon_i = \varepsilon$ for all i and that the coupling matrix $A(x)$ was coercive. In the case of $m = 2$, Shishkin [9] subdivided the problem class into

S. Margenov, L.G. Vulkov, and J. Waśniewski (Eds.): NAA 2008, LNCS 5434, pp. 104–115, 2009.

three subclasses corresponding to $(i)\,\varepsilon_1 = \varepsilon_2 = \varepsilon$, $(ii)\,\varepsilon_1 = \varepsilon$, $\varepsilon_2 = 1$, the general case $(iii)\,\varepsilon_1 = \varepsilon$, $\varepsilon_2 = \mu$ and established uniform convergence for each of these subclasses. When $m = 2$, in the case of general ε_1 and ε_2, Madden and Stynes [5] established almost first-order convergence which was later improved to second-order convergence by Linß and Madden [4]. More recently, Gracia et al. [10] analysed the case of m equations. Most of these papers impose conditions on the system matrix A such as an M-matrix structure so that the resulting operator is of inverse-monotone type, viz., if $\mathbf{z}(0) \geq \mathbf{0}, \mathbf{z}(1) \geq \mathbf{0}$ and $L_E\mathbf{z}(x) \geq \mathbf{0},\ x \in (0,1)$ then $\mathbf{z}(x) \geq \mathbf{0},\ x \in (0,1)$.

Assuming $B \equiv 0$ and strong diagonal dominance of A, Linß and Madden [11] use sharp stability inequalities from Andreev [12] for the scalar reaction-diffusion equation to analyse (1) without imposing any M-matrix structure onto the matrix A. Moreover, under the same assumptions as in [8] of coercivity and $\varepsilon_i = \varepsilon$, Kellogg et al. [1] extended the domain to two space dimensions and obtained almost second-order convergence. Shishkin [7], Shishkina and Shishkin [13] also obtained second-order convergence (ignoring logarithmic factors) for the two-dimensional problem, with $m = 2$, $\varepsilon_1 = \varepsilon_2$ and assuming that the coupling matrix A is strictly diagonally dominant with positive diagonal elements.

If B is a non-zero diagonal matrix, then (1) is said to be weakly coupled [14,15]. If B has a non-zero off diagonal entry on each row (i.e., for each $i = 1, \ldots, m$ there exists a $j \neq i$ such that $b_{ij} \neq 0$), we say that (1) is *strongly coupled.* Although the analysis in some earlier publications does not rely on inverse-monotonicity of the differential operator, it is not obvious how to extend these analytical techniques to the case of a strongly coupled convection-diffusion system. Note in particular that a strongly coupled system will not in general satisfy a maximum principle.

In contrast to the reaction-diffusion case, there have been relatively few publications on systems involving strongly coupled convection-diffusion equations. In [16], a first-order parameter-uniform method is analysed, under the assumptions that $m = 2$ and that the coupling matrix B of the first-order derivatives is assumed to be a strictly diagonally dominant M-matrix. It is difficult to see how to generalize this analysis to the case of $m > 2$. The papers [17,18] employ scalar stability inequalities from Andreev [19] to derive parameter-uniform error bounds. This analysis requires B to be strongly diagonally dominant. Our analysis in the present paper draws on that of [18], but an iterative Jacobi algorithm (cf. [2]) is employed to analyse the continuous solution and to generate the numerical solution. Under some assumptions on the data, we establish a parameter-uniform asymptotic error bound for our numerical solution. Some numerical results are presented to illustrate the performance of this iterative method.

Notation. Throughout the paper C denotes a generic constant that is independent of ε and any mesh, and can take on different values at different points in the argument. Write $\|\cdot\|_\infty$ for the norm on $L_\infty[0,1]$. Set $\|\mathbf{g}\|_\infty = \max\{\|g_1\|_\infty, \ldots, \|g_m\|_\infty\}$ for any vector-valued function $\boldsymbol{g} := (g_1, \ldots, g_m)^T$ having $g_i \in L_\infty(0,1)$ for all i. Set $\|\mathbf{g}\|_{n,\infty} := \max_i \|g_i^{(n)}\|_\infty$ when the right-hand side is defined. For each $w \in W^{-1,\infty}$ define the norm

$$\|w\|_{-1,\infty} := \inf\{\|W\|_\infty : W' = w\}.$$

We shall also use the usual $L_1[0,1]$ norm $\|\cdot\|_{L_1}$.

2 Analysis of the Continuous Problem

In this paper, we consider the strongly coupled system

$$\boldsymbol{L}\boldsymbol{u} := (-\varepsilon \boldsymbol{u}'' - \boldsymbol{B}\boldsymbol{u}' + \boldsymbol{A}\boldsymbol{u})(x) = \boldsymbol{f}(x), \quad x \in (0,1) \tag{2}$$

where $\varepsilon > 0$ is the only perturbation parameter present in the system. For $i = 1, \ldots, m$, set

$$\beta_i = \min_{x\in[0,1]} b_{ii}(x), \quad \underline{\beta} = \min_i \beta_i.$$

The following hypotheses are placed on the problem:

Assumption 1

(i) $\underline{\beta} > \beta > 0$ and $a_{ii}(x) \geq 0$ for $x \in [0,1]$;

(ii) $0 < \gamma < \min_{i=1,\ldots,m} \gamma_i$,

$$\textit{where } \gamma_i := \sum_{\substack{j=1\\ j\neq i}}^{m} \left(\beta_i^{-1}\|b'_{ij} + a_{ij}\|_{L_1} + \tilde{R}_i\|b_{ij}\|_\infty\right) \quad \textit{for each } i$$

$$\textit{and } \quad \tilde{R}_i := \left(1 + \frac{\|a_{ii}\|_\infty}{\beta_i}\right)\left(R_i^* + \frac{2}{\beta_i}\right), \qquad R_i^* := \int_{x=0}^1 \left|\left(\frac{1}{b_{ii}(x)}\right)'\right| dx.$$

Before examining the system (2) we first consider the scalar convection-diffusion two-point boundary value problem

$$-\varepsilon v''(x) - r(x)v'(x) + q(x)v(x) = p(x) \quad \text{on } (0,1), \qquad v(0) = v(1) = 0, \tag{3}$$

where $0 < \underline{r} \leq r(x) \leq R$ and $0 \leq q(x) \leq Q$ on $[0,1]$. Set

$$R^* := \int_{x=0}^1 \left|\left(\frac{1}{r(x)}\right)'\right| dx \quad \text{and} \quad \tilde{R} := \left(1 + \frac{Q}{\underline{r}}\right)\left(R^* + \frac{2}{\underline{r}}\right).$$

Then by the stability theory of Andreev [19, Theorem 3.1], which is based on a careful analysis of Green's functions, one has

$$\|v\|_\infty \leq \frac{1}{\underline{r}}\|p\|_{L_1}, \quad \|v\|_\infty \leq \tilde{R}\,\|p\|_{-1,\infty}. \tag{4}$$

If one introduces general Dirichlet boundary conditions $v(0)$ and $v(1)$ into (3), then a maximum principle yields

$$\|v\|_\infty \leq \frac{\|p\|_\infty}{\underline{r}} + \max\{|v(0)|, |v(1)|\}. \tag{5}$$

Define the decoupled operators

$$L_i v(x) := -\varepsilon v''(x) - b_{ii} v'(x) + a_{ii}(x)v(x) \quad \text{for } i = 1, \dots, m.$$

The proof of the next result resembles [2, Lemma 2.2]. It combines (4) and (5) with arguments from [18, Lemma 2.1].

Lemma 1. *Let $\boldsymbol{u}$ be a solution of (2). Let $\boldsymbol{u}^{[0]}$ be any function in $(C^1[0,1])^m$. For $k = 1, 2, \dots$, define the sequence $\boldsymbol{u}^{[k]} = (u_1^{[k]}, u_2^{[k]}, \dots, u_m^{[k]})$ by*

$$L_i u_i^{[k]} = f_i + \sum_{\substack{j=1 \\ j \neq i}}^{m} \left(b_{ij} \frac{du_j^{[k-1]}}{dx} - a_{ij} u_j^{[k-1]} \right) \quad \text{on } (0,1) \quad \text{for } i = 1, \dots, m, \qquad (6)$$

with $\boldsymbol{u}^{[k]}(0) = \boldsymbol{u}(0)$, $\boldsymbol{u}^{[k]}(1) = \boldsymbol{u}(1)$. Then $\lim_{k \to \infty} \|\boldsymbol{u} - \boldsymbol{u}^{[k]}\|_\infty = 0$. Furthermore we have

$$\|\boldsymbol{u}\|_\infty \leq \frac{1}{1-\gamma} \left(\frac{1}{\underline{\beta}} \|\boldsymbol{f}\|_\infty + \max\{|\boldsymbol{u}(0)|, |\boldsymbol{u}(1)|\} \right). \qquad (7)$$

Proof. Set $\boldsymbol{n}^{[k]} = \boldsymbol{u} - \boldsymbol{u}^{[k]}$ for $k = 0, 1, \dots$. For $k \geq 1$, we have

$$L_i n_i^{[k]} = \sum_{j \neq i} \left(b_{ij} \frac{dn_j^{[k-1]}}{dx} - a_{ij} n_j^{[k-1]} \right) \quad \text{on } (0,1), \quad n_i^{[k]}(0) = n_i^{[k]}(1) = 0, \quad \text{for } i = 1, \dots, m.$$

By splitting the right-hand side as in [18, Lemma 2.1] then invoking (4) for each i, one gets

$$\begin{aligned} \|n_i^{[k]}\|_\infty &\leq \sum_{\substack{j=1 \\ j \neq i}}^{m} \left(\beta_i^{-1} \|(b'_{ij} + a_{ij}) n_j^{[k-1]}\|_{L_1} + \tilde{R}_i \|b_{ij} n_j^{[k-1]}\|_\infty \right) \\ &\leq \|\boldsymbol{n}^{[k-1]}\|_\infty \sum_{\substack{j=1 \\ j \neq i}}^{m} \left(\beta_i^{-1} \|b'_{ij} + a_{ij}\|_{L_1} + \tilde{R}_i \|b_{ij}\|_\infty \right) = \gamma_i \|\boldsymbol{n}^{[k-1]}\|_\infty. \end{aligned}$$

Hence $\|\boldsymbol{n}^{[k]}\|_\infty \leq \gamma \|\boldsymbol{n}^{[k-1]}\|_\infty \leq \gamma^k \|\boldsymbol{n}^{[0]}\|_\infty$. It follows that $\|\boldsymbol{n}^{[k]}\|_\infty \to 0$.

Choose $\boldsymbol{u}^{[0]} = \boldsymbol{0}$. Set $\boldsymbol{z}^{[k]} = \boldsymbol{u}^{[k]} - \boldsymbol{u}^{[k-1]}$ for $k = 1, 2 \dots$. Then for $k \geq 2$, equation (6) implies that $L_i z_i^{[k]} = \sum_{j \neq i} \left(b_{ij} \frac{dz_j^{[k-1]}}{dx} - a_{ij} z_j^{[k-1]} \right)$ for $i = 1, \dots, m$, with $\boldsymbol{z}^{[k]}(0) = \boldsymbol{z}^{[k]}(1) = \boldsymbol{0}$. By the same argument as in the first part of the lemma, $\|\boldsymbol{z}^{[k]}\|_\infty \leq \gamma \|\boldsymbol{z}^{[k-1]}\|_\infty$ for $k = 2, 3, \dots$. Consequently $\|\boldsymbol{z}^{[k]}\|_\infty \leq \gamma^{k-1} \|\boldsymbol{z}^{[1]}\|_\infty$ for $k = 1, 2, 3, \dots$. Now

$$\|\boldsymbol{z}^{[1]}\|_\infty = \|\boldsymbol{u}^{[1]}\|_\infty \leq \frac{1}{\underline{\beta}} \|\boldsymbol{f}\|_\infty + \max\{\|\boldsymbol{u}(0)\|_\infty, \|\boldsymbol{u}(1)\|_\infty\}$$

by (5) applied to $L_i z_i^{[1]} = f_i$ on $(0,1)$, $\boldsymbol{z}(0) = \boldsymbol{u}(0)$, $\boldsymbol{z}(1) = \boldsymbol{u}(1)$. Thus

$$\|\boldsymbol{z}^{[k]}\|_\infty \leq \gamma^{k-1} \left(\frac{1}{\underline{\beta}} \|\boldsymbol{f}\|_\infty + \max\{\|\boldsymbol{u}(0)\|_\infty, \|\boldsymbol{u}(1)\|_\infty\} \right) \quad \text{for } k = 1, 2, \dots \qquad (8)$$

Finally,

$$\begin{aligned}\|\boldsymbol{u}\|_\infty &= \lim_{j\to\infty}\left\|\sum_{k=1}^{j}\boldsymbol{z}^{[k]}\right\|_\infty \le \lim_{j\to\infty}\sum_{k=1}^{j}\|\boldsymbol{z}^{[k]}\|_\infty \\ &\le \lim_{j\to\infty}\sum_{k=1}^{j}\gamma^{k-1}\left(\frac{1}{\underline{\beta}}\|\boldsymbol{f}\|_\infty + \max\{\|\boldsymbol{u}(0)\|_\infty, \|\boldsymbol{u}(1)\|_\infty\}\right) \\ &= \frac{1}{1-\gamma}\left(\frac{1}{\underline{\beta}}\|\boldsymbol{f}\|_\infty + \max\{\|\boldsymbol{u}(0)\|_\infty, \|\boldsymbol{u}(1)\|_\infty\}\right),\end{aligned}$$

where we used (8) in the calculation.

Corollary 1. *The system* (2) *has a solution and this solution is unique.*

Proof. If $\boldsymbol{f} = \boldsymbol{u}(0) = \boldsymbol{u}(1) = \boldsymbol{0}$, then (7) implies that $\boldsymbol{u} = \boldsymbol{0}$ is the only solution of (2). Now the standard theory for solutions of systems of linear boundary-value problems [20, §7.5] implies the desired result.

3 Discrete Problem

We use a piecewise-uniform Shishkin mesh $\bar{\Omega}_\sigma^N$, which is constructed as follows. Let $N \ge 4$ be an even positive integer. Partition the domain $[0,1]$ into two subintervals $[0,\sigma]$ and $[\sigma,1]$ where the *transition point* σ is

$$\sigma := \min\left\{\frac{1}{4}, C_*\varepsilon\ln N\right\}, \quad C_* > \frac{1}{\underline{\beta}}. \tag{9}$$

The subdomains $[0,\sigma]$ and $[\sigma,1]$ are each subdivided into $N/2$ intervals.

We introduce the finite difference operators

$$D^+v_i := \frac{v_{i+1}-v_i}{\bar{h}_i},\ D^-v_i := \frac{v_i - v_{i-1}}{\bar{h}_i}, \quad \delta^2 v_i := \frac{1}{\bar{h}_i}\left(\frac{v_{i+1}-v_i}{h_{i+1}} - \frac{v_i - v_{i-1}}{h_i}\right).$$

Here $h_i := x_i - x_{i-1}$ and $\bar{h}_i := (h_i + h_{i+1})/2$ for each i. The operator $a_i D^+ v_i$ is an upwinded approximation of $av'(x_i)$ and $\delta^2 v_i$ is the standard central difference approximation of $v''(x_i)$. Define the finite difference operator L_i^N by

$$L_i^N Z_i := \left(-\varepsilon\delta^2 Z_i - b_{ii}D^+Z_i + a_{ii}Z_i\right).$$

Our discretization of (2) uses a Jacobi iteration. For all $x_i \in \bar{\Omega}_\sigma^N$, let $\boldsymbol{U}^{[0]}(x_i) = \boldsymbol{u}(0) + x_i[\boldsymbol{u}(1) - \boldsymbol{u}(0)]$ be the initial guess. For $k = 1, 2, \ldots$, set

$$L_i^N U_i^{[k]}(x_j) = f_i(x_j) + \sum_{\substack{l=1\\ l\ne i}}^{m}\left(b_{il}D^+U_l^{[k-1]} - a_{il}U_l^{[k-1]}\right)(x_j) \quad \text{for } x_j \in \Omega_\sigma^N, \tag{10a}$$

$$U_i^{[k]}(0) = u_i(0), \quad U_i^{[k]}(1) = u_i(1), \tag{10b}$$

The sequence of discrete solutions $U^{[k]}$ is well defined since Assumption 1 implies that the matrix associated with L^N is invertible for all sufficiently large N, independently of ε.

Lemma 2. *Let $\boldsymbol{U}^{[k]}$ be the solution of* (10)*. Then* $\lim_{k\to\infty} \|\boldsymbol{U}-\boldsymbol{U}^{[k]}\|_{\infty,\bar{\Omega}_\sigma^N} = 0$. *Furthermore, we have*

$$\|\boldsymbol{U}^{[k]} - \boldsymbol{U}\|_\infty \leq C\gamma^k. \tag{11}$$

Proof. This is a discrete analogue of the proof of Lemma 1.

Combining this result with [18, Theorem 4.1] we arrive at our main result:

Theorem 1. *Let N be sufficiently large, independently of ε. Let K be any integer satisfying*

$$K \geq -\frac{\ln N}{\ln \gamma}. \tag{12}$$

Under the assumptions 1.1, 2.1, 3.1 and 3.2 of [18],

$$\|\boldsymbol{U}^{[K]} - \boldsymbol{u}\|_{\infty,d} \leq CN^{-1}(\ln N)^2, \tag{13}$$

where $\boldsymbol{U}^{[K]}$ is the solution of (10) *with $k = K$ and $\boldsymbol{u}$ is the solution of (1).*

4 Numerical Results

In this section numerical results are presented for the iterative numerical method (10) applied to some test problems of the form (1). The numerical results in this section can be compared to the corresponding results in [18], where the same problems were solved using a non-iterative numerical method. The data in these problems do not satisfy hypothesis (ii) of Assumption 1. Nevertheless, the iterative method (10) is applied and, motivated by the stopping criterion (12), the iteration is halted when

$$K > K^* \ln N, \tag{14}$$

where the value of K^* was determined experimentally as follows: each test problem was solved over an extensive range of ε, ($\varepsilon = 10^{-j}, j = 0, 1, 2, \ldots, 7$) and we examined approximations to the uniform rates of convergence for values of $K^* = 1, 2, 4, 8$ to see at what value of K^* the computed orders of convergence were unchanged by further doubling of K^*.

Example 1

$$B = \begin{pmatrix} 3 & -1 & -1 \\ -1 & 4 & -2 \\ -1 & -2 & 4 \end{pmatrix}, \ A = 0 \quad \text{and} \quad \boldsymbol{f} = \begin{pmatrix} -4 \\ 11 \\ -7 \end{pmatrix}.$$

The boundary conditions are $\boldsymbol{u}(0) = (-1, 4, -1)^T, \boldsymbol{u}(1) = (e^{-1/\varepsilon} - 2e^{-4/\varepsilon} + 1, e^{-1/\varepsilon} + e^{-4/\varepsilon} + 2e^{-6/\varepsilon} - 2, e^{-1/\varepsilon} + e^{-4/\varepsilon} - 2e^{-6/\varepsilon})^T$. The exact solution of this problem is

$$\boldsymbol{u} = (1,1,1)^T e^{-x/\varepsilon} + (-2,1,1)^T e^{-4x/\varepsilon} + (0,2,-2)^T e^{-6x/\varepsilon} + (x, -2x, x-1)^T.$$

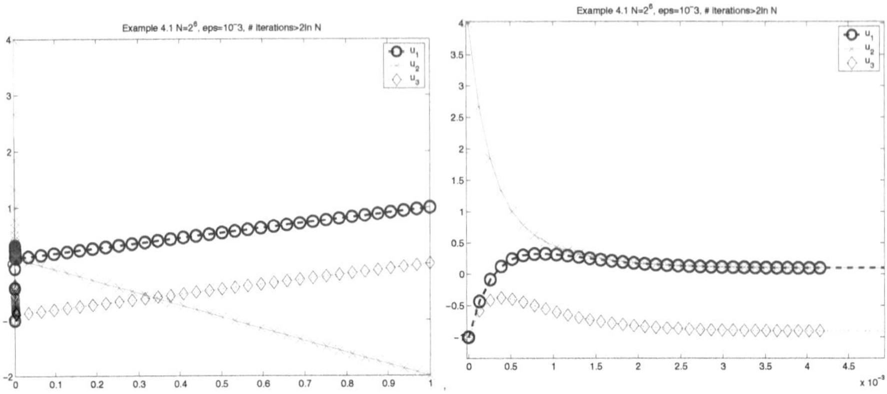

Fig. 1. Example 1 when $\varepsilon = 10^{-3}$, $N = 64$, $C_* = 1$ in (9) and $K > 2\ln N$; computed solution for $x \in [0,1]$ with a blow-up of the layer region for $0 \le x \le 5\varepsilon$

Table 1. The ε-uniform orders $q^{K,N}$ of(15) for Example 1 with $C_* = 1$ in (9) and $K^* = 1,2,4,8$ in (14); the final row gives q^N computed when the iterative process terminates after the maximum pointwise difference in the iterates is less than 10^{-6}

	N						
ε	8	16	32	64	128	256	512
$K^* = 1$	-0.214	1.036	0.778	-0.297	0.714	0.649	0.000
$K^* = 2$	0.410	0.626	0.808	0.823	0.896	0.929	0.963
$K^* = 4$	0.351	0.637	0.810	0.829	0.899	0.932	0.964
$K^* = 8$	0.366	0.639	0.811	0.830	0.899	0.932	0.964
$Tol = 10^{-6}$	0.366	0.639	0.811	0.830	0.899	0.932	0.964

Table 1 shows that the uniform rates of convergence $q^{K,N}$ (defined in (15) below) do not change significantly for $K^* \geq 2$. These rates were also compared with the rates computed when the iterations were continued until

$$\|\mathbf{U}^{K+1,N} - \mathbf{U}^{K,N}\| < 10^{-6}.$$

Table 2 displays computational results for Example 1 with $C_* = 1$ in (9), $N = 8, 16, 32, \ldots, 1024$ and over a more detailed range of the singular perturbation parameter $\varepsilon = 1, 2^{-1}, \ldots, 2^{-27}$. As the exact solution of the example is known, we compute the maximum pointwise error

$$E_\varepsilon^{K,N} := \|\mathbf{U}^{K,N} - \mathbf{u}\|, \quad K > 2\ln N$$

and the ε-uniform maximum pointwise error $E^{K,N}$

$$E^{K,N} := \max_{\varepsilon = 1, 2^{-1}, \ldots, 2^{-27}} E_\varepsilon^{K,N}, \quad K > 2\ln N.$$

Table 2. Maximum pointwise errors $E_\varepsilon^{K,N}$, ε-uniform errors $E^{K,N}$, ε-uniform orders $q^{K,N}$ of (15) and computed error constants $\tilde{C}_1^N, \tilde{C}_2^N$ for Example 1 with $C_* = 1$ in (9) and $K > 2\ln N$

	N							
ε	8	16	32	64	128	256	512	1024
2^0	2.630e-1	1.521e-1	8.142e-2	4.199e-2	2.141e-2	1.081e-2	5.430e-3	2.721e-3
2^{-1}	4.872e-1	2.895e-1	1.620e-1	8.696e-2	4.476e-2	2.258e-2	1.136e-2	5.701e-3
2^{-2}	6.491e-1	4.989e-1	2.852e-1	1.621e-1	8.651e-2	4.517e-2	2.317e-2	1.183e-2
2^{-3}	6.837e-1	5.989e-1	4.512e-1	2.914e-1	1.628e-1	8.638e-2	4.439e-2	2.246e-2
2^{-4}	7.042e-1	6.055e-1	4.543e-1	3.007e-1	1.933e-1	1.172e-1	6.861e-2	3.893e-2
2^{-5}	7.147e-1	6.095e-1	4.559e-1	3.017e-1	1.938e-1	1.174e-1	6.879e-2	3.905e-2
2^{-6}	7.200e-1	6.117e-1	4.568e-1	3.023e-1	1.941e-1	1.176e-1	6.898e-2	3.917e-2
2^{-7}	7.227e-1	6.129e-1	4.572e-1	3.026e-1	1.942e-1	1.177e-1	6.898e-2	3.917e-2
2^{-8}	7.240e-1	6.135e-1	4.574e-1	3.027e-1	1.943e-1	1.177e-1	6.898e-2	3.917e-2
2^{-9}	7.247e-1	6.139e-1	4.575e-1	3.028e-1	1.943e-1	1.177e-1	6.898e-2	3.917e-2
2^{-10}	7.250e-1	6.140e-1	4.576e-1	3.029e-1	1.944e-1	1.177e-1	6.898e-2	3.917e-2
2^{-11}	7.252e-1	6.141e-1	4.576e-1	3.029e-1	1.944e-1	1.177e-1	6.898e-2	3.917e-2
2^{-12}	7.253e-1	6.141e-1	4.576e-1	3.029e-1	1.944e-1	1.177e-1	6.898e-2	3.917e-2
$E^{K,N}$	7.254e-1	6.142e-1	4.576e-1	3.029e-1	1.944e-1	1.177e-1	6.898e-2	3.917e-2
$q^{K,N}$	0.410	0.626	0.808	0.823	0.896	0.929	0.963	
$\tilde{C}_1^N$	2.791	3.544	4.226	4.661	5.128	5.436	5.661	5.787
$\tilde{C}_2^N$	1.342	1.278	1.219	1.121	1.057	0.980	0.908	0.835

Assuming that one has a theoretical rate of convergence of the form $\mathcal{O}((N^{-1}\ln N)^p)$, the computed rate of convergence q^N is given by

$$q^{K,N} := \frac{\ln E^{K,N} - \ln E^{K,2N}}{\ln(2\ln N) - \ln(\ln(2N))} \tag{15}$$

and estimates of the associated error constants by

$$\tilde{C}_1^N := E^{K,N} N(\ln N)^{-1}, \quad \tilde{C}_2^N := E^{K,N} N(\ln N)^{-2}.$$

The results in Table 2 indicate that the numerical approximations $\boldsymbol{U}^{K,N}$ generated by (10) converge uniformly to the exact solution $\boldsymbol{u}$ of Example 1. The numerical results in this table can be compared to the results in [18], where the same problem was solved using a non-iterative scheme on a similar Shishkin mesh (with $C_* = 0.4$).

Our second test problem is a variable coefficient problem.

Example 2

$$B = \begin{pmatrix} 5+2x & 1+3x^2 & 3-x \\ 1+2e^{-4x} & 5-x^2 & x^3 \\ 1 & 2(2+x)/(1+x) & 6 \end{pmatrix}, \; A = 0 \quad \text{and} \quad \boldsymbol{f} = \begin{pmatrix} 1 \\ -4-4x \\ -12+2x^2 \end{pmatrix}.$$

The boundary conditions are $\boldsymbol{u}(0) = \boldsymbol{u}(1) = 0$. A representative computed solution is displayed in Figure 2. As the exact solution of the example is unknown,

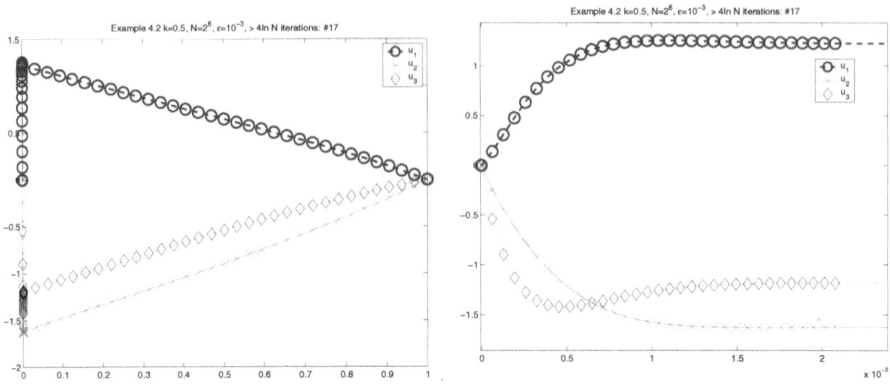

Fig. 2. Example 2 computed solution for $x \in [0,1]$ and blow-up of the layer region for $x \in [0, 2.5\varepsilon]$ when $\varepsilon = 10^{-3}$, $N = 64$, $C_* = 0.5$ in (9) and $K > 4 \ln N$

we follow the standard approach [21] by computing, for each N and ε, the two-mesh difference $D_\varepsilon^{N,K(N)}$ defined by

$$D_\varepsilon^{N,K} = \|\tilde{\mathbf{U}}^{2N,K(N)} - \mathbf{U}^{N,K(N)}\|, \qquad K > 2\ln N$$

where $\tilde{\mathbf{U}}^{2N,K(N)}$ is computed on the mesh obtained by bisecting $\bar{\Omega}^N$. Then the ε-uniform two-mesh difference is defined to be

$$D^{N,K} = \max_{\varepsilon=1,10^{-1},\dots,10^{-7}} D_\varepsilon^{N,K}.$$

Assuming that one has a theoretical rate of convergence of the form $\mathcal{O}((N^{-1}\ln N)^p)$, an estimate of the computed rate of convergence is given by

$$p^{N,K} := \frac{\ln D^{N,K} - \ln D^{2N,K}}{\ln(2\ln N) - \ln(\ln(2N))}. \tag{16}$$

and estimates of the associated error constants by

$$C_1^N := D^{K,N} N(\ln N)^{-1}, \quad C_2^N := D^{K,N} N(\ln N)^{-2}.$$

In Example 2 one sees that $\beta < 4$. Thus one should choose $C_* > 0.25$. In Table 3 the uniform rates of convergence $p^{K,N}$ do not change significantly for $K* \geq 4$ (see (14). Table 4 displays computational results for Example 2 with $C_* = 0.5$ in (9), $\varepsilon = 1, 2^{-1}, \dots, 2^{-27}$ and $N = 16, 32, \dots, 1024$. The results in Table 4 indicate that the numerical approximations $\boldsymbol{U}^{K,N}$ generated by (10) converge uniformly to the exact solution $\boldsymbol{u}$ of Example 2.

Our final example examines how the numerical method performs when it is applied to a problem where the coefficient matrix B may not be row diagonally dominant.

Table 3. The ε-uniform orders $p^{K,N}$ for Example 2 with $C_* = 0.5$ in (9) and $K^* = 1, 2, 4, 8$ in (14)

ε	N 8	16	32	64	128	256	512
$K^* = 1$	0.311	0.727	0.426	0.810	0.892	0.936	0.962
$K^* = 2$	0.887	0.619	0.685	0.785	0.894	0.942	0.956
$K^* = 4$	0.723	0.640	0.688	0.797	0.889	0.937	0.963
$K^* = 8$	0.716	0.637	0.688	0.797	0.889	0.937	0.963

Table 4. Maximum pointwise differences $D_\varepsilon^{K,N}$, ε-uniform differences $D^{K,N}$, ε-uniform orders $p^{K,N}$ and computed error constants C_1^N, C_2^N for Example 2 with $C_* = 0.5$ in (9) and $K > 4 \ln N$

ε	N 8	16	32	64	128	256	512	1024
2^0	8.647e-2	5.534e-1	3.154e-2	1.681e-2	8.693e-3	4.422e-3	2.230e-3	1.120e-3
2^{-1}	1.394e-1	9.056e-2	5.763e-2	3.251e-2	1.724e-2	8.906e-3	4.526e-3	2.282e-3
2^{-2}	1.425e-1	1.174e-1	8.659e-2	5.861e-2	3.296e-2	1.748e-2	9.020e-3	4.582e-3
2^{-3}	1.448e-1	1.168e-1	8.649e-2	6.045e-2	3.924e-2	2.386e-2	1.392e-2	7.911e-3
2^{-4}	1.483e-1	1.164e-1	8.621e-2	6.059e-2	3.935e-2	2.392e-2	1.396e-2	7.929e-3
2^{-5}	1.514e-1	1.165e-1	8.607e-2	6.072e-2	3.944e-2	2.397e-2	1.398e-2	7.942e-3
2^{-6}	1.536e-1	1.167e-1	8.603e-2	6.079e-2	3.950e-2	2.401e-2	1.400e-2	7.951e-3
2^{-7}	1.550e-1	1.167e-1	8.597e-2	6.078e-2	3.953e-2	2.404e-2	1.402e-2	7.957e-3
2^{-8}	1.557e-1	1.167e-1	8.590e-2	6.075e-2	3.955e-2	2.405e-2	1.403e-2	7.960e-3
2^{-9}	1.561e-1	1.167e-1	8.585e-2	6.074e-2	3.956e-2	2.405e-2	1.403e-2	7.960e-3
2^{-10}	1.563e-1	1.167e-1	8.581e-2	6.074e-2	3.956e-2	2.405e-2	1.403e-2	7.961e-3
2^{-11}	1.564e-1	1.167e-1	8.579e-2	6.074e-2	3.956e-2	2.405e-2	1.403e-2	7.962e-3
2^{-12}	1.564e-1	1.167e-1	8.578e-2	6.074e-2	3.956e-2	2.405e-2	1.403e-2	7.963e-3
$D^{K,N}$	1.565e-1	1.174e-1	8.659e-2	6.079e-2	3.956e-2	2.405e-2	1.403e-2	7.964e-3
$p^{K,N}$	0.709	0.647	0.693	0.797	0.889	0.937	0.963	
C_1^N	0.602	0.677	0.800	0.935	1.044	1.110	1.151	1.177
C_2^N	0.289	0.244	0.231	0.225	0.215	0.200	0.185	0.170

Example 3

$$B = \begin{pmatrix} 5 & 3 & \eta \\ \eta & 5 & 3 \\ \eta & 3 & 6 \end{pmatrix}, \qquad A = 0 \qquad \text{and} \qquad \boldsymbol{f} = \begin{pmatrix} 1 \\ -4 - 4x \\ -12 + 2x^2 \end{pmatrix}.$$

The parameter $\eta \geq 0$ will be varied in our numerical experiments. The boundary conditions are $\boldsymbol{u}(0) = \boldsymbol{u}(1) = 0$.

The data of this problem do not satisfy hypothesis (ii) of Assumption 1 for any value of η. Here $\beta = 5$ and strict diagonal dominance requires $\eta < 2$. The orders of convergence of the numerical method for a range of $\eta \geq 0$ are given in Table 5. The method apparently fails to converge only when $\eta \geq 2$. For the particular case $\eta = 2$, Table 6 shows that the iterative method appears to converge when the

Table 5. Computed values of $p^{K,N}$ with $C_* = 1$ in (9) and $K > 4\ln N$, for various values of η, for Example 3

	N						
η	8	16	32	64	128	256	512
0	0.176	0.372	0.706	0.633	0.805	0.864	0.917
1	0.248	0.420	0.744	0.635	0.818	0.865	0.923
2	-0.821	0.116	0.530	-0.798	-0.590	0.472	-0.107
3	-8.213	1.16	0.530	-0.798	-0.590	0.472	-0.107

Table 6. The ε-uniform orders $p^{K,N}$ for Example 3 for $\eta = 2$ with $C_* = 2$ in (9) and $K^* = 1, 2, 4, 8$ in (14)

	N						
ε	8	16	32	64	128	256	512
$K^* = 1$	0.378	0.462	0.361	0.714	0.836	0.764	0.880
$K^* = 2$	0.364	0.166	0.526	0.732	0.667	0.815	0.877
$K^* = 4$	0.196	0.541	0.421	0.699	0.669	0.815	0.877
$K^* = 8$	0.216	0.128	0.438	0.700	0.669	0.814	0.877

transition point is moved so that $C_* = 2$. For values of $\eta > 2$ in further numerical experiments, it appeared that the stopping criterion (14) was not satisfied for $C_* = 2$ and $K^* \leq 8$.

In conclusion, all of the above numerical experiments suggest that the numerical method (10) may yield uniformly convergent numerical approximations to the solutions of a wider class of problems of the form (1) than is covered by our current theory.

References

1. Kellogg, R., Linß, T., Stynes, M.: A finite difference method on layer-adapted meshes for an elliptic reaction-diffusion system in two dimensions. Math. Comp. 77, 2085–2096 (2008)
2. Kellogg, R., Madden, N., Stynes, M.: A parameter-robust numerical method for a system of reaction-diffusion equations in two dimensions. Numer. Methods. Partial Diff. Eqns. 24, 312–334 (2008)
3. Linß, T., Madden, N.: An improved error estimate for a numerical method for a system of coupled singularly perturbed reaction–diffusion equations. Comp. Meth. Appl. Math. 3, 417–423 (2003)
4. Linß, T., Madden, N.: Accurate solution of a system of coupled singularly perturbed reaction-diffusion equations. Computing 73, 121–133 (2004)
5. Madden, N., Stynes, M.: A uniformly convergent numerical method for a coupled system of two singularly perturbed linear reaction-diffusion problems. IMA J. Numer. Anal. 23, 627–644 (2003)
6. Matthews, S., O'Riordan, E., Shishkin, G.: A numerical method for a system of singularly perturbed reaction-diffusion equations. J. Comput. Appl. Math. 145, 151–166 (2002)

7. Shishkin, G.I.: Approximation of a system of singularly perturbed elliptic reaction–diffusion equations on a rectangle. In: Farago, I., Vabishchevich, P., Vulkov, L. (eds.) Proceedings of Fourth International Conference on Finite Difference Methods: Theory and Applications, Lozenetz, Bulgaria, Rousse University, Bulgaria, August 2006, pp. 125–133 (2007)
8. Bakhvalov, N.S.: On the optimization of methods for boundary-value problems with boundary layers. J. Numer. Meth. Math. Phys. 9(4), 841–859 (1969) (in Russian)
9. Shishkin, G.: Mesh approximation of singularly perturbed boundary-value problems for systems of elliptic and parabolic equations. Comp. Maths. Math. Phy 35, 429–446 (1995)
10. Gracia, J., Lisbona, F., O'Riordan, E.: A system of singularly perturbed reaction-diffusion equations. Preprint MS-07-10, Dublin City University (2007)
11. Linß, T., Madden, N.: Layer-adapted meshes for a system of coupled singularly perturbed reaction-diffusion problem. IMA Journal of Numerical Analysis 29, 109–125 (2009)
12. Andreev, V.: On the uniform convergence of a classical difference scheme on a nonuniform mesh for the one-dimensional singularly perturbed reaction-diffusion equation. Comp. Math. Math. Phys. 44, 449–464 (2004)
13. Shishkina, L., Shishkin, G.: Robust numerical method for a system of singularly perturbed parabolic reaction–diffusion equations on a rectangle. Math. Model. Anal. 13, 251–261 (2008)
14. Cen, Z.: Parameter-uniform finite difference scheme for a system of coupled singularly perturbed convection-diffusion equations. J. Syst. Sci. Complex. 18, 498–510 (2005)
15. Linß, T.: Analysis of an upwind finite-difference scheme for a system of coupled singularly perturbed convection-diffusion equations. Computing 79, 23–32 (2007)
16. O'Riordan, E., Stynes, M.: Numerical analysis of a strongly coupled system of two singularly perturbed convection-diffusion problems. Adv. Comput. Math. 30, 101–121 (2009)
17. Linß, T.: Analysis of a system of singularly perturbed convection-diffusion equations with strong coupling. Preprint MATH-NM-02-2007, Institut für Numerische Mathematik, Technische Universität Dresden (2007)
18. O'Riordan, E., Stynes, J., Stynes, M.: A parameter-uniform finite difference method for a coupled system of convection-diffusion two-point boundary value problems. Numerical Mathematics: Theory, Methods and Applications 1, 176–197 (2008)
19. Andreev, V.: A priori estimates for solutions of singularly perturbed two-point boundary value problems. Mat. Model. 14, 5–16 (2002); Second International Conference OFEA 2001, Optimization of Finite Element Approximation, Splines and Wavelets (St. Petersburg, 2001) (Russian)
20. Ladyzhenskaya, O., Uraltseva, N.: Linear and quasilinear elliptic equations. Translated from the Russian by Scripta Technica, Inc. Translation editor: Leon Ehrenpreis. Academic Press, New York (1968)
21. Farrell, P.A., Hegarty, A.F., Miller, J.J., O'Riordan, E., Shishkin, G.I.: Robust Computational Techniques for Boundary Layers. Chapman & Hall/CRC, Boca Raton (2000)

Improved Difference Scheme for a Singularly Perturbed Parabolic Reaction-Diffusion Equation with Discontinuous Initial Condition

Grigory Shishkin

Institute of Mathematics and Mechanics,
Ural Branch of Russian Academy of Sciences, Ekaterinburg 620219, Russia
shishkin@imm.uran.ru

Abstract. A Dirichlet problem is considered for a singularly perturbed parabolic reaction-diffusion equation with a piecewise-continuous initial condition. For this problem, using the method of additive splitting of singularities (generated by discontinuities of the initial function and its lowest derivatives) and the Richardson method, a finite difference scheme on piecewise-uniform meshes is constructed that converges ε-uniformly with the third and second order of accuracy in x and t, respectively.

1 Introduction

A decrease in the smoothness of the solution, for example, in the case when either initial or boundary condition is piecewise-smooth or is discontinuous, causes a lowering of the rate of ε-uniform convergence of constructed schemes and even a loss of convergence.

Special difference schemes for singularly perturbed parabolic reaction-diffusion and convection-diffusion equations with weak singularities (piecewise-smooth initial-boundary conditions) were considered in [1,2,3,4]. In these papers, for the construction of ε-uniformly convergent difference schemes the condensing mesh method was applied, i.e., monotone classical grid approximations of a problem (see, e.g., [5]) on piecewise-uniform meshes that condense in a neighbourhood of the boundary layer.

Problems for singularly perturbed parabolic reaction-diffusion equations with a strong singularity (a discontinuous initial condition) were considered in [6,7,8,9,10,11,12]. In these problems, when constructing special schemes, in addition to the condensing mesh method, a specific technique was used, such as the fitted operator method in [7,8,9,10] or the method of additive splitting of singularities in [6,11,12] (in a neighbourhood of the points of discontinuity to the initial function). In the fitted operator method, the singular components of the solution (or some of them) are solutions of a difference scheme, and in the method of additive splitting of singularities, these components are included in the approximate solution additively (for a description of this method in the case of a regular problem with nonsmooth data, see, e.g., [13] and the references therein). As a rule, order of ε-uniform convergence for difference schemes does not exceed 2 in spatial variables and 1 in the temporal variable.

S. Margenov, L.G. Vulkov, and J. Waśniewski (Eds.): NAA 2008, LNCS 5434, pp. 116–127, 2009.

In the present paper, for a singularly perturbed parabolic reaction-diffusion equations with a piecewise-continuous initial function, a special finite difference scheme is constructed that converges ε-uniformly at the rate $\mathcal{O}\big(N^{-2}\ln^2 N + N_0^{-1}\big)$, where $N+1$ and N_0+1 define the number of mesh points in x and t, respectively. When constructing the scheme, the method of additive splitting of singularities (generated by discontinuities of the initial function and its low-order derivatives) is used and also the condensing mesh method (based on piecewise-uniform meshes that condense in a neighbourhood of the boundary layer).

To improve the convergence order of the scheme, an approach based on the Richardson technique is applied. As a result, a difference scheme is also constructed that converges ε-uniformly at the rate $\mathcal{O}\big(N^{-3} + N_0^{-2}\big)$.

2 Problem Formulation. The Aim of the Research

1. In the domain $\overline{G}$ with the boundary S, where

$$\overline{G} = G \bigcup S, \quad G = D \times (0,\, T], \quad D = \{x:\ x \in (-d,\, d)\}, \tag{1}$$

we consider the Dirichlet problem for the singularly perturbed parabolic equation with constant coeficients[1]

$$\begin{aligned} L_{(2)}\, u(x,t) \equiv \Big\{\varepsilon^2\, a\, \frac{\partial^2}{\partial x^2} - c - p\, \frac{\partial}{\partial t}\Big\}\, u(x,t) &= f(x,t), \quad (x,t) \in G, \\ u(x,t) = \varphi(x,t), \quad (x,t) \in S \setminus S^d. & \end{aligned} \tag{2}$$

Here $a,\, p > 0$, $c \geq 0$, and the right-hand side $f(x,t)$ is sufficiently smooth on the set $\overline{G}$; the parameter ε takes arbitrary values in the open-closed interval $(0,\, 1]$.

The initial-boundary function $\varphi(x,t)$ is continuous on $S \setminus S^d$ and has jump on the set S^d, where $S^d = \{(0,\, 0)\}$. The function $\varphi(x,t)$ is assumed to be sufficiently smooth on the closure of $\overline{S}^L$ (the lateral part of the boundary S) and on the closure of those segments of the boundary S_0 (the lower part of the boundary S), where $\varphi(x,t)$ is continuous; continuations of $\varphi(x,t)$ to the closures are also denoted by $\varphi(x,t)$. Assume that on the set S^c of corner points compatibility conditions [14] are fulfilled that guarantee the required smoothness of the solution in a neighbourhood of the set. Here $S = S_0 \bigcup S^L$, $S_0 = \overline{S}_0$ and S^L are the lower and lateral parts of the boundary S, $S^L = \Gamma \times (0,\, T]$, $\Gamma = \overline{D} \setminus D$; $S^c = S_0 \bigcap \overline{S}^L$.

In the case of the condition

$$S^d \neq \emptyset, \tag{3}$$

the solution of the problem (2) means a function $u \in C(\overline{G}^*) \bigcap C^{2,1}(G)$ which is bounded on $\overline{G}$ and satisfies the differential equation on G and the boundary condition on S^*, where $\overline{G}^* = \overline{G} \setminus S^d$, $S^* = S \setminus S^d$.

[1] The notation $L_{(j.k)}$ ($m_{(j.k)}$, $M_{(j.k)}$, $G_{h(j.k)}$)) means that these operators (constants, grids) are introduced in formula $(j.k)$.

2. For fixed values of the parameter ε, the function $u(x,t)$ is discontinuous on S^d; the derivatives of the function $u(x,t)$ grow without bound as $(x,t) \to S^d$.

We discuss the behavior of the solution for small values of the parameter ε. Let $S^{\gamma_0} = \{(x,t) :\ x = \gamma_0(t) \equiv 0,\ t \in [0, T]\}$ be the characteristic of the reduced equation passing through S^d. When the parameter ε tends to zero, parabolic boundary and interior layers with typical scale ε arise in a neighbourhood of the sets S^L and S^γ (see the estimates in Section 3).

It is known that solutions of classical difference schemes do not converge ε-uniformly even in the case of singularly perturbed problems with sufficiently smooth data. For singularly perturbed equations with the parabolic boundary layer, there are no schemes based on the fitted operator method that converge ε-uniformly. When the function $\varphi(x,t)$ is discontinuous, difference schemes based on classical approximations of the boundary value problem do not converge in the maximum norm even for fixed values of ε. (see, e.g., [7]).

Our aim is for problem (3), (1) to construct an ε-uniformly convergent difference scheme of the method of additive splitting of singularities generated by discontinuities of the function $\varphi(x,t)$, $(x,t) \in S^d$ and its low-order derivatives. Besides, using the Richardson technique (applied for the construction of an improved grid approximation of the solution to the problem (2), (1), (3), which does not include an additively splitted solution component), we construct a high-order accurate difference scheme that converges ε-uniformly with the convergence order higher than two and one, respectively, with respect to x and t.

3 *A Priori* Estimates of Solutions for the Problem (1)

We give some estimates for the solution of the boundary value problem (1) and its derivatives. To derive the estimates, we apply a technique that can be found in [3,12,14], where the decomposition of the solution into its regular ("sufficiently smooth") and singular components is used. We assume that the functions $f(x,t)$ and $\varphi(x,t)$ are sufficiently smooth on the sets $\overline{G}$ and $\overline{S}^L$, S_0^+, S_0^-, respectively; $\varphi(x,t) \in C(S^*)$. Here $S_0 = S_0^- \bigcup S_0^+$, $S_0^- = S_0 \bigcap \{x \le 0\}$, $S_0^+ = S_0 \bigcap \{x \ge 0\}$.

1. The set $\overline{G}$ is decomposed into the sum of overlapping subsets[2]

$$\overline{G} = \bigcup_j \overline{G}^j, \quad j = 1, 2, 3, \quad \overline{G}^2 = \overline{G}^{2l} \bigcup \overline{G}^{2r}, \tag{4}$$

$$G^1 = G^1(m^1) = \{(x,t) :\ |x| < m^1, \quad t \in (0,T]\},$$

$$G^2 = G^2(m^2) = \{(x,t) :\ r(x, \Gamma) < m^2, \quad (x,t) \in G\},$$

$$G^3 = G^3(m^3) = G \setminus \{G^1(m^3) \bigcup G^2(m^3)\}, \quad m^3 < m^1, m^2;$$

$$G^{2l} = G^2 \bigcap \{x < 0\}, \quad G^{2r} = G^2 \bigcap \{x > 0\}.$$

---

[2] By m, m^i, m_i (by M, M^i, M_i) we denote sufficiently small (large) positive constants independent of the value of the parameter ε and of the discretization parameters.

Here $r(x,\,\Gamma)$ is the distance from the point x to the set Γ; G^1 and G^2 are the neighbourhoods of the interior and boundary layers, respectively; set $\overline{G}^1 \bigcap \overline{G}^2 = \emptyset$. We denote the solution of the problem (2), (1) considered on the set $\overline{G}^j$ by $u^j(x,t)$, $j = 1,\,2,\,3$. On $\overline{G}^3$ we have the estimate

$$\left| \frac{\partial^{k+k_0}}{\partial x^k \partial t^{k_0}}\, u(x,t) \right| \le M, \quad (x,t) \in \overline{G}^3, \quad k + 2k_0 \le K. \tag{5}$$

The value K is defined by the problem data; $K \ge 4$.

2. Let us study the solution of the problem on the set $\overline{G}^1$.
The function $u(x,t)$, $(x,t) \in \overline{G}^{1*}$ is the solution of the problem

$$L_{(6)} u(x,\,t) = f(x,\,t), \quad (x,\,t) \in G^1, \tag{6}$$

$$u(x,\,t) = \begin{cases} u^3(x,\,t),\; (x,\,t) \in S^1 \setminus S, \\ \varphi(x,\,t), \quad (x,\,t) \in S^{1*} \bigcap S. \end{cases}$$

Here S^1 is the boundary of the set G^1, $u^3(x,t) = u(x,t)$, $(x,t) \in \overline{G}^3$.
Write the function $u(x,t)$, $(x,t) \in \overline{G}^{1*}$, as the sum of the functions [12]:

$$u(x,\,t) = U(x,\,t) + W(x,\,t), \quad (x,\,t) \in \overline{G}^{1*}, \tag{7a}$$

where $U(x,t)$ and $W(x,t)$ are the regular ("sufficiently smooth") and singular parts of the solution; $W(x,t)$ is the function of the interior layer generated by discontinuity of the initial function $\varphi(x,\,t)$ and its lowest derivatives. The function $U(x,t)$ has ε-uniformly bounded derivatives in x and t up to the orders K and $2^{-1}K$, respectively. The function $W(x,t)$ is the solution of the Cauchy problem

$$\begin{aligned} L_{(6)} W(x,\,t) &= 0, \qquad (x,t) \in G^\infty, \quad G^\infty = I\!\!R \times (0,T], \\ W(x,\,t) &= \Phi_W(x), \;\; x \in I\!\!R \setminus \{0\}, \quad t = 0; \end{aligned} \tag{8}$$

$$\Phi_W(x) = 2^{-1}\,[\varphi(0,\,0)]\; sign\; x + 2^{-1} \sum_{k=1}^{K-1} (k!)^{-1} \left[\frac{\partial^k}{\partial x^k}\, \varphi(0,\,0) \right] |x|\, x^{k-1},$$

$\left[\frac{\partial^k}{\partial x^k}\, \varphi(0,\,0) \right] = \frac{\partial^k}{\partial x^k}\, \varphi(+0,\,0) - \frac{\partial^k}{\partial x^k}\, \varphi(-0,\,0)$ is the jump of the k-order derivative $\varphi(x,\,t)$ at the point $(0,\,0)$, $k = 0, 1, \ldots, K-1$. The function $\Phi_W(x)$ extended to the closures of the sets where it is continuous is denoted also by $\Phi_W(x)$.
The function $U(x,\,t)$ is the solution of the problem

$$L_{(6)} U(x,\,t) = f(x,\,t), \quad (x,\,t) \in G^1,$$

$$U(x,\,t) = \begin{cases} u^3(x,\,t) - W(x,\,t),\; (x,\,t) \in S^1, \;\; t > 0, \\ \varphi(x,\,t) - \Phi_W(x), \quad (x,\,t) \in S^{1*},\, t = 0. \end{cases}$$

It is convenient to represent the function $W(x,\,t)$ in the form

$$W(x,\,t)=\sum_{k=0}^{K-1} W_k(x,\,t), \quad (x,\,t)\in \overline{G}^{\infty}, \quad (x,\,t)\notin S^d; \tag{7b}$$

here $W_k(x,\,t)$ is the solution of the problem (8), where $\Phi_W(x)$ is

$\Phi_0(x)=2^{-1}\,[\varphi(0,\,0)]\ sign\ x\ \ for\ \ k=0,$

$\Phi_k(x)=2^{-1}\,(k!)^{-1}\left[\dfrac{\partial^k}{\partial x^k}\,\varphi(0,\,0)\right]\,|x|\,x^{k-1}, \quad x\in I\!R\setminus\{0\}, \quad k=1,2,...,K-1.$

The functions $W_k(x,\,t)$ can be written explicitly; for example, for $W_0(x,\,t)$ and $W_1(x,\,t)$ we have the representations

$$\begin{aligned}
&W_0(x,\,t)=2^{-1}\,[\varphi(0,0)]\,\exp(-\alpha t)\,w_0(x,\,t),\\
&w_0(x,\,t)=v\big(2^{-1}\,\varepsilon^{-1}\,a^{-1/2}\,p^{1/2}\,x\,t^{-1/2}\big), \quad (x,\,t)\notin S^d,\\
&W_1(x,\,t)=2^{-1}\left[\frac{\partial}{\partial x}\,\varphi(0,0)\right]\exp(-\alpha t)\,w_1(x,\,t),\\
&w_1(x,\,t)=x\,v(2^{-1}\,\varepsilon^{-1}a^{-1/2}\,p^{1/2}\,x\,t^{-1/2})+
\end{aligned} \tag{7c}$$

$+2\,\pi^{-1/2}\,\varepsilon\,a^{1/2}\,p^{-1/2}\,t^{1/2}\,\exp(-4^{-1}\,\varepsilon^{-2}\,a^{-1}\,p\,x^2\,t^{-1}), \quad (x,t)\in\overline{G}^{\infty}, \quad \alpha=c\,p^{-1},$
$v(x)=erf(x)=2\,\pi^{-1/2}\int_0^x \exp(-\xi^2)\,d\xi$, $x\in I\!R$, is the error function.

Write the function $u(x,t)$, $(x,t)\in\overline{G}^{1*}$, as the decomposition

$$u(x,t)=U^{\,1}(x,t)+W^{\,1}(x,t), \quad (x,t)\in\overline{G}^{1*}, \tag{9}$$

where $U^{\,1}(x,t)$ and $W^{\,1}(x,t)$ are the regular and singular parts of the solution,

$$U^{\,1}(x,t)=U^{\,1}(x,t;\,i)\equiv U(x,t)+\sum_{k=i+1}^{K-1} W_k(x,t), \quad (x,t)\in\overline{G}^{1};$$

$$W^{\,1}(x,t)=W^{\,1}(x,t;\,i)\equiv\sum_{k=0}^{i} W_k(x,t), \quad (x,t)\in\overline{G}^{*}, \quad i\geq 1.$$

Here the functions $U(x,t)$ and $W_k(x,t)$ correspond to the components in the representations (7a) and (7b).

For the components in the representation (9) in the case of the condition

$$\left[\frac{\partial^k}{\partial x^k}\,\varphi(0,0)\right]\neq 0, \;\; k=j \;\; and \;\; \left[\frac{\partial^k}{\partial x^k}\,\varphi(0,0)\right]=0, \;\; k\leq j-1 \;\; for \;\; j\neq 0, \tag{10}$$

where $j\geq 0$, we find the following estimates

$$\left|\frac{\partial^{k+k_0}}{\partial x^k\partial t^{k_0}}U^1(x,t)\right|\leq M\big[1+\varepsilon^{\,i+1-k}\rho^{i+1-k-2k_0}\exp(-m\varepsilon^{-1}|x|)\big], \; (x,t)\in\overline{G}^{1*},$$

$$\left|\frac{\partial^{k+k_0}}{\partial x^k\partial t^{k_0}}\,W^1(x,t)\right|\leq M\,\big[1+\varepsilon^{\,j-k}\,\rho^{j-k-2k_0}\,\exp(-m\,\varepsilon^{-1}\,|x|)\big], \quad (x,t)\in\overline{G}^{*},$$
$$k+2k_0\leq K, \quad \rho=\rho(t)=t^{1/2}, \tag{11}$$

where m is an arbitrary constant, and $i\geq 1$, $j=j_{(10)}$, $i\geq j$, $j\geq 0$.

3. Consider the problem (2), (1) on the set $\overline{G}^2$, assuming that the solution of the problem on $\overline{G}^2$ is sufficiently smooth. Write the solution of the problem as the sum of the functions

$$u(x,t) = U(x,t) + V(x,t), \quad (x,t) \in \overline{G}^2, \tag{12}$$

where $U(x,t)$ and $V(x,t)$ are the regular and singular components of the solution.

For simplicity, we assume that compatibility conditions are fulfilled on the set S^c that guarantee the local smoothness of the solution for fixed values ε [14]; suppose that on the set $\overline{G}^2$ the following condition is valid:

$$u \in C^{l+\alpha,(l+\alpha)/2}(\overline{G}^2), \quad l \geq 2, \quad \alpha \in (0,\,1). \tag{13}$$

In this case, for the functions $U(x,t)$ and $V(x,t)$ we have the estimates

$$\left| \frac{\partial^{k+k_0}}{\partial x^k \partial t^{k_0}} U(x,t) \right| \leq M, \tag{14}$$

$$\left| \frac{\partial^{k+k_0}}{\partial x^k \partial t^{k_0}} V(x,t) \right| \leq M\varepsilon^{-k} \exp\left(-m\varepsilon^{-1} r(x,\Gamma)\right), \; (x,t) \in \overline{G}^2, \; k + 2k_0 \leq K.$$

Here $K \leq l$, and $m = m_{(11)}$.

Theorem 1. *Let for the data of the boundary value problem (2), (1), the condition (10) hold and also the condition* $f \in C^{l_1,l_1/2}(\overline{G})$, $\varphi \in C(S^*) \bigcap \{C^{l_1}(S_0^-) \bigcap C^{l_1}(S_0^+) \bigcap C^{l_1/2}(\overline{S}^L)\}$, *for* $l_1 = l + \alpha$, *where* $l = K$ *and* $\alpha \in (0,1)$, *be fulfilled, and let the solution of the problem satisfy the condition (13). Then the solution of the boundary value problem and its components in the representations (9) and (12) satisfy the estimates (5), (11) and (14).*

Remark 1. Under the condition

$$[\varphi(0,0)] = 0, \quad \left[\frac{\partial}{\partial x} \varphi(0,0) \right] \neq 0, \tag{15}$$

the interior layer is weak; its first derivative in x in a nearest neighbourhood of S^{γ_0} is ε-uniformly bounded. □

4 Grid Approximations of the Problem (2), (1) for $u \in C(\overline{G})$

It is known that in the case of a discontinuous initial function classical difference schemes do not converge in the maximum grid norm even for fixed values of ε [7,12]. In this section, we consider a classical difference scheme for the problem (2), (1) under the condition

$$u \in C(\overline{G}). \tag{16}$$

1. Based on classical approximations of the problem (2), (1), in the case of a piecewise-smooth initial and sufficiently smooth boundary functions satisfying compatibility conditions on S^c, we construct a difference scheme that converges ε-uniformly.

On the set $\overline{G}_{(1)}$ we introduce the rectangular grid

$$\overline{G}_h = \overline{D}_h \times \overline{\omega}_0 = \overline{\omega} \times \overline{\omega}_0, \tag{17}$$

where $\overline{\omega}$ and $\overline{\omega}_0$ are meshes on the intervals $[-d,\, d]$ and $[0,\, T]$, respectively, the mesh $\overline{\omega}$ has an arbitrary distribution of its nodes satisfying only the condition $h \le MN^{-1}$, where $h = \max_i h^i$, $h^i = x^{i+1} - x^i$, x^i, $x^{i+1} \in \overline{\omega}$, the mesh $\overline{\omega}_0$ is uniform with step-size $h_0 = TN_0^{-1}$. Here $N+1$ and N_0+1 are the numbers of nodes in the meshes $\overline{\omega}$ and $\overline{\omega}_0$, respectively.

We approximate the boundary value problem (2), (1), (16) by the finite difference scheme [5]

$$\begin{aligned} \Lambda_{(18)}\, z(x,t) &\equiv \{\varepsilon^2\, a\, \delta_{\overline{x}\,\widehat{x}} - c - p\,\delta_{\overline{t}}\}\, z(x,t) = f(x,t), \quad (x,t) \in G_h, \\ z(x,t) &= \varphi(x,t), \quad (x,t) \in S_h. \end{aligned} \tag{18}$$

Here $\delta_{\overline{x}\,\widehat{x}}\, z(x,t) = z_{\overline{x}\,\widehat{x}}(x,t) = 2(h^i + h^{i-1})^{-1}[\delta_x\, z(x,t) - \delta_{\overline{x}}\, z(x,t)]$, $(x,t) = (x^i,t) \in G_h$, is the central difference derivative on a nonuniform mesh, where $\delta_x z(x,t)$ and $\delta_{\overline{x}} z(x,t)$, $\delta_{\overline{t}}\, z(x,t)$, are the first (forward and backward) difference derivatives, $\delta_x z(x,t) = \left(h^i\right)^{-1} \left(z(x^{i+1},t) - z(x^i,t)\right)$, $\delta_{\overline{x}} z(x,t) = \left(h^{i\;1}\right)^{-1} \left(z(x^i,t) - z(x^{i-1},t)\right)$, $\delta_{\overline{t}}\, z(x,t) = h_0^{-1} \left(z(x^i,t) - z(x^i,t-h_0)\right)$.

The difference scheme (18), (17) is monotone ε-uniformly [5].

We consider the behaviour of solution errors to the difference scheme (18), (17) when piecewise-uniform grids are used for different types of nonsmoothness of the initial function $\varphi(x,t)$.

2. On the set $\overline{G}$ we construct a grid condensing in a neighborhood of the boundary layer; this grid is similar to that used, e.g., in [3]

$$\overline{G}_h = \overline{D}_h \times \overline{\omega}_0 = \overline{\omega}^{\,s} \times \overline{\omega}_0, \tag{19}$$

where $\overline{\omega}_0 = \overline{\omega}_{0(17)}$, $\overline{\omega}^{\,s} = \overline{\omega}^{\,s}(\sigma)$ is a piecewise-uniform mesh on $[-d,\, d]$, and σ is a parameter depending on ε and N. The value σ is specified by $\sigma = \sigma(N,\, \varepsilon) = \min\,[\,\beta,\, 2\,m^{-1}\,\varepsilon\, \ln N\,]$, where β is an arbitrary number in the interval $(0,\, d)$, and $m = m_{(11)}$. The interval $[-d,\, d]$ is divided into three parts: $[-d,\, -d+\sigma]$, $[-d+\sigma,\, d-\sigma]$ and $[d-\sigma,\, d]$; in each part, the mesh-size is constant and equals $h^{(1)} = 2\,d\,\sigma\,\beta^{-1}\,N^{-1}$ on the intervals $[-d,\, -d+\sigma]$, $[d-\sigma,\, d]$ and $h^{(2)} = 2\,d\,(\,d - \sigma)\,(\,d-\beta)^{-1}\,N^{-1}$ on the interval $[-d+\sigma,\, d-\sigma]$.

We call the difference scheme (18) on the piecewise-uniform grid (19) the *basic scheme* for the problem (2), (1), (16).

3. We consider estimates of the grid solutions depending on order of the derivative of the initial function that has a discontinuity on S^d. Assume that the solution of the boundary value problem satisfies the condition (13) for $l = 4$. In the case

when the function $\varphi(x,t)$ satisfies the condition (15), and taking into account the a priori estimates of the solution of the boundary value problem (2), we obtain the following estimate for the solution of the basic scheme:

$$| u(x,t) - z(x,t) | \leq M \left[N^{-1} + N_0^{-1/2} \right], \quad (x,t) \in \overline{G}_h. \tag{20}$$

In the case when the function $\varphi(x,t)$ satisfies the condition

$$[\varphi(0,0)] = \left[\frac{\partial}{\partial x} \varphi(0,0) \right] = 0, \tag{21}$$

i.e., $\varphi(x,t)$ and the derivative $(\partial/\partial x)\,\varphi(x,t)$ are continuous on S_0 (under these conditions, in general, $u \notin C^{2,1}(\overline{G}^1)$), we obtain the ε-uniform estimate

$$| u(x,t) - z(x,t) | \leq M \left[N^{-2} \ln^2 N + N_0^{-1} \ln N_0 \right], \quad (x,t) \in \overline{G}_h. \tag{22}$$

But if the condition

$$\left[\frac{\partial^k}{\partial x^k} \varphi(0,0) \right] = 0, \quad k = 0, 1, 2, \tag{23}$$

holds (in this case $u \notin C^{3,3/2}(\overline{G}^1)$), we have the estimate

$$| u(x,t) - z(x,t) | \leq M \left[N^{-2} \ln^2 N + N_0^{-1} \right], \quad (x,t) \in \overline{G}_h. \tag{24}$$

Theorem 2. *Let for the solution of the problem (2), (1) and its components in the representations (9), (12), the condition (16) hold, the inclusion (13) for $l = 4$ hold and the a priori estimates (5), (11) for $i = j$, (14) for $K = 4$, and also either the condition (15), (21) or (23) in the case $i_{(11)} = 1, 2, 3$, respectively, be fulfilled. Then the difference scheme (18), (19) converges ε-uniformly. The grid solutions satisfy the estimates (20), (22) and (24) under the conditions (15), (21) and (23), respectively.*

Remark 2. Comparing the estimates (20) and (22), we conclude that the presence of discontinuities of the first-order derivative in x of the initial function on S_0 leads to an essential decrease of the ε-uniform convergence rate of the basic scheme (18), (19). Under the condition (15), the order of the ε-uniform convergence rate of the scheme drops twice up to logarithmic factors in N and N_0 in comparison with the scheme under the condition either (21) or (23). □

5 Difference Scheme for the Problem (2), (1), (3)

When constructing an ε-uniformly convergent difference scheme for the problem (2), (1) with a discontinuous initial function, we apply the method of additive splitting of singularities.

1. Consider the problem (2), (1) under the condition $S^d \neq \emptyset$.

Write the solution of the problem as the sum of the functions

$$u(x,\, t) = u_1(x,\, t) + u_2(x,\, t), \quad (x,\, t) \in \overline{G}, \tag{25a}$$

$$u_2(x,\, t) = W(x,t) \equiv W^{\,1}(x,t), \quad (x,\, t) \in \overline{G}^{\,*}, \quad W^{\,1}(x,t) = W^{\,1}_{(9)}(x,t;\, i), \quad i \geq 0.$$

The function $u_1(x,\, t)$ is the solution of the problem

$$L_{(2)}\, u_1(x,\, t) = f(x,\, t), \quad (x,\, t) \in G, \quad u_1(x,\, t) = \varphi_1(x,\, t), \quad (x,\, t) \in S, \tag{25b}$$

where $\varphi_1(x,\, t) = \varphi(x,\, t) - W(x,t)$; the function $\varphi_1(x,\, t)$ is continuous on S.

On the grid (17), we approximate the problem (25b) by the difference scheme

$$\Lambda_{(18)}\, z_1(x,\, t) = f(x,\, t), \quad (x,\, t) \in G_h, \quad z_1(x,\, t) = \varphi_1(x,\, t), \quad (x,\, t) \in S_h. \tag{26a}$$

For the function $z_1(x,\, t)$, $(x,\, t) \in \overline{G}_h$, we construct the interpolant $\overline{z}_1(x,\, t)$, $(x,\, t) \in \overline{G}$, (linear on triangular elements) and further the function

$$u_0^h(x,\, t) = \overline{z}_1(x,\, t) + W(x,\, t), \quad (x,\, t) \in \overline{G}^{\,*}, \tag{26b}$$

where $W(x,\, t) = W_{(25)}(x,\, t)$.

We call the function $u_0^h(x,\, t)$, $(x,\, t) \in \overline{G}^{\,*}$, the *solution* of the difference scheme (26), (17), i.e., the *scheme of the method of additive splitting of singularities* generated by discontinuities of the initial function and its low-order derivatives defined by the value $i_{(25)}$.

2. Consider the difference scheme (26) on piecewise-uniform grid (19) for different values of $i_{(25)}$ that define the component $W_{(26)}(x,t) = W^{\,1}_{(9)}(x,t;\, i)$.

Let for the value i that defines the additive component $W(x,t)$ in (26b) the following condition hold

$$i_{(25)} = 0. \tag{27}$$

In this case, the first-order derivative in x of the function $\varphi_1(x,\, t)$ is discontinuous on the set S^d.

Taking into account the *a priori* estimates of the function $u_1(x,\, t)$ (the solution of the problem (25b), (1)), for the function $u_0^h(x,\, t)$, $(x,\, t) \in \overline{G}$ (the solution of the difference scheme (26), (27), (19)), we obtain the estimate

$$|u(x,\, t) - u_0^h(x,\, t)| \leq M\,[N^{-1} + N_0^{-1/2}], \quad (x,\, t) \in \overline{G}. \tag{28}$$

In a similar way, for the solution of the difference scheme (26), (19) under the condition

$$i_{(25)} = 1 \tag{29}$$

(in this case, the second-order derivative in x of the function $\varphi_1(x,\, t)$ is discontinuous on S^d), we have the estimate

$$|u(x,\, t) - u_0^h(x,\, t)| \leq M\,[N^{-2} \ln^2 N + N_0^{-1} \ln N_0], \quad (x,\, t) \in \overline{G}. \tag{30}$$

But if for the function $W(x,t)$ the condition

$$i_{(25)} = 2, \tag{31}$$

holds, we obtain the following estimate for the solution of the difference scheme (26), (19):

$$|u(x,\, t) - u_0^h(x,\, t)| \le M\,[N^{-2}\ln^2 N + N_0^{-1}], \quad (x,\, t) \in \overline{G}. \tag{32}$$

Theorem 3. *Let for the data of the boundary value problem (2), (1), (3), the condition (13) for $l = 4$ hold and also the condition $f \in C^{l_1,l_1/2}(\overline{G})$, $\varphi \in C(S^*)\bigcap\{C^{l_1}(S_0^-)\bigcap C^{l_1}(S_0^+)\bigcap C^{l_1/2}(\overline{S}^L)\}$ for $l_1 = l+\alpha$, , where $l = K$, $K = 4$, and $\alpha \in (0,1)$, be fulfilled. Then the difference scheme (26) on the grid (19) in the case of the conditions (27), (29) and (31) converges ε-uniformly. The solutions of the difference scheme under these conditions satisfy the estimates (28), (30), (32), respectively.*

6 Difference Scheme of Improved Accuracy for the Problem (2), (1), (3)

In the case of the problem (2), (1), (3) the difference scheme of the method of additive splitting of singularities (26), (31), (19) converges ε-uniformly at the rate $\mathcal{O}\big(N^{-2}\ln^2 N + N_0^{-1}\big)$. For this problem, on the basis of the method of additive splitting of singularities and using a Richardson technique (applied for increase of accuracy of the component $z_{1(26a)}(x,t)$), we construct a difference scheme that converges ε-uniformly at the rate $\mathcal{O}\big(N^{-3} + N_0^{-2}\big)$.

1. In the difference scheme (26), (19), we choose the value $i_{(25)}$ in $W(x,t;i)$ sufficiently large ($i = i^0$, $i^0 > 2$). Then, for sufficiently smooth data of the problem (2), (1), (3) on S_0^-, S_0^+, $\overline{S}^L$ and appropriate compatibility conditions on S^c, the component $u_1(x,t)$ in the representation (25a) is sufficiently smooth. To improve accuracy of the difference scheme (26a), (19) we use the Richardson difference scheme.

Let $z_1^i(x,t)$, $(x,t) \in \overline{G}_h^i$, $i = 1,2$, be solutions of two difference schemes on the embedded grids $\overline{G}_h^1$ and $\overline{G}_h^2$ that approximate the problem (25b), (1):

$$\begin{aligned} \Lambda_{(18)}\, z_1^i(x,t) &= f(x,t),\ (x,t) \in G_h^i,\\ z_1^i(x,t) &= \varphi_1(x,t),\ (x,t) \in S_h^i, \quad i = 1,2. \end{aligned} \tag{33a}$$

Here $\overline{G}_h^2 = \overline{G}_{h(19)}$, and $\overline{G}_h^1$ is a "coarsened" grid obtained from $\overline{G}_h^2$; the step-sizes in the mesh $\overline{G}_h^1$, on the intervals of uniformity of the meshes $\overline{\omega}^{s1}$, are k times larger than the step-sizes in the meshes $\overline{\omega}^{s2}$ in the grid $\overline{G}_h^2$, and the step-size of the mesh $\overline{\omega}_0^1$ is k^2 times larger than the step-size in the mesh $\overline{\omega}_0^2$. The function

$$z_1^0(x,t) = \gamma z_1^1(x,t) + (1-\gamma) z_1^2(x,t), \quad (x,t) \in \overline{G}_h^0, \quad \overline{G}_h^0 = \overline{G}_h^1 \bigcap \overline{G}_h^2, \tag{33b}$$

where $\gamma = -(k^2 - 1)^{-1}$, is the solution of the Richardson scheme (33) for the problem (25b), (1).

For the function $z_1^0(x,t)$, $(x,t) \in \overline{G}_h^0$, we construct its extension on $\overline{G}$, i.e., the interpolant $\overline{z}_1^0(x,t)$

$$\overline{z}_1^0(x,t) = \overline{z}_1^0(x,t;\, z_1^0(\cdot)), \quad (x,t) \in \overline{G}. \tag{34a}$$

In the case when the Richardson scheme (33) converges on $\overline{G}_h^0$ ε-uniformly at the rate $\mathcal{O}(N^{-2}\ln^2 N + N_0^{-2})$, the function $\overline{z}_1^0(x,t)$ is a bilinear interpolant on the elementary rectangles generated by straight lines passing through the nodes in the grid $\overline{G}_h$ in parallel to the coordinate axes. Let the difference scheme converge on $\overline{G}_h^0$ at the rate $\mathcal{O}(N^{-k} + N_0^{-2})$, where $k > 2$; in this case the function $\overline{z}_1^0(x,t)$ is constructed in the following way. On straight lines passing through the nodes in the mesh $\overline{\omega}_0^0$, we construct interpolants $\widetilde{z}_1^0(x,t)$, $t \in \overline{\omega}_0^0$, $x \in [-d,d]$. The function $\widetilde{z}_1^0(x,t)$ on the interval $[x^j,\, x^{j+1}]$, where $x^j,\, x^{j+1} \in \overline{\omega}^{s0}$ and $t \in \overline{\omega}_0^0$, is the Lagrange interpolation polynomial that is constructed using the values of $z_1^0(x,t)$ in sequential k nodes $x^{j_1}, x^{j_1+1}, \ldots, x^{j_1+k-1} \in \overline{\omega}^{s0}$, which belong entirely to one from the intervals $[-d, -d+\sigma]$, $[-d+\sigma, d-\sigma]$, $[d-\sigma, d]$, including the nodes x^j, x^{j+1} [15]. The interpolant $\overline{z}_1^0(x,t)$ on $\overline{G}$ is constructed using the function $\widetilde{z}_1^0(x,t)$ by a linear interpolation in t.

Further, we construct the function

$$u_0^{0h}(x,t) = \overline{z}_1^0(x,t) + W(x,t;i^0), \quad (x,t) \in \overline{G}^*, \quad i^0 > 2, \tag{34b}$$

which we call the *solution* of the difference scheme $\{(26), (33), (34), (19)\}$, i.e., the *scheme of the method of additive splitting of singularities with improved accuracy based on the Richardson technique.*

2. Choosing $i^0 = 5$ in the function $W(x,t;i^0)$, we find the following estimate for the solution of the difference scheme $\{(26), (33), (34), (19)\}$:

$$|u(x,t) - u_0^{0h}(x,\, t)| \le M\ \left[N^{-3} + N_0^{-2}\right], \quad (x,t) \in \overline{G}, \tag{35}$$

i.e., the difference scheme converges ε-uniformly at the rate $\mathcal{O}(N^{-3} + N_0^{-2})$.

Theorem 4. *Let for the data of the boundary value problem (2), (1), (3) and its solutions, the hypotheses of Theorem 1 hold, where $l = 6$. Then the solution of the difference scheme $\{(26), (33), (34), (19)\}$, under the condition $i^0 = 5$ in $W(x,t;i^0)$ from* (34b)*, satisfies the estimate (35).*

Remark 3. The decrease of the value i^0 leads, in general, to lowering of the rate of the ε-uniform convergence of the improved scheme. The schemes $\{(26), (33), (34), (19)\}$ and (18), (19) converge ε-uniformly at the rate $\mathcal{O}(N^{-2}\ln^2 N + N_0^{-1})$ for $i^0 = 2$, and only $\mathcal{O}(N^{-1} + N_0^{-1/2})$ for $i^0 = 0$.

Acknowledgments

This research was supported by the Russian Foundation for Basic Research under grants Nos. 07–01–00729, the Boole Centre for Research in Informatics at the National University of Ireland, Cork, and by the Mathematics Applications Consortium for Science and Industry in Ireland (MACSI) under the Science Foundation Ireland (SFI) Mathematics Initiative.

References

1. Li, S., Shishkin, G.I., Shishkina, L.P.: Approximation of the solution and its derivative for the singularly perturbed Black-Scholes equation with nonsmooth initial data. Comp. Math. Math. Phys. 47(3), 442–462 (2007)
2. Shishkin, G.I.: Grid approximations of parabolic convection-diffusion equations with piecewise-smooth initial data. Doklady Mathematics 72(3), 850–853 (2005)
3. Shishkin, G.I.: Grid approximation of singularly perturbed parabolic convection-diffusion equations subject to a piecewise smooth initial condition. Comp. Math. Math. Phys. 46(1), 49–72 (2006)
4. Shishkin, G.I.: Grid approximation of singularly perturbed parabolic reaction-diffusion equations with piecewise smooth initial-boundary conditions. Mathematical Modelling and Analysis 12(2), 235–254 (2007)
5. Samarskii, A.A.: The Theory of Difference Schemes. Marcel Dekker, New York (2001)
6. Kolmogorov, V.L., Shishkin, G.I.: Numerical methods for singularly perturbed boundary value problems modelling diffusion processes. In: Miller, J.J.H. (ed.) Singular Perturbation Problems in Chemical Physics. Advances in Chemical Physics Series, vol. XCVII, pp. 181–362. J. Wiley & Sons, Chichester (1997)
7. Shishkin, G.I.: A difference scheme for a singularly perturbed equation of parabolic type with a discontinuous boundary condition. USSR Comput. Math. Math. Phys. 28(11), 1649–1662 (1988)
8. Shishkin, G.I.: A difference scheme for a singularly perturbed equation of parabolic type with a discontinuous initial condition. Soviet Math. Dokl. 37(3), 792–796 (1988)
9. Hemker, P.W., Shishkin, G.I.: Discrete approximation of singularly perturbed parabolic PDEs with a discontinuous initial condition. Computational Fluid Dynamics Journal 2(4), 375–392 (1994)
10. Shishkin, G.I.: Approximation of solutions and diffusion flows in the case of singularly perturbed boundary value problems with discontinuous initial conditions. Comp. Math. Math. Phys. 36(9), 1233–1250 (1996)
11. Shishkin, G.I.: Singularly perturbed boundary value problems with concentrated sources and discontinuous initial conditions. Comp. Math. Math. Phys. 37(4), 417–434 (1997)
12. Shishkin, G.I.: Grid approximation of singularly perturbed parabolic equations with piecewise continuous initial-boundary conditions. Proc. Steklov Inst. Math. 2, 213–230 (2007)
13. Marchuk, G.I., Shaidurov, V.V.: Difference Methods and Their Interpolations. Springer, New York (1983)
14. Ladyzhenskaya, O.A., Solonnikov, V.A., Ural'tseva, N.N.: Linear and Quasilinear Equations of Parabolic Type. In: Translations of Mathematical Monographs, vol. 23. American Mathematical Society, Providence (1967)
15. Bakhvalov, N.S.: Numerical Methods. Mir Publishers, Moscow (1976)

The Numerical Spherically Symmetric Modeling of Deep-Seated Geodynamics

A.V. Vyatkin, V.V. Shaidurov, and G.I. Shchepanovskaya

Institute of Computational Modeling SB RAS, Krasnoyarsk, Russia
{shidurov,gi}@icm.krasn.ru

Abstract. In this paper a computer model is proposed which allows one to consider geodynamics processes of the Earth's expansion, contraction, heating and cooling. Geosphere dynamics is studied in the framework of a viscous heat-conducting compressible medium where medium density and viscosity vary with time and space. This model includes the Earth's crust, mantle, and the core as well.

Research in the deep-seated structure, the composition, and geodynamics of the continent and ocean litosphere allows one to recognize systems relating to global processes of the development of the Earth (rifts, deep uncompensated trenches, continents, oceans) and to local phenomena within continents and oceans (folded areas, moving zones, cratons, and others). The study of these phenomena shows that the interference of these systems leads to complex processes of the geodynamic development and states within the Earth.

The interior structure of the Earth is determined from the known mass, the moment of inertia of the Earth, and the study of elastic waves caused by earthquakes. It is known that density of the substance at the center of the Earth is $\rho \geq 12.2$ gram per cm^3 and at a depth of 2900 km the Earth's core is separated from the upper layers by an abrupt jump of density of the order of 4 gram per cm^3. Thus, within the Earth elastic properties vary very rapidly at some specific depthes and vary gradually within the layers separated by these boundaries. The Mohorovičič discontinuity at a depth of 10–70 km and the Wichert-Gutenberg discontinuity at a depth of 2900 km are the most important boundaries. The Earth is subdivided by these boundaries into three main zones: the crust, the mantle, and the core [1]. The crust has maximal stiffness, the mantle is characterized by high viscosity, and the liquid external core responds to longitudinal waves only. There are less well-defined boundaries inside three main zones of the Earth. In the Table 1 dimensional values of pressure, temperature, density, and the coefficient of dynamic viscosity are given depending on depth [2].

The model proposed in this paper includes the Earth's crust, mantle and the core as well. The model is spherically symmetric, so when specifying properties of the layer of the Earth, the Earth's structure is assumed to be the same in various directions of a radius vector emanating from the Earth's center. We consider a nonstationary formulation of the problem with dimensionless equations [3] where linear dimensions are taken in the ratio to the Earth's radius and dynamic quantities does to corresponding typical values at the Earth's surface.

S. Margenov, L.G. Vulkov, and J. Waśniewski (Eds.): NAA 2008, LNCS 5434, pp. 128–138, 2009.

Table 1. Physical parameters of Earth's layers

h, km	15	60	100	150	300	400	600	1600	2700	2870	2900	4700	6371
P, kbar	3.5	14	29	46.5	99	130	190	530	1240	1340	1350	3040	3632
T, K	673	873	1103	1343	1623	1673	1723	2700			3500	4000	
ρ, g/cm^3	2.85	3.34	3.37	3.37	3.48	3.54	4.13	5.03	5.55	5.68	9.89	12.3	13.01
μ	0.35	0.72	0.72	0.72	0.75	0.82	1.3	2.2	2.9	2.95	0	0	

To describe the Earth spherically symmetric expansion, contraction, heating, and cooling, the Navier-Stokes equations are used. The spherical coordinate system (r, θ, φ) is related to the Cartesian coordinates in the following way: $x = r\,\cos\varphi \sin\theta,\ y = r\,\sin\varphi\sin\theta,\quad z = r\,\cos\theta,\ 0 \leq r < +\infty,\ 0 \leq \theta \leq \pi,\ 0 \leq \varphi \leq 2\pi$.

For the Navier-Stokes equations in the spherical coordinate system [2] spherical symmetry leads to the fact that the derivatives with respect to φ and θ and the velocity components in φ and θ vanish. As a result, we arrive at the equations of continuity, pulse, and energy in the dimensionless form [4]:

$$\frac{\partial\sigma}{\partial t} + \frac{1}{2r^2}\frac{\partial}{\partial r}(r^2\sigma u) + \frac{1}{2}u\frac{\partial\sigma}{\partial r} = 0, \quad \sigma(t,\,r) \;=\; \sqrt{\rho(t,\,r)}, \tag{1}$$

$$\frac{\partial}{\partial t}(\rho u) + \frac{1}{r^2}\frac{\partial}{\partial r}(r^2\rho u^2) \;=\; -\frac{\partial P}{\partial r} + \frac{1}{r^2}\frac{\partial}{\partial r}(r^2\tau_{rr}) + F_{em}(r,\rho), \tag{2}$$

$$\frac{\partial}{\partial t}(\rho e) + \frac{1}{r^2}\frac{\partial}{\partial r}(r^2 u\rho e) + P\frac{1}{r^2}\frac{\partial}{\partial r}(r^2 u) = -\frac{1}{r^2}\frac{\partial}{\partial r}(r^2 q_r) + \tau_{rr}\frac{\partial u}{\partial r}. \tag{3}$$

Density $\rho = \sigma^2$, velocity u, and internal energy e of a mass unit are unknown functions. Pressure P and a dynamic coefficient μ of viscosity can be expressed in terms of these unknowns from the state equations. $F_{em}(r,\rho)$ is potential of the gravitation forces. The state equations represent pressure and temperature complicated algebraic dependence of density where the phase transition and the metamorphism of the main components are taken into account. Notice that due to the fact that the first equation is written in terms of σ instead of ρ we get the law of mass conservation in the L_2-norm instead of the L_1-norm. Later this significantly simplifies justification of stability and convergence [5, 6].

The flow relations between the stress tensor τ_{rr} and the deformation velocity tensor which determine the heat flow q_r in terms of dimensionless parameters (the Reynolds number Re and the Prandtle number Pr) are given [7] by the formulae:

$$\tau_{rr} = \frac{4}{3\mathrm{Re}}\mu\frac{\partial u}{\partial r} - \frac{4}{3r\mathrm{Re}}\mu u, \qquad q_r = -\frac{\gamma}{\mathrm{Re\,Pr}}\mu\frac{\partial e}{\partial r}. \tag{4}$$

In this case potential of the gravitation forces is expressed in the following form [8]:

$$F_{em} = -4\pi F_r \kappa_l \cdot \frac{\rho}{r^2}\int\limits_0^r \rho x^2 dx, \tag{5}$$

where F_r is the Froude number, k_l is a dimensionless quantity which depends on the gravitation constant, the acceleration of the free fall, and density at the Earth's surface.

The values of density and the dynamic coefficient of viscosity given in the table 1 are taken as initial conditions [2].

Introduce a uniform grid with mesh size $h = 2/(2n+1)$. Introduce also $\bar{\Omega}_h = \{r_i = (i+1/2)h, i = -1, \ldots, n\}$ and a set of interior nodes $\Omega_h = \{r_i, i = 0, 1, \ldots, n-1\}$. As a result, a computational domain $\bar{\Omega}$ is subdivided into $n+1$ intervals (r_i, r_{i+1}), $i = -1, 0, \ldots, n-1$. With each node $r_i \in \bar{\Omega}_h$ we associate the basis function φ_i which equals one ar r_i, vanishes at all other nodes of $\bar{\Omega}_h$, and is linear on each interval (r_i, r_{i+1}) (Fig. 1).

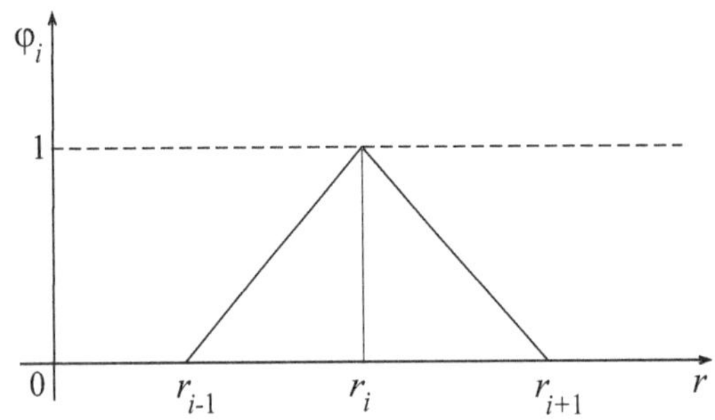

Fig. 1. Basis function φ_i

$$\varphi_i(r) = \begin{cases} \dfrac{r - r_{i-1}}{h}, & r \in [r_{i-1},\ r_i]; \\ \dfrac{r_{i+1} - r}{h}, & r \in [r_i,\ r_{i+1}]; \\ 0, & r \notin [r_{i-1},\ r_{i+1}]. \end{cases} \tag{6}$$

With these notations, we formulate the Bubnov-Galerkin method. We seek an approximation $\sigma_h(t, r)$ to the function $\sigma(t, r)$ in the form

$$\sigma_h(t, r) = \sum_{i=0}^{n} \sigma_i(t)\varphi_i(r) \tag{7}$$

with time-dependent coefficients $\sigma_i(t)$. Substitute the test solution σ_h into the equation (1) in place of an exact solution and multiply the equation by φ_l:

$$(R,\ \varphi_l) = 0, \quad l = 0,\ \ldots,\ n, \tag{8}$$

where

$$R = \frac{\partial \sigma_h}{\partial t} + \frac{1}{2r^2}\frac{\partial}{\partial r}(r^2\sigma_h u) + \frac{1}{2}u\frac{\partial \sigma_h}{\partial r}. \tag{9}$$

Due to spherical symmetry the inner product in R^3 is reduced to that in R^1, so we obtain the following expression:

$$\int_0^1 r^2 \varphi_l \left(\frac{\partial \sigma}{\partial t} + \frac{1}{2r^2} \frac{\partial}{\partial r}(r^2 \sigma u) + \frac{1}{2} u \frac{\partial \sigma}{\partial r} \right) dr = 0, \qquad l = 0, \ ..., \ n. \tag{10}$$

Applying the integration by parts to (10), we get the equality:

$$\int_0^1 r^2 \varphi_l \frac{\partial \sigma}{\partial t} dr + \left(\frac{1}{2}\varphi_l \sigma u\right)\Big|_{r=1} - \int_0^1 \frac{1}{2} r^2 \sigma u \frac{\partial \varphi_l}{\partial r} dr + \int_0^1 \frac{1}{2} r^2 \varphi_l u \frac{\partial \sigma}{\partial r} dr = 0, \tag{11}$$

$$l = 0, \ ..., \ n.$$

Substituting the approximate solution (7) into (11) in place of σ yields

$$\sum_{i=0}^{n} \Big[\int_0^1 \left(\frac{\partial \sigma_i(t)}{\partial t} r^2 \varphi_l \varphi_i - \frac{1}{2} r^2 \sigma_i u \frac{\partial \varphi_l}{\partial r} \varphi_i + \frac{1}{2} r^2 \sigma_i u \varphi_l \frac{\partial \varphi_i}{\partial r} \right) dr +$$

$$\frac{1}{2}(u \sigma_i \varphi_l \varphi_i)\Big|_{r=1} \Big] = 0, \qquad l = 0, \ ..., \ n. \tag{12}$$

From the form of the basis functions (6) it follows that only three terms in (12) do not vanish:

$$\sum_{i=l-1}^{l+1} \Big[\int_0^1 \left(\frac{\partial \sigma_i(t)}{\partial t} r^2 \varphi_l \varphi_i - \frac{1}{2} r^2 \sigma_i u \frac{\partial \varphi_l}{\partial r} \varphi_i + \frac{1}{2} r^2 \sigma_i u \varphi_l \frac{\partial \varphi_i}{\partial r} \right) dr +$$

$$\frac{1}{2}(u \sigma_i \varphi_l \varphi_i)\Big|_{r=1} \Big] = 0, \quad l = 0, \ ..., \ n. \tag{13}$$

To calculate the integrals in (13), we apply the trapezoid quadrature formula and the equalities $\sigma_{-1} = \sigma_0$, $u_{-1} = -u_0$, which follow from symmetry of the problem. Approximating the time derivative with the help of the forward difference, we arrive at the difference equations

$$\frac{h}{\tau} r_0^2 \sigma_0^{k+1} + \left(\frac{1}{4} r_1^2 u_1^k + \frac{1}{4} r_0^2 u_0^k \right) \sigma_1^{k+1} = \frac{h}{\tau} r_0^2 \sigma_0^k, \qquad l = 0,$$

$$\left(-\frac{1}{4} r_{l-1}^2 u_{l-1}^k - \frac{1}{4} r_l^2 u_l^k \right) \sigma_{l-1}^{k+1} + \frac{1}{\tau} r_l^2 h \sigma_l^{k+1} +$$

$$\left(\frac{1}{4} r_{l+1}^2 u_{l+1}^k + \frac{1}{4} r_l^2 u_l^k \right) \sigma_{l+1}^{k+1} = \frac{1}{\tau} r_l^2 h \sigma_l^k, \quad l = 1, \ ..., \ n-1,$$

$$\left(-\frac{1}{4} r_{n-1}^2 u_{n-1}^k - \frac{1}{4} r_n^2 u_n^k \right) \sigma_{n-1}^{k+1} + \left(\frac{h}{2\tau} r_n^2 + \frac{1}{2} r_n^2 u_n^k \right) \sigma_n^{k+1} = \frac{h}{2\tau} r_n^2 \sigma_n^k, \quad l = n. \tag{14}$$

Here in nonlinear terms the coefficients known from the previous time level are "frozen".

We seek the function $u_h(t,r)$ similar to (7). It approximates the function $u(t,r)$ in the form

$$u_h(t,r) = \sum_{i=0}^{n} u_i(t)\,\varphi_i(r) \tag{15}$$

with time-dependent coefficients $u_i(t)$. Again, reducing the inner product in R^3 to that in R^1, in the case of spherical symmetry we obtain the following expression:

$$\int_0^1 r^2\,\varphi_l\Big[\Big(\rho\frac{\partial u}{\partial t} + \frac{u}{2}\frac{\partial \rho}{\partial t}\Big) + \frac{1}{r^2}\frac{\partial}{\partial r}(r^2\rho u^2) - \frac{u}{2r^2}\frac{\partial}{\partial r}(r^2\rho u) + \frac{\partial P}{\partial r} -$$

$$\frac{1}{r^2}\frac{\partial}{\partial r}\Big(\frac{4}{3\mathrm{Re}}r^2\mu\,\frac{\partial u}{\partial r} - \frac{4}{3\mathrm{Re}}r\mu u\Big) - F_{em}\Big]\,dr = 0, \qquad l = 0,\ ...,\ n. \tag{16}$$

Notice that here from the original equation (2) we subtracted the equation of continuity multiplied by $u/2$. In continuous and discrete cases this results in stability in the L_2-norm instead of the L_1-norm for the original equation (2). Applying the integration by parts, we obtain

$$\int_0^1 r^2\,\varphi_l\Big(\rho\frac{\partial u}{\partial t} + \frac{u}{2}\frac{\partial \rho}{\partial t}\Big)\,dr + (\varphi_l\rho\,u^2)\Big|_{r=1} - \int_0^1 r^2\rho\,u^2\frac{\partial \varphi_l}{\partial r}dr - \frac{1}{2}(\varphi_l\rho\,u^2)\Big|_{r=1} +$$

$$\int_0^1 \frac{1}{2}r^2\rho\,u\frac{\partial}{\partial r}(\varphi_l\,u)dr + (r^2\varphi_l P)\Big|_{r=1} - \int_0^1 P\Big(2r\varphi_l + r^2\frac{\partial \varphi_l}{\partial r}\Big)\,dr -$$

$$\Big(\varphi_l\frac{4}{3\mathrm{Re}}\mu\frac{\partial u}{\partial r} - \varphi_l\frac{4}{3\mathrm{Re}}\mu u\Big)\Big|_{r=1} + \int_0^1 \frac{\partial \varphi_l}{\partial r}\Big(\frac{4}{3\mathrm{Re}}r^2\mu\frac{\partial u}{\partial r} - \frac{4}{3\mathrm{Re}}r\mu u\Big)\,dr - \tag{17}$$

$$\int_0^1 r^2\varphi_l F_{em}\,dr = 0, \qquad l = 0,\ ...,\ n.$$

Substitute the approximate solution (15) into (17) in place of u. Due to the basis functions (6), only the following terms do not vanish:

$$\sum_{i=l-1}^{l+1}\Big[\int_0^1\Big(r^2\varphi_l\varphi_i\Big(\rho\frac{\partial u_i}{\partial t} + \frac{u_i}{2}\frac{\partial \rho}{\partial t}\Big) - \frac{1}{2}r^2\rho u u_i\varphi_i\frac{\partial \varphi_l}{\partial r} + \frac{1}{2}r^2\rho u u_i\frac{\partial \varphi_i}{\partial r}\varphi_l +$$

$$\frac{4}{3\mathrm{Re}}r^2\mu u_i\frac{\partial \varphi_l}{\partial r}\frac{\partial \varphi_i}{\partial r} - \frac{4}{3\mathrm{Re}}r\mu u_i\varphi_i\frac{\partial \varphi_l}{\partial r}\Big)\,dr +$$

$$\Big(\frac{1}{2}\rho u u_i\varphi_l\varphi_i - \varphi_l\frac{4}{3\mathrm{Re}}\mu u_i\frac{\partial \varphi_i}{\partial r} + \varphi_l\frac{4}{3\mathrm{Re}}\mu u_i\varphi_i\Big)\Big|_{r=1}\Big] =$$

$$\int_0^1 P\Big(2r\varphi_l + r^2\frac{\partial \varphi_l}{\partial r}\Big)\,dr - (\varphi_l P)\Big|_{r=1} + \int_0^1 r^2\varphi_l F_{em}\,dr, \qquad l=0,\,...,\,n. \quad (18)$$

To calculate the integrals in (18), we apply the trapezoid formula and the equalities $u_{-1} = -u_0, \mu_{-1} = \mu_0$ which follow from symmetry of the problem. The time derivative is approximated in the following way:

$$\rho_l\frac{\partial u_i}{\partial t} + \frac{u_i}{2}\frac{\partial \rho_l}{\partial t} = \sqrt{\rho_l}\frac{\partial}{\partial t}(\sqrt{\rho_l}u_i) \approx \frac{\rho_l^{k+1}u_l^{k+1} - \sqrt{\rho_l^{k+1}}\sqrt{\rho_l^k}u_l^k}{\tau}.$$

As a result we obtain

$$\Big(\frac{h}{\tau}r_0^2\rho_0^{k+1} + \frac{2}{h\,\mathrm{Re}}r_0^2\mu_0^k + \frac{2}{3\,h\,\mathrm{Re}}r_1^2\mu_1^k\Big)\,u_0^{k+1} +$$
$$\Big(\frac{1}{4}r_1^2\rho_1^{k+1}u_1^k - \frac{2}{3\,h\,\mathrm{Re}}r_1^2\mu_1^k + \frac{2}{3\,\mathrm{Re}}r_1\mu_1^k +$$
$$\frac{1}{4}r_0^2\rho_0^{k+1}u_0^k - \frac{2}{3\,h\,\mathrm{Re}}r_0^2\mu_0^k\Big)\,u_1^{k+1} =$$
$$\frac{\sqrt{\rho_0^{k+1}}\sqrt{\rho_0^k}\,u_0^k}{\tau}h\,r_0^2 + \frac{1}{2}r_0^2P_0^k + 2hr_0P_0^k - \frac{1}{2}r_1^2P_1^k + hr_0^2F_{em,0}^k, \quad l=0,$$

$$\Big(-\frac{1}{4}r_{l-1}^2\rho_{l-1}^{k+1}u_{l-1}^k - \frac{2}{3\,h\,\mathrm{Re}}r_{l-1}^2\mu_{l-1}^k - \frac{2}{3\,\mathrm{Re}}r_{l-1}\mu_{l-1}^k -$$
$$\frac{1}{4}r_l^2\rho_l^{k+1}u_l^k - \frac{2}{3\,h\,\mathrm{Re}}r_l^2\mu_l^k\Big)\,u_{l-1}^{k+1} +$$
$$\Big(\frac{h}{\tau}r_l^2\rho_l^{k+1} + \frac{2}{3\,h\,\mathrm{Re}}r_{l-1}^2\mu_{l-1}^k + \frac{4}{3\,h\,\mathrm{Re}}r_l^2\mu_l^k + \frac{2}{3\,h\,\mathrm{Re}}r_{l+1}^2\mu_{l+1}^k\Big)\,u_l^{k+1} +$$
$$\Big(\frac{1}{4}r_{l+1}^2\rho_{l+1}^{k+1}u_{l+1}^k - \frac{2}{3\,h\,\mathrm{Re}}r_{l+1}^2\mu_{l+1}^k + \frac{2}{3\,\mathrm{Re}}r_{l+1}\mu_{l+1}^k +$$
$$\frac{1}{4}r_l^2\rho_l^{k+1}u_l^k - \frac{2}{3\,h\,\mathrm{Re}}r_l^2\mu_l^k\Big)\,u_{l+1}^{k+1} =$$
$$\frac{\sqrt{\rho_l^{k+1}}\sqrt{\rho_l^k}\,u_l^k}{\tau}h\,r_l^2 + \frac{1}{2}r_{l-1}^2P_{l-1}^k + 2hr_lP_l^k - \frac{1}{2}r_{l+1}^2P_{l+1}^k + hr_l^2F_{em,l}^k,$$
$$l=1,\,...,\,n-1,$$

$$\Big(-\frac{1}{4}r_{n-1}^2\rho_{n-1}^{k+1}u_{n-1}^k - \frac{2}{3\,h\,\mathrm{Re}}r_{n-1}^2\mu_{n-1}^k - \frac{2}{3\,\mathrm{Re}}r_{n-1}\mu_{n-1}^k -$$
$$\frac{1}{4}r_n^2\rho_n^{k+1}u_n^k - \frac{2}{3\,h\,\mathrm{Re}}r_n^2\mu_n^k\Big)\,u_{n-1}^{k+1} +$$
$$\Big(\frac{h}{2\tau}r_n^2\rho_n^{k+1} + \frac{2}{3\,h\,\mathrm{Re}}r_{n-1}^2\mu_{n-1}^k - \frac{2}{3\,h\,\mathrm{Re}}r_n^2\mu_n^k + \frac{2}{3\,\mathrm{Re}}r_n\mu_n^k + \frac{1}{2}r_n^2\rho_n^{k+1}u_n^k\Big)\,u_n^{k+1} =$$

$$\frac{\sqrt{\rho_n^{k+1}}\sqrt{\rho_n^k}\,u_n^k}{2\tau}h\,r_n^2+\frac{1}{2}r_{n-1}^2P_{n-1}^k+h\,r_nP_n^k-\frac{1}{2}r_n^2P_n^k+h\,r_n^2F_{em,n}^k, \qquad l=n. \quad (19)$$

We seek an approximation $e_h(t,r)$ to the function $e(t,r)$ in the form

$$e_h(t,r)=\sum_{i=0}^{n}e_i(t)\varphi_i(r) \quad (20)$$

with time-dependent coefficients $e_i(t)$.

In the case of spherical symmetry, reducing the inner product in R^3 to that in R^1 we get the following expression:

$$\int_0^1\left(r^2\varphi_l\frac{\partial}{\partial t}(\rho e)+\varphi_lP\frac{\partial}{\partial r}(r^2u)+\varphi_l\frac{\partial}{\partial r}(r^2u\rho e)-\frac{\gamma}{\mathrm{Re\,Pr}}\frac{\partial}{\partial r}\left(r^2\mu\frac{\partial e}{\partial r}\right)\varphi_l-\right.$$
$$\left.\frac{4}{3\mathrm{Re}}r^2\varphi_l\mu\left(\frac{\partial u}{\partial r}\right)^2+\frac{4}{3\mathrm{Re}}r\varphi_l\mu u\frac{\partial u}{\partial r}\right)dr=0,\ \ l=0,\,...,\,n. \quad (21)$$

Applying the integration by parts to (21) gives

$$\int_0^1\left(r^2\varphi_l\frac{\partial}{\partial t}(\rho e)+2r\varphi_luP+r^2\varphi_lP\frac{\partial u}{\partial r}-r^2u\rho e\frac{\partial\varphi_l}{\partial r}+\frac{\gamma}{\mathrm{Re\ Pr}}r^2\mu\frac{\partial e}{\partial r}\frac{\partial\varphi_l}{\partial r}-\right.$$
$$\left.\frac{4}{3\mathrm{Re}}r^2\varphi_l\mu\left(\frac{\partial u}{\partial r}\right)^2+\frac{4}{3\mathrm{Re}}r\varphi_l\mu u\frac{\partial u}{\partial r}\right)dr+(u\rho e\varphi_l)\Big|_{r=1}-$$
$$\left(\frac{\gamma}{\mathrm{Re\,Pr}}\varphi_l\mu\frac{\partial e}{\partial r}\right)\Big|_{r=1}=0, \qquad l=0,\,...,\,n. \quad (22)$$

Substitute the approximate solution (20) into (22) in place of e. Considering the form of the basis functions, only the following terms do not vanish:

$$\sum_{i=l-1}^{l+1}\left(\int_0^1\left(r^2\varphi_l\varphi_i\frac{\partial}{\partial t}(\rho e_i)-r^2u\rho\frac{\partial\varphi_l}{\partial r}e_i\varphi_i+\frac{\gamma}{\mathrm{Re\,Pr}}\mu r^2e_i\frac{\partial\varphi_i}{\partial r}\frac{\partial\varphi_l}{\partial r}\right)dr+\right.$$
$$\left.\left(u\rho\varphi_le_i\varphi_i-\frac{\gamma}{\mathrm{Re\,Pr}}\mu\,\varphi_le_i\frac{\partial\varphi_i}{\partial r}\right)\Big|_{r=1}\right)=-\int_0^1\left(2r\varphi_luP+r^2\varphi_lP\frac{\partial u}{\partial r}-\right.$$
$$\left.\frac{4}{3\mathrm{Re}}r^2\varphi_l\mu\left(\frac{\partial u}{\partial r}\right)^2+\frac{4}{3\mathrm{Re}}r\varphi_l\mu u\frac{\partial u}{\partial r}\right)dr, \quad l=0,\,...,\,n. \quad (23)$$

To calculate integrals in (23), we apply the trapezoid formula and the equalities $e_{-1}=e_0$, $\mu_{-1}=\mu_0$, which follows from symmetry of the problem. The time derivative is approximated with the help of the forward difference. As a result, we obtain the equations

$$\left(\frac{h}{\tau}r_0^2\rho_0^{k+1} + \frac{3\gamma}{2h\mathrm{Re\,Pr}}r_0^2\mu_0^k + \frac{\gamma}{2h\mathrm{Re\,Pr}}r_1^2\mu_1^k\right)e_0^{k+1} +$$
$$\left(-\frac{\gamma}{2h\mathrm{Re\,Pr}}r_0^2\mu_0^k + \frac{1}{2}r_1^2u_1^{k+1}\rho_1^{k+1} + \frac{\gamma}{2h\mathrm{Re\,Pr}}r_1^2\mu_1^k\right)e_1^{k+1} =$$
$$\frac{h}{\tau}r_0^2\rho_0^k e_0^k - 2hr_0u_0^{k+1}P_0^k - \frac{1}{2}r_0^2P_0^k(u_1^{k+1} - u_0^{k+1}) + \frac{1}{3h\mathrm{Re}}r_0^2\mu_0^k(u_1^{k+1} - u_0^{k+1})^2 -$$
$$\frac{2}{3\mathrm{Re}}r_0\mu_0^k u_0^{k+1}(u_1^{k+1} - u_0^{k+1}), \quad l = 0,$$

$$\left(-\frac{1}{2}r_{l-1}^2u_{l-1}^{k+1}\rho_{l-1}^{k+1} - \frac{\gamma}{2h\mathrm{Re\,Pr}}r_{l-1}^2\mu_{l-1}^k - \frac{\gamma}{2h\mathrm{Re\,Pr}}r_l^2\mu_l^k\right)e_{l-1}^{k+1} +$$
$$\left(\frac{h}{\tau}r_l^2\rho_l^{k+1} + \frac{\gamma}{2h\mathrm{Re\,Pr}}r_{l-1}^2\mu_{l-1}^k + \frac{\gamma}{h\mathrm{Re\,Pr}}r_l^2\mu_l^k + \frac{\gamma}{2h\mathrm{Re\,Pr}}r_{l+1}^2\mu_{l+1}^k\right)e_l^{k+1} +$$
$$\left(-\frac{\gamma}{2h\mathrm{Re\,Pr}}r_l^2\mu_l^k + \frac{1}{2}r_{l+1}^2u_{l+1}^{k+1}\rho_{l+1}^{k+1} + \frac{\gamma}{2h\mathrm{Re\,Pr}}r_{l+1}^2\mu_{l+1}^k\right)e_{l+1}^{k+1} =$$
$$\frac{h}{\tau}r_l^2\rho_l^k e_l^k - 2hr_lu_l^{k+1}P_l - \frac{1}{2}r_l^2P_l^k(u_{l+1}^{k+1} - u_{l-1}^{k+1}) + \frac{1}{3h\mathrm{Re}}r_l^2\mu_l^k(u_{l+1}^{k+1} - u_{l-1}^{k+1})^2 -$$
$$\frac{2}{3\mathrm{Re}}r_l\mu_l^k u_l^{k+1}(u_{l+1}^{k+1} - u_{l-1}^{k+1}), \qquad l = 1, ..., n-1, \qquad (24)$$

$$\left(-\frac{1}{2}r_{n-1}^2u_{n-1}^{k+1}\rho_{n-1}^{k+1} - \frac{\gamma}{2h\mathrm{Re\,Pr}}r_{n-1}^2\mu_{n-1}^k - \frac{\gamma}{2h\mathrm{Re\,Pr}}r_n^2\mu_n^k\right)e_{n-1}^{k+1} +$$
$$\left(\frac{h}{2\tau}r_n^2\rho_n^{k+1} + \frac{\gamma}{2h\mathrm{Re\,Pr}}r_{n-1}^2\mu_{n-1}^k + \frac{1}{2}r_n^2u_n^{k+1}\rho_n^{k+1} - \frac{\gamma}{2h\mathrm{Re\,Pr}}r_n^2\mu_n^k\right)e_n^{k+1} =$$
$$\frac{h}{2\tau}r_n^2\rho_n^k e_n^k - hr_nu_n^{k+1}P_n^k - \frac{1}{2}r_n^2P_n^k(u_n^{k+1} - u_{n-1}^{k+1}) +$$
$$\frac{2}{3h\mathrm{Re}}r_n^2\mu_n^k(u_n^{k+1} - u_{n-1}^{k+1})^2 - \frac{2}{3\mathrm{Re}}r_n\mu_n^k u_n^{k+1}(u_n^{k+1} - u_{n-1}^{k+1}), \quad l = n.$$

Thus, we get a first-order consistent conservative variational difference scheme. To solve a large system of quasilinear algebraic equations of a special form with a tridiagonal matrix, the nonmonotone sweep method is applied with iteration with respect to nonlinearity.

We give the formulae [9] of the nonmonotone sweep method for a problem of the form

$$c_0y_0 - b_0y_1 = f_0, \qquad i = 0,$$
$$-a_iy_{i-1} + c_iy_i - b_iy_{i+1} = f_i, \qquad i = 1, 2, ..., N-1,$$
$$-a_Ny_{N-1} + c_Ny_N = f_N, \qquad i = N.$$

First the initial values $C = c_0$, $A = a_1$, $F = f_0$, $\varPhi = f_1$, and $æ_1 = 0$ are specified. Then for $i = 0, 1, \ldots, N-1$ forward sweep is performed. As the situation requires, the following calculations are carried out:

if $|C| \geq |b_i|$ *then*

$$\alpha_{i+1} = b_i/C, \qquad \beta_{i+1} = F/C,$$

$$C = c_{i+1} - A\alpha_{i+1}, \qquad F = \Phi + A\beta_{i+1},$$

$$\Theta_{i+1} = æ_i, \qquad æ_{i+1} = i + 1,$$

$$A = a_{i+2}, \qquad \Phi = f_{i+2};$$

if $|C| < |b_i|$ *then*

$$\alpha_{i+1} = C/b_i, \qquad \beta_{i+1} = -F/b_i,$$

$$C = c_{i+1}\alpha_{i+1} - A, \qquad F = \Phi - c_{i+2}\beta_{i+1},$$

$$\Theta_{i+1} = i + 1, \qquad æ_{i+1} = æ_i,$$

$$A = a_{i+2}\,\alpha_{i+1}, \qquad \Phi = f_{i+2} + a_{i+2}\beta_{i+1}.$$

In both cases formulae for A and Φ for $i = N - 1$ are not cited.

Once forward sweep is completed, backward sweep for determining the unknowns in the reverse order is performed. First y_n where $n = æ_N$ is calculated by the formula

$$y_n = F/C.$$

Then for $i = N - 1,\ N - 2, \ldots, 0$ the remaining unknowns are calculated

$$y_m = \alpha_{i+1} y_n + \beta_{i+1},$$

where $m = \Theta_{i+1},\ n = æ_{i+1},\ i = N - 1,\ N - 2, \ldots, 0$.

This algorithm is implemented in a C^{++} program. The values of density, the dynamic coefficient of viscosity, pressure, and temperature given in the table which is copied from [2] are taken as initial conditions. In calculations density, the dynamic coefficient of viscosity, and temperature are taken in the ratio to the corresponding typical values at the Earth's surface.

A numerical experiment shows stability and convergence of the algorithm with decreasing a discretization parameter h. Besides, the algorithm is found to be sensitive to the state equation. For example, when variations in energy in phase and metamorphic transitions are not taken into account in the model, this immediate results in smoothing a density jump at the Mohorovičič and Wichert-Gutenberg discontinuities. Different forms of the state equation lead to some qualitative distinctions.

For example, for some state equations the Earth contracts with time under the action of the gravitation forces, its radius decreases and density of its substance increases in all layers (Fig. 2). The main geodynamic zones are retained but become more smoothed. In other cases pressure within the Earth increases, as a result the Earth's layers expand and its radius increases for the time being.

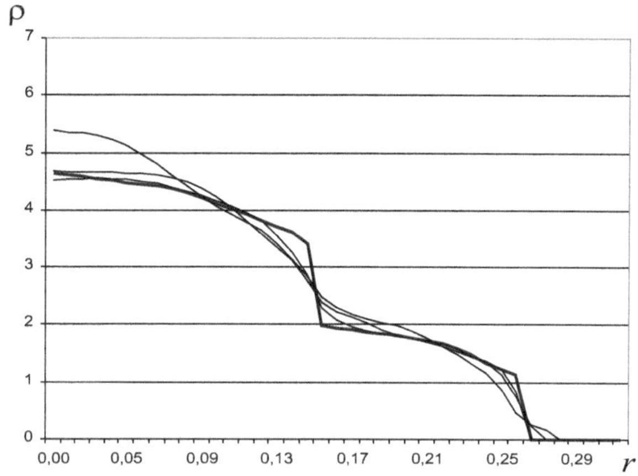

Fig. 2. Density function $\rho(r)$ with various times under the action of the gravitation forces

Conclusion

Thus, in this paper dynamics of the inner structure of the Earth is described by a model of a viscous compressible heat-conducting medium in the form of the Navier-Stokes equations. For the discretization of the spherically symmetric model the finite element method is used. The method is shown turned out to be highly sensitive to the state equation used. This results in decreasing or increasing the radius of the Earth, smoothing the boundaries of phase, chemical, or metamorphic transitions of the Earth's geodynamic layers. Therefore, in addition to the solution of the mathematical problems, a correct and accurate formulation of the state equation on the basis of the modern notion of the substance of the Earth and its physical and chemical properties at corresponding temperatures and pressures is of major importance.

Acknowledgement

The work was supported by the program № 16 of the Presidium of RAS (Project № 9).

References

1. Jeffreys, H.: The Earth: Its Origin, History and Physical Constitution. Cambridge University Press, Cambridge (1976)
2. Dobretsov, N.L., Kirdyashkin, A.G., Kirdyashkin, A.A.: Deep-seated Geodynamics, "GEO" branch. SB RAS Publisher, Novosibirsk (2001) (in Russian)

3. Shaidurov, V.V., Shchepanovskaya, G.I.: The mathematical and numerical modeling of the nonstationary propagation of a pulce of high-power energy in a viscous heat conducting gas. Part 1. Mathematical formulation of the problem. Registered in VINITI 24.10.03, 1860–B2003. Krasnoyarsk: ICM SB RAS, 50 p. (2003) (in Russian)
4. Karepova, E.D., Malyshev, A.V., Shaidurov, V.V., Shchepanovskaya, G.I.: The finite element method for the navier-stokes equations for a viscous heat conducting gas. In: Li, Z., Vulkov, L.G., Waśniewski, J. (eds.) NAA 2004. LNCS, vol. 3401, pp. 56–65. Springer, Heidelberg (2005)
5. Karepova, E.D., Malyshev, A.V., Shaidurov, V.V., Shchepanovskaya, G.I.: The parallel realization of the finite element method for the Navier-Stokes equations for a viscous heat conducting gas. In: Notes on Numerical Fluid Mechanics and Multidisciplinary Design, vol. 91, pp. 41–54. Springer, Heidelberg (2006)
6. Malyshev, V.A., Shaidurov, V.V., Shchepanovskaya, G.I.: The mathematical and numerical modeling of a supersonic interaction of heat pulses with a multiprocessor computational system. Krasn. State Univ. Bulletin. 4, 109–116 (2006) (in Russian)
7. Ushakova, O.A., Shaidurov, V.V., Shchepanovskaya, G.I.: The finite element method for the Navier-Stokes equations in spherical coordinates. Krasn. State Univ. Bulletin. 4, 151–156 (2006)
8. Shaidurov, V.V., Shchepanovskaya, G.I.: A gas dynamic model of the inner structure of the Earth II Conference in honour of Reshetnev. In: Transactions of the XI-th international scient. conference, pp. 268–269. Siberian state aerospace university, Krasnoyarsk (2007) (in Russian)
9. Samarskii, A.A., Nikolaev, E.S.: Methods for the Solution of Grid Equations. Moscow, Nauka (1978) (in Russian)

Temporally-Periodic Solitons of the Parametrically Driven Damped Nonlinear Schrödinger Equation

E.V. Zemlyanaya[1], I.V. Barashenkov[2], and N.V. Alexeeva[2]

[1] Laboratory for Information Technologies, Joint Institute for Nuclear Research, Dubna 141980, Russia
elena@jinr.ru
[2] Department of Mathematics, University of Cape Town, Rondebosch 7701, South Africa
{Igor.Barashenkov,Nora.Alexeeva}@uct.ac.za

Abstract. Time-periodic solitons of the parametrically driven damped nonlinear Schrödinger equation are determined as solutions of the boundary-value problem on a two-dimensional domain. We classify stability and bifurcations of single and double-periodic solutions.

1 Introduction

An efficient way of compensating dissipative losses in soliton-bearing systems consists in applying a resonant driving force. When the damping coefficient and the driving strength are weak, the amplitude of the arising localised solution is typically described by a directly or parametrically driven, damped nonlinear Schrödinger equation. The damped-driven nonlinear Schrödinger equations exhibit a wide variety of soliton solutions. For small driving amplitudes, the soliton is typically stationary and stable; as the driving strength is increased, the stationary soliton loses its stability to a time-periodic spatially localised solution. On further increase of the driver's strength, the periodic soliton may undergo a period-doubling transition to temporal chaos, followed by the decay and dissipation, or give way to spatio-temporal chaotic states.

Stability and bifurcation of stationary and steadily travelling solitons can be analysed using a number of analytical and numerical techniques. However, most of these are inapplicable to the time-periodic solutions. So far, the direct numerical simulation has been the only way to obtain periodic solutions and study their stability. The shortcoming of this method is that simulations capture only *stable* regimes. This means that the actual mechanisms and details of the transformations (which are bifurcations involving both stable and unstable solutions) remain unaccessible. Neither can simulations be used to identify alternative attractors in cases of bi- or multi-stability.

In this paper we propose a new approach to the analysis of these hidden mechanisms. Instead of direct numerical simulations, the time-periodic solitons are studied as solutions to a boundary-value problem formulated on a two-dimensional cylinder. The advantage of this approach is that it furnishes both stable and unstable solutions.

S. Margenov, L.G. Vulkov, and J. Waśniewski (Eds.): NAA 2008, LNCS 5434, pp. 139–150, 2009.

The particular equation that we will be concerned with here is the parametrically driven damped nonlinear Schrödinger equation,

$$i\psi_t + \psi_{xx} + 2|\psi|^2\psi - \psi = h\psi^* - i\gamma\psi. \tag{1}$$

Here $\gamma > 0$ is the damping coefficient, and $h > 0$ the amplitude of the parametric driver. Equation (1) describes a number of resonant phenomena in nonlinear media, in particular the nonlinear Faraday resonance in a vertically oscillating water trough [1,2,3] and the effect of phase-sensitive amplifiers on solitons in optical fibers [4,5,6]. The same equation controls the magnetization waves in an easy-plane ferromagnet exposed to a combination of a static and microwave field [7] and the amplitude of synchronized oscillations in vertically vibrated pendula lattices [8,9,10].

The stationary soliton solution of Eq.(1) exists for all $h \geq \gamma$ [7]:

$$\psi(x) = Ae^{-i\theta}\mathrm{sech}(Ax), \quad A = \sqrt{1+\sqrt{h^2-\gamma^2}}, \quad \theta = \frac{1}{2}\arcsin\frac{\gamma}{h}. \tag{2}$$

The soliton (2) is stable for $h \leq h_{\mathrm{Hopf}}(\gamma)$. As we increase h past $h_{\mathrm{Hopf}}(\gamma)$ keeping γ fixed, the stationary soliton loses its stability to a time-periodic soliton [7,11]. The transformation scenario arising as h is increased further, depends on the value of γ. According to the numerical simulations of [12], for γ smaller than 0.27 the periodic soliton follows a period-doubling route to temporal chaos. For h above the chaotic domain, the equation does not support any stable spatially-localised solutions. In a wide region above the chaotic domain, the only attractor determined in direct numerical simulations, was the trivial one, $\psi = 0$. Finally, for even larger values of h, the unstable soliton ignites spatio-temporal chaos [12].

For γ greater than 0.27, the soliton follows a different transformation scenario. Here, the period-doubling sequence does not arise and the soliton death does not occur. The periodic soliton remains stable until it yields directly to the spatio-temporal chaotic state [12].

The purpose of the present work is to follow the transformations of temporally periodic solutions of equation (1) as its parameters are varied, identify the arising bifurcations and eventually explain the attractor chart for this equation which was compiled using direct numerical simulations in Ref.[12]. We will also add missing details to this chart such as coexisting attractors in cases of bistability.

2 Periodic Solitons as Solutions of a Boundary-Value Problem in 2D

Instead of solving equation (1) with some initial condition and determining the resulting attractor by running the computation for a sufficiently long time, we will be looking for periodic solutions by solving Eq.(1) as a boundary-value problem on a two-dimensional domain $(-\infty, \infty) \times (0, T)$. The boundary conditions on the vertical sides of the domain will be set as

$$\psi(x,t) = 0 \quad \text{as } x \to \pm\infty, \tag{3}$$

while on the horizontal sides, we impose the periodic conditions

$$\psi(x, t+T) = \psi(x, t). \tag{4}$$

The period T is not known beforehand and is regarded as an unknown, along with the solution $\psi(x)$.

The periodic solutions will be path-followed (continued) in h for the fixed γ. We employ a predictor-corrector algorithm [14] with a fourth-order Newtonian iteration at each h. A finite-difference discretization with the stepsize $\Delta x = 0.05$ is used on the interval $(-L, L) = (-50, 50)$.

For the graphical representation of solutions, we will be using the averaged energy, defined by

$$E = \frac{1}{T}\int_0^T dt \int_{-\infty}^{\infty} dx\, \mathcal{E}(x,t), \tag{5}$$

$$\mathcal{E}(x,t) = |\psi_x|^2 + |\psi|^2 - |\psi|^4 + h\mathrm{Re}(\psi^2). \tag{6}$$

Note that the energy $\int \mathcal{E} dx$ is not an integral of motion for $\gamma \neq 0$.

3 Existence and Stability of the Stationary Two-Soliton Complexes

We start with the stationary solutions to eq.(1). Apart from the one-soliton solution (2), Eq.(1) supports several stationary complexes of solitons [13].

The two-soliton complex was found to exist for all $\gamma \geq 1.5 \times 10^{-8}$. Its domain of existence on the (γ, h)-plane is not bounded from above; however for h greater than $h_C = \sqrt{1+\gamma^2}$, the complex is unstable w.r.t. the continuous spectrum (as any other solution decaying to zero at the infinities [7]). Therefore for all practical purposes the value h_C can be regarded as the upper boundary of its existence

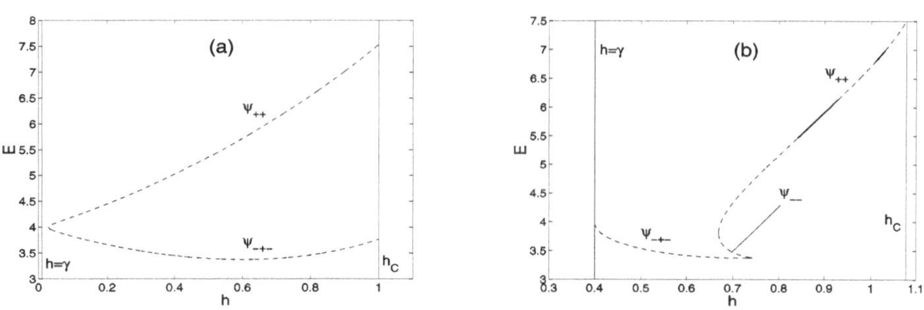

Fig. 1. Energy of the stationary two-soliton complex and stationary multisoliton solutions obtained from this complex by continuation in h for the fixed γ. (a) $\gamma = 0.01$; (b) $\gamma = 0.4$. Solid curves show stable and the dashed ones unstable solutions. Note two stable intervals of the ψ_{++} complex in (b).

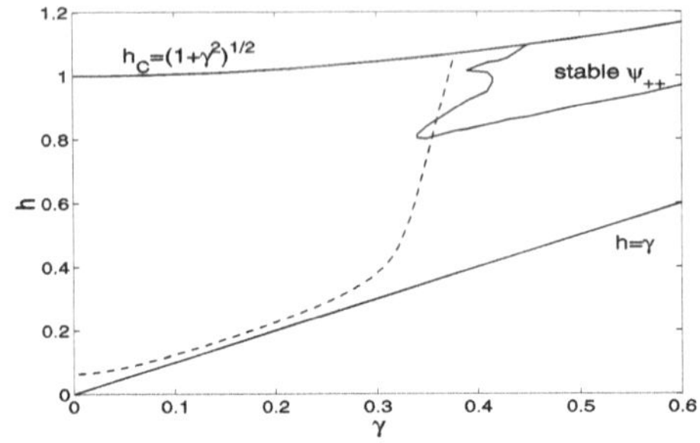

Fig. 2. The existence and stability chart of the stationary two-soliton complexes. For each γ, the region of existence of the complexes extends in the direction of larger h, beyond the value $h_C = \sqrt{1+\gamma^2}$.

domain. Reducing h from h_C for the fixed γ, we obtain one of two possible types of bifurcation diagrams on the (h, E) plane. Here E is given by

$$E = \int \left[|\psi_x|^2 + |\psi|^2 - |\psi|^4 + h\mathrm{Re}(\psi^2)\right] dx.$$

The diagram of the first type [Fig.1(a)] arises when h is decreased for a fixed *small* γ ($\gamma \lesssim 0.292$). In this case, there is only one turning point, $h = h_{\rm sn}$, with $h_{\rm sn} = h_{\rm sn}(\gamma) > \gamma$. (For $\gamma = 0.01$, $h_{\rm sn} = 0.02972$.) As h approaches $h_{\rm sn}$ along the top branch, the two-soliton complex $\psi_{(++)}$ transforms into a symmetric three-hump solution, with the distance between the lateral humps remaining approximately the same. The complex obtained by the continuation of this solution to the bottom branch can be identified as a three-soliton bound state $\psi_{(-+-)}$. As we continue away from $h_{\rm sn}$ along the bottom branch in Fig.1(a), the ψ_- solitons bound in this complex (the two side solitons) diverge to the infinities on the x axis. A more complicated diagram arises for larger values of γ ($\gamma \gtrsim 0.292$), see Fig.1(b). This bifurcation diagram has been described in [13] for a particular value of γ ($\gamma = 0.565$); here, we reproduce it for convenience. Reducing h from h_C for the fixed γ, the branch resulting from the two-soliton solution $\psi_{(++)}$ develops a sequence of turning points. As we pass the first turning point, the $\psi_{(++)}$ complex transforms into the $\psi_{(--)}$ solution. Moving away from the point along the bottom branch, the $\psi_{(--)}$ complex develops a third hump. This branch does not continue all the way to h_C but turns again, into a branch with even a lower energy. On this branch, the three-hump solution can be identified as $\psi_{(-+-)}$. The lowest branch continues to the point $h = \gamma$. This point defines the lower boundary of the domain of existence of the stationary complexes which result from the continuous transformation of the two-soliton solution $\psi_{(++)}$. As we approach the point $h = \gamma$, the distance between the two side solitons in the $\psi_{(-+-)}$ complex tends to infinity.

We were not able to obtain a symmetric two-soliton or three-soliton complex for $\gamma = 0$. As we continue in γ towards $\gamma = 0$ for a fixed h, the separation distance between the solitons in the complex increases without bounds; hence we conjecture that symmetric multisoliton complexes do not exist for $\gamma = 0$. (There are *nonsymmetric* complexes with $\gamma = 0$ though; see [15].)

The shape of the $E(h)$ curve pertaining to $\gamma = 0.4$ [Fig.1(b)] looks similar to the $E(h)$ curve for $\gamma = 0.565$ [13]. The main difference between the diagrams pertaining to $\gamma = 0.565$ and $\gamma = 0.4$ is that when $\gamma = 0.565$, the stability region of the two-soliton solution is seamless, i.e. does not have instability gaps in it, whereas in the $\gamma = 0.4$ case, the stability region consists of two segments of the curve separated by an interval of instability. The stability domain for the two-soliton complex ($\psi_{(++)}$) on the (γ, h)-plane is shown in Fig.2.

4 Stability of Periodic Solutions: The Method

Assume we have found a periodic solution $\psi_0(x,t) = \mathcal{R}(x,t) + i\mathcal{I}(x,t)$ satisfying the above conditions. Letting $\psi(x,t) = \psi_0(x,t) + u(x,t) + iv(x,t)$ and linearising (1) in the small perturbation $u + iv$, we obtain

$$J\mathbf{w}_t = \mathcal{H}\mathbf{w}, \tag{7}$$

where $\mathbf{w}$ is a two-component column-vector $\mathbf{w} = \begin{pmatrix} u \\ v \end{pmatrix}$, $\mathcal{H}$ is a matrix-differential operator

$$\mathcal{H} = \begin{pmatrix} -\partial_x^2 + 1 + h - 6\mathcal{R}^2 - 2\mathcal{I}^2 & -4\mathcal{R}\mathcal{I} + \gamma \\ -4\mathcal{R}\mathcal{I} - \gamma & -\partial_x^2 + 1 - h - 2\mathcal{R}^2 - 6\mathcal{I}^2 \end{pmatrix} \tag{8}$$

and J is a skew-symmetric matrix $J = \begin{pmatrix} 0 & -1 \\ 1 & 0 \end{pmatrix}$.

The solution to Eq.(7) with an initial condition $\mathbf{w}(x,0)$ can be written, formally, as $\mathbf{w}(x,t) = \mathcal{M}_t\mathbf{w}(x,0)$, where the evolution operator $\mathcal{M}_t$ acts on (vector-) functions of x but depends, parametrically, on t. The fundamental role is played by eigenvalues of the monodromy operator $\mathcal{M}_T$:

$$\mathcal{M}_T\mathbf{v}(x) = \mu\mathbf{v}(x). \tag{9}$$

Here $\mathcal{M}_T$ is the evolution operator $\mathcal{M}_t$ evaluated for $t = T$. The eigenvalues μ are usually referred to as the Floquet multipliers and the exponents λ, where $\mu = e^{\lambda T}$, as the Floquet exponents. According to the Floquet theory, for each λ there is a solution $\mathbf{w}(x,t)$ such that

$$\mathbf{w}(x,t) = e^{\lambda t}\mathbf{p}(x,t), \tag{10}$$

where $\mathbf{p}(x,t)$ is a periodic function: $\mathbf{p}(x,t+T) = \mathbf{p}(x,t)$ for all t.

To find out whether equation (7) admits solutions of the form (10) with Re$\lambda > 0$, we expand $u(x,t)$ and $v(x,t)$ in the Fourier series on the interval $(-L, L)$, and truncate the series as follows:

$$\mathbf{w}(x,t) = \sum_{n=-N}^{N} \mathbf{w}_n e^{-iq_n x} = \sum_{n=-N}^{N} \begin{pmatrix} u_n \\ v_n \end{pmatrix} e^{-iq_n x}.$$

Here $\mathbf{w}_n = \mathbf{w}_n(t)$ and $q_n = \pi n/L$. Substituting into (7) gives

$$J\dot{\mathbf{w}}_m = \sum_{n=-N}^{N} H_{mn}(t)\mathbf{w}_n \tag{11}$$

where H_{mn} $(m, n = -N, ..., N)$ is a 2×2 matrix of the form

$$H_{mn}(t) = \frac{1}{2L}\int_{-L}^{L} e^{i(q_m - q_n)x} \begin{pmatrix} q_n^2 + 1 + h - 6\mathcal{R}^2 - 2\mathcal{I}^2 & -4\mathcal{R}\mathcal{I} + \gamma \\ -4\mathcal{R}\mathcal{I} - \gamma & q_n^2 + 1 - h - 2\mathcal{R}^2 - 6\mathcal{I}^2 \end{pmatrix} dx. \tag{12}$$

The $2N+1$ vector equations (11) constitute a system of $4N+2$ equations for the unknowns $u_n(t)$ and $v_n(t)$ with periodic coefficients. Solutions of this system are of the form

$$u_n(t) = e^{\lambda t} f_n(t), \quad v_n(t) = e^{\lambda t} g_n(t), \tag{13}$$

where f_n and g_n are periodic functions: $f_n(t+T) = f_n(t)$, $g_n(t+T) = g_n(t)$ for all t. The Floquet exponents λ in (13) give approximations for the Floquet exponents in (10).

It is convenient to rearrange components of the two-dimensional vectors $\mathbf{w}_n$ $(n = -N, ..., N)$ to form a $(4N+2)$-dimensional vector $\boldsymbol{w}$:

$$\boldsymbol{w} = (u_{-N}, v_{-N}, u_{-N+1}, v_{-N+1}, ..., u_N, v_N).$$

In terms of $\boldsymbol{w}$, the system (11) can be written as

$$\boldsymbol{w}_t = \hat{H}\boldsymbol{w}, \tag{14}$$

where $\hat{H}$ is a matrix of dimension $(4N+2) \times (4N+2)$ whose elements are 2×2 blocks $J^{-1}H_{mn}$.

To find the Floquet exponents for the system (14), we form its principal fundamental matrix

$$M_t = \left(\boldsymbol{w}_{-N}^{(1)}, \boldsymbol{w}_{-N}^{(2)}, ..., \boldsymbol{w}_N^{(1)}, \boldsymbol{w}_N^{(2)}\right). \tag{15}$$

Here $\boldsymbol{w}_\alpha^{(1)}(t)$ (with $\alpha = -N, ..., N$) is the solution of the system (14) with the initial condition

$$u_n(0) = \delta_{n\alpha}, \quad v_n(0) = 0 \quad (n = -N, ..., N), \tag{16}$$

and $\boldsymbol{w}_\alpha^{(2)}(t)$ (where $\alpha = -N, ..., N$) is the solution of the same system with the initial condition

$$u_n(0) = 0, \quad v_n(0) = \delta_{n\alpha} \quad (n = -N, ..., N). \tag{17}$$

In (16)-(17), $\delta_{n\alpha}$ is Kronecker's delta: $\delta_{n\alpha} = 0$ for $n \neq \alpha$ whereas $\delta_{\alpha\alpha} = 1$. Letting $t = T$ in (15), we obtain the monodromy matrix for the system (14). The Floquet multipliers for this system are found as eigenvalues of its monodromy matrix:

$$M_T\boldsymbol{w} = \mu\boldsymbol{w}.$$

The bulk of our eigenvalue calculations were done with $N = 100$ but we went up to $N = 250$ when the eigenfunctions have shown variations on a small scale.

5 Results of Numerical Study

5.1 Moderate Damping: $\gamma = 0.3$

The leftmost point in figure 3 ($h = 0.385$) corresponds to the stationary single-soliton solution ψ_+. At this value of h the stable stationary soliton undergoes the Hopf bifurcation and a stable periodic soliton (shown by the solid curve in the figure) is born. At this point, the spectrum includes a positive real eigenvalue $\mu_4 < 1$, a complex-conjugate pair $\mu_5, \mu_6 = \mu_5^*$ with $|\mu_{5,6}| < 1$ and three unit eigenvalues $\mu_{1,2,3} = 1$. As h is increased, two of these unit eigenvalues remain in the spectrum whereas the third one moves inside the unit circle. The complex-conjugate eigenvalues grow in absolute value and cross through the unit circle at the turning point (where $h = 0.8761$). This is where the periodic solution looses its stability.

The end point of the dashed curve ($h = 0.61$) corresponds to the stationary three-soliton complex $\psi_{(+++)}$. As the stationary complex is continued in h through the value $h = 0.61$ (this stationary branch is not shown in Fig.3), two complex-conjugate eigenvalues in the linearised spectrum change the sign of their real parts and a periodic solution is born in the Hopf bifurcation. (This does not imply the change of stability of the solution $\psi_{(+++)}$ as there are pure

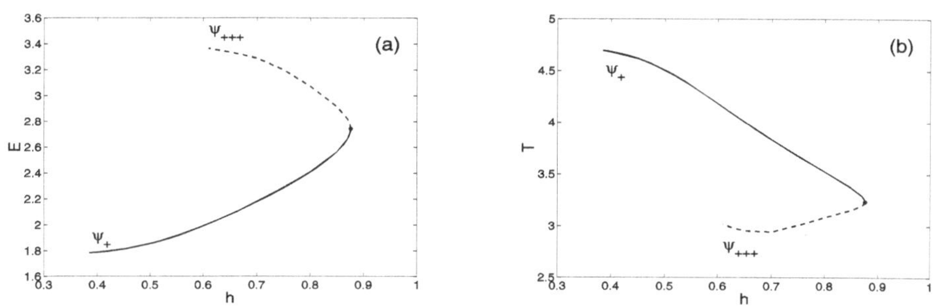

Fig. 3. The energy (a) and the period (b) of the periodic solution for $\gamma = 0.3$. The solid curve shows the stable and the dashed one unstable branch; the black dot marks the turning point.

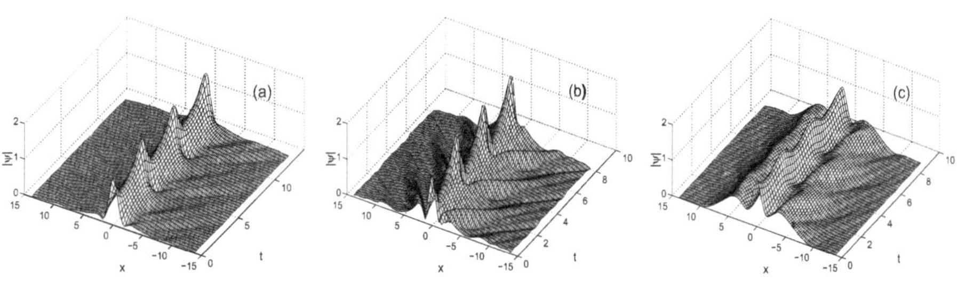

Fig. 4. The absolute value of the periodic solution with $\gamma = 0.3$: (a) $h = 0.55$, $T = 4.356$; (b) $h = 0.876$, $T = 2.743$; (c) $h = 0.64$, $T = 2.967$. In each case several periods of oscillation are shown.

real unstable eigenvalues on both sides of the point $h = 0.61$. The detaching periodic branch is unstable as well.)

On the dashed branch, we have $|\mu_{5,6}| > 1$. As we continue from the turning point to the value $h = 0.61$ along the dashed branch, the complex-conjugate eigenvalues $\mu_{5,6}$ converge and move on to the real axis (remaining greater than 1). As we approach the endpoint of the curve, the eigenvalue μ_3 approaches 1. At the endpoint we have three unit eigenvalues.

Representative solutions are shown at figure 4. Near the leftmost point in Fig.3, the periodic solution looks like a single soliton with a periodically oscillating amplitude and width [Fig.4(a)]. During its oscillations, the soliton emits radiation waves. As we move further along the curve in Fig.3, the amplitude of oscillations as well as the intensity of radiation increases [Fig.4(b)]. As we continue moving along the curve, the shape of the oscillating solution evolves into a three-hump structure which may be interpreted as a triplet of solitons. Near the end point of the curve, the amplitude of oscillations decreases [Fig.4(c)] and we arrive at the stationary three-soliton complex.

5.2 Weak Damping: $\gamma = 0.265$

The bifurcation diagram consists of three branches, two of which arise as a result of the period-doubling bifurcation of the solution on the third branch. The transformation of the solution as we move along this period-one branch [the bottom branch in Fig.5(b), denoted $T1$] is similar to the transformation of the solution along the curve shown in Fig.3. The starting point of the curve (its leftmost point) corresponds to the stationary solution ψ_+; the end point corresponds to the stationary complex $\psi_{(+++)}$. The only difference from the case $\gamma = 0.3$ occurs when a real eigenvalue crosses through -1 as h is increased through $h = 0.44$. As h is increased past 0.44, this negative eigenvalue continues to grow in absolute value, reaches a maximum, and then starts to decrease. As h is increased through $h = 0.75$, the negative eigenvalue crosses through -1 once again, this time in the direction of decreasing modulus. For larger h the solution is stable — all the way to the turning point. The described behaviour of the

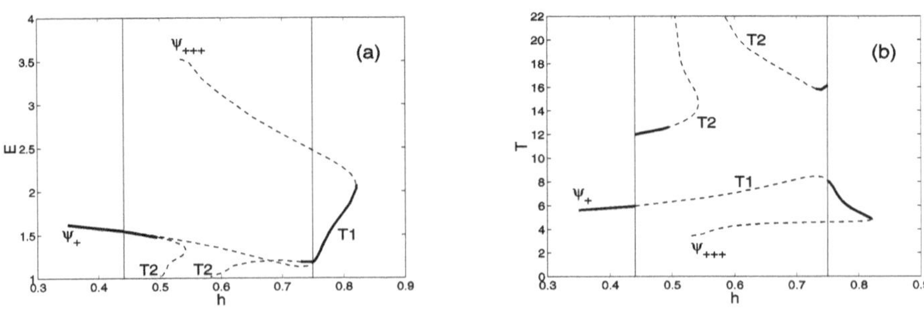

Fig. 5. The energy (a) and the period (b) of the periodic solution for $\gamma = 0.265$. Solid curves show the stable and the dashed ones unstable branches.

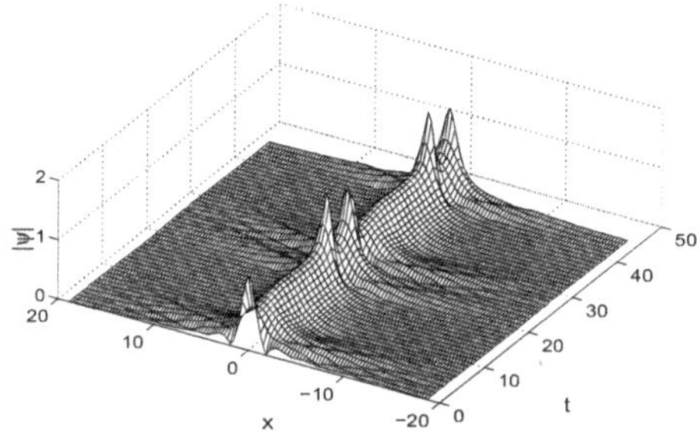

Fig. 6. The absolute value of the double-periodic solution found on the left upper branch in Fig. 5 at $h = 0.502$ and $T = 22.985$. A rapid oscillation is followed by a long quasistationary epoch.

negative eigenvalue corresponds to period-doubling bifurcations at $h = 0.44$ and $h = 0.75$, where periodic solutions with double period are detached.

The spectrum of the first double-periodic solution (detaching at $h = 0.44$, denoted $T2$) includes two unit eigenvalues, $\mu_{1,2} = 1$. For h close to 0.44, there is also a negative eigenvalue μ_3 with $|\mu_3| < 1$. As h increases from 0.44, the negative real eigenvalue μ_3 moves inside the unit circle from $\mu_3 = -1$, then reverses and crosses through $\mu_3 = -1$ once again (at $h = 0.495$). At this point, a new periodic solution is born, with the period equal to double the period of the double-periodic solution $T2$. (That is, the period of the newly born solution is approximately four times the period of $T1$.) This "period-four" solution is not shown in Fig.5.

In addition to the eigenvalues $\mu_{1,2,3}$, there is a pair of complex-conjugate eigenvalues μ_4, $\mu_5 = \mu_4^*$. As h approaches the turning point $h = 0.52$, the complex-conjugate pair crosses through the unit circle so that $|\mu_{4,5}| > 1$ on the top branch in Fig.5. As we continue further along the upper branch, the period grows because the solution develops a long epoch where it is very close to the stationary soliton ψ_-. Each period now consists of two phases: the solution performs a rapid oscillation with spatio-temporal profile very close to that of the $T1$ solution which is followed by a long quasistationary epoch [see Fig.6].

The second $T2$ solution detaches at $h = 0.75$. It is stable for $0.731 \leq h \leq 0.75$. As with the first double-periodic solution, the period of this solution grows as we move along the branch. [See Fig.5(b)]. The solution changes similarly to what we have described in the previous paragraph: a rapid oscillation is followed by a long quasistationary epoch when the solution is very close to the stationary soliton ψ_-.

5.3 Strong Damping: $\gamma = 0.565$

In this case, the stationary soliton ψ_+ is stable for all h between γ and $\sqrt{1+\gamma^2}$ and does not undergo a Hopf bifurcation [7]; hence no periodic solutions can detach from this stationary branch. However, the stationary *two*-soliton solution does exhibit a Hopf bifurcation at $h = 0.94$ [13] and we have used this point

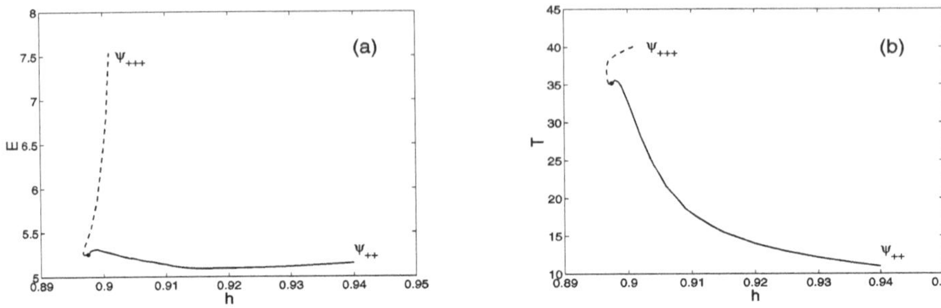

Fig. 7. The energy (a) and the period (b) of the periodic solution for $\gamma = 0.565$. The solid curve shows the stable and the dashed one unstable branch. The stability boundary is marked by a black blob.

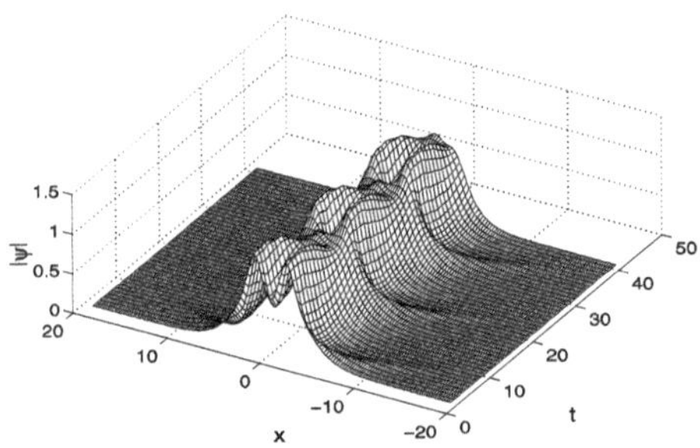

Fig. 8. A representative solution on the lower branch in Fig. 7. (Here $h = 0.92$, $T = 13.973$; shown is the absolute value of ψ.) Clearly visible is the two-soliton spatial structure of the solution.

as a starting point in our continuation process. The corresponding bifurcation diagrams showing the energy and period of the resulting solution are in Fig.7.

At the starting point $h = 0.94$, the spectrum includes three unit eigenvalues and two complex-conjugate pairs with modulus smaller than one. As we move along the curve away from $h = 0.94$, two unit eigenvalues remain in the spectrum while the third one moves inside the unit circle along the real axis. This eigenvalue decreases in modulus, crosses to the negative semiaxis at $h = 0.92$ and starts growing in the absolute value after that. For $h = 0.897$, the negative real eigenvalue crosses through $\mu = -1$. A period-doubling bifurcation occurs at this point; for $h < 0.897$, the periodic solution becomes unstable but a stable double-periodic solution arises. Note that the destabilization occurs *not* at the turning point of the $E(h)$ curve but to the right of this point. [The $E(h)$ curve turns at $h = 0.89665$.] Fig.8 shows a representative solution on the lower branch.

As for the two complex pairs, eigenvalues constituting one of these grow in absolute value as we move along the lower branch of the $E(h)$ curve towards smaller h. They cross through the unit circle as the curve turns back at $h = 0.89665$. As we move away from the turning point along the upper branch, the two complex-conjugate eigenvalues converge on the real axis but remain outside the unit circle. The other complex pair remains inside the unit circle; as we move along the lower branch from $h = 0.94$ to $h = 0.89665$, the imaginary parts of these eigenvalues decrease in absolute value and, at $h = 0.905$, the complex pair converges on the real axis. The two eigenvalues remain on the real axis as we continue along the upper branch.

The whole of the upper branch of the $E(h)$ curve is unstable. The curve ends at the stationary $\psi_{(+++)}$ solution (here $h = 0.901$). Near the endpoint of the upper branch, the solution looks similar to the one shown in Fig.4(c). As we approach the endpoint of the upper branch, the two real eigenvalues with $|\mu| < 1$ tend to 1. At the endpoint, the spectrum includes four unit eigenvalues and two positive real eigenvalues with $|\mu| > 1$.

6 Conclusions

Our analysis provides clear answers to a number of questions raised by the attractor chart of Ref.[12]. One such question is whether the transitions to chaos observed as h is increased for the fixed $\gamma < 0.27$ and $\gamma > 0.27$ do follow different routes (as the numerical simulations seem to indicate), or is this simply an effect caused by numerical approximations. The answer to this question is provided by diagrams in Fig.3 and 5. In the region $\gamma < 0.27$ (represented by the value $\gamma = 0.265$, Fig.5) the increase of h results in a sequence of period-doubling bifurcations of the periodic soliton. On the contrary, no period-doubling bifurcations occur as h is increased for the fixed $\gamma > 0.27$. This is established by the numerical continuation of solutions to the boundary-value problem on the two-dimensional (x, t)-domain and corroborated by the analysis of the eigenvalues of the linearised operator.

Another question is why no periodic solitons are observed in numerical simulations with sufficiently large h (not even as short-lived transients). Is there a well-defined boundary of the domain of existence of stable periodic solitons? We have shown that for $\gamma > 0.27$, the region of existence of periodic solitons on the h axis is bounded by the saddle-node bifurcation. No solitons, stable or unstable, exist above this turning point. For small γ, $\gamma < 0.27$, the situation is more complicated. The stable periodic soliton does indeed cease to exist after a period-doubling cascade; this accounts for the results of the simulations [12]. Furthermore, we have found that the period-one branch restabilises as h is increased above the region occupied by the higher-periodic solitons but hits a saddle-node bifurcation for even higher h. No periodic one-soliton solutions exist above the saddle-node bifurcation point which therefore defines the boundary of the periodic soliton existence domain.

The third issue clarified by our analysis pertains to the shape of the region on the (γ, h)-plane where periodic solitons are observed in numerical simulations. The question here is why does the $h(\gamma)$ curve bounding this region folds on itself for γ between 0.25 and 0.27. This phenomenon is accounted for by the restabilisation of the periodic soliton as h is increased for the fixed γ, $0.25 < \gamma < 0.27$. This restabilisation is exemplified by Fig.5 which corresponds to $\gamma = 0.265$.

Finally, the continuation algorithm furnishes a new stable two-soliton bound state which was not observed in numerical simulations (see Fig.7).

EZ was supported by the DST grant in the frame of JINR/RSA Research Collaboration Programme and by the RFBR (grant No. 06-01-00228).

References

1. Elphick, C., Meron, E.: Localized structures in surface waves. Phys. Rev. A 40, 3226–3229 (1989)
2. Chen, X.N., Wei, R.J.: Dynamic behaviour of a nonpropagating soliton under a periodically modulated oscillation. J. Fluid Mech. 259, 291–303 (1994)
3. Miao, G., Wei, R.: Parametrically excited hydrodynamic solitons. Phys. Rev. E 59, 4075–4078 (1999)
4. Mecozzi, A., et al.: Long-term storage of a soliton bit stream by use of phase-sensitive amplification. Opt. Lett. 19, 2050–2052 (1994)
5. Longhi, S.: Ultrashort-pulse generation in degenerate optical parametric oscillators. Opt. Lett. 20, 695–697 (1995)
6. Sánchez-Morcillo, V.J., et al.: Vectorial Kerr-cavity solitons. Opt. Lett. 25, 957–959 (2000)
7. Barashenkov, I.V., Bogdan, M.M., Korobov, V.I.: Stability Diagram of the Phase-Locked Solitons in the Parametrically Driven, Damped Non- Linear Schrodinger Equation. Europhys. Lett. 15, 113–118 (1991)
8. Denardo, B., et al.: Observations of localized structures in nonlinear lattices – domain walls and kinks. Phys. Rev. Lett. 68, 1730–1733 (1992)
9. Alexeeva, N.V., Barashenkov, I.V., Tsironis, G.P.: Impurity-induced stabilisation of solitons in arrays of parametrically driven nonlinear oscillators. Phys. Rev. Lett. 84, 3053–3056 (2000)
10. Chen, W., Hu, B., Zhang, H.: Interactions between impurities and nonlinear waves in a driven nonlinear pendulum chain. Phys. Rev. B 65, 134302(1-11) (2002)
11. Alexeeva, N.V., Barashenkov, I.V., Pelinovsky, D.E.: Dynamics of the parametrically driven NLS solitons beyond the onset of the oscillatory instability. Nonlinearity 12, 103–140 (1999)
12. Bondila, M., Barashenkov, I.V., Bogdan, M.M.: Topography of attractors of the parametrically driven damped nonlinear Schrödinger equation. Physica D 87, 314–320 (1995)
13. Barashenkov, I.V., Zemlyanaya, E.V.: Stable complexes of parametrically driven, damped Nonlinear Schrödinger solitons. Phys. Rev. Lett. 83, 2568–2571 (1999)
14. Zemlyanaya, E.V., Barashenkov, I.V.: Numerical study of the multisoliton complexes in the damped-driven NLS. Math. Modelling 16(3), 3–14 (2004)
15. Barashenkov, I.V., Zemlyanaya, E.V., Bär, M.: Travelling solitons in the parametrically driven nonlinear Schrödinger equation. Phys. Rev. E. 64, 016603(1-12) (2001)

Optimal Order FEM for a Coupled Eigenvalue Problem on 2D Overlapping Domains

A.B. Andreev[1] and M.R. Racheva[2]

[1] Department of Informatics, Technical University of Gabrovo
5300 Gabrovo, Bulgaria
[2] Department of Mathematics, Technical University of Gabrovo
5300 Gabrovo, Bulgaria

Abstract. In this paper we present a numerical approach to a nonstandard second-order elliptic eigenvalue problem defined on two overlapping rectangular domains with a nonlocal (integral) boundary condition. Usually, for this class of problems error estimates are suboptimal. By introducing suitable degrees of freedom and a corresponding interpolation operator we derive optimal order finite element approximation. Numerical results illustrate the efficiency of the proposed method.

1 Introduction

We consider a coupling eigenvalue problem on a two-component domain, with a nonlocal condition on the overlapping part of the subdomains. The study of such kind of contact problems is motivated by its applications in many engineering and physical disciplines. Problems with nonstandard boundary and coupling conditions appear, for instance, in heat conduction, fluid structure interaction and semi-conductors.

Recently there have been some studies of applying the finite element method (FEM) to overlapping grids [2,3,4]. Usually for eigenvalue problems, using nonstandard FE Lagrange interpolant the approximation of eigenpairs is found to be suboptimal (see, e. g. [2]). Our aim is to derive optimal order error estimates by means of a new finite element approach. For this purpose we introduce suitable modified degrees of freedom and a corresponding interpolation operator. Another advantage of this approach is that no additional requirements, compared to the standard ones, of the smoothness of the exact solution are needed (see also Remark 3).

The rest of the paper is organized as follows. In Section 2 we give a brief description for continuous eigenvalue problem defined on overlapping domains. In Section 3, a technique of finite element discretization is presented. Section 4 is devoted to the main result. Here the error estimates for approximate eigenpairs are derived. Finally, a numerical example is given to verify the validity of the analytic results.

2 Formulation of the Problem

Let Ω_1 and Ω_2 be overlapping rectangles with respective boundaries $\partial\Omega_1$ and $\partial\Omega_2$ and Ω be two-component rectangular domain, i.e. $\overline{\Omega} = \overline{\Omega}_1 \cup \overline{\Omega}_2$ (see

S. Margenov, L.G. Vulkov, and J. Waśniewski (Eds.): NAA 2008, LNCS 5434, pp. 151–158, 2009.

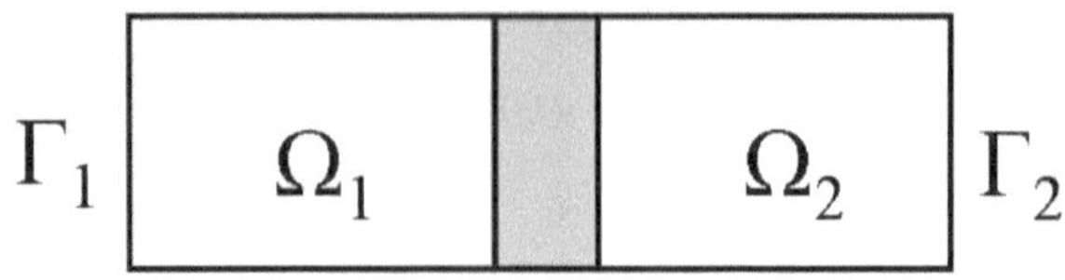

Fig. 1. Overlapping rectangular domains

Figure 1). Let also $H^m(\Omega_i)$ be the usual $m-$th order Sobolev space on $\Omega_i,\ i = 1,2$ with norm $\|\cdot\|_{m,\Omega_i}$.

Consider the following second-order two-dimensional elliptic operators:

$$L^{(i)} \equiv -\sum_{k,m=1}^{2} \frac{\partial}{\partial x_k}\left(a_{km}^{(i)} \frac{\partial}{\partial x_m}\right) + a_0^{(i)}, \quad i = 1,2,$$

where $a_{km}^{(i)}(x) > 0$ and $a_0^{(i)} \geq 0$ are bounded functions on $\Omega_i,\ i = 1,2$. For notational convenience we shall often drop the argument $x = (x_1, x_2)$. The eigenvalue problem is defined by: Find $(\lambda, u_1, u_2) \in \mathbf{R} \times H^2(\Omega_1) \times H^2(\Omega_2)$ such that

$$L^{(i)} u_i + (-1)^i \chi_{\Omega_1 \cap \Omega_2}.K = \lambda u_i \quad \text{in} \quad \Omega_i,\ i = 1,2, \tag{1}$$

and the classical Robin/Neumann and Dirichlet boundary conditions

$$\begin{aligned} &\frac{\partial u_i}{\partial \nu_{a^i}} + \sigma^{(i)} u_i = 0 \quad \text{on} \quad \Gamma_i', \\ &u_i = 0 \quad \text{on} \quad \Gamma_i,\ i = 1,2, \end{aligned} \tag{2}$$

as well as the following nonlocal coupling condition:

$$\int_{\Omega_1 \cap \Omega_2} [u_1(x) - u_2(x)]\, dx = 0. \tag{3}$$

Therein, the following notations are used: $\frac{\partial u_i}{\partial \nu_{a^i}}$ is the conormal derivative of u_i with respect to the coefficient matrix $a_{km}^{(i)},\ k,m,i = 1,2$; $\sigma^{(i)} \geq 0$; $\chi_{\Omega_1 \cap \Omega_2}$ denotes the characteristic function of $\Omega_1 \cap \Omega_2$ and $\Gamma_i' = \partial\Omega_i \setminus \Gamma_i$.

Remark 1. K is a real number depending on unknown function $u = (u_1, u_2)$. Its explicit representation could be obtained by integrating the equation (1) over $\Omega_1 \cap \Omega_2$ and using Green's formula (see [2], Proposition 3):

$$K = \frac{\int_{\Omega_1 \cap \Omega_2} (a_0^{(1)} u_1 - a_0^{(2)} u_2)\, dx - \int_{\partial(\Omega_1 \cap \Omega_2)} \left(\frac{\partial u_1}{\partial \nu_{a^1}} - \frac{\partial u_2}{\partial \nu_{a^2}}\right) ds}{2\text{meas}(\Omega_1 \cap \Omega_2)}.$$

Let us introduce the spaces

$$V_i = \left\{v_i \in H^1(\Omega_i) :\ v_i = 0 \text{ on } \Gamma_i\right\},\ i = 1,2 \quad \text{and} \quad \widetilde{V} = V_1 \times V_2.$$

Thus, the closed subspace of $\widetilde{V}$ which incorporates the coupling condition (3) is:

$$V = \left\{ v \in \widetilde{V} : \int_{\Omega_1 \cap \Omega_2} [v_1(x) - v_2(x)] \; dx = 0 \right\}.$$

Consider the variational eigenvalue problem: Find $(\lambda, u) \in \mathbf{R} \times V$ such that for all $v \in V$

$$a(u, v) = \lambda(u, v), \tag{4}$$

where

$$a(u,v) = \sum_{i=1}^{2} \left[\int_{\Omega_i} \left(\sum_{k,m=1}^{2} a_{km}^{(i)} \frac{\partial u_i}{\partial x_k} . \frac{\partial v_i}{\partial x_m} + a_0^{(i)} u_i v_i \right) dx + \int_{\Gamma_i'} \sigma^{(i)} u_i v_i \, ds \right].$$

Using the properties of coefficient functions as well as that $a_{12}^{(i)} = a_{21}^{(i)}$ a.e. in Ω_i, $i = 1, 2$ it is easy to see that:

- $a(\cdot, \cdot)$ is bounded, symmetric and strongly coersive on $V \times V$;
- V is a closed subspace of $H^1(\Omega_1) \times H^1(\Omega_2)$.

Thus, the considered problem (4) could be referred to the theory of abstract elliptic eigenvalue problems in Hilbert space [5] and the following result (see [2], Theorem 6) is valid:

Theorem 1. *The problems (1)-(3) and (4) are formally equivalent. Both problems have a countable infinite set of eigenvalues λ_l, all being strictly positive and having finite multiplicity, without a finite accumulation point. The corresponding eigenfunctions u_l can be chosen to be a Hilbert basis of V, orthonormal with respect to $(\cdot, \cdot)$.*

3 Finite Element Approximation

We approximate the eigenpairs of (4) by the finite element method. Consider a family of rectangulations $\tau_{h_i}^{(i)}$ of Ω_i, $i = 1, 2$, i.e. the partitions $\tau_{h_i}^{(i)}$ consist of rectangles $T_j^{(i)}$. Herein mesh parameters are denoted by h_i, $i = 1, 2$ and $h = \max_{i=1,2} h_i$.

The rectangles are constructed in such a way that for each element $T^{(i)}$ either $T^{(i)} \subset \Omega_1 \cap \Omega_2$ or $T^{(i)} \cap \Omega_1 \cap \Omega_2 = \emptyset$.

We introduce the FE spaces connected with the partitions $\tau_h^{(i)}$ for $i = 1, 2$:

$$X_{h_i}^{(i)} = \left\{ v_i \in C(\overline{\Omega}_i) : \; v_{i|_{T^{(i)}}} \in Q_2(T^{(i)}) \forall T^{(i)} \in \tau_h^{(i)} \right\},$$

$$X_h = X_{h_1}^{(1)} \times X_{h_2}^{(2)},$$

$$X_{h,0} = \left\{ v = (v_1, v_2) \in X_h : \; v_{i|_{\Gamma_i}} = 0 \right\},$$

where Q_2 denotes the set of polynomials of degree ≥ 2 for each variable.

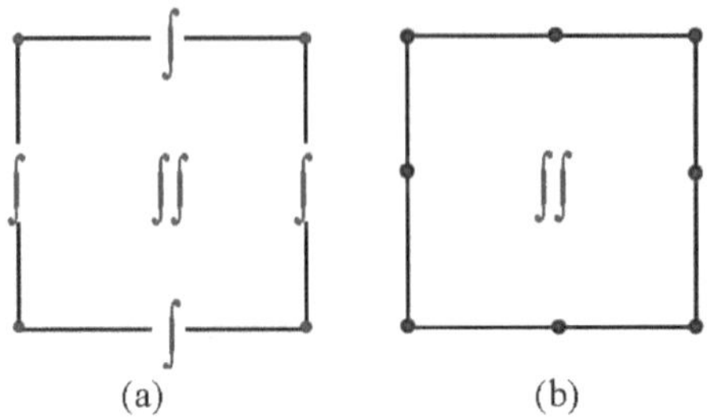

Fig. 2. (a) - vertex-edge conditions; (b) - integral value on $T^{(i)}$

Remark 2. Our presentation is restricted to polynomials of second degree. Nevertheless, the approach we will present could be applied when $v_i|_{T^{(i)}} \in Q_k(T^{(i)})$ where $k > 2$. The case $k = 1$ is not applicable to our method.

The FE space related to the nonstandard boundary condition (3) is:

$$V_h = \left\{ v \in X_{h,0} : \int_{\Omega_1 \cap \Omega_2} [v_1(x) - v_2(x)]\, dx = 0 \right\}, \quad V_h \subset V.$$

In order to derive optimal order error estimates we introduce suitable modified degrees of freedom and corresponding interpolation operator (cf. [1]). Let the vertices and the edges of any element $T^{(i)}$ be noted by $M_k^{(i)}$ and $l_k^{(i)}$, $k = 1, \ldots, 4$, respectively. For this element $T^{(i)}$ we choose its degrees of freedom in such a way that every polynomial $p(x) \in Q_2(T^{(i)})$ is determined by: the values at $M_k^{(i)}$; the integral value $\int_T^{(i)} p(x)\, dx$; the integral values $\int_{l_k^{(i)}} p(s)\, ds$ for case (a) and the values at the midpoints of $l_k^{(i)}$, $k = 1, \ldots, 4$ for case (b), respectively. It is to be mentioned here that all considerations into the next section are devoted to case (a), but case (b) is more simple and could be considered in a similar way.

According to the degrees of freedom, we define the interpolation operator $\pi_h : C(\Omega_1) \times C(\Omega_2) \to X_{h,0}$, where $\pi_h = (\pi_{h_1}, \pi_{h_2})$ on the analogy (in a sense) with the usual Lagrange quadratic interpolation operator $\Pi_{h_i} \in X_{h,0}$, $\Pi_{h_i} = (\Pi_{h_1}, \Pi_{h_2})$, $i = 1, 2$. Let us emphasize that for any $v \in V$ $\Pi_h v \notin V_h$, whereas $\pi_h v \in V_h$.

4 Main Result

The crucial point of our approach is to estimate the difference between both interpolants Π_h and π_h. The next theorem contains the main result:

Theorem 2. *Let the function $v = (v_1, v_2)$ belong to $V \cap H^3(\Omega)$, $\overline{\Omega} = \overline{\Omega}_1 \cup \overline{\Omega}_2$. Then there exists a constant $C = C(\Omega) > 0$, independent of h, such that*

$$\|v - \pi_h v\|_{m,\Omega} \le C h^{3-m} \|v\|_{3,\Omega}, \quad m = 0, 1. \tag{5}$$

Proof. First, we shall estimate $\Pi_{h_i} v_i - \pi_{h_i} v_i$ on each finite element $T_j^{(i)}$, $j = 1, \cdots, k_i$, $i = 1, 2$. For this purpose we transform any rectangle $T_j^{(i)}$ to the unit quadrilateral element T, called reference element. Let the coordinates of the left bottom vertex of $T_j^{(i)}$ be $(a_{j,1}^{(i)}, a_{j,2}^{(i)})$, $i = 1, 2$. Then we denote: $t_1 = (x_1 - a_{j,1}^{(i)})/h_1^{(i)}$, $t_2 = (x_2 - a_{j,2}^{(i)})/h_2^{(i)}$.

Obviously, $t_i \in [0, 1]$, $i = 1, 2$. The basis functions of the Lagrange interpolant Π_h, related to this interval, are $\{\psi_i(t_1).\psi_j(t_2)\}_{i,j=1}^3$, where:

$$\psi_1(t) = 2t^2 - 3t + 1;\ \psi_2 = -4t^2 + 4t;\ \psi_3 = 2t^2 - t,\ t \in [0, 1].$$

At the same time, the basis functions corresponding to the interpolant π_h are $\{\varphi_i(t_1).\varphi_j(t_2)\}_{i,j=1}^3$, where:

$$\varphi_1(t) = 3t^2 - 4t + 1;\ \varphi_2 = -6t^2 + 6t;\ \varphi_3 = 3t^2 - 2t,\ t \in [0, 1].$$

Let a_i, $i = 1, \ldots, 9$ denote classical biquadratic nodes of the reference element T. Then

$$(\Pi_{h_i} v_i - \pi_{h_i} v_i)_{|T} = v_i(a_1)\left[\psi_1(t_1)\psi_1(t_2) - \varphi_1(t_1)\varphi_1(t_2)\right]$$

$$+v_i(a_2)\left[\psi_3(t_1)\psi_1(t_2) - \varphi_3(t_1)\varphi_1(t_2)\right] + v_i(a_3)\left[\psi_3(t_1)\psi_3(t_2) - \varphi_3(t_1)\varphi_3(t_2)\right]$$

$$+v_i(a_4)\left[\psi_1(t_1)\psi_3(t_2) - \varphi_1(t_1)\varphi_3(t_2)\right] + v_i(a_5)\psi_2(t_1)\psi_1(t_2) + v_i(a_6)\psi_3(t_1)\psi_2(t_2)$$

$$+v_i(a_7)\psi_2(t_1)\psi_3(t_2) + v_i(a_8)\psi_1(t_1)\psi_2(t_2) + v_i(a_9)\psi_2(t_1)\psi_2(t_2)$$

$$-\int_{l_1} v_i(s)\,ds.\varphi_2(t_1)\varphi_1(t_2) - \int_{l_2} v_i(s)\,ds.\varphi_3(t_1)\varphi_2(t_2) - \int_{l_3} v_i(s)\,ds.\varphi_2(t_1)\varphi_3(t_2)$$

$$-\int_{l_4} v_i(s)\,ds.\varphi_1(t_1)\varphi_2(t_2) - \int_T v_i(t)\,dt.\varphi_2(t_1)\varphi_2(t_2)$$

$$= \varphi_2(t_1)\varphi_1(t_2)\left[\frac{1}{6}v_i(a_1) + \frac{4}{6}v_i(a_5) + \frac{1}{6}v_i(a_2) - \int_{l_1} v_i(s)\,ds\right]$$

$$+\varphi_3(t_1)\varphi_2(t_2)\left[\frac{1}{6}v_i(a_2) + \frac{4}{6}v_i(a_6) + \frac{1}{6}v_i(a_3) - \int_{l_2} v_i(s)\,ds\right]$$

$$+\varphi_2(t_1)\varphi_3(t_2)\left[\frac{1}{6}v_i(a_4) + \frac{4}{6}v_i(a_7) + \frac{1}{6}v_i(a_3) - \int_{l_3} v_i(s)\,ds\right]$$

$$+\varphi_1(t_1)\varphi_2(t_2)\left[\frac{1}{6}v_i(a_1) + \frac{4}{6}v_i(a_8) + \frac{1}{6}v_i(a_8) - \int_{l_4} v_i(s)\,ds\right]$$

$$+\varphi_2(t_1)\varphi_2(t_2)\left[\frac{1}{36}\sum_{k=1}^{4} v_i(a_k) + \frac{4}{36}\sum_{k=5}^{8} v_i(a_k) + \frac{16}{36}v_i(a_9) - \int_T v_i(t)\,dt\right].$$

The last expression could be presented by using error functionals of quadrature formulas (see e.q. [6]):

$$(\Pi_{h_i} v_i - \pi_{h_i} v_i)_{|T} = \varphi_2(t_1)\varphi_1(t_2)E_T^1(v_i) + \varphi_3(t_1)\varphi_1(t_2)E_T^2(v_i)$$

$$+\varphi_2(t_1)\varphi_3(t_2)E_T^3(v_i) + \varphi_1(t_1)\varphi_2(t_2)E_T^4(v_i) + \varphi_2(t_1)\varphi_2(t_2)E_T(v_i),$$

where 1-dimensional error functionals $E_T^k(v_i)$, $k = 1, \ldots, 4$ and 2-dimensional one $E_T(v_i)$ have the same order of convergence. It is easy to verify, that $E_T^k(v_i) = E_T(v_i) = 0$, $k = 1, \ldots, 4$ for all $v_i \in Q_2(T)$. However, the Simpson quadrature formula is more accurate, i.e. $E_T(v_i) = 0$ for all $v_i \in Q_3(T)$, but to make use of this property a higher order of regularity for the function v_i is needed. Obviously, in case (b) the 11-dimensional error functionals $E_T^K(v_i)$, $k = 1, \ldots, 4$ disappear.

On the other hand, the functions φ_i, $i = 1, 2, 3$ are bounded, namely $\varphi_1(t_i) \le 1$, $\varphi_2(t_i) \le 3/2$, $\varphi_3(t_i) \le 1$, $i = 1, 2$.

Using standard argument of the Bramble-Hilbert lemma [1,6], for any element T_j^i we have

$$|\Pi_{h_i} v_i - \pi_{h_i} v_i|_{|T_j^i}| \le C h_i^3 |v_i|_{3,T_j^i},$$

where $|\cdot|_{3,T_j^i}$ is third order Sobolev seminorm.

Then we obtain the following L_2-norm error estimate:

$$\|\Pi_h v - \pi_h v\|_{0,\Omega} = \left(\sum_{i=1}^{2} \sum_{T_j^i \in \tau_{h_i}^{(i)}} \int_{T_j^i} |\Pi_{h_i} v_i - \pi_{h_i} v_i|^2 \, dx \right)^{1/2} \le C h^3 \|v\|_{3,\Omega}. \quad (6)$$

By the same way, applying explicit calculations, we obtain:

$$|\frac{\partial}{\partial x} (\Pi_{h_i} v_i - \pi_{h_i} v_i) |_{|T_j^i} \le C h_i^2 |v_i|_{3,T_j}, \quad i, m = 1, 2.$$

Then, the H^1-norm error estimate is

$$\|\Pi_h v - \pi_h v\|_{1,\Omega} \le C h^2 \|v\|_{3,\Omega}.$$

This inequality and (6) give

$$\|\Pi_h v - \pi_h v\|_{m,\Omega} \le C h^{3-m} \|v\|_{3,\Omega}, \quad m = 0, 1. \quad (7)$$

Finally, we use the fact, that the order of $\|v - \Pi_h v\|_{m,\Omega}$, $m = 0, 1$ is an optimal one [5,6]. From (7) and applying

$$\|v - \pi_h v\|_{m,\Omega} \le \|v - \Pi_h v\|_{m,\Omega} + \|\Pi_h v - \pi_h v\|_{m,\Omega},$$

we prove the estimate (5).

As a consequence of this theorem we give the following proposition:

Proposition 1. *The finite element space $V_h \subset V$ satisfies the following approximation property:*

$$\inf_{v_h \in V_h} \{\|v - v_h\|_{0,\Omega} + h|v - v_h|_{1,\Omega}\} \le C h^3 \|v\|_{3,\Omega},$$
$$\|v - \mathcal{R}_h v\|_{1,\Omega} \le C h^2 \|v\|_{3,\Omega}, \quad \forall v \in V \cap H^3(\Omega), \quad (8)$$

where $\mathcal{R}_h : V \to V_h$ is the elliptic projector defined by

$$a(u - \mathcal{R}_h u, v_h) = 0, \quad \forall u \in V, \ v_h \in V_h.$$

Let us define finite element approximation of the eigenvalue problem (4): Find $(\lambda_h, u_h) \in \mathbf{R} \times V_h$ such that

$$a(u_h, v_h) = \lambda_h (u_h, v_h) \quad \forall v_h \in V_h. \tag{9}$$

The estimate (8) enables us to adapt the theory of the FE error analysis [5]. Using quadratic finite elements to solve (9) and presented finite elements, we get optimal order error estimate. If (λ, u) is an exact eigenpair of (4) and (λ_h, u_h) is the corresponding approximate solution of (9), then

$$\|u - u_h\|_{1,\Omega} \leq Ch^2 \|u\|_{3,\Omega},$$

$$|\lambda - \lambda_h| \leq Ch^4 \|u\|_{3,\Omega}^2,$$

where $C = C(\Omega)$ is independent of the mesh parameters.

Remark 3. Our approach has another advantage. It could be adapted to so-called patch-recovery postprocessing technique [7] in order to obtain considerable more accurate FE solutions.

5 Numerical Results

To illustrate our theoretical results we shall refer to the example on related two-dimensional eigenvalue problem. Let Ω be two-component domain with corresponding overlapping subdomains

$$\Omega_1 = \{(x_1, x_2) : -5/2 \leq x_1 \leq 1/2, -1/2 \leq x_2 \leq 1/2\},$$

$$\Omega_2 = \{(x_1, x_2) : -1/2 \leq x_1 \leq 5/2, -1/2 \leq x_2 \leq 1/2\}$$

as it can be seen in Figure 3. Let also $\Gamma_1 = \{(-5/2, x_2) : -1/2 \leq x_2 \leq 1/2\}$, $\Gamma_2 = \{(5/2, x_2) : -1/2 \leq x_2 \leq 1/2\}$.

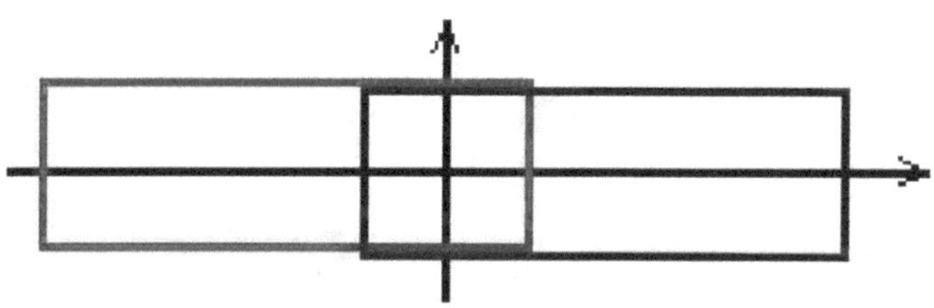

Fig. 3. The domain Ω, consisting of two overlapping parts

Consider the following model problem: Find $(\lambda, u_1, u_2) \in \mathbf{R} \times H^2(\Omega_1) \times H^2(\Omega_2)$ which obeys the differential equations

$$-\Delta u_i + (-1)^i \chi_{\Omega_1 \cap \Omega_2} K = \lambda u_i \text{ in } \Omega_i, \; i = 1, 2,$$

Table 1. The eigenvalues computed by the quadratic mesh

λ_h / N	6	24	96
$\lambda_{1,h}$	6.854597357	6.853936503	6.853127610
$\lambda_{2,h}$	44.75071381	44.59346577	44.563871082
$\lambda_{3,h}$	62.14904248	61.71661903	60.90145381
$\lambda_{4,h}$	206.8546291	197.1437513	186.9372563

and the following boundary conditions

$$u_i = 0 \text{ on } \Gamma_i, \quad \frac{\partial u_i}{\partial \nu} = 0 \text{ on } \partial\Omega_i \setminus \Gamma_i, \; i = 1, 2,$$

as well as the following coupling condition

$$\int_{\Omega_1 \cap \Omega_2} [u_1(x) - u_2(x)] \, dx = 0.$$

The numerical results for the first four eigenvalues computed by means of the the presented approach are given in Table 1. For both subdomains Ω_1 and Ω_2 are used N finite elements such as in case (b).

The exact eigenvalues for this problem are not known. The reason for this choice is to obtain veritable realization of our method. One-dimensional numerical experiments concerning a problem whose exact solution are known can be seen, e.q. in [1].

Acknowledgement. This work is supported by the Bulgarian Ministry of Science under grant VU-MI 202/2006.

References

1. Andreev, A.B., Racheva, M.R.: Optimal order finite element method for coupled eigenvalue problem on overlappingdomains. In: Dimov, I.T., Lirkov, I., Margenov, S., Zlatev, Z. (eds.) NMA 2002. LNCS, vol. 2542, pp. 637–644. Springer, Heidelberg (2003)
2. De Shepper, H.: Finite element analysis of a coupling eigenvalue problem on overlapping domains. J. Comput. Appl. Math. 132, 141–153 (2001)
3. Huang, Y.Q., Xu, J.C.: A conforming finite element method for overlapping and nonmatching grids. Math. Comp. 72(243), 1057–1066 (2003)
4. Cai, X.C., Dryja, M., Sarkis, M.: Overlapping nonmatchinggrids mortar element methods for elliptic problems. SIAM J. Numer. Anal. 36(2), 581–606 (1999)
5. Raviart, P.A., Thomas, J.M.: Introduction a l'Analyse Numerique des Equations aux Derivees Partielles. Masson Paris (1983)
6. Ciarlet, P.G.: The Finite Element Method for Elliptic Problems. North-Holland, Amsterdam (1978)
7. Zienkiewich, O.C., Zhu, J.Z.: The superconvergence patch-recovery (SPR) and adaptive finite element refinement. Comp. Methods Appl. Mech. Eng. 101, 207–224 (1992)

New Approach of FEM for Eigenvalue Problems with Non-local Transition Conditions

A.B. Andreev[1] and M.R. Racheva[2]

[1] Department of Informatics, Technical University of Gabrovo
5300 Gabrovo, Bulgaria
[2] Department of Mathematics, Technical University of Gabrovo
5300 Gabrovo, Bulgaria

Abstract. This paper is considered with the finite element method (FEM) for second order eigenvalue problems on a bounded multi-component domain in the plane. Non-local transition conditions on the interfaces between any two adjacent subdomains are imposed. A new finite element approach is proposed based on much more comprehensible theoretical proofs obtained under lower regularity requirements. The utility of this strategy when superconvergent postprocessing procedure is used as well as the numerical implementation are discussed. Finally, some numerical results are given.

1 Introduction and Problem Setting

Let $\Omega \subset \mathbf{R}^2$ be a bounded convex polygonal domain with boundary $\partial\Omega = \overline{\Gamma^{(1)} \cup \Gamma^{(2)}}$. Here $\Gamma^{(1)}$ and $\Gamma^{(2)}$ are disjoint parts of $\partial\Omega$, each consisting of an integer number of sides of Ω Let Ω be divided in M nonoverlapping open convex polygonal subdomains Ω_i, $i = 1, \ldots, M$ (see Fig. 1). Then, for $j = 1, 2$ and $i, k = 1, \ldots, M$ the following notations are introduced:

$$\Gamma_i^{(j)} = \Gamma^{(j)} \cap \partial\Omega_i, \text{ if } \operatorname{meas}(\Gamma^{(j)} \cap \partial\Omega_i) > 0;$$

$$\Gamma_{i,k} = \partial\Omega_i \cap \partial\Omega_k, \text{ if } \operatorname{meas}(\partial\Omega_i \cap \partial\Omega_k) > 0.$$

The model problem is: Find $\lambda \in \mathbf{R}$ and M functions u_i, $i = 1, \ldots, M$, which obey the differential equations

$$-\sum_{l,m=1}^{2} \frac{\partial}{\partial x_l}\left(a_{lm}^{(i)} \frac{\partial u_i}{\partial x_m}\right) + a_0^{(i)} u_i = \lambda u_i, \quad \text{in } \Omega_i,\ i = 1, \ldots, M, \tag{1}$$

and the classical Robin and Dirichlet boundary conditions

$$\frac{\partial u_i}{\partial \nu^{(i)}} + \sigma^{(i)} u_i = 0 \quad \text{on } \Gamma_i^{(1)}, \tag{2}$$

$$u_i = 0 \quad \text{on } \Gamma_i^{(2)},\ i = 1, \ldots, M, \tag{3}$$

S. Margenov, L.G. Vulkov, and J. Waśniewski (Eds.): NAA 2008, LNCS 5434, pp. 159–167, 2009.

as well as the following nonlocal Dirichlet transition conditions

$$\int_{\Gamma_{i,k}} [u_i(x) - u_k(x)]\, ds = 0, \quad i,k \in \{1,2,\dots,M\} \tag{4}$$

completed with

$$\frac{\partial u_i}{\partial \nu^{(i)}} = -\frac{\partial u_k}{\partial \nu^{(k)}} \quad \text{is constant on } \Gamma_{i,k}. \tag{5}$$

Here, $\partial u_i/\partial \nu^{(i)}$ stands for the conormal derivative of u_i, defined by

$$\frac{\partial u_i}{\partial \nu^{(i)}} \equiv \sum_{l,m=1}^{2} a_{lm}^{(i)} \frac{\partial u_i}{\partial x_m} \nu_l^{(i)},$$

where $\nu_l^{(i)}$ denotes the $l-$th component of the outward unit normal vector.

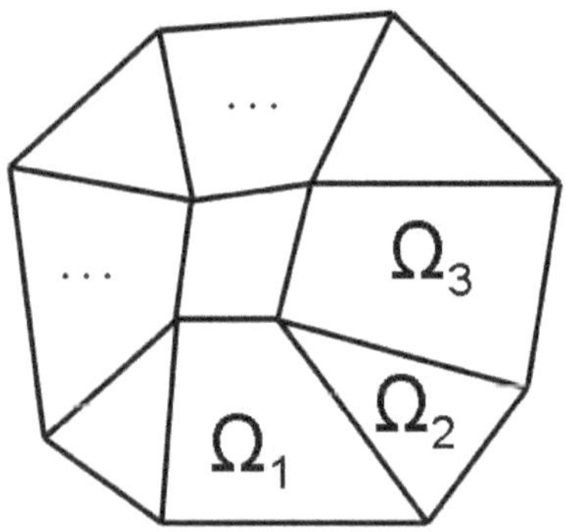

Fig. 1. Multi-component (composite) polygonal domain

Remark 1. The constant value of the conormal derivative on $\Gamma_{i,k}$ in (5) is not given a priori and must be determined as part of the problem.

For $i = 1,\dots,M$, the coefficient functions of the equations (1), (3) obey the usual regularity, symmetry and ellipticity conditions:

$$a_{lm}^{(i)} \in L_\infty(\Omega_i),\ a_{12}^{(i)} = a_{21}^{(i)} \text{ a.e. in } \Omega_i;\ \ a_0^{(i)} \in L_\infty(\Omega_i),\ a_0^{(i)} \geq 0 \text{ a.e. in } \Omega_i;$$

$$\sigma^{(i)}(x) \in L_\infty(\Gamma_i^{(1)}),\ \sigma_{(i)} > 0 \text{ a.e. in } \Gamma_i^{(1)};$$

$$\exists\, \alpha > 0 : \sum_{l,m=1}^{2} a_{lm}^{(i)} \xi_l \xi_m \geq \alpha(\xi_1^2 + \xi_2^2),\ \forall(\xi_1,\xi_2) \in \mathbf{R}^2 \text{ a.e. in } \Omega_i.$$

The problem (1)-(5) is considered and studied by De Schepper and Van Keer [1,2]. They recast this problem into the framework of abstract variational eigenvalue problems in Hilbert spaces. The error analysis in [1,2] is based on a modification of the Lagrange interpolant (so-called imperfect interpolant).

The aim of this paper is motivated by some results. First, due to the geometry of the multi-component domain Ω and to the nonlocal transition conditions on the interfaces between any two adjacent subdomains, the use of standard finite elements leads to involved considerations in theoretical aspects. Also, the numerical implementation is much more complicated than for this concerning 1-component domains.

Here, we propose a method which presents more naturally the character of the problem. On one hand, this approach expands the possibilities for solving problems with transition conditions. On the other hand, the finite elements by means of integral degrees of freedom (see [3]) which are used herein could be used to construct superconvergent patch-recovery technique [4,5]. This procedure can be obtained simultaneously with the finite element approximate solution.

2 Variational Eigenvalue Problem

Let $H^m(\Omega_i)$ be the usual $m-$th order Sobolev space on Ω_i (see [6]) with norm $\|\cdot\|_{m,\Omega_i}$, $i=1,\ldots,M$.

Consider the spaces $V_i = \left\{v_i \in H^1(\Omega_i):\ v_i = 0 \text{ on } \Gamma_i^{(2)}\right\}$, $i=1,\ldots,M$ and $\widetilde{V} = V_1 \times V_2 \times \ldots \times V_M$.

The space which incorporates the transition condition (4) is defined by

$$V = \left\{ v \in \widetilde{V}:\ \int_{\Gamma_{i,k}} [v_i(x) - v_k(x)]\, ds = 0,\quad i,k \in \{1,2,\ldots,M\} \right\}.$$

It is clear that V is a closed subspace of $\widetilde{V}$.

By means of vector space V as the space of trial and test functions, the eigenvalue problem (1)-(5) can be reformulated in variational form as follows (see [2]): Find $(\lambda, u) \in \mathbf{R} \times V$ such that

$$a(u,v) = \lambda(u,v),\quad \forall v \in V, \tag{6}$$

where

$$a(u,v) = \sum_{i=1}^{M} \left[\int_{\Omega_i} \left(\sum_{l,m=1}^{2} a_{lm}^{(i)} \frac{\partial u_i}{\partial x_l} \frac{\partial v_i}{\partial x_m} + a_0^{(i)} u_i v_i \right) dx + \int_{\Gamma_i^{(1)}} \sigma^{(i)} u_i v_i\, ds \right]$$

and the inner product in the Lebesgue space $H \equiv L_2(\Omega_1) \times \ldots \times L_2(\Omega_M)$ is

$$(u,v) = \sum_{i=1}^{M} \int_{\Omega_i} u_i v_i\, dx.$$

The formal equivalence of the variational problem (6) to the problem (1)-(5) is proved in [1] (see Theorem 2.1).

Clearly, $a(\cdot,\cdot)$ is bounded, symmetric and strongly coercive on $\widetilde{V} \times \widetilde{V}$, while V is densily and compactly embedded in H. The next theorem proves the existence of the exact eigenpairs of (6).

Theorem 1. *([1], Theorem 2.1) The problem (6) has an infinite number of eigenvalues λ_l, all being strictly positive, having finite multiplicity and showing no finite accumulation point. We arrange them as $0 < \lambda_1 \leq \lambda_2 \leq \ldots \to +\infty$. The corresponding eigenfunctions u_l can be chosen to be orthonormal in $L_2(\Omega)$, the sequence $\{u_l/\sqrt{\lambda_l}\}$ then being orthonormal w.r.t. $a(\cdot,\cdot)$. They constitute a Hilbert basis for V as well as for $L_2(\Omega)$.*

3 Finite Element Approximation

Consider families $\tau_{h_i}^{(i)}$ of regular finite element partitions of Ω_i, $i = 1,\ldots,M$ which fulfil standard assumptions (see [6], Chapter 3). Herein h_i, $i = 1,\ldots,M$ are mesh parameters and $h = \max_i h_i$. It is important to emphasize that the nodes of any adjacent elements coincide on its element interface.

With triangulation $\tau_{h_i}^{(i)}$ and a fixed natural number n we associate the following functional spaces:

$$X_{h_i}^{(i)} = \left\{v_i \in C(\overline{\Omega}_i) : v_{i|_{T^{(i)}}} \in P(T^{(i)})\ \forall T^{(i)} \in \tau_{h_i}^{(i)}\right\},\ i = 1,\ldots,M,$$

$$X_{h_i,0}^{(i)} = \left\{v_i \in X_{h_i}^{(i)} : v_i|_{\Gamma_i^{(2)}} = 0\right\},\ i = 1,\ldots,M,$$

where $P(T^{(i)})$ is either $Q_n(T^{(i)})$, or $P_n(T^{(i)})$ in the case of rectangular or triangular elements, respectively. Here $P_n(T^{(i)})$ stands for the set of polynomials of degree $\leq n$ and $Q_n(T^{(i)})$ - for the set of polynomials of degree $\leq n$ in each variable. We introduce the product space: $X_{h,0} = X_{h_1,0}^{(1)} \times X_{h_2,0}^{(2)} \times \ldots \times X_{h_M,0}^{(M)}$.

Then the FE space V_h, defined by

$$V_h = \left\{v \in X_{h,0} : \int_{\Gamma_{i,k}} [u_i(x) - u_k(x)]\, ds = 0,\ i,k \in \{1,2,\ldots,M\}\right\}$$

is a finite-dimensional subspace of V.

For $v \in V \cap H^2(\Omega) \subset C(\overline{\Omega})$, if $\Pi_h v \in X_{h,0}$ denotes the piecewise Lagrange interpolant of v on the global mesh, then $\Pi_h v \notin V_h$. Supposing the problem will be considered by means of usual finite elements, this makes it necessary to construct an imperfect interpolant $\widetilde{\Pi}_h v$ by suitably modifying the value in a selected number of nodes per subdomains Ω_i in such a way that $\widetilde{\Pi}_h v \in V_h$ [1,2].

In order to avoid construction of an imperfect interpolant we introduce suitable modified degrees of freedom and corresponding interpolation operator (cf. [3,4]). Let the vertices and the edges of any triangular element $T^{(i)}$ be noted by $a_k^{(i)}$ and $l_k^{(i)}$, $k = 1,2,3$, $i = 1,\ldots,M$, respectively. For this element $T^{(i)}$ we choose its degrees of freedom in such a way that every polynomial $p(x) \in P_n(T^{(i)})$ is determined by: the values at $a_k^{(i)}$ and the integral values $\int_{l_k} p(s)\, ds$. (see Fig.2). The "vertices-edges" conditions as degrees of freedom for any rectangular element are: the values at $a_k^{(i)}$; the integral values $\int_{l_k} p(s)\, ds$, $k = 1,2,3,4$, $i = 1,\ldots,M$ where $p(x) \in Q_n(T^{(i)})$ and it is not obligatory to add either the value at the element center or the integral value $\int_{T^{(i)}} p(x)\, dx$ (Fig. 2).

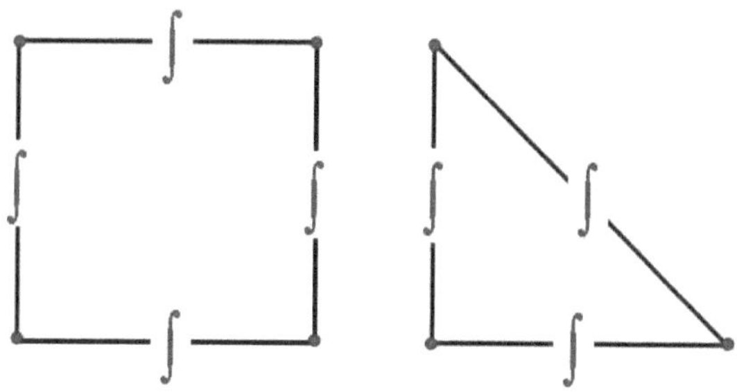

Fig. 2. Rectangular and triangular element by means of integral degrees of freedom

Remark 2. We restrict ourselves to the case of second degree polynomials, but the proposed method could be generalized to higher degree. The linear case (nonconforming elements) concerning our method should be discussed separately.

Next, using the introduced elements, by analogy with the usual Lagrange quadratic interpolation operator $\Pi_h : C(\Omega_1) \times \ldots \times C(\Omega_M) \to X_{h,0}$, $\Pi_h = (\Pi_{h_1}, \ldots, \Pi_{h_M})$, we define the interpolation operator $\pi_h : C(\Omega_1) \times \ldots \times C(\Omega_M) \to X_{h,0}$, where $\pi_h = (\pi_{h_1}, \ldots, \pi_{h_M})$.

In view of non-local transition conditions it is obvious that in general if $v \in V$ then $\Pi_h v \notin V_h$. At the same time, due to the appropriate chosen degrees of freedom on the edges of any finite element, it is easy to see that $\pi_h v \in V_h$ will be satisfied.

4 Main Result

The considerations here are devoted to triangular finite elements. The case of quadrilateral FEM could be investigated in a similar way (cf. [3]). First, we estimate the difference between both interpolants Π_h and π_h.

Theorem 2. *Let the function $v = (v_1, v_2, \ldots, v_M)$ belong to $V \cap H^3(\Omega)$, $\overline{\Omega} = \overline{\Omega}_1 \cup \overline{\Omega}_2 \cup \ldots \cup \overline{\Omega}_M$. Then there exists a constant $C = C(\Omega) > 0$, independent of h, such that*

$$\|v - \pi_h v\|_{m,\Omega} \leq Ch^{3-m} \|v\|_{3,\Omega}, \quad m = 0, 1. \tag{7}$$

Proof. First, we shall estimate $\Pi_{h_i} v_i - \pi_{h_i} v_i$ on each finite element $T_j^{(i)}$, $j = 1, \cdots, k_i$, $i = 1, \ldots, M$.

Let us introduce reference element T: $\{(t_1, t_2) \in \mathbf{R}^2 : \ t_1 \geq 0, \ t_2 \geq 0, \ t_1 + t_2 \leq 1\}$. We denote the vertices of T by a_k and its side midpoints by a_{3+k}, $k = 1, 2, 3$. The sides of this unit triangle are denoted by l_k where any l_k is opposite to the corresponding vertex a_k, $k = 1, 2, 3$.

Any triangle $T_j^{(i)}$ is transformed to T by means of linear functions, i.e.

$$t_1 = L_{1,j}^{(i)}(x_1, x_2); \ t_2 = L_{2,j}^{(i)}(x_1, x_2), \ j = 1, \ldots, k_i; \ i = 1, \ldots, M.$$

Obviously, $|L_{s,j}^{(i)}(x_1,x_2)| = \mathcal{O}(1/h_j^{(i)})$, where $h_j^{(i)}$ is diameter of $T_j^{(i)}$. The basis functions of the Lagrange interpolant Π_h, related to T are:

$$\psi_1(t_1,t_2) = 2t_1^2 + 2t_2^2 + 4t_1t_2 - 3t_1 - 3t_2 + 1;$$

$$\psi_2(t_1,t_2) = 2t_1^2 - t_1; \quad \psi_3(t_1,t_2) = 2t_2^2 - t_2; \quad \psi_4(t_1,t_2) = 4t_1t_2;$$

$$\psi_5(t_1,t_2) = -4t_2^2 - 4t_1t_2 + 4t_2; \quad \psi_6(t_1,t_2) = -4t_1^2 - 4t_1t_2 + 4t_1.$$

Accordingly, the basis functions corresponding to the interpolant π_h are:

$$\varphi_1(t_1,t_2) = 3t_1^2 + 3t_2^2 + 6t_1t_2 - 4t_1 - 4t_2 + 1;$$

$$\varphi_2(t_1,t_2) = 3t_1^2 - 2t_1; \quad \varphi_3(t_1,t_2) = 3t_2^2 - 2t_2; \quad \varphi_4(t_1,t_2) = 6t_1t_2;$$

$$\varphi_5 = -6t_2^2 - 6t_1t_2 + 6t_2; \quad \varphi_6(t_1,t_2) = -6t_1^2 - 6t_1t_2 + 6t_1.$$

Then for $i = 1,\ldots,M$ we obtain:

$$(\Pi_{h_i}v_i - \pi_{h_i}v_i)_{|T} = \sum_{k=1}^{3} v_i(a_k)\left[\psi_k(t_1,t_2) - \varphi_k(t_1,t_2)\right]$$

$$+\sum_{k=1}^{3} v_i(a_{3+k})\psi_{3+k}(t_1,t_2) - \sum_{k=1}^{3}\int_{l_k} v_i(s)\,ds.\varphi_{3+k}(t_1,t_2)$$

$$= \varphi_4(t_1,t_2)\left[\frac{1}{6}v_i(a_2) + \frac{4}{6}v_i(a_4) + \frac{1}{6}v_i(a_3) - \int_{l_1} v_i(s)\,ds\right]$$

$$+\varphi_5(t_1,t_2)\left[\frac{1}{6}v_i(a_1) + \frac{4}{6}v_i(a_5) + \frac{1}{6}v_i(a_3) - \int_{l_2} v_i(s)\,ds\right]$$

$$+\varphi_6(t_1,t_2)\left[\frac{1}{6}v_i(a_1) + \frac{4}{6}v_i(a_6) + \frac{1}{6}v_i(a_2) - \int_{l_3} v_i(s)\,ds\right].$$

The expression in the brackets could be presented by using error functionals of quadrature formulas [6]. Namely,

$$(\Pi_{h_i}v_i - \pi_{h_i}v_i)_{|T} = \sum_{k=1}^{3} \varphi_{3+k}(t_1,t_2)E_T^k(v_i), \tag{8}$$

where error functionals E_T^k, $k = 1,2,3$ represent the error of the Simpson quadrature formula on the sides of reference element T. It is easy to verify that $E_T^k(v_i) = 0$, $k = 1,2,3$ for all $v_i \in P_2(T)$. On the other hand $|\varphi_{3+k}(t_1,t_2)| \leq 3/2$, $(t_1,t_2) \in T$.

From (8), using standard argument of the Bramble-Hilbert lemma [3,6], for any element $T_j^{(i)}$, $j = 1,\ldots,k_i$; $i = 1,\ldots,M$ we obtain

$$|\Pi_{h_i}v_i - \pi_{h_i}v_i|_{|_{T_j^{(i)}}} \leq Ch_i^3|v_i|_{3,T_j^{(i)}},$$

where $|\cdot|_{3,T_j^{(i)}}$ is third order Sobolev seminorm.

From this inequality it follows the L_2-norm error estimate:

$$\|\Pi_{h_i}v_i - \pi_{h_i}v_i\|_{0,\Omega} = \left(\sum_{i=1}^{M} \sum_{T_j^{(i)} \in \tau_{h_i}^{(i)}} \int_{T_j^{(i)}} |\Pi_{h_i}v_i - \pi_{h_i}v_i|^2 \, dx \right)^{1/2} \leq Ch^3 \|v\|_{3,\Omega} \tag{9}$$

Applying explicit calculations we get

$$\left| \frac{\partial}{\partial x_s} (\Pi_{h_i}v_i - \pi_{h_i}v_i) \right|_{T_j^{(i)}} \leq Ch_i^2 |v_i|_{3,T_j^{(i)}}, \; s = 1, 2.$$

Consequently, the H^1-norm error estimate is $\|\Pi_h v - \pi_h v\|_{1,\Omega} \leq Ch^2 \|v\|_{3,\Omega}$. This inequality and (9) give

$$\|\Pi_h v - \pi_h v\|_{m,\Omega} \leq Ch^{3-m} \|v\|_{3,\Omega}, \; m = 0, 1. \tag{10}$$

The Lagrange interpolant has one and the same order of convergence (see [6]). Then the estimate (7) follows from (10), taking into account that

$$\|v - \pi_h v\|_{m,\Omega} \leq \|v - \Pi_h v\|_{m,\Omega} + \|\Pi_h v - \pi_h v\|_{m,\Omega}, \; m = 0, 1.$$

Let us define the elliptic projector $R_h : \; V \to V_h$ such that

$$a(u - R_h u, v_h) = 0, \quad \forall u \in V, \; v_h \in V_h.$$

From Theorem 2 it follows that the finite element space $V_h \subset V$ satisfies the approximation property:

$$\begin{aligned} &\inf_{v_h \in V_h} \{\|v - v_h\|_{0,\Omega} + h|v - v_h|_{1,\Omega}\} \leq Ch^3 \|v\|_{3,\Omega}, \\ &\|v - R_h v\|_{1,\Omega} \leq Ch^2 \|v\|_{3,\Omega}, \quad \forall v \in V \cap H^3(\Omega). \end{aligned} \tag{11}$$

We define finite element approximation of the eigenvalue problem (6): Find $(\lambda_h, u_h) \in \mathbf{R} \times V_h$ such that

$$a(u_h, v_h) = \lambda_h (u_h, v_h), \quad \forall v_h \in V_h. \tag{12}$$

The theory of the FE error analysis could be adapted to this problem using the estimate (11) (see [1,2,3]). Let (λ, u) be an exact eigenpair of (6). If (λ_h, u_h) is the corresponding approximate solution of (12), then

$$\|u - u_h\|_{1,\Omega} \leq Ch^2 \|u\|_{3,\Omega},$$

$$|\lambda - \lambda_h| \leq Ch^4 \|u\|_{3,\Omega}.$$

5 Numerical Results

For purpose of illustration our theoretical results we consider a two-dimensional eigenvalue problem for which the domain Ω is the unit square and its two components are two identical isosceles triangles (Fig. 3).

The model problem is:

$$-\Delta u_i = \lambda u_i \ \text{ in } \Omega_i, \ i = 1, 2,$$

$$u_i = 0 \ \text{ on } \Gamma_i, \ i = 1, 2, \qquad \int_{\Gamma_{1,2}} [u_1(x) - u_2(x)] \, ds = 0,$$

$$\frac{\partial u_1}{\partial \nu^{(1)}} = -\frac{\partial u_2}{\partial \nu^{(2)}} \ \text{ is constant on } \Gamma_{1,2}.$$

The numerical results for the first four eigenvalues obtained by means of the presented approach as well as the corresponding exact values are given in Table 1. For both subdomains Ω_1 and Ω_2 are used N triangular finite elements like this shown in Fig.2. Constructing mass and stiffness matrices, one should take into account the nonlocal transition conditions, which leads to transformation of rows

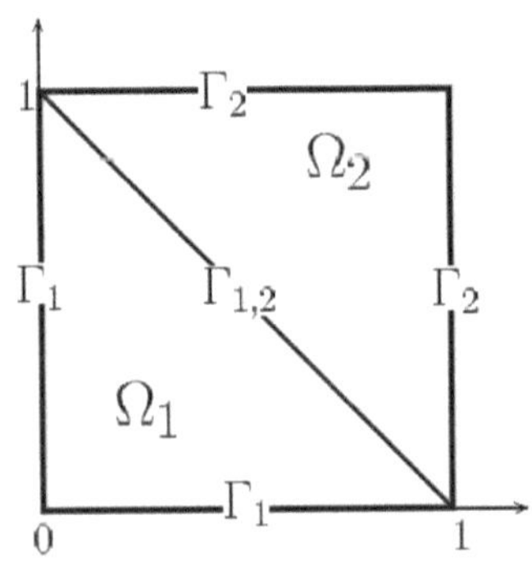

Fig. 3. The domain Ω and its subdomains

Table 1. The eigenvalues computed by the quadratic mesh

λ_h / N	8	32	128	exact
$\lambda_{1,h}$	21.082	19.814	19.739	19.739
$\lambda_{2,h}$	51.692	41.617	40.382	40.381
$\lambda_{3,h}$	70.565	49.454	49.350	49.348
$\lambda_{4,h}$	71.217	50.612	49.356	49.348

and columns of these matrices [1,2]. It is worth to observe that if proposed finite elements are used, much less rows and columns of the matrices become involved into the transformations.

Acknowledgement. This work is supported by the Bulgarian Ministry of Science under grant VU-MI 202/2006.

References

1. De Shepper, H., Van Keer, R.: A finite element method for elliptic eigenvalue problems in a multi-component domain in 2D with non-local Dirichlet transition conditions. J. Comput. Appl. Math. 111, 253–265 (1999)
2. De Shepper, H., Van Keer, R.: On a variational approximation method for 2nd order eigenvalue problems in a multi-component domain with nonlocal Dirichlet transition conditions. Numer. Func. Anal. Optim. 19(9&10), 971–994 (1998)
3. Andreev, A.B., Racheva, M.R.: Optimal order FEM for a coupled eigenvalue problem on 2D overlapping domains. In: Margenov, S., Vulkov, L.G., Waśniewski, J. (eds.) NAA 2008. LNCS. vol. 5434, Springer, Heidelberg (2008)
4. Andreev, A.B., Racheva, M.R.: Superconvergence of the interpolated quadratic finite elements on triangular meshes. Math. Balkanica, New Series 19, 3-4, 385–404 (2005)
5. Lin, Q., Yan, N., Zhou, A.: A rectangular test for interpolated finite elements. In: Proceedings of Systems Science & Systems Engineering, pp. 217–229. Culture Publish Co. (1991)
6. Ciarlet, P.G.: The Finite Element Method for Elliptic Problems. North-Holland, Amsterdam (1978)

Minimal Simplex for IFS Fractal Sets

Elena Babače[1] and Ljubiša Kocić[2]

[1] Ss Cyril and Methodius University, Faculty of Electrical Engineering and Information Technologies, P.O. Box 574, Skopje, Macedonia
elena.babace@feit.ukim.edu.mk

[2] University of Niš, Faculty of Electronic Engineering, P.O. Box 73, 18 000 Niš, Serbia
kocic@elfak.ni.ac.yu

Abstract. Fractal sets manipulation and modeling is a difficult task due to their complexity and unpredictability. One of the basic problems is to determine bounds of a fractal set given by some recursive definitions, for example by an Iterated Function System (IFS). Here we propose a method of bounding an IFS-generated fractal set by a minimal simplex that is affinely identical to the standard simplex. First, it will be proved that for a given IFS attractor, such simplex exists and it is unique. Such simplex is then used for definition of an Affine invariant Iterated Function System (AIFS) that then can be used for affine transformation of a given fractal set and for its modeling.

1 Introduction

One of the most suitable ways to define and generate fractal sets is by using (hyperbolic) Iterated Function Systems (IFSs). We will briefly introduce the basic terms and theorems concerning the IFSs.

The (hyperbolic) Iterated Function System (IFS) is the system

$$S = \{\mathbb{R}^m; w_1, w_2, \ldots, w_N\}, \quad N \geq 2$$

where w_i, $i = 1, 2, \ldots N$, are contractive mappings with corresponding contractive factors $s_i < 1$, $i = 1, 2, \ldots, N$, of the Euclidean metric space $(\mathbb{R}^m, d_E)$ $(m > 1)$ into itself.

Typical choice of w_i is affine mapping i.e.

$$w_i(\mathbf{x}) = A\mathbf{x} + \mathbf{b}, \quad \mathbf{x} \in \mathbb{R}^m, \quad i = 1, 2, \ldots, N,$$

where A is $m \times m$ real matrix and $\mathbf{b}$ is m-dimensional real vector.

$\mathcal{H}(\mathbb{R}^m)$ denotes the space whose points are the nonempty compact subsets of $\mathbb{R}^m$. h is Hausdorff metric induced by d_E, i.e.

$$h(A, B) = \max\left\{\max_{a\in A}\min_{b\in B} d_E(a, b),\ \max_{b\in B}\min_{a\in A} d_E(b, a)\right\},\ A, B \in \mathcal{H}(\mathbb{R}^m).$$

With such defined metrics, $(\mathcal{H}(\mathbb{R}^m), h)$ is a complete metric space, ([1]). Attached to S is the Hutchinson operator, $W_S : \mathcal{H}(\mathbb{R}^m) \to \mathcal{H}(\mathbb{R}^m)$ defined by

S. Margenov, L.G. Vulkov, and J. Waśniewski (Eds.): NAA 2008, LNCS 5434, pp. 168–175, 2009.

$W_S(B) = \bigcup_{i=1}^{N} w_i(B)$. If the IFS S is a contraction, i.e. if $s = \max_{i=1,2,\dots,N}\{|s_i|\} < 1$, then as a consequence, W_S is also a contraction of the complete metric space $(\mathcal{H}(\mathbb{R}^m), h)$. According to Banach's theorem, W_S has a unique fixed point A, $A = W_S(A)$, i.e.

$$A = \lim_{n\to\infty} W_S^{\circ n}(B), \quad \text{for any} \quad B \in \mathcal{H}(\mathbb{R}^m). \tag{1}$$

The fixed point A is called the *attractor* of the IFS, ([4]). We will call $W_S^{\circ n}(B)$ *n-th preattractor* of the IFS S.

Our interest is directed to attractors that have fractal structure and are generated by affine IFSs. Some algorithms for generating attractors by using IFS are described in [3], [5]. In spite of their simplicity, IFSs have essential disadvantages, mainly from the interactive modeling viewpoint. Namely, it is hard to predict the shape and the position of the attractor only knowing the IFS, and also, the attractor can be changed only if the IFS itself is changed by redefinition of all or some of the w_i-s.

Inspired by the subdivision technique used in CAGD for modeling by means of free-form curves, especially by the paper of Berger [2], Kocic and Simoncelli have offered an alternative concept of an affine invariant IFS, trying to trace a way for overcoming the mentioned shortcomings of IFS ([6], [7], [8], [9]).

Affine invariant Iterated Function System (AIFS) is an IFS, rewritten in barycentric coordinates, i.e., $\Omega(\mathbf{T}) = \{\mathbf{T}; S_1, S_2, ..., S_N\}$, where $\mathbf{T}$ is non-degenerate simplex in $\mathbb{R}^{m+1}$, and $S_i, i = 1, 2 \dots, N$, are real square non-singular row-stochastic matrices that define linear mappings $\mathcal{L}_i : \mathbf{x} \mapsto S_i\mathbf{x}$, $\mathbf{x} \in \mathbb{R}^{m+1}$.

The main and the most important property of the AIFS is affine invariance, the property that IFS does not possess. Besides, the AIFS has several other properties important for interactive modeling: continuity property, interpolation property, symmetry property, ability to generate smooth shapes, etc ([9]). These properties are common for many free-form models (Beziér, spline, NURBS), but they have one more essential property - *the convex hull property*. In order to supply fractal modeling schemes with this property, the problem, called *minimal simplex problem* is faced by authors. The minimal simplex problem is solved to the extent of having practical use, both for AIFS and IFS.

The reason of using minimal simplex in building AIFS for a given attractor lies in *stability*. It is straightforward that having a simplex that contains the entire attractor leads to the stable modeling process. This means that finite perturbation of simplex dimensions results in reduced changing of the dimensions of the attractor. More precisely, the AIFS $\Omega(\mathbf{T}) = \{\mathbf{T}; S_1, S_2, ..., S_N\}$ has *stability property* provided that $h(A, A') \le h(\mathbf{T}, \mathbf{T}')$, where $\mathbf{T}'$ is perturbation of $\mathbf{T}$ and A' is the attractor of the AIFS $\Omega(\mathbf{T}') = \{\mathbf{T}'; S_1, S_2, ..., S_N\}$.

It is obvious that any simplex that contains the whole attractor satisfies the mentioned implication. But, it is the most convenient to have minimal simplex by two reasons. First, the minimal simplex is the body that is close to the convex hull, which is in turn very difficult to determine, meaning that it locates the attractor in the optimal way by giving its closest bounds; Second, too large simplex will result in vast reduction of the modeling effect. Namely, for a tiny change of the attractor, major perturbation of the simplex is needed.

2 Minimal Simplex Problem

What does convex hull property in the theory of IFS really mean? It means that the limit set A (1) stays within the convex hull of the starting set (B). It is very convenient to have a simplex as a starting set. Further, it is desirable this simplex to be as close as possible to the minimal one, so that the position of the attractor can be predicted more precisely.

Before stating the problem, we need few notations and definitions. We denote by $\{\mathbf{e}_1, \mathbf{e}_2, \ldots, \mathbf{e}_m\}$ the standard orthonormal basis of $\mathbb{R}^m$, $\mathbf{e}_i = [\delta_{ij}]_{j=1}^m$, $i = 1, 2, \ldots, m$.

Definition 1. An *m-simplex* (or just *simplex*) is the convex hull of a set of $m+1$ affinely independent points (or vectors) in Euclidean space of dimension m or higher (a set of points such that no m-plane contains more than $m+1$ of them). The simplex defined by the points $\mathbf{N}_1, \mathbf{N}_2, \ldots, \mathbf{N}_{m+1}$ will be denoted by $\mathbf{N}_\triangle = \mathbf{N} = \mathrm{conv}\{\mathbf{N}_1, \mathbf{N}_2, \ldots, \mathbf{N}_{m+1}\}$

Definition 2. The simplex $\mathbf{E} = \mathrm{conv}\{\mathbf{e}_1, \mathbf{e}_2, \ldots, \mathbf{e}_{m+1}\}$ is called *canonical simplex.*

Definition 3. The (orthogonal) projection of $\mathbf{E}$ on $\mathbb{R}^m$, $\mathbf{E}_0 = \mathrm{proj}_\perp \mathbf{E}$, is called *standard simplex*, where $\mathrm{proj}_\perp : \mathbb{R}^n \to \mathbb{R}^m$, $n \geq m+1$, is defined by

$$\mathrm{proj}_\perp(x_1, x_2, \ldots, x_m, \ldots, x_n) = (x_1, x_2, \ldots, x_m).$$

In other words, $\mathbf{E}_0 = \mathrm{conv}\{\mathrm{proj}_\perp \mathbf{e}_1, \mathrm{proj}_\perp \mathbf{e}_2, \ldots, \mathrm{proj}_\perp \mathbf{e}_m, \ldots, \mathrm{proj}_\perp \mathbf{e}_n\}$

Definition 4. The *minimal simplex of a compact set* $C \subset \mathbb{R}^m$ is the simplex with the least $m-$volume of all simplices that contain C.

Remark 1. As the compact sets of the metric space are bounded, the minimal simplex of a given compact set always exists.

Now, let us state the problem of minimal simplex.

Minimal Simplex Problem. *Let the IFS* $S = \{\mathbb{R}^m; w_1, w_2, \ldots, w_N\}$, *be given. Find the minimal simplex of the attractor of the IFS* S.

The task of finding minimal simplex for a given IFS is a difficult problem. Anyway, the Minimal Simplex Problem can be simplified by using a subset of the set of simplices, the set of *standard-like* simplices.

Definition 5. A simplex is called *standard-like* if it can be mapped into standard simplex just by translation and non-homogenous scaling.

For the aims of CAGD, the minimal simplex can be replaced by minimal standard-like simplex or even a simplex that approximates the last one from above. This choice is acceptable by the AIFS method since it does not make difference which simplex we are using, as far as it is close enough to the minimal one. Hence, in the following lines a solution of the problem for a minimal standard-like simplex will be suggested.

As it is mentioned, the minimal simplex for a given attractor exists (Remark 1.), but in general it is not unique. On the other hand, the minimal standard-like simplex is unique.

Given the IFS $S = \{\mathbb{R}^m; w_1, w_2, \ldots, w_N\}$ and the associated Hutchinson operator W_S. Let $B = \{\mathbf{x}_0\}$, $\mathbf{x}_0$ be an arbitrary point from $\mathbb{R}^m$. Then, the attractor of S is defined by

$$A = \lim_{n\to\infty} W^{\circ n}(B) = \lim_{n\to\infty} W^{\circ n}(\{\mathbf{x}_0\}).$$

Theorem 1. *Let $\mathbf{S}^{(n)}$ be the minimal standard-like simplex of the preattractor $A^{(n)} = W^{\circ n}(\{\mathbf{x}_0\})$. Then, there exists $\lim_{n\to\infty} \mathbf{S}^{(n)}$ and it is the minimal standard-like simplex of the attractor $A = \lim_{n\to\infty} A^{(n)}$.*

Proof. The proof proceeds in three steps:

(i) Construction of $\mathbf{S}^{(n)}$, the minimal standard-like simplex of the preattractor $A^{(n)}$;

(ii) Showing that the sequence $\{\mathbf{S}^{(n)}\}$ is a Cauchy sequence in $(\mathcal{H}(\mathbb{R}^m), h)$;

(iii) Conclusion: $\lim_{n\to\infty} \mathbf{S}^{(n)}$ is the minimal standard-like simplex of the attractor A.

(i) The n-th preattractor of A, $A^{(n)} = W^{\circ n}(\{\mathbf{x}_0\})$ is a finite set (it contains N^n points)

$$A^{(n)} = \{\mathbf{x}_i^{(n)} = (x_{1,i}^{(n)}; x_{2,i}^{(n)}; \ldots; x_{m,i}^{(n)}) \,|\, i = 1, 2, \ldots, N^n\}. \tag{2}$$

Therefore, for $n \in \mathbb{N}$, there exist $\alpha_1 = \alpha_1(n)$, $\alpha_2 = \alpha_2(n)$, $\ldots, \alpha_m = \alpha_m(n)$, $\beta = \beta(n)$ – all belonging to $\{1, 2, \ldots, N^n\}$, such that:

$$\begin{aligned}
x_{1,\alpha_1}^{(n)} &= \min\{x_{1,i}^{(n)} \,|\, i = 1, 2, \ldots, N^n\},\\
x_{2,\alpha_2}^{(n)} &= \min\{x_{2,i}^{(n)} \,|\, i = 1, 2, \ldots, N^n\},\\
&\ldots\ldots\ldots\ldots\ldots \\
x_{m,\alpha_m}^{(n)} &= \min\{x_{m,i}^{(n)} \,|\, i = 1, 2, \ldots, N^n\},\\
x_{1,\beta}^{(n)} + x_{2,\beta}^{(n)} + \ldots + x_{m,\beta}^{(n)} &= \max\{x_{1,i}^{(n)} + x_{2,i}^{(n)} + \ldots + x_{m,i}^{(n)} \,|\, i = 1, 2, \ldots N^n\},
\end{aligned} \tag{3}$$

where $x_{1,i}^{(n)}, x_{2,i}^{(n)}, \ldots, x_{m,i}^{(n)}$ are the first, second, i.e. m-th coordinate of the i-th point of $A^{(n)}$.

For $n \in \mathbb{N}$, we denote by $\mathbf{S}^{(n)}$ the following set from $\mathbb{R}^m$

$$\mathbf{S}^{(n)} = \Big\{(x_1, x_2, \ldots, x_m) \,|\, x_1 \geq x^{(n)}_{1,\alpha_1};\ x_2 \geq x^{(n)}_{2,\alpha_2};\ \ldots\ ; x_m \geq x^{(n)}_{m,\alpha_m};$$
$$x_1 + x_2 + \ldots + x_m \leq x^{(n)}_{1,\beta} + x^{(n)}_{2,\beta} + \ldots + x^{(n)}_{m,\beta}\Big\}. \tag{4}$$

Knowing that for $A \subset \mathbb{R}^m$ and $\varepsilon > 0$, $A + \varepsilon = \{y \in \mathbb{R}^m : \exists x \in A,\ d(x, y) \leq \varepsilon\}$, we just note that the set $\mathbf{S}^{(n)} + \varepsilon$ will be

$$\mathbf{S}^{(n)} + \varepsilon = \Big\{(x_1, x_2, \ldots, x_m) \,|\, x_1 \geq x^{(n)}_{1,\alpha_1} - \varepsilon;\ x_2 \geq x^{(n)}_{2,\alpha_2} - \varepsilon;\ x_m \geq x^{(n)}_{m,\alpha_m} - \varepsilon;$$
$$x_1 + x_2 + \ldots + x_m \leq x^{(n)}_{1,\beta} + x^{(n)}_{2,\beta} + \ldots + x^{(n)}_{m,\beta} + \varepsilon\Big\}. \tag{5}$$

$\mathbf{S}^{(n)}$ is the minimal standard like simplex of the compact set $A^{(n)}$, as it is constructed such that $A^{(n)} \subseteq \mathbf{S}^{(n)}$ ((3), (4)); and also, $\mathbf{S}^{(n)}$ is minimal for $A^{(n)}$, because every edge of $\mathbf{S}^{(n)}$ contains at least one point from $A^{(n)}$. (for example, for $m = 2$, the points $(x^{(n)}_{1,\alpha_1}; x^{(n)}_{2,\alpha_1}), (x^{(n)}_{1,\alpha_2}; x^{(n)}_{2,\alpha_2})$ and $(x^{(n)}_{1,\beta}; x^{(n)}_{2,\beta})$, are positioned on the edges $x_1 = x^{(n)}_{1,\alpha_1}$, $x_2 = x^{(n)}_{2,\alpha_2}$ and $x_1 + x_2 = x^{(n)}_{1,\beta} + x^{(n)}_{2,\beta}$, respectively.)

(ii) The sequence $\{A^{(n)}\}$ is a Cauchy sequence, because it is convergent in the metric space $(\mathcal{H}(\mathbb{R}^m), h)$. Let $\varepsilon > 0$. Then there exists $n_0 \in \mathbb{N}$ such that $\forall k, p \geq n_0$, $k, p \in \mathbb{N}$,

$$h(A^{(k)}, A^{(p)}) \leq \varepsilon,$$

which is equivalent with ([1])

$$A^{(k)} \subseteq A^{(p)} + \varepsilon \quad \text{and} \quad A^{(p)} \subseteq A^{(k)} + \varepsilon.$$

Taking into account (3), the relation $A^{(k)} \subseteq A^{(p)} + \varepsilon$ actually means that all of the following inequalities are satisfied:

$$x^{(k)}_{1,\alpha_1} \geq x^{(p)}_{1,\alpha_1} - \varepsilon, \quad x^{(k)}_{2,\alpha_2} \geq x^{(p)}_{2,\alpha_2} - \varepsilon, \quad \ldots, \quad x^{(k)}_{m,\alpha_m} \geq x^{(p)}_{m,\alpha_m} - \varepsilon,$$
$$x^{(k)}_{1,\beta} + x^{(k)}_{2,\beta} + \ldots + x^{(k)}_{m,\beta} \leq x^{(p)}_{1,\beta} + x^{(p)}_{2,\beta} + \ldots + x^{(p)}_{m,\beta} + \varepsilon \tag{6}$$

Similar conclusion can be derived from the relation $A^{(p)} \subseteq A^{(k)} + \varepsilon$.

We will show that

$$\mathbf{S}^{(k)} \subseteq \mathbf{S}^{(p)} + \varepsilon \quad \text{and} \quad \mathbf{S}^{(p)} \subseteq \mathbf{S}^{(k)} + \varepsilon$$

take place.

Suppose that $\mathbf{S}^{(k)} \not\subseteq \left(\mathbf{S}^{(p)} + \varepsilon\right)$, i.e. $\exists \mathbf{a} = (a_1, a_2, \ldots, a_m) \in \mathbf{S}^{(k)} \backslash \left(\mathbf{S}^{(p)} + \varepsilon\right)$. Then, because of (4) and (5), at least one of the following $m + 1$ cases occur:

1° $x^{(k)}_{1,\alpha_1} \leq a_1 < x^{(p)}_{1,\alpha_1} - \varepsilon$;

2° $x^{(k)}_{2,\alpha_2} \leq a_2 < x^{(p)}_{2,\alpha_2} - \varepsilon$;

. . . .

m^o $x^{(k)}_{m,\alpha_m} \le a_m < x^{(p)}_{m,\alpha_m} - \varepsilon;$

$m+1^o$ $x^{(k)}_{1,\beta} + x^{(k)}_{2,\beta} + \ldots + x^{(k)}_{m,\beta} \ge a_1 + a_2 + \ldots + a_m >$
$> x^{(p)}_{1,\beta} + x^{(p)}_{2,\beta} + \ldots + x^{(p)}_{m,\beta} + \varepsilon.$

All cases lead to a contradiction with the relations (6), i.e. with the fact that $A^{(k)} \subseteq A^{(p)} + \varepsilon$. Therefore, it has to be $\mathbf{S}^{(k)} \subseteq \mathbf{S}^{(p)} + \varepsilon$. Similarly, it can be shown that $\mathbf{S}^{(p)} \subseteq \mathbf{S}^{(k)} + \varepsilon$. Hence,

$$h(\mathbf{S}^{(p)}, \mathbf{S}^{(k)}) \le \varepsilon,$$

i.e. the sequence $\{\mathbf{S}^{(n)}\}$ is a Cauchy sequence in $(\mathcal{H}(\mathbb{R}^m), h)$, so there exists $\lim\limits_{n\to\infty} \mathbf{S}^{(n)} = \mathbf{S}$.

(iii) From the construction of $\mathbf{S}$ itself it is clear that $\mathbf{S}$ is the unique minimal standard-like simplex of the attractor $A = \lim\limits_{n\to\infty} A^{(n)}$. The uniqueness of the minimal standard-like simplex is not only intuitive conclusion, but it follows from the uniqueness of the limit point $\lim\limits_{n\to\infty} \mathbf{S}^{(n)}$ in the complete metric space $(\mathcal{H}(\mathbb{R}^m), h)$.

Remark 2. Note that the Random Iteration Algorithm is more efficient from the point of view of modeling, especially in comparison with above described Deterministic Algorithm. The reason is the number of points which in every step grows exponentially when using the last one. If the Random Iteration Algorithm is used with enough number of points $(5 \cdot 10^4 - 10^5)$, a standard-like simplex $\mathbf{S}^{(n)}$ can be obtained close enough to the minimal standard-like simplex $\mathbf{S}$. We call this simplex quasi-minimal standard-like simplex; more precisely, $\mathbf{M}'_\triangle$ is *quasi-minimal* if its volume approximates the volume of the minimal simplex $\mathbf{M}_\triangle$, but such that $\mathbf{M}_\triangle \subseteq \mathbf{M}'_\triangle$; and is *quasi-minimal standard-like* if it is standard like and quasi minimal.

For the examples given bellow a Random Iteration Algorithm is used and quasi-minimal standard-like simplex is found.

Example 1. The quasi-minimal standard-like simplex for the famous "seahorse", defined by the IFS

$$w_1: \quad A_1 = \begin{bmatrix} 0.824074 & 0.281482 \\ -0.212346 & 0.864198 \end{bmatrix}, \quad \mathbf{b}_1 = \begin{bmatrix} -1.88229 \\ -0.110607 \end{bmatrix},$$

$$w_2: \quad A_2 = \begin{bmatrix} 0.088272 & 0.520988 \\ -0.463889 & -0.377778 \end{bmatrix}, \quad \mathbf{b}_2 = \begin{bmatrix} 0.78536 \\ 8.095795 \end{bmatrix}.$$

is given on the left on the Figure 1.

Example 2. The IFS for the fractal set known as "galaxy" is given with:

$$w_1: \quad A_1 = \begin{bmatrix} -0.08 & 0.06 \\ 0.00 & -0.28 \end{bmatrix}, \quad \mathbf{b}_1 = \begin{bmatrix} -2.02 \\ -2.40 \end{bmatrix},$$

$$w_2: \quad A_2 = \begin{bmatrix} -0.05 & 0.20 \\ -0.03 & -0.21 \end{bmatrix}, \quad \mathbf{b}_2 = \begin{bmatrix} -1.51 \\ -3.46 \end{bmatrix},$$

$$w_3: \quad A_3 = \begin{bmatrix} -0.05 & 0.23 \\ -0.05 & -0.20 \end{bmatrix}, \quad \mathbf{b}_3 = \begin{bmatrix} 2.04 \\ 2.47 \end{bmatrix},$$

$$w_4: \quad A_4 = \begin{bmatrix} -0.01 & 0.30 \\ -0.05 & -0.04 \end{bmatrix}, \quad \mathbf{b}_4 = \begin{bmatrix} 0.67 \\ 3.18 \end{bmatrix},$$

$$w_5: \quad A_5 = \begin{bmatrix} 0.70 & 0.52 \\ -0.53 & 0.69 \end{bmatrix}, \quad \mathbf{b}_5 = \begin{bmatrix} 0.54 \\ 0.14 \end{bmatrix},$$

and its quasi-minimal standard-like simplex is on the right on the figure 1.

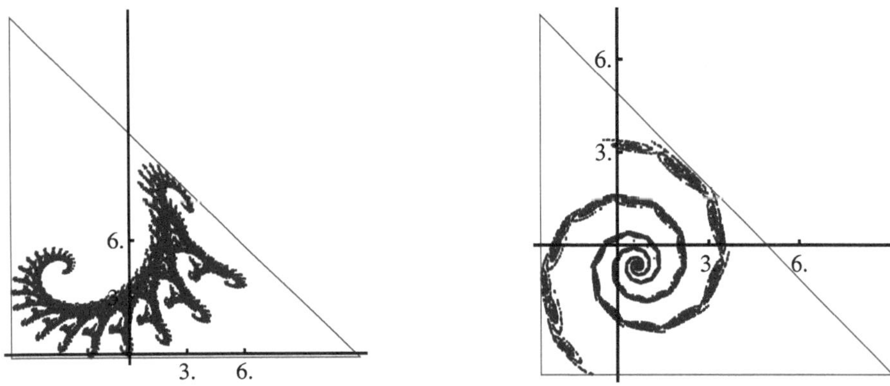

Fig. 1. The quasi-minimal standard-like simplices for the "seahorse" and "galaxy"

This result for the minimal simplex problem can be easily converted in the theory of AIFS, thus obtaining an opportunity to affinly interchange the attractor ([10]) by means of a minimal standard-like simplex.

3 Conclusion

A minimal standard-like simplex that bounds m-dimensional IFS attractor is constructed, thus endowing both IFS and AIFS with convex hull property. As a consequence, AIFS defined on such minimal simplex, possesses stability property in the sense that perturbation of simplex causes smaller in amount perturbation of the attractor. Examples are given for 2-dimensional attractors. This result, adapted for AIFS schemes, offers stable manipulation and control over complex and "wild" fractal sets.

References

1. Barnsley, M.F.: Fractals everywhere. Academic Press, London (1988)
2. Berger, M.A.: Random affine iterated function systems: curve generation and wavelets. SIAM Review 34, 361–385 (1992)
3. Dubuc, S., Elqortobi, A.: Approximation of fractal sets. J. Comput. Appl. Math. 29, 79–89 (1990)
4. Hutchinson, J.E.: Fractals and self-similarity. Indiana Univ. Math. J. 30(5) (1981)
5. Kocić, L.M.: Descrete methods for visualising fractal sets. FILOMAT (Niš) 9(3), 753–764 (1995)
6. Kocić, L.M., Simoncelli, A.C.: Towards free-form fractal modelling. In: Daehlen, M., Lyche, T., Schumaker, L.L. (eds.) Mathematical Methods for Curves and Surfaces II, pp. 287–294. Vanderbilt University Press, Nashville (1998)
7. Kocić, L.M., Simoncelli, A.C.: Stochastic approach to affine invariant IFS. Prague Stochastic 1998 theory, Statistical Decision Functions and Random Processes 2, 317–320 (1998)
8. Kocić, L.M., Simoncelli, A.C.: Cantor Dust by AIFS. FILOMAT (Niš) 15, 265–276 (2001); Math. Subj. Class. 28A80 (65D17) (2000)
9. Kocić, L.M., Simoncelli, A.C.: Shape predictable IFS representations. In: Novak, M.M. (ed.) Emergent Nature, pp. 435–436. World Scientific, Singapore (2002); Math. Subj. Class. 65D17, 28A80 (1991)
10. Kocić, L.M., Stefanovska, L., Babače, E.: Affine invariant iterated function system and minimal simplex problem. In: Proceedings of the International Conference DGDS-2007, pp. 119–128. Geometry Balkan Press, Bucharest (2008)
11. Lawlor, O.S., Hart, J.C.: Bounding Recursive Procedural Models using Convex Optimization. In: Proc. Pacific Graphics 2003 (October 2003)
12. Mandelbrot, B.: The Fractal Geometry of Nature. Freeman and Company, New York (1977)

Computational Analysis of Expected Climate Change in the Carpathian Basin Using a Dynamical Climate Model

Judit Bartholy, Rita Pongrácz, Ildikó Pieczka,
Péter Kardos, and Adrienn Hunyady

Department of Meteorology, Eötvös Loránd University,
Pázmány st. 1/a, H-1117 Budapest, Hungary
bari@ludens.elte.hu, prita@nimbus.elte.hu, pieczka@nimbus.elte.hu,
peter.kardos@metnet.hu, adrienn.hunyady@metnet.hu
http://nimbus.elte.hu/

Abstract. For analyzing the possible regional climate change in the Carpathian Basin, model PRECIS has been adapted, which is the hydrostatic regional climate model HadRM3P developed at the UK Met Office, Hadley Centre, and nested in HadCM3 GCM. First, control run simulations (1961-1990) of the PRECIS model (with two different sets of boundary conditions) are analyzed. In the validation, seasonal temperature and precipitation mean values from the CRU datasets are used. According to the results, model PRECIS slightly overestimates the temperature and underestimates the precipitation. Then, model results for the periods 2071-2100 (using SRES A2 scenario) and 1961-1990 (as the reference period) are compared. The results suggest that the temperature increase expected in the Carpathian Basin may considerably exceed the global warming rate. The climate of this region is expected to become wetter in winter and drier in the other seasons.

Keywords: Regional climate modeling, Carpathian Basin, temperature, precipitation.

1 Introduction

Global climate models (GCM) are widely used for estimating the possible global warming due to increased anthropogenic influences. The results from these coarse resolution (about 300 km) models can only be considered as a first-guess of regional climate change consequences of the global warming. Regional climate models (RCM) are dynamical models nested in GCMs and they may lead to a much better estimation of future climate conditions in the European subregions since the horizontal resolution of these RCMs is much finer (maybe as fine as 10-25 km) than the GCMs [8]. Moreover, high resolution model results are essential for the generation of national climate change scenarios, as it is recommended by the United Nations Development Programme (UNDP). The expected regional

S. Margenov, L.G. Vulkov, and J. Waśniewski (Eds.): NAA 2008, LNCS 5434, pp. 176–183, 2009.

climate change in the Carpathian Basin (located in Central/Eastern Europe) is modelled by four different RCMs. Two of them (RegCM and PRECIS) are run by the Department of Meteorology, Eötvös Loránd University, Budapest [3]. The other two RCMs are run by the Hungarian Meteorological Service: ALADIN (developed by the Météo-France) and REMO (developed by the Max Planck Institute, Hamburg). The present paper discusses the climate modeling experiments using the model PRECIS. First, we evaluate the model capability of reconstructing the present climate (1961-1990) using two different sets of boundary conditions, (i) from the European Centre for Medium Range Weather Forecast (ECMWF) ERA-40 reanalysis database, (ii) from the HadCM3 GCM output data. In order to fulfill the validation task, the results of the different model experiments are compared to the monthly climatological datasets of the Climatic Research Unit (CRU) of the University of East Anglia as a reference. Then, results of A2 scenario run (for 2071-2100) of model PRECIS are presented and discussed. Finally, the main conclusions of the paper are summarized in the last section.

2 Regional Climate Model PRECIS

The installation and the adaptation of the regional climate model PRECIS at the Department of Meteorology, Eötvös Loránd University (Budapest, Hungary) has started in 2004. At the beginning of our studies, version 1.3 was used, but the results presented in this paper are from an updated model version (1.4.8). PRECIS is a high resolution limited area model with both atmospheric and land surface modules. The model was developed at the Hadley Centre of the UK Met Office [16], and it can be used over any part of the globe (e.g., [12], [1]). The PRECIS regional climate model is based on the atmospheric component of HadCM3 [7] with substantial modifications to the model physics [9]. The atmospheric component of PRECIS is a hydrostatic version of the full primitive equations (1–5):

$$\frac{\partial \boldsymbol{v}}{\partial t} = -\boldsymbol{v} \cdot \nabla \boldsymbol{v} - \omega \frac{\partial \boldsymbol{v}}{\partial p} + f\boldsymbol{k} \times \boldsymbol{v} - \nabla \Phi + \boldsymbol{D}_M \tag{1}$$

$$\frac{\partial T}{\partial t} = -\boldsymbol{v} \cdot \nabla T + \omega \left(\frac{\kappa T}{p} - \frac{\partial T}{\partial p} \right) + \frac{\widetilde{Q}_{rad}}{c_p} + \frac{\widetilde{Q}_{con}}{c_p} + \boldsymbol{D}_H \tag{2}$$

$$\frac{\partial q}{\partial t} = -\boldsymbol{v} \cdot \nabla q - \omega \frac{\partial q}{\partial p} + E - C + D_q \tag{3}$$

$$\frac{\partial \omega}{\partial p} = -\nabla \cdot \boldsymbol{v} \tag{4}$$

$$\frac{\partial \Phi}{\partial p} = -\frac{RT}{p} \tag{5}$$

Table 1. Configuration and properties of PRECIS

Prognostic variables	Surface pressure, zonal and meridional wind components, water vapour, potential temperature [9]
Horizontal coordinate-system	Spherical polar coordinates rotated to the equator [9]
Horizontal resolution	$0.44\,^\circ \times 0.44\,^\circ$ or $0.22\,^\circ \times 0.22\,^\circ$ [9]
Horizontal grid	Arakawa B grid [2]
Horizontal discretization	Finite differences (split-explicit) [9]
Vertical coordinate-system	Hybrid: terrain-following + pressure [14]
Coupling (LBC treatment)	Davies relaxation scheme [9]
Linear term	Explicit scheme [9]
Advection	Heun-scheme [13]
Timestep	5 min [9]

where $D_M = (D_\lambda, D_\phi)$ are dissipation terms for momentum and D_H and D_q are diffusion terms for heat and moisture, respectively, q is the specific humidity, E and C are the rates of evaporation and condensation due to cloud processes, and Φ is the geopotential [15].

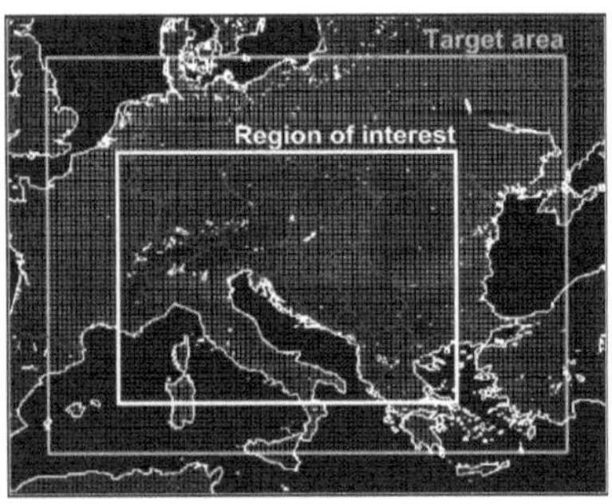

Fig. 1. Topography of the selected Central European domain used in model PRECIS

There are 19 vertical levels in the model, the lowest at ~ 50 m and the highest at 0.5 hPa [4]. The vertical coordinate η is defined by the following implicit relationship:

$$p(x, y, \eta, t) = A(\eta) + B(\eta)p_s(x, y, t) \tag{6}$$

where p is the pressure, $p_s(x, y, t)$ is the surface pressure, and A and B are constants depending on the vertical levels. The η varies between 0 (at the top) and 1 (at the bottom) and the following conditions are fulfilled:

$$A(1) = 0, B(1) = 1, B(0) = 0, \partial\eta/\partial p > 0. \tag{7}$$

Splitting techniques are commonly used when large-scale models - like climate models - are treated numerically [5]. In our model during the integration the geostrophic adjustment is separated from the advection part: adjustment is iterated three times per 5 minutes advection timestep. Averaged velocities over the three adjustment timesteps are used for advection, which is integrated in time using the Heun scheme [13]. This finite difference scheme is 4th order accurate except at high wind speeds when it is reduced to 2nd order accuracy for stability.

In our studies, we used the finest possible horizontal resolution (0.22 °) for modeling the Central European climate. Hence, the target region contains 123 × 96 grid points, with special emphasis on the Carpathian Basin and its Mediterranean vicinity containing 105 × 49 grid points (Fig. 1). In the post-processing of the RCM outputs, daily mean values are used. In case of the control period (1961-1990), the initial and the lateral boundary conditions for the regional model are taken from (i) the ERA-40 reanalysis database [6] using 1 ° horizontal resolution, compiled by the ECMWF, and (ii) the HadCM3 ocean-atmosphere coupled GCM using 150 km as a horizontal resolution. For the validation of the PRECIS results CRU TS 1.0 ([10],[11]) datasets are used.

3 Results of the Control Experiments of PRECIS

During the validation process, we analyzed monthly, seasonal, and annual temperature mean values and precipitation amounts for the reference period. Fig. 2 and Fig. 3 summarize the seasonal differences between the simulated (PRECIS outputs) and the observed (CRU data) values in case of temperature and precipitation, respectively.

On the basis of the maps shown in Fig. 2, the seasonal mean temperature is well reproduced by the regional model in the Carpathian Basin, the spatially averaged overestimation is less than 2.2 °C. Large overestimation can be seen in the high-elevated regions, especially, in the Alps (in winter and spring, the simulated mean temperature can be larger than the observed value by more than 5 °C). In Hungary, the largest overestimation is found in summer (+2.2 °C on average). The smallest difference values close to 0 °C are determined in spring (0.2 °C average overestimation), but also, the winter and the autumn daily temperature is overestimated by less than 1 °C in Hungary on average.

Precipitation is far more variable both in time and in space than temperature. The spatially averaged precipitation is overestimated in the entire model domain, especially, in spring and winter. According to the maps shown in Fig. 3, the precipitation of the high-elevated regions is overestimated (by more than 50% in each season), while the overestimation of the seasonal precipitation occurring in the plain regions is much less in spring than in the mountains. In summer and in autumn the precipitation is underestimated in the lowlands. The underestimation is larger in the southern subregions than in the northern part of the domain. In case of Hungary, the spring precipitation is overestimated (by 33% on average), while in the other three seasons the precipitation is slightly underestimated (by less than 10% on average) in the country.

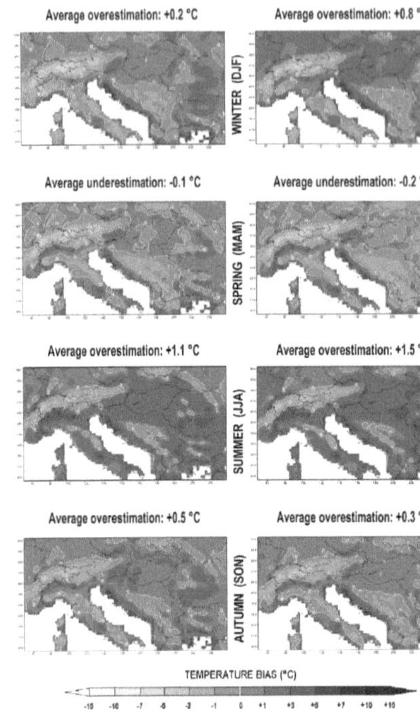

Fig. 2. Results of control runs (1961-1990) for Hungary: difference between PRECIS and CRU seasonal mean temperature data. Maps shown on the left/right are generated using the ERA-40/HadCM3 driving data.

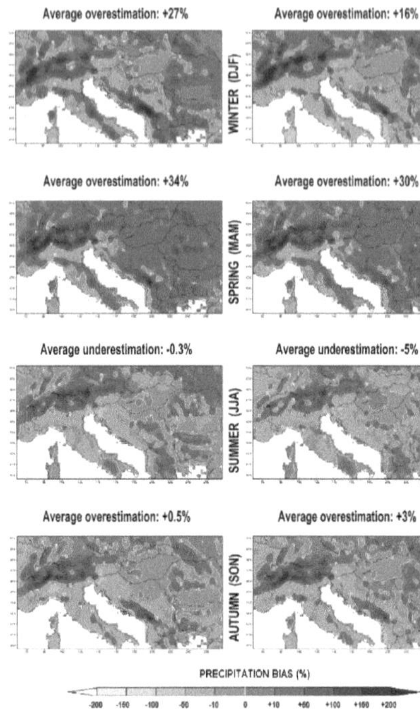

Fig. 3. Results of control runs (1961-1990) for Hungary: difference between PRECIS and CRU seasonal mean precipitation data. Maps shown on the left/right are generated using the ERA-40/HadCM3 driving data.

4 Results of the Scenario Experiments of PRECIS

In our research plan several future scenario runs are scheduled using the PRECIS model for the 2071-2100 period. Here, the results of the first accomplished PRECIS experiments are summarized using the SRES A2 emission scenario [8]. Fig. 4 shows the expected mean seasonal temperature and precipitation change in the target domain by the end of the 21st century. In all the four seasons considerable warming is projected (both in the mountains and the lowlands), the largest warming can be expected in summer. Over the sea, the expected warming is certainly less than over the continent. The largest temperature increase (by more than 8 °C) is expected in the plain region of the river Po, and the southeastern region of the target domain including Hungary. In case of precipitation, winter is expected to become considerably wetter, especially, in the western mountainous subregions. Seasonal precipitation amounts in spring, summer and autumn are projected to decrease by the end of the 21st century. The largest decrease can be expected in autumn.

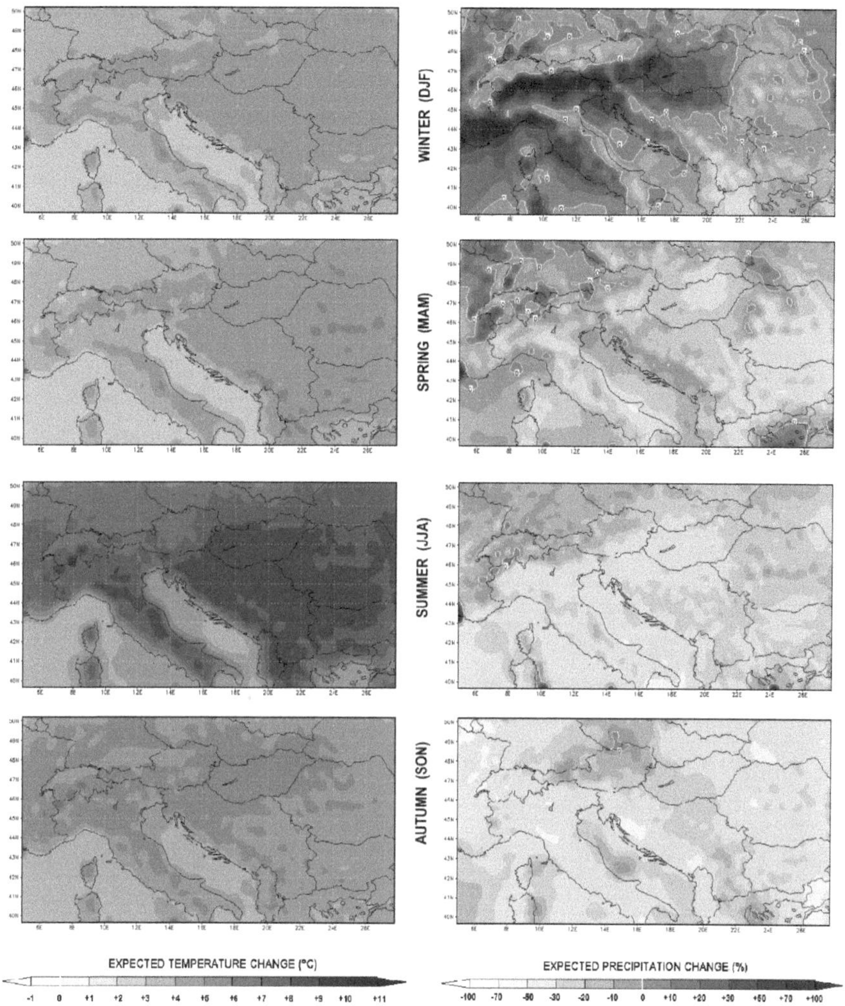

Fig. 4. Expected climate change for the Carpathian Basin and its Mediterranean vicinity in case of A2 scenario runs (2071-2100)

5 Conclusions

In this paper, results of the regional climate model PRECIS are discussed and compared for the Carpathian Basin and its vicinity in the 1961-1990 reference period using the ERA-40 and the HadCM3 driving data. In addition to the control experiments, the A2 scenario run is also completed for 2071-2100. Based on the results presented here, the following main conclusions can be drawn.

1. Only slight differences are found between the two control run experiments, both in case of temperature and precipitation.

2. In general, the seasonal mean temperature fields are overestimated by the PRECIS simulation. The largest bias values are found in summer, when the average overestimation of PRECIS is about 0 – 3 °C.
3. The seasonal precipitation is generally underestimated by the PRECIS simulation, except spring when the precipitation is overestimated.
4. According to the PRECIS projections for the SRES A2 scenario, the climate of the Carpathian Basin is expected to warm (especially, in summer by about 7 – 9 °C). The winter precipitation in Hungary is expected to increase, and the other three seasons are expected to become drier by the end of the 21st century.

Acknowledgments. Research leading to this paper has been supported by the following sources: the Hungarian Academy of Sciences under the program 2006/TKI/246 titled Adaptation to climate change, the Hungarian National Research Development Program under grants NKFP-3A/082/2004 and NKFP-6/079/2005, the Hungarian National Science Research Foundation under grants T-049824, K-67626, and K-69164, the Hungarian Ministry of Environment and Water, and the CECILIA project of the European Union Nr. 6 program (contract no. GOCE-037005).

MEC Albert Apponyi programme

Established by the support of the National Office for Research and Technology.

NKTH

National Office for Research and Technology

References

1. Akhtar, M., Ahmad, N., Booij, M.J.: The impact of climate change on the water resources of Hindukush-Karakorum-Himalaya region under different glacier coverage scenarios. Journal of Hydrology (2008) doi: 10.1016/j.jhydrol.2008.03.015
2. Arakawa, A., Lamb, V.R.: Computational design of the basic dynamical processes of the UCLA general circulation model. In: Chang, J. (ed.) Methods in Computational Physics, vol. 17, pp. 173–265. Academic Press, New York (1977)
3. Bartholy, J., Pongrácz, R., Torma, C., Hunyady, A.: Regional climate projections for the Carpathian Basin. In: Láng, I., Faragó, T., Iványi, Z. (eds.) Proceedings of the International Conference on Climate Change: Impacts and Responses in Central and Eastern European Countries, pp. 55–62. Hungarian Academy of Sciences, Hungarian Ministry of Environment and Water, Regional Environment Center for Central and Eastern Europe, Budapest (2006)
4. Cullen, M.J.P.: The unifed forecast/climate model. Meteorological Magazine 122, 81–94 (1993)
5. Dimov, I., Faragó, I., Havasi, Á., Zlatev, Z.: Different splitting techniques with application to air pollution models. Int. J. Environment and Pollution. 32(2), 174–199 (2008)

6. Gibson, J.K., Kallberg, P., Uppala, S., Nomura, A., Hernandez, A., Serrano, A.. ERA description. Reading: ECMWF Reanalysis Project Report Series 1, 77 (1997)
7. Gordon, C., Cooper, C., Senior, C.A., Banks, H., Gregory, J.M., Johns, T.C., Mitchell, J.F.B., Wood, R.A.: The simulation of SST, sea ice extents and ocean heat transports in a version of the Hadley Centre coupled model without flux adjustments. Climate Dynamics 16, 147–168 (2000)
8. Solomon, S., Qin, D., Manning, M., Chen, Z., Marquis, M., Averyt, K.B., Tignor, M., Miller, H.L. (eds.): IPCC: Climate Change 2007: The Physical Science Basis. Contribution of Working Group I to the Fourth Assessment Report of the IPCC, p. 996. Cambridge University Press, Cambridge (2007)
9. Jones, R.G., Noguer, M., Hassell, D.C., Hudson, D., Wilson, S.S., Jenkins, G.J., Mitchell, J.F.B.: Generating high resolution climate change scenarios using PRECIS, 40p. UK Met Office Hadley Centre, Exeter (2004)
10. New, M., Hulme, M., Jones, P.: Representing twentieth-century space-time climate variability. Part I: Development of a 1961-90 mean monthly terrestrial climatology. Journal of Climate 12, 829–856 (1999)
11. New, M., Hulme, M., Jones, P.: Representing twentieth-century space-time climate variability. Part 2: Development of 1901-96 monthly grids of terrestrial surface climate. Journal of Climate 13, 2217–2238 (2000)
12. Rupa Kumar, K., Sahai, A.K., Krishna Kumar, K., Patwardhan, S.K., Mishra, P.K., Revadekar, J.V., Kamala, K., Pant, G.B.: High-resolution climate change scenarios for India for the 21st century. Current Science 90, 334–345 (2006)
13. Mesinger, F.: Horizontal Advection Schemes of a Staggered Grid - An Enstrophy and Energy-Conserving Model. Monthly Weather Review 109, 467–478 (1981)
14. Simmons, A.J., Burridge, D.M.: An energy and angular-momentum conserving vertical finite difference scheme and hybrid vertical coordinates. Monthly Weather Review 109, 758–766 (1981)
15. Trenberth, K.E. (ed.): Climate System Modeling, Cambridge University Press, 788p. Cambridge University Press, Cambridge (1992)
16. Wilson, S., Hassell, D., Hein, D., Jones, R., Taylor, R.: Installing and using the Hadley Centre regional climate modelling system, PRECIS. Version 1.5.1, 157p. UK Met Office Hadley Centre, Exeter (2007)

An Efficient Computational Technique for a System of Singularly Perturbed Initial Value Problems

Rajesh K. Bawa[1] and Vinod Kumar[2]

[1] Department of Computer Science, Punjabi University, Patiala - 147 002, India
[2] Chitkara Institute of Engineering and Technology Jansla, Rajpura, Patiala - 140401, India

Abstract. In this paper, a parameter-uniformly convergent computational technique for a system of singularly perturbed initial value problems, which is applied on a piecewise uniform Shishkin mesh is presented. Numerical experiments are carried out on some test problems which shows almost second order uniform convergence, confirming the efficiency of the proposed technique.

1 Introduction

Singularly perturbed problems have received a significant amount of attention in the past and recent years. To solve these type of problems, mainly there are two approaches, namely fitted operators and fitted mesh methods. More details can be found in the books by Farrell et al. [1] and Roos et al. [2]. Matthews et al. [3] have suggested parameter robust numerical methods for second order system of singularly perturbed ordinary differential equations with one or two small parameters. In [4], a computational method is presented for a system of first order singularly perturbed ordinary differential equations. Here, we present a computational technique for a system of first order singularly perturbed ordinary differential equations of the form:

$$L_\epsilon \boldsymbol{u_\epsilon}(x) = \begin{pmatrix} \epsilon D & 0 & . . & 0 \\ 0 & \epsilon D & . . & 0 \\ . & . & . . & . \\ . & . & . . & . \\ 0 & 0 & . . & \epsilon D \end{pmatrix} \boldsymbol{u_\epsilon}(x) + A(x)\boldsymbol{u_\epsilon}(x) = \boldsymbol{f}(x) \tag{1}$$

$$\boldsymbol{u_\epsilon}(0) = (u_{\epsilon,1}(0), u_{\epsilon,2}(0), ...u_{\epsilon,n}(0))^T, x \in (0,1]$$

where $\boldsymbol{u_\epsilon} = (u_{\epsilon,1}, u_{\epsilon,2}, ...u_{\epsilon,n})^T$, $A(x) = \begin{pmatrix} a_{11}(x) & a_{12}(x) & . . & a_{1n}(x) \\ a_{21}(x) & a_{22}(x) & . . & a_{2n}(x) \\ . & . & . . & . \\ . & . & . . & . \\ a_{n1}(x) & a_{n2}(x) & . . & a_{nn}(x) \end{pmatrix}$

$\boldsymbol{f}(x) = (f_1(x), f_2(x), ..., f_n(x))^T$, D denotes $\frac{d}{dx}$ and $u_\epsilon \in C^{(1)}(\Omega), \Omega = (0,1)$.

S. Margenov, L.G. Vulkov, and J. Waśniewski (Eds.): NAA 2008, LNCS 5434, pp. 184–191, 2009.

The singular perturbation parameter ϵ satisfies $0 < \epsilon \leq 1$. The functions $a_{ij}, f_i \in C^2(\overline{\Omega}), \overline{\Omega} = [0,1]$, $i, j = 1, 2, ..., n$ satisfies the following inequalities:

$$(i) \quad a_{ii} > \sum_{j=1, j \neq i}^{n} |a_{ij}(x)|, \quad i = 1(1)n$$

$$(ii) \quad a_{ij} < 0, \quad i, j = 1(1)n, i \neq j$$

We introduce the positive number, $\beta = min\{\sum_{j=1}^{n} a_{1,j}, \sum_{j=1}^{n} a_{2,j}, ..., \sum_{j=1}^{n} a_{n,j}\}$.

In the present work, we propose a hybrid scheme for (1) on variable meshes and apply it on piece-wise uniform Shishkin mesh. The basic idea behind the Shishkin mesh is to divide the domain $\overline{\Omega} = [0, \sigma] \cup (\sigma, 1]$, where σ is a transition point, a function of N and ϵ, and place $N/2$ mesh points in the region $[0, \sigma]$ known as "inner region " where the solution varies fast and place remaining $N/2$ mesh points in the region $(\sigma, 1]$ called "outer region" where the solution varies slowly. In this hybrid scheme, we are taking Trapezoidal scheme in the inner region and Euler scheme in the outer region. A transition parameter σ_0 (defined later) is suitably chosen, so that we can retain the oscillation free behavior of Euler scheme and higher order convergence of Trapezoidal scheme.

The outline of this paper is as follows: The piecewise uniform Shishkin mesh and the Hybrid scheme are given in Sect. 2.1. Convergence results of the scheme are obtained in Sect. 3. Finally, some numerical examples are presented in Sect. 4. The paper ends with some conclusions.

2 Proposed Hybrid Scheme

Let the mesh points of $\overline{\Omega} = [0, 1]$ be

$$x_0 = 0, x_i = \sum_{k=0}^{i-1} h_k, h_k = x_{k+1} - x_k, x_N = 1, i = 1, 2, ..., N-1.$$

We define the scheme as

$$L_\epsilon^N \boldsymbol{U_\epsilon}(x_j) \equiv \begin{cases} Q\boldsymbol{U_\epsilon}(x_j) + \frac{A(x_{j-1})\boldsymbol{U_\epsilon}(x_{j-1}) + A(x_j)\boldsymbol{U_\epsilon}(x_j)}{2} = \frac{\boldsymbol{f}(x_{j-1}) + \boldsymbol{f}(x_j)}{2}, 0 < j \leq \frac{N}{2} \\ Q\boldsymbol{U_\epsilon}(x_j) + A(x_j)\boldsymbol{U_\epsilon}(x_j) = \boldsymbol{f}(x_j), \quad \frac{N}{2} < j \leq N \end{cases} \tag{2}$$

where $\boldsymbol{U_\epsilon} = (U_{\epsilon,1}, U_{\epsilon,2}, ...U_{\epsilon,n})^T$, $Q = \begin{pmatrix} \epsilon D^- & 0 & .. & 0 \\ 0 & \epsilon D^- & .. & 0 \\ . & . & .. & . \\ . & . & .. & . \\ 0 & 0 & .. & \epsilon D^- \end{pmatrix}$,

$D^- U_{\epsilon,i}(x_j) = \frac{U_{\epsilon,i}(x_j) - U_{\epsilon,i}(x_{j-1})}{h_{j-1}}$, j=1(1)N, i= 1(1)n.

2.1 Piece-Wise Uniform Shishkin Mesh

As mentioned earlier, in this mesh, the domain $\overline{\Omega}$ is divided into two subintervals as $\overline{\Omega} = [0, \sigma] \cup (\sigma, 1]$ for some σ such that $0 \leq \sigma \leq \frac{1}{2}$, defined as

$$\sigma = min\{\frac{1}{2}, \sigma_0 \epsilon \ln N\}$$

where $\sigma_0 \geq \frac{2}{\sqrt{\beta}}$. Further, we denote the mesh size in the region $[0, \sigma]$ by $h = 2\frac{\sigma}{N}$ and in $(\sigma, 1]$ by $H = 2\frac{(1-\sigma)}{N}$.

3 Convergence Analysis

Here, we present some analytical results which include a maximum principle, uniform stability and estimates of the derivatives of the solution.

Lemma 1. *Consider the system of initial value problems (1). Then* $u_{\epsilon,i}(0) \geq 0, i = 1, ..., n$ *and* $(L_\epsilon u_\epsilon)_i(x) \geq 0$ *for all* $x \in (0, 1], i = 1, ..., n$ *imply that* $u_{\epsilon,i}(x) \geq 0, i = 1, ..., n$ *for all* $x \in \overline{\Omega}$.

Proof. See Ref. [4].

Lemma 2. *If* $u_\epsilon(x)$ *is the solution of system of initial value problems (1), then*

$$\| u_\epsilon(x) \| \leq Cmax\{\| u_\epsilon(x) \|, \| L_\epsilon u_\epsilon(x) \|\} for\ all x \in \overline{\Omega}.$$

where C *is a constant independent of* x *and* ϵ

Proof. See Ref. [4].

Lemma 3. *If* $u_\epsilon(x)$ *is the solution of system of initial value problems (1), then* $u_\epsilon(x)$ *for i= 1,2,...,n satisfy*

$$| u^k_{\epsilon,i}(x) | \leq C(1 + \epsilon^{-k} e^{-\beta x/\epsilon}) \quad for i = 1, 2, ..., n$$

for $0 \leq k \leq 2$, $x \in \overline{\Omega}$ *and* C *is a constant independent of* x *and* ϵ

Proof. See Ref. [4].

Lemma 4. *Consider the discrete IVP (2). Then* $U_{\epsilon,i}(0) \geq 0, i = 1, ..., n$ *and* $(L^N_\epsilon U_\epsilon)_i(x_j) \geq 0$ *for all* $x \in (0, 1]$, *i=1,...,n, j=1,2,...,N imply that* $U_{\epsilon,i}(x_j) \geq 0, i = 1, ..., n, j = 0, 1, ..., N$.

Proof. This can be proved by using the idea presented in [4].

Lemma 5. *If* $U_\epsilon(x_j)$ *is any mesh function, then*

$$\| U_\epsilon(x_j) \| \leq Cmax\{\| U_\epsilon(0) \|, \| L^N_\epsilon U_\epsilon(x_j) \|\} j = 1, 2, ..., N.$$

where C *is a constant independent of* x *and* ϵ

Proof. This can be proved by using the idea presented in [4].

Now, we give a parameter uniform error estimate. For simplicity of proof, we are considering the following scalar case which can be generalized easily for system of singularly perturbed problems. Let us consider the following singularly perturbed initial value problem (IVP):

$$L_\epsilon u(x) \equiv \varepsilon u'(x) + b(x)u(x) = f(x), \quad x \in \Omega = (0,1) \tag{3}$$

$$u(0) = A \tag{4}$$

where A is a constant and $\varepsilon > 0$ is a small parameter, b and f are sufficiently smooth functions, such that $b(x) \geq \beta > 0$ on $\overline{\Omega} = [0,1]$. Under these assumptions, the IVP possesses a unique solution $u(x)$ [6]. For this case the hybrid scheme becomes:

$$L_\epsilon^N U_i \equiv \begin{cases} \varepsilon D^- U_i + \frac{b_{i-1}U_{i-1}+b_iU_i}{2} = \frac{f_{i-1}+f_i}{2} & 0 < i \leq N/2, \\ \varepsilon D^- U_i + b_i U_i = f_i, \quad N/2 < i \leq N \end{cases} \tag{5}$$

where $D^- U_i = \frac{U_i - U_{i-1}}{h_{i-1}}$ and $b_i = b(x_i), f_i = f(x_i)$.

Proposition 1. *Let $u(x)$ and U_i be respectively the solutions of (3-4) and (5). Then, the local truncation error satisfies the following bounds:*

$$|L_\varepsilon^N(U_i - u(x_i))| \leq CN^{-2}\sigma_0^2 \ln^2 N, \quad \textit{for} \quad 0 < i \leq N/2,$$
$$|L_\varepsilon^N(U_i - u(x_i))| \leq C(N^{-1}\varepsilon + N^{-\beta\sigma_0}), \quad \textit{for} \quad N/2 < i \leq N, \quad \textit{and} \quad H \leq \varepsilon,$$
$$|L_\varepsilon^N(U_i - u(x_i))| \leq C(N^{-2} + N^{-\beta\sigma_0}), \quad \textit{for} \quad N/2 < i \leq N, \quad \textit{and} \quad H > \varepsilon.$$

Proof. We distinguish several cases depending on the location of the mesh points. Firstly, we state the bound for the derivatives of the continuous solution,i.e., the solution $u(x)$ of the IVP (3-4) satisfies the following bound

$$|u^{(k)}(x)| \leq C\left[1 + \varepsilon^{-k}\exp(-\beta x/\varepsilon)\right] \tag{6}$$

For $x_i \in (0, \sigma]$, by using the usual Taylor series expansion, we get

$$|L_\varepsilon^N(U_i - u(x_i))| \leq C\varepsilon h^2 |u'''(\xi)|, \quad \text{for} \quad 0 < i \leq N/2 \tag{7}$$

for some point ξ, $x_{i-1} \leq \xi \leq x_i$.

Using $h = 2N^{-1}\sigma_0\varepsilon lnN$ on the above bound and bounding the exponential function by a constant, we have

$$|L_\varepsilon^N(U_i - u(x_i))| \leq CN^{-2}\sigma_0^2 ln^2 N, \quad \text{for} \quad 0 < i \leq N/2, \tag{8}$$

For $x_i \in (\sigma, 1]$, by using the usual Taylor series expansion, we get

$$|L_\varepsilon^N(U_i - u(x_i))| \leq C\varepsilon H |u''(\xi)|, \quad \text{for} \quad N/2 < i \leq N, \tag{9}$$

Note that the above expression for truncation error in outer region can also be represented as

$$|L_\varepsilon^N(U_i - u(x_i))| = \frac{\varepsilon}{h_i} R_1(x_i, x_{i+1}, u), \tag{10}$$

where $R_n(a,p,g) = \frac{1}{n!}\int_a^p (p-\xi)^n g^{(n+1)}(\xi)d\xi$ denotes the remainder obtained from Taylors expansion in integral form.

We discuss two cases: First, if $H < \varepsilon$, from (6), we obtain

$$|L_\varepsilon^N(U_i - u(x_i))| \le C\varepsilon H|u''(\xi)|, \tag{11}$$

$$\le C[H\varepsilon + H\varepsilon^{-1}\exp(-\beta x_i/\varepsilon)] \tag{12}$$

$$\le C[N^{-1}\varepsilon + N^{-\beta\sigma_0}] \tag{13}$$

Secondly, if $H \ge \varepsilon$, then using the bounds of the derivatives of $u(x)$ from (6), one can obtain the following

$$|L_\varepsilon^N(U_i - u(x_i))| \le C\left(H\varepsilon + \int_{x_{i-1}}^{x_i}(x_i-\xi)\varepsilon^{-2}\exp(-\beta\xi/\varepsilon)d\xi\right). \tag{14}$$

Integrating by parts, we get

$$\int_{x_{i-1}}^{x_i}(x_i-\xi)\varepsilon^{-2}\exp(-\beta\xi/\varepsilon)d\xi \le C\left(H\varepsilon + \int_{x_{i-1}}^{x_i}\varepsilon^{-1}\exp(-\beta\xi/\varepsilon)d\xi\right)$$

$$\le C\left[H\varepsilon + N^{-\beta\sigma_0}\right]$$

Using that $H < 2N^{-1}$ and $\varepsilon \le H$, we get

$$|L_\varepsilon^N(U_i - u(x_i))| \le C\left(N^{-2} + N^{-\beta\sigma_0}\right). \tag{15}$$

Combining all the previous results, we obtain the required truncation error. Hence, we obtain the required result.

Theorem 1. *Let $u(x)$ be the solution of the IVP (3-4) and U_i be the numerical solution obtained from the difference scheme as given in (5). Then, for sufficiently large N, and $N^{-1}\sigma_0 \ln N\beta^* < 1$, where $\beta^* = \max_{0\le i\le N} b(x_i)$, we have,*

$$|U_i - u(x_i)| \le C\left[N^{-2}\ln^2 N + N^{-1}\varepsilon + N^{-\sqrt{\beta}\sigma_0}\right], \quad \forall x_i \in \overline{\Omega}.$$

Proof. Let $B_i^- = (2-\rho_i b_i)$, $B_i^+ = (2+\rho_{i-1}b_i)$ and $b_i^+ = (1+\rho_i b_i)$, where $\rho_i = \frac{h_i}{\varepsilon}$ The solution of the scheme (5) can be expressed as: For $0 < i \le N/2$

$$U_i = \frac{\Pi_{j=0}^{i-1}B_j^-}{\Pi_{j=1}^{i}B_j^+}U_0 + \frac{\rho_i\Pi_{j=1}^{i-1}B_j^-}{\Pi_{j=1}^{i}B_j^+}(f_0+f_1) + \frac{\rho_i\Pi_{j=2}^{i-1}B_j^-}{\Pi_{j=2}^{i}B_j^+}(f_1+f_2) + \ldots + \frac{\rho_i}{B_i^+}(f_{i-1}+f_i)$$

and for $N/2 < i \leq N$

$$U_i = \frac{1}{\Pi_{j=N/2+1}^{i} b_j^+} U_{N/2} + \frac{\rho_i}{\Pi_{j=N/2+1}^{i} b_j^+} f_{N/2+1} + \frac{\rho_i}{\Pi_{j=N/2+2}^{i} b_j^+} f_{N/2+2} + \ldots + \frac{\rho_i}{b_i^+} f_i$$

Clearly, $B_i^+ 's$ and $b_i^+ 's$ are non-negative. Also, $B_i^- 's$ are non-negative for $0 < i \leq N/2$ due to $N^{-1}\sigma_0 \ln N\beta^* < 1$. So, the solution satisfies the Discrete Maximum Principle and hence there are no oscillations.

Defining the discrete barrier function

$$\phi_i = C\left[N^{-2}\ln^2 N + N^{-1}\varepsilon + N^{-\beta\sigma_0}\right].$$

Now, Choosing C sufficiently large, and using discrete maximum principle, it is easy to see that,

$$L_\epsilon^N(\phi_i \pm (U_i - u(x_i))) \geq 0$$

equivalently,

$$L_\epsilon^N(\phi_i) \geq |U_i - u(x_i)|$$

Therefore, it follows that

$$|U_i - u(x_i)| \leq |\phi_i|, \quad \forall x_i \in \overline{\Omega}.$$

Thus, we have the required ε-uniform error bound.

Remark 1. In Theorem 1, we obtain the error bound of order $N^{-1}\varepsilon$ only in outer region for the case $H < \varepsilon$, which is not the practical case. Therefore, we conclude that the order of convergence is almost two (up to a logarithmic factor). Our numerical results given in Sect 4 reveal the same behavior.

4 Numerical Experiments and Discussions

To show the accuracy of the present scheme, it has been applied on two test problems.

Example 1. Consider the following Problem:

$$\begin{aligned}
\epsilon u_1'(x) + (2+x)u_1(x) - u_2(x) &= 1 + x,\\
\epsilon u_2'(x) - (1+x)u_1(x) + (2+x)u_2(x) &= x,\ x \in (0,1]\\
u_1(0) = 1, \quad u_2(0) &= 0.5.
\end{aligned}$$

The exact solution of this example is not available. Therefor, to obtain the maximum point-wise errors and rates of convergence, we use double mesh principle. By following the idea of Sun and Styne [5], we modify Shishkin mesh by altering σ. Let $\overline{\Omega}^N$ be a Shishkin mesh with the parameter σ altered slightly to $\overline{\sigma} = min\{\frac{1}{2}, \sigma_0\epsilon \ln\frac{N}{2}\}$. Then, for i=0,1...N the ith point of the mesh $\overline{\Omega}^N$ with the (2i)th point of the mesh $\overline{\Omega}^{2N}$. As a result, the transition point does

Table 1. Maximum point wise errors and rates of convergence by Proposed method for Example 1

ε	Number of mesh points					
	32	64	128	256	512	1024
2−2	3.49825E-04 0.989	1.76265E-04 0.992	8.86340E-05 0.995	4.44634E-05 0.997	2.22709E-05 0.999	1.11456E-05
2^{-4}	4.63986E-03 1.815	1.31901E-03 2.018	3.25680E-04 2.016	8.05314E-05 2.002	2.01092E-05 1.728	6.06851E-06
2^{-8}	4.78790E-03 1.719	1.45453E-03 1.574	4.88392E-04 1.619	1.58993E-04 1.662	5.02567E-05 1.698	1.54884E-05
2^{-14}	4.79784E-03 1.718	1.45843E-03 1.576	4.89084E-04 1.618	1.59310E-04 1.662	5.03439E-05 1.698	1.55162E-05
2^{-18}	4.79799E-03 1.718	1.45849E-03 1.576	4.89094E-04 1.618	1.59314E-04 1.662	5.03452E-05 1.698	1.55166E-05
⋮	⋮	⋮	⋮	⋮	⋮	⋮
2^{-30}	4.79800E-03 1.718	1.45849E-03 1.576	4.89095E-04 1.618	1.59315E-04 1.662	5.03452E-05 1.698	1.55166E-05
2^{-32}	4.79800E-03 1.718	1.45849E-03 1.576	4.89095E-04 1.618	1.59315E-04 1.662	5.03452E-05 1.698	1.55166E-05

Table 2. Maximum point wise errors and rates of convergence by Proposed method for Example 2

ε	Number of mesh points					
	32	64	128	256	512	1024
2^{-2}	2.69486E-04 1.019	1.32957E-04 1.008	6.61174E-05 1.003	3.29792E-05 1.002	1.64712E-05 1.001	8.23115E-06
2^{-4}	4.47551E-03 1.818	1.26928E-03 2.016	3.13923E-04 2.016	7.76156E-05 2.003	1.93698E-05 2.001	4.84031E-06
2^{-8}	4.77780E-03 1.719	1.45169E-03 1.575	4.87296E-04 1.619	1.58655E-04 1.662	5.01471E-05 1.698	1.54548E-05
2^{-14}	4.79768E-03 1.718	1.45839E-03 1.576	4.89067E-04 1.618	1.59304E-04 1.662	5.03422E-05 1.698	1.55157E-05
2^{-18}	4.79798E-03 1.718	1.45849E-03 1.576	4.89093E-04 1.618	1.59314E-04 1.662	5.03451E-05 1.698	1.55166E-05
⋮	⋮	⋮	⋮	⋮	⋮	⋮
2^{-30}	4.79800E-03 1.718	1.45849E-03 1.576	4.89095E-04 1.618	1.59315E-04 1.662	5.03452E-05 1.698	1.55166E-05
2^{-32}	4.79800E-03 1.718	1.45849E-03 1.576	4.89095E-04 1.618	1.59315E-04 1.662	5.03452E-05 1.698	1.55166E-05

not move, when N is changed to 2N. Hence, use of interpolation for double mesh principle can be avoided. The double mesh difference is defined as $E_\epsilon^N = \max_{x_i \varepsilon \overline{\Omega}^N} \{|U_i^N - U_i^{2m}|\}$, where U_i^N and U_i^{2N} respectively denoted the numerical solutions obtained using N and 2N mesh intervals. The rates of convergence are calculated as:

$$p_\epsilon^N = \frac{lnE_\epsilon^N - lnE_\epsilon^{2m}}{ln2}$$

Numerical results by proposed scheme are given in Table 1.

Example 2. Consider the following Problem:

$$\epsilon u_1'(x) + 2u_1(x) - (1 + \tfrac{x}{2})u_2(x) = 1,$$
$$\epsilon u_2'(x) - u_1(x) + (2 + 2x)u_2(x) = x + 2,\ x \in (0, 1]$$
$$u_1(0) = 1, \quad u_2(0) = 1.5.$$

Numerical results are given in Table 2.

5 Conclusions

It is observed that although Implicit Euler Method satisfy discrete maximum principle in whole domain $[0, 1]$, but its order is almost one even for values of $\sigma_0 \geq 1$. We can get order two(up to a logarithmic factor) by applying Trapezoidal scheme in $[0, 1]$ with $\sigma_0 \geq 2$, but it results in small oscillations, hence the solution is not stable unless mesh size is very small even in outer region, where a coarse mesh is enough to give satisfactory result.

We have overcome this problem in the proposed scheme by proper combination of Euler and Trapezoidal schemes which gives second order for $\sigma_0 \geq 2$. The advantages of this scheme are several-fold. This scheme can be applied to non-linear problems. Also, the results of many boundary value and initial value technique's (IVT's and BVT's) can be improved and applied on Shishkin's mesh for obtaining higher order convergence, as these techniques mostly use either first order exponentially fitted operator methods or first order Implicit Euler method.

References

1. Farrell, P.A., Hegarty, A.F., Miller, J.J.H., O'Riordan, E., Shishkin, G.I.: Robust Computational Techniques for Boundary Layers. Chapman & Hall/CRC Press (2000)
2. Roos, H.-G., Stynes, M., Tobiska, L.: Numerical Methods for Singularly Perturbed Differential Equations. Springer, Berlin (1996)
3. Matthews, S., Miller, J.J.H., O'Riordan, E., Shishkin, G.I.: A Parameter Robust Numerical Method for a System of Singularly Pertyrbed Ordinary Differential Equations. Nova Science Publishers, New York (2000)
4. Hemavathi, S., Bhuvaneswari, T., Valarmathi, S., Miller, J.J.H.: A Parameter Uniform Numerical Method for a System of Singularly Perturbed Ordinary Differential Equations. Appl. Math. Comput. 191, 1–11 (2007)
5. Sun, G., Stynes, M.: An almost fourth order uniformly convergent difference scheme for a semilinear singularly perturbed reaction-diffusion problem. Numerische Mathematik 70, 487–500 (1995)
6. Doolan, E.P., Miller, J.J.H., Schilders, W.H.A.: Uniform Numerical Methods For Problems With Initial And Boundary Layers. Boole Press, Dublin (1980)
7. Miller, J.J.H., O'Riordan, E., Shishkin, G.I.: Fitted numerical methods for singular perturbation problems. World Scientific, Singapore (1996)

Model Predictive Control – Numerical Methods for the Invariant Sets Approximation

H. Benlaoukli and S. Olaru

SUPELEC, Automatic Control Department, 91192 Gif sur Yvette, France

Abstract. This paper deals with the computational issues encountered in the construction of invariant sets for LTI (Linear Time Invariant) systems subject to linear constraints. Three algorithms to compute or approximate the invariant set are presented. Two of theme are based on expansive and contractive strategy, while the third one uses the transition graph over the partition of the closed loop piecewise affine system.

1 Introduction

Model Predictive Control (MPC) is an optimization based control technique which applies in a receding horizon manner the solution of an open loop optimal control problem. The feasibility turns out to be a crucial demand in MPC synthesis as long as it represents the main ingredient for the stability of the entire closed loop system. In this context, the positive invariance is an important concept to be exploited for assuring the feasibility at all future instants and by consequence the viability of the control law.

This paper deals with the computational issues encountered in the construction of invariant sets for LTI (Linear Time Invariant) systems subject to linear constraints. It is shown that explicit MPC laws transform the closed loop dynamics in a piecewise affine (PWA) system. The main contribution is the comparison of two effective numerical procedures for the outer approximations of the maximal positively invariant (MPI) set for such PWA systems. Indeed, when iterative construction procedures are employed (especially if no finite-time algorithms exists to construct the exact MPI set), consistent approximations are to be found within a predefined precision.

The main tools employed here are the polyhedral computations but in order to decrease the computational complexity, the interval analysis procedure and transition graph constructions are used to avoid the treatment of the regions which do not meet the neighboring properties. This proves that computational geometry provides versatile set theoretic methods for engineering problems involving constraints, bounded-disturbances or construction of domain of attraction for stabilizing compensators.

The rest of the paper is organized as follows: in the next section, the predictive control generalities are recalled. Section 3 describes the invariant set construction, further illustrated on a numerical example in section 4.

S. Margenov, L.G. Vulkov, and J. Waśniewski (Eds.): NAA 2008, LNCS 5434, pp. 192–199, 2009.

2 Model Predictive Control – Piecewise Affine System

Considered a discrete time LTI system defined by:

$$x_{k+1} = Ax_k + Bu_k \tag{1}$$

In Model Predictive Control, a constrained optimal control problem over a finite receding horizon is solved at each sampling time. Generally the cost function to be minimized has the form:

$$\begin{aligned} &\min_{u_k, u_{k+1}, \ldots, u_{k+H_u-1}} \quad \textstyle\sum_{i=1}^{H_p} \|x_{k+i}\|_{Q_i}^2 + \sum_{i=0}^{H_u-1} \|\Delta u_{k+i}\|_{R_i}^2 \\ &\text{subject to: } \begin{cases} x_{k+i+1} = Ax_{k+i} + Bu_{k+i} \\ Cx_{k+i} + Du_{k+i} \leq E \end{cases} \end{aligned} \tag{2}$$

where H_p is the prediction horizon, H_u is the control horizon and for $i \geqslant H_u$ a pre-defined stabilizing control law $u_{k+i} = K_{LQ}x_{k+i}$ is used in order to construct the predictions. $Q_i = Q_i^T$ and $R_i = R_i^T$ are the weighting factors for the states trajectory and the control variation [1]. The success of the MPC strategies resides in the constraints handling capabilities.

The methodology has real-time limitations for fast dynamical system and the construction of explicit solution [2] may overcome the on-line computational burden. This is equivalent to the resolution of multi-parametric quadratic programs (mp-QP) [3] where the parameters are the components of the state vector and the future references.

The feedback control law obtained for linear systems (1) with linear constraints is a piecewise affine function of the current state (regulation study):

$$u_k = F_i x_k + G_i \tag{3}$$

and thus the closed loop dynamics will be given by:

$$x_{k+1} = f_{PWA}^i(x_k) = (A + BF_i)x_k + BG_i \tag{4}$$

defined over a set of polytopic regions $P_{set} = \bigcup_{i=1}^{N} P_i$ where the intersections $P_i \cap P_j$ are not full dimensional $\forall i \neq j$.

We define *image* and *preImage* operators as: $Image(P_i) = f_{PWA}^i(P_i)$ and $preImage_i(R) = \{x \in P_i | \quad f_{PWA}^i(x) \in R\}$.

The stability of MPC scheme (4) is a well understood topic [4], guarantees can be obtained through the use of terminal constraints that assure the positive invariance of the feasible domain with respect to the closed loop dynamics. An alternative method is to use receding horizon optimization as in (2) without terminal constraints, build the explicit solution and *a posteriori* determine the invariant set for the piecewise linear system (4). By using this former technique one has to obtain the invariant sets in an efficient way and in the case when the finite-time determination is not possible, to obtain upper and lower bounds within a given precision. In the next section two algorithms are proposed in this direction.

The results can be extended to the reference tracking case where the construction of invariant sets provides meaningful information about the stability for arbitrary reference. It should be mentioned that the existing stability results on this problems are restricted to specific class of signals [5], or make use of auxiliary concepts as *reference governors* to assure the global stability [6].

3 Algorithms for Invariant Set Computation

The invariance is an important concept in control theory and practice, mainly due to the fact that it can be used as a versatile stability tool [7]. Basically, the positive invariance is understood in the sense that if the system state enters in some subspace it will remain within at all future instants. The two algorithms proposed to compute or approximate the invariant set are based on expansive and contractive iterative construction. The difficulty in this construction is to decide when iteration must be stopped for a given precision computation, see [8]. Another problem is the heavy computation when system has large dimension. To overcome this problem, interval analysis is proposed here to diminish the computation burden. In section (3.4) an alternative procedure for the invariant set will be proposed by using the *transition graph* of the PWA model.

3.1 Contractive Iterations

Let a piecewise affine autonomous dynamic system defined by (4). The following algorithm computes an upper bound $\Phi \supseteq \Phi^{MPI}$ where Φ^{MPI} is the maximal positively invariant set (MPI) of the piecewise affine system (4).

Algorithm 1. *Contractive set construction.*
Input: PWA system defined over polyhedron regions P_{set}.
Output: The (approximate) maximal positive invariant set $\Phi \supseteq \Phi^{MPI}$.

```
while ("precision condition" not true) ∧ (max. no. of iterations)
    N = cardinal(P_set)
    i = 1
    while (i ≤ N)
        po = ∅
        for j = 1, N
            po = po ⋃ preImag(Image(P_i) ∩ P_j)
        end
        m = cardinal(po)
        P_set = {P_1, ..., P_{i-1}, po, P_{i+1}, ..., P_N}
        i = i + m
        N = cardinal(P_set)
    end
end
Φ = P_set
```

The algorithm eliminates at each iteration the subset of each P_i, $i = 1, ..., N$ that leaves P_{set} in one step with the forward dynamic (4). If P_{set} do not change at the update (step 10), then the construction stops and the result is the exact invariant set. Otherwise an alternative stopping criterion has to be used (see the complementarity with the *Expansive algorithm* to approximate Φ^{MPI} in [9]).

Remark 1. *If there exist a limit cycle phenomena, at the construction of the invariant set with contractive procedure, the algorithm is locked and terminate in a finite number of iterations. The use of expansive algorithm can overcame this problem.*

3.2 Expansive Iterations

The expansive algorithm proceeds in reverse way comparing to algorithm 1. A lower bound for the invariant set $\Psi \subseteq \Psi^{MPI} = \Phi^{MPI} \subseteq \Phi$ will be computed starting from an initial positive invariant set (for example the maximal output admissible set [10] for the region containing the origin). Alternatively, by choosing P_1, the region in P_{set} that contain the origin and by applying algorithm 1 one gets the initial invariant set Ψ_1. The result is stored in Ψ. subsequently, the idea is to compute all regions of P_{set} that transit in one step to Ψ. Iteratively Ψ is updated to contain all the sets reaching the invariant set in one step. It can be shown that Ψ expands and thus a lower bound for the maximal positively invariant set is obtained.

Algorithm 2. *Expansive set construction.*
Input: PWA system defined over polyhedron regions P_{set}.
Output: The (approximate) maximal invariant set $\Psi \subseteq \Psi^{MPI}$.

1: Let $P_1 \in P_{set}$ be the polytope that contains origin
2: Use Algorithm 1 to compute the invariant set of P_1. Let the result be Ψ_1
3: $StepSet = \Psi_1$
4: **while**("precision condition" not true) $\wedge$ (max. no. of iterations)
5: $\quad N = cardinal(P_{set})$
6: $\quad Interm = \emptyset$
7: $\quad L = cardinal(StepSet)$
8: $\quad$ **for** $i = 1, N$
9: $\quad\quad po = \emptyset$
10: $\quad\quad$ **for** $j = 1, L$
11: $\quad\quad\quad po = po \bigcup preImag(Image(P_i) \cap StepSet_j)$
12: $\quad\quad$ **end**
13: $\quad\quad Interm = Interm \bigcup po$
14: $\quad$ **end**
15: $\quad \Psi = \Psi \bigcup Interm$
16: $\quad StepSet = Interm$
17: **end**

If the exact invariant set cannot be obtained in finite time, the algorithm 1-2 provide an inner approximation of the invariant set see [9].

Remark 2. *If* $\Psi_1 = \emptyset$ *in (step 2), then the expansive algorithm can not compute any invariant set as long as the initialization step cannot be performed.*

3.3 Interval Analysis

Interval tree search (see [11] for basic definitions and procedures) is an algorithm which allows to efficiently identify all the intervals, in a predefined collection, that overlap a given point or interval. Figure 1 illustrates the principle: instead to intersect the black ploytope with all set P_{set} we intersect only with candidate ones (in gray color) [9] which have at least one projection with a non-empty intersection.

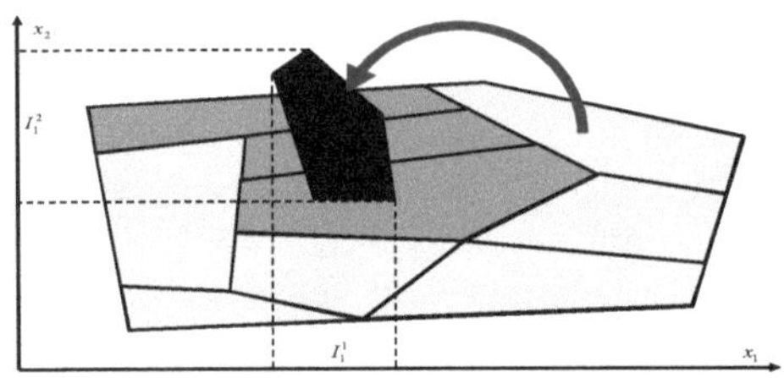

Fig. 1. In gray are represented the candidate polytopes for the intersection with the black one

3.4 Using Transition Graph

The third method for the invariant set approximation uses the graph structure associated to the state evolution for a given piecewise affine system defined over a polytopic regions $P_{set} = \bigcup_{i=1}^{N} P_i$.

Definition 1. *A transition graph associated to a PWA system* $x_{k+1} = f^i_{PWA}(x_k)$ *will be described by the pair* (Ω, Θ) *where* Ω *is the set of nodes and* Θ *the set of arches. Each region* P_i *is associated with a node* $n_i \in \Omega$. *By default a node* n_0 *is associated to the region* $D = \mathbb{R}^n \setminus P_{set}$. *The existence of an oriented arch* t_{ij} *between the node* n_i *and* n_j *is equivalent with the existence of at least one state trajectory starting in* $x \in P_i$ *and satisfying* $f^i_{PWA}(x) \in P_j$ *(It is clear that there is no arch from* n_0 *to* $n_i \in \Omega$*).*

The numerical procedure to construct the graph, has to check the intersection of the set dynamics with the original regions of the PWA models using the *image* and *preImage* operators. The arches Θ can be stored in a *non-symmetric* matrix representation - $M \in \{0,1\}^{(N+1)\times(N+1)}$. Each row and column of the matrix M corresponds to a node of transition graph such that if node n_i is connected to the node n_j, $M(n_i, n_j) = 1$ otherwise $M(n_i, n_j) = 0$. The transition matrix M, is offering a neighboring information that can not be obtained from the PWA description without a costly dynamics evaluation. Indeed, if the reachability analysis can be carried out by following the path in the transition graph. The following algorithm computes the (approximation of the) maximal positive invariant set for a given piecewise affine system.

Algorithm 3. *Invariant set construction.*
Input: PWA system defined over polyhedron regions P_{set}.
Output: The (approximate) maximal positive invariant set.

1: Compute the transition graph (Ω, Θ) corresponding to PWA system. The results are the matrix M and a pair of sets $(P_i, S_i), i = 1, \ldots, N$
2: **while** ($\exists i$ s.t. $t_{i0} \in \Theta$) $\wedge$ (max. no. of iterations)
3: Select the maximal subset $\mathbf{J} \subset \{1, \ldots, N\}$ such that $t_{i0} \in \Theta, \forall i \in \mathbf{J}$
4: $invP_i = S_i$, $outP_i = P_i \setminus S_i$ for all $i \in \mathbf{J}$
5: **for** $j \in \mathbf{J}$
6: **while** $(IP \neq \emptyset)$
7: $IP = preImag_j(Image(invP_j) \cap outP_j)$
8: $outP_j = outP_j \cup IP$;
9: $invP_j = invP_j \setminus IP$
10: **end**
11: $P_j = invP_j$
12: $S_j = invP_j$
13: **end**
14: Reconstruct the transition graph to take into account the changes in the PWA system definition
15: **end**

For each node connected to the *escape node* n_0 the corresponding region is decomposed as $P_j = invP_j \cup outP_j$, where $outP_j$ is the subset which force the transition to n_0 exclusively upon the dynamics defined for the region P_i. $invP_j$ the complement with respect to P_j. The problem consists in the determination of the number of sampling times before the effective transition to the node n_0. The finite determinedness of the algorithm 3 is not guaranteed.[1] Indeed the external loop with the condition mentioned at the (step 6) may not be satisfied mainly due to the graph reconstruction performed in (step 14) (There is no need to reconstruct the entire transition graph, but only the nodes and arches neighboring the nodes in $\mathbf{J}$). If the maximal number of iterations is reached, then the resulting set $P = \cup P_i$ is an external approximation of the maximal positive invariant set. Even for this case, the previous algorithm provides a useful information: the exact description of a region which have to be adapted $P_i \setminus S_i$ in order to reinforce the invariance.[2]

4 Example

The following example shows how the **Algorithm 3** calculates the invariant set for a given PWA system, and a comparison is done with the contractive algorithm Algo. 1.

[1] It is known that in the general case, even for a linear system, the finite determinedness is not guaranteed [10], [12].
[2] The algorithm 3 can be used in conjunction with an expansive invariant set approximation in order to get an ϵ inner approximation of the maximal positive invariant set.

Let a double integrator defined by:

$$\begin{cases} x_{k+1} = \begin{bmatrix} 1 & 0 \\ 1 & 1 \end{bmatrix} x_k + \begin{bmatrix} 1 \\ 0.5 \end{bmatrix} u_k \\ y_k = [0\ 1] x_k \end{cases} \tag{5}$$

In order to illustrate the construction of the invariant set using algorithm 3, a simple optimal control problem with prediction horizon $N = 2$, $Q = I_2$ and $R = 1$ is considered. The constraint to be satisfied are:

$$\begin{cases} -20 \leqslant y_k \leqslant 20 \\ -1 \leqslant u_k \leqslant 1 \end{cases} \tag{6}$$

Using multiparametric toolbox [13], an explicit controller is obtained with 11 regions. The iterations of the invariant set construction are illustrated in Fig. 2.

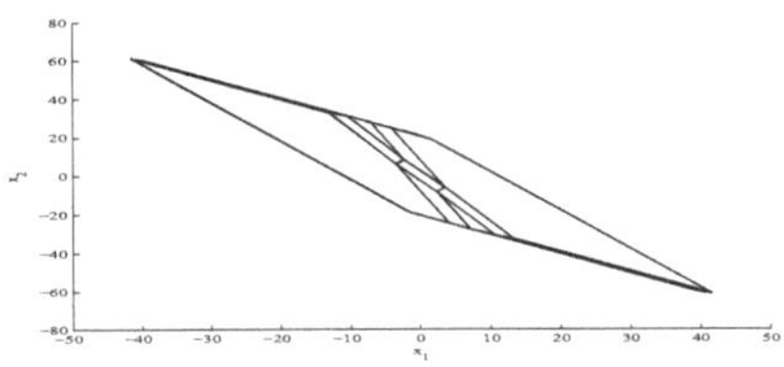

(a)

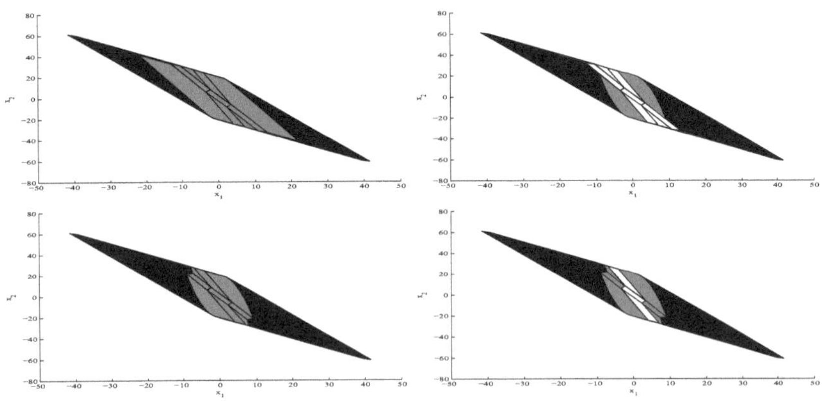

(b)

Fig. 2. Explicit solution (a). Two iterations are sufficient to obtain the invariant set (gray and white colors) using the transition graph (b) right column. The white regions are subregions not treated being inside the MPI set. Eight iterations (only two are shown) are needed to obtain the MPI set (gray color) using the contractive algorithm (b) left column.

5 Conclusion

In this paper three algorithms was presented to compute or to approximate the maximal positively invariant set. Algorithms 1 and 2 were presented in previous work [9] were completed here with the use of graph transition. This gives us *a priori* informations concerning the interconnection between all regions constituting the piecewise controller, so the number of geometric calculation may be reduced and only the regions containing an escape subregion may be processed. Another idea can be developed in conjunction with this result; it consist on forcing the invariance by changing the control law of escape regions.

References

1. Maciejowski, J.M.: Predictive Control with Constraints. Prentice-Hall, Englewood Cliffs (2002)
2. Bemporad, A., Morari, M., Dua, V., Pistikopoulos, E.N.: The explicit linear quadratic regulator for constrained systems. Automatica 38(1), 3–20 (2002)
3. Tøndel, P., Johansen, T.A., Bemporad, A.: An algorithm for mpqp and explicit mpc solutions. Automatica 39(3), 489–497 (2003)
4. Mayne, D., Rawlings, J., Rao, C., Scokaert, P.: Constrained model predictive control: Stability and optimality. Automatica 36, 789–814 (2000)
5. Limon, D., Alvarado, I., Alamo, T., Camacho, E.F.: Mpc for tracking of piece-wise constant references for constrained linear systems. In: Proceeding 16th IFAC World Congress, Prague (2005)
6. Olaru, S., Dumur, D.: Compact explicit mpc with guarantee of feasibility for tracking. In: Proceedings of the IEEE Conference on Decision and Control and European Control Conference, pp. 969–974 (2005)
7. Blanchini, F.: Set invariance in control. Automatica 35(11), 1747–1767 (1999)
8. Rakovic, S.V., Grieder, P., Kvasnica, M., Mayne, D.Q., Morari, M.: Computation of invariant sets for piecewise affine discrete time systems subject to bounded disturbances. In: Proceeding 43th IEEE Conference on Decision and Control (December 2004)
9. Benlaoukli, H., Olaru, S.: Computation and bounding of robust invariant sets for uncertain systems. In: Proceedings of the IFAC World Congress, Seoul, Korea (2008)
10. Gilbert, E., Tan, K.: Linear systems with state and control constraints, the theory and application of maximal output admissible sets. IEEE Transaction on Automatic Control 36, 1008–1020 (1991)
11. de Berg, M., van Kreveld, M., Overmars, M., Schwarzkopf, O.: Computational Geometry, Algorithms and Applications. Springer, Heidelberg (2000)
12. Rakovic, S.V., Kerrigan, E.C., Kouramas, K.I., Mayne, D.Q.: Invariant approximations of the minimal robust positively invariant set. IEEE Transaction on Automatic Control 50, 406–410 (2005)
13. Kvasnica, M., Grieder, P., Baotić, M.: Multi-parametric toolbox (mpt) (2004)

Quartic Spline of Interpolation with Minimal Quadratic Oscillation

Alexandru Mihai Bica

Department of Mathematics and Informatics,
University of Oradea,
Str. Universitatii no.1, 410087, Oradea, Romania
abica@uoradea.ro, smbica@yahoo.com

Abstract. The quartic spline of interpolation generated by initial conditions is constructed. The initial values corresponding to the first, second and third derivative of spline in the first knot are uniquely determined such that the quadratic oscillation in average of the quartic spline to be minimal (this notion was recently introduced by the author for any spline of interpolation function).

1 Introduction

The most used splines are cubic splines. Afterthere quartic splines was defined and constructed (see [1], [4], [5], [10], [12] and [14]). Some of the properties of quartic splines can be viewed in [6], [7], [9], [11], [13] and [15]. Here we present a natural way to obtain quartic splines of interpolation generated by initial conditions. With classical notations, let $[a,b]$ be a compact interval of $\mathbb{R}$, Δ a partition of $[a,b]$ given by

$$\Delta : a = x_0 < x_1 < ... < x_{n-1} < x_n = b,$$

$y = (y_0, ..., y_n) \in \mathbb{R}^{n+1}$ and the quartic spline $s : [a,b] \to \mathbb{R}$, $s \in C^3[a,b]$ for which we denote $s(x_i) = y_i, s'(x_i) = m_i, s''(x_i) = M_i, s'''(x_i) = V_i, \ \ i = \overline{0,n}$. Let $h_i = x_i - x_{i-1}, i = \overline{1,n}$.

From the smoothness conditions, the values of the parameters m_i, M_i, V_i, $i = \overline{1,n}$ are recurrently determined by $y_0, ..., y_n, m_0, M_0, V_0$. For given values $y_0, ..., y_n$ (obtained by practical measurements), the parameters m_0, M_0, V_0 remain free. In this paper will be determined such that the obtained quartic spline to have an optimal property: minimal quadratic oscillation in average. This notion was firstly introduced in [2] being proper for any interpolation function and was used for cubic splines generated by initial conditions in [3] (such cubic splines was introduced in [8]). We recall the notion of quadratic oscillation in average from [2].

Consider the line segments

$$D_i(x) = y_{i-1} + \frac{y_i - y_{i-1}}{h_i} \cdot (x - x_{i-1}), \quad \forall x \in [x_{i-1}, x_i], \quad \forall i = \overline{1,n} \qquad (1)$$

S. Margenov, L.G. Vulkov, and J. Waśniewski (Eds.): NAA 2008, LNCS 5434, pp. 200–207, 2009.

that together form the polygonal line joining the points (x_i, y_i), $i = \overline{0,n}$, $D(y) : [a,b] \longrightarrow \mathbb{R}$, given by

$$D(y)(x) = D_i(x),\ \forall x \in [x_{i-1}, x_i],\ \text{for all } i = \overline{1,n}.$$

Let $f : [a,b] \longrightarrow \mathbb{R}$, be a continuous function such that $f(x_i) = y_i,\ \forall i = \overline{0,n}$ and denote by f_i, the restriction of f to the interval $[x_{i-1}, x_i]$, for $i = \overline{1,n}$. For any $i = \overline{1,n}$ we define $\overline{f_i} : [a,b] \longrightarrow \mathbb{R}$, and $\overline{D_i}(y) : [a,b] \longrightarrow \mathbb{R}$, by :

$$\overline{f_i}(x) = \begin{cases} 0, & x < x_{i-1} \\ f_i(x), & x \in [x_{i-1}, x_i] \\ 0, & x > x_i \end{cases},\quad \overline{D_i}(y)(x) = \begin{cases} 0, & x < x_{i-1} \\ D_i(x), & x \in [x_{i-1}, x_i] \\ 0, & x > x_i. \end{cases}$$

Definition 1. *Let $k(\Delta) = n + 1$ the dimension of the division Δ and*

$$C([a,b], \Delta, y) = \{f \in C[a,b] : f(x_i) = y_i,\quad \forall\, i = \overline{0,n}\}.$$

An oscillation of interpolation type is a function $\rho : C([a,b], \Delta, y) \times \mathbb{R}^{k(\Delta)} \longrightarrow \mathbb{R}$, having the properties:

(i) (positivity): $\rho(f, y; \Delta) \geq 0,\ \forall f \in C([a,b], \Delta, y), \forall y \in \mathbb{R}^{k(\Delta)}$ and $\rho(f, y; \Delta) = 0 \Longleftrightarrow f = D(y)$,
(ii) (absolute homogeneity)

$$\rho(\alpha \cdot f, \alpha \cdot y; \Delta) = |\alpha| \cdot \rho(f, y; \Delta),\quad \forall \alpha \in \mathbb{R}^*,\quad \forall f \in C([a,b], \Delta, y), \forall y \in \mathbb{R}^{k(\Delta)},$$

(iii) (monotonicity): considering $y = (y_0,\ y_1, ...,\ y_n) \in \mathbb{R}^{k(\Delta)}$ be fixed, for $f, g \in C([a,b], \Delta, y)$ we have the implication:

$$\left\|\overline{f_i} - \overline{D_i}(y)\right\|_C \leq \left\|\overline{g_i} - \overline{D_i}(y)\right\|_C,\quad \forall i = \overline{1,n} \Longrightarrow \rho(f, y; \Delta) \leq \rho(g, y; \Delta),$$

where $\|\cdot\|_C$ is the sup-norm of uniform convergence. Here, $\rho(f, y; \Delta)$ is named the oscillation of interpolation type of the function f corresponding to the division Δ and to the vector y.

Definition 2. *(in [2]) The quadratic oscillation in average of f corresponding to Δ and to y is the function $\rho_2 : C([a,b], \Delta, y) \times \mathbb{R}^{k(\Delta)} \longrightarrow \mathbb{R}$, defined by*

$$\rho_2(f, y; \Delta) = \sqrt{\int_a^b \left(\sum_{i=1}^n [\overline{f_i}(x) - \overline{D_i}(y)(x)]^2\right) dx}. \tag{2}$$

Remark 1. It is easy to see that the quadratic oscillation in average is an oscillation of interpolation type. Moreover, we have $\rho_2(f, y; \Delta) \leq \sqrt{b-a} \cdot \|f - D(y)\|_C$ and after elementary calculus, follows

$$[\rho_2(f, y; \Delta)]^2 = \sum_{i=1}^n \int_{x_{i-1}}^{x_i} [f_i(x) - D_i(x)]^2 dx. \tag{3}$$

Here is the geometric interpretation : In the plane, there exists a set between the graph of f and the polygonal line joining the points (x_i, y_i), $i = \overline{0,n}$. If we rotate this set round about the x-axis, we obtain a body having the volume equal with the value of $[\rho_2(f, y; \Delta)]^2$ multiplied with π. Minimizing $\rho_2(f, y; \Delta)$, the volume of this body will be minimized.

2 The Quartic Spline Generated by Initial Conditions

Let the partition Δ of $[a, b]$,

$$\Delta : a = x_0 < x_1 < ... < x_{n-1} < x_n = b, \tag{4}$$

and $y_0, ..., y_n \in \mathbb{R}$. Consider the quartic spline of interpolation $s : [a, b] \to \mathbb{R}$, $s \in C^3[a, b]$ with $s(x_i) = y_i, \forall\ i = \overline{0,n}$. The restrictions of s to the intervals $[x_{i-1}, x_i]$ will be denoted by $s_i, i = \overline{1,n}$. These restrictions are fourth order polynomials and will be obtained solving the initial value problems:

$$\begin{cases} s_i'''(x) = V_{i-1} + \frac{V_i - V_{i-1}}{h_i} \cdot (x - x_{i-1}), \qquad x \in [x_{i-1}, x_i] \\ s_i(x_{i-1}) = y_{i-1} \\ s_i'(x_{i-1}) = m_{i-1} \\ s_i''(x_{i-1}) = M_{i-1} \end{cases}, i = \overline{1,n}. \tag{5}$$

We get,

$$s_i(x) = \frac{V_i - V_{i-1}}{24h_i} \cdot (x - x_{i-1})^4 + \frac{V_{i-1}}{6} \cdot (x - x_{i-1})^3 + \frac{M_{i-1}}{2} \cdot (x - x_{i-1})^2 + \tag{6}$$

$$+ m_{i-1} \cdot (x - x_{i-1}) + y_{i-1}, \qquad \forall x \in [x_{i-1}, x_i], i = \overline{1,n}.$$

We see that $s_i'''(x_{i-1}) = V_{i-1}$, $s_i'''(x_i) = V_i$ and from the smoothness requirements $s \in C[a, b], s \in C^1[a, b], s \in C^2[a, b]$ follows the conditions: $s_i(x_i) = y_i$, $s_i'(x_i) = m_i$, respectively $s_i''(x_i) = M_i$, $\forall i = \overline{1,n}$. These conditions lead to the relations:

$$\begin{cases} \frac{V_i - V_{i-1}}{24} \cdot h_i^3 + \frac{V_{i-1}}{6} \cdot h_i^3 + \frac{M_{i-1}}{2} \cdot h_i^2 + m_{i-1} \cdot h_i = y_i - y_{i-1} \\ \frac{V_i - V_{i-1}}{6} \cdot h_i^2 + \frac{V_{i-1}}{2} \cdot h_i^2 + M_{i-1} \cdot h_i = m_i - m_{i-1} \\ \frac{V_i - V_{i-1}}{2} \cdot h_i + V_{i-1} \cdot h_i = M_i - M_{i-1}, \qquad i = \overline{1,n}, \end{cases}$$

which can be written in the recurrent form:

$$\begin{cases} m_i = \frac{4(y_i - y_{i-1})}{h_i} - 3m_{i-1} - M_{i-1} \cdot h_i - V_{i-1} \cdot \frac{h_i^2}{6} \\ M_i = \frac{12(y_i - y_{i-1})}{h_i^2} - \frac{12m_{i-1}}{h_i} - 5M_{i-1} - V_{i-1} \cdot h_i \quad , i = \overline{1,n}. \\ V_i = \frac{24(y_i - y_{i-1})}{h_i^3} - \frac{24m_{i-1}}{h_i^2} - \frac{12M_{i-1}}{h_i} - 3V_{i-1} \end{cases} \tag{7}$$

The matrices form of (7) is:

$$\begin{pmatrix} m_i \\ M_i \\ V_i \end{pmatrix} = \begin{pmatrix} -3 & -h_i - & \frac{h_i^2}{6} \\ -\frac{12}{h_i} & -5 & -h_i \\ -\frac{24}{h_i^2} & -\frac{12}{h_i} & -3 \end{pmatrix} \cdot \begin{pmatrix} m_{i-1} \\ M_{i-1} \\ V_{i-1} \end{pmatrix} + (y_i - y_{i-1}) \cdot \begin{pmatrix} \frac{4}{h_i} \\ \frac{12}{h_i^2} \\ \frac{24}{h_i^3} \end{pmatrix}, \ i = \overline{1,n}. \tag{8}$$

The recurrent relations (7) lead to the following existence result:

Theorem 1. *For given $y_0, ..., y_n, m_0, M_0, V_0$ there exists an unique quartic spline of interpolation $s : [a, b] \to \mathbb{R}$, generated by initial conditions and determined by the interpolation conditions:*

$$\begin{cases} s(x_i) = y_i, \quad i = \overline{0,n} \\ s'(x_0) = m_0, \quad s''(x_0) = M_0, \quad s'''(x_0) = V_0. \end{cases}$$

Proof. The function s_i given in (6) is the unique solution of the problem (5) and the function given by the expression $V_{i-1} + \frac{V_i - V_{i-1}}{h_i} \cdot (x - x_{i-1})$ is the unique first order polynomial s_i''' for which $s_i'''(x_{i-1}) = V_{i-1}$ and $s_i'''(x_i) = V_i$ for all $i = \overline{1,n}$. The result follows observing that the parameters $m_i, M_i, V_i,\ i = \overline{1,n}$ are uniquely obtained in (7) starting from $y_0, ..., y_n, m_0, M_0, V_0$.

3 The Main Results

By induction, we obtain

$$\begin{pmatrix} m_1 \\ M_1 \\ V_1 \end{pmatrix} = (y_1 - y_0) \cdot \begin{pmatrix} \frac{4}{h_1} \\ \frac{12}{h_1^2} \\ \frac{24}{h_1^3} \end{pmatrix} - \begin{pmatrix} -3 & -h_1 - & \frac{h_1^2}{6} \\ -\frac{12}{h_1} & -5 & -h_1 \\ -\frac{24}{h_1^2} & -\frac{12}{h_1} & -3 \end{pmatrix} \cdot \begin{pmatrix} m_0 \\ M_0 \\ V_0 \end{pmatrix}$$

and for $k = \overline{2,n}$ follows,

$$\begin{pmatrix} m_k \\ M_k \\ V_k \end{pmatrix} = (y_k - y_{k-1}) \cdot \begin{pmatrix} \frac{4}{h_k} \\ \frac{12}{h_k^2} \\ \frac{24}{h_k^3} \end{pmatrix} + \sum_{j=1}^{k-1} (-1)^j \cdot (y_{k-j} - y_{k-j-1}) \cdot \tag{9}$$

$$\cdot \prod_{i=1}^{j} \begin{pmatrix} 3 & h_{k+1-i} & \frac{1}{6} h_{k+1-i}^2 \\ 12/h_{k+1-i} & 5 & h_{k+1-i} \\ 24/h_{k+1-i}^2 & 12/h_{k+1-i} & 3 \end{pmatrix} + (-1)^k \cdot$$

$$\cdot \prod_{j=1}^{k} \begin{pmatrix} 3 & h_{k+1-j} & \frac{1}{6} h_{k+1-j}^2 \\ 12/h_{k+1-j} & 5 & h_{k+1-j} \\ 24/h_{k+1-j}^2 & 12/h_{k+1-j} & 3 \end{pmatrix} \cdot \begin{pmatrix} m_0 \\ M_0 \\ V_0 \end{pmatrix}.$$

The relations (9) can be written in the form:

$$\begin{pmatrix} m_k \\ M_k \\ V_k \end{pmatrix} = \begin{pmatrix} g_1^k(y;h) \\ g_2^k(y;h) \\ g_3^k(y;h) \end{pmatrix} + A_k(h) \cdot \begin{pmatrix} m_0 \\ M_0 \\ V_0 \end{pmatrix}, \quad \forall k = \overline{1,n}, \tag{10}$$

where $A_k(h) \in M_3(\mathbb{R})$, $h = (h_1, ..., h_n)$. In this way, we obtain:

$$\begin{cases} m_k = a_k(h) \cdot m_0 + b_k(h) \cdot M_0 + c_k(h) \cdot V_0 + g_1^k(y;h) \\ M_k = a_k'(h) \cdot m_0 + b_k'(h) \cdot M_0 + c_k'(h) \cdot V_0 + g_2^k(y;h) \\ V_k = a_k''(h) \cdot m_0 + b_k''(h) \cdot M_0 + c_k''(h) \cdot V_0 + g_3^k(y;h) \end{cases}, \quad \forall k = \overline{1,n}. \tag{11}$$

Putting these expressions in (6) we obtain:

$$s_k(x) = A_k(x) \cdot m_0 + B_k(x) \cdot M_0 + C_k(x) \cdot V_0 + E_k(x), \quad \forall x \in [x_{k-1}, x_k], \quad (12)$$

where $A_k, B_k, C_k, E_k, k = \overline{1,n}$ are fourth degree polynomials:

$$A_k(x) = \frac{a_k''(h) - a_{k-1}''(h)}{24h_k} \cdot (x - x_{k-1})^4 + \frac{a_{k-1}''(h)}{6} \cdot (x - x_{k-1})^3 + \quad (13)$$
$$+ \frac{a_{k-1}'(h)}{2} \cdot (x - x_{k-1})^2 + a_{k-1}(h) \cdot (x - x_{k-1})$$

$$B_k(x) = \frac{b_k''(h) - b_{k-1}''(h)}{24h_k} \cdot (x - x_{k-1})^4 + \frac{b_{k-1}''(h)}{6} \cdot (x - x_{k-1})^3 + \quad (14)$$
$$+ \frac{b_{k-1}'(h)}{2} \cdot (x - x_{k-1})^2 + b_{k-1}(h) \cdot (x - x_{k-1})$$

$$C_k(x) = \frac{c_k''(h) - a_{k-1}''(h)}{24h_k} \cdot (x - x_{k-1})^4 + \frac{c_{k-1}''(h)}{6} \cdot (x - x_{k-1})^3 + \quad (15)$$
$$+ \frac{c_{k-1}'(h)}{2} \cdot (x - x_{k-1})^2 + c_{k-1}(h) \cdot (x - x_{k-1})$$

$$E_k(x) = \frac{g_3^k(y;h) - g_3^{k-1}(y;h)}{24h_k} \cdot (x - x_{k-1})^4 + \frac{g_3^{k-1}(y;h)}{6} \cdot (x - x_{k-1})^3 + \quad (16)$$
$$+ \frac{g_2^{k-1}(y;h)}{2} \cdot (x - x_{k-1})^2 + g_1^{k-1}(y;h) + y_{k-1}, \quad k = \overline{1,n}.$$

From (12) we infer that can be written: $s_k(x; m_{k-1}, M_{k-1}, V_{k-1}, V_k) = s_k(x; m_0, M_0, V_0)$, for all $x \in [x_{k-1}, x_k]$ and $k = \overline{1,n}$. Now, we define the residual:

$$R(m_0, M_0, V_0) = \sum_{i=1}^{n} \int_{x_{i-1}}^{x_i} [s_i(x; m_0, M_0, V_0) - D_i(x)]^2 dx = \quad (17)$$

$$= \sum_{i=1}^{n} \int_{x_{i-1}}^{x_i} [A_i(x) \cdot m_0 + B_i(x) \cdot M_0 + C_i(x) \cdot V_0 + E_i(x) - D_i(x)]^2 dx,$$

where A_i, B_i, C_i, E_i are given in (13)-(16) and $D_i, i = \overline{1,n}$ are given in (1).

Theorem 2. *For given $y_0, ..., y_n$, there exists unique $(\overline{m_0}, \overline{M_0}, \overline{V_0}) \in \mathbb{R}^3$ such that*

$$R(\overline{m_0}, \overline{M_0}, \overline{V_0}) = \min\{R(m_0, M_0, V_0) : (m_0, M_0, V_0) \in \mathbb{R}^3\}.$$

Proof. To find the minimum point of $R(m_0, M_0, V_0)$ we firstly solve the system:

$$\begin{cases} \frac{\partial R}{\partial m_0}(m_0, M_0, V_0) = 0 \\ \frac{\partial R}{\partial M_0}(m_0, M_0, V_0) = 0 \\ \frac{\partial R}{\partial V_0}(m_0, M_0, V_0) = 0. \end{cases} \quad (18)$$

The system (18) has the form:

$$\left\{\begin{array}{l}
m_0 \cdot \left(\sum_{i=1}^{n} \int_{x_{i-1}}^{x_i} [A_i(x)]^2 dx \right) + M_0 \cdot \left(\sum_{i=1}^{n} \int_{x_{i-1}}^{x_i} [A_i(x) \cdot B_i(x)] dx \right) + \\
V_0 \cdot \left(\sum_{i=1}^{n} \int_{x_{i-1}}^{x_i} [A_i(x) \cdot C_i(x)] dx \right) = \sum_{i=1}^{n} \int_{x_{i-1}}^{x_i} [D_i(x) - E_i(x)] \cdot A_i(x)\, dx \\
m_0 \cdot \left(\sum_{i=1}^{n} \int_{x_{i-1}}^{x_i} [A_i(x) \cdot B_i(x)] dx \right) + M_0 \cdot \left(\sum_{i=1}^{n} \int_{x_{i-1}}^{x_i} [B_i(x)]^2 dx \right) + \\
+V_0 \cdot \left(\sum_{i=1}^{n} \int_{x_{i-1}}^{x_i} [B_i(x) \cdot C_i(x)] dx \right) = \sum_{i=1}^{n} \int_{x_{i-1}}^{x_i} [D_i(x) - E_i(x)] \cdot B_i(x)\, dx \\
m_0 \cdot \left(\sum_{i=1}^{n} \int_{x_{i-1}}^{x_i} [A_i(x) \cdot C_i(x)] dx \right) + M_0 \cdot \left(\sum_{i=1}^{n} \int_{x_{i-1}}^{x_i} [B_i(x) \cdot C_i(x)] dx \right) + \\
+V_0 \cdot \left(\sum_{i=1}^{n} \int_{x_{i-1}}^{x_i} [C_i(x)]^2 dx \right) = \sum_{i=1}^{n} \int_{x_{i-1}}^{x_i} [D_i(x) - E_i(x)] \cdot C_i(x)\, dx.
\end{array}\right. \tag{19}$$

We see that

$$\frac{\partial^2 R}{\partial m_0^2} = 2 \cdot \left(\sum_{i=1}^{n} \int_{x_{i-1}}^{x_i} [A_i(x)]^2 dx \right) > 0$$

and

$$\delta_R = \begin{vmatrix} \frac{\partial^2 R}{\partial m_0^2} & \frac{\partial^2 R}{\partial m_0 \partial M_0} \\ \frac{\partial^2 R}{\partial m_0 \partial M_0} & \frac{\partial^2 R}{\partial M_0^2} \end{vmatrix} = 4 \left(\sum_{i=1}^{n} \int_{x_{i-1}}^{x_i} [A_i(x)]^2 dx \right) \cdot$$

$$\cdot \left(\sum_{i=1}^{n} \int_{x_{i-1}}^{x_i} [B_i(x)]^2 dx \right) - 4 \left(\sum_{i=1}^{n} \int_{x_{i-1}}^{x_i} [A_i(x) \cdot B_i(x)] dx \right)^2 > 0,$$

according to the Cauchy-Schwarz inequality. The Hessian of R has the determinant $8 \cdot \Delta_R$, where Δ_R is the determinant of the system (19). After elementary calculus, follows $\Delta_R > 0$. So, the system (19) has unique solution $\left(\overline{m_0}, \overline{M_0}, \overline{V_0}\right)$ which minimize the residual R.

Let $s\left(\cdot; \overline{m_0}, \overline{M_0}, \overline{V_0}\right)$ be the quartic spline having the restrictions $s_i\left(\cdot; \overline{m_0}, \overline{M_0}, \overline{V_0}\right)$ to the intervals $[x_{i-1}, x_i]$, $i = \overline{1, n}$.

Corollary 1. *The quartic spline* $\overline{s} = s\left(\cdot; \overline{m_0}, \overline{M_0}, \overline{V_0}\right)$ *has minimal quadratic oscillation in average.*

Proof. Follows observing that $\rho_2(\overline{s}, y; \Delta) = \sqrt{R\left(\overline{m_0}, \overline{M_0}, \overline{V_0}\right)}$.

As can be viewed in [7], if $f \in C^5[a,b]$ then the error estimation in approximation with quartic spline has order $O(h^4)$. Here, we present the error estimation in the less restrictive case for f to obtain the order $O(h^3)$.

Theorem 3. *If $f \in C^1[a,b]$, Δ is a partition as in (4) and f is piecewise two times differentiable with Lipschitzian second derivatives in this sense, then the error estimation is:*

$$\|f - \overline{s}\|_C \leq 2\left(L' + \max\{|V_i| : i = \overline{0,n}\}\right) \cdot h^3 \tag{20}$$

and

$$\|f' - \overline{s}'\|_C \leq 2\left(L' + \max\{|V_i| : i = \overline{0,n}\}\right) \cdot h^2 \tag{21}$$

where $L' > 0$ is the greatest Lipschitz constant of the second derivatives of the corresponding restrictions of f and $h = \max\{h_i : i = \overline{1,n}\}$.

Proof. Since $f(x_i) = \overline{s}(x_i)$, $\forall i = \overline{0,n}$, then for $\varphi = f - \overline{s}$ there exist $\xi_i \in (x_{i-1}, x_i)$ such that $\varphi'(\xi_i) = 0$, $\forall\, i = \overline{1,n}$, that is $f'(\xi_i) = \overline{s}'(\xi_i)$, $\forall\, i = \overline{1,n}$. With similar reason, there exists $u_i \in (\xi_i, \xi_{i+1})$ such that $f''(u_i) = \overline{s}''(u_i)$. Let arbitrary $x \in [a,b]$. Then there exists an $i \in \{1,...,n\}$ such that $x \in [x_{i-1}, x_i]$. We have,

$$|f'(x) - \overline{s}'(x)| \leq \int_{\xi_i}^{x} |f''(t) - \overline{s}''(t)|\, dt$$

$$\leq \int_{\xi_i}^{x} [|f''(t) - f''(u_i)| + |\overline{s}''(u_i) - \overline{s}''(t)|]dt \leq$$

$$\leq \int_{\xi_i}^{x} [L'|t - u_i| + \|\overline{s}'''\|_C \cdot |t - u_i|]dt \leq \int_{\xi_i}^{x} (L' + \|\overline{s}'''\|_C) \cdot 2h dt \leq 2(L' + \|\overline{s}'''\|_C) \cdot h^2$$

and

$$|f(x) - \overline{s}(x)| \leq \int_{x_{i-1}}^{x} |f'(t) - \overline{s}'(t)|\, dt$$

$$\leq \int_{x_{i-1}}^{x} 2(L' + \|\overline{s}'''\|_C) \cdot h^2 dt \leq 2(L' + \|\overline{s}'''\|_C) \cdot h^3.$$

Since $\|\overline{s}'''\|_C = \max\{\|\overline{s_i}'''\|_C : i = \overline{1,n}\}$ and $\|\overline{s_i}'''\|_C = \max\{|V_i|, |V_{i-1}|\}$, we obtain (20) and (21).

Remark 2. In the above theorem, the " second derivatives of the restrictions of f " are taken corresponding to the partition of $[a,b]$ realized by the points $\xi_i, i = \overline{1,n}$. The aim of this paper is to prove the existence and uniqueness of the quartic spline of interpolation generated by initial conditions having minimal quadratic oscillation in average and to determine its error estimation, but not to compute effectively them. The effective computation of this quartic spline and the corresponding algorithm will be the content of a future work.

Acknowledgement. The research on this paper is supported by the grant 2Cex-06-11-96/19.09.2006 of the National Authority for Scientific Research from the Minister of Education and Research, Romanian Government.

References

1. Ahlberg, J.H., Nilson, E.N., Walsh, J.L.: The theory of splines and their applications. Academic Press, New York (1967)
2. Bica, A.M.: Iterative numerical methods for operatorial equations. University of Oradea Press (2006)
3. Bica, A.M., Căuş, V.A., Fechete, I., Mureşan, S.: Application of the Cauchy-Buniakovski-Schwarz's inequality to an optimal property for cubic splines. J. of Computational Analysis and Applications 9(1), 43–53 (2007)
4. De Boor, C.: A practical guide to splines. Applied Math. Sciences, vol. 27. Springer, Berlin (1978)
5. Gao, X., Shu, S., Fu, K.: Quartic spline on spline interpolation. J. Comput. Appl. Math. 71(2), 213–223 (1996)
6. Grandine, T.A., Hogan, T.A.: A parametric quartic spline interpolant to position, tangent and curvature. Computing 72(1-2), 65–78 (2004)
7. Howell, G., Varma, A.K.: Best error bounds for quartic spline interpolation. J. Approximation Theory 58(1), 58–67 (1989)
8. Iancu, C.: On the cubic spline of interpolation. Semin. of Funct. Anal. and Num. Meth. 4, 52–71 (1981) (preprint)
9. Karaballi, A.A., Sallam, S.: Quartic spline interpolation on uniform meshes with application to quadratures. J. Math. Res. Expo. 19(3), 533–538 (1999)
10. Micula, G., Micula, S.: Handbook of splines. Mathematics and its Applications, vol. 462. Kluwer Academic Publishers, Dordrecht (1999)
11. Rana, S.S., Dubey, Y.P.: Best error bounds for deficient quartic spline interpolation. Indian J. Pure Appl. Math. 30(4), 385–393 (1999)
12. Rana, S.S., Gupta, R.: Deficient discrete quartic spline interpolation. Rocky Mt. J. Math. 35(4), 1369–1379 (2005)
13. Usmani, R.A.: Error bounds in periodic quartic spline interpolation. Approx. Theory Appl. 12(3), 1–9 (1996)
14. Usmani, R.A.: On nonperiodic quartic spline interpolation. Int. J. Comput. Math. 57(3-4), 197–211 (1995)
15. Volkov, Y.S.: Best error bounds for the derivative of a quartic interpolation spline. Sib. Adv. Math. 9(2), 140–150 (1999)

Expressions of Solutions of Linear Partial Differential Equations Using Algebraic Operators and Algebraic Convolution

Liepa Bikulčienė[1] and Zenonas Navickas[2]

[1] Department of Applied Mathematics, Kaunas University of Technology, Studentu 50-323, Lithuania
liepa.bikulciene@ktu.lt

[2] Department of Applied Mathematics, Kaunas University of Technology, Studentu 50-325c, Lithuania
zenonas.navickas@ktu.lt

Abstract. An operator – algebraic algorithm of solution of linear partial differential equations suitable to be realized using computers is presented in this article; furthermore, examples of application are given as well.

Keywords: Algebraic operator, Cauchy problem, perfect operator, algebraic convolution, Appell polynomials.

Introduction

Usually digital methods are used for solving differential equations, besides a lot of articles have been written and exhaustive studies have been performed on this topic. However, in this article only the operator method will be discussed. Usually the operator expressions of solutions (for instance, described in [1]) are not convenient to be realized using computers.

As the potential of the electronic calculation has been developing, a possibility to apply other, the so called algebraic methods for the solution of differential equations occurs. The substance of these methods is that a solution can be recorded in a conformation of power series. The above mentioned power series can be derived by using the operator – algebraic method and specially determined algebraic convolution. Furthermore, the series consist of algebraic symbols (parameters, specifying the series), and when digital values are inserted instead of these symbols, a digital series – an exact solution of the differential equation - are obtained. It is possible to get quite exact approximations of solutions from the series at our request and according to our opportunities.

L. Euler and G.W. Leibnitz can be regarded as the originators of the methods of algebraic calculation. Mathematicians of subsequent ages S. Lie, E. Cartan, G. Weil and others noticeably improved the methods of algebraic calculation, creating indefectibly coherent exhaustive theory of algebraic operators. Mathematicians of the present times, such as: O. Viskov, M. Rahula, V. Maslov, Ph. Feinsilver and R. Schott, added esthetic – logical shine to the method of algebraic

S. Margenov, L.G. Vulkov, and J. Waśniewski (Eds.): NAA 2008, LNCS 5434, pp. 208–215, 2009.

calculations, for instancc, showing that any linear operator can be expressed only by two operators, namely differential (operator "Down") and multiplication (operator "Up"), using algebraic expressions of these operators. In this article the authors show the way to record solutions of linear partial differential equations by algebraic operators. The authors have published several articles on the application of these operators for solution of ordinary differential equations [2], [3]. It must be noted that it is expedient to use the calculation methods presented in this article in parallel with the digital method.

Features of algebraic operators used in this paper are presented in section 1. The algebraic-operator algorithm of solution for differential equations is presented in the section 2, whereas applications of this algorithm are shown in sections 3 and 4.

1 The Auxiliary Results

The main concepts, related with algebraic operators and some their features, described in the [2], will by represented in this section.

Solutions of the differential equations will be expressed by power series:

$$f := \sum_{k,l,r=0}^{+\infty} a_{k,l,r} \left(x - x_0\right)^k s^l t^r, \tag{1}$$

where $a_{k,l,r}, x_0 \in R$ are the constants and $x, s, t \in R$ are the variables, which get values in the domain E_{xst}(in this domain the power series (1) converge absolutely). Now domain of convergence E_{xst} is not indicated, but in problems of application this domain can be found easily.

The set of power series (1) is noted by symbol $F_{x-x_0,s,t}$, the set of series without variable t is noted as $F_{x-x_0,s}$, etc.

The ordinary addition and the product with scalar operations can be performed with power series (1) from set $F_{x-x_0,s,t}$. So, the set $F_{x-x_0,s,t}$(or $F_{x-x_0,s,\ldots}$) with following operations form the linear space, noted as symbol $lsp_{x-x_0,s,t}$ (or $lsp_{x-x_0,s,\ldots}$).

The linear operators defined in linear space $lsp_{x-x_0,s,t}$ are called as algebraic and are noted by the letters $A, B, \ldots$, i.e. $A, B, \ldots$: $lsp_{x-x_0,s,t} \to lsp_{x-x_0,s,t}$. If operator A has inverse operator A^{-1}, then notation $A : lsp_{x-x_0,s,t} \leftrightarrow lsp_{x-x_0,s,t} : A^{-1}$ can be used, etc.

The elementary operators L_{x-x_0}, D_x, D_s, D_t will be defined by expressions: $L_{x-x_0} \left(x - x_0\right)^n := \frac{(x-x_0)^{n+1}}{n+1}$, $D_x \left(x - x_0\right)^n := n \left(x - x_0\right)^{n-1}$, $D_s s^n := n s^{n-1}$, $D_t t^n := n t^{n-1}$, where $n = 0, 1, 2, \ldots$. It is observed that $L_{x-x_0} 1 := x - x_0$, $D_x 1 := 0$, etc.

The algebraic combinations of the elementary operators are used below. In this case the ordinary addition and the product with scalar operations are known, besides, $L_{x-x_0} D_s = D_s L_{x-x_0}$, $D_s D_t := D_t D_s$ etc., (but $D_x L_{x-x_0} \neq L_{x-x_0} D_x$).

Let us suppose that using linear operator $A : lsp_{x-x_0,s,t} \to lsp_{x-x_0,s,t}$ it is possible to form the linear operator $g(A) := \sum_{n=0}^{+\infty} A^n$, if $A^0 := 1$ and this operator is algebraic also. Besides, representing any series $f \in F_{x-x_0,s,t}$ using operator $g(A)$, only the finite number of arithmetic operations with the real numbers is executing. This means that in calculation of coefficients of fixed degrees $(x-x_0)^k s^l t^r$ is needless to sum an infinite number of components or multiply the infinite number of factors. Such operator A is called the perfect operator in the linear space $lsp_{x-x_0,s,t}$.

The operators $g(A)$ will be used below in the case of perfect operator A, because in inverse case the calculations with algebraic operators will be incorrect.

Let $A : lsp_{x-x_0,s,t} \to lsp_{x-x_0,s,t}$, then

$KerA := \{f| \, f \in F_{x-x_0,s,t}, Af = 0\}$, $ImA := \left\{\hat{f} \middle| \, Af = \hat{f}; f, \hat{f} \in F_{x-x_0,s,t}\right\}$.

Let A is perfect operator. Then following relationships hold true:

$(g(A))^{-1} = 1 - A$.

Let $A = A_0 - A_1$ and, besides, it is possible to find the operator B_0, which satisfy conditions:

a) $A_0 B_0 = 1$, b) operator $B_0 A_1$ is perfect.

Then $(A_0 - A_1)\, g(B_0 A_1) = A_0$.
Let operator $A = A_0 - A$ satisfy conditions. Then

$$KerA = g(B_0 A_1)\, KerA_0, KerA_0 = (1 - B_0 A_1)\, KerA. \tag{2}$$

Let operator $A = A_0 - A_1$ satisfy the conditions and, besides, it is possible to relate linear operators T and R with linear space $\langle Q; +|\, R\rangle$ (further this space is called the linear space of Cauchy conditions). The following operators satisfy the conditions:

$$T : lsp_{x-x_0,s,t} \to \langle Q; +|\, R\rangle, T : \langle KerA; +|\, R\rangle \leftrightarrow \langle Q; +|\, R\rangle : R \tag{3}$$

i.e. $TR|_Q = 1|_Q$, $RT|_{KerA} = 1|_{KerA}$.
Then the system of operator equations (the operator Cauchy problem)

$$\begin{cases} Af = \hat{f}, \hat{f} \in \mathrm{Im}A, \\ Tf = \hat{q}, \hat{q} \in Q \end{cases} \tag{4}$$

has the unique solution $f \in F_{x-x_0,s,t}$, specified by expression

$$f = (1 - RT)\, g(B_0 A_1)\, B_0 \hat{f} + R\hat{q}. \tag{5}$$

Thus, the relationship (5) shows the expedience of composition of the linear algebraic operator $g(A)$.

The generalized expressions of the features are presented in [2].

Let the convergent for all x, s, t series be given:

$P := P(x-x_0, s, t)$, $Q := Q(x-x_0, s, t)$, $f_k := f(x-x_0, s, t)$, $f_r := f_r(s, t)$, $r = 0, 1, \ldots, k-1; k, m, n \in N$, i.e. $p, q, f_k \in F_{x-x_0,s,t}, f_r \in F_{s,t}$. Then the perfect linear operator $L_x(PD_x^m + QD_t^n)$ in the linear space $lsp_{x-x_0,s,t}$, i.e. $G :=$

$\sum_{j=0}^{+\infty}\left(L_{x-x_0}^{j}\left(PD_x^m+QD_t^n\right)^j\right): lsp_{x-x_0,s,t}\leftrightarrow lsp_{x-x_0,s,t}: 1-L_{x-x_0}\left(PD_x^m+QD_t^n\right)$ is algebraic, besides, the solution of differential equation with boundary conditions

$$\begin{cases} \frac{\partial^k w}{\partial x^k}-P\frac{\partial^m w}{\partial s^m}-Q\frac{\partial^n w}{\partial t^n}=f_k\left(x-x_0;s,t\right) \\ \frac{\partial^r w}{\partial x^r}\big|_{x=x_0}=f_r\left(s,t\right), r=0,1,\ldots,k-1; \end{cases} \tag{6}$$

is expressed by relationship

$$w=Gf, \tag{7}$$

if $f:=f_0\left(s,t\right)+f_1\left(s,t\right)\frac{(x-x_0)}{1!}+\cdots+f_{k-1}\left(s,t\right)\frac{(x-x_0)^{k-1}}{(k-1)!}+L_{x-x_0}^{k}f_k\left(x;s,t\right)$.

2 Algebraic Algorithm of Solution

It is possible to observe that relationship (7) is powerful when $P,Q,f_r;r=0,1,\ldots,k$ are the conditional simple functions (like in the Example 1), but it is necessary to restructure relationship (7) if following functions are complicated.

In this section one modification of relationship (7) is presented. This modification allows realizing algebraic algorithm, using computer.

The principle assumption is made:

$$P=P\left(s\right),Q=Q\left(t\right),\rho=\rho\left(s\right),\psi=\psi\left(t\right). \tag{8}$$

Then linear operators PD_s^m and QD_t^n become commutative and after the notation $\lambda_j\left(s\right):=\left(PD_s^m\right)^j\rho\left(s\right), \mu_r\left(t\right):=\left(QD_t^n\right)^r\psi\left(t\right)$ it is possible to get that

$\left(PD_s^m\right)^j\left(QD_t^n\right)^r\rho\left(s\right)\psi\left(t\right)=\lambda_j\left(s\right)\mu_r\left(t\right)$, if $j,r=0,1,\ldots$, and $\left(PD_s^m\right)^0=\left(QD_t^n\right)^0=1, \lambda_0\left(s\right)=\rho\left(s\right), \mu_0\left(t\right)=\psi\left(t\right)$.

After choosing $k=1,2,\ldots$, it is possible to make the series

$\tilde{\rho}\left(x,s\right):=\sum_{j=0}^{+\infty}\lambda_j\left(s\right)\frac{x^{kj}}{(kj)!};\tilde{\psi}\left(x,t\right)=\sum_{j=0}^{+\infty}\mu_r\left(t\right)\frac{x^{kj}}{(kj)!}$.

The correctness of the following identity can be examined for $a,b\in R$:

$\left(a\rho_1\left(s\right)\tilde{+}b\rho_2\left(s\right)\right)=a\tilde{\rho}_1\left(x,s\right)+b\tilde{\rho}_2\left(x,s\right)$. Besides, such property is also hold true.

Property 1. $D_x^k\tilde{\rho}\left(x,s\right)=\left(PD_x^m\right)\tilde{\rho}\left(x,s\right);D_x^k\tilde{\psi}\left(x,t\right)=\left(QD_t^n\right)\tilde{\psi}\left(x,t\right)$.

The algebraic operation of convolution $*$ can be described with series

$f_1:=\sum_{j=0}^{+\infty}a_j\frac{x^{kj}}{(kj)!}$ and $f_2:=\sum_{j=0}^{+\infty}b_j\frac{x^{kj}}{(kj)!}$, if $\begin{pmatrix} j \\ l \end{pmatrix}:=\frac{j!}{l!(j-l)!}, l=0,1,\ldots,j$, i.e.,

$$f_1*f_2:=\sum_{j=0}^{+\infty}\left(\sum_{j=0}^{l}\begin{pmatrix} j \\ l \end{pmatrix}a_lb_{j-l}\right)\frac{x^{kj}}{(kj)!}. \tag{9}$$

It is observed that $f_1 * f_2 = f_1 \cdot f_2$, if $k = 1$. Thus, the operation of algebraic convolution introduced there is generalization of the ordinary product of series.

The following identities hold true.

$f_1 * f_2 = f_2 * f_1$, because $\binom{j}{l} = \binom{j}{l-j}$.

The series $e := 1$ exist, that $e * f = f * e = f$.

$(f_1 + f_2) * f_3 = f_1 * f_3 + f_2 * f_3$.
$\eta (f_1 * f_2) = (\eta f_1) * f_2 = f_1 * (\eta f_2)$, if $\eta \in R$.

There and further, the sum, product and product with scalar of series are interpreted in classical sense.

Property 2. $(f_1 * f_2) * f_3 = f_1 * (f_2 * f_3)$.

The properties, presented below, and the theorem for algebraic convolution with series $\hat{\rho}(x, s)$ and $\hat{\psi}(x, t)$ are hold true.

Property 3. $\tilde{\rho}(x, s) = g\left(L_x^k P(s) D_s^m\right) \rho(s)$.

Property 4. $\tilde{\rho} * \tilde{\psi} = g\left(L_x^k \left(PD_s^m + QD_t^n\right)\right) \rho\psi$.

It is possible to write the expressions of solutions of Cauchy problems as algebraic convolution, using formulation of Cauchy problem (6), solution (7) and theorem.

Relationships between Three Solutions of the Cauchy Problem
Let the differential equations be given:

$$\frac{\partial^k u}{\partial x^k} - P\frac{\partial^m u}{\partial s^m} = \rho_k(s),\ \frac{\partial^k v}{\partial x^k} - Q\frac{\partial^n v}{\partial t^n} = \psi_k(t),$$
$$\frac{\partial^k w}{\partial x^k} - \left(P\frac{\partial^m w}{\partial s^m} + Q\frac{\partial^n w}{\partial t^n}\right) = \rho_k(s)\,\psi_k(t),$$

if $u = u(x, s), v = v(x, t), w = w(x, s, t)$, and, besides,

$$\left.\frac{\partial^r u}{\partial x^r}\right|_{x=0} = \rho_r,\ \left.\frac{\partial^r v}{\partial x^r}\right|_{x=0} = \psi_r,\ \left.\frac{\partial^r w}{\partial x^r}\right|_{x=0} = \rho_r\psi_r, r = 0, 1, \ldots, k-1.$$

Then their solutions are written as follow expressions:

$$u = \sum_{r=0}^{k} L_x^r \hat{\rho}_r(x, s), v = \sum_{r=0}^{k} L_x^r \hat{\psi}_r(x, t), w = \sum_{r=0}^{k} L_x^r \left(\hat{\rho}_r(x, s) * \hat{\psi}_r(x, t)\right). \quad (10)$$

Properly, using (7) relationship it is observed that:

$$u(x, s) = g\left(L_x^k PD_s^m\right) \sum_{r=1}^{k} L_x^r \rho_r(s) = \sum_{r=1}^{k} L_x^r g\left(L_x^k PD_s^m\right) \rho_r(s) = \sum_{r=1}^{m} L_x^r \hat{\rho}_r(x, s).$$

3 The Applications (Theoretical Part)

The presented above with definition of convolution (9) and expression of solution method (10) allows easily solve partial differential equations with constant coefficients.

Let $P := a^m; Q := b^n; a, b \in R$ be given. Then it is possible to define auxiliary functions:

$\lambda_j^{(l)} := (aD_s)^{mj} \frac{s^l}{l!} = \frac{a^{mj} s^{l-mj}}{(l-mj)!}; \mu_j^{(l)} := (bD_t)^{nj} \frac{t^l}{l!} = \frac{b^{mj} s^{l-nj}}{(l-nj)!}$, if $j, l = 0, 1, 2, \ldots$, and, besides, $(-n)! = \pm\infty$, when $n \in N$.

The series $\overline{\left(\frac{s^l}{l!}\right)} := \sum\limits_{j=0}^{+\infty} \lambda_j^{(l)} \frac{x^{kj}}{(kj)!} = \sum\limits_{j=0}^{\left[\frac{l}{m}\right]} \frac{a^{mj} s^{l-mj}}{(l-mj)!} \frac{x^{kj}}{(kj)!}$,

$\overline{\left(\frac{t^l}{l!}\right)} := \sum\limits_{j=0}^{+\infty} \mu_j^{(l)} \frac{x^{kj}}{(kj)!} = \sum\limits_{j=0}^{\left[\frac{l}{n}\right]} \frac{b^{nj} s^{l-nj}}{(l-nj)!} \frac{x^{kj}}{(kj)!}$

become to polynomials of Appell. The definition of polynomials of Appell by O.Viskov is presented in [4].

There $\left[\frac{l}{n}\right]$ is a integer part of real number $\frac{l}{n}$, i.e., $\left[\frac{l}{n}\right] \in N$ and, besides, $\frac{l}{n} - 1 < \left[\frac{l}{n}\right] \leq \frac{l}{n}$. The convolution of the functions $\overline{\left(\frac{s^\alpha}{\alpha!}\right)}$ and $\overline{\left(\frac{t^\beta}{\beta!}\right)}$

$\overline{\left(\frac{s^\alpha}{\alpha!}\right)} * \overline{\left(\frac{t^\beta}{\beta!}\right)} = \sum\limits_{j=0}^{\left[\frac{\alpha}{m}\right]+\left[\frac{\beta}{n}\right]} \left(\sum\limits_{l=0}^{j} \left(\binom{j}{l} \lambda_l^{(\alpha)}(s) \mu_{j-l}^{(\beta)}(t) \right) \frac{x^{kl}}{(kj)!} \right)$ is a polynomial also.

For instance, if $P = -1, Q = 2, k = 1, m = 1, n = 2$, then

$\overline{\left(\frac{s^2}{2!}\right)} = \frac{s^2}{2} - sx + \frac{x^2}{2}, \hat{t} = t, \overline{\left(\frac{s^2}{2!}\right)} * t = \frac{s^2 t}{2} - stx + t\frac{x^2}{2}$.

It is observed that series $\tilde{\rho}(x, s)$ and $\tilde{\psi}(x, t)$ are the generalization of polynomials of Appell. It is possible to express the solutions of following differential equations, using substitutions $\overline{\left(\frac{s^\alpha}{\alpha!}\right)}, \overline{\left(\frac{t^\beta}{\beta!}\right)}$ of the functions $\tilde{\rho}$and $\tilde{\psi}$, and substitution $\overline{\left(\frac{s^\alpha}{\alpha!}\right)} * \overline{\left(\frac{t^\beta}{\beta!}\right)}$ of theirs convolution $\tilde{\rho} * \tilde{\psi}$.

Problem 1. Let a differential equation

$$\frac{\partial^k u}{\partial x^k} - a^m \frac{\partial^m u}{\partial s^m} = \sum_{l=0}^{+\infty} \nu_{kl} \frac{s^l}{l!} \tag{11}$$

with boundary conditions

$$\left. \frac{\partial^r u}{\partial x^r} \right|_{x=0} = \sum_{l=0}^{+\infty} \nu_{rl} \frac{s^l}{l!}, r = 0, 1, \ldots, k-1 \tag{12}$$

be given. Then its solution is

$$u = \sum_{r=0}^{k} L_x^r \sum_{l=0}^{+\infty} \nu_{rl} \overline{\left(\frac{s^l}{l!}\right)} = \sum_{r=0}^{k} \sum_{\alpha=0}^{+\infty} \nu_{r\alpha} \sum_{j=0}^{+\infty} \frac{a^{mj} s^{\alpha-mj}}{(\alpha-mj)!} \cdot \frac{x^{kj+r}}{(kj+r)!}, \tag{13}$$

where $\frac{1}{(-n)!} := 0, n = 1, 2, 3, \ldots$.

Problem 2. Let a differential equation

$$\frac{\partial^k w}{\partial x^k} - a^m \frac{\partial^m w}{\partial s^m} - b^n \frac{\partial^n w}{\partial t^n} = \sum_{\alpha,\beta=0}^{+\infty} \gamma_{k\alpha\beta} \frac{s^\alpha}{\alpha!} \frac{t^\beta}{\beta!} \tag{14}$$

with boundary conditions

$$\left.\frac{\partial^r w}{\partial x^r}\right|_{x=0} = \sum_{\alpha,\beta=0}^{k} \gamma_{r\alpha\beta} \frac{s^\alpha}{\alpha!} \frac{t^\beta}{\beta!}, r = 0, 1, \ldots, k-1 \tag{15}$$

be given. Then its solution is expressed by convolutions of polynomials of Appell:

$$\begin{aligned} w &= \sum_{r=0}^{k} L_x^r \sum_{\alpha,\beta=0}^{+\infty} \gamma_{r\alpha\beta} \overline{\left(\tfrac{s^\alpha}{\alpha!}\right)} * \overline{\left(\tfrac{t^\beta}{\beta!}\right)} \\ &= \sum_{r=0}^{k} \sum_{\alpha,\beta=0}^{+\infty} \gamma_{r\alpha\beta} \sum_{j,l=0}^{+\infty} \binom{j}{l} \frac{a^{ml} s^{\alpha-ml}}{(\alpha-ml)!} \cdot \frac{b^{n(j-l)} t^{\beta-n(l-l)}}{(\beta-n(j-l))!} \cdot \frac{x^{kj+r}}{(kj+r)!}. \end{aligned} \tag{16}$$

4 The Applications (Practical Part)

The examples of calculations, not associated with exact problem, are proposed in this section. The aim is demonstrate the possibilities of described above method.

Let the differential operator $D_{xst} := \frac{\partial^3}{\partial x^3} - \frac{\partial^4}{\partial s^4} - 4\frac{\partial^2}{\partial t^2}$ be given. The some problems of Cauchy

1) $D_{xst} w_1 = \sin(st)$; $\left.\frac{\partial^r w_1}{\partial x^r}\right|_{x=0} = 0; r = 0, 1, 2,$
2) $D_{xst} w_2 = 0$; $w_2(0, s, t) = t \sin s \cos t$; $\left.\frac{\partial^r w_2}{\partial x^r}\right|_{x=0} = 0; r = 1, 2,$
3) $D_{xst} w_3 = 0$; $\left.\frac{\partial^r w_3}{\partial x^r}\right|_{x=0} = 0; r = 0, 1$; $\left.\frac{\partial^2 w_3}{\partial x^2}\right|_{x=0} = te^{-s}$

are composed. Using solution (16) of problem of Cauchy (14), the expressions of solutions w_1, w_2, w_3 by power series are obtained:

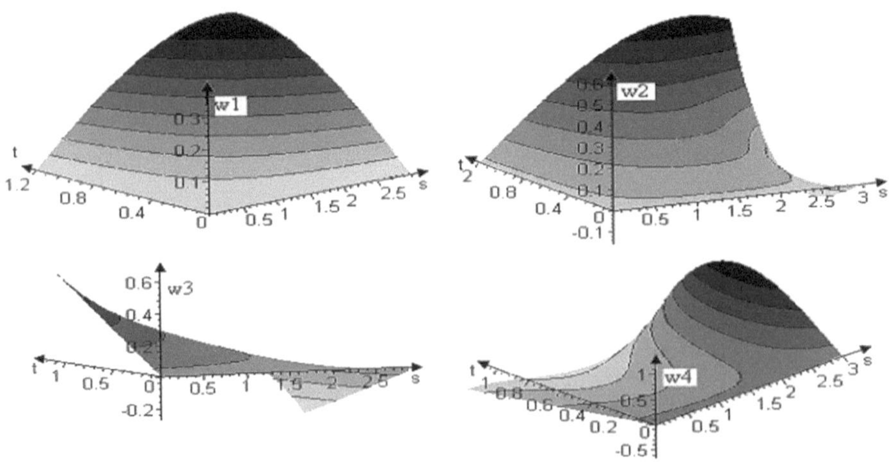

Fig. 1. Approximations of solutions ($x = 1, s = 0..3, t = 0..1.2$)

$$w_1 = \sum_{k=0}^{+\infty} \Psi_{1k}(s,t)x^k, w_2 = \sum_{k=0}^{+\infty} \Psi_{2k}(s,t)x^k, w_3 = \sum_{k=0}^{+\infty} \Psi_{3k}(s,t)x^k.$$

Using the partial sum of series, the approximations $\tilde{w}_1, \tilde{w}_2, \tilde{w}_3$ of solutions w_1, w_2, w_3 are calculated.

It is possible to make the approximation $\tilde{w}_4 = 2\tilde{w}_1 - 3\tilde{w}_2 - \tilde{w}_3$ for solution of problem of Cauchy

4) $D_{xst}w_4 = 2\sin(st)$; $w_4(0,s,t) = -3\sin s\cos t$; $\frac{\partial w_4}{\partial x}|_{x=0} = 0$; $\frac{\partial^2 w_4}{\partial x^2}|_{x=0} = -te^{-s}$.

The pictures of approximations are shown in the Fig 1. There the variable x is fixed, the variable s vary from 0 to 3, t vary from 0 to 1,2.

Computer realization is done using software Maple. The graphical information are realized using the standard function *contourplot3d*.

In the paper, the accuracy of approximations and errors is not analyzed.

5 Conclusions

It is expedient to change an operator expression of a differential equation solution $w = Gf(x,s,t)$ into operator – algebraic expression of a solution $w = \sum_{r=0}^{+\infty} L_x^r\left(\hat{\rho}_r(x,s) * \hat{\psi}(x,t)\right)$ as the electronic calculation technology has been developing. This allows utilize the resources of electronic calculation in a more optimal way.

The algebraic – operator method of solution of differential equations presented in his study, can be successfully applied for quite a broad type of linear differential equations.

Obtained solutions in the form of algebraic series and polynomials can also be successfully used for further procedure of formation of topological solutions.

It is possible to perform the research of analytical and other characteristics of both solutions and differential equations using obtained algebraic – operator expressions of solutions.

From the algebraic – operator algorithms here presented other algorithms can be formed as well that later can be quite effectively utilized creating computer calculation algorithms.

References

1. Weil, J.-A.: Recent algorithms for Solving second – order differential equations. In: Chyzak, F. (ed.) Algorithms Seminar, 2001-2002, pp. 43–46. INRIA (2003)
2. Navickas, Z.: Constructive solution of the Cauchy problem for a special class of partial differential equations with constant coefficients. Lithuanian Mathematical Journal 34(4), 404–414 (1994)
3. Bikulčienė, L., Marcinkevičius, R., Navickas, Z.: Computer Realization of the Operator Method for Solving of Differential Equations. In: Li, Z., Vulkov, L.G., Waśniewski, J. (eds.) NAA 2004. LNCS, vol. 3401, pp. 182–189. Springer, Heidelberg (2005)
4. Viskov, O.V.: Operator characterization for generalized Appell polynomials. Report Akad. Nauk SSSR 225(4) (mathematics), 749–752 (1975)

Multilevel Splitting of Weighted Graph-Laplacian Arising in Non-conforming Mixed FEM Elliptic Problems

P.T. Boyanova and S.D. Margenov

Institute for Parallel Processing, Bulgarian Academy of Sciences, Bulgaria
{petia,margenov}@parallel.bas.bg

Abstract. We consider a second-order elliptic problem in mixed form that has to be solved as a part of a projection algorithm for unsteady Navier-Stokes equations. The use of Crouzeix-Raviart non-conforming elements for the velocities and piece-wise constants for the pressure provides a locally mass-conservative algorithm. Then, the Crouzeix-Raviart mass matrix is diagonal, and the velocity unknowns can be eliminated exactly. The reduced matrix for the pressure is referred to as weighted graph-Laplacian. In this paper we study the construction of optimal order preconditioners based on algebraic multilevel iterations (AMLI). The weighted graph-Laplacian for the model 2-D problem is considered. We assume that the finest triangulation is obtained after recursive uniform refinement of a given coarse mesh. The introduced hierarchical splitting is the first important contribution of this article. The proposed construction allows for a local analysis of the constant in the strengthened Cauchy-Bunyakowski-Schwarz (CBS) inequality. This is an important characteristic of the splitting and is associated with the angle between the two hierarchical FEM subspaces. The estimates of the convergence rate and the computational cost at each iteration show that the related AMLI algorithm with acceleration polynomial of degree two or three is of optimal complexity.

1 Introduction

In this paper we consider the elliptic problem in mixed form:

$$\begin{aligned} \mathbf{u} + \nabla p = \mathbf{f} \quad & \text{in} \quad \bar{\Omega}, \\ \nabla \cdot \mathbf{u} = 0 \quad & \text{in} \quad \Omega, \\ \mathbf{u} \cdot \mathbf{n} = 0 \quad & \text{on} \quad \partial\Omega, \end{aligned} \tag{1}$$

where Ω is a convex polygonal domain in $\mathbb{R}^2$ and $\mathbf{f}$ is a given smooth vector valued function. In the paper we use the CFD terminology and refer to the vector $\mathbf{u}$ as a velocity and p as a pressure.

This problem has to be solved at the projection step of a composite time stepping method for unsteady Navier-Stokes equations. Crouzeix-Raviart linear finite elements (FEs) for the velocity and piece-wise constants for the pressure

S. Margenov, L.G. Vulkov, and J. Waśniewski (Eds.): NAA 2008, LNCS 5434, pp. 216–223, 2009.

are assumed, see e.g. [2] and the references therein. The resulting scheme is stable and locally mass conservative.

The construction of efficient methods for the solution of the discretized projection step has been addressed before. In [6] a MIC(0) preconditioning was proposed for the reduced Schur complement system. We follow the same approach of eliminating the velocity unknowns in the original system of the symmetric saddle point problem. As the Crouzeix-Raviart mass matrix is diagonal, this can be done locally and the derived system for the pressure has a symmetric and positive semidefinite matrix with structure similar to the one of a graph-Laplacian (see e.g.[7] for the motivation of the name). We refer to it as weighted graph-Laplacian. The values of the links between elements introduce weights. For the 2-D model problem and uniform mesh of right triangles the Schur complement matrix corresponds to the T-shaped four point stencil shown in Fig. 1.

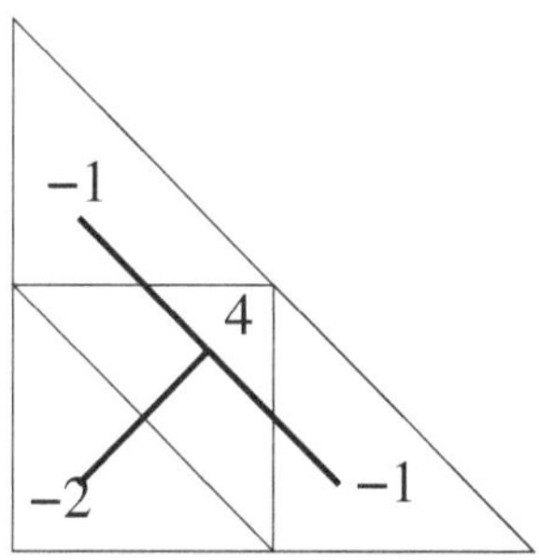

Fig. 1. Schur complement four point stencil for the pressure

The rest of the paper is organized as follows. In Section 2 we introduce the AMLI preconditioner and the main theorem concerning the estimate of its relative condition number with respect to the original system matrix. In Section 3 we present the framework for local analysis of the CBS constant and pivot block preconditioner. In the final Section 4 we propose a decomposition of the weighted graph-Laplacian for the model problem in 2-D based on a properly introduced hierarchical basis.

2 AMLI Preconditioner

The framework of the algebraic multilevel iterations (AMLI) method was originally proposed in [1] for the case of linear conforming FEs. More recently, AMLI methods for non-conforming FEs and discontinuous Galerkin (DG) systems were developed, see [3,4,5,7] and the references therein. This paper follows the notations and the spirit of [7] where the case of graph-Laplacian matrices arising in preconditioning of DG systems is studied.

We consider a sequence of nested triangulations $\mathcal{T}_m \subset \mathcal{T}_{m-1} \subset \cdots \subset \mathcal{T}_0$ of the domain Ω, constructed by recursive uniform refinement of a given initial mesh.

Next we define the corresponding spaces of piece-wise constant functions $\mathcal{V}^{(m)} \subset \mathcal{V}^{(m-1)} \subset \cdots \subset \mathcal{V}^{(0)}$, the spaces of degrees of freedom $\mathbb{V}^{(m)}, \mathbb{V}^{(m-1)}, \cdots, \mathbb{V}^{(0)}$, the numbers of degrees of freedom $n_m < n_{m-1} < \cdots < n_0$, and the weighted graph-Laplacians associated with each triangulation level $A^{(m)}, A^{(m-1)} \dots A^{(0)}$. We are interested in the solution of the discrete problem $A^{(0)}\mathbf{u} = \mathbf{b}$ at the finest mesh.

We denote by $\chi^{(k)} = \{\chi_i^{(k)}\}_{i=1}^{n_k}$ the set of standard piece-wise constant basis functions on level k and by $\widehat{\chi}^{(k)} = \{\widehat{\chi}_i^{(k)}\}_{i=1}^{n_k}$ the set of properly defined hierarchical basis (HB) functions. The hierarchical basis $\widehat{\chi}^{(k)}$ is determined by a nonsingular transformation matrix $J^{(k)}$, i.e., $\widehat{\chi}^{(k)} = J^{(k)}\chi^{(k)}$. Then the hierarchical basis stiffness matrix $\widehat{A}^{(k)}$ and hierarchical basis spaces of degrees of freedom $\widehat{\mathbb{V}}^k$ are expressed as follows:

$$\widehat{\mathbb{V}}^{(k)} = {J^{(k)}}^{-T}\mathbb{V}^{(k)} \text{ and } \widehat{A}^{(k)} = J^{(k)}A^{(k)}{J^{(k)}}^T. \tag{2}$$

On each level k the matrix $\widehat{A}^{(k)}$ is partitioned into a two-by-two block form

$$\widehat{A}^{(k)} = \begin{bmatrix} \widehat{A}_{11}^{(k)} & \widehat{A}_{12}^{(k)} \\ \widehat{A}_{21}^{(k)} & \widehat{A}_{22}^{(k)} \end{bmatrix} \begin{matrix} \}n_k - n_{k+1} \\ \}n_{k+1} \end{matrix}, \tag{3}$$

where n_{k+1} is the dimension of the space $\mathbb{V}^{(k+1)}$.

Assumption 1. *The hierarchical basis is locally constructed so that the transformation matrix is sparse. Moreover, the following relation holds*

$$\widehat{A}_{22}^{(k)} = A^{(k+1)}. \tag{4}$$

The splitting (3) generates a splitting in the space of the degrees of freedom. It is characterized by the constant $\gamma^{(k)}$ in the strengthened Cauchy-Bunyakowski-Schwarz (CBS) inequality, associated with the cosine between the two subspaces.

Assumption 2. *There is an absolute constant γ such that $\gamma^{(k)} \le \gamma < 1$ is valid for all $k \ge 0$.*

Regarding partitioning (3), the AMLI method is defined as follows:

Algorithm 3. *(AMLI method):*

$C^{(m)} = A^{(m)}$;
for $k = 0, 1, \dots, m-1$

$$C^{(k)} = {J^{(k)}}^{-1} \begin{bmatrix} \widehat{C}_{11}^{(k)} & 0 \\ \widehat{A}_{21}^{(k)} & \tilde{A}^{(k+1)} \end{bmatrix} \begin{bmatrix} I & {\widehat{C}_{11}^{(k)}}^{-1}\widehat{A}_{12}^{(k)} \\ 0 & I \end{bmatrix} {J^{(k)}}^{-T},$$

where the blocks $\widehat{C}_{11}^{(k)}$ are symmetric positive definite approximations of $\widehat{A}_{11}^{(k)}$, that satisfy

$$\mathbf{v}^t\widehat{A}_{11}^{(k)}\mathbf{v} \le \mathbf{v}^t\widehat{C}_{11}^{(k)}\mathbf{v} \le (1+b)\mathbf{v}^t\widehat{A}_{11}^{(k)}\mathbf{v}, \quad \textit{for all } \mathbf{v} \in \mathbb{R}^{n_k - n_{k+1}}, \tag{5}$$

and the Schur complement approximation is stabilized by

$$\tilde{A}^{(k+1)^{-1}} = \left[I - p_\beta \left(C^{(k+1)^{-1}} A^{(k+1)}\right)\right] A^{(k+1)^{-1}}.$$

The acceleration polynomial is explicitly defined by

$$p_\beta(t) = \left[1 + T_\beta \left(\frac{1+\alpha-2t}{1-\alpha}\right)\right] / \left[1 + T_\beta \left(\frac{1+\alpha}{1-\alpha}\right)\right], \tag{6}$$

where T_β stands for the Chebyshev polynomial of degree β with L^∞-norm 1 on $(-1,1)$. The choice of the parameter $\alpha \in (0,1)$, providing optimal convergence rate of the AMLI preconditioner is given in the next theorem, which is a straightforward reformulation of the basic result from [1].

Theorem 4. *Let the Assumptions 1 and 2 hold and let the integer β satisfy $\frac{1}{\sqrt{1-\gamma^2}} < \beta < \rho$, where $\rho = \min_k \frac{n_k}{n_{k+1}}$. Then there exists a positive root of the scalar equation*

$$1-\gamma^2 = tb\gamma^2 + \left[\frac{(1+t^{1/2})^\beta + (1-t^{1/2})^\beta}{2\sum_{s=1}^{\beta}(1+t^{1/2})^{\beta-s}(1-t^{1/2})^{s-1}}\right]^2$$

$$+\frac{tb^2(1-\gamma^2)}{b+\gamma^2+(4(1-t)^\beta/([(1+t^{1/2})^\beta-(1-t^{1/2})^\beta]^2))}$$

and if the parameter α in (6) is chosen to be the smallest such root, then the relative condition number of $C^{(0)}$ with respect to $A^{(0)}$ is bounded by

$$\kappa(C^{(0)^{-1}}A^{(0)}) \le \frac{1}{1-\gamma^2}\left\{b\gamma^2 + \left[\frac{(1+\alpha^{1/2})^\beta + (1-\alpha^{1/2})^\beta}{(1+\alpha^{1/2})^\beta - (1-\alpha^{1/2})^\beta}\right]\right\}$$

$$+\frac{b^2}{b+\gamma^2+(4(1-\alpha)^\beta/([(1+\alpha^{1/2})^\beta-(1-\alpha^{1/2})^\beta]^2))}$$

and the total computational complexity is $O(n_0)$.

3 Local Analysis of the CBS Constant and Pivot Block Preconditioner

The hierarchical splitting that will be proposed for the weighted graph-Laplacian in the next section is locally constructed. The analysis of the associated AMLI method can be done locally too as will be demonstrated here.

Let us assume that

$$\widehat{A}^{(k)} = \sum_{e\in\mathcal{F}} \widehat{A}_e^{(k)}, \qquad \mathbf{v} = \sum_{e\in\mathcal{F}} \mathbf{v}_e,$$

where $\widehat{A}_e^{(k)}$ are symmetric positive semidefinite local matrices, $\mathcal{F}$ is some set of indices, and the summation is understood as assembling. The global hierarchical basis splitting naturally induces the block two-by-two presentation of the local matrix $\widehat{A}_e^{(k)}$, namely,

$$\widehat{A}_e^{(k)} = \begin{bmatrix} \widehat{A}_{e:11}^{(k)} & \widehat{A}_{e:12}^{(k)} \\ \widehat{A}_{e:21}^{(k)} & \widehat{A}_{e:22}^{(k)} \end{bmatrix}, \quad \mathbf{v}_e = \begin{bmatrix} \mathbf{v}_{e,1} \\ \mathbf{v}_{e,2} \end{bmatrix}. \tag{7}$$

Let $\widehat{\mathbb{V}}_e^{(k)}$ be the restriction of $\widehat{\mathbb{V}}^{(k)}$, corresponding to the local matrix $\widehat{A}_e^{(k)}$, and let $\widehat{\mathbb{V}}_e^{(k)} = \widehat{\mathbb{V}}_{e:1}^{(k)} \times \widehat{\mathbb{V}}_{e:2}^{(k)}$ be the partitioning corresponding to (7).

Lemma 1. *[4,7] Assume that for all* $\mathbf{w} = \begin{bmatrix} \mathbf{v}_1 \\ \mathbf{v}_2 \end{bmatrix} \in ker(\widehat{A}_e^{(k)})$, $\mathbf{v}_1 \in \widehat{\mathbb{V}}_{e:1}^{(k)}$, $\mathbf{v}_2 \in \widehat{\mathbb{V}}_{e:2}^{(k)}$, *it holds that* $\mathbf{v}_2 \in ker(\widehat{A}_{e:22}^{(k)})$. *Then the local CBS constant* $\gamma_e^{(k)}$ *is determined by*

$$\gamma_e^{(k)} = \sup_{\mathbf{v}_2 \in \widehat{\mathbb{V}}_{e:2}^{(k)} \setminus ker(\widehat{A}_{e:22}^{(k)})} \sqrt{\frac{\mathbf{v}_2^T \widehat{A}_{e:21}^{(k)} \left(\widehat{A}_{e:11}^{(k)}\right)^{-1} \widehat{A}_{e:12}^{(k)} \mathbf{v}_2}{\mathbf{v}_2^T \widehat{A}_{e:22}^{(k)} \mathbf{v}_2}} < 1, \tag{8}$$

and the following estimate holds

$$\gamma^{(k)} \le \max_{e \in \mathcal{F}} \gamma_e^{(k)}\ . \tag{9}$$

Let us consider a local approximation $\widehat{D}_{e:11}^{(k)}$ of the pivot block $\widehat{A}_{e:11}^{(k)}$. We denote by $\lambda_{e,M}$ the maximal and by $\lambda_{e,m}$ the minimal eigenvalues of the generalized eigenvalue problem

$$\widehat{A}_{e:11}^{(k)} \mathbf{u}_{e,1} = \lambda_e \widehat{D}_{e:11}^{(k)} \mathbf{u}_{e,1}.$$

Then $\lambda_{e,m} \mathbf{v}_{e,1}^t \widehat{D}_{e:11}^{(k)} \mathbf{v}_{e,1} \le \mathbf{v}_{e,1}^t \widehat{A}_{e:11}^{(k)} \mathbf{v}_{e,1} \le \lambda_{e,M} \mathbf{v}_{e,1}^t \widehat{D}_{e:11}^{(k)} \mathbf{v}_{e,1}$ for every $\mathbf{v}_{e,1}$ and if we define $\widehat{C}_{e:11}^{(k)} = \lambda_{e,M} \widehat{D}_{e:11}^{(k)}$ the following relations can be readily seen:

$$\mathbf{v}_{e,1}^t \widehat{A}_{e:11}^{(k)} \mathbf{v}_{e,1} \le \mathbf{v}_{e,1}^t \widehat{C}_{e:11}^{(k)} \mathbf{v}_{e,1} \le \frac{\lambda_{e,M}}{\lambda_{e,m}} \mathbf{v}_{e,1}^t \widehat{A}_{e:11}^{(k)} \mathbf{v}_{e,1}$$

If we now assemble all local pivot block matrices and define $\widehat{C}_{11}^{(k)} = \sum_{e \in \mathcal{F}} \widehat{C}_{e:11}^{(k)}$ we will arrive at the estimate $\mathbf{v}_1^t \widehat{A}_{11}^{(k)} \mathbf{v}_1 \le \mathbf{v}_1^t \widehat{C}_{11}^{(k)} \mathbf{v}_1 \le \max_{e \in \mathcal{F}} \left(\frac{\lambda_{e,M}}{\lambda_{e,m}}\right) \mathbf{v}_1^t \widehat{A}_{11}^{(k)} \mathbf{v}_1$ which is summarized in the next lemma.

Lemma 2. *Let us assume the approximation of the first pivot block is defined by* $\widehat{C}_{11}^{(k)} = \sum_{e \in \mathcal{F}} \widehat{C}_{e:11}^{(k)}$, *where* $\widehat{C}_{e:11}^{(k)} = \lambda_{e,M} \widehat{D}_{e:11}^{(k)}$. *Then the parameter b in (5) is given by:* $b = \max_{e \in \mathcal{F}} \left(\frac{\lambda_{e,M}}{\lambda_{e,m}}\right) - 1$.

4 Model Analysis of 2-D Weighted Graph-Laplacian

The main problem in the derivation of efficient AMLI methods is to define a proper hierarchical decomposition of the FE spaces. In the case of nonconforming finite elements such a splitting is neither obvious nor unique. In what follows we present a construction of a HB for the weighted graph-Laplacian based on the ideas originally introduced in [7]. Let us consider two consecutive triangulations $\mathcal{T}_k \subset \mathcal{T}_{k-1}$ and a decomposition of the wighted graph-Laplacian:

$$A^{(k)} = \sum_{e \in \mathcal{E}} A_e^{(k)} , \quad \widehat{A}^{(k)} = \sum_{e \in \mathcal{E}} \widehat{A}_e^{(k)} ,$$

as a sum of local matrices associated with the set of edges $\mathcal{E}$ of the coarser one. We are interested in the model 2-D problem with rectangle polygonal domain Ω covered by a uniform mesh of right triangles $\mathcal{T}_m$. Let each refined mesh be obtained by dividing the current triangles in four congruent ones connecting the midpoints of the sides. Following the numbering from Fig. 2, we introduce the local (macroelement) matrix $A_e^{(k)}$, corresponding to a hypotenuse, in the form

$$A_e^{(k)} = A_{e;H}^{(k)} = \left[\begin{array}{cccc|cccc} t+1 & -2t & \frac{t-1}{2} & \frac{t-1}{2} & & & & \\ -2t & 2t & & & & & & \\ \frac{t-1}{2} & & \frac{5-t}{2} & & & & -2 & \\ \frac{t-1}{2} & & & \frac{5-t}{2} & & & & -2 \\ \hline & & & & t+1 & -2t & \frac{t-1}{2} & \frac{t-1}{2} \\ & & & & -2t & 2t & & \\ & & -2 & & \frac{t-1}{2} & & \frac{5-t}{2} & \\ & & & -2 & \frac{t-1}{2} & & & \frac{5-t}{2} \end{array}\right] . \qquad (10)$$

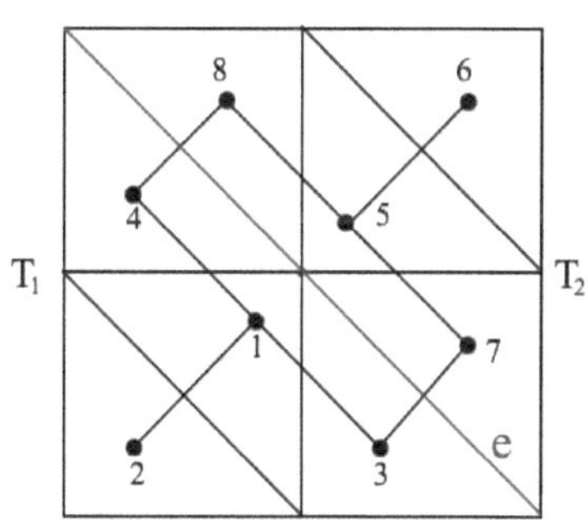

Fig. 2. Macroelement of two adjacent triangles from $\mathcal{T}_k$ with a common hypotenuse

The corresponding local matrix in the case of a common cathetus is as follows

$$A_e^{(k)} = A_{e;C}^{(k)} = \left[\begin{array}{cccc|cccc} \frac{3-1}{2} & t-1 & \frac{t-1}{2} & -t & & & & \\ t-1 & 2-t & & & & & & -1 \\ \frac{t-1}{2} & & \frac{3-1}{2} & & & -1 & & \\ -t & & & t & & & & \\ \hline & & & & \frac{3-1}{2} & t-1 & -t & \frac{t-1}{2} \\ & & -1 & & t-1 & 2-t & & \\ & & & & -t & & t & \\ & -1 & & & \frac{t-1}{2} & & & \frac{3-1}{2} \end{array}\right]. \tag{11}$$

The role of the weight parameter $t \in (0,1)$ is correctly to distribute the contribution of the links between the interior nodes among the (macroelement) edge matrices of the current coarse triangle.

We define now the (macroelement) local transformation matrix $J_e^{(k)}$ as

$$J_e^{(k)} = \left[\begin{array}{cccccccc} 1 & p & q & q & & & & \\ 1 & q & p & q & & & & \\ 1 & q & q & p & & & & \\ & & & & 1 & p & q & q \\ & & & & 1 & q & p & q \\ & & & & 1 & q & q & p \\ r & r & r & r & & & & \\ & & & & r & r & r & r \end{array}\right], \tag{12}$$

where p, q are parameters to be determined later, and r is a scaling factor. Then the hierarchical basis local matrices are obtained as follows:

$$\widehat{A}_e^{(k)} = J_e^{(k)} A_e^{(k)} J_e^{(k)T} = \begin{bmatrix} \widehat{A}_{e:11}^{(k)} & \widehat{A}_{e:12}^{(k)} \\ \widehat{A}_{e:21}^{(k)} & \widehat{A}_{e:22}^{(k)} \end{bmatrix}. \tag{13}$$

Lemma 3. *Consider the hierarchical basis defined by the transformation (12). Then*

$$\widehat{A}_{22}^{(k)} = A^{(k+1)} \quad \textit{if and only if} \quad r = \frac{\sqrt{2}}{2}.$$

Proof. The equivalence statements $\widehat{A}_{e:22}^{(k)} = \widehat{A}_{e;H:22}^{(k)} = \begin{bmatrix} 2 & -2 \\ -2 & 2 \end{bmatrix}$, $\widehat{A}_{e:22}^{(k)} = \widehat{A}_{e;C:22}^{(k)} = \begin{bmatrix} 1 & -1 \\ -1 & 1 \end{bmatrix}$ directly follow from (10), (11), (12), and (13), that is, the equations $\widehat{A}_{e:22}^{(k)}(1,1) = r^2 \sum_{i,j=1}^{4} A_e^{(k)}(i,j)$, $\widehat{A}_{e:22}^{(k)}(1,2) = r^2 \sum_{i=1}^{4} \sum_{j=5}^{8} A_e^{(k)}(i,j)$, $\widehat{A}_{e:22}^{(k)}(2,2) = r^2 \sum_{i,j=5}^{8} A_e^{(k)}(i,j)$, $\widehat{A}_{e:22}^{(k)}(2,1) = r^2 \sum_{i=5}^{8} \sum_{j=1}^{4} A_e^{(k)}(i,j)$.

From (8) it is seen that $(\gamma_e^{(k)})^2 = 1 - \lambda$ where λ is the eigenvalue (which is unique in this particular case) of the eigenproblem $\widehat{S}_e^{(k)}\mathbf{v} = \lambda \widehat{A}_{e:22}^{(k)}\mathbf{v}\,, \quad \mathbf{v} \neq [1,1]^t$, where $\widehat{S}_e^{(k)} = \widehat{A}_{e:22}^{(k)} - \widehat{A}_{e:21}^{(k)}(\widehat{A}_{e:11}^{(k)})^{-1}\widehat{A}_{e:12}^{(k)}$.

Varying the parameters (p, q, t) we get a family of hierarchical splittings. The first parameter setting we examine is $p = 1$, $q = -0.5$ and $t = 0.5$, see [7]. Using Lemma 1 and the fact that the construction of the hierarchical basis and all related matrices are independent of the current refinement level, we get the estimate

$$\gamma^2 \leq \max\{\gamma_{e;H},\ \gamma_{e;C}\} = 0.73 \qquad (14)$$

Let us assume now that the pivot blocks $\widehat{A}_{e:11}^{(k)}$ are approximated using their diagonal parts. Then according to Lemma 2 we get $b = 16$. At the end of this analysis Theorem 4 is applied to get the uniform condition number estimate of the AMLI preconditioner $\sqrt{\kappa(C^{(0)^{-1}}A^{(0)})} \leq 23.53$ which holds for the acceleration polynomial of degree $\beta = 2$.

We now modify the parameter setting aimed at the improvement of the local CBS constant estimate. We vary q and t when $p = 1$. The analysis shows that a local minimum is obtained for $q = -0.1$ and $t = 0.75$. Then

$$\gamma^2 \leq \max\{\gamma_{e;H},\ \gamma_{e;C}\} = 0.58. \qquad (15)$$

As above, approximating the pivot blocks $\widehat{A}_{e:11}^{(k)}$ using their properly (locally) scaled diagonal parts, and if $\beta = 2$ we obtain the improved uniform estimate of the AMLI preconditioner $\sqrt{\kappa(C^{(0)^{-1}}A^{(0)})} \leq 14.8$.

Acknowledgement. Research has been partially supported by the Bulgarian NSF Grant VU-MI-202/2006.

References

1. Axelsson, O., Vassilevski, P.S.: Algebraic multilevel preconditioning methods II. SIAM J. Numer. Anal. 27, 1569–1590 (1990)
2. Bejanov, B., Guermond, J., Minev, P.: A locally *div*-free projection scheme for incompressible flows based on non-conforming finite elements. Int. J. Numer. Meth. Fluids 49, 239–258 (2005)
3. Blaheta, R., Margenov, S., Neytcheva, M.: Robust optimal multilevel preconditioners for non-conforming finite element systems. Numer. Lin. Alg. Appl. 12(5-6), 495–514 (2005)
4. Kraus, J., Margenov, S.: Multilevel methods for anisotropic elliptic problems. Lectures on Advanced Computational Methods in Mechanics, Radon Series Comp. Appl. Math. 1, 47–87 (2007)
5. Kraus, J., Tomar, S.: A multilevel method for discontinuous Galerkin approximation of three-dimensional anisotropic elliptic problems. Num. Lin. Alg. Appl (2007) doi: 10.1002/nla.544
6. Margenov, S., Minev, P.: On a MIC(0) preconditioning of non-conforming mixed FEM elliptic problems. Math. Comput. Simul. (2007) doi:10.1016/j.matcom.2007.01.021
7. Lazarov, R., Margenov, S.: CBS constants for multilevel splitting of graph-Laplacian and application to preconditioning of discontinuous Galerkin systems. J. Complexity 23(4-6), 498–515 (2007)

Stability and Bifurcation of the Magnetic Flux Bound States in Stacked Josephson Junctions

Ivan Christov, Stefka Dimova, and Todor Boyadjiev

Faculty of Mathematics and Infromatics, University of Sofia,
5 James Bourchier Boulevard, 1164 Sofia, Bulgaria
{ivanh,dimova,todorlb}@fmi.uni-sofia.bg

Abstract. The static distributions of the magnetic flux in stacked Josephson junctions are investigated numerically. To solve the nonlinear boundary value problem an iterative algorithm, based on the Continuous analog of Newton method is constructed. The linearized problems at every iteration step are solved by the Galerkin finite element method. In order to study the stability of possible distributions a Sturm-Liouville problem is generated. A minimal eigenvalue equal to zero means a bifurcation of the corresponding solution. The subspace iteration method is used to find the smallest eigenvalues and the corresponding eigenvectors.

1 Statement of the Problem

In the last decades the propagation of electromagnetic waves in long Josephson Junctions (JJs) has been extensively studied both theoretically and experimentally (see, for example, [1]-[11]). Stacking the junctions may increase the usability of these devices especially for storage and transmission of information. Such structures make it possible to state and study new physical effects [4,9] that do not occur in single JJs. In this paper we propose numerical technique for investigation the properties of the solutions in the model of stacked JJs, based on inductively coupled layers [3]. The general case which takes into account both inductive and capacitive coupling is considered in [11].

A simple scheme of N-layered JJ is shown on Fig. 1, where black layers are barriers (insulators) with thickness D and white layers are superconductors with thickness d. In the symmetric case the electromagnetic interaction between junctions is represented only by a coupling constant s [3,11].

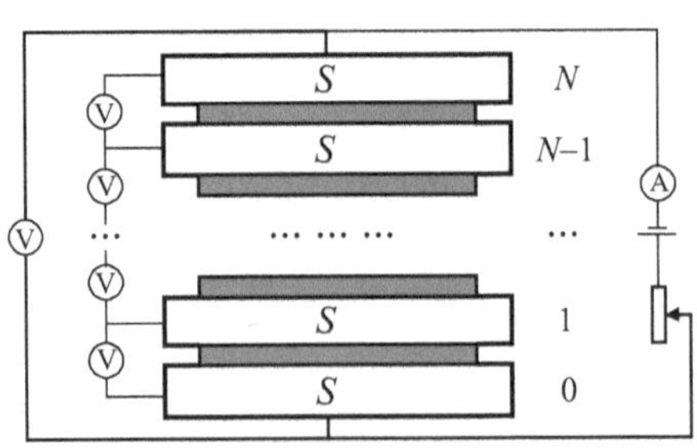

Fig. 1. N-layered JJ

S. Margenov, L.G. Vulkov, and J. Waśniewski (Eds.): NAA 2008, LNCS 5434, pp. 224–232, 2009.

The simplest generalizable model of stacked JJ is three-layered Josephson junction because it takes into account the difference in the behavior of the interior and exterior junctions. The first and the third junctions are coupled only to one neighboring junction while the second junction is coupled to its two neighbors below and above. The numerical results, presented in this paper, are obtained mainly for the particular case of three-layered JJs, but the method of investigation and its program realization are developed for the general N-layered case ($N \geq 1$).

The existence of Josephson current generates a specific magnetic flux [12]. When the external current is less then some critical value γ_{cr}, the junction layers are in superconductive regime [12]. The transitions from superconductive to resistive regime are mathematically interpreted as bifurcation of the static distributions of the magnetic flux under the change of the parameters [13].

The vector $\varphi(x) = (\varphi_1(x), \varphi_2(x), \ldots, \varphi_N(x))^T$ of static distributions of the magnetic flux satisfies [3], [11] the nonlinear boundary value problem

$$-\varphi_{xx} + L\,(J_z + \Gamma) = 0, \tag{1a}$$

$$\varphi_x(\pm l) = H, \tag{1b}$$

where $J_z = (\sin\varphi_1, \sin\varphi_2, \ldots, \sin\varphi_N)^T$ is the vector of the Josephson current density, $\Gamma = \gamma\,(1,1,\ldots,1)^T$ is the vector of the external current density, $H = h_e\,(1,1,\ldots,1)^T$ is the vector of the external magnetic field and $2l$ is the length of the layers (the subscript x means differentiation, the superscript T means transposition). The matrix L is a matrix of the inductive interaction ($-0.5 < s \leq 0$ for arbitrary N):

$$L_{(N\times N)} = \begin{pmatrix} 1\; s\; 0 \ldots 0\; 0 \\ s\; 1\; s \ldots 0\; 0 \\ \cdot\;\cdot\;\cdot\;\cdot\;\cdot\;\cdot\;\cdot\;\cdot \\ 0\; 0\; 0 \ldots 1\; s \\ 0\; 0\; 0 \ldots s\; 1 \end{pmatrix}.$$

2 Numerical Method

In order to solve the nonlinear boundary value problem (1) we use an iterative algorithm, based on the continuous analog of Newton's method (CAMN) [15]. As initial approximations for the iteration process we take combinations (for the different layers) of solutions which exist in the one-layered case and $h_e = 0$, $\gamma = 0$:

– Meissner solutions (denoted further by M) of the form $\varphi(x) = k\pi$, $k = 0, \pm 1, \pm 2, \ldots$,

– fluxon (antifluxon) solutions, for which there are exact analytical expressions in the case of infinite junctions ($l \to \infty$) [19]. The single fluxon/antifluxon solution has the well known form $\varphi(x) = 4\arctan\exp\{\pm x\} + 2k\pi$, $k = 0, \pm 1, \ldots$ Further for n-fluxon distributions we use the simple notation Φ^n, $n = \pm 1, \pm 2, \ldots$ For junctions of finite length objects of type Φ^n are not fluxons in a strong sense, but by analogy the same terminology is used.

CANM gives a linearized boundary value problem at each iteration step:

$$-v_{xx} + LJ'_z\, v = \varphi_{xx} - L\,(J_z + \Gamma), \tag{2a}$$

$$v_x(\pm l) = H - \varphi_x(\pm l), \tag{2b}$$

where $\varphi = (\varphi_1, \varphi_2, \ldots, \varphi_N)^T$ is the approximate solution found at the previous iteration step, $v = (v_1, v_2, \ldots, v_N)^T$ is the vector of the iteration corrections and $J'_z(\varphi)$ is the matrix $\mathrm{diag}(\cos\varphi_1, \cos\varphi_2, \ldots, \cos\varphi_N)$.

The linear boundary value problems (2) are solved by means of Galerkin finite element method [14] and quadratic elements. The matrices of the corresponding linear algebraic problems are nonsymmetric. They are stored and used in *sky – line* form [16]. The linear algebraic problems are solved using LU-decomposition.

In order to study the global stability of the possible distributions of the magnetic flux, a matrix Sturm-Liouville problem [20] is generated

$$-Au_{xx} + Q(x)u = \lambda\, u, \tag{3a}$$

$$u_x(\pm l) = 0, \tag{3b}$$

$$\int_{-l}^{l} \langle u, u\rangle\, dx - 1 = 0. \tag{3c}$$

Here the matrix $A = L^{-1}$, and the potential $Q(x) \equiv J'_z(\varphi(x))$. The minimal eigenvalue λ_{min} determines the stability of the distribution under consideration.

A minimal eigenvalue λ_{min} equal to zero means a bifurcation of this distribution caused by change of some parameter(s). For example, in infinite junction Meissner solution M is stable, while Φ^1 is quasistable (bifurcation solution with $\lambda_{min} = 0$) [17,18].

The finite element method is used to reduce the above matrix Sturm-Liouville problem (3) to a matrix algebraic problem, whose few smallest eigenvalues and the corresponding eigenfunctions are found by the subspace iteration method [16].

Table 1. Order of convergence $\sigma(x)$

x_i	$h = 0.125$	$h = 0.0625$	$h = 0.03125$	σ
-5	0.019669254	0.019669211	0.019669208	3.99
-4	0.027395953	0.027395900	0.027395896	3.99
-3	0.057938303	0.057938217	0.057938216	3.99
-2	0.139892955	0.139892807	0.139892798	3.99
-1	0.323331077	0.323330864	0.323330851	4.00
0	0.486766104	0.486766424	0.486766444	4.02
1	0.323331077	0.323330864	0.323330851	3.98
2	0.139892955	0.139892807	0.139892798	3.98
3	0.057938303	0.057938217	0.057938212	3.99
4	0.027395953	0.027395900	0.027395896	3.99
5	0.019669254	0.019669211	0.019669208	3.99
λ	-0.002815748	-0.002816608	-0.002816662	4.00

To test the accuracy of all realized methods we have used the method of Runge by computing the solutions on three embedded meshes. The numerous experiments made show a super-convergence of order four. As an example in the table below the computed orders of convergence σ

$$\sigma(x) = \ln \left| \frac{u_h(x) - u_{h/2}(x)}{u_{h/2}(x) - u_{h/4}(x)} \right| / \ln 2,$$

of the smallest eigenvalue, as well as of the eigenfunction, corresponding to the first component of solution type (Φ^1, M, Φ^1), $s = -0.3$, $2l = 10$, are shown.

3 Numerical Experiment

We briefly discuss some numerical results obtained by the developed algorithms. Especially we investigate numerically the static distributions of the magnetic flux and seek for critical values of the parameters h_e and γ where these distributions fail to exist.

3.1 Three-Layered Junctions

Because of big variety of different triplets of solutions, in the present paper we explain results for (M, M, M) and (Φ^1, M, Φ^{-1}) solutions only. On Fig. 2 and Fig. 3 the distributions of the internal magnetic field $\varphi_x(x)$ in the first layer of a three layered junction are graphically shown. The results are for three different values of the external magnetic field h_e (0 and the two bifurcation values) when $\gamma = 0$, $l = 5$ and $s = -0.3$. Changing the value of γ for given h_e when the geometrical parameters l and s are fixed, we get the region of existence of the corresponding solution on the plane $P \equiv (h_e, \gamma)$. We expected a symmetry of this region with respect to h_e and γ for solutions of types (M, M, M) and (Φ^1, M, Φ^{-1}), and this was confirmed by the numerical results. On Fig. 4 and Fig. 5 the regions of existence ($h_e \geq 0$) in the plane P for the same solutions are shown. Every point on these curves is a bifurcation point and, consequently, satisfies

$$\lambda_{min}(\gamma_{cr}, h_e) = 0.$$

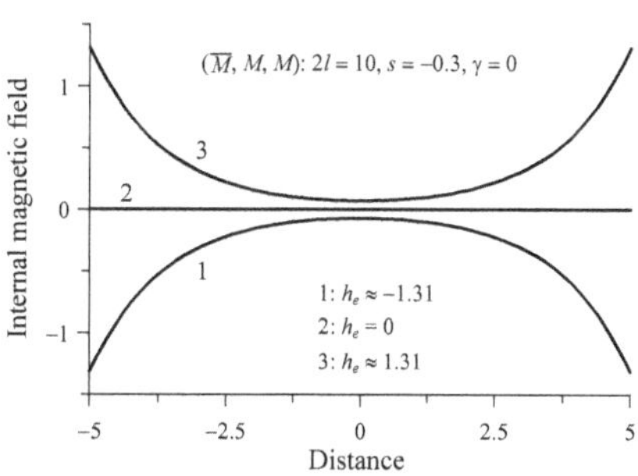

Fig. 2. Solution of type (M, M, M)

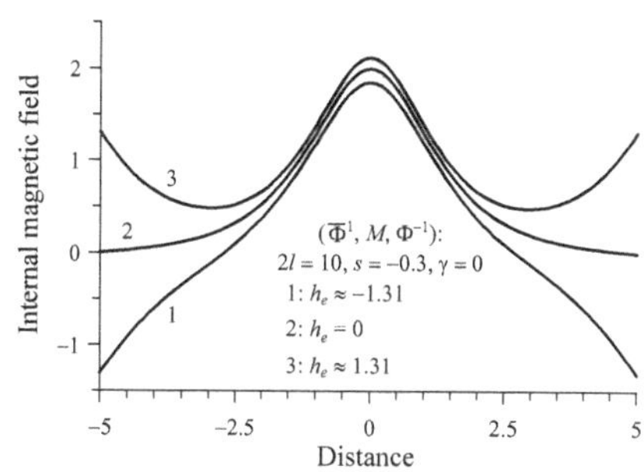

Fig. 3. Solution of type (Φ^1, M, Φ^{-1})

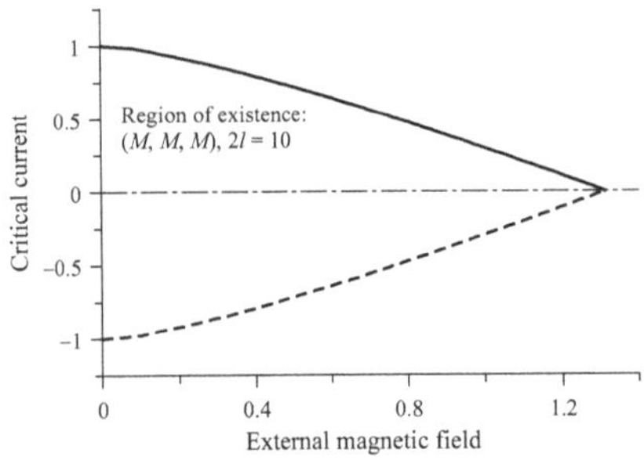

Fig. 4. Region of existence for (M, M, M)

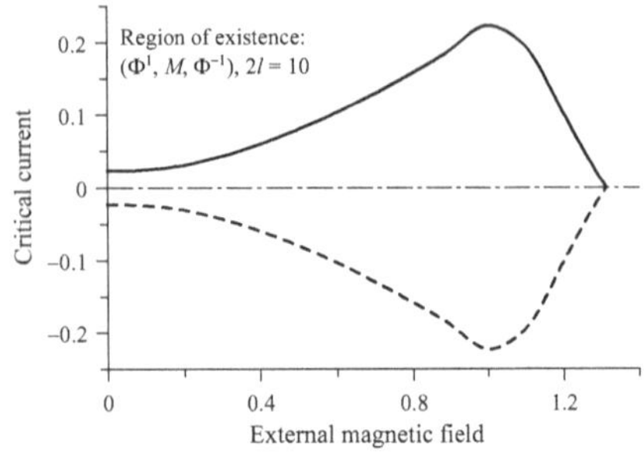

Fig. 5. Region of existence (Φ^1, M, Φ^{-1})

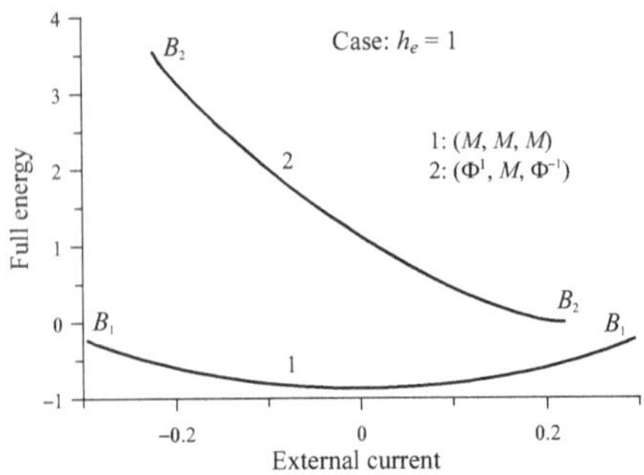

Fig. 6. Full energy $F(\gamma)$ for solutions (M, M, M) and (Φ^1, M, Φ^{-1})

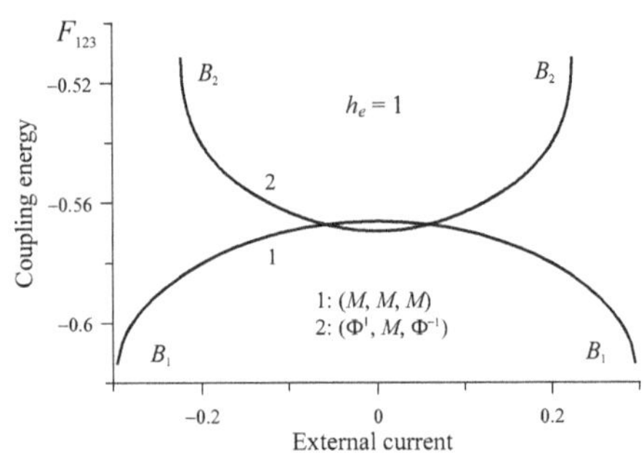

Fig. 7. Coupling energy $F_{123}(\gamma)$ for solutions (M, M, M) and (Φ^1, M, Φ^{-1})

The main physical characteristics of interest — energy, number of fluxons, full magnetic flux - are computed as well. The dependence of these characteristics on the geometric parameters of the junction is investigated and new interesting results are found. Let us denote $\Delta\varphi$ the vector of full magnetic flux through the junction:

$$\Delta\varphi = (\Delta\varphi_1, \ldots, \Delta\varphi_N)^T, \quad \Delta\varphi_i = \varphi_i(l) - \varphi_i(-l).$$

Equations (1) can be considered as necessary conditions for extremum of the functional of full energy $F[\varphi]$ of the junction:

$$F[\varphi] = \int_{-l}^{l} \left[\frac{1}{2}\langle \varphi_x, A\varphi_x\rangle + U(\varphi)\right] dx - \langle AH, \Delta\varphi\rangle \tag{4}$$

where the density $U = \sum_{i=1}^{N} U_i(\varphi_i)$, $U_i = 1 - \cos\varphi_i + \gamma\varphi_i$. Equation (1a) is the Euler-Lagrange equation for (4) and the boundary conditions (1b) follow from

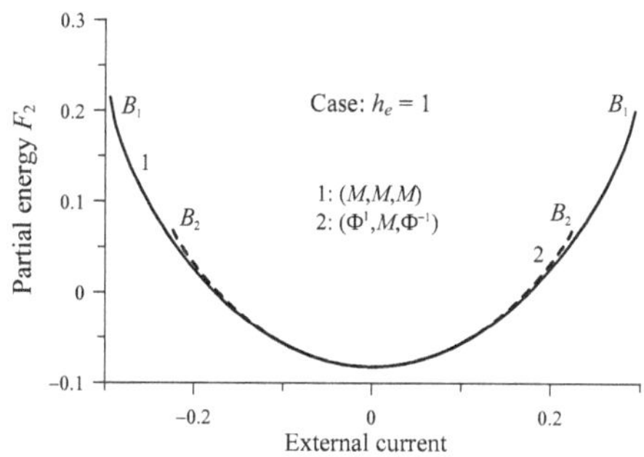

Fig. 8. Partial energies $F_2(\gamma)$ for (M, M, M) and (Φ^1, M, Φ^{-1})-type solutions

Weierstrass-Erdmann conditions. The full energy F is presented as a sum of the energies F_i, $i=1,\ldots,N$ of the non-interacting layers and the coupling energy F_{int}:

$$F = \sum_{i=1}^{N} F_i + F_{int}.$$

For this reason the matrix A is presented as a sum $A = I + B$, where I is the identity matrix. Then:

$$F[\varphi] = \int_{-l}^{l} \left[\frac{1}{2}\langle\varphi_x, \varphi_x\rangle + \frac{1}{2}\langle\varphi_x, B\varphi_x\rangle + U(\varphi)\right] dx - \langle IH, \Delta\varphi\rangle - \langle BH, \Delta\varphi\rangle, \quad (5)$$

$$F_i[\varphi_i] = \int_{-l}^{l} \left[\frac{1}{2}\varphi_{i,x}^2 + U_i(\varphi_i)\right] dx - h_e\Delta\varphi_i,$$

$$F_{int}[\varphi] = \frac{1}{2}\int_{-l}^{l} \langle\varphi_x, B\varphi_x\rangle dx - \langle BH, \Delta\varphi\rangle.$$

A comparison of the dependence $F(\gamma)$ for solutions of type (M, M, M) (see curve 1) and (Φ^1, M, Φ^{-1}) (curve 2) is made on Fig. 6. The values of $F(\gamma)$ are computed by formulae (5). The boundary points B_1 and B_2 are bifurcation points for the parameter γ. The curve (1) lies under curve (2), which means that for given $h_e = 1$ the distribution (M, M, M) has bigger critical current (by modulus) than the distribution (Φ^1, M, Φ^{-1}).

Let us mention the different behavior of the dependence $F_{123}(\gamma)$ for the two distributions (see Fig. 7). When $|\gamma|$ increases the coupling energy $F_{123}(\gamma)$ for (M, M, M)-distribution monotonically decreases in contrary of $F_{123}(\gamma)$ for (Φ^1, M, Φ^{-1}) which monotonically increases.

Finally, on Fig. 8 a comparison of the energy $F_2(\gamma)$ of the middle layer for (M, M, M) and (Φ^1, M, Φ^{-1}) solutions is shown. It is clearly seen that the changes $M \to \Phi^1$ and $M \to \Phi^{-1}$ in the external layers result mainly in decreasing of the region of existence with respect to external current γ.

3.2 Single Josephson Junction

In the process of investigation some interesting observations about the behavior of the solutions depending on the length of the junction were made. For their deeper understanding we analyzed the case of single junction, which corresponds to $s = 0$.

The numerical experiments have shown that when the length $2l$ grows, the supports of the bifurcation curves for fluxons $\Phi^1, \Phi^2, \Phi^3, \ldots$ tend consequently to the support of the bifurcation curve for Meissner solution. This fact is illustrated on Fig. 9 and 10. One can see that the support of the bifurcation curve for Φ^1 (Fig. 9, $2l = 14$) and for Φ^1, Φ^2, Φ^3 (Fig. 10, $2l = 30$) coincides within the plotting resolution with the support of the bifurcation curve for Meissner solution. And what is more, the maximal critical current in these cases lies on the same curve.

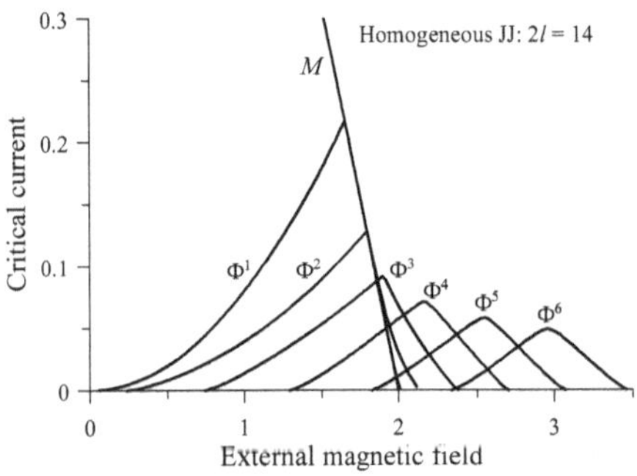

Fig. 9. Bifurcation curves for $2l = 14$

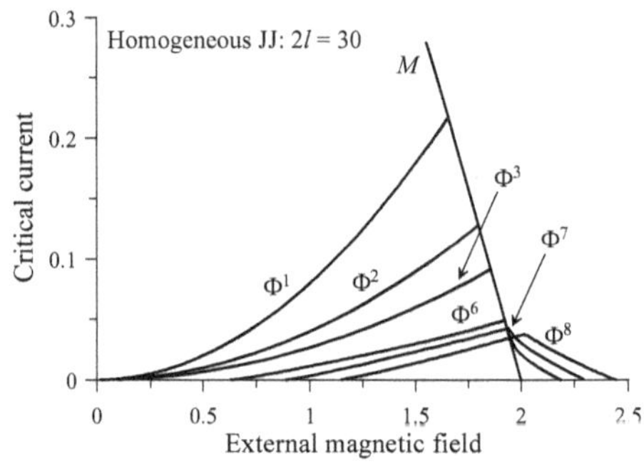

Fig. 10. Bifurcation curves for $2l = 30$

The numerical observations gave a hint to ask for analytical dependencies. The following ones are found:

1. Let $(\alpha(\Phi^k, l), \beta(\Phi^k, l))$ is the interval in h_e for $\gamma = 0$ and fixed halflength l, for which the solution Φ^k is stable. Then
$$\alpha(\Phi^k, l) \searrow 0 \quad \text{when} \quad l \to \infty, \quad \beta(\Phi^k, l) \searrow 2 \quad \text{when} \quad l \to \infty.$$
2. Let $(-\alpha(M, l), \alpha(M, l))$ is the interval in h_e for $\gamma = 0$ and fixed halflength l, for which the solution M is stable. Then
$$\alpha(M, l) \searrow 2 \quad \text{when} \quad l \to \infty.$$
3. For arbitrary l the following inequalities hold:
$$\alpha(\Phi^1, l) < \alpha(\Phi^2, l) < \alpha(\Phi^3, l) < \ldots$$
$$\alpha(M, l/2) = \beta(\Phi^1, l) < \beta(\Phi^2, l) < \beta(\Phi^3, l) < \ldots$$
$$\alpha(\Phi^m, l) = \alpha\left(\Phi^k, \frac{k}{m} l\right), \quad \beta(\Phi^m, l) = \beta\left(\Phi^k, \frac{k+1}{m+1} l\right), \ k \geq m.$$

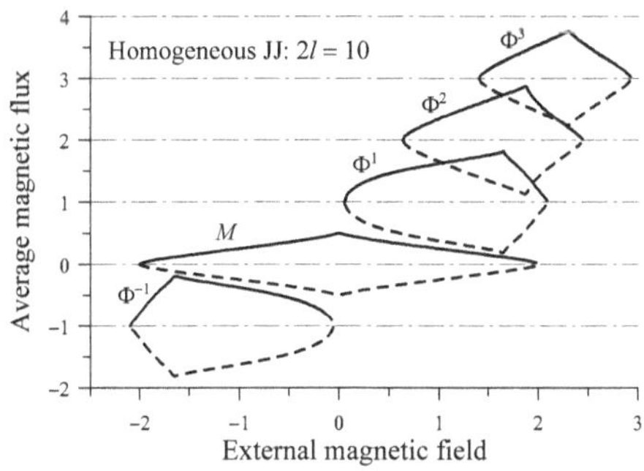

Fig. 11. The number of fluxons in single JJ

4. If for $h_e \leq 2$, $\gamma = 0$ and fixed l the solution Φ^k is stable, then Φ^1, $\Phi^2, \ldots,$ Φ^{k-1} are stable as well.

Let us denote by

$$N[\varphi] = \frac{1}{2l\pi} \int_{-l}^{l} \varphi(x)\, dx, \tag{6}$$

the average value of the bound state $\varphi(x)$ on the junction [15]. Functional (6) depends on all of the parameters of the model. On Fig. 11 the dependence $N(h_e)$ for the bifurcation solutions Φ^{-1}, M, Φ^1, Φ^2 and Φ^3 is presented in the case of single JJ of full length $2l = 10$. The solid lines correspond to $\gamma_{cr} \geq 0$, the dashed ones — to $\gamma_{cr} \leq 0$. Note that for $\gamma_{cr} = 0$ the values of $N[\varphi]$ are integer numbers: $N[\Phi^n] = n$, $N[M] = 0$. When $\gamma_{cr} \neq 0$ the following relations are fulfilled

$$N[\Phi^n_+] + N[\Phi^n_-] = 2n, \quad N[M_+] + N[M_-] = 0,$$

where the subscript $\pm$ correspond to the sign of γ_{cr}.

The above mentioned facts can be considered as conservation laws of the average magnetic flux. So formulae (6) determines the admissible number of fluxons in the solutions of problem (1), having fixed sign of $\varphi_x(x)$.

4 Conclusions

Effective numerical algorithms are worked out for solving the nonlinear system of ODE for finding the static distributions of the magnetic flux in $N-$stacked JJs, as well as the corresponding matrix Sturm-Liouville problem for studying their global stability. The developed technique gives a possibility for detailed investigation of these multiparametric problems. We illustrate its application to analyze the existence, stability, lack of stability and some physical characteristics of two types magnetic flux distributions in 3-layered JJs. In addition some new numerical and analytical results for 1-layered JJs are found.

Acknowledgement

This work is supported by Sofia University Scientific foundation under Grant No 135/2008.

References

1. Volkov, A.F.: Solitons in Josephson superlaticess. JETP Lett. 45(6), 299–301 (1987)
2. Boyadjiev, T.L., Pavlov, D.V., Puzynin, I.V.: Computation of Bifurcations of Stable States in Two-Layer Inhomogeneous Josephson Junctions. Comm. JINR, Dubna, P5-89-173 (1989)
3. Sakai, S., Bodin, P., Pedersen, N.F.: Fluxons in thin-film superconductor-insulator superlattices. J. Appl. Phys. 73(5), 2411–2418 (1993)
4. Nevirkovets, I.P., Evetts, J.E., Blamire, M.G.: Transition From Single Junction to Double Junction Behavior In Sisis- Type Nb-Based Devices. Phys. Lett. A 187(1), 119 (1994)
5. Bulaevskii, L.N., Zamora, M., Baeriswyl, D., Beck, H., Clem, J.R.: Time-dependent equations for phase differences and a collective mode in Josephson-coupled layered superconductors. Phys. Rev. B 50, 12831–12834 (1994)
6. Kleiner, R., Muller, P., Kohlstedt, H., Pedersen, N.F., Sakai, S.: Phys. Rev. B 50, 3942–3952 (1994)
7. Song, S.N., Auvil, P.R., Ulmer, M., Ketterson, J.B.: Vortex structure and Josephson supercurrent in stacked double Josephson junctions. Phys. Rev. B 53(10), R6018 (1996)
8. Krasnov, V.M., Winkler, D.: Static and dynamic properties of stacked Josephson junctions: Analytic solution. Phys. Rev. B 56, 9106–9115 (1997)
9. Goldobin, E., Ustinov, A.V.: Current locking in magnetically coupled long Josephson junctions. Phys. Rev. B 59(17), 11532–11538 (1999)
10. Atanasova, P.K., Bojadjiev, T.L., Dimova, S.N.: Numerical Simulation of Critical Dependences for Symmetric Two-Layered Josephson Junctions. Computational Mathematics and Mathematical Physics 46(4), 666–679 (2006)
11. Machida, M., Sakai, S.: Unifed theory for magnetic and electric field coupling in multistacked Josephson junctions. PRB 70, 144520 (2004)
12. Licharev, K.K.: Dynamics of Josephson Junctions and Circuits. Gordon and Breach, New York (1986)
13. Gal'pern, Y.S., Filippov, A.T.: Bound states of solitons in inhomogeneous Josephson junctions. Sov. Phys. JETP 59 (1984)
14. Thomee, V.: Galerkin finite element method for parabolic problems. Spinger, Berlin (1997)
15. Puzynin, I.V., et al.: Methods of computational physics for investigation of models of complex physical systems. Particals & Nucley 38(1) (2007)
16. Bathe, K.J., Wilson, E.: Numerical Methods in Finite Element Analisis. Prentice Hall, Englewood Cliffs (1976)
17. Rubinstein, J.: Sine-Gordon equation. J. Math. Phys. 16(96) (1970)
18. Fogel, M.B., et al.: Dinamics of sine-Gordon solitons in the presence of perturbations. Phys. Rev. B 15(3) (1977)
19. Iliev, I.D., Khristov, E.K., Kirchev, K.P.: Spectral methods in soliton equations. Longman Sci. & Techn. Wiley (1994)
20. Naimark, M.A.: Linear differential operators. Nauka, Moskow (1969)

On thc Numbcr of Spikes of Solutions for a Singularly Perturbed Boundary-Value Problem

Ognyan Christov

Faculty of Mathematics and Informatics, Sofia University,
5 J.Bouchier blvd., 1164 Sofia, Bulgaria
christov@fmi.uni-sofia.bg

Abstract. We study the stationary solutions of the famous Fisher - Kolmogorov - Petrovsky - Piscounov (FKPP) equation

$$u_t = Du_{xx} + \gamma u(1-u),$$

where the diffusion parameter is small - $D = \epsilon^2$ and γ is normalized to be 1. Next, we state a singularly perturbed boundary-value problem (BVP) on the interval $[0, 1]$.

$$\left| \begin{array}{l} \epsilon^2 \ddot{u} + u(1-u) = 0, \\ u(0) = a, \quad u(1) = a, \quad a \in (0,1). \end{array} \right.$$

Asymptotic formulas, as $\epsilon \to 0^+$ are obtained for the solutions of the above (BVP). The solutions of the (BVP) can have "spikes". Estimates for the number of spikes and number of solutions to the (BVP) are given.

1 Introduction and Motivation

The Fisher - Kolmogorov - Petrovsky - Piscounov (FKPP) equation [1,2] is one of the fundamental models in mathematical biology and ecology [3]. It describes a population $u(x,t)$ that evolves under the combined effects of spatial diffusion and local logistic growth and saturation. One dimensional FKPP equation can be written as

$$u_t = Du_{xx} + \gamma u(1-u), \tag{1}$$

with diffusion coefficient D, low - density growth rate γ and where the population has been normalized so that the stable saturation level is $u = 1$ $(0 \leq u(x,t) \leq 1)$. Actually, we are interested in the corresponding travelling wave fronts $u(x,t) = f(x-ct)$, with speed c. Insertion this expression into the FKPP equation (1) yields $Df'' + cf' + \gamma f(1-f) = 0$.

Usually a boundary - value problem (BVP) for the last equation is stated on an infinite interval (see [4]). Here we consider small diffusion $D = \epsilon^2$ and again denote the unknown function with u, t is the independent variable $t \in [0,1]$, hence we consider a BVP on a finite interval. In addition, we relax the restrictions on u. If $\gamma > 0$ without loss of generality, we put $\gamma = 1$ (for $\gamma < 0$ the things can

S. Margenov, L.G. Vulkov, and J. Waśniewski (Eds.): NAA 2008, LNCS 5434, pp. 233–240, 2009.

be treated analogously). Here, we study the case $c = 0$, the asymptotics for the case $c = O(1)$ are obtained in the standard way (see [5,6] for example). However, the case $c = O(\epsilon)$ needs more efforts, so it will be considered elsewhere. So, the (BVP) under consideration is

$$\left| \begin{array}{l} \epsilon^2 \ddot{u} + u(1-u) = 0, \quad t \in [0,1] \\ u(0) = a, \quad u(1) = a, \quad 0 < a < 1, \quad \epsilon << 1 \end{array} \right. \tag{2}$$

Our motivation is the following. In [7] Ou and Wong have studied the so called Carrier - Pearson BVP

$$\epsilon \ddot{u} + u^2 = 1, \quad u(-1) = u(1) = 0 \tag{3}$$

from rigorous point of view. Instead of Poincaré type perturbation series which lead to incorrect results, they expand the solutions in powers of exponentially small functions to locate the internal boundary layers, so called spikes and to evaluate their number in $(-1, 1)$.

Ward [8] points out some open problems in studying the spike solutions for one space dimension - among them to estimate number of spikes with various boundary conditions and to establish results for other nonlinearities. In the second paper of [7] using the same method as in the first article, the authors consider the cubic nonlinearity, subjected to certain conditions. Similar estimates on error terms and number of internal boundary layers are obtained.

2 Results

Here we follow the approach of Ou and Wong [7], based on the shooting method.

2.1 The Shooting Method

First let us investigate the associated initial-value problem

$$\left| \begin{array}{l} \epsilon^2 \ddot{u} + u(1-u) = 0, \\ u(0) = a, \quad \dot{u}(0) = k, \quad k \in \mathbb{R}. \end{array} \right. \tag{4}$$

It is known that this problem has a unique solution, which can be extended to infinity. Introduce the potential energy $F(u) = \frac{u^2}{2} - \frac{u^3}{3}$. Hence, the total energy $\epsilon^2 \frac{(\dot{u})^2}{2} + F(u) = E = \epsilon^2 k^2/2 + a^2/2 - a^3/3$ is a first integral for (4). The corresponding phase portrait is given on Fig. 1. Let $u(t,k)$ be the solution of the (IVP) (4) and denote $k_{max} := \sqrt{\frac{2}{\epsilon^2}(\frac{1}{6} - \frac{a^2}{2} + \frac{a^3}{3})}$.

Lemma 1

1) If $k > k_{max}$, then $u(t,k)$ is increasing in $(0,\infty)$ and $\lim_{t\to\infty} u(t,k) = \infty$;
2) If $k = \pm k_{max}$, then $u(t,k) = 1 - \frac{3}{2}\mathrm{sech}^2(\pm\frac{t}{2\epsilon} + d_1)$, where $d_1 : \cosh d_1 = \sqrt{\frac{3}{2(1-a)}}$;

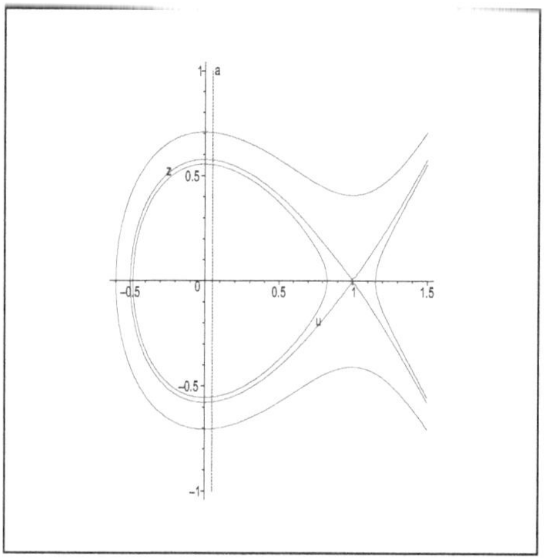

Fig. 1. The phase portrait

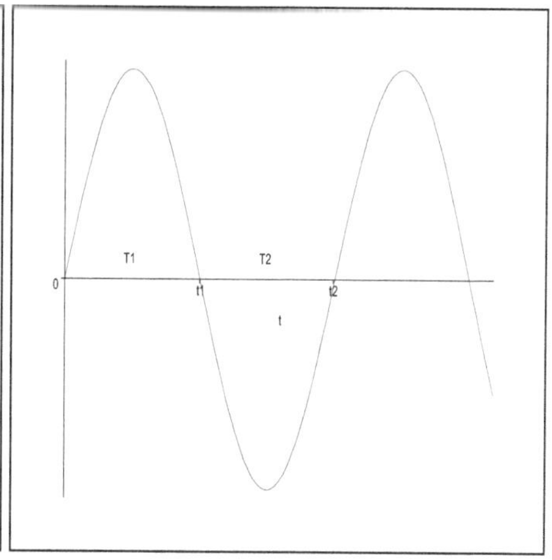

Fig. 2. The intervals T_1, T_2

3) If $k < -k_{max}$, then $u(t,k)$ first decreasing and increasing to infinity;

4) If $|k| < k_{max}$, then $u(t,k)$ is periodic and intersects t - axis infinitely many times.

Lemma 2. *There exists $\epsilon_0 = \epsilon_0(a)$ such that if $\epsilon \le \epsilon_0$ and $|k| \ge k_{max}$, then $u(1, \epsilon, k) \neq a$.*

It is clear from the last lemma that in order to achieve $u(1, \epsilon, k) = a$ we must choose $k \in (-k_{max}, k_{max})$. We consider only the case $k \ge 0$, because the other is similar. Let t_1 be the first point to the right of 0 such that $u(t_1, \epsilon, k) = a$ and $u > a$ in $(0, t_1)$, and t_2 be the first point to the right of t_1, such that $u(t_2, \epsilon, k) = a$ and $u < a$ in (t_1, t_2). We denote by T_1 and T_2, respectively, the lengths of the intervals $(0, t_1)$ and (t_1, t_2) (see Fig. 2).

In order to achieve more convenient parametrization, we define a one - to - one mapping from k to m in $(a, 1)$ such that

$$E = \epsilon^2 \frac{k^2}{2} + \frac{a^2}{2} - \frac{a^3}{3} = \frac{m^2}{2} - \frac{m^3}{3}.$$

From the energy integral we obtain

$$\epsilon^2 \frac{(\dot{u})^2}{2} = -\frac{u^2}{2} + \frac{u^3}{3} + \frac{m^2}{2} - \frac{m^3}{3} = \frac{1}{6}(u - u_1)(u - u_2)(u - u_3),$$

where $u_2 = m$ and

$$u_1 = \frac{3 - 2m - \sqrt{9 + 12m - 12m^2}}{4}, \quad u_2 = \frac{3 - 2m + \sqrt{9 + 12m - 12m^2}}{4}.$$

Now, it is straightforward that

$$T_1(m) = \epsilon\sqrt{2} \int_a^m (\frac{u^3}{3} - \frac{u^2}{2} + \frac{m^2}{2} - \frac{m^3}{3})^{-1/2} du,$$

$$T_2(m) = \epsilon\sqrt{2}\int_{u_1}^{a}(\frac{u^3}{3} - \frac{u^2}{2} + \frac{m^2}{2} - \frac{m^3}{3})^{-1/2}du.$$

Lemma 3. *$T_1(m)$ is increasing ($T_1(a) = 0, T_1(1) = \infty$) and $T_2(m)$ is decreasing in $a < m < 1$. Moreover, in the same interval*

$$\frac{d(T_1 + T_2)}{dm} > 0, \quad \frac{d^2(T_2)}{dm^2} > 0, \quad \frac{d^2(T_1 + T_2)}{dm^2} > 0.$$

Remark. For small values of ϵ, $T_2(m) \leq T_2(a) < 1$, that is (BVP) (2) does not have a solution which is below the line $u = a$ in the entire interval $(0, 1)$.

2.2 Number of Spikes

After this preparation, we are ready to investigate whether (BVP) (2) has solutions and to estimate the maximum number of spikes they have. Recall that by spikes, we mean those, which lie under line $u = a$. This is equivalent to finding suitable values of k so that

$$u(1, \epsilon, k) = a. \tag{5}$$

Since $u(t, \epsilon, k)$ is periodic and oscillatory about $u = a$, the equation (5) can be written in the form

$$MT_1(m) + NT_2(m) = 1, \tag{6}$$

where M, N are nonnegative integers representing, respectively, the number of intervals with lengths T_1 and T_2. Note that for every given N, then M can be one of the following values $N - 1, N, N + 1$. Let us define

$$N_\epsilon := \left[\frac{1}{T_2(a)}\right].$$

Lemma 4. *If $N > N_\epsilon + 1$, then (BVP) (2) has no solutions with N spikes. For any integer $0 \leq N \leq N_\epsilon$, there exists at least one solution with N spikes (including the boundary and internal spikes). If $N = N_\epsilon + 1$, then either case can happen.*

Denote by n_ϵ the number of solutions to the (BVP) (2).

Proposition 1. *If $N_\epsilon \neq 1/T_2(a)$, then n_ϵ satisfies*

$$4N_\epsilon \leq n_\epsilon \leq 4N_\epsilon + 2.$$

If $N_\epsilon = 1/T_2(a)$, then it satisfies

$$4(N_\epsilon - 1) + 1 \leq n_\epsilon \leq 4(N_\epsilon - 1) + 2.$$

The proof is practically the same as in [7].

2.3 Solutions without Internal Spikes

In order to study endpoint layers, first we put $u = 1 + v(\tau)$ in (2) where $\tau = t/\epsilon$ is the stretched time, we get

$$\left| \begin{array}{l} \frac{d^2v}{d\tau^2} - v(v+1) = 0, \\ v(0) = a - 1, \quad v(\infty) = 0 \end{array} \right. \tag{7}$$

The exact solution is $v(\tau) = -\frac{3}{2}\text{sech}^2(\kappa_1 \frac{\tau}{2} + d_1)$ where $\kappa_1 = \pm 1$. Similarly, if we put $u = 1 + w(\sigma)$, $\sigma = \frac{1-t}{\epsilon}$ we can get $w(\sigma) = -\frac{3}{2}\text{sech}^2(\kappa_2 \frac{\sigma}{2} + d_1)$ where $\kappa_2 = \pm 1$. Denote by $\tilde{u}(t) = 1 + v(\tau) + w(\sigma)$ the approximate solution. It is clear that there are four such approximate solutions (see Fig. 3).

$$\tilde{u}_1(t,\epsilon) = 1 - \frac{3}{2}\text{sech}^2(\frac{t}{2\epsilon} + d_1) - \frac{3}{2}\text{sech}^2(\frac{1-t}{2\epsilon} + d_1) \tag{8}$$

$$\tilde{u}_2(t,\epsilon) = 1 - \frac{3}{2}\text{sech}^2(-\frac{t}{2\epsilon} + d_1) - \frac{3}{2}\text{sech}^2(\frac{1-t}{2\epsilon} + d_1) \tag{9}$$

$$\tilde{u}_3(t,\epsilon) = 1 - \frac{3}{2}\text{sech}^2(\frac{t}{2\epsilon} + d_1) - \frac{3}{2}\text{sech}^2(-\frac{1-t}{2\epsilon} + d_1) \tag{10}$$

$$\tilde{u}_4(t,\epsilon) = 1 - \frac{3}{2}\text{sech}^2(-\frac{t}{2\epsilon} + d_1) - \frac{3}{2}\text{sech}^2(-\frac{1-t}{2\epsilon} + d_1) \tag{11}$$

We have the following estimate.

Theorem 1. *For ϵ small, there exist exactly four solutions $u_i(t,\epsilon)$, $i = 1, \ldots, 4$, to the BVP (2), which have no internal spikes. These solutions are uniformly approximated by the functions $\tilde{u}_i(t,\epsilon)$, given in (8) - (11) with error estimate of order $O(\exp(-1/2\epsilon)$, that is*

$$u_i(t,\epsilon) = \tilde{u}_i(t,\epsilon) + \mathrm{O}(e^{-\frac{1}{2\epsilon}}), \quad i = 1, \ldots, 4$$

as $\epsilon \to 0^+$, uniformly for $t \in [0,1]$.

2.4 Solutions with Spikes

Let us study the situation in which there are n internal boundary layers, that is n spikes. Let these spikes are located at

$$0 < \bar{t}_1 < \bar{t}_2 < \ldots < \bar{t}_n < 1.$$

Now, consider the expansion in the neighborhood of $\bar{t}_j$. Put $\tau_j = \frac{t-\bar{t}_j}{\epsilon}$ and $u = 1 + q(\tau_j)$. As above we obtain $q(\tau_j) = -\frac{3}{2}\text{sech}^2(\frac{\tau_j}{2})$. Then, the function $u = 1 + v(\tau) + w(\sigma) + \sum_{j=1}^{n} q(\tau_j)$ could be an approximate solution for BVP (2). Again we have the following estimate for the case with n internal spikes.

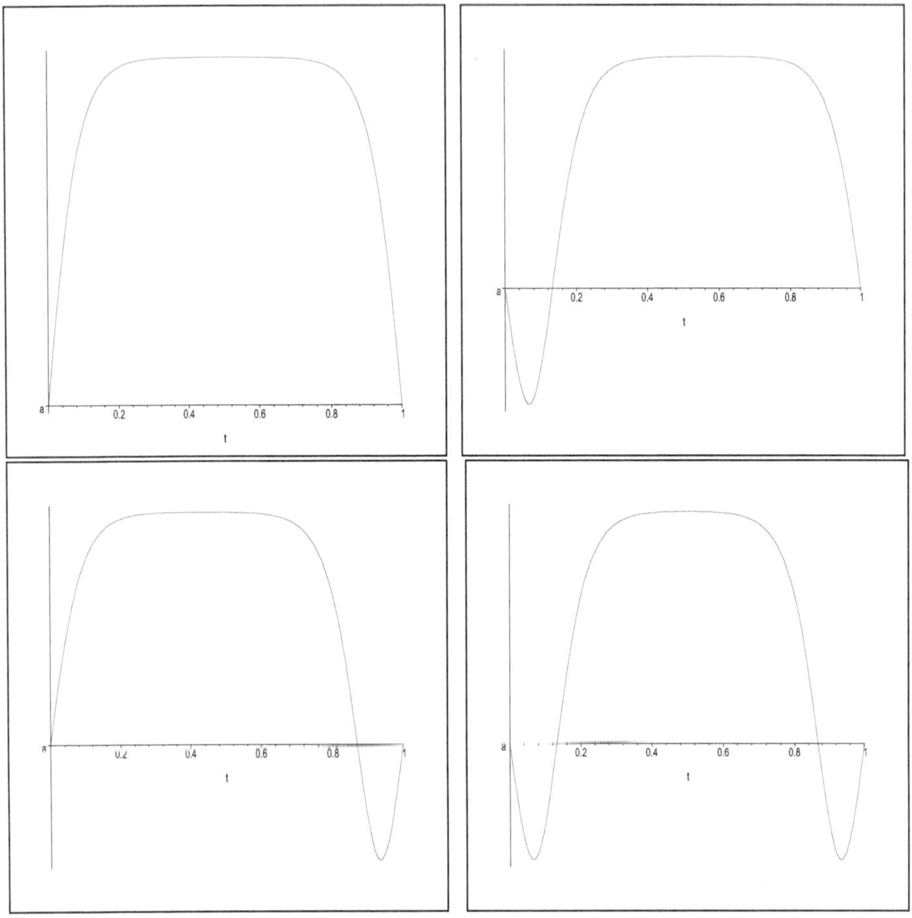

Fig. 3. Graphs of the approximate solutions

Theorem 2. *For $n \geq 1$ and each $i = 1, \ldots, 4$, there exists unique solution $u_{i,n}(t, \epsilon)$, satisfying*

$$u_{i,n}(t, \epsilon) = \tilde{u}_i(t, \epsilon) + \sum_{j=1}^{n} q\left(\frac{t - t_{i,j}}{2\epsilon}\right) + O\left(e^{-\frac{1}{2(n+1)\epsilon}}\right)$$

for all $t \in [0, 1]$, where the approximated positions $t_{i,j}$ of the spikes are

$$t_{1,j} = -2\epsilon d_1 + \frac{j}{n+1}(1 + 4\epsilon d_1)$$

$$t_{2,j} = 2\epsilon d_1 + \frac{j}{n+1}$$

$$t_{3,j} = -2\epsilon d_1 + \frac{j}{n+1}$$

$$t_{4,j} = 2\epsilon d_1 + \frac{j}{n+1}(1 - 4\epsilon d_1)$$

for $j = 1, \ldots, n$.

3 Conclusion

There is no doubt that it is essential for the numerical analysts to know the location of the spikes as well the approximate solutions with their asymptotics in such singularly perturbed problems.

The problem of finding asymptotic behaviour of the solutions to such kind conservative problems is studied by O'Malley [5] using phase - plane analysis. Although his approach provides qualitative information for the solutions with internal spikes, it does not give quantitative information such as asymptotic formulas for the solution.

Probably the best known approach to derive asymptotic formulas is the method of matched asymptotics. However, it is known that routine application of this method fails to determine the spike location.

Another approach has been introduced by Ward, known as the projection method. Ward's method does not require the knowledge of the explicit form of the internal layer solutions, but does not give the number of solutions to the (BVP) for small fixed ϵ.

The approach of Ou and Wong is based on the shooting method and we follow their ideas here.

Finally, one can find the explicit solution to the (IVP) in our problem (as well as in Carrier - Pearson's problem) and then to adjust it in order to obtain the solution of the (BVP). This solution is given by the $\wp$ - function of Weierstrass. This function is expressed via theta functions, which are defined as very fast convergent series of exponents. Knowing the asymptotic formulas, one can easily obtain the estimates. Moreover, the location of spikes are obtained explicitly as the minima (in our case) of the $\wp$ function. Again, it is difficult to obtain the number of solutions to the (BVP).

The common limitation of the last two approaches is that they rely heavily on analytical formulae that are available for very specific nonlinearities and hardly can be extended to treat problems with several space dimensions.

Acknowledgements

My thanks are to the anonymous referee which comments and suggestions have helped to improve the text.

References

1. Fisher, R.: The wave of advance of advantageous genes. Annals of Eugenics 7, 353 (1937)
2. Kolmogorov, A., Petrovsky, I., Piskounov, N.: Etude de l' equation de la diffusion avec croisance de la quantité de matière et son application a un probleme biologique. Moscow University, Bull. Math. 1, 1–25 (1937)

3. Murray, J.: Mathematical Biology. Springer, New York (2001)
4. Doering, C., Mueller, C., Smereka, P.: Interacting particles, the stochastic Fisher - Kolmogorov - Petrovsky - Piscounov equation, and duality. Phys. A 325, 243–259 (2003)
5. O'Malley, R.: Singular perturbation Methods for Ordinary Differential Equations, vol. 89. Springer, New York (1991); Phase - Plane Solutions to Some Singular Perturbation Problems. J. Math. Anal. Appl. 54, 449–466 (1976)
6. Vasil'eva, A., Butuzov, V.: Asymptotic Expansions of Solutions of Singularly Perturbed Equations, Nauka, Moscow (1973) (in Russian)
7. Ou, C., Wong, R.: On a Two - Point Boundary - Value Problem with Spurious Solutions. Stud. in Appl. Math. 111, 377–408 (2003); Shooting Method for Nonlinear Singularly Perturbed Boundary - Value Problems. Stud. in Appl. Math. 112, 161–200 (2004)
8. Ward, M.: Spikes for singularly perturbed reaction - diffusion systems and Carrier's problem. In: Hua, C., Wong, R. (eds.) Differential Equations and Asymptotic Theory in Mathematical Physics. Series in Analysis, vol. 2, pp. 100–188. World Scientific, Singapore (2005)

On a Class of Almost Orthogonal Polynomials

Bratislav Danković[1], Predrag Rajković[2], and Sladjana Marinković[1]

[1] Department of Automatic Control, Faculty of Electronic Engineering,
{dankovic,sladjana}@elfak.ni.ac.yu
[2] Department of Mathematics, Faculty of Mechanical Engineering,
University of Niš, A. Medvedeva 14, 18 000 Niš, Serbia
pecar@masfak.ni.ac.yu

Abstract. In this paper, we define a new class of almost orthogonal polynomials which can be used successfully for modelling of electronic systems which generate orthonormal basis. Especially, for the classical weight function, they can be considered like a generalization of the classical orthogonal polynomials (Legendre, Laguerre, Hermite, ...). They are very suitable for analysis and synthesis of imperfect technical systems which are projected to generate orthogonal polynomials, but in the reality generate almost orthogonal polynomials.

1 Introduction

It is known that orthogonal polynomials are useful tool in technical sciences. Here we will emphasize their role in signal approximation ([1],[2]) and design of the electronic systems which generate the orthogonal signals ([3], [4]). However, since the components of those systems can not be made quite exactly, the polynomials which are generated by these systems are not quite orthogonal, but rather *almost orthogonal*. The measure of nearness between the obtained and the regular orthogonal polynomials depends on the exactness of the manufacturing of the components. Until now, some classes of almost orthogonal functions are defined and investigated by other authors and their applications are considered too (see, for example, [5], [6] and [7]).

Let $\lambda(x)$ be a positive Borel measure on an interval $(a,b) \subset \mathbb{R}$ with infinite support and such that all moments

$$\lambda_n = \mathcal{L}[x^n] = \int_a^b x^n d\lambda(x) \tag{1}$$

exist. In this manner, we define linear functional $\mathcal{L}$ in the linear space of real polynomials $\mathcal{P}$. Also, we can introduce an inner product as follows (see [8]):

$$(f,\ g) = \mathcal{L}\big[f \cdot g\big] \qquad (f, g \in \mathcal{P}), \tag{2}$$

which is positive-definite because of the property $\|f\|^2 = (f,f) \geq 0$. Hence it follows that monic polynomials $\{P_n(x)\}$ orthogonal with respect to this inner product exist and they satisfy the three-term recurrence relation

$$P_{k+1}(x) = (x-\alpha_k)P_k(x) - \beta_k P_{k-1}(x) \quad (k \geq 0), \qquad P_{-1} \equiv P_0 \equiv 1. \tag{3}$$

S. Margenov, L.G. Vulkov, and J. Waśniewski (Eds.): NAA 2008, LNCS 5434, pp. 241–248, 2009.

For a very small positive real number ε, let us consider a functional

$$\mathcal{L}_\varepsilon[f] = \mathcal{L}[f] - \varepsilon \qquad (0 < \varepsilon \ll 1). \tag{4}$$

If it exists, the sequence of monic polynomials $\{P_n^{(\varepsilon)}(x)\}$ which satisfies the relations

$$\mathcal{L}_\varepsilon\Big[P_j^{(\varepsilon)} \cdot P_k^{(\varepsilon)}\Big] = 0 \ (j \neq k), \qquad |||P_k^{(\varepsilon)}|||_\varepsilon^2 = \mathcal{L}_\varepsilon\Big[\big(P_k^{(\varepsilon)}\big)^2\Big] > 0 \ (j,k \in \mathbb{N}_0) \tag{5}$$

will be called *orthogonal* with respect to $\mathcal{L}_\varepsilon$, or *almost orthogonal* w.r.t. $\mathcal{L}$.

Obviously, it is valid

$$\lim_{\varepsilon \to 0} P_n^{(\varepsilon)}(x) = P_n(x). \tag{6}$$

Notice that

$$\mathcal{L}_\varepsilon[cf(x)] = c\mathcal{L}_\varepsilon[f(x)] + (c-1)\varepsilon, \qquad \mathcal{L}_\varepsilon[f(x)+g(x)] = \mathcal{L}_\varepsilon[f(x)] + \mathcal{L}_\varepsilon[g(x)] + \varepsilon,$$

wherefrom the presence of additional ε abandons the distribution property. It means that a lot of simple procedures useful for the classical orthogonal polynomials cannot be applied here.

We have to use the next procedure for their construction. Denoting

$$\mu_{j,k} = \mathcal{L}\big[x^j P_k^{(\varepsilon)}(x)\big],$$

we can find the coefficients in

$$P_n^{(\varepsilon)}(x) = x^n + \sum_{k=0}^{n-1} a_{n,k}\, x^k$$

like solutions of the linear algebraic system

$$\begin{bmatrix} \mu_{0,0} & \mu_{1,0} & \cdots & \mu_{n-1,0} \\ \mu_{0,1} & \mu_{1,1} & & \mu_{n-1,1} \\ \vdots & & & \\ \mu_{0,n-1} & \mu_{1,n-1} & \cdots & \mu_{n-1,n-1} \end{bmatrix} \cdot \begin{bmatrix} a_{n,0} \\ a_{n,1} \\ \vdots \\ a_{n,n-1} \end{bmatrix} = \begin{bmatrix} \varepsilon - \mu_{n,0} \\ \varepsilon - \mu_{n,1} \\ \vdots \\ \varepsilon - \mu_{n,n-1} \end{bmatrix}.$$

2 The Connection between Almost Orthogonal and Orthogonal Polynomials

Since the monic polynomials $\{P_k(x)\}_{k=0}^n$ form a basis in the space of the polynomials $\mathcal{P}_n$ whose degree is not greater than n, we can write an expression as follows:

$$P_n^{(\varepsilon)}(x) = P_n(x) + \sum_{k=0}^{n-1} \frac{b_k^{(n)}(\varepsilon)}{\|P_k\|^2} P_k(x). \tag{7}$$

Theorem 1. *The coefficients* $b_j^{(n)}$ *do not depend on* n, *i.e.*

$$b_j^{(n)}(\varepsilon) = b_j(\varepsilon) \qquad (\forall n \in \mathbb{N}_0). \tag{8}$$

Proof. By taking the inner product of $P_n^{(\varepsilon)}(x)$ with P_j, we can write

$$(P_n^{(\varepsilon)}, P_j) = (P_n, P_j) + \sum_{k=0}^{n-1} \frac{b_k^{(n)}(\varepsilon)}{\|P_k\|^2}(P_k, P_j).$$

By the orthogonality relations for the polynomials $\{P_k(x)\}_{k=0}^{n}$, we get

$$(P_n^{(\varepsilon)}, P_j) = b_j^{(n)}(\varepsilon).$$

On the other hand, we can express

$$P_j(x) = \sum_{i=0}^{j} c_{j,i}(\varepsilon) P_i^{(\varepsilon)}(x).$$

Hence

$$b_j^{(n)}(\varepsilon) = \sum_{i=0}^{j} c_{j,i}(\varepsilon)(P_n^{(\varepsilon)}, P_i^{(\varepsilon)}) = \varepsilon \sum_{i=0}^{j} c_{j,i}(\varepsilon),$$

wherefrom we see that $b_j(\varepsilon)$ does not depend on n. ◊

Theorem 2. *The coefficients* $b_k = b_k(\varepsilon)$ $(k \in \mathbb{N}_0)$ *satisfy the next recurrence relation*

$$b_n + \frac{b_{n-1}^2}{\|P_{n-1}\|^2} + \cdots + \frac{b_0^2}{\|P_0\|^2} = \varepsilon \quad (n \in \mathbb{N}). \tag{9}$$

Proof. From the fact

$$\mathcal{L}_\varepsilon\left(P_{n+1}^{(\varepsilon)} \cdot P_n^{(\varepsilon)}\right) = 0 \quad \Leftrightarrow \quad (P_{n+1}^{(\varepsilon)}, P_n^{(\varepsilon)}) = \varepsilon,$$

and from the orthogonality of $\{P_k(x)\}$, we have

$$\begin{aligned}
\varepsilon &= \left(P_{n+1}(x) + \textstyle\sum_{k=0}^{n} \frac{b_k}{\|P_k\|^2} P_k(x)\ ,\ P_n(x) + \sum_{i=0}^{n-1} \frac{b_i}{\|P_i\|^2} P_i(x)\right) \\
&= \textstyle\sum_{k=0}^{n} \frac{b_k}{\|P_k\|^2} \left(P_k(x)\ ,\ P_n(x) + \sum_{i=0}^{n-1} \frac{b_i}{\|P_i\|^2} P_i(x)\right) \\
&= \textstyle\frac{b_n}{\|P_n\|^2}(P_n(x), P_n(x)) + \sum_{k=0}^{n-1} \frac{b_k}{\|P_k\|^2} \left(P_k(x)\ ,\ \frac{b_k}{\|P_k\|^2} P_k(x)\right),
\end{aligned}$$

wherefrom the formula follows. ◊

By mathematical induction and the previous theorem, we can prove the next recurrence relation.

Corollary 1. *The polynomials* $\{b_k(\varepsilon)\}$ $(\deg b_k(\varepsilon) = 2^k)$ *are related recurrently by*

$$b_0 = \varepsilon, \quad b_{k+1} = b_k\Big(1 - \frac{b_k}{\|P_k\|^2}\Big) \qquad (k \in \mathbb{N}). \tag{10}$$

Proof. We get it by subtracting the formula from the previous theorem for $n = k+1$ and $n = k$. ♢

Corollary 2. *The squared norm of the polynomials* $\{P_n^{(\varepsilon)}(x)\}$ *is*

$$\|P_n^{(\varepsilon)}\|^2 = \|P_n\|^2 + \sum_{k=0}^{n-1} \frac{b_k^2}{\|P_k\|^2}.$$

Proof. From the definition of norm and orthogonality of polynomials, we get

$$\begin{aligned}\|P_n^{(\varepsilon)}\|^2 &= (P_n^{(\varepsilon)}, P_n^{(\varepsilon)}) \\ &= \Big(P_n(x) + \textstyle\sum_{k=0}^{n-1} \frac{b_k}{\|P_k\|^2} P_k(x)\ ,\ P_n(x) + \sum_{i=0}^{n-1} \frac{b_i}{\|P_i\|^2} P_i(x)\Big) \\ &= (P_n, P_n) + \textstyle\sum_{k=0}^{n-1} \frac{b_k}{\|P_k\|^2}\Big(P_k(x)\ ,\ \sum_{i=0}^{n-1} \frac{b_i}{\|P_i\|^2} P_i(x)\Big) \\ &= \|P_n\|^2 + \textstyle\sum_{k=0}^{n-1} \frac{b_k}{\|P_k\|^2}\Big(P_k(x)\ ,\ \frac{b_k}{\|P_k\|^2} P_k(x)\Big),\end{aligned}$$

wherefrom the conclusion follows. ♢

Remark 1. Unfortunately, the recurrence relation for $\{P_n^{(\varepsilon)}(x)\}$ is not known. Comparing the first members of the sequence, we easily conclude that they do not satisfy recurrence relation of the form (3).

3 Some Classical Cases

The monic *shifted Legendre polynomials* $P_n(x)$ are orthogonal with respect to the weight $w(x) \equiv 1$ on $(0,1)$. They satisfy the three-term recurrence relation

$$P_{k+1}(x) = (x - \alpha_k)P_k(x) - \beta_k P_{k-1}(x) \quad (k \geq 0), \qquad P_{-1} \equiv P_0 \equiv 1,$$

with

$$\alpha_k = \frac{1}{2} \quad (k \geq 0), \qquad \beta_0 = 1, \quad \beta_k = \frac{k^2}{4(4k^2-1)} \quad (k > 0). \tag{11}$$

The square norm is

$$\|P_0\|^2 = \beta_0 = 1, \qquad \|P_n\|^2 = \prod_{k=0}^{n} \beta_k = \frac{1}{4^n} \frac{(n!)^2}{(2n-1)!!(2n+1)!!} \quad (n > 0). \tag{12}$$

and their explicit form is

$$P_n(x) = \frac{n!}{(2n)!} \sum_{j=0}^{n} (-1)^{n-j} \binom{n}{j} \frac{(n+j)!}{j!} x^j. \tag{13}$$

Let us consider the functional

$$\mathcal{L}_\varepsilon[f] = \int_0^1 f(x)dx - \varepsilon \qquad (0 < \varepsilon \ll 1) \tag{14}$$

and the sequence of monic polynomials $\{P_n^{(\varepsilon)}(x)\}$ which satisfies the relation of orthogonality w.r.t. $\mathcal{L}_\varepsilon$.

The first members of this sequence are:

$$\left\{1, \quad x - (\frac{1}{2} - \varepsilon), \quad x^2 - (1 - 12\varepsilon + 12\varepsilon^2)x + \frac{1}{6}(1 - 30\varepsilon + 36\varepsilon^2)\right\}.$$

Now, in the expansion (7), we get the coefficients b_k as follows:

$$b_0 = \varepsilon, \qquad b_1 = (1-\varepsilon)b_0, \qquad b_2 = (1 - 12\varepsilon + 12\varepsilon^2)b_1.$$

The monic *Chebyshev polynomials of the first kind* $T_n(x)$ are orthogonal with respect to the weight $w(x) = 1/\sqrt{1-x^2}$ on $(-1, 1)$. They satisfy the three-term recurrence relation

$$T_{k+1}(x) = xT_k(x) - \beta_k T_{k-1}(x) \quad (k \geq 0), \qquad T_{-1} \equiv T_0 \equiv 1,$$

with

$$\beta_1 = \frac{1}{2}, \qquad \beta_k = \frac{1}{4} \quad (k \geq 2). \tag{15}$$

Also, the norms are: $\|T_0\|^2 = \pi$, $\|T_n\|^2 = \pi/2^{2n-1}$ $(n \geq 1)$. The explicit form is

$$T_n(x) = n \sum_{k=0}^{[n/2]} \frac{(-1)^k (n-k-1)!}{k!(n-2k)!\, 4^k} x^{n-2k}. \tag{16}$$

If we define the functional

$$\mathcal{L}_\varepsilon[f] = \int_{-1}^1 f(x) \frac{dx}{\sqrt{1-x^2}} - \varepsilon \qquad (0 < \varepsilon \ll 1), \tag{17}$$

we can introduce the sequence of polynomials $\{T_n^{(\varepsilon)}(x)\}$ orthogonal with respect to $\mathcal{L}_\varepsilon$:

$$\left\{1, \quad x + \frac{\varepsilon}{\pi}, \quad x^2 + \frac{2\varepsilon(\pi - \varepsilon)}{\pi^2} x - \frac{\pi - 2\varepsilon}{2\pi}, \ldots\right\},$$

with

$$b_0 = \varepsilon, \qquad b_1 = \frac{\pi - \varepsilon}{\pi} b_0, \qquad b_2 = \frac{\pi^2 - 2\pi\varepsilon + 2\varepsilon^2}{\pi^2} b_1, \ldots .$$

The *Laguerre polynomials* $L_n(x)$ satisfy the three-term recurrence relation

$$L_{k+1}(x) = (x - \alpha_k)L_k(x) - \beta_k L_{k-1}(x) \quad (k \geq 0), \qquad L_{-1} \equiv L_0 \equiv 1,$$

with

$$\alpha_k = 2k+1 \quad (k \geq 0), \qquad \beta_0 = 1, \quad \beta_k = k^2 \quad (k > 0). \tag{18}$$

Their explicit expression and the norm are:

$$L_n(x) = n! \sum_{k=0}^{n} (-1)^{n-k} \binom{n}{k} \frac{x^k}{k!}, \qquad \|L_n\| = n! \qquad (n \in \mathbb{N}) .$$

We will denote by $\{L_n^{(\varepsilon)}(x)\}$ the sequence of polynomials which satisfies the relation of orthogonality with respect to the functional

$$\mathcal{L}_\varepsilon[f] = \int_0^{+\infty} f(x)\, \mathrm{e}^{-x} dx - \varepsilon \qquad (0 < \varepsilon \ll 1). \tag{19}$$

The first three members of this sequence are:

$$\{1, \quad x - (1-\varepsilon), \quad x^2 - (4 - \varepsilon + \varepsilon^2)x + (2 + \varepsilon^2)\} .$$

Now, the expansion (7) involves the following polynomials over ε:

$$b_0 = \varepsilon, \qquad b_1 = (1-\varepsilon)b_0, \qquad b_2 = (1 - \varepsilon + \varepsilon^2)b_1 .$$

4 The Least Square Approximation over Almost Orthogonal Polynomials

For an integrable function $f(x)$ we are considering the approximations over the polynomials $\{P_k^{(\varepsilon)}(x)\}$:

$$f(x) \approx \Phi_n(x) = \sum_{k=0}^{n} d_k P_k^{(\varepsilon)}(x) \qquad (n \in \mathbb{N}).$$

The error of the approximation can be estimated by

$$\mathcal{G}(d_0, d_1, \ldots, d_n) = \||\Phi_n(x) - f(x)\||_\varepsilon^2 .$$

The least square approximation w.r.t. such introduced norm will be reached in the point of minimum of the function $\mathcal{G}(d_0, d_1, \ldots, d_n)$. Hence

$$\frac{1}{2}\frac{\partial \mathcal{G}}{\partial d_j} \equiv \sum_{k=0}^{n} d_k (P_k^{(\varepsilon)}, P_j^{(\varepsilon)}) - (f, P_j^{(\varepsilon)}) = 0 \qquad (0 \le j \le n).$$

So we get the system of linear algebraic equations

$$\|P_j^{(\varepsilon)}\|^2\, d_j + \varepsilon \sum_{k:k\neq j} d_k = (f, P_j^{(\varepsilon)}) \qquad (j = 0, 1, \ldots, n).$$

Denoting by $s_k = \|P_k^{(\varepsilon)}\|^2$ and $f_k = (f, P_k^{(\varepsilon)})$ $(0 \le k \le n)$, we can write the matrix form of this system like

$$\begin{bmatrix} s_0 & \varepsilon & \cdots & \varepsilon \\ \varepsilon & s_1 & & \varepsilon \\ \vdots & & & \\ \varepsilon & \varepsilon & & s_n \end{bmatrix} \cdot \begin{bmatrix} d_0 \\ d_1 \\ \vdots \\ d_n \end{bmatrix} = \begin{bmatrix} f_0 \\ f_1 \\ \vdots \\ f_n \end{bmatrix} . \tag{20}$$

By Cramer's rule, we can evaluate d_k's. At first, we find that the determinant of the system is

$$\Delta(\varepsilon) = \prod_{i=0}^{n}(s_i - \varepsilon)\Big(1 + \varepsilon\sum_{i=0}^{n}\frac{1}{s_i - \varepsilon}\Big).$$

Including the Corollary 2, we see that the acceptable values for ε are:

$$0 < \varepsilon \ll \tau_n = \min_{0\le k\le n} \|P_k\|^2 \qquad (\forall n \in \mathbb{N}_0). \tag{21}$$

If we want to evaluate coefficients in approximation we have to find determinants $\Delta_k(\varepsilon)$ which we get from $\Delta(\varepsilon)$ putting free elements in the $(k+1)$th column. Now, we have

$$\Delta_k(\varepsilon) = \sum_{i=0}^{n} u_i f_i,$$

where

$$u_k = \prod_{\substack{i=0\\ i\neq k}}^{n}(s_i - \varepsilon)\left(1 + \varepsilon\sum_{\substack{i=0\\ i\neq k}}^{n}\frac{1}{s_i - \varepsilon}\right), \qquad u_i = \varepsilon\, g_i(\varepsilon)\ (i \neq k),$$

with $g_i(\varepsilon)$ a polynomial over ε. It is bounded function for $\varepsilon \in (0, \tau_n)$.

Obviously, for $\varepsilon = 0$, this approximation becomes the least square approximation over the orthogonal polynomials $\{P_n(x)\}$ and the system (20) is diagonal and well-coditioned.

But, for $\varepsilon \neq 0$, we must use almost orthogonal instead of orthogonal polynomials. Although theoretically we can achieve the approximation of the same quality, in practical computing it is not always true. The system (20) can be ill-conditioned for bigger ε, that is why it is important to use condition (21).

Example 1. The sequence $\{\|P_n\|^2\}$ of squared norms of shifted Legendre polynomials is decreasing rapidly to zero. Really, for $n = 5$, the exactness of expansion of the function $f(x) = \ln(1+x)\cos 5x$ over $\{P_j^{(\varepsilon)}(x)\}$ is holding on for $\varepsilon < \|P_5\|^2 = 1.43155(-6)$, but for larger it can decrease significantly. Numerical evaluating shows that for $\varepsilon = 0.01$, the exactness is still good $1.1883(-6)$, but for $\varepsilon = 0.02$, it is only $3.6331(-4)$.

If we compare approximations on $(-1, 1)$ with Chebyshev weight and on $(0, \infty)$ with Laguerre weight, the influence of $\varepsilon \in (0, 1)$ is negligible because the norm-sequence is slowly decreasing in Chebyshev case in spite of the increase in Laguerre case.

Conjecture 1. Under condition (21), the polynomials $\{P_n^{(\varepsilon)}(x)\}$ have all zeros on the support interval (a, b).

Acknowledgement

This research was supported by Ministry of Science of Republic Serbia through the Project NPEE 24-2006 and Project 144023.

References

1. MacInnes, C.S.: The Reconstruction of Discontinuous Piecewise Polynomial Signals. IEEE Transactions on Signal Processing 53(7), 2603–2607 (2005)
2. Karam, L.J., McClellan, J.H.: Complex Chebyshev Approximation for FIR Filter Design. IEEE Transactions on Circuits-II: Analog and Digital Signal Processing 42(3), 207–216 (1995)
3. Nie, X., Raghuramireddy, D., Unbehauen, R.: Orthonormal Expansion of Stable Rational Transfer Functions. Electronics Letters 27(16), 1492–1494 (1991)
4. Tseng, C.C.: Digital Differentiator Design Using Fractional Delay Filter and Limit Computation. IEEE Transactions on Circuits and Systems-4, Regular Papers 52(10), 2248–2259 (2005)
5. Benyi, A., Torres, R.H.: Almost Orthogonality and a Class of Bounded Bilinear Pseudodifferential Operators. Mathematical Research Letters 11, 1–11 (2004)
6. Ben-Yaacov, I., Wagner, F.O.: On Almost Orthogonality in Simple Theories. J. Symbolic Logic 69(2), 398–408 (2004)
7. Cotlar, M.: A Combinatorial Inequality and Its Applications to L^2-Spaces. Rev. Mat. Cuyana 1, 41–55 (1955)
8. Szegő, G.: Orthogonal Polynomials, 4th edn., vol. 23. Amer. Math. Soc. Colloq. Publ. (1975)

Numerical Experiments for Reaction-Diffusion Equations Using Exponential Integrators

Gabriel Dimitriu and Răzvan Ştefănescu

University of Medicine and Pharmacy "Gr. T. Popa"
Department of Mathematics and Informatics, 700115 Iaşi, Romania
dimitriu.gabriel@gmail.com, rastefanescu@yahoo.com

Abstract. In this study we focus on a comparative numerical approach of two reaction-diffusion models arising in biochemistry by using exponential integrators. The goal of exponential integrators is to treat exactly the linear part of the differential model and allow the remaining part of the integration to be integrated numerically using an explicit scheme. Numerical simulations including both the global error as a function of time step and error as a function of computational time are shown.

1 Introduction

Reaction-diffusion equations are frequently encountered in mathematical biology, ecology, physics and chemistry. This type of equations leads to interesting phenomena, such as, pattern formation far from equilibrium, pulse splitting and shedding, reactions and competitions in excitable systems, nonlinear waves and spatio-temporal chaos. The efficient and accurate simulation of such systems, however, represent a difficult task. This is because they couple a stiff diffusion term with a (typically) strongly nonlinear reaction term. When discretised this leads to large systems of strongly nonlinear, stiff ODEs.

In this work we perform a comparative numerical approach of two reaction-diffusion models arising in biochemistry by using exponential integrators. The paper is organized as follows. Section 2 briefly describes the exponential integrators and their features. In Section 3 the two reaction kinetics – Gierer-Meinhardt and Thomas models, respectively – are presented on the basis of which the numerical study is carried out. Section 4 is devoted to a short description of the numerical schemes applied to the models under study, together with results of the numerical simulations. Finally, some concluding remarks are drawn in the last section.

2 Exponential Integrators

The exponential integrators represent numerical schemes specifically constructed for solving differential equations (see for details [8]), where it is possible to split the problem into a linear and a nonlinear part

$$\dot{y} = Ly + N(y,t), \qquad y(t_{n-1}) = y_{n-1}\,, \tag{1}$$

S. Margenov, L.G. Vulkov, and J. Waśniewski (Eds.): NAA 2008, LNCS 5434, pp. 249–256, 2009.

where $y \in \mathbf{C}^d$, $L \in \mathbf{C}^{d \times d}$ and $N : \mathbf{C}^d \times \mathbf{R} \to \mathbf{C}^d$. In specific applications (discretizations of PDEs) the matrix L is unbounded. Generally, solving such problems requires an implicit scheme; the goal of the exponential integrators is to treat the linear term exactly and allow the remaining part of the integration to be integrated numerically using an explicit scheme.

An exponential integrator has two main characteristics: (i) If $L = 0$, then the scheme reduces to a standard general linear scheme. This is often called the underlying general linear scheme; (ii) If $N(y,t) = 0$ for all y and t, then the scheme reproduces the exact solution of (1). To satisfy (i) the exponential function must be used within the numerical scheme. Despite the fact that L is unbounded, typically the coefficients of the scheme will be bounded.

For an s-stage exponential integrator of Runge-Kutta type, we define the internal stages and output approximation:

$$Y_i = h \sum_{j=1}^{s} a_{ij}(hL) N(Y_j, t_{n-1} + c_j h) + u_{i1}(hL) y_{n-1}\,, \quad i = 1, \ldots, s\,,$$

$$y_n = h \sum_{i=1}^{s} b_i(hL) N(Y_j, t_{n-1} + c_j h) + v_1(hL) y_{n-1}\,. \tag{2}$$

The feature (i) above is satisfied if we require in (2) as $u_{i1}(0) = 1$, $a_{ij}(0) = a_{ij}$, $v_1(0) = 1$, and $b_i(0) = b_i$, where the real numbers a_{ij} and b_j represent the coefficients of the underlying Runge-Kutta scheme.

The extension to general linear schemes is carried out as follows. A step of length h in an exponential general linear scheme, requires to import r aproximations into the step, denoted as $y_i^{[n-1]}$, $i = 1, \ldots, r$. The internal stages (as in the Runge-Kutta case) are written as Y_i, $i = 1, \ldots, s$. After the step is completed, r updated approximations are computed. These are then used in the next step. Each step in an exponential general linear scheme can be written as

$$Y_i = h \sum_{j=1}^{s} a_{ij}(hL) N(Y_j, t_{n-1} + c_j h) + \sum_{j=1}^{r} u_{ij}(hL) y_j^{[n-1]}\,, \quad i = 1, \ldots, s\,,$$

$$y_i^{[n]} = h \sum_{j=1}^{s} b_{ij}(hL) N(Y_j, t_{n-1} + c_j h) + \sum_{j=1}^{r} v_{ij}(hL) y_j^{[n-1]}\,, \quad i = 1, \ldots, r. \tag{3}$$

The exponential integrators of Runge-Kutta type are easily seen to be a special case when $r = 1$ with $u_{i1}(z) = a_{i0}(z)$, $v_{11}(z) = b_0(z)$ and $b_{1j}(z) = b_j(z)$.

3 Model Equations

In this section we shortly describe the models governing different reaction kinetics arising in biochemistry, which will be solved numerically in Section 4.

Gierer-Meinhardt reaction kinetics. This represents a phenomenological model suggested by Gierer and Meinhardt ([5]), whereby reaction kinetics are chosen in such a way that one of the chemicals (termed activator) activates the production of the other chemical (the inhibitor) which, in turn, inhibits the production of the activator. The non-dimensionalised reaction-diffusion system is given by

$$\begin{aligned} \frac{\partial u}{\partial t} &= D_u \nabla^2 u + \gamma \left(a - bu + \frac{u^2}{v(1+\kappa u^2)} \right), \\ \frac{\partial v}{\partial t} &= D_v \nabla^2 v + \gamma (u^2 - v), \end{aligned} \tag{4}$$

where $u(x,t)$ is the concentration of the activator, $v(x,t)$ is the concentration of the inhibitor, t is time and ∇^2 is the 1-dimensional Laplacian. D_u, D_v, a, b and γ are all nondimensionalised positive parameters and k is a measure of the saturation concentration ([10]). The biological interpretation of the reaction kinetics in (4) is that u is produced at a constant rate γa and is degraded linearly at rate γb. The $\gamma \frac{u^2}{v(1+\kappa u^2)}$ term implies autocatalysis in u with saturation at high concentration values of u, and inhibition of u through the production of v. In the second equation of the system (4), v is activated (produced) by u and degraded linearly.

Thomas reaction kinetics. This model is based on a specific substrate-inhibition reaction involving the substrates oxygen $v(x,t)$, and uric acid $u(x,t)$, which react in the presence of the enzyme uricase. The reaction kinetics, derived by fitting the kinetics to experimental data ([12]), can be written in nondimensional form as

$$\begin{aligned} \frac{\partial u}{\partial t} &= D_u \nabla^2 u + \gamma \left(a - u - h(u,v) \right), \\ \frac{\partial v}{\partial t} &= D_v \nabla^2 v + \gamma \left(\alpha b - \alpha v - h(u,v) \right), \end{aligned} \tag{5}$$

with $h(u,v) = \frac{\rho u v}{1+u+Ku^2}$. Here D_u, D_v, a, α, b, γ and ρ are positive parameters. The term $h(u,v)$ indicates the rate at which u and v are used up, in particular $h(u,v)$ exhibits what is known as substrate-inhibition, that is, for small u, $h(u,v)$ increases with u, while it decreases with large u.

4 Description of the Numerical Schemes and Computational Issues

In what follows, we briefly describe the numerical schemes defining the exponential integrators that have been used in our comparative study. All these integrators belong to the package `EXPINT` written in Matlab ([1]). In this description we will use two terms of order. The non-stiff order refers to the case when the operator L is bounded, such conditions were derived in [8]. The stiff order refers to the case when L is unbounded ([3]), for various schemes. We remark here that the stiff order convergence analysis is performed for the parabolic case only.

The first scheme that has been applied to our models is named `Lawson4`. The scheme `Lawson4` belongs to the Lawson schemes constructed by applying the Lawson transformations ([7]) to the semi-linear problem. It is based on the

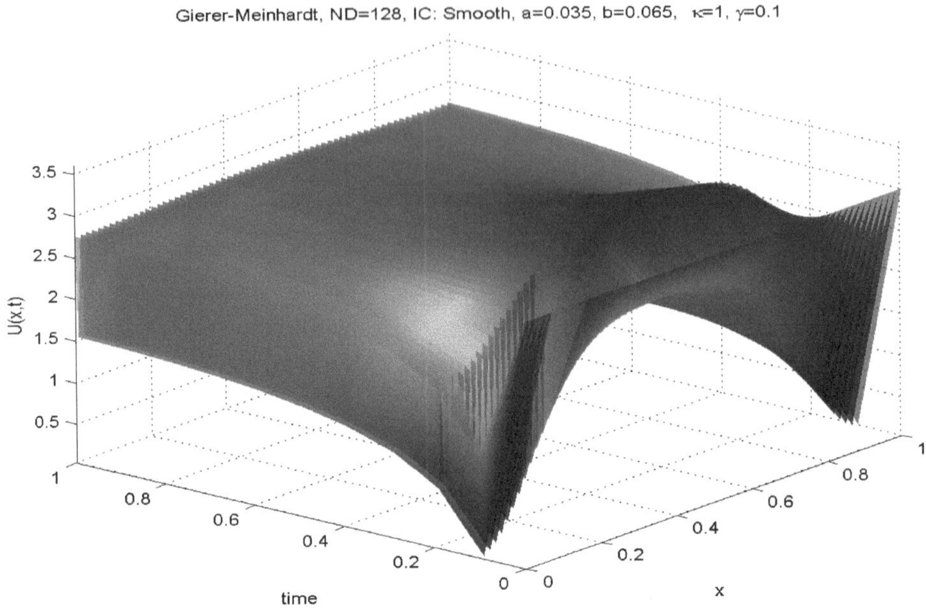

Fig. 1. The variation profile of the variable $u(x,t)$ in the model (4) representing the concentration of the activator which stimulates the production of the inhibitor denoted by $v(x,t)$ (Gierer-Meinhardt reaction kinetics)

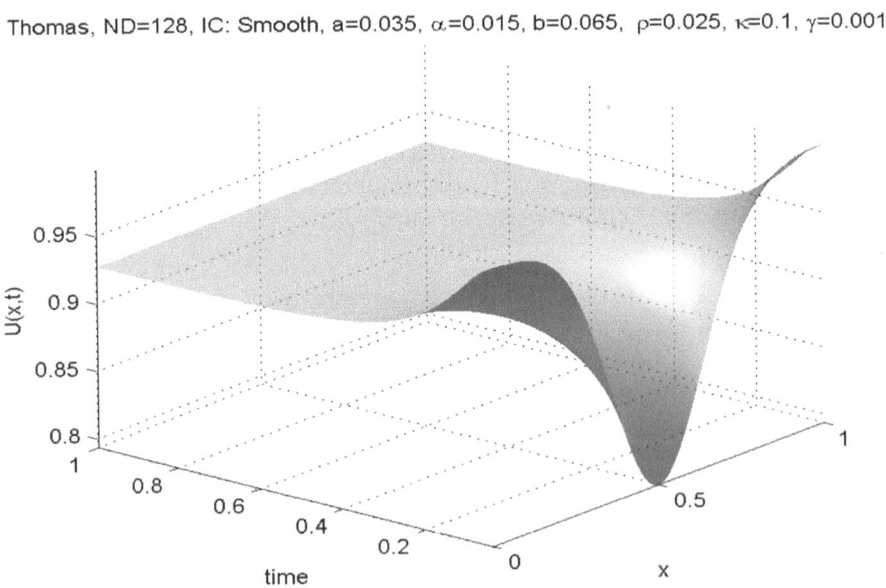

Fig. 2. The variation profile of the variable $u(x,t)$ in the model (5) representing the concentration of the uric acid in the substrate-inhibition reaction involving the substrate oxygen $v(x,t)$ (Thomas reaction kinetics)

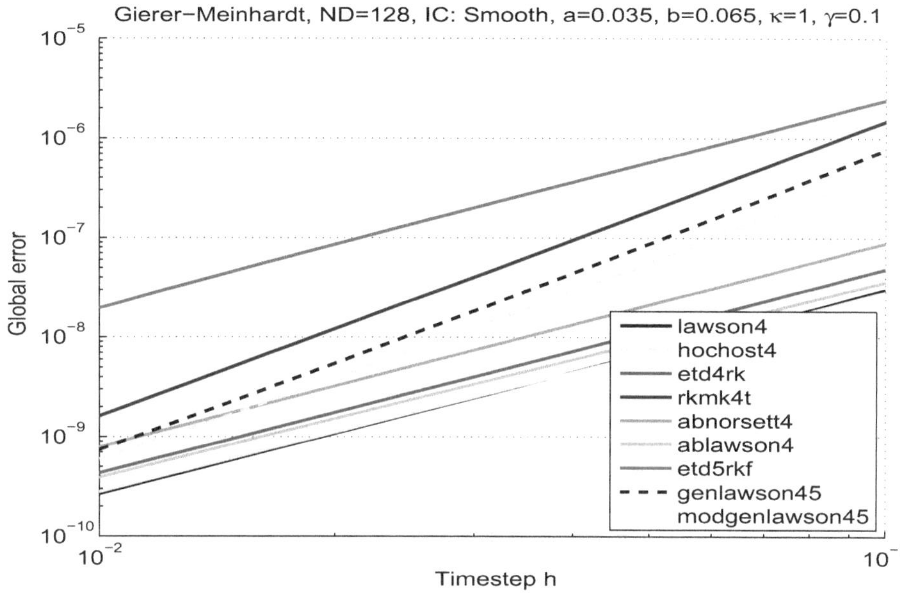

Fig. 3. Comparative results concerning the quality of the numerical schemes: the global error as a function of timestep h for the Gierer-Meinhardt reaction kinetics

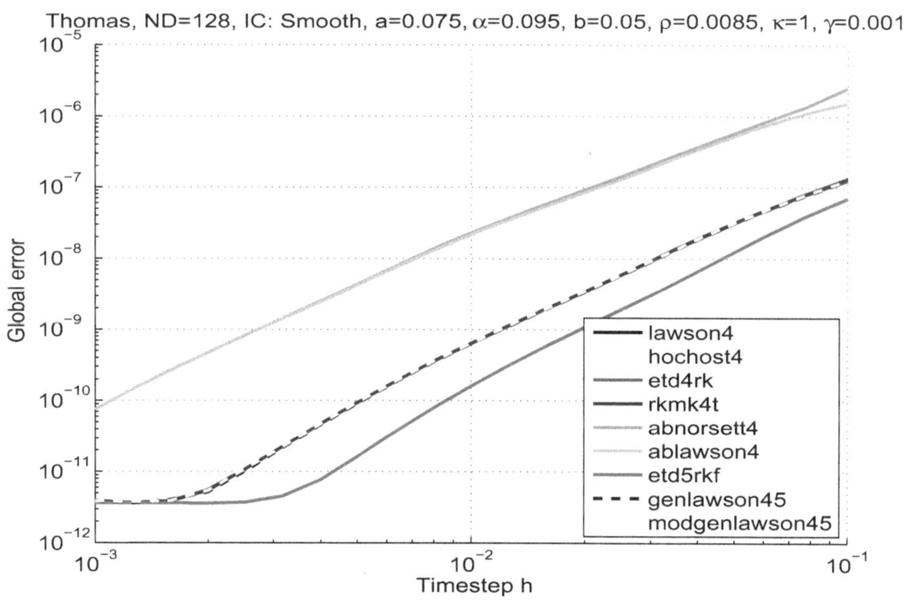

Fig. 4. Comparative results concerning the quality of the numerical schemes: the global error as a function of timestep h for the Thomas reaction kinetics

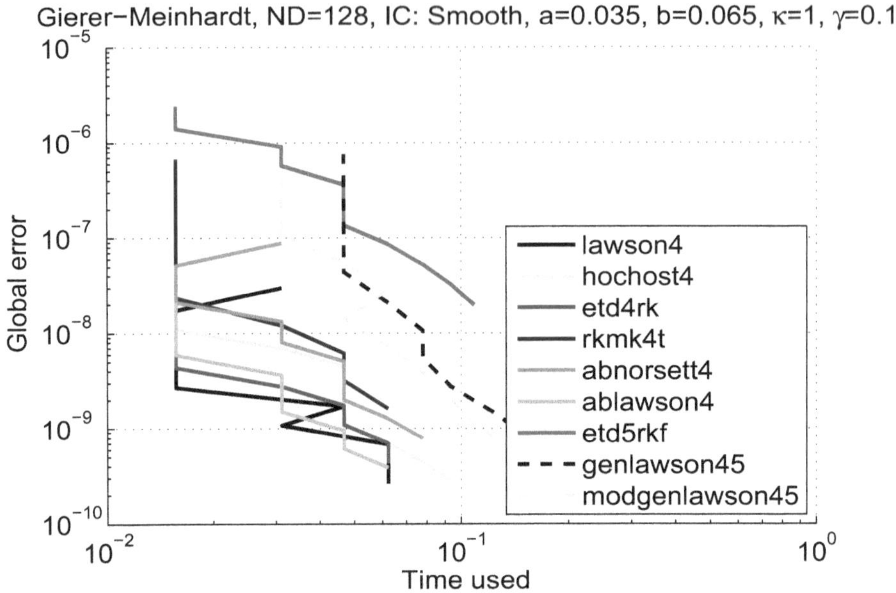

Fig. 5. Comparative results concerning the global error as a function of computational time for the Gierer-Meinhardt reaction kinetics

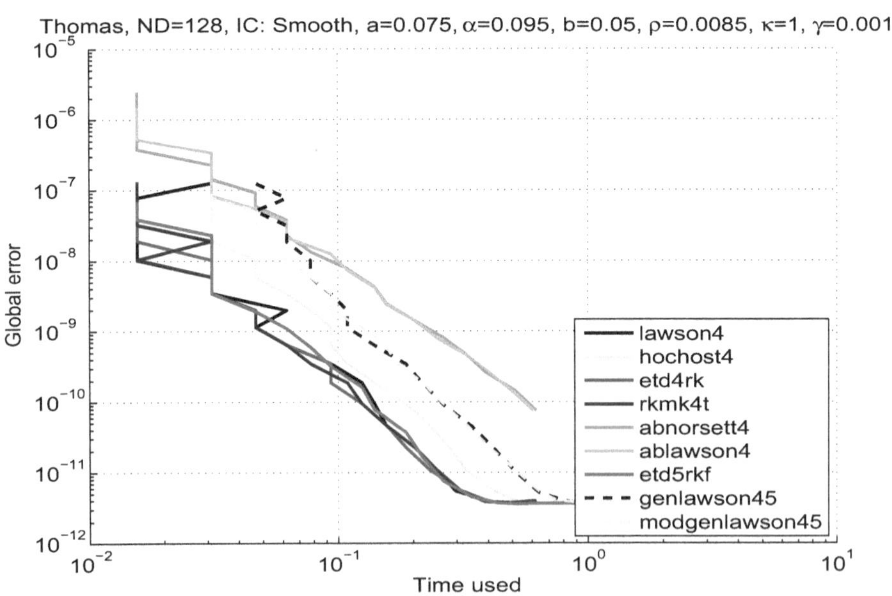

Fig. 6. Comparative results concerning the global error as a function of computational time for the Thomas reaction kinetics

clasical fourth order scheme of Kutta (see [2], Eq. (235i)), and this scheme has stiff order one.

The scheme denoted by `hochost4` was developed by Hochbruck and Ostermann. It has five-stages and is the only known exponential Runge-Kutta method with stiff order four.

Nørsett designed in [11] a class of schemes which reduced to the Adams-Bashforth methods when the linear part of the problem is zero.

`ABLawson4` has stiff order one and is based on the Adams-Bashforth scheme of order four and is represented in this way so that the incoming approximation has the form $y^{[n-1]} = [y_{n-1}, hN_{n-2}, hN_{n-3}, hN_{n-4}]^T$.

`ABNørsett4` is a stiff order four scheme of Norsett ([11]), which is implemented so that the incoming approximation has the same form as in `ABLawson4`.

`ETD` schemes are based on algebraic approximations to the nonlinear term in the variation of constants formula. `ETD` means "Exponential Time Differencing" and the name stems from [4]. The scheme `ETD4RK` due to Cox and Matthews in ([4], Eqs. (26)-(29)) started the recent focus on exponential integrators, unfortunately it has only stiff order two. `ETD5RKF` is a non-stiff fifth order scheme based on the six stage fifth order scheme of Fehlberg.

The scheme `RKMK4t` uses a convenient truncation of the $dexp^{-1}$ operator, leading to the method of Munthe-Kaas [9], which again is of stiff order two but suffers from instabilities, especially when non-periodic boundary conditions are used.

Krogstad [6] constructed the generalized Lawson schemes as a means of overcoming some of the undesirable properties of the Lawson schemes. This class of schemes uses approximations of the nonlinear term from previous steps, resulting in an exponential general linear method. The scheme `genlawson45` included in the package mentioned above is also used for our numerical study.

Figures 1 and 2 present the variation profiles for the concentration variable of $u(x,t)$ in the two reaction kinetics, Gierer-Meinhardt and Thomas, respectively.

Figures 3 and 4 illustrate comparison results concerning the quality of the numerical schemes that have been used in this analysis. There are shown relationships between the global error and the timestep h varying from 10^{-2} to 10^{-1}. We note that for Gierer-Meinhardt model good behaviours have had the schemes `lawson4`, `hochcost4`, `etd4rk`, `ablawson4`, while the schemes `rkmk4t`, and `etd5rkf` indicated a more significant increasing rate of the global errors with respect to the computed global error (see Fig. 3). In the case of Thomas reaction kinetics, the scheme `etd5rkf` has had the best behaviour (see Fig. 4).

Figures 5 and 6 give timing results in the sense that present dependencies of the global error as a function of computational time for the two models. In this respect, good results are obtained with the schemes `lawson4`, `etd4rk` and `ablawson4` for the Gierer-Meinhardt model (see Fig. 5), and `lawson4`, `rkmk4t`, `etd5rkf` for the Thomas model (see Fig. 6).

All the plots indicate in their title: "ND=128" and "IC: Smooth". This means that we have used 128 Fourier modes in the spatial direction (must be power of 2), and as initial condition for the model variables, we have chosen a set of values with a Gaussian distribution.

5 Conclusions

In this paper we focused on a comparative numerical study for two models arising in biochemsitry: Gierer-Meinhardt reaction kinetics and Thomas reaction kinetics. The numerical approach has been performed by using several exponential integrators belonging to Matlab package EXPINT ([1]). The numerical findings together with global error and timing results were presented in illustrative plots.

References

1. Berland, H., Skaflestad, B., Wright, W.: EXPINT – A Matlab package for exponential integrators. Numerics 4 (2005)
2. Butcher, J.C.: Numerical methods for ordinary differential equations. John Wiley & Sons, Chichester (2003)
3. Celledoni, E., Marthinsen, A., Owren, B.: Commutator-free Lie group methods. FGCS 19(3), 341–352 (2003)
4. Cox, S.M., Matthews, P.C.: Exponential time differencing for stiff systems. J. Comp. Phys. 176(2), 430–455 (2002)
5. Gierer, A., Meinhardt, H.: A theory of biological pattern formation. Kybernetik 12, 30–39 (1972)
6. Krogstad, S.: Generalized integrating factor methods for stiff PDEs. Journal of Computational Physics 203(1), 72–88 (2005)
7. Lawson, D.J.: Generalized Runge-Kutta processes for stable systems with large Lipschitz constants. SIAM J. Numer. Anal. 4, 372–380 (1967)
8. Minchev, B., Wright, W.M.: A review of exponential integrators for semilinear problems. Technical Report 2, The Norwegian University of Science and Technology (2005)
9. Munthe-Kaas, H.: High order Runge-Kutta methods on manifolds. Applied Numerical Mathematics 29, 115–127 (1999)
10. Murray, J.D.: Mathematical Biology, 2nd edn. Springer, New York (1993)
11. Nørsett, S.P.: An A-stable modification of the Adams-Bashforth methods. In: Conf. on Numerical solution of Differential Equations (Dundee, 1969), pp. 214–219. Springer, Berlin (1969)
12. Thomas, D.: Artificial enzyme membrane, transport, memory and oscillatory phenomena. In: Thomas, D., Kervenez, J.-P. (eds.) Analysis and Control of Immobilised Enzyme Systems, pp. 115–150. Springer, Heidelberg (1975)

Diaphony of Uniform Samples over Hemisphere and Sphere

I.T. Dimov[1], S.S. Stoilova[2], and N. Mitev

[1] Institute for Parallel Processing, Bulgarian Academy of Sciences
Acad. G. Bonchev Str., bl. 25 A, 1113 Sofia, Bulgaria
[2] Institute of Mathematics and Informatics, Bulgarian Academy of Sciences
Acad. G. Bonchev Str., bl. 8, 1113 Sofia, Bulgaria
I.T.Dimov@reading.ac.uk, stoilova@math.bas.bg

Abstract. The sampling of certain solid angle is a fundamental operation in realistic image synthesis, where the rendering equation describing the light propagation in closed domains is solved. In this work we consider the problem for generation of uniformly distributed random samples over hemisphere and sphere. Using two algorithms we obtain samples in orthogonal spherical triangle and spherical quadrangle. Our aim is to prove uniform distribution of the obtained samples. The importance of the uniformly distributed samples is determined by the effectiveness of the algorithms for numerical solution of the rendering equation. We use numerical characteristic for uniform distribution of points, called diaphony. The diaphony of these samples is calculated numerically. Analysis and comparison of the obtained results are made.

1 Introduction

1.1 Uniform Sampling of Hemisphere and Sphere

In computer graphics for realistic image synthesis the sampling of certain solid angle is a fundamental operation. For the numerical solving the rendering equation describing the light propagation in closed domains the Monte Carlo and Quasi-Monte Carlo methods [6] are used. The domain of the numerical integration is the solid angle subtended by unit hemisphere or unit sphere.

We consider the problem for generation of uniformly distributed samples over hemisphere and sphere. We construct and study the sampling scheme for hemisphere and sphere.The partitioning of hemisphere and sphere is made on the base of the property symmetry. We obtain the equal sub-domains. Each sub-domain represents solid angle subtended by orthogonal spherical triangle or spherical quadrangle.

Consider hemisphere and sphere with center in the origin of a Descartes coordinate system. It is obvious that the coordinate planes partition the hemisphere into 4 equal areas and the sphere into 8 equal areas. The partitioning of each one area into sub-domains can be continued by the three bisector planes. One can see

S. Margenov, L.G. Vulkov, and J. Waśniewski (Eds.): NAA 2008, LNCS 5434, pp. 257–264, 2009.

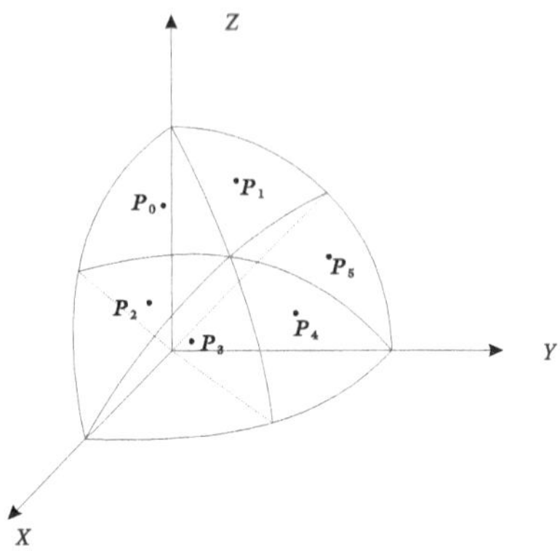

Fig. 1. Partition of a spherical area into 6 sub-domains

that the bisector planes to the dihedral angles $(\overrightarrow{X}, \overrightarrow{Y})$, $(\overrightarrow{X}, \overrightarrow{Z})$ and $(\overrightarrow{Z}, \overrightarrow{Y})$, partition each area into 6 equal sub-domains. In Fig. 1 we show the partitioning of the area with positive coordinate values of X, Y and Z into 6 equal sub-domains.

Two algorithms for sampling of orthogonal spherical triangles and spherical quadrangles are used. The first sampling algorithm generates a sample by mapping of the unit square onto orthogonal spherical triangle. The second algorithm directly computes the unit radius vector of a sampling point inside to the orthogonal spherical triangle.

Algorithm 1
Let us consider the solid angle subtended by the spherical triangle $\triangle ABC$ shown in Fig. 2.

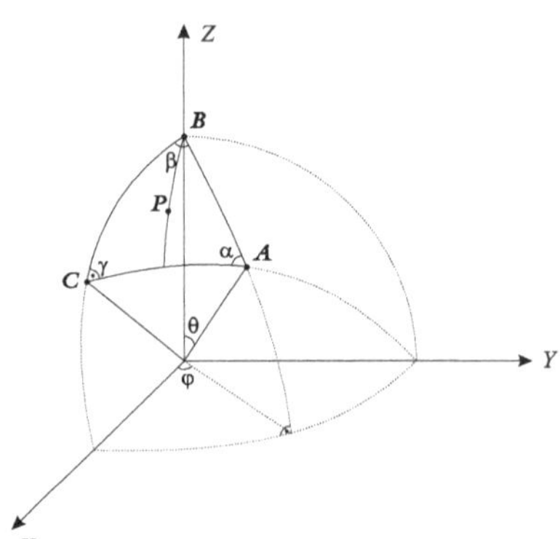

Fig. 2. Sampling point generation by Algorithm 1

For $u, v \in [0, 1)$ the coordinates of the point P are calculate by

▷ Generate Random Variables: (real u, real v)
▷ Calculate angles : $\varphi = \frac{u\pi}{4}$ and $\theta = \arctan \frac{v}{\cos \frac{u\pi}{4}}$
▷ Calculate the sampling point coordinates:
$P_x = \cos\varphi \cdot \sin\theta, \quad P_y = \sin\varphi \cdot \sin\theta \quad \textit{and} \quad P_z = \cos\theta$
▷ Return Sampling Point: $P(P_x, P_y, P_z)$.

Algorithm 2
This algorithm compute directly the unit radius vector of a sampling point inside to the orthogonal spherical triangle. Consider a point P inside for the spherical triangle $\triangle ABC$ shown in Fig. 3. The coordinates of an arbitrary sampling point P could be calculated by finding the intersection point of spherical triangle $\triangle ABC$ with the two coordinate planes having normal vectors $N_x(1, 0, 0)$ and $N_y(0, 1, 0)$, and rotated respectively on angle φ and angle θ, where $\theta, \varphi \in [0, \frac{\pi}{4}]$. It is clear that when $\theta = \varphi = \frac{\pi}{4}$, the point $P = A$ and when $\theta = \varphi = 0$, the point $P = B$.

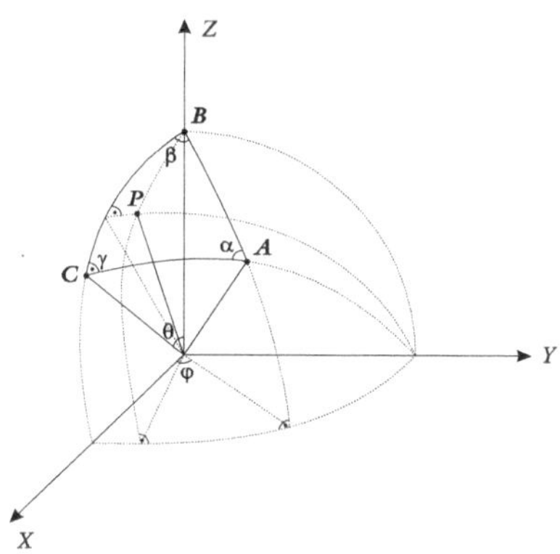

Fig. 3. Sampling point generation by Algorithm 2

The calculation of the coordinates of the point P for $u, v \in [0, 1)$ by the second algorithm is

▷ Generate Random Variables:(real u, real v)
▷ Calculate angles : $\varphi = \frac{u\pi}{4}$ and $\theta = \frac{v\pi}{4}$
▷ Calculate the sampling point coordinates:
$$P_x = \frac{\sin\theta \cdot \cos\varphi}{\sqrt{\sin^2\theta + \cos^2\theta\cos^2\varphi}}, \quad P_y = \frac{\sin\theta \cdot \sin\varphi}{\sqrt{\sin^2\theta + \cos^2\theta\cos^2\varphi}}$$
and $P_z = \frac{\cos\theta \cdot \cos\varphi}{\sqrt{\sin^2\theta + \cos^2\theta\cos^2\varphi}}$
▷ Return Sampling Point: $P(P_x, P_y, P_z)$.

The description in details of the algorithms and the using of the obtained spherical nets can be found in [1] and [2].

When we are talking about spherical quadrangle, we will consider it like a combination of two neighbour spherical triangles having a common arc. The exposed algorithms do not change by their essence. The difference is in the second step where the angles φ and θ are computed. For algorithm 1 we have $\varphi = \frac{u\pi}{4}, \quad \theta = \arctan \frac{v}{\sin \frac{u\pi}{4}}$ and for algorithm 2 we have $\varphi = \frac{u\pi}{4}, \quad \theta = \frac{v\pi}{2}$.

Using different generators of u and v, we obtain different sampling schemes. With inspection of the complexity of the numerical computations, is necessary to find sampling schemes with as possible least points and as possible best distributed points in the spherical triangle or spherical quadrangle. In this work we use the uniformly distributed sequences and nets for generators in the sampling schemes.

1.2 Uniform Distribution of Sequences

Let $\xi = (\mathbf{x}_j)_{j\geq 1}$ be sequence in $[0,1)^s$. For arbitrary integer N and $J = [\mathbf{a}, \mathbf{b}) \subseteq [0,1)^s$ $A_N(\xi; J)$ is the number of belonging to J points of ξ. We denote the Lebesgue measure on J with $\mu(J) = \prod_{i=1}^{s}(b_i - a_i)$.

Definition 1. [8] *The sequence $\xi = (\mathbf{x}_j)_{j\geq 1}$ is uniformly distributed mod 1 in $[0,1)^s$ if for every $J = [\mathbf{a}, \mathbf{b}) \subseteq [0,1)^s$ the equality*

$$\lim_{N\to\infty} \frac{A_N(\xi; J)}{N} = \mu(J)$$

is hold.

Definition 1 shows when a sequence of points in $[0,1)^s$ is uniformly distributed but it does not give opportunity to comparison of the distributions of two sequences. For that purpose to use measures of the distribution of the sequences. One of these measures is the discrepancy.

Definition 2. *Let $N \geq 1$ be an arbitrary fixed integer and $\xi_N = \{\mathbf{x}_0, \dots \mathbf{x}_{N-1}\}$ is a net of real numbers in $[0,1)^s$. The quantity $D(\xi_N)$ definite by the equality*

$$D(\xi_N) = \sup_{J\subseteq[0,1)^s} \left| \frac{A(\xi_N; J)}{N} - \mu(J) \right|$$

is called a discrepancy of the given net.

Another measure for uniform distribution of sequences and nets is the diaphony.

Definition 3. [9] *The diaphony $F_N(\mathcal{T}; \xi)$ of the first $N \geq 1$ elements of the sequence $\xi = (\mathbf{x})_{j\geq 1}$ in $[0,1)^s$ is defined as*

$$F_N(\mathcal{T}; \xi) = \left(\sum_{\mathbf{k}\in\mathbf{Z}^s\setminus\{\mathbf{0}\}} R^{-2}(\mathbf{k}) \left| \frac{1}{N} \sum_{j=1}^{N} e_{\mathbf{k}}(\mathbf{x}_j) \right|^2 \right)^{\frac{1}{2}},$$

where for $\mathbf{k} = (k_1, k_2, \ldots, k_s) \in \mathbf{Z}^s$ $R(\mathbf{k}) = \prod_{i=1}^{s} \max(|k_i|, 1)$ *and* $\mathcal{T} = \{e_{\mathbf{k}}(\mathbf{x}) = \prod_{j=1}^{s} \exp(2\pi \mathbf{i} k_j x_j) : \mathbf{k} \in \mathbf{Z}^s, \mathbf{x} \in [0,1)^s\}$ *is the trigonometric functional system.*

For the diaphony is known that

$$0 \leq F_N(\mathcal{T}; \xi) < 1.$$

There is other similar numerical characteristics for uniform distribution of sequences and nets as dyadic diaphony[5], b-adic diaphony[3] etc.

The fact that the discrepancy and the diaphony are measures for uniform distribution of sequences and nets is given in the next two theorems.

Theorem 1. *The sequence* $\xi = (\mathbf{x}_j)_{j \geq 1}$ *is uniformly distributed mod 1 in* $[0,1)^s$ *if and only if*

$$\lim_{N \to \infty} D_N(\xi) = 0.$$

By analogy, we have

Theorem 2. *The sequence* $\xi = (\mathbf{x}_j)_{j \geq 1}$ *is uniformly distributed mod 1 in* $[0,1)^s$ *if and only if*

$$\lim_{N \to \infty} F_N(\mathcal{T}; \xi) = 0.$$

In this work we will use the diaphony as a measure for uniform distribution of the nets obtaining by mentioned above algorithms. Take into account the complexity of the diaphony definition and the generation method of the point it is obvious that the analytically computations of the diaphony of the net are very hard, very expensive. For that reason we calculate numerically the diaphony of the generated above point nets.

2 Numerical Calculation of the Diaphony

The numerical computation problem of the diaphony given in the form of Definition 3 is very difficult. By this reason we represent the diaphony in the form

$$F_N(\mathcal{T}; \xi) = \left(\frac{1}{N^2} \sum_{i=1}^{N} \sum_{j=1}^{N} H(\mathbf{x}_i - \mathbf{x}_j) \right)^{\frac{1}{2}},$$

where for an arbitrary $\mathbf{x} = (x_1, x_2, \ldots, x_s) \in [0,1)s$

$$H(\mathbf{x}) = \prod_{j=1}^{s} g(x_j) - 1,$$

$$g(x) = 1 - \frac{\pi^2}{6} + \frac{\pi^2}{2}(1 - 2\{x\})^2$$

and $\{x\}$ is the fractional part of x.

From described above algorithms can be seen the dependence of generated points from u and v. It is very important to select by appropriate way these variables. The issue that stands here back is if we use uniformly distributed nets for the variables u and v then will we obtain uniformly distributed net of points in the spherical triangle, respectively spherical quadrangle? The answer of this question consists in the prove of the fact that the diaphony of the spherical net tends to zero with increasing the number of the generated by given algorithms points in the spherical triangle or quadrangle.

Here, we use the well known uniformly distributed net of Roth [7] and Halton[4]. We remember their definitions. Let the natural number $b \geq 2$ be the base and $i \geq 0$ is integer. We define the function ϕ as

$$\phi_b(i) = \sum_{j=0}^{\infty} a_j(i) b^{-j-1}, \quad \text{when} \quad i = \sum_{j=0}^{\infty} a_j(i) b^j.$$

For integer N the net of Roth is defined as

$$R_N(i) = \left\{ \left(\frac{i}{N}, \phi_b(i) \right), 0 \leq i < N \right\}$$

and the net of Halton is defined as

$$H_N(i) = \left\{ \left(\phi_{b_1}(i), \phi_{b_2}(i) \right), 0 \leq i < N \right\},$$

where the bases b_1 and b_2 are primes.

To prove that obtained spherical nets by Algorithms 1 and 2 are uniformly distributed in spherical triangle and quadrangle, we compute the diaphony of these nets.

The results from the numerical calculations of the diaphony are given in Table 1 and Table 2 respectively for Algorithm 1 and Algorithm 2.

Table 1. Diaphony of the points generated by Algorithm 1

Num. points	Sph. Tr. Roth gen.	Sph. Tr. Halton gen.	Sph. Quad. Roth gen.	Sph. Quad. Halton gen.
50	0.00361486	0.00362232	0.00259929	0.00275899
100	0.00087062	0.00087121	0.00062317	0.0006435
1000	$8.44683469.10^{-06}$	$8.44889617.10^{-06}$	$6.11011646.10^{-06}$	$6.13663460.10^{-06}$
10000	$8.42145014.10^{-08}$	$8.42155444.10^{-08}$	$6.10672324.10^{-08}$	$6.10933768.10^{-08}$

It may be remarked that the diaphony values for the shown nets are in the interval $[0, 1)$. It is proving that the nets are *uniformly distributed* in spherical triangle and quadrangle. It is making an impression that the diaphony values in the in spherical quadrangle are less from the diaphony values in the spherical triangle. This fact shows the point distribution in the spherical quadrangle is better from the distribution in the spherical triangle.

Table 2. Diaphony of the points generated by Algorithm 2

Num. points	Sph. Tr. Roth gen.	Sph. Tr. Halton gen.	Sph. Quad. Roth gen.	Sph. Quad. Halton gen.
50	0.00359156	0.003597	0.00296329	0.00283578
100	0.00086508	0.00086530	0.000747	0.00072903
1000	$8.39300849.10^{-06}$	$8.39523207.10^{-06}$	$7.30704654.10^{-06}$	$7.2844826.10^{-06}$
10000	$8.3677299.10^{-08}$	$8.36786114.10^{-08}$	$7.28309822.10^{-08}$	$7.2807439.10^{-08}$

Can be seen that even for small number of the points - 50, the diaphony order is 10^{-3} which goes to show that the distribution of the obtained nets is very good. The augmentation of the point number with 50 leads to diaphony order 10^{-4}, and for 10000 points the diaphony order becomes 10^{-8}. It is obvious that the diaphony tending to zero is very fast.

The generation of the points by special way on spherical triangle and spherical quadrangle and the demonstration of their uniform distribution are necessary on their application for numerical solution of the rendering equation in the computer graphics. Applying generated by algorithm 1 and algorithm 2 points for numerical integration of the rendering equation we have faster convergence of the quasi-Monte Carlo method. The fast diaphony convergence leads to using small number of points for the numerical integration. That gives opportunity for decrease the complexity of the calculations and increases the effectiveness of the method.

3 Conclusion

The considered schemes for uniform separation of hemisphere and sphere and the generation of uniformly distributed points on spherical triangle and quadrangle improve the effectiveness of the numerical solution of the rendering equation. The uniform distribution of using spherical nets by numerical calculation of a characteristic for distribution, called diaphony was proving. It is shown that the diaphony of the constructed nets fast comes to zero. This gives us reason to see that these nets are very good distributed in spherical triangle and quadrangle. Due to that is not necessary to use too large number of points for numerical solution of the rendering equation. Contrariwise, the small number of points is sufficiently for obtaining of good convergence of the numerical process.

Here, the question for analytically estimation of the diaphony of the so constructed nets is open. The answer of this question is subject of future consideration and research.

References

1. Dimov, I.T., Penzov, A.A., Stoilova, S.S.: Parallel Monte Carlo Sampling Scheme for Sphere and Hemisphere. In: Li, Z., Vulkov, L.G., Waśniewski, J. (eds.) NAA 2004. LNCS, vol. 3401, pp. 148–155. Springer, Heidelberg (2005)
2. Dimov, I.T., Penzov, A.A., Stoilova, S.S.: Parallel Monte Carlo Approach for Integration of the Rendering Equation. In: Li, Z., Vulkov, L.G., Waśniewski, J. (eds.) NAA 2004. LNCS, vol. 3401, pp. 140–147. Springer, Heidelberg (2005)

3. Grozdanov, V., Stoilova, S.: On the Theory of b-adic Diaphony. Comp. Ren. Akad. Bul. Sci. 54(3), 31–34 (2001)
4. Halton, J.H.: On the efficiency of certain quasi-random sequences of points in evaluating multi-dimensional integrals. Numer. Math. 2, 84–90 (1960)
5. Hellekalek, P., Leeb, H.: Dyadic Diaphony. Acta Arithmetica LXXX(2), 187–196
6. Keller, A.: Quasi-Monte Carlo Methods in Computer Graphics: The Global Illumination Problem. Lectures in Applied Mathematics 32, 455–469 (1996)
7. Roth, K.: On irregularities of distribution. Mathematika 1, 73–79 (1954)
8. Weyl, H.: Über die Gleichverteilung von Zahlen mod. Eins. Math. Ann. 77(S), 313–352 (1916)
9. Zinterhof, P.: Über einige Abschätzungen bei der Approximation von Funktionen mit Gleichverteilungsmethoden. Sitzungsber. Osterr. Akad. Wiss. Math.-Naturwiss. abt II 185, 121–132 (1976)

Tensor Product $q-$Bernstein Bézier Patches

Çetin Dişibüyük and Halil Oruç

Department of Mathematics, Dokuz Eylül University
Fen Edebiyat Fakültesi, Tınaztepe Kampüsü, Buca, 35160, İzmir, Turkey

Abstract. In this work we define a new de Casteljau type algorithm, which is in barycentric form, for the $q-$Bernstein Bézier curves. We express the intermediate points of the algorithm explicitly in two ways. Furthermore we define tensor product patches, based on this algorithm, depending on two parameters. Degree elevation procedure for the tensor product patch is studied. Finally, the matrix representation of tensor product patch is given and we find the transformation matrix between classical tensor product Bézier patch and tensor product $q-$Bernstein Bézier patch.

1 Introduction

One parameter family of Bernstein polynomials are defined in [7]. These polynomials are used in computer graphics as well as approximation theory. In [5] one parameter Bernstein Bézier curves are represented using these polynomials. These curves have nice geometric properties. Convergence properties of $q-$Bernstein polynomials are investigated and it is shown that when $q \geqslant 1$ the generalized Bernstein polynomials $B_n f$ converge to f as $n \to \infty$ if f is a polynomial in [6]. G.M. Phillips showed that these polynomials can be generated by a de Casteljau type algorithm (see, [8]).

A one parameter family of $q-$Bernstein Bézier curve of degree n is defined by

$$\mathsf{P}(t) = \sum_{i=0}^{n} \mathsf{b}_i B_i^{n,q}(t). \tag{1}$$

The points $\mathsf{b}_i \in \mathbb{R}^2$ or $\in \mathbb{R}^3$ are called control points and $B_i^{n,q}(t)$ is the $q-$Bernstein Bézier basis functions given by

$$B_i^{n,q}(t) = \begin{bmatrix} n \\ i \end{bmatrix} t^i \prod_{s=0}^{n-i-1} (1 - q^s t), \quad t \in [0,1], \quad 0 \leqslant i \leqslant n. \tag{2}$$

where an empty product denotes 1 and the parameter q is positive real number.

The $q-$binomial coefficient $\begin{bmatrix} n \\ i \end{bmatrix}$, or Gaussian polynomial (see [1]), is defined as

$$\begin{bmatrix} n \\ i \end{bmatrix} = \frac{[n][n-1]\cdots[n-i-1]}{[i][i-1]\cdots[1]} \tag{3}$$

S. Margenov, L.G. Vulkov, and J. Waśniewski (Eds.): NAA 2008, LNCS 5434, pp. 265–272, 2009.

for $0 < i \leqslant n$, and has the value 1 when $i = 0$ and the value 0 otherwise. Here $[i]$ denotes a $q-$integer, defined by

$$[i] = \begin{cases} (1-q^i)/(1-q), & q \neq 1 \\ i, & q = 1. \end{cases}$$

When $q = 1$ the $q-$binomial coefficients reduce to the usual binomial coefficients. It satisfies the following recurrence relations

$$\begin{bmatrix} n \\ i \end{bmatrix} = q^{n-i} \begin{bmatrix} n-1 \\ i-1 \end{bmatrix} + \begin{bmatrix} n-1 \\ i \end{bmatrix} \tag{4}$$

and

$$\begin{bmatrix} n \\ i \end{bmatrix} = \begin{bmatrix} n-1 \\ i-1 \end{bmatrix} + q^i \begin{bmatrix} n-1 \\ i \end{bmatrix}. \tag{5}$$

Using (5) one may show by induction on n that

$$(1-x)(1-qx)\cdots(1-q^{n-1}x) = \sum_{r=0}^{n} (-1)^r q^{r(r-1)/2} \begin{bmatrix} n \\ r \end{bmatrix} x^r. \tag{6}$$

It is shown in [8] that (1) may be evaluated by the following de Casteljau type algorithm:

Algorithm 1: For the given control points $\mathsf{b}_0, \mathsf{b}_1, \ldots, \mathsf{b}_n$ compute

$$\hat{\mathsf{b}}_i^r(t) = (q^i - q^{r-1}t)\hat{\mathsf{b}}_i^{r-1}(t) + t\hat{\mathsf{b}}_{i+1}^{r-1}(t) \begin{cases} r = 1, 2, \ldots, n \\ i = 0, 1, \ldots, n-r, \end{cases} \tag{7}$$

where $\hat{\mathsf{b}}_i^0(t) = \mathsf{b}_i$ for all i. Thus, $\hat{\mathsf{b}}_0^n(t) = \mathsf{P}(t)$.

The outline of this paper is as follows: Section 2 introduces a new de Casteljau type algorithm which is in barycentric form for $\mathsf{P}(t)$. This algorithm cannot be obtained by putting $w = 0$ in algorithm defined in [4, Lemma 2.7.]. The main difference is their intermediate points. In section 3 we define tensor product patch based on this algorithm and geometric properties of this patch are investigated. Finally, the matrix representation of tensor product patch is given and the transformation matrix between classical tensor product Bézier patch and tensor product $q-$Bernstein Bézier patch is found.

2 A New Algorithm

We now give a new de Casteljau type algorithm for computing the $q-$Bernstein Bézier curves.

Algorithm 2: For the given control points $\mathsf{b}_0, \mathsf{b}_1, \ldots, \mathsf{b}_n$, compute

$$\mathsf{b}_i^r(t) = (1 - q^{r-i-1}t)\mathsf{b}_i^{r-1}(t) + q^{r-i-1}t\mathsf{b}_{i+1}^{r-1}(t) \begin{cases} r = 1, 2, \ldots, n \\ i = 0, 1, \ldots, n-r. \end{cases} \tag{8}$$

The main difference of Algorithm 1 and Algorithm 2 is that each step of the latter is in barycentric form which eventually make up a curve that remain invariant under affine maps. Note that in CAGD systems it is desirable to express curves and surfaces in barycentric form and that $q = 1$ recovers standard de Casteljau algorithm for both of the above algorithms.

Theorem 1. *The explicit form of the intermediate points of the Algorithm 2 are*

$$\mathsf{b}_i^r(t) = \sum_{j=0}^{r} q^{-ri} \mathsf{b}_{i+j} \begin{bmatrix} r \\ j \end{bmatrix} t^j \prod_{s=0}^{r-j-1} (q^i - q^s t). \tag{9}$$

Proof. The proof is by induction on r and is analogous to the proof of Theorem 2.1. in [8]

Corollary 1. $\mathsf{b}_0^n(t)$ *is a point on the $q-$Bernstein Bézier curve with a value t. That is*

$$\mathsf{b}_0^n(t) = \mathsf{P}(t).$$

We can also express the intermediate points of the Algorithm 2 in terms of $q-$differences, where we define

$$\Delta_q^{k+1} \mathsf{b}_i = \Delta_q^k \mathsf{b}_{i+1} - q^k \Delta_q^k \mathsf{b}_i$$

recursively and $\Delta_q^0 \mathsf{b}_i = \mathsf{b}_i$ for $i = 0, 1, \ldots, n$.

Theorem 2. *The intermediate points of Algorithm 2 are*

$$\mathsf{b}_i^r(t) = \sum_{j=0}^{r} q^{-ij} \begin{bmatrix} r \\ j \end{bmatrix} \Delta^j \mathsf{b}_i t^j \tag{10}$$

Proof. The proof is by induction on r, using the recurrence relation (6) as in the proof of Theorem 2.3. in [8].

Note that although both algorithms consequently evaluate the same curve, their associated intermediate points are different. Namely, $\mathsf{b}_i^r(t) = q^{-ri} \hat{\mathsf{b}}_i^r(t)$.

3 Tensor Product $q-$Bernstein Bézier Surfaces

We define a two-parameter family of tensor product surface of degree (m, n) as follows:

$$\mathsf{S}(u, v) = \sum_{i=0}^{m} \sum_{j=0}^{n} \mathsf{b}_{i,j} B_i^{m,q_1}(u) B_j^{n,q_2}(v), \qquad (u, v) \in [0, 1] \times [0, 1] \tag{11}$$

where $\mathsf{b}_{i,j}$ are control net points and $B_i^{m,q_1}(u)$, $B_j^{n,q_2}(v)$ are $q-$Bernstein polynomials with parameters q_1, q_2 respectively for $i = 0, 1, \ldots, m$ and $j = 0, 1, \ldots, n$.

3.1 Properties

1. **Affine invariance property**
   Since $\sum_{i=0}^{m}\sum_{j=0}^{n} B_i^{m,q_1}(u)B_j^{n,q_2}(v) = 1$, $\mathsf{S}(u,v)$ is an affine combination of its control net points. Thus $\mathsf{S}(u,v)$ is affinely invariant.
2. **Convex hull property**
   When $0 < q_1, q_2 \leqslant 1$, the bases polynomials are nonnegative and their sum is 1. Hence $\mathsf{S}(u,v)$ is a convex combination of $\mathsf{b}_{i,j}$ and lies in the convex hull of its control net points.
3. **Boundary curves**
   Boundary curves of $\mathsf{S}(u,v)$ are evaluated by $\mathsf{S}(u,0), \mathsf{S}(u,1), \mathsf{S}(0,v)$ and $\mathsf{S}(1,v)$. The first two curves are $q-$Bernstein Bézier curves in u and the last two curves are $q-$Bernstein Bézier curves in v.
4. **Corner point interpolation**
   The control points of the boundary curves are the boundary points of control net of $\mathsf{S}(u,v)$. Thus it follows from end point interpolation property of the $q-$Bernstein Bézier curves that the corner control net points coincide with the four corners of the surface. Namely,

$$\mathsf{S}(0,0) = \mathsf{b}_{0,0}, \mathsf{S}(0,1) = \mathsf{b}_{0,n}, \mathsf{S}(1,0) = \mathsf{b}_{m,0}, \mathsf{S}(1,1) = \mathsf{b}_{m,n}.$$

As a result, $\mathsf{S}(u,v)$ mimics the shape of control net.

In what follows is the de Casteljau type algorithm to compute $\mathsf{S}(u,v)$.

Algorithm 3: Given the control net $\mathsf{b}_{i,j} \in \mathbb{R}^3; \quad i = 0,1,\dots,m,\ j = 0,1,\dots,n.$ Compute

$$\mathsf{b}_{i,j}^{r,r} = \left[1 - q_1^{r-i-1}u \;\; q_1^{r-i-1}u\right] \begin{bmatrix} \mathsf{b}_{i,j}^{r-1,r-1} & \mathsf{b}_{i,j+1}^{r-1,r-1} \\ \mathsf{b}_{i+1,j}^{r-1,r-1} & \mathsf{b}_{i+1,j+1}^{r-1,r-1} \end{bmatrix} \begin{bmatrix} 1 - q_2^{r-j-1}v \\ q_2^{r-j-1}v \end{bmatrix} \tag{12}$$

for $r = 1,2,\dots,k, \quad i,j = 0,1,\dots,k-r$, where $k = \min(m,n)$.

Since, in general $m \neq n$, performing the direct de Casteljau algorithm k times may not give a point on the surface. Then to get a point on the surface after k^{th} application of (12) we perform Algorithm 2 for the intermediate points $\mathsf{b}_{i,j}^{k,k}$ (see [3]).

3.2 Degree Elevation

Let $\mathsf{S}(u,v)$ be two parameter tensor product surface of degree (m,n). The aim is to obtain the same surface as a surface of degree $(m+1,n)$. Hence we need to find control points $\mathsf{b}_{i,j}^{(1,0)}$ such that

$$\begin{aligned}\mathsf{S}(u,v) &= \sum_{j=0}^{n}\left[\sum_{i=0}^{m} \mathsf{b}_{i,j} B_i^{m,q_1}(u)\right] B_j^{n,q_2}(v) \\ &= \sum_{j=0}^{n}\left[\sum_{i=0}^{m+1} \mathsf{b}_{i,j}^{(1,0)} B_i^{m+1,q_1}(u)\right] B_j^{n,q_2}(v).\end{aligned}$$

For this purpose, we first write tensor product Bézier patches in the form

$$\mathsf{S}(u,v) = \sum_{j=0}^{n} \mathsf{b}_j B_j^{n,q_2}(v) \tag{13}$$

where

$$\mathsf{b}_j = \sum_{i=0}^{m} \mathsf{b}_{i,j} B_i^{m,q_1}(u). \tag{14}$$

Thus, the problem is now reduced to express m^{th} degree $q-$Bernstein Bézier curve b_j as that of $(m+1)^{\text{th}}$ degree. From the degree elevation procedure (see, [5]) for b_j in the latter equation, we obtain

$$\mathsf{b}_j = \sum_{i=0}^{m} \mathsf{b}_{i,j} B_i^{m,q_1}(u) = \sum_{i=0}^{m+1} \mathsf{b}_{i,j}^{(1,0)} B_i^{m+1,q_1}(u),$$

where

$$\mathsf{b}_{i,j}^{(1,0)} = \left(1 - \frac{[m+1-i]_{q_1}}{[m+1]_{q_1}}\right)\mathsf{b}_{i-1,j} + \frac{[m+1-i]_{q_1}}{[m+1]_{q_1}}\mathsf{b}_{i,j}, \quad i = 0,1,\dots,m+1,$$

and $[i]_{q_1}$ denotes the $q-$integer $[i]$ with parameter value q_1 in the place of q. Similarly, to obtain the same surface as one of degree $(m, n+1)$ we need control points $\mathsf{b}_{i,j}^{(0,1)}$ such that

$$\mathsf{b}_{i,j}^{(0,1)} = \left(1 - \frac{[n+1-j]_{q_2}}{[n+1]_{q_2}}\right)\mathsf{b}_{i,j-1} + \frac{[n+1-j]_{q_2}}{[n+1]_{q_2}}\mathsf{b}_{i,j}, \quad j = 0,1,\dots,n+1.$$

Finally, to obtain $\mathsf{S}(u,v)$ as a surface of degree $(m+1, n+1)$ evaluate the new control net points from the product

$$\mathsf{b}_{i,j}^{(1,1)} = \left[1 - \frac{[m+1-i]_{q_1}}{[m+1]_{q_1}} \;\; \frac{[m+1-i]_{q_1}}{[m+1]_{q_1}}\right] \begin{bmatrix} \mathsf{b}_{i-1,j-1} & \mathsf{b}_{i-1,j} \\ \mathsf{b}_{i,j-1} & \mathsf{b}_{i,j} \end{bmatrix} \begin{bmatrix} 1 - \frac{[n+1-j]_{q_2}}{[n+1]_{q_2}} \\ \frac{[n+1-j]_{q_2}}{[n+1]_{q_2}} \end{bmatrix}.$$

Repeated degree elevation procedure can be obtained analogously.

3.3 Matrix Form and Change of Basis

The tensor product $q-$Bernstein Bézier patch can be written in matrix form as

$$\mathsf{S}(u,v) = [B_0^{m,q_1}(u), \dots, B_m^{m,q_1}(u)] \begin{bmatrix} \mathsf{b}_{0,0} & \cdots & \mathsf{b}_{0,n} \\ \vdots & & \vdots \\ \mathsf{b}_{m,0} & \cdots & \mathsf{b}_{m,n} \end{bmatrix} \begin{bmatrix} B_0^{n,q_2}(v) \\ \vdots \\ B_n^{n,q_2}(v) \end{bmatrix}$$

The basis of tensor product polynomial space $\mathbb{P}_m \otimes \mathbb{P}_n$ has dimension $(m+1)(n+1)$ and each basis element is in the form $u^i v^j, \quad i = 0,1,\dots,m, \quad j =$

$0, 1, \ldots, n$. For simplicity we take m and $q_1 = q_2 = q$. Let $C = [c_0, c_1, \ldots, c_n]^T$ be a $(n+1)^2 \times 1$ block vector with elements $c_i = [u^i, u^i v, \ldots, u^i v^n]^T$ and let $B^q = [b_0, b_1, \ldots, b_n]^T$ be a block matrix with block elements

$$b_i = [B_i^{n,q}(u)B_0^{n,q}(v), B_i^{n,q}(u)B_1^{n,q}(v), \ldots, B_i^{n,q}(u)B_n^{n,q}(v)]^T$$

for $i = 0, 1, \ldots, n$. Since the tensor product $q-$Bernstein Bézier patches span the space of tensor product polynomials, there exist a transformation matrix $M^{n,q}$ such that $B^q = M^{n,q}C$. The functions for the tensor product $q-$Bernstein Bézier patch are

$$B_i^{n,q}(u)B_j^{n,q}(v) = \begin{bmatrix} n \\ i \end{bmatrix} u^i \prod_{s=0}^{n-i-1} (1 - q^s u) \begin{bmatrix} n \\ j \end{bmatrix} v^j \prod_{s=0}^{n-j-1} (1 - q^s v).$$

By using the property (6) we deduce that

$$B_i^{n,q}(u)B_j^{n,q}(v) = \sum_{k=i}^{n} (-1)^{k-i} q^{\binom{k-i}{2}} \begin{bmatrix} n \\ i \end{bmatrix} \begin{bmatrix} n-i \\ k-i \end{bmatrix} u^k \sum_{l=j}^{n} (-1)^{l-j} q^{\binom{l-j}{2}} \begin{bmatrix} n \\ j \end{bmatrix} \begin{bmatrix} n-j \\ l-j \end{bmatrix} v^l.$$

Rearranging the terms using the definition (3) we have

$$B_i^{n,q}(u)B_j^{n,q}(v) = \sum_{k=0}^{n} \sum_{l=0}^{n} (-1)^{(k+l)-(i+j)} q^{\binom{k-i}{2} + \binom{l-j}{2}} \begin{bmatrix} n \\ i \end{bmatrix} \begin{bmatrix} n-i \\ k-i \end{bmatrix} \begin{bmatrix} n \\ j \end{bmatrix} \begin{bmatrix} n-j \\ l-j \end{bmatrix} u^k v^l.$$

Since

$$\begin{bmatrix} n-i \\ k-i \end{bmatrix} = \frac{\begin{bmatrix} n \\ k \end{bmatrix} \begin{bmatrix} k \\ i \end{bmatrix}}{\begin{bmatrix} n \\ i \end{bmatrix}} \tag{15}$$

we obtain

$$B_i^{n,q}(u)B_j^{n,q}(v) = \sum_{k=0}^{n} \sum_{l=0}^{n} (-1)^{(k+l)-(i+j)} q^{\binom{k-i}{2} + \binom{l-j}{2}} \begin{bmatrix} n \\ k \end{bmatrix} \begin{bmatrix} k \\ i \end{bmatrix} \begin{bmatrix} n \\ l \end{bmatrix} \begin{bmatrix} l \\ j \end{bmatrix} u^k v^l.$$

As a consequence, one may write $B^q = M^{n,q}C$ where $M^{n,q}$ is an upper triangular block matrix with a generic element

$$\left((M_{ij}^{n,q})_{k,l=0}^{n}\right)_{i,j=0}^{n} = (-1)^{(j+l)-(i+k)} q^{\binom{j-i}{2} + \binom{l-k}{2}} \begin{bmatrix} n \\ j \end{bmatrix} \begin{bmatrix} j \\ i \end{bmatrix} \begin{bmatrix} n \\ l \end{bmatrix} \begin{bmatrix} l \\ k \end{bmatrix}.$$

Conversely, to express the monomial (power) basis in terms of $q-$Bernstein basis we multiply the equation

$$1 = \sum_{k=0}^{n-i} B_k^{n-i,q}(u) \sum_{l=0}^{n-j} B_j^{n-l,q}(v)$$

by $u^i v^j$. Then we have

$$u^i v^j = \sum_{k=0}^{n-i} \begin{bmatrix} n-i \\ k \end{bmatrix} u^{i+k} \prod_{s=0}^{n-i-k-1} (1 - q^s u) \sum_{l=0}^{n-j} \begin{bmatrix} n-j \\ l \end{bmatrix} v^{j+l} \prod_{s=0}^{n-j-l-1} (1 - q^s v).$$

Shifting the limits of the sums and rearranging the terms using the equation (15) yields

$$u^i v^j = \sum_{k=i}^{n} \frac{\begin{bmatrix} k \\ i \end{bmatrix}}{\begin{bmatrix} n \\ i \end{bmatrix}} B_k^{n,q}(u) \sum_{l=j}^{n} \frac{\begin{bmatrix} l \\ j \end{bmatrix}}{\begin{bmatrix} n \\ j \end{bmatrix}} B_l^{n,q}(v).$$

From definition (3) one may write

$$u^i v^j = \sum_{k=0}^{n} \sum_{l=0}^{n} \frac{\begin{bmatrix} k \\ i \end{bmatrix} \begin{bmatrix} l \\ j \end{bmatrix}}{\begin{bmatrix} n \\ i \end{bmatrix} \begin{bmatrix} n \\ j \end{bmatrix}} B_k^{n,q}(u) B_l^{n,q}(v).$$

Using the above equation we can write $C = \tilde{M}^{n,q} B^q$ where $\tilde{M}^{n,q}$ is a block matrix with a generic element

$$\left((\tilde{M}_{i,j}^{n,q})_{k,l=0}^{n} \right)_{i,j=0}^{n} = \frac{\begin{bmatrix} j \\ i \end{bmatrix} \begin{bmatrix} l \\ k \end{bmatrix}}{\begin{bmatrix} n \\ i \end{bmatrix} \begin{bmatrix} n \\ k \end{bmatrix}}.$$

Note that $\tilde{M}^{n,q}$ is an upper triangular block matrix and $(\tilde{M}^{n,q})^{-1} = M^{n,q}$. Now, one can find a transformation matrix between $q-$Bernstein basis and standard Bernstein basis. Since $C = \tilde{M}^{n,q} B^q$, we can express the power basis in terms of standard Bernstein basis when $q = 1$. We have

$$C = \tilde{M}^{n,1} B^1,$$

where B^1 is the standard tensor product Bernstein basis (i.e. $q = 1$ in (2)) and the matrix $\tilde{M}^{n,1}$ is a block matrix with a generic element

$$\left((\tilde{M}_{i,j}^{n,1})_{k,l=0}^{n} \right)_{i,j=0}^{n} = \frac{\binom{j}{i} \binom{l}{k}}{\binom{n}{i} \binom{n}{k}}.$$

It follows that

$$\tilde{M}^{n,q} B^q = \tilde{M}^{n,1} B^1.$$

Multiplying both sides from left with the matrix $M^{n,q}$, we obtain

$$B^q = T^{n,q,1} B^1,$$

where $T^{n,q,1} = M^{n,q} \tilde{M}^{n,1}$. It is worth noting that the transformation matrix $T^{n,q,1}$ make it possible to exchange $q-$Bernstein and Bézier representations of the surface $\mathsf{S}(u, v)$.

References

1. Andrews, G.E.: The theory of partitions. Cambridge Mathematical Library. Cambridge University Press, Cambridge (1998); Reprint of the 1976 original
2. Dişibüyük, Ç., Oruç, H.: A generalization of rational Bernstein-Bézier curves. BIT 47(2), 313–323 (2007)

3. Farin, G.: Curves and surfaces for computer-aided geometric design. In: Computer Science and Scientific Computing, 5th edn. Academic Press Inc., San Diego (2002)
4. Lewanowicz, S., Woźny, P.: Generalized Bernstein polynomials. BIT 44(1), 63–78 (2004)
5. Oruç, H., Phillips, G.M.: q-Bernstein polynomials and Bézier curves. J. Comput. Appl. Math. 151(1), 1–12 (2003)
6. Oruç, H., Tuncer, N.: On the convergence and iterates of q-Bernstein polynomials. J. Approx. Theory 117(2), 301–313 (2002)
7. Phillips, G.M.: Bernstein polynomials based on the q-integers. Ann. Numer. Math. 4(1-4), 511–518 (1997); The heritage of P. L. Chebyshev: a Festschrift in honor of the 70th birthday of T. J. Rivlin
8. Phillips, G.M.: A de Casteljau algorithm for generalized Bernstein polynomials. BIT 37(1), 232–236 (1997)

Modeling of a Vertical Cavity Surface Emitting Laser Containing a Multi-QW Heterostructure

N.N. Elkin, A.P. Napartovich, V.N. Troshchieva, and D.V. Vysotsky

State Science Center Troitsk Institute for Innovation and Fusion Research(TRINITI), 142190, Troitsk, Moscow Region, Russia
elkin@triniti.ru

Abstract. A vertical cavity surface emitting laser (VCSEL) is an object of numerical study in the present paper.

The linear non-hermitian eigenvalue problem arises in the first stage when we consider a "cold" cavity and neglect changes of material characteristics induced by electromagnetic field. The round-trip operator technique and Krylov subspace methods were used for determination of eigenfunctions, which represent intra-cavity wave field distributions. Corresponding complex eigenvalues determine the wavelength shifts relative to reference value and threshold gains.

The next stage of study relates to a case of a loaded cavity when self-consistent solving of a wave field equation and material equations is required. The eigenvalue problem for a non-linear operator must be solved to find the lasing electromagnetic field spatial profile and its frequency. The gain element of a typical VCSEL device comprises several quantum wells (QW). The charge carriers distributions in each of QW obey non-linear diffusion equation. The round-trip operator is a non-linear operator in this case, and its evaluation needs an iteration procedure. We propose the iteration procedure, which is applicable for a set of QW of any size and has computational costs growing linearly with number of QW.

The computational procedures and results of calculations for a cylindrical VCSEL will be reported.

1 Introduction

A multi-QW heterostructure is of practical interest for VCSELs illuminated by a beam of high energy electrons. Modeling of a VCSEL containing a multi-QW heterostructure represents a very difficult computational problem because of complicated geometry and non-linear partial differential equations containing eigenvalues. The traditional for optical resonators Fox-Li iteration method [1] is inapplicable owing to dispersion effects. We propose an efficient computational algorithm using moderate computational resources. The algorithm is based on the bi-directorial beam propagation method (BiBPM) [2] and the modified round-trip operator technique [3]. The results of determination of a mode spatial profile, exact wavelength and attenuation are presented.

S. Margenov, L.G. Vulkov, and J. Waśniewski (Eds.): NAA 2008, LNCS 5434, pp. 273–280, 2009.

2 Numerical Model of the VCSEL

The VCSEL consisting of a multi-QW heterostructure placed between top and bottom distributed Bragg reflectors (DBR) is considered. Additionally, thin protective layers separate a multi-QW heterostructure from DBR's. Directing z-axis perpendicularly to the substrate surface we represent the VCSEL as a pile of plane layers: $\{[z_{k-1}, z_k], k = 1, \dots, m\}$, where m is the total number of layers, $\{z_k, k = 0, \dots, m\}$ is an ascending sequence of coordinates of layer interfaces, $h_k = z_k - z_{k-1}$ is thickness of the k-th layer. Moreover, account must be taken that the VCSEL is mounted on the substrate. We suppose that the substrate is unbounded and assign to it number $k = 0$. Similarly, an unbounded layer must be accounted on top of a VCSEL which has material characteristics of an output window. We assign to it number $k = m + 1$.

The index and absorption are constant in each layer except the active layers (QWs) where non-uniform distributions are controlled by electrical current and light intensity. To distinguish QWs from other layers we define the index array $\{\nu(j), j = 1, \dots, q\}$, where q is the number of QWs. If $k = \nu(j)$ then the layer $[z_{k-1}, z_k]$ is the j-th QW. For uniformity of expressions we define that $\nu(0) = 0$ and $\nu(q+1) = m$. The origin of the vertical coordinate was set according to condition $z_{\nu(q)} = 0$.

We start from Maxwell equations and assume that the polarization effects can be neglected and the scalar diffraction theory is applicable. Further, we assume that index and absorption (gain) distributions have circular symmetry. Therefore, we use cylindrical coordinates. Laser modes have a time dependence of the form $E(r, \varphi, z, t) = U(r, \varphi, z)\exp(-i\Omega t)$, $\Omega = \omega_0 + \Delta\omega - i\delta$, where ω_0 is the reference frequency, $\Delta\omega = \omega - \omega_0$ is the frequency shift and δ is the attenuation factor. The reference frequency ω_0, in other words, the frequency guess must be chosen close to the operation frequency of a laser. The reference wavenumber and reference wavelength are defined by standard relations: $\omega_0 = k_0 c$, $k_0 = 2\pi/\lambda_0$. We search solutions in a form $U(r, \varphi, z) = U_m(r, z)\exp(im\varphi)$, where φ is the azimuthal angle. Introducing new variables $g_t = 2\delta/c$, $\Delta k = \Delta\omega/c$, $\beta = g_t + i2\Delta k$, we have obtained the wave equation for $m-$th angular harmonic

$$\frac{\partial^2 U_m}{\partial z^2} + \frac{1}{r}\frac{\partial}{\partial r}\left(r\frac{\partial U_m}{\partial r}\right) - \frac{m^2}{r^2}U_m + (k_0^2 n^2 - ik_0 g)U_m - ik_0 n^2 \beta U_m = 0, \quad (1)$$

containing a complex eigenvalue β. Here n and g are index and gain respectively.

There is need to define the boundary condition at the interfaces between adjoint layers and the boundary condition at the lateral boundary $r = r_m$. We use condition of continuity for the wave field U_m and its normal derivative at the interfaces. Luckily there is no problem of boundary conditions at the lateral boundary because the active layers have strong attenuation in the absence of pump. The value of r_m is chosen large sufficiently to embrace the pump region and to secure a negligible value of a wave field at $r = r_m$.

At the first stage we neglect dependence of material characteristics on electromagnetic field intensity. The equation (1) with the boundary conditions described formes the linear non-hermitian eigenvalue problem.

The second problem consists of self-consistent solving of the wave field equation and material equations in order to find the spatial profile of a laser electromagnetic field and its frequency in steady-state mode of operation. We restrict our consideration with axisymmetric laser modes. According to this condition, we are to solve the axisymmetric ($m = 0$) equation (1) jointly with the set of non-linear diffusion equations [4]

$$\frac{1}{r}\frac{\partial}{\partial r}\left(r\frac{\partial Y_j}{\partial r}\right) - \frac{Y_j}{D\tau_{nr}} - \frac{B}{D}N_{tr}Y_j^2 - \frac{|U_0|^2 \ln(\chi(Y_j))}{D\tau_{nr}} = -\frac{J}{eDdN_{tr}},\ j = 1, \ldots, q \tag{2}$$

for normalized carrier density $Y_j = N_j/N_{tr}$ at the j-th active layers. Here N_j is the carrier density, D is the diffusion coefficient, τ_{nr} is a recombination time, B is a coefficient of nonlinearity, d is thickness of the QW, e is the elementary charge, $N_{tr} = \left(-1/\tau_{nr} + \sqrt{1/\tau_{nr}^2 + 4BJ_{tr}/(ed)}\right)/(2B)$ is the carrier density for conditions of transparency, J_{tr} is the injection current density for conditions of transparency, $|U_0|^2$ is the normalized light intensity, $J = \kappa J_{tr} f(r/r_0)$ is the effective current density which produces the same rate of carrier generation as the electron beam, κ is the pump level, $f(\rho)$ is the pump profile function, r_0 is the pump region raduis. Zero boundary conditions for $Y_j(r)$ are set at the lateral boundary of the active layer ($r = r_m$). The function $\chi(Y)$, gain and index at the active layers are approximated by the formulas

$$\chi(Y) = \begin{cases} \alpha + (1-\alpha)Y^{1/(1-\alpha)}, & Y < 1 \\ Y, & Y \geq 1 \end{cases},\ g_j = g_0 \ln(\chi(Y_j)),\ n_j = n_0 - \frac{R(g_0 + g_j)}{2k_0}, \tag{3}$$

where $\alpha = \exp(-1)$, g_0 is a gain parameter, n_0 is the refractive index in the absence of carriers, R is the line enhancement factor. The equation (1) at $m = 0$ jointly with the equations (2) and (3) supplemented with corresponding boundary conditions form the eigenvalue problem for a non-linear operator. The supplementary condition $\delta = 0$ ($\mathrm{Re}(\beta) = 0$) is required for steady-state operation.

To synthesize numerical algorithm we use representation of the wave field $U_m(r, z)$ in terms of m–th order Hankel transform over r. Introducing an appropriate numerical meshes $\{r_l,\ l = 0, 1, \ldots, N_r\}$, $\{\kappa_n,\ n = 0, 1, \ldots, N_r\}$ for radius r and transverse wave number κ and going on to discrete approximation of the Hankel transform we can define $\psi_{nm}(z) = \mathbf{H}_m\{U_m(r_l, z)\}$, where n is the number of radial harmonic and $\mathbf{H}_m$ is the discrete m–th order Hankel transform operator. We simulate the QW as a uniform layer containing a non-uniform phase screen with gain and phase according to equations (2) and (3). Suppose for certainty that a phase screen is placed to top boundary of the layer. Thus, the longitudinal structure of a VCSEL is a combination of sets of uniform layers separated by non-uniform phase screens. The set of uniform layers between j-th and $(j-1)$-th phase screens is assigned by number j, at that $j = 1$ corresponds to layers below 1-st phase screen and $j = q + 1$ corresponds to layers above the top phase screen.

As a result of Hankel transform, the wave equation (1) in the k-th layer can be expressed in a form: $d^2\psi_n/dz^2 + q_{kn}^2\psi_n = 0, q_{kn}^2 = k_0^2 n_k^2 - ik_0 g_k - ik_0 n_k^2 \beta - \kappa_n^2$, where index m is omitted. We define $q_{kn} = \sqrt{k_0^2 n_k^2 - ik_0 g_k - ik_0 n_k^2 \beta - \kappa_n^2}$ with the condition that $\mathbf{Re}(q_{kn}) \geq 0$. The general solution to the last equation has a form: $\psi_n(z) = E_{kn}^{j,+} \exp(iq_{kn}z) + E_{kn}^{j,-} \exp(-iq_{kn}z)$, where $z_{k-1} < z < z_k$, j is the number of set of layers which contains the k-th layer. Applying to $\psi_n(z)$ the operator $\mathbf{H}^{-1}$ we can represent the wave field in the k-th layer as a sum of upward propagating wave $V_{kl}^{j,+} = \mathbf{H}^{-1}\{E_{kn}^{j,+} \exp(iq_{kn}z)\}$ and downward propagating wave $V_{kl}^{j,-} = \mathbf{H}^{-1}\{E_{kn}^{j,-} \exp(-iq_{kn}z)\}$, where l is the number of node of the radial mesh. Using interface boundary conditions following relations between wave field amplitudes can be derived

$$\begin{pmatrix} E_{k+1}^{j,+} \\ E_{k+1}^{j,-} \end{pmatrix} = T_k \begin{pmatrix} E_k^{j,+} \\ E_k^{j,-} \end{pmatrix}, \quad T_k = \frac{1}{2q_{k+1}} \begin{pmatrix} q_k^+ \exp(-iq_k^- z_k) & q_k^- \exp(-iq_k^+ z_k) \\ q_k^- \exp(iq_k^+ z_k) & q_k^+ \exp(iq_k^- z_k) \end{pmatrix},$$

where $q_k^+ = q_{k+1} + q_k$ and $q_k^- = q_{k+1} - q_k$. Here the index n is omitted for short. Let us define T-matrices for sets of layers: $\{T^j = T_{\nu(j)-1} T_{\nu(j)-2} \cdots T_{\nu(j-1)},\ j = 1, \ldots, q\}$, $T^{q+1} = T_{\nu(q+1)} T_{\nu(q+1)-1} \cdots T_{\nu(q)}$. Combining described above procedures we can approximate the equation (1) by a set of discrete equations:

$$\begin{aligned}
V_{\nu(j)}^{j+1,-} &= \mathbf{H}^{-1}\{E_{\nu(j)}^{j+1,-} \exp(-iq_{\nu(j)} z_{\nu(j)})\}, \quad V_{\nu(j)}^{j,-} = V_{\nu(j)}^{j+1,-} \exp(G_j + i\varphi_j) \\
E_{\nu(j)}^{j,-} &= \mathbf{H}\{V_{\nu(j)}^{j,-} \exp(iq_{\nu(j)} z_{\nu(j)})\}, \quad V_{\nu(j)}^{j,+} = \mathbf{H}^{-1}\{E_{\nu(j)}^{j,+} \exp(iq_{\nu(j)} z_{\nu(j)})\}, \\
V_{\nu(j)}^{j+1,+} &= V_{\nu(j)}^{j,+} \exp(G_j + i\varphi_j), \quad E_{\nu(j)}^{j+1,+} = \mathbf{H}\{V_{\nu(j)}^{j+1,+} \exp(-iq_{\nu(j)} z_{\nu(j)})\}, \\
E_{\nu(j)}^{j,+} &= t_{11}^j E_{\nu(j-1)}^{j,+} + t_{12}^j E_{\nu(j-1)}^{j,-}, \quad E_{\nu(j)}^{j,-} = t_{21}^j E_{\nu(j-1)}^{j,+} + t_{22}^j E_{\nu(j-1)}^{j,-},
\end{aligned} \tag{4}$$

where t_{11}^j, t_{12}^j, t_{21}^j and t_{22}^j are elements of the matrix T^j, $G_j = g_j h_{\nu(j)}$, $\varphi_j = ik_0(n_j - n_0)h_{\nu(j)}$. The first six equations hold at $j = 1, \ldots, q$. The last two equations hold at $j = 1, \ldots, q+1$ with conditions $E_{\nu(0)}^{1,+} = 0$ and $E_{\nu(q+1)+1}^{q+1,-} = 0$. The set of equations (4) have non-trivial solutions only for discrete values of the complex eigenvalue β.

If index and gain distributions $\{(n_j, g_j),\ j = 1, \ldots, q\}$ do not depend on the wave field intensity, we consider the eigenvalue problem for a "cold" cavity. In contrast to equation (1) the eigenvalue β enters to equations (4) in non-linear way.

The second case we consider assumes that the index and gain distributions at the active layers depend on the wave field through diffusion equation (2). The axisymmetric ($m = 0$) equations (4) jointly with the finite-difference approximations of (2) and (3) were used to find a self-consistent solution. The wave field amplitude in the active layer can be approximated by the formula: $U_0 = (V_{\nu(j)}^{j,+} + V_{\nu(j)}^{j,-} + V_{\nu(j)}^{j+1,+} + V_{\nu(j)}^{j+1,-})/2$. We call this task as the eigenvalue problem for a "hot" cavity. The supplementary condition $\mathrm{Re}(\beta) = 0$ is required.

Similarly to [3], the round-trip operator was built up using equations (4). We specify as the start field $u = V_{\nu(q)}^{q+1,+}$, that is the upward propagating wave at the upper boundary of the top active layer. The required round-trip operator $\mathbf{P}(g, n, \beta)$ can be represented as a composition of four operators. The first one is evaluation of the wave $V_{\nu(q)}^{q+1,-}$ by means of u and T^{q+1} after reflection from the top DBR. The second operator evaluates transmission of wave through the top active layer: $V_{\nu(q)}^{q+1,-} \to V_{\nu(q)}^{q,-}$. The third operator is evaluation of the wave $V_{\nu(q)}^{q,+}$ by means of $V_{\nu(q)}^{q,-}$ after downward transmission through the multi-QW heterostructure except the top QW, reflection from the bottom DBR and upward transmission. Lastly, the fourth operator $V_{\nu(q)}^{q,+} \to V_{\nu(q)}^{q+1,+} \equiv \mathbf{P}(g,n,\beta)u$ evaluates transmission of wave through the top active layer. All the steps of round-trip evaluation can be easily performed except the third step, where severe difficulties are met. We can't go through the active layers sequentially because downward and upward wave amplitudes are not available simultaneously as the BiBPM requires. The third step can be completed by an iteration procedure only. We have developed the iteration procedure, which is applicable for an array of QWs of any size and has computational costs growing linearly with number of QWs. Each step of the iteration procedure consists of calculation of vectors $\{E_{\nu(j)}^{j+1,-},\ j=q-1,\ldots,1\}$ at the next iteration using current values. Calculations are performed using appropriate equations from (4) in sequence:

$$E_{\nu(j)}^{j+1,-} \Rightarrow V_{\nu(j)}^{j+1,-} \Rightarrow V_{\nu(j)}^{j,-} \Rightarrow E_{\nu(j)}^{j,-}, \quad j=q-1,\ldots,1, \quad E_{\nu(0)}^{1,-} = E_{\nu(1)}^{1,-}/t_{22}^{1}.$$

$$E_{\nu(j-1)}^{j,+} \Rightarrow E_{\nu(j)}^{j,+} \Rightarrow V_{\nu(j)}^{j,+} \Rightarrow V_{\nu(j)}^{j+1,+} \Rightarrow E_{\nu(j)}^{j+1,+} \Rightarrow E_{\nu(j)}^{j+1,-}, \quad j=1,\ldots,q-1.$$

All the vectors $\{E_{\nu(j)}^{j+1,-},\ j=q-1,\ldots,1\}$ have to be initialized before start of the iteration procedure.

In conclusion, we can note that use the Fast Hankel transform algorithm [5] and a T-matrix approach for spectral components within plane layers structure give us the very fast algorithm for round-trip evaluation.

Round-trip condition gives us the following equation:

$$\mathbf{P}(g,n,\beta)u = u \tag{5}$$

for a function u and eigenvalue β subject to finding. This formulation is similar to problem definition in [2], but authors of [2] restrict themselves to case of "cold" cavity. They calculate explicitly the matrix of the operator $\mathbf{A} = \mathbf{P}(g,n,\beta) - \mathbf{I}$ and apply singular value decomposition (SVD) to the matrix of $\mathbf{A}$. The value β is adjusted until the smallest singular value is zero within a certain tolerance. For some applications explicit calculation of the matrix of $\mathbf{A}$ and SVD procedure bring to enormous computational expenditures because of high dimension of $\mathbf{A}$. Our approach to problem (5) consists in solution of the auxiliary problem

$$\mathbf{P}(g,n,\beta)u = \gamma u \tag{6}$$

for a function u and eigenvalue γ to be found provided the value of β is specified. The value β is adjusted until $\gamma = 1$ within a certain tolerance. For case of "cold" cavity we have the linear non-hermitian eigenvalue problem (6). Only several eigenpairs (u,γ) are required. Such a problem can be solved efficiently by the Arnoldi method. Calculation of all elements of matrix of $\mathbf{P}(g,n,\beta)$ is not required. It is necessary to calculate elements of vector $\mathbf{P}(g,n,\beta)u$ only. For case of "hot" cavity the equation (6) represents the eigenvalue problem for a non-linear operator because gain g and index n are determined by equations (2), (3) and depend on u. This problem is solved by the Fox-Li iteration method [1]. In both cases of problem (6) the value of β is adjusted by the secant method.

3 Results and Discussion

We set the reference wavelength $\lambda_0 = 640\,\text{nm}$. The Bragg mirrors are made up of quarter-wave alternate layers of SiO_2 ($n = 1.465$, $h = 109.2$ nm) and TiO_2 ($n = 2.4$, $h = 66.7$ nm). The bottom DBR consists of 7.5 pairs $SiO_2 - TiO_2$ and the top DBR consists of 5 ones. The active region of VCSEL consists of $q = 25$ QWs ($Ga_{0.5}In_{0.5}P$, $n = n_0 = 3.62$, $h = d = 8$ nm) separated from each other and from outer regions by barrier layers ($Al_{0.35}Ga_{0.15}In_{0.5}P$, $n = 3.345$, $h = 182.67$ nm). So, the set of QW's form a finite periodical structure. The optical length of the one period is equal to λ_0 exactly, therefore, we have a resonant heterostructure. The active region is separated from DBRs by the thin protective layers having thickness $h = 6$ nm and index $n = 3.62$. At last, the VCSEL is mounted on the aluminium plate and finished at the top by the sapphire plate ($n = 1.716$). Both aluminium and sapphire plates are taken as unbounded. The other parameters were given as follows: $D = 0.5\,\text{cm}^2\text{s}^{-1}$, $\tau_{nr} = 10^{-9}\,\text{s}$, $B = 3.5 \times 10^{-10}\,\text{cm}^3\text{s}^{-1}$, $J_{tr} = 400\,\text{A}\,\text{cm}^{-2}$, $R = 2.5$, $r_0 = 13\,\mu\text{m}$, $g_0 = 3400\,\text{cm}^{-1}$. We have no precise data about the pump profile. The numerical results presented were calculated for the profile function $f(\rho) = (1+\rho^4)^{-1}$. The pump level κ is expressed through the total current I by the formula: $\kappa = I/(2\pi J_{tr} \int f(r/r_0) r dr)$.

We use standard notation TEM_{nm} for optical modes in a VCSEL where m is the angular quantum number responding to dependence $\sim \exp(im\varphi)$ and n is the radial quantum number. The pump profile in the axial plane is shown in Fig. 1. Number of mesh nodes $N_r = 1024$ for all results presented in Figs. 2-6. To estimate the error of discretization we have repeated calculations for $I = 6\,\text{mA}$ using $N_r = 2048$. We call the value $0.5 \times |x_1 - x_2|/|x_1 + x_2|$ as the relative change of x where x_1 and x_2 are the results of calculation of x at $N_r = 1024$ and $N_r = 2048$, correspondingly. For case of "hot" cavity we have calculated the

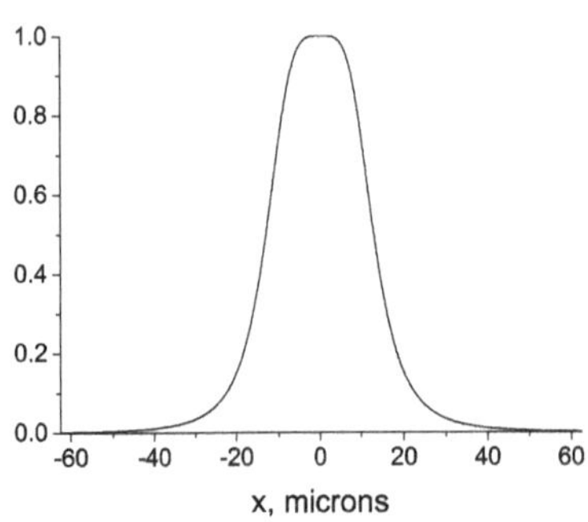

Fig. 1. Pump profile

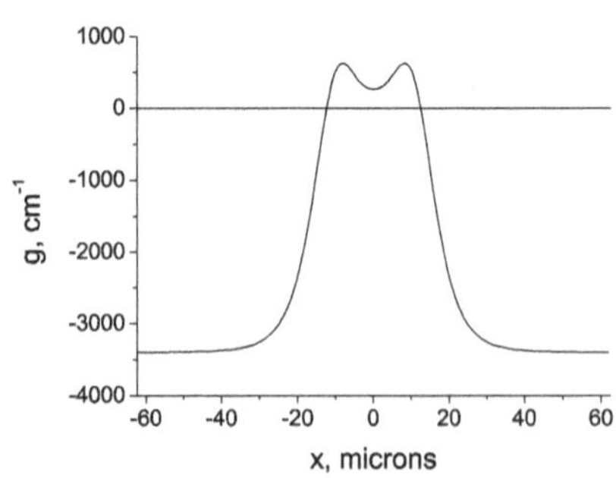

Fig. 2. Gain at the top QW. $I = 6\,\text{mA}$

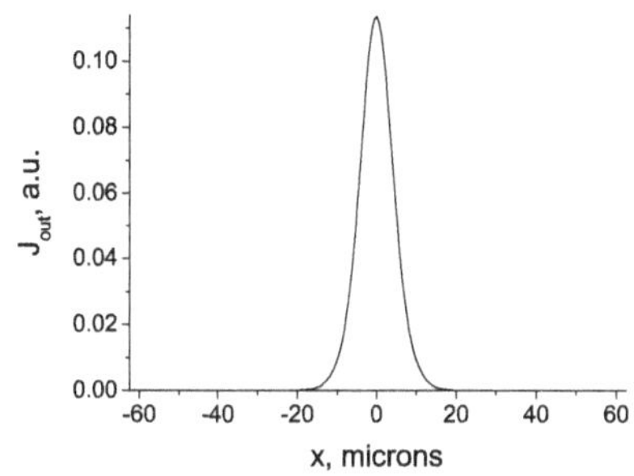

Fig. 3. TEM_{00} output field distribution. $I = 6\,\mathrm{mA}$.

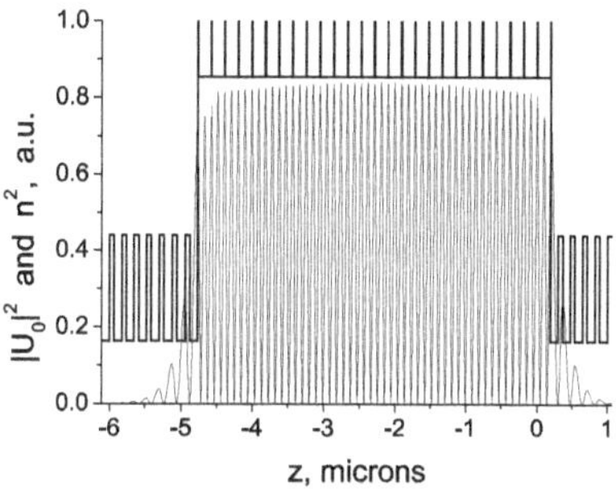

Fig. 4. Longitudinal field distribution at $r = 0$ (continuous function) and squared index (stepwise function). $I = 6\,\mathrm{mA}$.

TEM_{00} lasing mode. The relative change of Δk amounted to 0.8×10^{-5} and the relative change of output power P_{out} amounted to 3.2×10^{-3}. The last quantity was calculated by the formula $P_{out} = 2\pi \int J_{out} r dr$, where $J_{out} = |V_{m+1}^{q+1,+}|^2$ is the intensity distribution of the output electromagnetic field. For case of "cold" cavity we have calculated the TEM_{01} mode using $N_r = 1024$ and $N_r = 2048$. The relative change of β was equal to 3.4×10^{-5}. The results of error estimation seem to be quite satisfactory.

The results for pump current $I = 6\,\mathrm{mA}$ are presented in Figs. 2-4. The cross section of the gain distribution at the top QW ($z = z_{\nu(q)} = 0$) is shown in Fig. 2. The gain distributions in other QWs differ insignificantly. The cross section of the output field intensity is shown in Fig. 3. The longitudinal distribution of the field intensity is shown in Fig. 4. The calculated value $\Delta k = -405.36\,\mathrm{cm}^{-1}$ corresponds to the lasing wavelength $\lambda = 642.65\,\mathrm{nm}$. The final series of calculations was performed at varied total current I with the purpose of determining of the threshold current for lasing and the upper limit of single-mode operation. The threshold current I_t is determined from condition $P_{out}(I_t) = 0$. Extrapolating

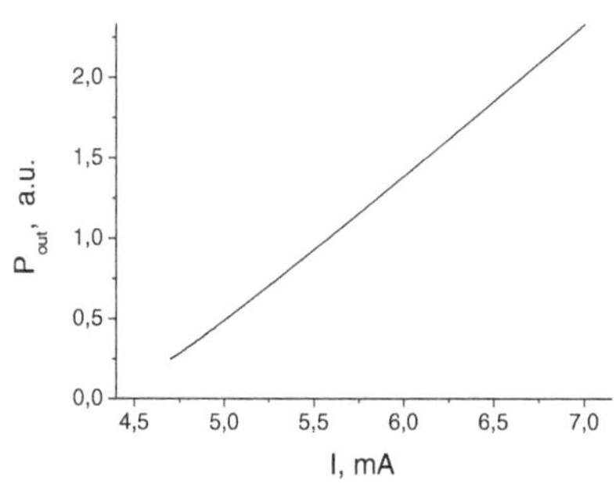

Fig. 5. TEM_{00} output power vs total current

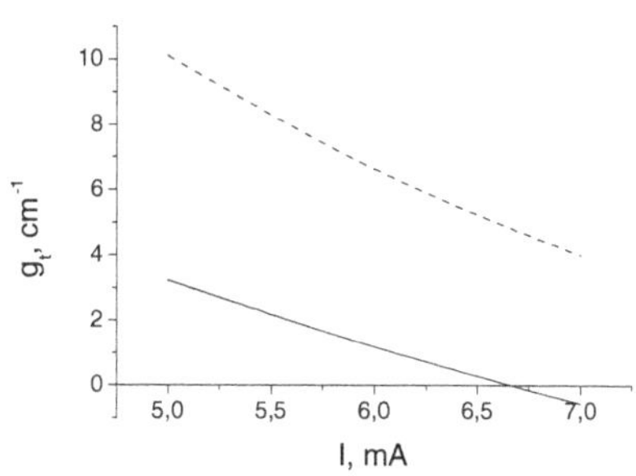

Fig. 6. Threshold gain for TEM_{01} mode (solid line) and for TEM_{10} mode (dashed line)

the graph in Fig. 5 we have found that $I_t \approx 4.43\,\mathrm{mA}$. To analyze stability of the operating mode we solve the eigenvalue problem (5) when gain and index are established by the operating mode and "frozen". Case of "cold" cavity is considered in this connection. The upper limit of single-mode operation is defined by the condition $g_t = 0$ for any mode except the lasing mode. As one can see from Fig. 6 the single-mode operation became unstable at $I \approx 6.7\,\mathrm{mA}$ because $g_t = 0$ for TEM_{01} mode at this pump current.

4 Conclusion

The given numerical algorithm allows us to calculate the mode spatial profile, output power, exact wavelength and other characteristics of an oscillating mode. Typical computational time for one variant amounts to several hours on IBM PC. Varying pump current we can determine the threshold current for lasing. Applying linear mode stability analysis we can find stability limit of single-mode lasing.

Acknowledgments

Work is partially supported by the RFBR project No. 08-02-00796-a. The authors appreciate helpful discussions with V.I. Kozlovsky.

References

1. Fox, A.G., Li, T.: Effect of gain saturation on the oscillating modes of optical masers. IEEE Journal of Quantum Electronics QE-2, 774–783 (1966)
2. Rao, H., Steel, M.J., Scarmozzino, R., Osgood Jr., R.M.: VCSEL design using the bidirectorial beam-propagating method. IEEE J. Quantum Electron. 37, 1435–1440 (2001)
3. Elkin, N.N., Napartovich, A.P., Troshchieva, V.N., Vysotsky, D.V.: Round-trip operator technique applied for optical resonators with dispersion elements. In: Boyanov, T., Dimova, S., Georgiev, K., Nikolov, G. (eds.) NMA 2006. LNCS, vol. 4310, pp. 542–549. Springer, Heidelberg (2007)
4. Hadley, G.R.: Modeling of diode laser arrays. In: Botez, D., Scifres, D.R. (eds.) Diode Laser Arrays, pp. 1–72. Cambridge Univ. Press, Cambridge (1994)
5. Siegman, A.E.: Quasi fast Hankel transform. Optics Letters 1, 13–15 (1977)

Memetic Simulated Annealing for the GPS Surveying Problem

Stefka Fidanova[1], Enrique Alba[2], and Guillermo Molina[2]

[1] IPP – Bulgarian Academy of Sciences, Acad. G. Bonchev str. bl.25A, 1113 Sofia, Bulgaria
stefka@parallel.bas.bg

[2] Universidad de Málaga, E.T.S.I. Informática, Grupo GISUM (NEO) Málaga, España
{eat,guillermo}@lcc.uma.es

Abstract. In designing Global Positioning System (GPS) surveying network, a given set of earth points must be observed consecutively (schedule). The cost of the schedule is the sum of the time needed to go from one point to another. The problem is to search for the best order in which this observation is executed. Minimizing the cost of this schedule is the goal of this work. Solving the problem for large networks to optimality requires impractical computational times. In this paper, several Simulated Annealing (SA) algorithms are developed to provide near-optimal solutions for large networks with bounded computational effort.

1 Introduction

A Global Positioning System (GPS) is a satellite-based radio-navigation system that permits land, sea, and airborne users to determine their three dimensional position, velocity, and time. This service is always available at any time and under any weather condition. In addition, satellite navigation systems have an impact in geoscience, in particular on surveying work in quick and effective determining positions and changes in positions networks. Measuring requires that the survey crew physically passes through all the intervening terrain to measure the distance between any two adjacent points. The most widely known space systems are: the American NAVSTAR global positioning system, the Russian GLObal Navigation Satellite System (GLONASS), and the forthcoming European satellite navigation system (GALILEO).

In this paper, we investigate the use of GPS to establish surveying networks. GPS satellites continuously transmit radio signals to the Earth while orbiting it. A receiver, with unknown position on Earth, has to detect and convert the signals received from all of the satellites into useful measurements. These measurements would allow a user to compute a three-dimensional coordinate position: the location of the receiver.

Solving this large problem to optimality requires a very high computational time. Therefore, new methods are needed to provide near-optimal solutions for

S. Margenov, L.G. Vulkov, and J. Waśniewski (Eds.): NAA 2008, LNCS 5434, pp. 281–288, 2009.

large networks within an acceptable amount of computational effort. These techniques are usually based on structured metaheuristics [3,5,10]. In this paper, several Simulated Annealing (SA) algorithms are introduced and applied on test problems with different dimension.

The organization of the paper is as follows. The general framework for a GPS surveying network problem as a combinatorial optimization problem is described in Section 2. Then, different search strategies for SA are explained in Section 3. The presentation of the initial temperature as a function of the initial solution cost is investigated in Section 4. The numerical results are presented and discussed in Section 5. The paper ends with conclusions and directions for future research.

2 Problem Description

The GPS network can be defined as set of stations $(a_1, a_2, \ldots a_n)$, which are co-ordinated by placing receivers $(X1, X2, \ldots)$ on them to determine sessions $(a_1a_2, a_1a_3, a_2a_3, \ldots)$ between them. The GPS surveying problem consists in searching for the best order in which these sessions can be organized to give the best schedule. Thus, the schedule can be defined as a sequence of sessions to be observed consecutively. The solution is represented by a linear graph with weighted edges. The nodes represent the stations and the edges represent the moving cost. The objective function of the problem is the cost of the solution, which is the sum of the costs (time) to move from one point to another one, $C(V) = \sum C(a_i, a_j)$, where a_ia_j is a session in solution $V = (a_1, a_2, \ldots, a_n)$.

The initial data is a cost matrix, which represents the cost of moving a receiver from one point to another. The cost could be evaluated purely upon the time or purely upon the distance; for more details see Dare [3]. This problem resembles the Traveling Salesman Problem (TSP). The main difference is that the TSP requires a closed path through all the nodes, where the initial and final nodes are the same, whereas the GPS problem finds an open path. Thus the strategies to solve GPS surveying problem can be different from these for TSP.

3 Simulated Annealing for the GPS Surveying Problem

SA is a heuristic method that has been implemented to obtain good solutions of an objective function defined on a number of discrete optimization problems [7]. This method has proved to be a flexible local search method and can be successfully applied to the majority of real-life problems [1,4,9]. The fundamental idea is to allow moves resulting in solutions of worse quality than the current solution in order to escape from local minima.

Simulated Annealing is a stochastic heuristic method which explores the solution space using a stochastic hill-climbing process. SA is inspired by the Metropolis scheme [8]. An initial state of a thermodynamic system was chosen with energy E and temperature T, holding T constant the initial configuration is perturbed and the change of energy dE is computed. The current state of the thermodynamic system is analogous to the current solution to the combinatorial problem,

```
Simulated Annealing

t:=0;
Initialize(V) - initial solution;
Initialize(T) - temperature;
Initialize(F) - cooling rate;
while not end_condition(t,V) do
      while not cooling_condition(t)
            V' := Choose_neighbor(V);
            Δ = C(V')-C(V);
            if Δ < 0 then
               V := V';
            end if
            else
               Generate_random number Θ
               if e^(-Δ/T) > Θ
                     V:=V';
               end if
            end else
            t := t+1;
      end while
      Cooldown(T);
end while
```

Fig. 1. Pseudocode for SA

the energy equation for the thermodynamic system is analogous to the objective function, and the ground state is analogous to the global minima. The algorithm starts by generating an initial solution and by initializing the so-called temperature parameter T. The temperature is decreased during the search process, thus at the beginning of the search the probability of accepting uphill moves is high and it gradually decreases. The structure of the SA algorithm is shown in Figure 1. The key objective of this paper is to find an effective solution in a short period of time with close to lowest cost for a given GPS network using Simulated Annealing.

The basic parts of SA are the initial solution construction and current solution perturbation. Two kinds of initial solutions, greedy and random, are used in this paper. To construct the greedy initial solution we start from a random node. The next node is the closest (cheapest) node to the current, which has not yet been included in the solution.

SA is a Local Search (LS) based method. The main concept of LS is searching the local neighborhood of the current solution [11]. In general, neighborhood for large-size problems can be much complicated to search. Therefore, LS attempts to improve a current schedule V to a GPS network by a small series of local improvements. A move generation is a transition from a schedule V to another one $V' \in I(V)$ in one step (iteration). The returned schedule V' may not be optimal, but it is the best schedule in its local neighborhood $I(V)$. A local optimal schedule is a schedule with the local minimal cost value.

In this paper several local search procedures $L(k, l)$ are applied, where k is the number of generated neighbor solutions and l is the number of used perturbation method. This raises a concept that can be called *memetic SA* because of the potential benefits of focusing on an internal local search step instead of the (usually) light perturbation of the canonical SA. Also, including problem knowledge in the initial generation of the solution is in the philosophy of memetic approaches [6]. As a consequence, for regular SA $k = 1$. The prepared by us perturbations are as follows:

1. Nodes Sequential Swaps: **for** $i = 1$ **to** $n - 1$; **for** $j = i + 1$ **to** n; swap a_i and a_j; [10];
2. Nodes Random Swaps: two nodes are chosen randomly and are swapped;
3. Randomly Delete an Edge: let the current solution is $(a_1 , a_2 , \dots , a_i , a_{i+1} , \dots , a_n)$. The edge $(i, i+1)$ is randomly chosen and deleted. The new solution is $(a_{i+1} , \dots , a_n , a_1 , \dots , a_i)$;
4. Greedy Delete an Edge: The longest (most expensive) edge is deleted. The new solution is constructed as in upper case;
5. Randomly Delete 2 Edges: Let the current solution is $(a_1 , a_2 , \dots , a_i , a_{i+1} , \dots , a_j , a_{j+1} , \dots , a_n)$. The edges $(i, i + 1)$ and $(j, j + 1)$ are randomly chosen and deleted. The new solutions are $(a_{i+1} , \dots , a_j , a_1 , \dots , a_i , a_{j+1} , \dots ,a_n)$, $(a_{j+1} , \dots , a_n , a_{i+1} , \dots , a_j , a_1 , \dots , a_i)$, $(a_1 , \dots , a_i , a_{j+1} , \dots , a_n , a_{i+1} , \dots , a_j)$;
6. Greedy Delete 2 Edges: The two longest edges are deleted. The new solutions are prepared as in the upper case.

When the cost of the best neighbor solution is equal to the cost of the current solution, we choose randomly the new current solution from the set of neighbors. The aim is to prevent $\Delta = 0$ and thus to prevent repetition of the same solutions. It is a way for diversification of the search process.

4 Parameter Settings

The parameter settings play a crucial role in the behavior of SA. The main SA parameters are initial temperature T_0 and cooling rate F. Most authors use a constant value for T_0, which is not related to the solved problem [9,10]. In [2] the temperature starts at T_0 close to ∞ and decreases very quickly. The goal of the mentioned work is to quicken the search process.

Our idea is to set the initial temperature T_0 to be a linear function of the cost of the initial solution, $T_0 = K * C(V_0)$, where K is a parameter. We decided T_0 to be a linear function of $C(V_0)$, because Δ is a linear function of $C(V)$. The acceptance probability depends on the difference Δ between the costs of the current and candidate solutions. Thus, if Δ is large, the probability of accepting the candidate solution is higher. The main problem is what large means. For example: Let $C(V) = 100$ and $\Delta = 20$ thus Δ is 20% of the value of $C(V)$ and Δ is large with respect to $C(V)$. Let $C(V) = 100\,000$ and $\Delta = 20$, thus Δ is 0.02% of the value of $C(V)$ and Δ is not large with respect to $C(V)$. For most problems the expected cost of the optimal solution is unknown, so our proposal is to automatically set the initial temperature to be proportional to the initial solution cost.

$$\frac{\Delta}{T} = \frac{C(V') - C(V)}{K * C(V_0)} \tag{1}$$

The other important parameter is the cooling parameter F. In previous works [2,7] it is recommended to use a value for the cooling parameter of 0.85 or higher in order to guarantee the theoretical convergence of SA to the global optima. In the next section the values of parameters K and F are investigated to understand their influence on the SA algorithm applied on the GSP surveying problem.

5 Experimental Results

In this section we analyze the experimental results obtained using the various SA algorithms described in previous sections. As a test problems we use real data from Malta and Seychelles GPS networks. The Malta GPS network is composed of 38 sessions and the Seychelles GPS network is composed of 71 sessions. We use 6 larger test problems too, from http://www.informatik.uni-heidelberg.de/ groups/ comopt/ software/ TSLIB95/ ATSP.html. These test problems range from 100 to 443 sessions.

For every experiment, the results are obtained by performing 30 independent runs, then averaging the fitness values obtained in order to ensure statistical confidence of the observed difference. Analysis of the data using ANOVA/Kruskal-Wallis test plus Multicompare function has been used to get statistical confidence of the results with a confidence level of 95% .

First the SA algorithms with random initial solution and various kinds of current solution perturbation are applied to all test problems with the same parameters as follows: initial temperature is 198, temperature decay is 0.85, Markov's chain is 1200, number of evaluations is 116000.

Comparing all the perturbation methods (Table 1), the best results (cost of the schedule, in bold) are obtained by $L(1,5)$, except for the Seychelles problem. With $L(1,5)$ perturbation the diversification is larger than with the rest of the techniques. We decided to further analyze the effects of using a greedy initial solution with only $L(1,5)$, since it achieves the best results.

Table 1. Comparison of simulated annealing algorithms with random initial solution, applied to various types of GPS networks

Test	sessions	LS(1,2)	LS(1,3)	LS(1,4)	LS(1,5)	LS(1,6)
Malta	38	1405	1285	1285	**1021**	1285
Seychelles	71	994	994	994	1052	**949**
rro124	100	206653	205643	205643	**48604**	200724
ftv170	170	7101	6908	6908	**6043**	7001
rbg323	323	6412	6396	6396	**5599**	6363
rbg358	358	7068	7050	7050	**6205**	7035
rbg403	403	7945	7923	7923	**6996**	7858
rbg443	443	8694	8486	8486	**7592**	8685

Table 2. Influence of the initial solution on the achieved results using L(1,5)

Tests stations	Malta 38	Seychelles 71	rro124 100	ftv170 170	rbg323 323	rbg358 358	rbg403 403	rbg443 443
random	1021	1052	48604	6043	5599	6205	6996	7592
greedy	**960**	**1037**	**43836**	**3980**	**1731**	**1820**	**3533**	**3891**

Comparing the influence of the initial solution generation (Table 2), we can conclude that starting from a greedy initial solution achieves lower cost results than starting from a random initial solution. The greedy initial solution is much better for most of the problems and gives the possibility for the algorithm to start from a solution which is closer to the optimal one, therefore our memetic algorithm approach represents a step forward in the techniques for solving this problem.

Next, we include in this study various kinds of local search procedures to improve the algorithm performance. To keep the running time low, the neighbor set consists of as many solutions as the number (n) of sessions. The number of iterations is equal to $10 \times n$.

Table 3. Simulated annealing algorithms plus local search applied to various types of GPS networks, n is equal to the number of sessions

Test	sessions	L(n,1)	L(n,2)	L(n,3)	L(n,4)	L(n,5)	L(n,6)
Malta	38	1345	**903**	1285	1285	**903**	1285
Seychelles	71	986	**930**	994	994	937	949
rro124	100	125606	56023	205463	205463	**38230**	191931
ftv170	170	6942	4994	6908	6908	**4599**	6605
rbg323	323	3855	1898	6396	6396	**1429**	5808
rbg358	358	3354	1858	7050	7050	**1335**	4996
rbg403	403	4845	2929	7923	7923	**2528**	5747
rbg443	443	5715	3242	8486	8486	**2844**	6065

Comparing SA with different local search procedures (Table 3) the best results are obtained by $L(n,5)$, except for the Malta and Seychelles test problems. For these test instances there are no significant differences between the results obtained by $L(n,2)$ and $L(n,5)$, but the standard deviations of $L(n,5)$ are three times smaller than using $L(n,2)$, which suggests its higher robustness.

We test the algorithm for all pairs (K, F),where K is from the set $\{0.002, 0.01, 0.1, 0.25, 0.50, 0.75, 1\}$ and F is from the set $\{0.85, 0.90, 0.95, 0.99\}$, for

Table 4. Influence of the initial temperature parameter K

Test	sessions	K=0.002	K=0.01	K=0.1	K=0.25	K=0.50	K=0.75	K=1
Malta	38	**939**	974	1030	1035	1035	1035	1035
Seychelles	71	**965**	1021	1036	1038	1040	1040	1040
rro124	100	**46622**	47106	47879	47879	47879	47879	47879
ftv170	170	**3706**	3926	3980	3980	3980	3980	3980
rbg323	323	**1726**	1731	1738	1738	1738	1738	1738
rbg358	358	**1808**	1818	1820	1820	1820	1820	1820
rbg403	403	**3530**	3532	3533	3533	3533	3533	3533
rbg443	443	**3888**	3891	3891	3892	3892	3892	3892

all test problems. We observed that the achieved costs are statistically similar went we fixed the value of K and changed the value of F.

In Table 4 we show the achieved costs for every test problem for different values of the temperature parameter K and $F = 0.85$. Analyzing the results, we conclude that, the algorithm in which $K = 0.002$ outperforms the others. Minimal costs are obtained by $K = 0.002$ for most of the test problems except tests rbg323, rbg403 and rbg443. For these three test problems there is no significant difference between the costs obtained by $K = 0.002$ and $K = 0.01$. For $K \geq 0.1$ the achieved costs are statistically similar. When the parameter $K \geq 0.1$, the initial temperature is too large and the probability to accept an inferior solution is too high. In this case the algorithm operates like random search. When the parameter $K < 0.002$, the value of the initial temperature is too small and it is difficult for the algorithm to climb the hills. In this case the algorithm proceeds in a greedy way. Therefore, we decided not to decreases the value of K.

6 Conclusion

The GPS surveying problem is addressed in this paper. Instances containing from 38 to 443 sessions have been solved using Simulated Annealing algorithms with various kinds of current solution perturbation and two kinds of initial solution construction. For solution improvement several local search procedures are developed and applied. The solution perturbation that deletes two randomly chosen edges and includes two new outperforms others. The results obtained using a greedy initial solution outperform those obtained using a random one. The combination of the SA with local search procedures has shown improvements of the results. The best results are achieved by using LS deleting two edges randomly and including two new ones, $L(n, 5)$ where n is equal to the number of sessions. The initial temperature is represented as a linear function of the initial solution cost and the best results are achieved when $T_0 = 0.002 * C(V_0)$. The cooling parameter F has vary low rate of influence.

In future work, other metaheuristics algorithms will be developed, applied and compared.

Acknowledgments. This work has been partially funded by the Spanish Ministry of Science and Technology and FEDER under contract TIN2005-08818-C04-01 (the OPLINK project). We also acknowledge partial funding from the Spanish Ministry of Industry and European EUREKA-CELTIC under contract FIT-330225-2007-1, label CP3-005 (the CARLINK project). Guillermo Molina is supported by grant AP2005-0014 from the Spanish government.

References

1. Arts, E., Van Loarhoven, P.: Statistical Cooling: A General Approach to Combinatorial Optimization Problems. Phillips Journal of Research 40, 193–226 (1985)
2. Chen, T.-C., Chang, Y.-W.: Modern Floorplaning Based on B* Fast Simulated Annealing. IEEE Trans. on Computer Aided Design 25(4), 637–650 (2006)
3. Dare, P.J., Saleh, H.A.: GPS Network Design:Logistics Solution Using Optimal and Near-Optimal Methods. J. of Geodesy 74, 467–478 (2000)
4. Dowsland, K., Thomson, J.: Variants of Simulated Annealing for the Examination Timetabling Problem. Annals of OR 63, 105–128 (1996)
5. Fidanova, S.: An Heuristic Method for GPS Surveying Problem. In: Shi, Y., van Albada, G.D., Dongarra, J., Sloot, P.M.A. (eds.) ICCS 2007. LNCS, vol. 4490, pp. 1084–1090. Springer, Heidelberg (2007)
6. Hart, W.E., Krasnogor, N., Smith, J.E. (eds.): Recent Advances in Memetic Algorithms. Studies in Fuzziness and Soft Computing, vol. 166 (2005)
7. Kirkpatrick, S., Gellat, C.D., Vecchi, P.M.: Optimization by Simulated Annealing. Science 220, 671–680 (1983)
8. Metropolis, N., Rosenbluth, A., Rosenbluth, M., Teller, A., Teller, E.: Equation of State Calculations by Fast Computing Machines. J. of Chem Phys. 21(6), 1087–1092 (1953)
9. Rene, V.V.: Applied Simulated Annealing. Springer, Berlin (1993)
10. Saleh, H.A., Dare, P.: Effective Heuristics for the GPS Survey Network of Malta: Simulated Annealing and Tabu Search Techniques. Journal of Heuristics 7(6), 533–549 (2001)
11. Schaffer, A.A., Yannakakis, M.: Simple Local Search Problems that are Hard to Solve. Society for Industrial Applied Mathematics Journal on Computing 20, 56–87 (1991)

On the Numerical Solution of a Transmission Eigenvalue Problem

S. Gegovska-Zajkova[1], Boško S. Jovanović[2], and Irena M. Jovanović[2]

[1] University "St. St. Cyril and Methody", Faculty of Electrical Engineering and Information Technologies, P.O. Box 574, 1000 Skopje, Macedonia
szajkova@feit.ukim.edu.mk
[2] University of Belgrade, Faculty of Mathematics
Studentski trg 16, 11000 Belgrade, Serbia
bosko@matf.bg.ac.yu, irenaire@gmail.com

Abstract. A transmission eigenvalue problem in disjoint intervals is examined. Distribution of the eigenvalues is obtained. The corresponding difference scheme is proposed and tested on few numerical examples.

1 Introduction

It happens frequently in applications that the domain under consideration is occupied by several materials with different physical properties. Problems of such type are usually called transmission (or diffraction, or interface) problems. For example, let us consider two thin rods, each of which is clamped at one end but which may contact at their free ends. We assume that the vibration process is independent of all but horizontal variable so that we can describe the reference configuration of the left rod as $a_1 \le x \le b_1$ and the right rod as $a_2 \le x \le b_2$. We let $u_1 = u_1(x,t)$ and $u_2 = u_2(x,t)$ denote the displacement of the cross-sections of the rods. The fixed ends occur at $x = a_1$ and $x = b_2$ while the ends $x = b_1, x = a_2$ must satisfy $a_1 < b_1 \le a_2 < b_2$. The hyperbolic system modelling the vibration contact of two rods is as follows:

$$\frac{\partial^2 u_1}{\partial t^2} - \frac{\partial}{\partial x}\left(p_1(x)\frac{\partial u_1}{\partial x}\right) + q_1(x)u_1(x) = f_1(x,t), \quad x \in (a_1,b_1), \quad t > 0 \qquad (1)$$

$$\frac{\partial^2 u_2}{\partial t^2} - \frac{\partial}{\partial x}\left(p_2(x)\frac{\partial u_2}{\partial x}\right) + q_2(x)u_2(x) = f_2(x,t), \quad x \in (a_2,b_2), \quad t > 0 \qquad (2)$$

where $p_i(x)$ are the modules of elasticity of the rods, $q_i(x)$ are the springs of the support at x and $f_i(x,t)$ are external loads. We have assumed that the area of the cross section and the specific mass to be constants. Note that since the rods are clamped at $x = a_1$ and $x = b_2$ we have

$$u_1(a_1,t) = 0, \quad u_2(b_2,t) = 0. \qquad (3)$$

If we suppose a non-perfect contact between the rods we have

$$p_1(b_1)\frac{\partial u_1}{\partial x}(b_1,t) + \alpha_1 u_1(b_1,t) = \beta_1 u_2(a_2,t), \qquad (4)$$

S. Margenov, L.G. Vulkov, and J. Waśniewski (Eds.): NAA 2008, LNCS 5434, pp. 289–296, 2009.

$$-p_2(a_2)\frac{\partial u_2}{\partial x}(a_2,t)+\alpha_2 u_2(a_2,t)=\beta_2 u_1(b_1,t). \tag{5}$$

Here α_1, β_1 are spring constants of springs attached to the left and right rods at $x=b_1$ and $x=a_2$, respectively. The second boundary condition has a similar interpretation.

Finally, we need to impose initial conditions. We specify the initial positions and velocities

$$u_1(x,0)=\varphi_1(x), \quad \frac{\partial u_1}{\partial t}(x,0)=\psi_1(x), \quad x\in(a_1,b_1) \tag{6}$$

$$u_2(x,0)=\varphi_2(x), \quad \frac{\partial u_2}{\partial t}(x,0)=\psi_2(x), \quad x\in(a_2,b_2) \tag{7}$$

Problem (1)-(7) is investigated in [1]. Analogous parabolic problem is treated in [2]. Problems with nonstandard boundary and/or conjugation conditions are treated also in [8,9].

2 Transmission Eigenvalue Problem

Let us now consider the corresponding eigenvalue problem for (1)-(7):

$$-\left(p_1(x)v_1'\right)'+q_1(x)v_1(x)=\lambda v_1(x), \quad x\in(a_1,b_1), \tag{8}$$

$$-\left(p_2(x)v_2'\right)'+q_2(x)v_2(x)=\lambda v_2(x), \quad x\in(a_2,b_2), \tag{9}$$

$$v_1(a_1)=0, \quad v_2(b_2)=0, \tag{10}$$

$$p_1(b_1)\,v_1'(b_1)+\alpha_1 v_1(b_1)=\beta_1 v_2(a_2), \tag{11}$$

$$-p_1(a_2)\,v_2'(a_2)+\alpha_2 v_2(a_2)=\beta_2 v_1(b_1). \tag{12}$$

Assuming that conditions

$$\alpha_i>0, \quad \beta_i>0, \quad i=1,2 \tag{13}$$

hold, consider the product space $L=L_2(a_1,b_1)\times L_2(a_2,b_2)$, endowed with the inner product and associated norm

$$(u,v)_L=\beta_2(u_1,v_1)_{L_2(a_1,b_1)}+\beta_1(u_2,v_2)_{L_2(a_2,b_2)}, \quad \|v\|_L=(v,v)_L^{1/2},$$

where $(u_i,v_i)_{L_2(a_i,b_i)}=\int_{a_i}^{b_i}u_i v_i\,dx,\ i=1,2$. We also introduce the space

$$H^1=\{v=(v_1,v_2)\,|\,v_i\in H^1(a_i,b_i)\ \text{ and }\ v_1(a_1)=0,\ \ v_2(b_2)=0\}$$

endowed with the inner product and norm

$$(u,v)_{H^1}=\beta_2(u_1,v_1)_{H^1(a_1,b_1)}+\beta_1(u_2,v_2)_{H^1(a_2,b_2)}, \quad \|v\|_{H^1}=(v,v)_{H^1}^{1/2},$$

where $(u_i,v_i)_{H^1(a_i,b_i)}=(u_i,v_i)_{L_2(a_i,b_i)}+(u_i',v_i')_{L_2(a_i,b_i)},\ i=1,2.$

We have the following assertion.

Lemma 1. *Under the regularity conditions*

$$p_i(x),\, q_i(x) \in L_\infty(a_i,b_i), \quad i=1,2 \tag{14}$$

$$0 < p_{i0} \le p_i(x), \quad 0 \le q_i(x) \quad a.\,e.\ in \quad (a_i,b_i), \quad i=1,2 \tag{15}$$

and the sign conditions (13), the spectral problem (8)-(12) is formally equivalent to the following variational problem: find $(\lambda, v) \in R \times H^1$ *such that*

$$A(v,w) = \lambda\,(v,w)_L, \quad \forall\, w \in H^1, \tag{16}$$

where

$$\begin{aligned} A(u,v) = \beta_2 \int_{a_1}^{b_1} p_1\, u_1'\, v_1'\, dx + \beta_1 \int_{a_2}^{b_2} p_2\, u_2'\, v_2'\, dx + \beta_2\alpha_1 u_1(b_1)v_1(b_1) \\ +\beta_1\alpha_2 u_2(a_2)v_2(a_2) - \beta_1\beta_2\big[u_1(b_1)v_2(a_2) + u_2(a_2)v_1(b_1)\big]. \end{aligned} \tag{17}$$

The proof is given in [3].

We state the following important properties of the spaces H^1 and L.

(i) H^1 and L are Hilbert spaces,
(ii) H^1 is compactly embedded in L.

In the following lemma we deal with some properties of the bilinear form $A(u,v)$.

Lemma 2. *Under the conditions (13), (14), (15) and*

$$\beta_1\beta_2 \le \alpha_1\alpha_2 \tag{18}$$

the bilinear form A*, defined by (17), is symmetric and bounded on* $H^1 \times H^1$*. Moreover, this form is also coercive, i.e. there exist a constant* $c_0 > 0$ *such that*

$$A(v,v) \ge c_0 \|v\|^2_{H^1}, \quad \forall\, v \in H^1.$$

For the proof see [1].

The Lemmas 1, 2 and the properties (i), (ii), allow us to recast the problem (8)-(12) into the general theory of abstract eigenvalue problems for bilinear forms in Hilbert spaces, see e.g. [5]. This ensures the existence of exact eigenpairs as stated in the following theorem.

Theorem 1. *The problem (8)-(12) has a countable sequence of real eigenvalues*

$$0 < \lambda_1 < \lambda_2 < \ldots \to \infty.$$

The corresponding eigenfunctions $v^n \equiv (v_1^n, v_2^n)$*,* $n = 1, 2\ldots$*, can be chosen to be orthonormal in* L*. They constitute a Hilbert basis for* H^1 *as well as for* L*.*

The λ_n can be characterized by the minimum principle of the Rayligh quotient $R(v) = A(v,v)/\|v\|_L^2$:

$$\lambda_1 = \min_{v\in H^1} R(v) = R(v_1)$$

$$\lambda_n = \min_{v\in H^1,\ A(v,v_i)=0,\ i=1,\ldots,n-1} R(v) = R(v_n), \qquad n = 2, 3, \ldots$$

3 Structure of the Spectrum

We need some additional properties of the problem (8)-(12). Since he eigenfunction $v = (v_1, v_2)$ is defined up to constant multiplyer we normalize v by

$$\|v\|_L = 1\,, \quad v_1'(a_1) > 0.$$

Theorem 2. *The eigenvalues λ_n of the problem (8)-(12) are given as $n \to \infty$ by the asymptotic formula*

$$\sqrt{\lambda_n} = C\,n + \frac{O(1)}{n}\,.$$

Proof. Following the theory of spectral problems with classical boundary conditions [4] one can prove that the asymptotic distribution of the eigenvalues of all regular problems (8)-(12) satisfying (13) and (18) is the same as for the problem with constant coefficients

$$-v_1'' = \lambda v_1\,, \qquad x \in (a_1, b_1), \tag{19}$$

$$-v_2'' = \lambda v_2\,, \qquad x \in (a_2, b_2), \tag{20}$$

$$v_1(a_1) = 0, \qquad v_2(b_2) = 0. \tag{21}$$

$$v_1'(b_1) + \alpha_1 v_1(b_1) = \beta_1 v_2(a_2), \tag{22}$$

$$-v_2'(a_2) + \alpha_2 v_2(a_2) = \beta_2 v_1(b_1), \tag{23}$$

We seek the eigenvector $v = (v_1, v_2)$ in the form

$$v_1(x) = A \sin \mu(x - a_1)\,, \qquad x \in (a_1, b_1),$$

$$v_2(x) = B \sin \mu(b_2 - x)\,, \qquad x \in (a_2, b_2).$$

The boundary conditions (21) are automatically satisfied. The equations (19) and (20) are also satisfied if one sets $\lambda = \mu^2$. The unknown parameters A, B we will find from the boundary conditions (22) and (23):

$$\begin{aligned} &A\,[\mu \cos \mu(b_1 - a_1) + \alpha_1 \sin \mu(b_1 - a_1)] - B\beta_1 \sin \mu(b_2 - a_2) = 0 \\ &-A\beta_2 \sin \mu(b_1 - a_1) + B\,[\mu \cos \mu(b_2 - a_2) + \alpha_2 \sin \mu(b_2 - a_2)] = 0 \end{aligned} \tag{24}$$

In order the homogeneous system (24) with respect to A, B to have nontrivial solutions its determinant must to be zero. After some algebra we get

$$\begin{aligned} &\mu^2 + \mu\,[\alpha_1\,\tan \mu(b_1 - a_1) + \alpha_2\,\tan \mu(b_2 - a_2)] \\ &+(\alpha_1\,\alpha_2 - \beta_1\,\beta_2) \tan \mu(b_1 - a_1)\,\tan \mu(b_2 - a_2) = 0 \end{aligned} \tag{25}$$

The equation (25) can be written in the form

$$\begin{aligned} \mu = &-\frac{1}{2}\,[\alpha_1\,\tan \mu(b_1 - a_1) + \alpha_2\,\tan \mu(b_2 - a_2)] \pm \frac{1}{2}\Big\{\,[\alpha_1\,\tan \mu(b_1 - a_1) \\ &+ \alpha_2\,\tan \mu(b_2 - a_2)]^2 - 4(\alpha_1\,\alpha_2 - \beta_1\,\beta_2) \tan \mu(b_1 - a_1)\,\tan \mu(b_2 - a_2)\Big\}^{1/2} \end{aligned} \tag{26}$$

The expression under the root is always positive. In fact setting

$$t_1 = \tan \mu(b_1 - a_1), \quad t_2 = \tan \mu(b_2 - a_2), \quad s = t_1/t_2$$

we get

$$\begin{aligned} &[\alpha_1 \tan \mu(b_1 - a_1) + \alpha_2 \tan \mu(b_2 - a_2)]^2 \\ &-4(\alpha_1 \alpha_2 - \beta_1 \beta_2) \tan \mu(b_1 - a_1) \tan \mu(b_2 - a_2) \\ &= t_2^2 \left[\alpha_1^2 s^2 + (4\beta_1 \beta_2 - 2\alpha_1 \alpha_2)s + \alpha_2^2\right] \end{aligned}$$

For the discriminant of the quadratic form $\alpha_1^2 s^2 + (4\beta_1 \beta_2 - 2\alpha_1 \alpha_2)s + \alpha_2^2$ we find

$$D = 16\beta_1 \beta_2 (\beta_1 \beta_2 - \alpha_1 \alpha_2) \leq 0$$

This implies

$$\alpha_1^2 s^2 + (4\beta_1 \beta_2 - 2\alpha_1 \alpha_2)s + \alpha_2^2 \geq 0$$

and

$$\begin{aligned} &[\alpha_1 \tan \mu(b_1 - a_1) + \alpha_2 \tan \mu(b_2 - a_2)]^2 \\ &-4(\alpha_1 \alpha_2 - \beta_1 \beta_2) \tan \mu(b_1 - a_1) \tan \mu(b_2 - a_2) \geq 0. \end{aligned}$$

Using properties of function $\tan x$, in particular its behavior in the neighborhood of asymptotes, we easily conclude that there exist two families of solutions of equation (26) – $\mu_{11}, \mu_{12}, \mu_{13}, \ldots$ and $\mu_{21}, \mu_{22}, \mu_{23}, \ldots$ such that

$$\mu_{in} = C\, n + \frac{O(1)}{n}, \quad i = 1, 2, \quad n = 1, 2, 3, \ldots \qquad \square$$

In the case $\alpha_1 \alpha_2 = \beta_1 \beta_2$ the equation (25) takes the simpler form:

$$\mu = -\left[\alpha_1 \tan \mu(b_1 - a_1) + \alpha_2 \tan \mu(b_2 - a_2)\right].$$

Also, if $b_2 - a_2 = b_1 - a_1 = d$, from (25) one gets:

$$\mu = -\frac{1}{2}\left[\alpha_1 + \alpha_2 \mp \sqrt{(\alpha_1 - \alpha_2)^2 + 4\beta_1 \beta_2}\right] \tan \mu d\,.$$

4 Alternative Formulation

Problem (8)-(12) can be transformed into analogous one containing Dirac distribution. Suppose that $p_1(b_1) = p_2(a_2)$. (This condition always can be satisfied by suitable change of variable). Letting

$$v(x) = \begin{cases} v_1(x + a_1), & x \in (0, \xi), \quad \xi = b_1 - a_1, \\ v_2(x - \xi + a_2), & x \in (\xi, l), \quad l = b_1 - a_1 + b_2 - a_2, \end{cases}$$

and analogously defining $p(x)$ and $q(x)$ problem (8)-(12) reduces to

$$-\left(p(x)v'\right)' + q(x)v(x) = \lambda v(x), \quad x \in (0, \xi) \cup (\xi, l), \tag{27}$$

$$v(0) = 0, \quad v(l) = 0, \tag{28}$$

$$p(\xi-0)\,v'(\xi-0)+\alpha_1 v(\xi-0)=\beta_1 v(\xi+0), \tag{29}$$

$$-p(\xi+0)\,v'(\xi+0)+\alpha_2 v(\xi+0)=\beta_2 v(\xi-0). \tag{30}$$

Let us denote

$$[v]_x=v(x+0)-v(x-0)\,.$$

For $\alpha_1\alpha_2>\beta_1\beta_2$ conditions (29)-(30) are equivalent to

$$[pv']_\xi=p(\xi)\,[v']_\xi=(\alpha_2-\beta_1)\,v(\xi+0)+(\alpha_1-\beta_2)\,v(\xi-0), \tag{31}$$

$$[v]_\xi=p(\xi)\,\frac{(\alpha_1-\beta_1)\,v'(\xi+0)+(\alpha_2-\beta_2)\,v'(\xi-0)}{\alpha_1\alpha_2-\beta_1\beta_2}\,. \tag{32}$$

Using properties of Dirac distribution $\delta(x)$ we easily conclude that (27), (28), (31), (32) is equivalent to boundary value problem

$$\begin{aligned}
&-(p(x)v')'+q(x)v(x)+\Big((\alpha_2-\beta_1)v(\xi+0)+(\alpha_1-\beta_2)v(\xi-0)\Big)\delta(x-\xi)\\
&\qquad+p^2(\xi)\,\frac{(\alpha_1-\beta_1)\,v'(\xi+0)+(\alpha_2-\beta_2)\,v'(\xi-0)}{\alpha_1\alpha_2-\beta_1\beta_2}\,\delta'(x-\xi)\\
&\qquad=\lambda v(x),\qquad x\in(0,l),\\
&\qquad v(0)=v(l)=0,
\end{aligned} \tag{33}$$

where differentiation and equality are assumed in the sense of distributions [7].

5 Finite Difference Approximation

Suppose p_i, $q_i\in C([a_i,b_i])$. Take an integer $N\ge 2$ and define uniform meshes

$$\bar\omega_i=\{x=x_{ij}=a_i+jh_i\,|\,j=0,1,\dots,N,\ h_i=(b_i-a_i)/N\},\quad \omega_i=\bar\omega_i\cap(a_i,b_i).$$

We introduce finite differences $v_{i,x}$ and $v_{i,\bar x}$ in the standard manner [6] and approximate spectral problem (8)-(12) with the following finite difference scheme

$$\begin{aligned}
&-(r_1\,y_{1,\bar x})_x+d_1\,y_1=\lambda^h\,y_1\,,\quad x\in\omega_1\,,\\
&-(r_2\,y_{2,\bar x})_x+d_2\,y_2=\lambda^h\,y_2\,,\quad x\in\omega_2\,,\\
&y_1(a_1)=y_2(b_2)=0,\\
&\frac{2}{h_1}\Big[r_1(b_1)\,y_{1,\bar x}(b_1)+\alpha_1\,y_1(b_1)-\beta_1\,y_2(a_2)\Big]+d_1(b_1)\,y_1(b_1)=\lambda^h\,y_1(b_1),\\
&\frac{2}{h_2}\Big[-r_2(x_{21})y_{2,x}(a_2)+\alpha_2 y_2(a_2)-\beta_2 y_1(b_1)\Big]+d_2(a_2)y_2(a_2)=\lambda^h y_2(a_2),
\end{aligned} \tag{34}$$

where $r_i(x)=p_i(x-h_i/2)$ and $d_i(x)=q_i(x)$.

6 Numerical Experiments

Eigenvalues and eigenfunctions of the problem (8)-(12) are determined for different values of input data: $p_1 = p_2 = 1$, $q_1 = q_2 = 0$ and

a) $a_1 = 0, b_1 = 3/8, a_2 = 5/8, b_2 = 1, \ \alpha_1 = 2, \alpha_2 = 4, \beta_1 = 1, \beta_2 = 2;$
b) $a_1 = 0, b_1 = 1/2, a_2 = 3/4, b_2 = 1, \ \alpha_1 = 2, \alpha_2 = 4, \beta_1 = 1, \beta_2 = 2;$

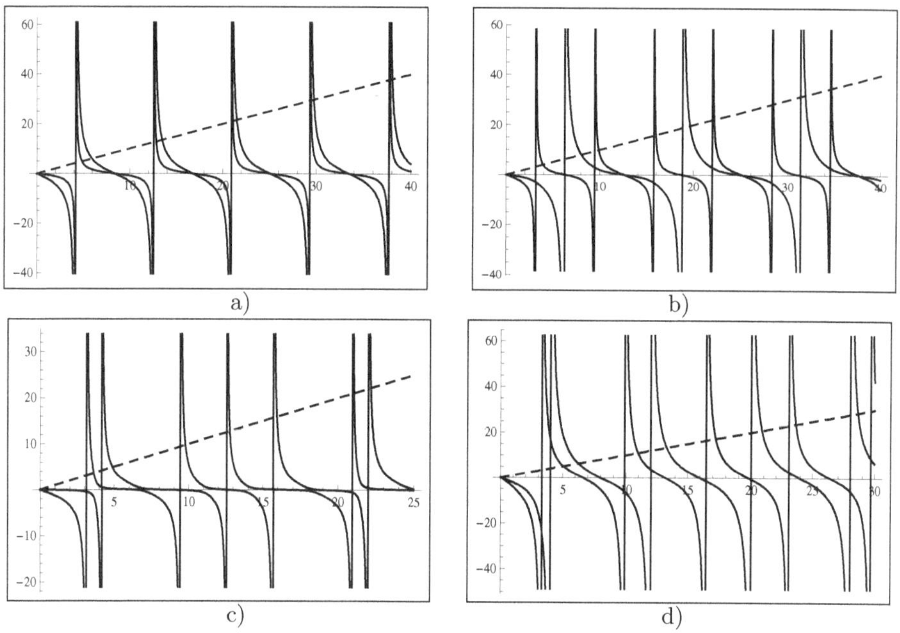

Fig. 1. Solutions of equation (26)

Table 1. First three eigenvalues of (8)-(12) and (34)

		$n = 1$	$n = 2$	$n = 3$
a	λ_n	23.699736	35.700620	164.585162
	λ_n^h	23.664991	35.657893	161.685623
	error (%)	0.147	0.120	1.762
b	λ_n	15.624021	64.599901	97.435160
	λ_n^h	15.604264	64.454206	95.860128
	error (%)	0.126	0.226	1.617
c	λ_n	13.255155	24.006976	95.267203
	λ_n^h	13.232531	23.971214	93.635879
	error (%)	0.171	0.149	1.712
d	λ_n	25.458130	39.134361	120.042078
	λ_n^h	25.420106	39.055711	118.134114
	error (%)	0.149	0.201	1.589

c) $a_1 = 0,\, b_1 = 4/8,\, a_2 = 5/8,\, b_2 = 1, \quad \alpha_1 = \sqrt{3},\, \alpha_2 = 1,\, \beta_1 = 1,\, \beta_2 = \sqrt{2}$;
d) $a_1 = 0,\, b_1 = \pi/8,\, a_2 = \pi/6,\, b_2 = 1, \quad \alpha_1 = 7.5,\, \alpha_2 = 6.5,\, \beta_1 = 4.2,\, \beta_2 = 1.2$.

Solutions of equation (26) are presented in Fig. 1. Comparison between exact eigenvalues and eigenvalues of difference scheme (34) for $N = 10$ are presented in Table 1.

Acknowledgement

The research of the second author was supported by MS of Republic of Serbia (project 144005A) and Bulgarian NFS (project HS-MI-106/2005).

References

1. Jovanović, B.S., Vulkov, L.G.: Numerical solution of a hyperbolic transmission problem. Comput. Method. Appl. Math. 8(4) (2008) (to appear)
2. Jovanović, B.S., Vulkov, L.G.: Finite difference approximation of strong solutions of a parabolic interface problem on disconected domains. Publ. Inst. Math. 83 (2008) (to appear)
3. Jovanović, B.S., Vulkov, L.G.: Formulation and analysis of parabolic interface problems on disjoint intervals (submitted)
4. Mikhlin, S.G.: Linear PDE. Vysshaya shkola, Moscow (1977) (Russian)
5. Renardy, M., Rogers, R.C.: An Introduction to PDE. Springer, Berlin (1993)
6. Samarskiĭ, A.A.: Theory of Difference Schemes. Nauka, Moscow (1989) (Russian)
7. Vladimirov, V.S.: Equations of Math. Physics. Nauka, Moscow (1988) (Russian)
8. Vulkov, L.G.: Applications of Steklov-type eigenvalue problems to convergence of difference schemes for parabolic and hyperbolic equations with dynamical boundary conditions. In: Vulkov, L.G., Yalamov, P., Waśniewski, J. (eds.) WNAA 1996. LNCS, vol. 1196, pp. 557–564. Springer, Heidelberg (1997)
9. Vulkov, L.G.: Well posedness and a monotone iterative method for a nonlinear interface problem on disjoint intervals. Amer. Inst. Phys. Proc. Ser. 946 (2007)

On Weakening Conditions for Discrete Maximum Principles for Linear Finite Element Schemes

Antti Hannukainen[1], Sergey Korotov[1], and Tomáš Vejchodský[2]

[1] Institute of Mathematics, Helsinki University of Technology, P.O. Box 1100, FIN–02015 Espoo, Finland
{antti.hannukainen,sergey.korotov}@hut.fi
[2] Institute of Mathematics, Czech Academy of Sciences, Žitná 25, CZ–115 67 Prague 1, Czech Republic
vejchod@math.cas.cz

Abstract. In this work we discuss weakening requirements on the set of sufficient conditions due to Ph. Ciarlet [4,5] for matrices associated to linear finite element schemes, which is commonly used for proving validity of discrete maximum principles (DMPs) for the second order elliptic problems.

1 Model Problem and Maximum Principle

We consider the following test problem: Find a function u such that

$$-\operatorname{div}(\mathcal{A}\nabla u) + cu = f \quad \text{in} \quad \Omega, \tag{1}$$

$$u = g \quad \text{on} \quad \partial\Omega, \tag{2}$$

where $\Omega \subset \mathbf{R}^d$ is a bounded polytopic domain with Lipschitz boundary $\partial\Omega$. The diffusive tensor $\mathcal{A}$ is assumed to be a symmetric and uniformly positive definite matrix. The reactive coefficient c is assumed to be nonnegative in Ω.

The classical solutions of elliptic problems of the second order are known to satisfy the so-called maximum principles (MPs), see e.g. [11,7]. For our test problem the corresponding MP is the following implication:

$$f \leq 0 \qquad \Longrightarrow \qquad \max_{x\in\overline{\Omega}} u(x) \leq \max\{0\,,\ \max_{s\in\partial\Omega} g(s)\}. \tag{3}$$

To the authors' knowledge the first reasonable DMP and conditions providing its validity were formulated in 1966 by R. Varga [14] for the finite difference method. Later, in 1970 in [4] (and [5]), Ph. Ciarlet presented a more general form of DMP suitable for finite element (FE) and finite difference types of discretizations. He also proposed a practical set of (sufficient) conditions on matrices involved, providing a validity of his DMP. Since that time these conditions became popular in numerical community, see e.g. [7,8,9,10] and references therein, for proving various DMPs for problems of elliptic type. In this work we consider the issue of weakening the conditions proposed by Ciarlet.

S. Margenov, L.G. Vulkov, and J. Waśniewski (Eds.): NAA 2008, LNCS 5434, pp. 297–304, 2009.

2 Finite Element Discretization

Standard (linear) schemes for construction of FE approximations for the (unknown) solution u of (1)–(2) are based on the so-called weak formulation: Find $u \in g + H_0^1(\Omega)$ such that

$$a(u, v) = \mathcal{F}(v) \quad \forall v \in H_0^1(\Omega),$$

where

$$a(u, v) = \int_\Omega \mathcal{A}\nabla u \cdot \nabla v \, \mathrm{d}x + \int_\Omega cuv \, \mathrm{d}x \quad \text{and} \quad \mathcal{F}(v) = \int_\Omega fv \, \mathrm{d}x.$$

Here, the matrix $\mathcal{A}$ is assumed to be in $[L^\infty(\Omega)]^{d\times d}$, $c \in L^\infty(\Omega)$, $g \in H^1(\Omega)$, and $f \in L^2(\Omega)$. The existence and uniqueness of the weak solution u is provided by the standard Lax-Milgram lemma.

Let $\mathcal{T}_h$ be a FE partition (mesh) of $\overline{\Omega}$ with interior nodes $B_1, \dots, B_N$ lying in Ω and boundary nodes $B_{N+1}, \dots, B_{N+N^\partial}$ lying on $\partial\Omega$. Further, let V_h be a finite-dimensional subspace of $H^1(\Omega)$, associated with $\mathcal{T}_h$ and its nodes, being spanned by the basis functions $\phi_1, \phi_2, \dots, \phi_{N+N^\partial}$ with the following properties: $\phi_i \geq 0$ in $\overline{\Omega}$, $i = 1, \dots, N + N^\partial$, and $\sum_{i=1}^{N+N^\partial} \phi_i \equiv 1$ in $\overline{\Omega}$.

We also assume that the basis functions $\phi_1, \phi_2, \dots, \phi_N$ vanish on the boundary $\partial\Omega$. Thus, they span a finite-dimensional subspace V_h^0 of $H_0^1(\Omega)$. Let, in addition, $g_h = \sum_{i=N+1}^{N+N^\partial} g_i\phi_i \in V_h$ be a suitable approximation of the function g, for example its nodal interpolant.

The FE approximation is defined as a function $u_h \in g_h + V_h^0$ such that

$$a(u_h, v_h) = \mathcal{F}(v_h) \quad \forall v_h \in V_h^0,$$

whose existence and uniqueness are also provided by the Lax-Milgram lemma.

Algorithmically, $u_h = \sum_{i=1}^{N+N^\partial} y_i\phi_i$, where the coefficients y_i are the entries of the solution $\bar{\mathbf{y}} = [y_1, \dots, y_{N+N^\partial}]^\top$ of the following square system of $N + N^\partial$ linear algebraic equations

$$\bar{\mathbf{A}}\bar{\mathbf{y}} = \bar{\mathbf{F}}, \tag{4}$$

where

$$\bar{\mathbf{A}} = \begin{bmatrix} \mathbf{A} & \mathbf{A}^\partial \\ \mathbf{0} & \mathbf{I} \end{bmatrix}, \quad \bar{\mathbf{F}} = \begin{bmatrix} \mathbf{F} \\ \mathbf{F}^\partial \end{bmatrix}, \quad \text{and} \quad \bar{\mathbf{A}}^{-1} = \begin{bmatrix} \mathbf{A}^{-1} & -\mathbf{A}^{-1}\mathbf{A}^\partial \\ \mathbf{0} & \mathbf{I} \end{bmatrix}. \tag{5}$$

In the above, blocks $\mathbf{A}$ and $\mathbf{A}^\partial$ are matrices of size $N \times N$ and $N \times N^\partial$, respectively, $\mathbf{I}$ stands for the unit matrix, and $\mathbf{0}$ – for the zero matrix. The entries of $\bar{\mathbf{A}}$ are denoted by $a_{ij} = a(\phi_j, \phi_i)$, $i = 1, \dots, N$, $j = 1, \dots, N + N^\partial$. The block-vector $\mathbf{F}$ consists of entries $f_i = \mathcal{F}(\phi_i)$, $i = 1, \dots, N$, and the block-vector $\mathbf{F}^\partial$ has entries $f_i = g_i$, $i = N + 1, \dots, N + N^\partial$, given by the boundary data. For the later reference we also include the formula for $\bar{\mathbf{A}}^{-1}$ in (5). Notice that $\bar{\mathbf{A}}$ is nonsingular if and only if $\mathbf{A}$ is nonsingular.

3 Sufficient Algebraic Conditions of Ph. Ciarlet

We will distinguish two essentially different types of DMPs.

Algebraic DMP: A natural algebraic analogue of (3) is as follows (cf. (4)):

$$\mathbf{F} \leq \mathbf{0} \qquad \Longrightarrow \qquad \max_{i=1,\ldots,N+N^{\partial}} y_i \leq \max\{0\,, \max_{j=N+1,\ldots,N+N^{\partial}} y_j\}.$$

Functional DMP: A natural functional imitation of (3) is as follows:

$$f \leq 0 \qquad \Longrightarrow \qquad \max_{\overline{\Omega}} u_h \leq \max\{0\,, \max_{\partial\Omega} u_h\}.$$

Remark 1. It is easy to see that the above types of DMPs are equivalent in the case of linear and multilinear finite elements. However, these DMPs are not equivalent, in general, for higher-order finite elements, see [15].

In [4], Ciarlet formulated and proved the following theorem:

Theorem 1. *The algebraic DMP is satisfied if and only if*

(A) $\bar{\mathbf{A}}$ *is monotone (i.e.,* $\bar{\mathbf{A}}$ *nonsingular and* $\bar{\mathbf{A}}^{-1} \geq 0$*)*
(B) $\xi + \mathbf{A}^{-1}\mathbf{A}^{\partial}\xi^{\partial} \geq 0$*, where* ξ *and* ξ^{∂} *are vectors of all ones of sizes* N *and* N^{∂}*, respectively.*

Since conditions (A) and (B) are difficult to verify, Ciarlet proposed in [4] the following *standard set of sufficient conditions* which is more practical.

Theorem 2. *The algebraic DMP is valid provided the matrix* $\bar{\mathbf{A}}$ *satisfies*

(a) $a_{ii} > 0,\ i = 1,\ldots,N,$
(b) $a_{ij} \leq 0,\ i \neq j,\ i = 1,\ldots,N,\ j = 1,\ldots,N+N^{\partial},$
(c) $\sum_{j=1}^{N+N^{\partial}} a_{ij} \geq 0,\ i = 1,\ldots,N,$
(d) $\mathbf{A}$ *is irreducibly diagonally dominant.*

Ciarlet essentially proposed the above conditions in order to utilize the following result of Varga [13, p. 85]:

Lemma 1. *If* $\mathbf{A} \in \mathbf{R}^{N\times N}$ *is an irreducibly diagonally dominant matrix with strictly positive diagonal and nonpositive off-diagonal entries then* $\mathbf{A}^{-1} > 0$.

Now, we can easily demonstrate the proof of Theorem 2. We follow the steps of Ciarlet [4]. Conditions (a), (b), and (d) together with Lemma 1 imply $\mathbf{A}^{-1} \geq 0$ and, hence, condition (A). Further, condition (c) is equivalent to $\mathbf{A}\xi + \mathbf{A}^{\partial}\xi^{\partial} \geq 0$ and since $\mathbf{A}^{-1} \geq 0$ we conclude that (B) is valid as well. Theorem 1 then guarantees the algebraic DMP.

Remark 2. In the case of homogeneous Dirichlet boundary conditions system (4) reduces to a simpler form $\mathbf{A}\mathbf{y} = \mathbf{F}$ (cf. [10]). Then the algebraic DMP holds if and only if $\mathbf{A}^{-1} \geq \mathbf{0}$, i.e., if and only if $\mathbf{A}$ is monotone.

4 Associated Geometrical Conditions on FE Meshes

For some types of finite elements, the entries of $\bar{\mathbf{A}}$ can be computed explicitly, therefore condition (b) can often be guaranteed a priori by imposing suitable geometrical requirements on the shape (and size) of FE meshes employed.

For example, if $\mathcal{A}$ is a diagonal matrix then there exist the following popular geometrical conditions providing (b):

(i) for simplicial finite elements ($d \geq 2$) – all dihedral angles between facets of simplices have to be nonobtuse or acute [5,2,7,8,10];
(ii) for bilinear elements – all rectangular elements have to be nonnarrow (i.e. $\sqrt{2}/2 \leq b_1/b_2 \leq \sqrt{2}$, where b_1, b_2 are the edges of the rectangle), trilinear elements have to be cubes, see [8];
(iii) for 3D meshes consiting of right triangular prisms the altitudes of prisms are limited from both sides by certain quantities dependent on the area and angles of the triangular base and on the magnitude of the reaction coefficient c, see [6].

5 Typical Problems with Standard Conditions

Not only condition (b) but also the other conditions (a), (c), and (d) have to be addressed. The positivity of diagonal entries (a) is trivially satisfied for elliptic problems. Also the row sums (c) are nonnegative automatically for problem (1)–(2), because the basis functions form a partition of unity:

$$\sum_{j=1}^{N+N^{\partial}} a_{ij} = a\left(\sum_{j=1}^{N+N^{\partial}} \phi_j, \phi_i\right) = a(1, \phi_i) = \int_{\Omega} c\phi_i \geq 0, \quad i = 1, \ldots, N. \qquad (6)$$

On the other hand, the irreducibility of $\mathbf{A}$ required in (d) is not always obvious. For illustration, we present three examples of triangulations which lead to reducible matrices in Figure 1.

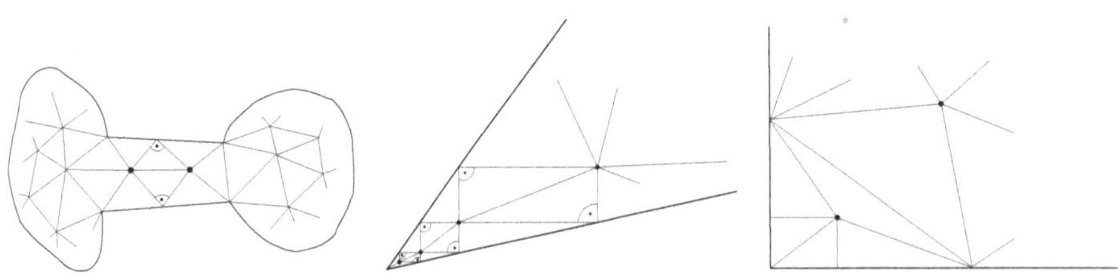

Fig. 1. Examples of meshes leading to reducible matrix $\mathbf{A}$ for the Poisson problem with Dirichlet boundary conditions. The angles with dots are right.

It might be a difficult task to satisfy all conditions (a)–(d) practically, especially in 3D. For example, the existence of a face-to-face partition of a cube into acute tetrahedra is still an open problem, see (i). Another practical problem in

3D is to keep the desired geometrical limitations on the elements during global and local refinements of meshes. Condition (b) leads to severe limitation in the case of 3D rectangular blocks (only cubes are allowed), see the point (ii) above. Moreover, if the diffusive tensor $\mathcal{A}$ is not diagonal then proving the irreducibility (d) could be a nontrivial task.

6 Less Severe Conditions: Stieltjes Matrices

Conditions (a)–(d) can be weakened using the concept of M-matrices and Stieltjes matrices [13]. A real square matrix $\mathbf{A}$ is an *M-matrix* if all its off-diagonal entries are nonpositive and if it is nonsingular and $\mathbf{A}^{-1} \geq 0$. A real square matrix $\mathbf{A}$ is a *Stieltjes matrix* if all its off-diagonal entries are nonpositive and if it is symmetric and positive definite. The following lemma [13, p. 85] enables to eliminate conditions (a), (c), and (d) from the standard set as we state in Theorem 3 below.

Lemma 2. *If $\mathbf{A}$ is a Stieltjes matrix then it is also an M-matrix.*

Theorem 3. *If the finite element matrix $\bar{\mathbf{A}}$, associated to (1)–(2), satisfies condition* (b) *from Theorem 2 then the algebraic DMP is valid.*

Proof. We verify conditions (A) and (B) of Theorem 1. (A) For problem (1)–(2), the FE matrix $\mathbf{A}$ is always symmetric and positive definite and hence if condition (b) is satisfied then $\mathbf{A}^{-1} \geq 0$ by Lemma 2. Further, since $\mathbf{A}^{\partial} \leq 0$ by (b), we obtain $-\mathbf{A}^{-1}\mathbf{A}^{\partial} \geq 0$ and therefore $\bar{\mathbf{A}}^{-1} \geq 0$, see (5). (B) Condition (c) is satisfied for problem (1)–(2) due to (6) and, as we already mentioned, (c) implies (B).

7 Testing the Sharpness of Theoretical Conditions

The standard DMP results [2,5,7,8] provide conditions which guarantee condition (b) and consequently that $\mathbf{A}$ is a Stieltjes matrix. However, the Stieltjes matrices form only a certain subclass of monotone matrices. In this section we test how sharp the conditions based on the Stieltjes matrix concept are, i.e., we try to test how wide is the class of meshes which lead to $\mathbf{A}$ being monotone but not Stieltjes. For this purpose we solve the 2D Poisson problem with homogenous Dirichlet boundary conditions on various domains using various triangulations.

We present three tests. In each test we construct a simple triangulation which is characterized by two parameters (angles) $0 < \alpha < \pi$ and $0 < \gamma < \pi$, see Figure 2. We prepare N sampling points and we go through all values $\alpha_i = i\pi/(N+1)$, $\gamma_j = j\pi/(N+1)$, $i, j = 1, 2, \ldots, N$. For each pair α_i, γ_j, we construct the basic mesh as indicated in Figure 2, provided it is possible. Then we refine each triangle in the basic mesh into 100 similar subtriangles (each edge in the original mesh is divided into 10 segments). Then we assemble the stiffness matrix $\mathbf{A}$ on the refined mesh and we compute the inverse $\mathbf{A}^{-1}$. Finally, we investigate the entries of $\mathbf{A}$ and $\mathbf{A}^{-1}$. If all off-diagonal entries of $\mathbf{A}$ are nonpositive we

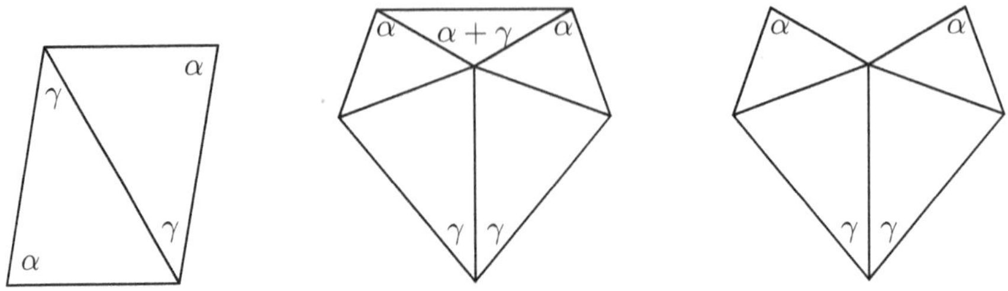

Fig. 2. The basic meshes for the three tests. The meshes used for the actual computations are 10-fold refined basic meshes.

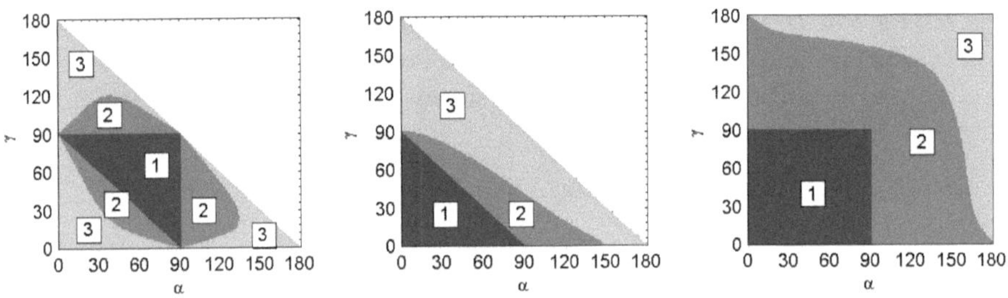

Fig. 3. Results of three DMP tests. Domain 1: triangulations with non-obtuse maximal angle (matrix **A** is a Stieltjes matrix). Domain 2: triangulations with obtuse maximal angle providing the DMP (matrix **A** is monotone but not Stieltjes). Domain 3: triangulations with obtuse maximal angle, DMP is not valid (matrix **A** is not monotone).

mark the pair α_i, γ_j by 1. Otherwise, we check the nonnegativity of $\mathbf{A}^{-1}$. We mark the pair α_i, γ_j by 2 if $\mathbf{A}^{-1} \geq 0$ and by 3 if it is not.

The results of the computations are visualized in Figure 3, where we used $N = 200$ sampling points for each of the angles α and γ. The white areas correspond to the angles α_i, γ_j for which the indicated triangulation does not exist.

The stiffness matrix **A** for the Poisson equation in 2D is well-known to be Stieltjes matrix if and only if the sum of the two angles opposite to each interior edge in the mesh is at most π. For the investigated meshes this sufficient and necessary condition reduces to the requirement of non-obtuseness of the greatest angle in the triangulation, see the point (i) above. If we compare the sizes of domains 1 with domains 2 in Figure 3 we may conclude that the standard sufficient condition (b) is not very sharp. There is a wide space for its generalization. However, any generalization have to utilize more general criteria for the monotonicity of a matrix. These criteria, see e.g. [1], are more complicated and their application for the DMP is not straightforward. Certain success in this respect was reported in [3,9,12].

8 Conclusions

We have analyzed the standard approach for proving the DMP for elliptic problems and showed that the positivity of the diagonal entries (a), the nonnegativity of the row sums (c), and the irreducibility and diagonal dominance (d) are, in fact, not needed as sufficient conditions. Moreover, the presented numerical experiments indicate that the known geometric conditions guaranteeing $\mathbf{A}^{-1} \geq 0$ are not very sharp and that there is a space for possible generalization and further research.

Acknowledgement

The first author was supported by Project no. 124619 from the Academy of Finland.

The second author was supported by the Academy Research Fellowship no. 208628 and Grant no. 121283 from the Academy of Finland.

The third author was supported by grant no. IAA100760702 of the Grant Agency of the Czech Academy of Sciences, grant no. 102/07/0496 of the Czech Science Foundation, and by the institutional research plan no. AV0Z10190503 of the Czech Academy of Sciences.

References

1. Bouchon, F.: Monotonicity of some perturbations of irreducibly diagonally dominant M-matrices. Numer. Math. 105, 591–601 (2007)
2. Brandts, J., Korotov, S., Křížek, M.: The discrete maximum principle for linear simplicial finite element approximations of a reaction-diffusion problem. Linear Algebra Appl. 429, 2344–2357 (2008)
3. Bramble, J.H., Hubbard, B.E.: On a finite difference analogue of an elliptic boundary problem which is neither diagonally dominant nor of non-negative type. J. Math. and Phys. 43, 117–132 (1964)
4. Ciarlet, P.G.: Discrete maximum principle for finite-difference operators. Aequationes Math. 4, 338–352 (1970)
5. Ciarlet, P.G., Raviart, P.-A.: Maximum principle and uniform convergence for the finite element method. Comput. Methods Appl. Mech. Engrg. 2, 17–31 (1973)
6. Hannukainen, A., Korotov, S., Vejchodský, T.: Discrete maximum principle for FE-solutions of the diffusion-reaction problem on prismatic meshes. J. Comput. Appl. Math. (to appear)
7. Karátson, J., Korotov, S.: Discrete maximum principles for finite element solutions of nonlinear elliptic problems with mixed boundary conditions. Numer. Math. 99, 669–698 (2005)
8. Karátson, J., Korotov, S., Křížek, M.: On discrete maximum principles for nonlinear elliptic problems. Math. Comput. Simulation 76, 99–108 (2007)
9. Korotov, S., Křížek, M., Neittaanmäki, P.: Weakened acute type condition for tetrahedral triangulations and the discrete maximum principle. Math. Comp. 70, 107–119 (2001)
10. Křížek, M., Qun, L.: On diagonal dominance of stiffness matrices in 3D. East-West J. Numer. Math. 3, 59–69 (1995)
11. Ladyzhenskaya, O.A., Ural'tseva, N.N.: Linear and quasilinear elliptic equations. Leon Ehrenpreis Academic Press, New York (1968)

12. Ruas Santos, V.: On the strong maximum principle for some piecewise linear finite element approximate problems of non-positive type. J. Fac. Sci. Univ. Tokyo Sect. IA Math. 29, 473–491 (1982)
13. Varga, R.: Matrix Iterative Analysis. Prentice Hall, New Jersey (1962)
14. Varga, R.: On discrete maximum principle. J. SIAM Numer. Anal. 3, 355–359 (1966)
15. Vejchodský, T., Šolín, P.: Discrete maximum principle for higher-order finite elements in 1D. Math. Comp. 76, 1833–1846 (2007)

On the Sign-Stability of Finite Difference Solutions of Semilinear Parabolic Problems

Róbert Horváth

Budapest University of Technology and Economics,
Egry J. u. 1, 1111 Budapest, Hungary
rhorvath@math.bme.hu

Abstract. The sign-stability property is one of the important qualitative properties of the one-dimensional heat conduction equation, or more generally, of one-dimensional parabolic problems. This property means that the number of the spatial sign-changes of the solution function cannot increase in time. In this paper, sufficient conditions will be given that guarantee the fulfillment of a numerical analogue of the sign-stability for the finite difference solution of a semilinear parabolic problem. The results are demonstrated on a numerical test problem.

1 Introduction

Previous results regarding the sign-stability considered the initial boundary value problem

$$\frac{\partial v}{\partial t} - \frac{\partial}{\partial x}\left(\kappa \frac{\partial v}{\partial x}\right) + \gamma v = 0, \quad (x,t) \in (0,1) \times (0,T) =: Q_T, \tag{1}$$

$$v(x,0) = v_0(x), \quad x \in (0,1), \tag{2}$$

$$v(0,t) = v(1,t) = 0, \quad 0 \le t \le T, \tag{3}$$

where $T > 0$, and $\gamma = \gamma(x)$ and $\kappa = \kappa(x)$ are continuous functions. A function $v : \bar{Q}_T \to \mathbb{R}$ is called the solution of problem (1)-(3) if it is sufficiently smooth and satisfies the equalities in (1)-(3).

Parabolic problems can be considered as mathematical models of heat conduction problems. From the laws of thermodynamics, a number of characteristic properties of the heat conduction process can be deduced. These properties can be formulated for the mathematical models, too. Some of the important qualitative properties are the nonnegativity preservation, maximum-minimum principle and the maximum norm contractivity.

The sign-stability property fits well into the above series of qualitative properties. This property of problem (1)-(3) expresses the fact that the number of the sign-changes of the functions $x \mapsto v(x,t)$ does not grow in t.

S. Margenov, L.G. Vulkov, and J. Waśniewski (Eds.): NAA 2008, LNCS 5434, pp. 305–313, 2009.

In the nineteen-thirties, Sturm and Pólya showed [11,13] the sign-stability of equation (1). The parameters γ and κ were positive constants. Pólya's proof is based on some elementary properties of periodic differentiable functions.

It is a natural requirement that the numerical equivalents of the above properties must be valid also for numerical solutions of problem (1)-(3) (e.g., [2,3,4,7,8]). The sign-stability property is the strongest and most important one among the qualitative properties in that sense that the validity of the sign-stability property guarantees the validity of the other qualitative properties in the numerical models [7].

The discrete analogue of the sign-stability property can be formulated as follows. A discretization

$$0 = x_0 < x_1 < \cdots < x_{n+1} = 1 \tag{4}$$

of the interval $[0, 1]$ and a positive time step τ defines a space-time mesh $\Omega_{h,\tau}$ on $\bar{Q}_T$. In the case of finite difference and finite element methods, the numerical solution can be constructed from a real valued mesh function defined on $\Omega_{h,\tau}$. This mesh function can be generally produced with a one-step vector iteration in the form

$$\mathbf{v}^{k+1} = \mathbf{A}^{(k)}\mathbf{v}^k, \; k = 0, 1, \ldots, \tag{5}$$

where $\mathbf{v}^k = [v_1^k, \ldots, v_n^k]^T \in \mathbb{R}^n$ $(k = 0, 1, \ldots)$, v_i^k is the approximation of the solution v at the point $(x_i, k\tau)$, and the iteration matrix $\mathbf{A}^{(k)} \in \mathbb{R}^{n\times n}$ may depend on the iteration step k.

For a given vector $\mathbf{y} = [y_1, \ldots, y_n]^T \in \mathbb{R}^n$, let us denote the number of the sign-changes in the sequence $y_1, y_2, \ldots, y_n$, where we leave out the occurrent zero values, by $\mathcal{S}(\mathbf{y})$. If $\mathbf{y}$ is the zero vector, then we set $\mathcal{S}(\mathbf{y}) = -1$. Naturally, the trivial relations $-1 \le \mathcal{S}(\mathbf{y}) \le n - 1$ and $\mathcal{S}(\mathbf{y}) = \mathcal{S}(-\mathbf{y})$ hold. We notice that the number of the sign-changes of a vector $\mathbf{y}$ is equal to the number of the sign-changes of that continuous piecewise linear scalar function f defined on $[0, 1]$ for which the equalities $f(x_i) = y_i$ $(i = 1, \ldots, n)$ and $f(0) = f(1) = 0$ hold at the mesh points.

Now we are able to define the sign-stability of a numerical method.

Definition 1. *A numerical method is said to have the sign-stability property on a fixed space-time mesh $\Omega_{h,\tau}$ if the sequence $\{\mathcal{S}(\mathbf{v}^k)\}$ defined with the vectors of the one-step vector iteration (5) is a nonincreasing sequence for any initial vector $\mathbf{v}^0$.*

Definition 2. *A matrix $\mathbf{A} \in \mathbb{R}^{n\times n}$ is said to be sign-stable (resp. sign-unstable) if the condition $\mathcal{S}(\mathbf{Ax}) \le \mathcal{S}(\mathbf{x})$ (resp. $\mathcal{S}(\mathbf{Ax}) \ge \mathcal{S}(\mathbf{x})$) is fulfilled for any vector $\mathbf{x} \in \mathbb{R}^n$.*

Clearly, the inverse of a regular sign-stable matrix is sign-unstable, and vice versa. Moreover, the product of sign-stable matrices are sign-stable and the same is true also for sign-unstable matrices. It is known that totally nonnegative matrices (all minors are nonnegative) or minordefinite matrices (no two minors with the same order that have different sign) are sign-stable [5,12].

The next theorems give sufficient conditions of the sign-stability and the sign-unstability of tridiagonal matrices. A direct proof of the theorems can be found in [9].

Theorem 1. *Let* $\mathbf{A} \in \mathbb{R}^{n\times n}$ *be a tridiagonal matrix with the properties*
(P1) $a_{i,i}, a_{i,i\pm1} > 0$ $(i = 1, \ldots n)$,
(P2) $a_{i,i} \geq a_{i,i-1} + a_{i,i+1}$ $(i = 1, \ldots n)$ *(weak row-diagonal dominance).*
Then the matrix $\mathbf{A}$ *is sign-stable.*

Theorem 2. *Let* $\mathbf{A} \in \mathbb{R}^{n\times n}$ *be a tridiagonal matrix with the properties*
(Q1) $a_{i,i} > 0$ $(i = 1, \ldots n)$,
(Q2) $a_{i,i\pm1} < 0$ $(i = 1, \ldots n)$,
(Q3) $a_{i,i} \geq -a_{i,i-1} - a_{i,i+1}$ $(i = 1, \ldots n)$.
Then the matrix $\mathbf{A}$ *is sign-unstable.*

It is evident that if all iteration matrices $\mathbf{A}^{(k)}$ in (5) are sign-stable then the numerical solution will possess the sign-stability property on the space-time mesh $\Omega_{h,\tau}$.

It was shown in [6] that the finite difference method with the uniform spatial step-size h and with the θ time-discretization method is sign-stable for problems with $\gamma = 0$, $\kappa = 1$ if the relation

$$\frac{\tau}{h^2} \leq \frac{1}{4(1-\theta)} \tag{6}$$

is satisfied. If $\theta = 1$, then there is no upper bound for the quotient τ/h^2. In [8], we showed that condition (6) is the necessary and sufficient condition of the uniform (independent of h) sign-stability. We extended the result also for finite element solutions. In paper [10], sufficient conditions are given that guarantee the sign-stability of the finite difference and finite element solutions of problem (1)-(3).

In [14], it is shown that the number of the "peaks" and the "valleys" of the solution function of a certain semilinear problem cannot grow in time. This raised the idea to investigate the sign stability of the same problem.

2 The Investigated Semilinear Problem and Its Discretization

We are going to give sufficient conditions in order to guarantee the sign-stability of the finite difference solution of the semilinear problem

$$\begin{aligned} \frac{\partial v}{\partial t} &= a(x,t)\frac{\partial^2 v}{\partial x^2} + b(x,t)\frac{\partial v}{\partial x} + f(t,v)v \quad in\ Q_T, \\ v(x,0) &= v_0(x), \\ v(0,t) &= v(1,t) = 0, \end{aligned} \tag{7}$$

where the coefficient functions are supposed to satisfy the following conditions:

i) The functions a, a'_x and a'_t are continuous on $\bar{Q}_T$. There exists a positive constant $a_0 > 0$ with the property $a \geq a_0$ in $\bar{Q}_T$.
ii) The functions b and b'_x are continuous on $\bar{Q}_T$.
iii) The functions f and f'_v are continuous in $[0,T] \times \mathbb{R}$. There esists a number M_0 with the property $|f(t,v)| \leq M_0$ in $[0,T] \times \mathbb{R}$.

Remark 1. Existence and uniqueness results about the solution of the problem can be found in [14]. With the technique that was used to show the decrease of the number of the peaks and valleys it can be showed that the number of the sign-changes also decreases.

Let us define a space time mesh $\Omega_{h,\tau}$ on $\bar{Q}_T$, that is a positive time step τ and a spatial discretization (4), and introduce the notations $h_i = x_i - x_{i-1}$ $(i = 1, \ldots, n+1)$ for the spatial grid sizes. Introducing the notations v_i^k for the approximations of the exact values of the solution v at the points $(x_i, k\tau)$, equation (7) can be discretized as follows

$$\begin{aligned} \frac{v_i^{k+1} - v_i^k}{\tau} &= \theta \left(a_i^{k+1} \Delta_h v_i^{k+1} + b_i^{k+1} D_h v_i^{k+1} + f(k\tau, v_i^k) v_i^{k+1} \right) \\ &+ (1-\theta) \left(a_i^k \Delta_h v_i^k + b_i^k D_h v_i^k + f(k\tau, v_i^k) v_i^k \right), \end{aligned} \tag{8}$$

where the known values a_i^k and b_i^k are the approximations of the functions a and b at the point $(x_i, k\tau)$, respectively. The constant θ is an arbitrary time discretization parameter form the interval $[0,1]$. The second and first order difference operators Δ_h and D_h (in the centered difference case) are defined, respectively, as

$$\Delta_h v_i^k = q_{i,i-1} v_{i-1}^k + q_{i,i} v_i^k + q_{i,i+1} v_{i+1}^k \tag{9}$$

and

$$D_h v_i^k = d_{i,i+1} v_{i+1}^k - d_{i,i-1} v_{i-1}^k, \tag{10}$$

where

$$\begin{aligned} q_{i,i-1} &= \frac{2}{h_{i-1}(h_{i-1}+h_{i+1})}, \quad q_{i,i+1} = \frac{2}{h_{i+1}(h_{i-1}+h_{i+1})}, \\ q_{i,i} &= \frac{-2}{h_{i+1}(h_{i-1}+h_{i+1})} + \frac{-2}{h_{i-1}(h_{i-1}+h_{i+1})}, \\ d_{i,i-1} &= \frac{-1}{h_{i-1}+h_{i+1}}, \quad d_{i,i+1} = \frac{1}{h_{i-1}+h_{i+1}}. \end{aligned} \tag{11}$$

Taking into the account the homogeneous Dirichlet boundary condition, we define the above coefficients to be zero if one of the indices equals zero. The finite difference equations (8) $(i = 1, \ldots, n)$ can be written in a one-step vector iteration form

$$\mathbf{A}_1^{(k)} \mathbf{v}^{k+1} = \mathbf{A}_2^{(k)} \mathbf{v}^k, \quad k = 0, 1, \ldots, [T/\tau] - 1. \tag{12}$$

With the help of this iteration, the vectors $\mathbf{v}^k = [v_1^k, \ldots, v_n^k]^T \in \mathbb{R}^n$ of the approximate values can be determined starting from the known initial vector $\mathbf{v}^0 = [v_0(x_1), \ldots, v_0(x_n)]^T$. The iteration matrices $\mathbf{A}_1^{(k)}$ and $\mathbf{A}_2^{(k)}$ have the form

$$\begin{aligned} \mathbf{A}_1^{(k)} &= \mathbf{I} - \theta\left(\tau\langle a_i^{k+1}\rangle\mathbf{Q} + \tau\langle b_i^{k+1}\rangle\mathbf{D} + \tau\langle f(k\tau, v_i^k)\rangle\right), \\ \mathbf{A}_2^{(k)} &= \mathbf{I} + (1-\theta)\left(\tau\langle a_i^k\rangle\mathbf{Q} + \tau\langle b_i^k\rangle\mathbf{D} + \tau\langle f(k\tau, v_i^k)\rangle\right). \end{aligned} \tag{13}$$

Notice that the above matrices are dependent on the number of the iteration step. This fact is indicated in the superscript $()^{(k)}$. The matrix $\mathbf{I} \in \mathbb{R}^{n\times n}$ is the unit matrix and the other $n \times n$ matrices are defined as

$$\langle a_i^k\rangle = \mathrm{diag}\,(a_1^k, \ldots, a_n^k), \quad \langle b_i^k\rangle = \mathrm{diag}\,(b_1^k, \ldots, b_n^k), \tag{14}$$

$$\langle f(k\tau, v_i^k)\rangle = \mathrm{diag}\,(f(k\tau, v_1^k), \ldots, f(k\tau, v_n^k)), \tag{15}$$

$$\mathbf{Q} = \mathrm{tridiag}\,(q_{i,i-1}, q_{i,i}, q_{i,i+1}), \quad \mathbf{D} = \mathrm{tridiag}\,(d_{i,i-1}, 0, d_{i,i+1}). \tag{16}$$

3 Sign-Stability of the Numerical Solution

The sign-stability of the numerical solution on a fixed mesh $\Omega_{h,\tau}$ means that the sequence $\{\mathcal{S}(\mathbf{v}^k)\}$ is a non-increasing sequence (Definition 1). A sufficient condition of the sign-stability can be given applying the results of Theorem 1 and Theorem 2. Let us introduce the notations $\|a\| = \max_{\bar{Q}_T}\{|a(x,t)|\}$ and $\|b\| = \max_{\bar{Q}_T}\{|b(x,t)|\}$ for the maximum norms of the coefficient function a and b, respectively. Moreover, let $h_{min} = \min_i\{h_i\}$ and $h_{max} = \max_i\{h_i\}$.

Theorem 3. *The simultaneous fulfillment of the conditions below guarantee the sign-stability of the numerical solution* (8) *of the semilinear parabolic problem* (7) *on a fixed space time mesh $\Omega_{h,\tau}$.*

(C1) The spatial discretization has the property that

$$c_0 := \frac{\|b\| h_{max}^2}{2 h_{min} a_0} < 1,$$

(C2) The time step fulfills the following conditions:

- (C2a) If $\theta = 0$, then

$$\tau \le \frac{1}{(4\|a\|/h_{\min}^2 + M_0)};$$

- (C2b) if $\theta = 1$, then either $M_0 = 0$ or $M_0 \neq 0$ and $\tau < 1/M_0$;

- (C2c) if $0 < \theta < 1$, then

$$\tau \le \frac{1}{(1-\theta)(4\|a\|/h_{\min}^2 + M_0)}$$

and if $M_0 \neq 0$ then the condition $\tau < 1/(M_0\theta)$ also must be valid.

Proof. It is enough to show that under the above conditions the matrices $\mathbf{A}_1^{(k)}$ are regular and sing-unstable and the matrices $\mathbf{A}_2^{(k)}$ are sign-stable for all $k = 0, 1, \ldots, [T/\tau] - 1$.

The matrix $\mathbf{A}_1^{(k)}$ is a tridiagonal matrix with the elements

$$\begin{aligned}(\mathbf{A}_1^{(k)})_{i,i\pm1} &= -\theta\tau(a_i^{k+1} q_{i,i\pm1} + b_i^{k+1} d_{i,i\pm1}),\\ (\mathbf{A}_1^{(k)})_{i,i} &= 1 - \theta\tau(a_i^{k+1} q_{i,i} + f(k\tau, v_i^k)).\end{aligned} \tag{17}$$

If $\theta = 0$, then this matrix is a unit matrix that has both the sign-stability and the sign-instability property. Let us suppose that $\theta \neq 0$. It can be seen that Condition (C1) implies the negativity of the sub- and superdiagonal elements of the matrices. Indeed, for the subdiagonal elements we obtain the estimation

$$\begin{aligned}(\mathbf{A}_1^{(k)})_{i,i-1} &= -\theta\tau(a_i^{k+1} q_{i,i-1} + b_i^{k+1} d_{i,i-1}) \le -\theta\tau\left(a_0 \tfrac{1}{h_{\max}^2} - \|b\| \tfrac{1}{2h_{\min}}\right)\\ &\le \tfrac{-\theta\tau a_0}{h_{\max}^2}\left(1 - \tfrac{\|b\| h_{\max}^2}{2a_0 h_{\min}}\right) = \tfrac{-\theta\tau a_0}{h_{\max}^2}(1 - c_0) < 0.\end{aligned}$$

Moreover, in the case $i = 2, \ldots, n-1$ we have

$$\begin{aligned}&(\mathbf{A}_1^{(k)})_{i,i-1} + (\mathbf{A}_1^{(k)})_{i,i} + (\mathbf{A}_1^{(k)})_{i,i+1}\\ &= 1 - \theta\tau(a_i^{k+1}(q_{i,i-1} + q_{i,i} + q_{i,i+1}) + b_i^{k+1}(d_{i,i-1} + d_{i,i+1}) + f(k\tau, v_i^k))\\ &- 1 - \theta\tau f(k\tau, v_i^k) \ge 1 - \theta\tau M_0 > 0.\end{aligned}$$

Now let $i = 1$ (the case $i = n$ can be handled similarly). Then we have

$$\begin{aligned}&(\mathbf{A}_1^{(k)})_{1,1} + (\mathbf{A}_1^{(k)})_{1,2}\\ &= 1 - \theta\tau(a_1^{k+1}(q_{1,1} + q_{1,2}) + b_1^{k+1} d_{1,2} + f(k\tau, v_1^k))\\ &\ge 1 - \theta\tau\left(a_0 \frac{-1}{h_{\max}^2} + \|b\| \frac{1}{2h_{\min}} + M_0\right) = 1 - \frac{\theta\tau a_0}{h_{\max}^2}\left(-1 + \frac{\|b\| h_{\max}^2}{2a_0 h_{\min}}\right) - \theta\tau M_0\\ &\ge 1 - \frac{\theta\tau a_0}{h_{\max}^2}(-1 + c_0) - 1 = \frac{\theta\tau a_0}{h_{\max}^2}(1 - c_0) > 0.\end{aligned}$$

In the above estimations, which show the strict diagonal dominance of the matrices, we used conditions (C2b) and (C2c) and we assumed that $M_0 \neq 0$. If $M_0 = 0$ (that is $f(t, v) \equiv 0$), then the strict diagonal dominance of the matrices is evident. The strict diagonal dominance also means that the main diagonal of the matrices are positive. Thus, based on Theorem 2, the matrices $\mathbf{A}_1^{(k)}$ are sign-unstable. Moreover, combining the sign structure with the strict diagonal dominance of the matrices, it can be seen that the matrices $\mathbf{A}_1^{(k)}$ are so-called regular M-matrices [1].

Let us turn to the investigation of the matrices $\mathbf{A}_2^{(k)}$. These matrices are tridiagonal matrices again with the nonzero elements

$$\begin{aligned}(\mathbf{A}_2^{(k)})_{i,i\pm1} &= (1-\theta)\tau(a_i^k q_{i,i\pm1} + b_i^k d_{i,i\pm1}),\\ (\mathbf{A}_2^{(k)})_{i,i} &= 1 + (1-\theta)\tau(a_i^k q_{i,i} + f(k\tau, v_i^k)).\end{aligned} \tag{18}$$

If $\theta = 1$, then we obtain the unit matrix, which matrix is trivially sign-stable. Let us suppose that $\theta \neq 1$. The positivity of the sub- and superdiagonal elements is guaranteed by the condition (C1). Moreover, in view of conditions (C2a) and (C2c), we have for the indices $i = 2, \ldots, n-1$ that

$$\begin{aligned}(\mathbf{A}_2)_{i,i}^{(k)} - (\mathbf{A}_2)_{i,i-1}^{(k)} - (\mathbf{A}_2)_{i,i+1}^{(k)} &= 1 + (1-\theta)\tau(a_i^k(q_{i,i} - q_{i,i-1} - q_{i,i+1}) + f(k\tau, v_i^k)) \\ &\geq 1 + (1-\theta)\tau\left(a_i^k \frac{-4}{h_{\min}^2} - M_0\right) \\ &\geq 1 + (1-\theta)\frac{1}{(1-\theta)(4\|a\|/h_{\min}^2 + M_0)}\left(a_i^k \frac{-4}{h_{\min}^2} - M_0\right) \geq 0.\end{aligned}$$

If $i = 1$ (the case $i = n$ can be handled similarly) we obtain the sum

$$\begin{aligned}(\mathbf{A}_2)_{i,i}^{(k)} - (\mathbf{A}_2)_{i,i+1}^{(k)} &= 1 + (1-\theta)\tau(a_i^k(q_{i,i} - q_{i,i+1}) - b_i^k d_{i,i+1} + f(k\tau, v_i^k)) \\ &\geq 1 + (1-\theta)\tau\left(a_0 \frac{-3}{h_{\min}^2} - \|b\| \frac{1}{2h_{\min}} - M_0\right) \\ &\geq 1 - (1-\theta)\frac{1}{(1-\theta)(4\|a\|/h_{\min}^2 + M_0)}\left(\frac{4a_0}{h_{\min}^2} + M_0\right) \geq 0.\end{aligned}$$

The above estimates implies the positivity of the diagonal and the weak diagonal dominance of the matrices. Thus, based on Theorem 1, the matrices $\mathbf{A}_2^{(k)}$ are sign-stable. This completes the proof of the theorem.

Remark 2. For uniform spatial meshes and for problems with $a \equiv 1$, $b \equiv 0$ and $M_0 = 0$, the conditions of Theorem 3 reduce to the condition (6). For fixed number of spatial grid points, the uniform mesh produces the largest possible time step.

Remark 3. Condition (C1) means for uniform meshes that the spatial discretization step must be sufficiently small. For nonuniform meshes we obtain additionally that the maximal step size cannot be arbitrarily large compared to the minimal step size.

Remark 4. We stress that Theorem 3 provides only an a priori sufficient condition for the numerical sign-stability.

4 Numerical Verification

Let us consider problem (7) with the choice $a \equiv 1$, $b \equiv 1$, $f(t, v) = \sin v$ and the initial function is chosen as $v_0(x) = x(x-1)(x-1/2)$, $x \in [0, 1]$. Then the coefficients of the problem trivially satisfies conditions i)-iii). Moreover, we have $a_0 = 1$, $\|a\| = 1$, $\|b\| = 1$ and $M_0 = 1$. The initial function has only one sign-change, namely at $x = 1/2$.

In the numerical solution we use the Crank-Nicolson method, that is we set $\theta = 1/2$ and in the uniform space partition we apply the step size $h = 1/13$. Applying the time step $\tau = 1/10$, the solution at the 19th time level has five sign-changes (Figure 1, left panel). Thus, the numerical solution is not sign-stable. Indeed the sufficient condition in Theorem 3 results in the upper bound

$$\tau \leq \frac{2h^2}{4+h^2}$$

for the time step. With $h = 1/13$, this yields the condition $\tau \leq 0.003$. Applying this time step, we get a sign-stable numerical solution (Figure 1, right panel).

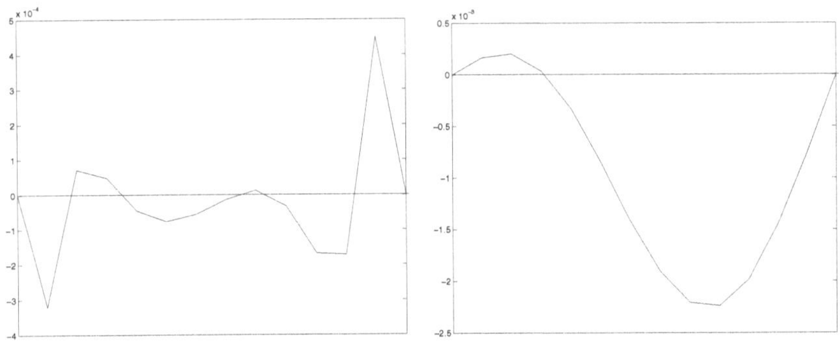

Fig. 1. Left: Numerical solution at the 19th time level with five sign-changes ($\tau = 1/10$). Right: the numerical solution at the 32nd time level with one sign-change ($\tau = 3/1000$).

Acknowledgement

The author of the paper was supported by the National Scientific Research Fund (OTKA) N. T043765, K61800. Moreover, the author is a grantee of the János Bolyai Research Scholarship.

References

1. Berman, A., Plemmons, R.J.: Nonnegative Matrices in the Mathematical Sciences. Academic Press, New York (1979)
2. Faragó, I.: Nonnegativity of the Difference Schemes. Pure Math. Appl. 6, 38–56 (1996)
3. Faragó, I., Horváth, R.: On the Connections Between the Qualitative Properties of the Numerical Solutions of Linear Parabolic Problems. SIAM J. Sci. Comput. 28, 2316–2336 (2006)
4. Faragó, I., Horváth, R., Korotov, S.: Discrete Maximum Principle for Linear Parabolic Problems Solved on Hybrid Meshes. Appl. Num. Math. 53(2-4), 249–264 (2005)
5. Gantmacher, F.R., Krein, M.G.: Oszillationsmatrizen, Oszillationskerne und kleine Schwingungen mechanischer Systeme. Akademie-Verlag, Berlin (1960)
6. Glashoff, K., Kreth, H.: Vorzeichenstabile Differenzenverfaren für parabolische Anfangsrandwertaufgaben. Numerische Mathematik 35, 343–354 (1980)
7. Horváth, R.: Maximum Norm Contractivity in the Numerical Solution of the One-Dimensional Heat Equation. Appl. Num. Math. 31, 451–462 (1999)

8. Horváth, R.: On the Sign-Stability of the Numerical Solution of the Heat Equation. Pure Math. Appl. 11, 281–291 (2000)
9. Horváth, R.: On the Sign-Stability of the Finite Difference Solutions of One-Dimensional Parabolic Problems. In: Boyanov, T., Dimova, S., Georgiev, K., Nikolov, G. (eds.) NMA 2006. LNCS, vol. 4310, pp. 458–465. Springer, Heidelberg (2007)
10. Horváth, R.: On the Sign-Stability of Numerical Solutions of One-Dimensional Parabolic Problems. Appl. Math. Model. 32(8), 1570–1578 (2008)
11. Pólya, G.: Qualitatives über Wärmeausgleich. Ztschr. f. angew. Math. und Mech. 13, 125–128 (1933)
12. Schoenberg, I.J.: Über variationsvermindernde lineare Transformationen. Math. Z. 32 (1930)
13. Sturm, Ch.: Journal de Math. pures et appliquées. 1, 373–444 (1836)
14. Tabata, M.: A Finite Difference Approach to the Number of Peaks of Solutions for Semilinear Parabolic Problems. J. Math. Soc. Japan 32, 171–192 (1980)

Numerical Solution of the Discrete-Time Coupled Algebraic Riccati Equations

Ivan Ganchev Ivanov

Faculty of Economics and Business Administration,
Sofia University "St. Kliment Ohridski",
Sofia 1113, Bulgaria
i_ivanov@feb.uni-sofia.bg

Abstract. We consider the numerical solution of a set of discrete-time coupled algebraic Riccati equations that arise in quadratic optimal control. Several iterations for computing a symmetric solution of this system are investigated and compared. New iterations are based on the properties of a Stein equation. It is necessary to solve a Stein equation at each step of considered algorithms. We will compare the corresponding solvers in regard of accuracy, number of iterations and time of executing. Several sets of test examples are used to demonstrate the performance.

1 Introduction

Systems of coupled Riccati equations arise in several classes of optimal control problems such as jump linear control systems in the literature. We will study the numerical solution of the following set of coupled algebraic Riccati equations $(k = 1, \ldots, N)$ with unknown matrices $X_1, \ldots, X_N$:

$$\begin{aligned} X_k = \mathcal{R}_k(\mathbf{X}) \ :=\ & A_k^T \mathcal{E}_k(\mathbf{X}) A_k + C_k^T C_k \\ & - A_k^T \mathcal{E}_k(\mathbf{X}) B_k \left(D_k^T D_k + B_k^T \mathcal{E}_k(\mathbf{X}) B_k \right)^{-1} B_k^T \mathcal{E}_k(\mathbf{X}) A_k \,, \end{aligned} \tag{1}$$

where $\mathcal{E}(\mathbf{X}) = (\mathcal{E}_1(\mathbf{X}), \ldots, \mathcal{E}_N(\mathbf{X}))$ with $\mathbf{X} = (X_1, \ldots, X_N)$ and

$$\mathcal{E}_k(\mathbf{X}) = \sum_{j=1}^{N} \lambda_{kj} X_j \,, \qquad X_j \ \text{is an}\ m \times m \ \text{symmetric matrix}.$$

Here $\lambda_{kj} \geq 0$ for all k, j and $\Lambda = (\lambda_{kj})$. We will say that $\mathbf{X}$ will be a hermitian solution for (1) if $\left(D_k^T D_k + B_k^T \mathcal{E}_k(\mathbf{X}) B_k \right)$ is invertible and $\mathbf{X}$ satisfies the set of equations above and $X_k = X_k^T$ for all k.

For equations (1) matrices A_k, B_k, D_k and C_k are given ones of sizes $m \times m, m \times p, s \times p$ and $r \times m$, respectively, for $k = 1, \ldots, N$.

Some iterations and numerical algorithms based on Riccati equations [9], Lyapunov equations [4,5], and convex programming [3,8] for analyzing Markovian jump linear systems have been proposed in the literature. A convex optimization problem for finding the optimal solution $\mathbf{P}$ to (1) and an iterative algorithm for

S. Margenov, L.G. Vulkov, and J. Waśniewski (Eds.): NAA 2008, LNCS 5434, pp. 314–321, 2009.

deriving a stabilizing solution, if it exists, of H_∞-control of discrete-time Markovian jump linear systems are obtained in [3]. The temporal difference method for receiving the optimal solution to (1) is analyzed in [1]. Moreover, all these approaches for solving a set of equations (1) are derived in terms of the concept of mean square stabilizability and a convex set not being empty.

In [3] Costa and Marques have considered a set of recurrence equations based on the linear matrix equations

$$Y_k - \Phi_k^T\, \mathcal{E}_k(\mathbf{Y})\, \Phi_k = S_k\,, \qquad k = 1, \ldots, N$$

for solving a set of equations (1). However, the solution of the last system needs many computational work.

We consider several iterative procedures to solve a set of Riccati equations (1) and compare these new iterations based on the solution a system of matrix equations of the type $Z_k - A_k^T Z_k A_k = C_k$. For this purpose we make some matrix manipulations on system (1) to derive new recurrence equations.

The notation $\mathcal{H}^m$ stands for the linear space of symmetric matrices of size m over the field of real numbers. For any $X, Y \in \mathcal{H}^m$, we write $X > Y$ or $X \geq Y$ if $X - Y$ is positive definite or $X - Y$ is positive semidefinite. The spectrum for any square real $m \times m$ matrix A will be denoted by $\sigma(A)$. A matrix A is said to be d-stable if the all eigenvalues of A lie in the open unit disk, i.e. $|\lambda_s(A)| < 1$ for $s = 1, \ldots, m$, and almost d-stable if $|\lambda_s(A)| \leq 1$ for $s = 1, \ldots, m$. The notations $\mathbf{X} \in \mathcal{H}^m$ and the inequality $\mathbf{Y} \geq \mathbf{Z}$ mean that for $k = 1, \ldots, N$, $X_k \in \mathcal{H}^m$ and $Y_k \geq Z_k$, respectively. The linear space $\mathcal{H}^m$ is a Hilbert space with the Frobenius inner product $< X, Y >= trace(XY)$. For a linear operator $\mathcal{L}$ on $\mathcal{H}^m$, let $r_\sigma(\mathcal{L})$ denote the spectral radius.

2 Iterative Algorithms

The following theorem can be found in [3].

Theorem 1. *[3, Theorem 1] Suppose that* $(\mathbf{A}, \mathbf{B})$ *is mean square stabilizable and there exists* $\hat{\mathbf{X}} \in \mathcal{H}^m$ *such that* $\mathcal{R}_k(\hat{\mathbf{X}}) \geq \hat{X}_k$. *Then for* $l = 0, 1, 2, \ldots$ *there exists* $\mathbf{X}^{(l)} = (X_1^{(l)}, \ldots, X_N^{(l)})$ *which satisfies the following properties:*

a) $\mathbf{X}^{(0)} \geq \mathbf{X}^{(1)} \geq \ldots \geq \mathbf{X}^{(l)} \geq \hat{\mathbf{X}}$;

b) $r_\sigma(\mathcal{L}^{(l)}) < 1$, *where* $\mathcal{L}^{(l)}(.) = (\mathcal{L}_1^{(l)}(.), \ldots, \mathcal{L}_N^{(l)}(.))$ *and for* $k = 1, \ldots, N$:

$$\mathcal{L}_k^{(l)}(.) = A_k^{(l)T}\, \mathcal{E}_k(.)\, A_k^{(l)}\,, \qquad A_k^{(l)} = A_k + B_k \mathcal{F}_k^{(l)}\,, \text{ and}$$

$$\mathcal{F}_k^{(l)} = -(R_k + B_k^T \mathcal{E}_k(\mathbf{X}^{(l-1)}) B_k)^{-1} (B_k^T \mathcal{E}_k(\mathbf{X}^{(l-1)}) A_k) \text{ for } l = 1, 2, \ldots$$

c) $X_k^{(l)} - A_k^{(l)T}\, \mathcal{E}_k(\mathbf{X}^{(l)})\, A_k^{(l)} = T_{k,\mathbf{X}^{(l-1)}}$ *for* $k = 1, \ldots, N$.

Moreover there exists a solution $\mathbf{X}^+ = (X_1^+, \ldots, X_N^+)$ *to (1) such that* $\mathbf{X}^+ \geq \hat{\mathbf{X}}$ *for any* $\hat{\mathbf{X}}$ *and* $\mathbf{X}^{(l)} \to \mathbf{X}^+$ *as* $l \to \infty$. *Furthermore* $r_\sigma(\mathcal{L}^+) \leq 1$, *where* $\mathcal{L}^+(.) = (\mathcal{L}^+{}_1(.), \ldots, \mathcal{L}^+{}_N(.))$ *is defined as* $\mathcal{L}^+{}_k(.) = A_k^{+T}\, \mathcal{E}_k(.)\, A_k^+$ *for* $k = 1, \ldots, N$ *and*

$$A_k^+ = A_k + B_k \mathcal{F}_k^+\,, \qquad \mathcal{F}_k^+ = -(R_k + B_k^T \mathcal{E}_k(\mathbf{X}^+) B_k)^{-1} (B_k^T \mathcal{E}_k(\mathbf{X}^+) A_k)\,.$$

We introduce the following algorithm realized iteration c) defined in Theorem 1:

Algorithm CM. (It is considered in detail in [6]).

1. We compute the initial point $\mathbf{X}^{(0)}$*:*

1a. We choose a matrix $\mathbf{G}^{(0)}$ *such that* $r_\sigma(\mathcal{L}^{(0)}) < 1$*, where*

$$\mathcal{L}^{(0)}(.) = (\mathcal{L}_1^{(0)}(.), \ldots, \mathcal{L}_N^{(0)}(.)) \text{ and } \mathcal{L}_k^{(0)}(.) = A_k^{(0)T}\, \mathcal{E}_k(.)\, A_k^{(0)}$$

with $A_k^{(0)} = (A_k + B_k\, G_k^{(0)})$ *for* $k = 1, \ldots, N$*;*

1b. We define the matrices $X_k^{(0)}$ *as a solution of the following system*

$$X_k^{(0)} - A_k^{(0)T}\, \mathcal{E}_k(\mathbf{X}^{(0)})\, A_k^{(0)} = Q_k + G_k^{(0)T} R_k G_k^{(0)} \quad \text{for } k = 1, \ldots, N\,.$$

2. For $l = 1, 2, \ldots$

2a. We compute $\mathcal{F}_k^{(l)}, A_k^{(l)}$ *as defined in Theorem 1 and* $T_{k,\mathbf{X}^{(l-1)}} = Q_k + \mathcal{F}_k^{(l)T} D_k^T D_k \mathcal{F}_k^{(l)}$*.*

2b. We define

$$\mathcal{X}^{(l)} = \operatorname{diag}(X_1^{(l)}, \ldots, X_N^{(l)}), \qquad \Gamma^{(l)} = \operatorname{diag}(A_1^{(l)}, \ldots, A_N^{(l)}),$$

$$\mathcal{S}^{(l-1)} = \operatorname{diag}(T_{1,\mathbf{X}^{(l-1)}}, \ldots, T_{N,\mathbf{X}^{(l-1)}})\,.$$

2c. We solve the linear system $\hat{W}x = s$ *with* $s = vec(\mathcal{S}^{(l-1)})$ *and*

$$\hat{W} = I_{Nm} \otimes I_{Nm} - \sum_{k=1}^{N} \sum_{j=k}^{N} (\Gamma^{(l)T}\, W_{kj}) \otimes (\Gamma^{(l)T}\, W_{kj})\,.$$

Note that the matrices W_{kj} *do not depend on the iteration number* l *and the condition* $r_\sigma(\mathcal{L}^{(l)}) < 1$ *is equivalent to* $\rho\left(\sum_{k=1}^{N} \sum_{j=k}^{N} (\Gamma^{(l)T}\, W_{kj}) \otimes (\Gamma^{(l)T}\, W_{kj})\right) < 1$.

2d. By the relation $x = vec(\mathcal{X}^{(l)})$ *we obtain* $\mathcal{X}^{(l)}$ *which is a solution to the system of matrix equations* $X_k^{(l)} - A_k^{(l)T}\, \mathcal{E}_k(\mathbf{X}^{(l)})\, A_k^{(l)} = T_{k,\mathbf{X}^{(l-1)}}$ *for* $l = 1, 2, \ldots$*.*

2e. If $\left\| \mathcal{R}_k\left(\mathbf{X}^{(l)}\right) - X_k^{(l)} \right\|_2 \leq \varepsilon$ *for* $\varepsilon > 0$*, then STOP.*

3. $\mathbf{X}^{(l)} \to \tilde{\mathbf{X}}$ *and* $\tilde{\mathbf{X}}$ *is the maximal solution to (1).*

End.

Note that the realization of this algorithm requires the solution the system $\hat{W}x = s$ where $\hat{W}$ is a $Nm \times Nm$ matrix. The solution of this system can be very expensive when Nm is large. On the other hand a matrix equation of the form

$$Z - A^T Z A = C\,, \tag{2}$$

with $m \times m$ unknown matrix Z can be solved via known algorithms in $O(m^3)$ operations when it has a unique solution. This observation lead us to investigate a new iteration where the solution of a linear matrix equation of the form (2) is required in each step.

After some matrix manipulations on system (1) it is received the following iteration:

$$X_k^{(i)} - \lambda_{kk}\,\tilde{A}^T_{k,\mathbf{X}^{(i-1)}}\,X_k^{(i)}\,\tilde{A}_{k,\mathbf{X}^{(i-1)}} = T_{k,\mathbf{X}^{(i-1)}} + \tilde{A}^T_{k,\mathbf{X}^{(i-1)}} \sum_{j=1, j\neq k}^{N} \lambda_{kj}\,X_j^{(i-1)}\,\tilde{A}_{k,\mathbf{X}^{(i-1)}}\,. \tag{3}$$

A variant of the above iteration regarding to $\mathbf{X}^{(i)}$ of carrying new updates would be the following

$$\begin{aligned} X_k^{(i)} - \lambda_{kk}\,\tilde{A}^T_{k,\mathbf{X}^{(i-1)}}\,X_k^{(i)}\,\tilde{A}_{k,\mathbf{X}^{(i-1)}} &= T_{k,\mathbf{X}^{(i-1)}} \\ &+ \tilde{A}^T_{k,\mathbf{X}^{(i-1)}}\mathcal{E}_{k1}(\mathbf{X}^{(i)})\tilde{A}_{k,\mathbf{X}^{(i-1)}} + \tilde{A}^T_{k,\mathbf{X}^{(i-1)}}\mathcal{E}_{k2}(\mathbf{X}^{(i-1)})\tilde{A}_{k,\mathbf{X}^{(i-1)}} \end{aligned} \tag{4}$$

where

$$\tilde{A}_{k,\mathbf{Z}} = A_k + B_k F_{k,\mathbf{Z}} = A_k - B_k \left(R_k + B_k^T \mathcal{E}_k(\mathbf{Z}) B_k\right)^{-1} B_k^T \mathcal{E}_k(\mathbf{Z}) A_k\,,$$

$$T_{k,\mathbf{Z}} = Q_k + F^T_{k,\mathbf{Z}} R_k F_{k,\mathbf{Z}}\,,$$

$$\mathcal{E}_{k1}(\mathbf{X}^{(i)}) = \sum_{j=1}^{k-1} \lambda_{kj}\,X_j^{(i)}, \quad \text{and} \quad \mathcal{E}_{k2}(\mathbf{X}^{(i-1)}) = \sum_{j=k+1}^{N} \lambda_{kj}\,X_j^{(i-1)}\,.$$

The above iterations are studied in detail in [6]. It is proved a sufficient condition for convergence of these iterations.

Further on, we present two iterations proposed and investigated in [7]. Using the fact that $\delta_{kk} < 1$ (see [7]) and notations $\delta_{kp} = \sum_{s\neq k} \lambda_{ks}\mu_{sp}$, $(k,p = 1,\dots,N)\,,(\mu_{kj}) = \Lambda^{-1}$,

$$\begin{cases} \mathbf{Y} = (Y_1,\dots,Y_N)\,, \text{ where } Y_k = \mathcal{E}_k(\mathbf{X})\,, \\ \tilde{A}_k = \sqrt{\frac{\lambda_{kk}}{1-\delta_{kk}}}\,A_k\,, \quad \tilde{C}_k = \sqrt{\frac{\lambda_{kk}}{1-\delta_{kk}}}\,C_k\,, \quad \tilde{S}_{k,Y_k} = \tilde{A}_k^T Y_k B_k\,, \\ (\gamma_{kp})_1^N = \left\{ \begin{array}{l} \gamma_{kk} = 0 \\ \gamma_{kp} = \frac{\delta_{kp}}{1-\delta_{kk}}, \text{ if } k\neq p \end{array} \right\}, \quad \mathcal{G}_k(\mathbf{Y}) = \sum_{j\neq k} \gamma_{kj}\,Y_j\,, \\ K_{k,Y_k} = D_k^T D_k + B_k^T Y_k B_k\,, \quad F_{k,Y_k} = -K^{-1}_{k,Y_k}\,\tilde{S}^T_{k,Y_k}\,. \end{cases}$$

we obtain the set of Riccati equations

$$Y_k = \mathcal{P}_k\left(\mathbf{Y}\right) := \tilde{A}_k^T Y_k \tilde{A}_k + \tilde{C}_k^T \tilde{C}_k - \tilde{S}_{k,Y_k}\,K^{-1}_{k,Y_k}\,\tilde{S}^T_{k,Y_k} + \mathcal{G}_k(\mathbf{Y})\,. \tag{5}$$

By presentation (5) with $F_k^{(i-1)} = F_{k,Y_k^{(i-1)}}$ and $\hat{A}_k^{(i-1)} = \hat{A}_{k,Y_k^{(i-1)}} = \tilde{A}_k + B_k F_k^{(i-1)}$ we derive a recurrence equation as the unknown matrix $Y_k^{(i)}$ is its solution:

$$
\begin{aligned}
Y_k^{(i)} = & \left(\hat{A}_k^{(i-1)}\right)^T Y_k^{(i)} \hat{A}_k^{(i-1)} \\
& + \tilde{C}_k^T \tilde{C}_k + \left(F_k^{(i-1)}\right)^T D_k^T D_k F_k^{(i-1)} + \mathcal{G}_k(\mathbf{Y}^{(i-1)}),
\end{aligned} \tag{6}
$$

for $k = 1, \ldots, N$ and the initial matrices $Y_k^{(0)}$ are properly chosen. Naturally generalization of the last iteration is the following

$$
\begin{aligned}
Y_k^{(i)} = & \left(\hat{A}_k^{(i-1)}\right)^T Y_k^{(i)} \hat{A}_k^{(i-1)} + \tilde{C}_k^T \tilde{C}_k + \left(F_k^{(i-1)}\right)^T D_k^T D_k F_k^{(i-1)} \\
& + \mathcal{G}_{k1}(\mathbf{Y}^{(i)}) + \mathcal{G}_{k2}(\mathbf{Y}^{(i-1)}), \qquad \text{for } k = 1, \ldots, N,
\end{aligned} \tag{7}
$$

where $\mathcal{G}_{k1}(\mathbf{Z}) = \sum_{j=1}^{k-1} \gamma_{kj} Z_j$, and $\mathcal{G}_{k2}(\mathbf{Z}) = \sum_{j=k+1}^{N} \gamma_{kj} Z_j$.

We name iterations (3) and (6) the Stein iterations and corresponding iterations (4) and (7) the accelerated Stein iterations. In these iterations we have to solve a Stein matrix equation at each iterative step.

The equations (5) have the following advantage. For solving this system the standard Newton method can be applied as described in [7]. The Newton iteration is

$$
\mathcal{Y}_{i+1} = (A + BF_{\mathcal{Y}_i})^T \mathcal{Y}_{i+1} (A + BF_{\mathcal{Y}_i}) + C^T C + F_{\mathcal{Y}_i}^T D^T D F_{\mathcal{Y}_i} + \Pi_1(\mathcal{Y}_{i+1}) \tag{8}
$$

where

$$
\begin{cases}
\mathcal{Y} = \operatorname{diag}[Y_1, \ldots, Y_N], \quad (\mathcal{Y} \neq \mathbf{Y}), \quad A = \operatorname{diag}\left[\tilde{A}_1, \ldots, \tilde{A}_N\right], \\
C = \left[\tilde{C}_1, \ldots, \tilde{C}_N\right], \quad B = \begin{bmatrix} B_1 \\ \vdots \\ B_N \end{bmatrix}, \quad D = \begin{bmatrix} D_1 \\ \vdots \\ D_N \end{bmatrix},
\end{cases}
$$

and

$$
\begin{cases}
F_{\mathcal{Y}_i} = -(D^T D + B^T \mathcal{Y}_i B)^{-1} B^T \mathcal{Y}_i A, \\
\Pi_1(\mathcal{Y}) = \sum_{k,j=1, j\neq k}^{N} P_{kj}\left(\sqrt{\gamma_{kj}}\right) \mathcal{Y} P_{kj}^T \left(\sqrt{\gamma_{kj}}\right), \\
P_{kj}(\alpha) = (p_{rs})_{r,s=1}^{N} = \begin{cases} p_{rs} = 0, r \neq k, s \neq j \\ p_{kj} = \alpha I \end{cases}
\end{cases}
$$

for $i = 0, 1, 2, \ldots$. The matrix $\mathcal{Y}_i$ is a solution to linear matrix equation (8) on the each iteration step.

3 Numerical Experiments

Our aim is to study considered iterations for finding a symmetric solution of (1). We will carry out experiments for numerical solving a set of coupled algebraic Riccati equations (1) with introduced iterations: Stein iterations (StI) (3) and (6), accelerated Stein iterations (AStI) (4) and (7), and **Algorithm CM** and Newton's iteration (8) for different initial points.

Solutions of iterations (3) and (4) can be found in terms of the solutions of N algebraic Stein equations at each step. For this purpose the MATLAB procedure *dlyap* is applied and the flops are $N\,\frac{27}{2}m^3$ per one iteration. Moreover, we have to solve an $Nm \times Nm$ linear system at each step of **Algorithm CM**.

Our experiments are executed in MATLAB on an 1,81GHz PENTIUM(R) Dual CPU computer. We denote *tol*- a small positive real number denoting the accuracy of computation; $Error_s = \max_{k=1,\dots,N} \left\| X_k^{(s)} - \mathcal{R}_k\left(\mathbf{X}^{(s)}\right)\right\|_2$ or $Error_s = \max_{k=1,\dots,N} \left\| Y_k^{(s)} - \mathcal{P}_k\left(\mathbf{Y}^{(s)}\right)\right\|_2$; *It*- number of iterations for which the inequality $Error_{It} \leq tol$ holds. The last inequality is used as a practical stopping criterion. For all experiments we take $N = 3$ and

$$(\lambda_{ij})_{i,j=1}^{3} = \begin{pmatrix} 0.67 & 0.17 & 0.16 \\ 0.3 & 0.47 & 0.23 \\ 0.26 & 0.10 & 0.64 \end{pmatrix}.$$

We have executed hundred examples of size $m = 3, 6, \dots, 10$. We compare all iterations introducing the following parameters: m It - the biggest number of iterations; av It - the average number of iterations. For initial points we choose $X_1^{(0)} = X_2^{(0)} = X_3^{(0)} = 20\,I$ for iterations (3) and (4). The matrices $Y_1^{(0)} = Y_2^{(0)} = Y_3^{(0)} = 10\,I$ are initial ones for iterations (6) and (7). Newton's method (8) starts with $Y_1^{(0)} = Y_2^{(0)} = Y_3^{(0)} = 40\,I$.

Test 1. For the first set of examples A_k, B_k, Q_k and $R_k (k = 1, 2, 3)$ were constructed as follows using the MATLAB notation:

$$\begin{array}{lll} A_1 = randn(m,m)/2; & A_2 = randn(m,m)/2; & A_3 = randn(m,m)/2; \\ B_1 = B_2 = B_3 = randn(m,m)/m; & & \\ Q_1 = 2 * eye(m,m); & Q_2 = 0.25 * eye(m,m); & Q_3 = Q_1; \\ R_1 = 0.8 * eye(m,m); & R_2 = 0.75 * eye(m,m); & R_3 = 0.5 * eye(m,m). \end{array}$$

We present the results from experiments in Tables 1 and 2.

Test 2. We generate a different set of test matrices. They were constructed as follows:

$$\begin{array}{lll} A_1 = randn(m,m)/4; & A_2 = randn(m,m)/4; & A_3 = randn(m,m)/4; \\ B_1 = B_2 = B_3 = randn(m,m)/2; & & \\ R_1 = 0.1 * eye(m,m); & R_2 = 0.75 * eye(m,m); & R_3 = 0.5 * eye(m,m); \\ Q_1 = 10 * eye(m,m); & Q_2 = 0.25 * eye(m,m); & Q_3 = Q_1. \end{array}$$

We present the results from experiments in Tables 3 and 4.

Tables 1 and 3 report the maximal number of iterations m It of each size for all examples needed for achieving the stoping criterion. Moreover, an average number av It of iterations is determined.

Table 1. Test 1: 100 runs, $tol = 10^{-12}$

	StI (3)		AStI (4)		Algor CM		StI (6)		AStI (7)		Newton (8)	
m	m It	av It	m It	av It	m It	av It	m It	av It	m It	av It	m It	av It
3	26	15.5	15	10.0	17	7.7	75	56.1	44	32.4	9	5.3
4	29	17.7	17	11.1	11	7.9	79	59.2	45	34.2	6	5.4
5	42	21.2	23	13.0	17	9.2	120	64.9	68	37.3	9	5.8
6	55	23.7	30	14.4	16	9.6	118	68.2	60	38.9	7	5.9
7	46	25.4	36	15.7	21	9.6	120	72.1	62	41.0	11	6.2
8	42	27.2	25	16.3	17	10.0	143	75.1	59	42.9	17	6.3
9	44	30.0	27	18.1	16	10.7	163	82.3	73	47.0	7	6.1
10	44	30.7	31	18.8	18	10.5	130	82.1	86	48.1	17	6.6

Table 2. CPU time in seconds for all iterations in Test 1

Runs	m	StI (3)	AStI (4)	Algor CM	StI (6)	AStI (7)	Newton (8)
$tol = 10^{-12}$							
10	10	0.8	0.54	56	2.45	1.32	26
10	20	1.98	1.3	2556	10.43	3.93	1666
$tol = 10^{-11}$							
100	20	19.25	12.82	slowly	78	41	slowly

Table 3. Test 2: 100 runs, $tol = 10^{-12}$

	StI (3)		AStI (4)		Algor CM		StI (6)		AStI (7)		Newton (8)	
m	m It	av It	m It	av It	m It	av It	m It	av It	m It	av It	m It	av It
3	13	6.5	8	5.3	7	5.1	50	43.8	29	25.7	5	4.0
4	10	6.7	7	5.5	8	5.5	47	44.2	27	25.8	5	4.1
5	11	7.1	7	5.7	10	5.7	47	44.5	28	26.1	5	4.2
6	10	7.3	7	5.8	8	5.9	48	44.8	28	26.3	6	4.3
7	10	7.2	7	5.8	8	6.0	47	45.0	28	26.4	5	4.4
8	11	7.4	8	5.9	9	6.1	48	45.1	28	26.6	5	4.4
9	12	7.9	9	6.2	8	6.3	49	45.6	29	26.8	5	4.5
10	17	8.0	11	6.3	9	6.4	56	45.7	33	26.9	5	4.5

Table 4. CPU time in seconds for all iterations in Test 2, $tol = 10^{-11}$

Runs	m	StI (3)	AStI (4)	Algor CM	StI (6)	AStI (7)	Newton (8)
10	10	1.04	0.38	52	1.71	1	33
10	20	1.5	1.3	2010	7.35	3.87	1315
20	20	1.05	0.87	very slowly	5.4	3.1	very slowly
50	20	2.65	2.18	very slowly	13	7.9	very slowly

4 Conclusion

We have made numerical experiments for computing a Hermitian solution to a system of stochastic Riccati equations with several iterative methods. Our numerical experiments confirm that the iterations based on the solution a Stein equation at each step are very effective for solving (1). We have compared the results from these experiments in regard of number of iterations and time for executing. All iterations achieve the same accuracy for different number of iterations and different time. All examples retain the average number of iterations av independently of the value of m. Newton's method has quadratic convergence, it executes the least of all number of iterations, it achieves the best accuracy and it is not the fastest because it performs huge computational work per one iteration. It is assumed for the other methods that they have the linear rate of convergence, however that is not proved. The **Algorithm CM** carries out many flops per one iteration and thus it is included for completeness and illustration of the results from the experiments. Examples show that iterations (4) and (7) are faster than iterations (3) and (6), however we do not have a theoretical proof for this fact.

References

1. Costa, O.L.V., Aya, J.C.C.: Temporal Difference Methods for the Maximal Solution of Discrete-Time Coupled Algebraic Ricacti Equations. Journal of Optimization Theory and Application 109, 289–309 (2001)
2. Costa, O.L.V., Fragoso, M.D.: Stability Results for Discrete-Time Linear Systems with markovian jumping parameters. J. Math. Analysis and Applic. 179, 154–178 (1993)
3. Costa, O.L.V., Marques, R.P.: Maximal and Stabilizing Hermitian Solutions for Discrete-Time Coupled Algebraic Ricacti Equations. Mathematics of Control, Signals and Systems 12, 167–195 (1999)
4. Gajic, Z., Borno, I.: Lyapunov iterations for optimal control of jump linear systems at steady state. IEEE Transaction on Authomatic Control 40, 1971–1975 (1995)
5. Gajic, Z., Losada, R.: Monotonicity of algebraic Lyapunov iterations for optimal control of jump parameter linear systems. Systems & Control Letters 41, 175–181 (2000)
6. Ivanov, I.: Stein Iterations for the Coupled Discrete-Time Riccati Equations (submitted)
7. Ivanov, I.: A method to solve the discrete-time coupled algebraic Riccati equations. Applied Mathematics and Computation (accepted)
8. Rami, M., Ghaoui, L.: LMI Optimization for Nonstandard Riccati Equations Arising in Stochastic Control. IEEE Transactions on Automatic Control 41, 1666–1671 (1996)
9. do Val, J.B., Geromel, J.C., Costa, O.L.V.: Solutions for the linear quadratic control problem of Markov jump linear systems. J. Optimization Theory and Applications 103(2), 283–311 (1999)

The Weierstrass Canonical Form of a Regular Matrix Pencil: Numerical Issues and Computational Techniques

Grigorios Kalogeropoulos, Marilena Mitrouli, Athanasios Pantelous, and Dimitrios Triantafyllou

Department of Mathematics, University of Athens, Panepistimiopolis 15784, Athens, Greece

Abstract. In the present paper, we study the derivation of the Weierstrass Canonical Form (WCF) of a regular matrix pencil. In order to compute the WCF, we use two important computational tools: a) the QZ algorithm to specify the required root range of the pencil and b) the updating technique to compute the index of annihilation. The proposed updating technique takes advantages of the already computed rank of the sequences of matrices that appears during our procedure reducing significantly the required floating-point operations.

The algorithm is implemented in a numerical stable manner, giving efficient results. Error analysis and the required complexity of the algorithm are included.

1 Introduction - Preliminary Results

Linear generalized (or descriptor) differential (difference) systems of type

$$S(F,G) : F\dot{\underline{x}}(t) = G\underline{x}(t) \ (Fx_{k+1} = G\underline{x}_k), F, G \in C^{n\times n} \tag{1}$$

with $detF = 0$, where $\underline{x} \in C^n$ and $\underline{x}_0$ an initial value. Such systems are required in modelling of many physical, electrical, mechanical, as well as, actuarial and financial problems. For instance, in Economics, the Leontief multi input-multi output descriptor (continuous or discrete) model is one of the most famous, see [9]. Recently, a generalized dynamic system for modelling actuarial pricing-claim processes is also derived to be descriptor whenever there is a further hope of success, an effective operation on the market, a further achievement of maximum profits etc, see for further details [6]. Systems of type (1) are related to matrix pencil theory, since the algebraic geometric, and dynamic properties stem from the structure by the associated pencil $sF - G$.

Definition 1. *Given $F, G \in C^{m\times n}$ and an indeterminate s, the matrix pencil $sF - G$ is called regular when $m = n$ and $det(sF - G) \neq 0$. In any other case, the pencil will be called singular.*

The pencil $sF - G$ is said to be strictly equivalent to the pencil $s\widetilde{F} - \widetilde{G}$ if and only if there exist $P \in C^{n\times n}$ and $Q \in C^{n\times n}$ such as $P(sF - G)Q = s\widetilde{F} - \widetilde{G}$, where $detP$, $detQ \neq 0$.

S. Margenov, L.G. Vulkov, and J. Waśniewski (Eds.): NAA 2008, LNCS 5434, pp. 322–329, 2009.

The class of strict equivalent matrix pencils is characterized by a uniquely defined element, known as a complex Weierstrass canonical form, $sF_w - Q_w$, see [3], specified by the complete set of invariants.

This is the set of elementary divisors (e.d.) obtained by factorizing the invariant polynomials $f_i(s, \hat{s})$ into powers of homogeneous polynomials irreducible over C.

In the case where $sF - G$ is a regular pencil and $detF = 0$, we have elementary divisors of the following type:

(a) e.d. of the type $(s-a)^p$ are called finite elementary divisors (f.e.d.).
(b) e.d. of the type $\hat{s}^q$ are called infinite elementary divisors (i.e.d).

Then, the complex Weierstrass form $sF_w - Q_w$ of the regular pencil is defined by

$$sF_w - Q_w = block\, diag\{sI_p - J_p, sH_q - I_q\} \tag{2}$$

where the first normal Jordan type block $sI_p - J_p$ is uniquely defined by the set of f.e.d.

$$(s-\alpha_1)^{p_1}, \ldots, (s-\alpha_\nu)^{p_\nu}, \sum_{j=1}^{\nu} p_j = p \tag{3}$$

of $sF - G$ and has the form

$$sI_p - J_p = block\, diag\{sI_{p_1}(\alpha_1), \ldots, sI_{p_\nu}(\alpha_\nu)\}. \tag{4}$$

And also the q blocks of the second uniquely defined block correspond to the i.e.d.

$$(\hat{s})^{q_1}, \ldots, (\hat{s})^{q_\sigma}, \sum_{j=1}^{\sigma} q_j = q \tag{5}$$

of $sF - G$ has the form

$$sH_q - I_q = block\, diag\{sH_{q_1} - I_{q_1}, \ldots, sH_{q_\sigma} - I_{q_\sigma}\}. \tag{6}$$

Thus the H_q is a nilpotent matrix of index $\widetilde{q} = max\{q_j : j = 1, 2, \ldots, \sigma\}$, where

$$H_q^{\widetilde{q}} = 0 \tag{7}$$

$I_{p_j}, J_{p_j}(\alpha_j), H_{q_j}$ are the matrices

$$I_{p_j} = \begin{bmatrix} 1 & 0 & \ldots & 0 \\ 0 & 1 & \ldots & 0 \\ \vdots & \vdots & \ddots & \vdots \\ 0 & 0 & \ldots & 1 \end{bmatrix}, \; J_{p_j}(\alpha_j) = \begin{bmatrix} \alpha_j & 1 & 0 & \ldots & 0 \\ 0 & \alpha_j & 1 & \ldots & 0 \\ \vdots & \vdots & \ddots & \ddots & \vdots \\ 0 & 0 & & \alpha_j & 1 \\ 0 & 0 & \ldots & 0 & \alpha_j \end{bmatrix} \in C^{p_j \times p_j},$$

$$H_{q_j} = \begin{bmatrix} 0 & 1 & 0 & \ldots & 0 \\ 0 & 0 & 1 & \ldots & 0 \\ \vdots & \vdots & \ddots & \ddots & \vdots \\ 0 & 0 & \ldots & 0 & 1 \\ 0 & 0 & \ldots & 0 & 0 \end{bmatrix} \in R^{q_j \times q_j}. \tag{8}$$

Now, the following Theorem is derived.

Theorem 1. [3] *From the regularity of $sF - G$, there exist nonsingular $C^{n\times n}$ matrices P and Q such that*

$$PFQ = F_w = \begin{bmatrix} I_p & 0_{p,q} \\ 0_{q,p} & H_q \end{bmatrix}, \tag{9}$$

$$PGQ = G_w = \begin{bmatrix} J_p & 0_{p,q} \\ 0_{q,p} & I_q \end{bmatrix}, \tag{10}$$

where I_p, J_p, H_q are given by (8).

Consequently, the original pencil $sF - G$ has been transferred to a canonical form, $sF_w - G_w$. Thus, system (1) is divided into two subsystems

$$\dot{\underline{x}}_p(t) = J_p(\alpha)\underline{x}_p(t)\ (\underline{x}_{p,k+1} = J_p(\alpha)\underline{x}_{p,k}), \tag{11}$$

$$H_q\dot{\underline{x}}_q(t) = \underline{x}_q(t)\ (H_q\underline{x}_{q,k+1} = \underline{x}_{q,k}). \tag{12}$$

In this paper, our main aim is to propose a new more efficient and computationally more stable methodology for the WCF of a regular matrix pencil with known set of root range, i.e. for a fixed set of roots of $det(sF - G)$.

Following [4], for every $(s - a_j)^{p_j}$ of $sF - G$, there exists a maximal length, linearly independent chain of vectors $\{\underline{\nu}_1, \underline{\nu}_2, \dots, \underline{\nu}_{p_j}\}$ such that

$$G\underline{\nu}_i = \alpha_j F\underline{\nu}_{i-1}, \tag{13}$$

for $i = 1, 2, \dots, p_j$ and $j = 1, 2, \dots, \nu$, $\underline{\nu}_0 = \underline{0}$.

Otherwise, when we have i.e.d., i.e. $\alpha = \infty$, expression (13) is reduced to

$$F\underline{\nu}_i = G\underline{\nu}_{i-1}, \tag{14}$$

for $i = 1, 2, \dots, q_{j-1}$ and $j = 1, 2, \dots, \sigma$, $\underline{\nu}_0 = \underline{0}$. Note that $\{\underline{\nu}_1, \underline{\nu}_2, \dots, \underline{\nu}_{q_j}\}$ is the maximal length, linearly independent chain of vectors for each of the i. e. d.. The expressions (13) and (14) may be rewritten in a matrix form and this leads to the definition of the following sequence of matrices.

a) For expression (13)

$$P^1_{\alpha_j}(F, G) = G - a_j F \in C^{n\times n},$$

$$P^2_{\alpha_j}(F, G) = \begin{bmatrix} G - \alpha_j F & 0 \\ -F & G - \alpha_j F \end{bmatrix} \in C^{2n\times 2n} \dots,$$

$$P^i_{\alpha_j}(F, G) = \begin{bmatrix} G - \alpha_j F & 0 & \dots & 0 & 0 \\ -F & G - \alpha_j F & \dots & 0 & 0 \\ \vdots & \vdots & \ddots & \vdots & \vdots \\ 0 & 0 & \dots & G - \alpha_j F & 0 \\ 0 & 0 & \dots & -F & G - \alpha_j F \end{bmatrix} \in C^{in\times in},\ for\ i = 1, \dots \tag{15}$$

b) For expression (14)

$$P^1_\infty(F,G) = F \in C^{n\times n},\ P^2_\infty(F,G) = \left[\begin{smallmatrix} F & 0 \\ -G & F \end{smallmatrix}\right] \in C^{2n\times 2n},\ldots,$$

$$P^i_\infty(F,G) \begin{bmatrix} F & 0\ldots & 0 & 0 & \\ -G & F & \ldots & 0 & 0 \\ \vdots & \vdots & \ddots & \vdots & \vdots \\ 0 & 0 & \ldots & F & 0 \\ 0 & 0 & \ldots & -G & F \end{bmatrix} \in C^{in\times in},\ for\ i = 1,\ldots. \quad (16)$$

Obviously, sequences (15) and (16) are called i^{th} order α_j - and ∞-block Toeplitz matrices of the pair (F,G). Furthermore, we denote the rank and the right nullspace of the sequence $P^i_{\alpha_j}(F,G)$, as follows:

$$\rho^i_{\alpha_j} = \rho_{C[\alpha_j]}\{P^i_{\alpha_j}(F,G), |\alpha_j \in C,\ j = 1,\ldots,\nu\}, for\, i = 1,2,\ldots \quad (17)$$

and

$$\mathcal{N}^i_{\alpha_j} = \mathcal{N}_r\{P^i_{\alpha_j}(F,G)|\,\alpha_j \in C,\ j = 1,\ldots,\nu\},\ for\, i = 1,2,\ldots. \quad (18)$$

The sequence

$$\mathcal{J}_{\alpha_j}(F,G) = \{\eta_0 = 0, \eta_i = \eta i - \rho^i_{\alpha_j} = dim\mathcal{N}_r,\ i = 1,2,\ldots\}, \quad (19)$$

is referred to as the α_j-characteristic sequence of matrix pair (F,G), for $j = 1,2,\ldots,\nu$. The properties of $J_{\alpha_j}(F,G)$ have been extensively studied in [7] and they are summarized below.

Proposition 1. *Let $F,G \in C^{n\times n}, \alpha_j \in C$ belong to the root range of the pencil, and $\mathcal{T}_{\alpha_j}(F,G) = \{(d_r,\sigma_r):\ r = 1,2,\ldots,\rho,\ 0 < d_1 < \ldots < d_\rho\}$, where σ_i is the multiplicity of e.d. $(s-\alpha)^d_i$. The sequence $\mathcal{J}_{\alpha_j}(F,G)$ is defined by*

$$\eta_k = \begin{cases} k\sum_{r=1}^{\rho}\sigma_r & ,if\, k < d_1 \\ \sum_{r=1}^{i}\sigma_r d_r + k\sum_{r=i+1}^{\rho}\sigma_r & ,if\, d_i \le k < d_{i+1} \\ k\sum_{r=1}^{\rho}\sigma_r d_r & ,if\, k \ge d_\rho \end{cases} \quad (20)$$

Theorem 2. *Let $F,G \in C^{n\times n}, \alpha_i \in C$ belong to the root range of the pencil, and $\mathcal{T}_{\alpha_j}(F,G) = \{(d_r,\sigma_r):\ r = 1,2,\ldots,\rho,\ 0 < d_1 < \ldots < d_\rho\}$ and $\mathcal{L}_{\alpha_j}(F,G) = \{d_r:\ r = 1,2,\ldots,\rho,\ 0 < d_1 < \ldots < d_\rho\}$.*

The sequence $\mathcal{J}_{\alpha_j}(F,G)$ is non-decreasing and satisfies the following condition

$$\eta_k \ge \frac{\eta_{k-1} + \eta_{k+1}}{2},\ for\, k = 1,2,\ldots \quad (21)$$

In particular, we have that

i) strict inequality holds if and only if $k = d_r \in \mathcal{L}_{\alpha_j}(F,G)$. In this case $\sigma_r = 2\eta_k - \eta_{k-1} - \eta_{k+1}$.
ii) Equality holds if and only if $k \notin \mathcal{L}_{\alpha_j}(F,G)$.

The sequence $\mathcal{J}_{\alpha_j}(F,G)$ satisfying condition (21) is known as the α_j Piecewise Arithmetic Progression Sequence (PAPS) and its properties define $\mathcal{T}_{\alpha_j}(F,G)$ by developing Ferrers type diagrams, see for further details [7].

Definition 2. *For every $\alpha_i \in C$ in root range of the matrix pair (F,G), the smallest integer τ_{α_j} for which $\eta_{\tau_{\alpha_j}} = \eta_{\tau_{\alpha_j}+1}, \eta_k \in \mathcal{J}_{\alpha_j}(F,G)$ is called the index of annihilation of (F,G) at $s = \alpha_j$.*

Finally, in the present paper, in order to compute the WCF, a new updating technique to compute the index of annihilation is analytically presented in the third section. The proposed updating takes advantages of the already computed rank of the sequence of matrices that appears during our procedure reducing significantly the required floating-point operations.

2 Computational Tools

2.1 The QZ Algorithm

In this section we briefly describe the QZ algorithm, which concerns the generalized eigenvalue problem. Given a matrix pencil $sF - G$ (or equivalently (F,G)), the QZ procedure applies the QR iteration to the matrix $F^{-1}G$ or FG^{-1} without forming the previous product and computes another pair $(F^{'}, G^{'})$ which has the same eigenvalues with the initial pair (F,G), where $F^{'}$ and $G^{'}$ is an upper real Schur and an upper triangular matrix, respectively. Since the eigenvalues of $F^{-1}G$ or FG^{-1} are the same with those of the pencil $sF - G$, by applying the QZ algorithm to (F,G) we compute at the same time and the root range of the pencil, see [2]. An analytical evaluation of the required operations of the QZ algorithm for $n \times n$ matrices F and G is $15n^3$ which results to a complexity of order $O(n^3)$.

2.2 The Updating Technique

In order to compute the WCF of a regular matrix pencil, firstly we need to calculate the rank of matrices, which generally can be of huge dimensions. According to the literature, the most stable and efficient way to compute them is the Singular Value Decomposition method (SVD); see [2] or the Partial Value Decomposition method (PSVD); see [10,11]. However, considering greater dimensions for the matrices, the first step of SVD or PSVD, which is the bidiagonalization, becomes quickly inefficient, since the classical techniques, such as the QR factorization etc, are applied. However, our matrices have a very special structure, i.e. they are Toeplitz matrices, see (15) and (16). In more details, the updating technique takes advantage of the already computed rank of the previous term in order to exploit the next one. This methodology reduces significantly the required complexity of the algorithm and it is as stable as the QR iteration algorithm. The updating algorithm follows.

The Updating Algorithm

Step 1: We set $A = G - \alpha_j F$ and $B = -F$. **Triangularize** matrix A (using LU or QR factorization) zeroing and discarding any linear dependent rows of A: $A \rightarrow A^{(0)}$.

Step 2: **Triangularize** matrix B zeroing any linear dependent rows of B and **update** the entries of the bottom right matrix A: $[B\ A] \rightarrow [B^{(0)} A^{'}]$.

Step 3: **Check** the row linear dependence of $B^{(0)}$ with $A^{(0)}$ by performing QR or LU factorization with partial pivoting [1] to the columns of $\begin{bmatrix} A^{(0)} \\ B^{(0)} \end{bmatrix}$, **zero** and **move** to the end the linear dependent rows resulting to $\begin{bmatrix} A^{(0)} & 0 \\ B^{(1)} & A^{(1)} \end{bmatrix}$.

Step 4: **Compute** the $rank[B^{(1)}\ A^{(1)}] = r_1$.

Step 5: For $k = 1,\ldots,n$

Check the row linear dependence of $B^{(0)}$ with $A^{(k)}$ after the first $k-1$ zero columns by performing QR or LU factorization with partial pivoting, see for instance [1], to the columns of $\begin{bmatrix} A^{(k)} \\ B^{(0)} \end{bmatrix}$,

zero and **move** to the end the linear dependent rows resulting to $\begin{bmatrix} 0\ldots 0\ B^{(k)} & A^{(k)} & 0 \\ 0\ldots 0\ \ 0 & B^{(0)} & A^{'} \end{bmatrix} \rightarrow \begin{bmatrix} 0\ldots 0\ B^{(k)} & A^{(k)} & 0 \\ 0\ldots 0\ \ 0 & B^{(k+1)} & A^{(k+1)} \end{bmatrix}$.

Compute the $rank[B^{(k+1)}\ A^{(k+1)}] = r_{k+1}$.

The total complexity of the updating algorithm is

$$\frac{2n^3}{3} + \sum_{i=1}^{k-1} [n^2(p_i - \frac{1}{6}n) + (\frac{p_i^2}{2}(n - \frac{p_i}{3}))]$$

flops, by using either the LU factorization or the double QR. The check of row linear dependence numerically is highly required. Denote p_i to be the dimension of $B^{(i)}$ without the zero linear dependent rows in the corresponding sub-matrix, $p_i \leq n$.

3 The Weierstrass Canonical Form

If $sF - G$ is a given regular matrix pencil, $F, G \in R^{n\times n}$, then the following algorithm computes its Weierstrass Canonical form:

The Weierstrass Algorithm

Step 1: **Specify** the root range of $sF - G$ using the QZ-algorithm.

Step 2: **If** $\deg(\det(sF - G)){=}m < n$ then $|sF - G|$ has a root at $s = \infty$. **Specify** the degrees q_j of the i.e.d. $\hat{s}^{q_j}$, using the updating algorithm

Step 3: For every $a_i \in C, i = 1, \ldots, r \leq n$
Specify the degrees d_i and the multiplicities σ_i of the f.e.d. $(s - a_j)^{d_i}$

Step4: Form the matrices:
$F_w := diag\{H_{q_j}, \ldots, I_{d_i}, \ldots\},$
$G_w := diag\{I_{q_j}, \ldots, J_{d_i}(a_j), \ldots\}$
Specify the WCF, i.e.
$$sF_w - G_w = diag\{sH_{q_j} - I_{q_j}, \ldots, sI_{d_i} - J_{d_i}(a_i), \ldots\}$$

3.1 Complexity

In this sub-section, the complexity of the whole procedure is provided. Analytically, in Step1, the call of the QZ algorithm requires $15n^3$ flops. Moreover, Step2 and Step3 are called $(\sigma + \nu)$ times. Thus, they demand $O((\sigma + \nu)(\frac{2n^3}{3} + \sum_{i=1}^{k-1}[n^2(p_i - \frac{1}{6}n) + \frac{p_i^2}{2}(n - \frac{p_i}{3})]))$ flops. Step4 is implemented symbolically, so the total complexity of the algorithm does not increase at all.

3.2 Error Analysis

Theoretically, the QZ algorithm is stable. If $\hat{R}$ and $\hat{S}$ are the computed quasi-triangular and triangular outputs respectively, the following equalities are satisfied:
$$Q_0^T(A + E_1)Z_0 = \hat{R}, \quad Q_0^T(B + E_2)Z_0 = \hat{S},$$
where Q_0 and Z_0 are orthogonal with $||E_1|| \cong \mu||A||$ and $||E_2|| \cong \mu||B||$, μ is the machine precision, see [2].

Moreover, in the proposed updating technique, each evaluation of the Gaussian elimination computes the exact factorization of a nearby matrix of $\Gamma_k + E$: $L^k U^k = \Gamma_k + E$, with $||E^k|| \leq p_k^2 \rho u ||\Gamma_k + E^{k-1}||_\infty$, where ρ is the growth factor and u the unit round off error.

4 Conclusions

In this paper, an efficient computational technique for the determination of Weierstrass Canonical Form (WCF) of a regular matrix pencil is presented. In that direction, the QZ algorithm which specifies the required root range of the pencil and the updating technique which computes the index of annihilation are proposed. The algorithm is implemented in a numerical stable manner, giving efficient results. Error analysis and the required complexity of the algorithm are included.

It should be stressed out that the present paper might be considered as a bird-sight view of WCF, useful for several other extensions and connections with other fields of numerical linear algebra and control theory. For instance, a numerical algorithm for the efficient calculation of the Drazin inverse through the WCF approach of a regular matrix pencil is an immediate, but not trivial, extension. In this direction, some work is in progress.

Acknowledgement

This research was financially supported by PENED 03ED740 of the Greek Secretariat for Research and Technology.

References

1. Chen, C.T.: Linear System Theory and Design, pp. 546–550. CBS College Publishing (1984)
2. Datta, B.N.: Numerical Linear Algebra and Applications, 2nd edn. Brooks/Cole Publishing Company, United States of America (1995)
3. Gantmacher, R.F.: The theory of matrices, Chelsea, New York., U.S.A, vol. I, II (1959)
4. Kalogeropoulos, G.: Matrix Pencils and linear systems theory. PhD thesis. Control Engineering Centre, City University, London (1985)
5. Kalogeropoulos, G., Mitrouli, M.: On the computation of the Weierstrass Canonical form of a Regular Matrix Pencil. Control and Computers 20(3), 61–67 (1992)
6. Kalogeropoulos, G.I., Pantelous, A.A.: Can be linear difference descriptor systems appeared in insurance? In: Proceedings of the 7th International Conference APLIMAT 2008, Bratislava, Slovakia, pp. 467–478 (2008)
7. Karcanias, N., Kalogeropoulos, G.: On the Segre$'$, Weyl characteristics of right (left) regular pencils. International Journal of Control 44, 991–1015 (1986)
8. Karcanias, N., Kalogeropoulos, G.: The prime and generalized nullspaces of right regular pencils. Circuits Systems Signal Processes 14(4), 495–524 (1995)
9. Leontief, W.: Input-Output Economics, 2nd edn. Oxford Univ. Press, N. Y (1986)
10. Van Huffel, S., Vandewalle, J.: An efficient and reliable algorithm for computing the singular subspace of a matrix, associated with its smallest singular values. Journal of Computers and Applied Mathematics 19, 313–330 (1987)
11. Van Huffel, S.: Partial singular value decomposition algorithm. Journal of Computers and Applied Mathematics 33, 105–112 (1990)
12. Yalamov, P.Y., Mitrouli, M.: A fast Algorithm for Index Annihilation Computations. Journal of Computers and Applied Mathematics 108, 99–111 (1999)

A Coupling Interface Method for a Nonlinear Parabolic-Elliptic Problem

Juri D. Kandilarov

Faculty of Natural Sciences and Education,
University of Rousse,7017 Rousse, Bulgaria
ukandilarov@ru.acad.bg

Abstract. A coupling interface method (CIM) on uniform cartesian grid for solving parabolic-elliptic problems is proposed. The domain Ω is divided in two subregions Ω^- and Ω^+. On Ω^- we consider an elliptic equation with nonlinear term and on Ω^+ - a linear parabolic equation. On the interface Γ the standard transmission conditions are stated: the jumps of the solution and the flux are zero. A method of upper and lower solutions is used to deal with the nonlinearity. Numerical experiments are discussed.

1 Introduction

Let Ω be a bounded domain in $\mathbf{R}^1$ ($\mathbf{R}^2$) with a piecewise smooth boundary $\partial\Omega$ and $\Gamma \subset \Omega$ be a point (a smooth curve), which splits the domain Ω into two separate subregions Ω^-, Ω^+, $\Omega = \Omega^+ \cup \Omega^- \cup \Gamma$. We consider the following parabolic-elliptic interface problem

$$0 = \beta \Delta u + g(u(x,y,t)) + f(x,y,t) \quad in \quad \Omega_T^- \equiv \Omega^- \times (0,T], \tag{1}$$

$$u_t = \beta \Delta u + f(x,y,t) \quad in \quad \Omega_T^+ \equiv \Omega^+ \times (0,T], \tag{2}$$

$$u(x,y,0) = u_0(x,y), \qquad (x,y) \quad in \quad \bar{\Omega}^-, \tag{3}$$

$$u(x,y,t) = u_B(x,y,t), \quad (x,y,t) \quad in \quad \partial\Omega \times [0,T], \tag{4}$$

with conjugation conditions on the interface $\Gamma_T = \Gamma \times [0,T]$

$$[u]_\Gamma := u^+(x(s),y(s),t) - u^-(x(s),y(s),t) = 0, \tag{5}$$

$$\left[\beta \frac{\partial u}{\partial \mathbf{n}}\right]_\Gamma = 0, \tag{6}$$

where s is a parameter of Γ, the superscripts $+$ and $-$ denote the limiting values of a function from one side in Ω^+ and another side in Ω^- respectively, and $\mathbf{n}$ is the normal vector at the point $(x(s),y(s)) \in \Gamma$, directed from Ω^- to Ω^+.

S. Margenov, L.G. Vulkov, and J. Waśniewski (Eds.): NAA 2008, LNCS 5434, pp. 330–337, 2009.

For the initial data β, f, u_0, u_B and the solution u we assume to be sufficiently smooth everywhere with exception of β, f and the derivatives of u, that may have discontinuity on Γ.

The problem (1)-(6) arises, when we study the enzyme fermentation reactor, described in [4] and also as a special case - in electromagnetism [3]. Existence and uniqueness of the solution has been studied by J. Henry [4]. Another important application of parabolic-elliptic problems appears in the heat-mass transfer when we consider the heat flux through a domain with subregions, in which the thermal capacity approaches zero [5,6].

Numerous methods have been developed for interface problems [12,16]. The IIM proposed by LeVeque and Li [12] solves elliptic equations with jump relations on Γ, which are known functions, defined on the interface. It has been successfully implemented for $1D$ and $2D$ linear and nonlinear elliptic and parabolic equations. Some $2D$ problems with jump conditions, that depend on the solution on the interface are considered by Kandilarov and Vulkov [7,9,10,15]. In [1] the IIM for parabolic-elliptic problem was studied. The main goal of this work is the application of the CIM, develop by I. Chern and Yu-Chen Shu [2] to the proposed nonlinear parabolic-elliptic problem.

This paper is organized as follows. In the next section we discuss the CIM. Then we present the difference scheme for the problem. The method of upper and lower solutions [14] for the nonlinear algebraic system is used. In the last section numerical results are discussed.

2 The Numerical Method

We use a uniform grid $x_i = ih$, $i = 0, 1, ..., M$, where $h = 1/M$ is the step size in space and $t^n = n\tau$, $n = 0, 1, ..., N$, where $\tau = T/N$ is the step size in time. Let $x_I \leq \xi < x_{I+1}$. We call the points (x_I, t^n) and (x_{I+1}, t^n) **irregular**. The approximation consists in two steps: approximation in space, using CIM1 or CIM2; approximation in time, using Euler method.

The attention here is on the first stage. At a regular point x_i the standard central finite difference approximation is adopted:

$$(\beta u_x)_x(x_i) \approx \frac{1}{h^2}(\beta_{i+1/2}(x_{i+1} - x_i) - \beta_{i-1/2}(x_i - x_{i-1})). \tag{7}$$

At irregular grid points the CIM1 and CIM2 [2] is used. Let the jump conditions (5)-(6) are of more general form:

$$[u]_{x=\xi} = \varphi, \quad [\beta u_x]_{x=\xi} = \psi. \tag{8}$$

2.1 CIM 1 in 1D Case

We assume, that there is at most one interface point in each mesh cell. Let $\xi \in [x_I, x_{I+1})$. Let x_I be located in Ω^-. Also, for simplicity we omit the indexes for the time variable t. Let $\rho_I := (x_I - \xi)/h$ and $\rho_{I+1} := (x_{I+1} - \xi)/h$. From the jump conditions (8) we get

$$(u_{i+1} - \rho_{I+1} h(u')^+_{i+1/2}) - (u_i + \rho_I h(u')^-_{i+1/2}) \approx \varphi,$$
$$\beta^+ (u')^+_{i+1/2} - \beta^- (u')^-_{i+1/2} \approx \psi.$$

This yields

$$(u')^-_{i+1/2} = \frac{1}{h}\left(\widehat{\beta}^+ (u_{i+1} - u_i) - \widehat{\beta}^+ \varphi - \rho_{I+1} h \frac{\psi}{\widehat{\beta}}\right) + O(h),$$
$$(u')^+_{i+1/2} = \frac{1}{h}\left(\widehat{\beta}^- (u_{i+1} - u_i) - \widehat{\beta}^- \varphi - \rho_I h \frac{\psi}{\widehat{\beta}}\right) + O(h),$$

where $\widehat{\beta} = \rho_{I+1}\beta^- + \rho_I \beta^+$ and $\widehat{\beta}^{\pm} = \beta^{\pm}/\widehat{\beta}$. Similarly we work, if there is another interface point. Otherwise, $(u')^-_{i-1/2} = (u_i - u_{i-1})/h$. With this we have

$$(\beta u_x)_x(x_i) \approx \frac{1}{h}\beta_i((u_x)^-_{i+1/2} - (u_x)^-_{i-1/2}) + O(1) \tag{9}$$

2.2 CIM2 in 1D Case

At x_I we use the following semi-discretization:

$$(\beta u_x)_x(x_i) \approx \frac{\beta_I - \beta_{I-1}}{h}\frac{u_I - u_{I-1}}{h} + \beta_I u''_I; \tag{10}$$

To obtain $O(h)$ truncation error at x_I, we need a first order approximation for the term u''_I. The idea is to approximate u by quadratic functions on both sides of ξ. By Taylor expansion we have:

$$u^-(\xi) = u_I + \left(\frac{u_I - u_{I-1}}{h} + \frac{1}{2} h u''_I\right)(\xi - x_I) + \frac{1}{2} u''_I (\xi - x_I)^2 + O(h^3),$$
$$u^+(\xi) = u_{I+1} + \left(\frac{u_{I+2} - u_{I+1}}{h} - \frac{1}{2} h u''_{I+1}\right)(\xi - x_{I+1}) + \frac{1}{2} u''_{I+1}(\xi - x_{I+1})^2 + O(h^3).$$

From here, using the jump conditions

$$u^+(\xi) - u^-(\xi) = \varphi, \qquad (\beta u')^+(\xi) - (\beta u')^-(\xi) = \psi$$

we find for u''_I and u''_{I+1} the following expressions:

$$u''_I = \frac{1}{h^2}(L u_I + J_I) + O(h), \tag{11}$$
$$u''_{I+1} = \frac{1}{h^2}(L u_{I+1} + J_{I+1}) + O(h).$$

The terms Lu_I, Lu_{I+1} J_I J_{I+1} are

$$Lu_I = a_{I,-1} u_{I-1} + a_{I,0} u_I + a_{I,1} u_{I+1} + a_{I,2} u_{I+2}$$
$$Lu_{I+1} = a_{I+1,-1} u_{I-1} + a_{I+1,0} u_I + a_{I+1,1} u_{I+1} + a_{I+1,2} u_{I+2}$$
$$J_I = -(1 + 2\rho_{I+1})\widetilde{\beta}^+ \varphi - (\rho_{I+1} + \rho^2_{I+1})\frac{h}{\widetilde{\beta}}\psi$$
$$J_{I+1} = (1 + 2\rho_I)\widetilde{\beta}^- \varphi - (\rho_{I+1} + \rho^2_{I+1})\frac{h}{\widetilde{\beta}}\psi$$

where

$$\widetilde{\beta} = (\rho_{I+1} + \rho_{I+1}^2)(\frac{1}{2} + \rho_I)\beta^- + (\rho_I + \rho_I^2)(\frac{1}{2} + \rho_{I+1})\beta^+ \quad \widetilde{\beta}^{\pm} = \frac{\beta^{\pm}}{\widetilde{\beta}}$$

$$a_{I,-1} = (\rho_{I+1} + \rho_{I+1}^2)\widetilde{\beta}^- + \rho_I(1 + 2\rho_{I+1})\widetilde{\beta}^+ \qquad a_{I+1,-1} = -\rho_I^2\widetilde{\beta}^-$$

$$a_{I,0} = -(\rho_{I+1} + \rho_{I+1}^2)\widetilde{\beta}^- - (1 + \rho_I)(1 + 2\rho_{I+1})\widetilde{\beta}^+ \qquad a_{I+1,0} = (1 + \rho_I)^2\widetilde{\beta}^-$$

$$a_{I,1} = (1 + \rho_{I+1})^2\widetilde{\beta}^+ \qquad a_{I+1,1} = -(\rho_I + \rho_I^2)\widetilde{\beta}^+ - (1 + \rho_{I+1})(1 + 2\rho_I)\widetilde{\beta}^-$$

$$a_{I,2} = -\rho_{I+1}^2\widetilde{\beta}^+ \qquad a_{I+1,2} = (\rho_I + \rho_I^2)\widetilde{\beta}^+ + \rho_{I+1}(1 + 2\rho_I)\widetilde{\beta}^-$$

From (10) and (11) we obtain

$$(\beta u_x)_x(x_I) \approx \frac{\beta_I - \beta_{I-1}}{h}\frac{u_I - u_{I-1}}{h} + (\frac{1}{h^2}\beta_I(Lu_I + J_I)); \tag{12}$$

In a similar way we get

$$(\beta u_x)_x(x_{I+1}) \approx \frac{\beta_{I+2} - \beta_{I+1}}{h}\frac{u_{I+2} - u_{I+1}}{h} + (\frac{1}{h^2}\beta_{I+1}(Lu_{I+1} + J_{I+1})). \tag{13}$$

2.3 CIM 1 in 2D Case

Let $\overline{\Omega}$ be a unit square and let us introduce on $\overline{\Omega}_T = \overline{\Omega} \times [0, T]$ the uniform mesh $\overline{\omega}_{h,\tau} = \overline{\omega}_h \times \overline{\omega}_\tau$, where $\overline{\omega}_h = \overline{\omega}_{h_1} \times \overline{\omega}_{h_2}$ and

$$\begin{aligned} \overline{\omega}_{h_1} &= \{x_i = ih_1,\ i = 0, 1, ..., M_1,\ h_1 = 1/M_1\}, \\ \overline{\omega}_{h_2} &= \{y_j = jh_2,\ j = 0, 1, ..., M_2,\ h_2 = 1/M_2\}, \\ \overline{\omega}_\tau &= \{t_n = n\tau,\ n = 0, 1, ..., N,\ \tau = T/N\}. \end{aligned}$$

Let ω_h and $\partial\omega_h$ be the sets of mesh points of Ω and $\partial\Omega$ respectively. Let also ω_h^+ and ω_h^- be the sets of mesh points of Ω^+ and Ω^-. With γ_h we denote the points of ω_h, that lie on the interface curve Γ. Then $\omega_h = \omega_h^+ \cup \omega_h^- \cup \gamma_h$.

If the interface curve Γ does not intersect the arms of the standard 5-point stencil at some mesh point, we call such point **regular** and the following approximation is used:

$$\begin{aligned} ((\beta u_x)_x + (\beta u_y)_y)_{(x_i,y_j)} \approx{}& \beta_{i+\frac{1}{2},j}\frac{u_{i+1,j} - u_{i,j}}{h_1^2} - \beta_{i-\frac{1}{2},j}\frac{u_{i,j} - u_{i-1,j}}{h_1^2} \\ &+ \beta_{i,j+\frac{1}{2}}\frac{u_{i,j+1} - u_{i,j}}{h_2^2} - \beta_{i,j-\frac{1}{2}}\frac{u_{i,j} - u_{i,j-1}}{h_2^2} \end{aligned} \tag{14}$$

In other cases the point is **irregular**. Let $(x_i, y_j) \in \Omega^-$. Then we approximate

$$(\beta u_x)_x + (\beta u_y)_y \approx \frac{\beta_{i,j}}{h}\left((u_x^-)_{i+\frac{1}{2},j} - (u_x^-)_{i-\frac{1}{2},j} + (u_y^-)_{i,j+\frac{1}{2}} - (u_x^-)_{i,j-\frac{1}{2}}\right) \tag{15}$$

We will illustrate the approximation to $(u_x^-)_{i+\frac{1}{2},j}$. Let the interface curve Γ intersects the right arm of the stencil (i.e. the segment $[x_i, x_{i+1}) \times y_j$) at point (ξ_r, y_j). Then

$$(u_x^-)_{i+\frac{1}{2},j} \approx \frac{1}{h}\left(\widehat{\beta}^+(u_{i+1,j} - u_{i,j}) - \widehat{\beta}^+[u]_{(\xi_r,y_j)} - \rho_{I+1}h\frac{[\beta u_x]_{(\xi_r,y_j)}}{\widehat{\beta}}\right). \tag{16}$$

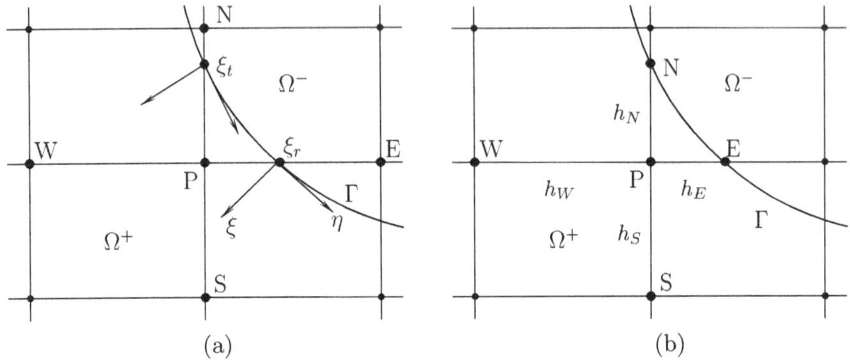

Fig. 1. (a)The stencil for the CIM1 in 1D and local coordinate system at points (ξ_r, y_j), (x_i, ξ_t), when the interface curve Γ intersects the right and top arm of the standard 5-point stencil, centered at $P = (x_i, y_j)$; (b) The stencil at irregular point $P = (x_i, y_j)$ for the numerical solution in Ω^+

Here ρ_{I+1}, $\widehat{\beta}$, $\widehat{\beta}^+$ and etc. are calculated at point (ξ_r, y_j) by the same formulas as in CIM1 for 1D.

Next, we describe a decomposition of jump data. In (16) the jump $[\beta u_x]_{(\xi_r,y_j)}$ is not available from the prescribed jump conditions. Therefore we decompose it into tangential and normal directions of Γ at point (ξ_r, y_j), where (ξ_1, ξ_2) and (η_1, η_2) are unit normal and tangential vectors at this point:

$$\begin{aligned}[][\beta u_x]_{(\xi_r,y_j)} &= [\beta u_\xi]_{(\xi_r,y_j)}\xi_1 + [\beta u_\eta]_{(\xi_r,y_j)}\eta_1 \\ &= [\beta u_\xi]_{(\xi_r,y_j)}\xi_1 + \left(\beta^+[u_\eta]_{(\xi_r,y_j)} + (\beta^+ - \beta^-)(u_\eta^-)\right)\eta_1 .\end{aligned} \tag{17}$$

Here $[\beta u_\xi]$ and $[u_\eta]$ are obtained from the prescribed data, but u_η^- is not available. We use $u_\eta^- = (u_x^-)\eta_1 + (u_y^-)\eta_2$ and approximate (u_x^-) and (u_y^-) at point (ξ_r, y_j):

$$(u_x^-)_{(\xi_r,y_j)} \approx (u_x)_{i+1/2,j},\ (u_y^-)_{(\xi_r,y_j)} \approx \begin{cases} (u_y)_{i,j+1/2} \text{ if } \Gamma \text{ intersects } x_i \times [y_j, y_{j+1}), \\ (u_{i,j+1} - u_{i,j})/h \ \text{ otherwise.} \end{cases}$$

We plug this in (15):

$$(1 - \bar{b}_r\xi_{1,r}\xi_{1,r})(u_x^-)_{i+1/2,j} \approx \frac{1}{h}\left(\hat{\beta}_r^+(u_{i+1,j} - u_{i,j}) + \bar{b}_r\xi_{1,r}T_x u_{i,j} + \bar{J}_r\right), \tag{18}$$

where $\bar{b}_r = -\rho_{i+1,r}(\hat{\beta}_r^+ - \hat{\beta}_r^-)$, $\bar{J}_r = -\hat{\beta}_r^+[u]_{\xi_r,y_j} - h\rho_{i+1,r}([\beta u_x]_{(\xi_r,y_j)}\xi_{1,r}/\hat{\beta}_r + \hat{\beta}_r^+[u_\eta]\eta_{1,r})$ and

$$T_x u_{i,j} = \begin{cases} 0 & \text{if } \Gamma \text{ intersects } x_i \times [y_j, y_{j+1}), \\ (u_{i,j+1} - u_{i,j})\eta_{2,r} & \text{otherwise.} \end{cases}$$

Similarly we obtain a coupling equations for $(u_x^-)_{i+1/2,j}$ and $(u_x^-)_{i,j+1/2}$:

$$M\begin{pmatrix} (u_x^-)_{i+1/2,j} \\ (u_x^-)_{i,j+1/2} \end{pmatrix} = \frac{1}{h}\begin{pmatrix} \hat{\beta}_r^+(u_{i+1,j} - u_{i,j}) + \bar{b}_r\eta_{1,r}T_x u_{i,j} + \bar{J}_r \\ \hat{\beta}_t^+(u_{i,j+1} - u_{i,j}) + \bar{b}_t\eta_{2,t}T_x u_{i,j} + \bar{J}_t \end{pmatrix}$$

with matrix

$$M = \begin{pmatrix} 1 - \bar{b}_r \eta_{1,r} \eta_{1,r} & \bar{b}_r \eta_{1,r} \eta_{2,r} \\ \bar{b}_t \eta_{1,t} \eta_{2,t} & 1 - \bar{b}_t \eta_{2,t} \eta_{2,t} \end{pmatrix}.$$

2.4 Algorithm Description

We will describe the algorithm in 1D case. On the first stage we solve the non-linear elliptic problem

$$0 = (\beta u')' + g(u(x,0)) + f(x,0) \quad in \quad \Omega^-, \tag{19}$$

with boundary conditions on $\partial\Omega^+$ and Γ. For this we use nonuniform grid near the interface Γ, i.e. we add to the grid points x_i, $i = 0, 1, ..., I$ also $x = \xi$ as an end grid point. So we find an approximated initial value U_i^0 on the elliptic region Ω^-. From the initial condition (3) we find the values of U_i^0 on the parabolic region Ω^+.

Next, for the discretization in time we use implicit method of Euler and for discretization in space we use the described coupling interface method:

$$\sigma \frac{U_i^{n+1} - U_i^n}{\tau} = (\beta U_x^{n+1})_{x,i} + (1-\sigma) g(U^{n+1}, t^{n+1}) + f(x_i, t^{n+1}), \tag{20}$$

where $\sigma = 1$ if $x_i \in \Omega^+$ and $\sigma = 0$ if $x_i \in \Omega^-$. The terms $(\beta U_x^{n+1})_{x,i}$ at irregular points x_I and x_{I+1} are defined by (9) for CIM1 and respectively by (12), (13) for CIM2. We obtain a linear-nonlinear difference problem, which we solve using the monotone method of upper and lower solutions, described in [8,11].

Theorem 1. *Let the initial and boundary data be sufficiently smooth and appropriate compatibility conditions are fulfilled, so that the classical solution of the 1D problem (1)-(6)* $u \in C(\overline{\Omega}_T) \bigcap C^{3,2}(\Omega_T^-) \bigcap C^{3,2}(\Omega_T^+)$. *Let also the nonlinear function* $g(u)$ *be monotone and continuous function. Then, the local truncation error of the CIM1 (CIM2) is* $\mathcal{O}(h^2 + \tau)$ *at the regular grid points and* $\mathcal{O}(1 + \tau)$ $(\mathcal{O}(h + \tau))$ *at the irregular grid points. The error in maximum norm is respectively* $\mathcal{O}(h + \tau)$ $(\mathcal{O}(h^2 + \tau))$.

Remark 1. For the 2D problem on the first stage we use the nonuniform grid, as is shown on Fig. 1b. This approximation is standard at regular grid points and then $h_E = h_W = h_1$, $h_N = h_S = h_2$. At irregular grid points, which are closed to the interior boundary Γ we must know the intersection points of the interface Γ with the grid lines through the point P, see for an example the points E and N in Fig. 1(b), when Γ crosses the right and top arm of the stencil. For more details see also [13], Section 6.3 and [1].

3 Numerical Experiments

We consider the problem

$$\begin{aligned} 0 &= \beta_l u'' + g(u(x,t)) + f_l(x,t) \quad && in \quad [-1, \xi] \times (0, T], \\ u_t &= \beta_r u'' + f_r(x,t) && in \quad [\xi, 1] \times (0, T] \end{aligned}$$

with exact solution

$$u(x,t) = \begin{cases} D + A\exp(-\alpha t)(c_{l1}\cos(\mu_l x) + c_{l2}\sin(\mu_l x)) & x \leq \xi, \\ D + A\exp(-\alpha t)(c_{r1}\cos(\mu_r x) + c_{r2}\sin(\mu_r x)) & x > \xi, \end{cases}$$

The nonlinear term is $g(u) = u^3$, $\xi = 0$, $\mu_r = \sqrt{2}/10$, $\alpha = 0.01$, and μ_l is the solution of the following nonlinear equation

$$\tan(\mu_l) = \tan(\mu_r)\frac{\beta_r}{\beta_l}\frac{1}{\mu_l}.$$

The corresponding coefficients are: $D = 1.2$, $A = 1$, $c_{l1} = 1$, $c_{l2} = -\tan(\mu_l)$, $c_{r1} = 1$, $c_{r2} = \tan(\mu_r)$. As initial guess for the first stage we use the solution in Ω^-: $k * u(x,0)$, where $k \leq 0.5$ for down solutions and $k \geq 1.5$ for upper solutions. The stabilized coefficient γ, using in the iterations (see also [8,11]) is taken to be $\gamma = 50$.

Table 1. Grid refinement analysis for CIM1 and CIM2

M	N	CIM1		CIM2	
		$\|E\|_\infty$	m	$\|E\|_\infty$	m
39	40	1.6141e-01	-	2.3341e-02	-
79	80	7.6764e-02	1.07	6.2242e-03	1.90
159	160	3.4517e-02	1.15	1.6083e-03	1.95
319	320	1.7842e-02	0.96	4.1029e-04	1.97
639	640	9.0826e-03	0.97	1.1361e-04	1.85

The results of the mesh refinement analysis are presented in Table 1. With $|E|_\infty$ we denote the error of the numerical solution in maximum norm and with m we denote the rate of convergence. The experiments confirm first order of convergence for CIM1 and second order for CIM2. The number of iterations vary from 67 to 98 on the first stage, when we are on the first stage of the algorithm. Having U^0, on the next time layer we use as initial guess the solution on the previous layer. So the number of iterations decreases extraordinary to 2 or 1.

4 Conclusions

In this paper we study a finite difference schemes based on the Coupling Interface Method for elliptic-parabolic problem with nonlinear term in the elliptic domain. Using method of upper and lower solutions we deal with the nonlinearity. A comparison of the numerical results against the exact solution shows that the method is first order accurate for CIM1 and second order accurate for CIM2. Our future plans are the implementation of CIM1 and CIM2 for 2D case and their theoretical validation.

Acknowledgements

This work was supported by the National Science Fund of Bulgaria under contract HS-MI-106/2005.

References

1. Brayanov, I.A., Kandilarov, J.D., Koleva, M.N.: Immersed Interface Difference Schemes for a Parabolic-Elliptic Interface Problem. In: Lirkov, I., Margenov, S., Waśniewski, J. (eds.) LSSC 2007. LNCS, vol. 4818, pp. 661–669. Springer, Heidelberg (2008)
2. Chern, I.-L., Shu, Y.-C.: A coupling interface method for elliptic interface problems. J. of Comput. Phys. 225(2), 2138–2174 (2007)
3. Al-Droubi, A., Renardy, M.: Energy methods for a parabolic-hyperbolic interface problem arising in electromagnetism. J. Appl. Math. Phys. 39, 931–936 (1988)
4. Henry, J.: Optimization of fermentation reactor with non-moving layer. In: Num. Meth. in Appl. Math., Science, Siberian Branch, Novosibirsk, pp. 163–173 (1982) (in Russian)
5. Jovanovic, B., Vulkov, L.: On the convergence of difference schemes for the heat equation with concentrated capacity. Numer. Math. 89(04), 715–734 (2002)
6. Jovanovic, B., Vulkov, L.: On the convergence of difference schemes for parabolic problems with concentrated data. IJNAM 5(03), 386–407 (2008)
7. Kandilarov, J.D.: Immersed-boundary level set approach for numerical solution of elliptic interface problems. In: Lirkov, I., Margenov, S., Waśniewski, J., Yalamov, P. (eds.) LSSC 2003. LNCS, vol. 2907, pp. 456–464. Springer, Heidelberg (2004)
8. Kandilarov, J.: A monotone iterative method for numerical solution of diffusion equations with nonlinear localized chemical reactions. In: Boyanov, T., Dimova, S., Georgiev, K., Nikolov, G. (eds.) NMA 2006. LNCS, vol. 4310, pp. 615–622. Springer, Heidelberg (2007)
9. Kandilarov, J., Vulkov, L.: The immersed interface method for a nonlinear chemical diffusion equation with local sites of reactions. Numer. Algor. 36, 285–307 (2004)
10. Kandilarov, J., Vulkov, L.: The immersed interface method for two-dimensional heat-diffusion equations with singular own sources. Appl. Num. Math. 57, 486–497 (2007)
11. Li, Z., Pao, C.V., Qiao, Z.: A finite difference method and analysis for 2D nonlinear Poisson-Boltzmann equations. J. Sci. Comput. 30, 61–81 (2007)
12. Li, Z., Ito, K.: The Immersed Interface Method: Numerical Solutions of PDEs Involving Interfaces and Irregular Domains. SIAM, Philadelphia (2006)
13. Morton, K.W., Mayers, D.F.: Numerical Solution of Partial Differential Equations. Cambridge University Press, Cambridge (1995)
14. Pao, C.V.: Nonlinear Parabolic and Elliptic Equations. Plenum Press, New York (1992)
15. Vulkov, L., Kandilarov, J.: Construction and implementation of finite-difference schemes for systems of diffusion equations with localized nonlinear chemical reactions. Comp. Math. Math. Phys. 40, 705–717 (2000)
16. Zhou, Y.C., Zhao, S., Feig, M., Wei, G.W.: High order matched interface and boundary method for elliptic equations with discontinuous coefficients and singular sources. J. Comput. Phys. 213, 1–30 (2005)

A New Method for Solving Transient Lossy Transmission Line Problem

Turhan Karaguler

Beykent University, Department of Mathematics and Computing,
Sisli-Ayazaga Campus, 34396 Istanbul, Turkey
turank@beykent.edu.tr

Abstract. This paper presents a new technique for the transient analysis of lossy transmission lines. The proposed method is based on discretization of Telegrapher's equation via the auxiliary problem equations to which well known numerical methods can be applied easily. The new method also lets simple and well structured algorithm be developed. A SPICE model is used to verify the results obtained from the new method.

1 Introduction

Transferring electrical power and signals from the source to the load all require transmission lines. The simplest transmission line consists of two or more parallel conductors connecting a source to a load. Pairs of lands on printed circuits board carrying the digital signals are also considered as the transmission lines. Transmission line problem can be described as either finding the current and voltage values or obtaining the electric and magnetic field distribution along the line. The problem in nature possess the wave phenomena due to the fact that the large distance between the source and the load leads to unavoidable time delay effects. Many text books provide the solution of the problem under steady state conditions in which voltage and current vary sinusoidally at a single frequency or transient but under the assumption of lossless line [1,2]. However it is rather difficult to solve the transmission line problem when both transient and lossy states exist in the model. It may not even be possible to get an analytical solution for the simplest line therefore only numerical solutions are attempted. These solutions are usually obtained either by direct approximation of the telegraphers equations in the time domain [3] or via the frequency domain solutions [4]. In this work the direct approximation in the time domain is used.

The following well known telegraphers equations are considered as the governing equations of the problem which represents a typical transmission line depicted in figure 1.

$$L\frac{\partial i(x,t)}{\partial t} + \frac{\partial v(x,t)}{\partial x} + R\, i(x,t) = 0 \tag{1}$$

$$C\frac{\partial v(x,t)}{\partial t} + \frac{\partial i(x,t)}{\partial x} + G\, v(x,t) = 0 \tag{2}$$

S. Margenov, L.G. Vulkov, and J. Waśniewski (Eds.): NAA 2008, LNCS 5434, pp. 338–344, 2009.

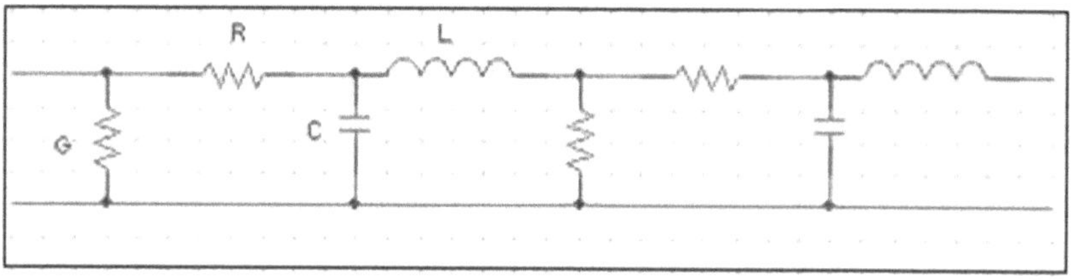

Fig. 1. Distributed Parameters of a Transmission Line

In the telegrapher's equation, $v(x,t)$ and $i(x,t)$ are potential and current functions at any point x along the transmission line at a given time t respectively. The line parameters R, G, L, C are not discrete but distributed therefore per unit length convention is used. For just sake of simplicity, the medium is assumed to be linear and homogeneous therefore all the line parameters are taken constant. In general, in order to define the problem completely, together with telegraphers equations (1),(2), the following initial- boundary conditions are also required.

$$i(x,0) = i_0(x), \tag{3}$$

$$v(x,0) = v_0(x), \tag{4}$$

$$i(0,t) = i_1(t), \tag{5}$$

$$v(0,t) = v_1(t), \tag{6}$$

The problem defined by these equations is named as the main problem throughout the paper. The conditions (3), (4) indicate that at t = 0, the line can be supplied by any type of geometry dependent continuous or piecewise continuous voltage and current functions. The conditions (5),(6) in fact express the initial current and voltage profiles of the line.

The governing equations (1),(2) possess the wave characteristics. This is obvious, such that, while trying to eliminate either $v(x,t)$ or $i(x,t)$ from the equations (1),(2) to obtain an equation involving only one quantity, these telegrapher's equations (1),(2) are first differentiated with respect to x and t respectively and then substituted one into other.

$$\frac{\partial^2 v(x,t)}{\partial x^2} = a_0 \frac{\partial^2 v(x,t)}{\partial t^2} + b_0 \frac{\partial v(x,t)}{\partial t} + c_0 v(x,t) \tag{7}$$

where $a_0 = LC$, $b_0 = (RC + GL)$, $c_0 = GR$.

As outlined in reference [2], Heaviside showed that if the $\frac{R}{L} = \frac{G}{C}$ relation exists between the parameters of the line, the equation (7) in fact becomes the familiar second order wave equation and the line is said to be distortionless.

$$\frac{\partial^2 u(x,t)}{\partial x^2} = a^2 \frac{\partial^2 u(x,t)}{\partial t^2}. \tag{8}$$

In the equation above, u is used as generic function instead of v or i, and $a = \frac{1}{\sqrt{LC}}$.

There are various difference schemes currently available in literature to find the solution of the problem. Due to the nature of these methods, it is common that the calculations need to be carried out also at fictitious points placed beyond the problem interval of $(0, L)$. Additionally, it is also known that the solution of the main problem may have weak discontinuity on the characteristics therefore the solution is on the characteristics and continuously differentiable but its first order derivatives are piecewise continuously differentiable. This in practice prevents the application of well known numerical methods to the main equations. Besides, the initial functions may possess the singular points which even make the application of well known numerical techniques further complicated. The mathematical details of these can be found in [5].

2 The Auxiliary Problem

In order to overcome the difficulties described above for solving the telegrapher's equations, a new approach to the transmission line problem is introduced. This method is first outlined by Rasulov and his co-authors in reference [6]. An application of the method to the fields of hydrodynamics and gas dynamics are given in references [7-8]. This new method is based on replacing the actual unknown functions, $i(x,t)$ and $v(x,t)$, with the new $I(x,t)$ and $V(x,t)$ auxiliary functions. For this aim, the following modification on telegrapher's equations (1), (2) are applied.

$$L \int_0^x \frac{\partial i(\xi,t)}{\partial t} d\xi + \int_0^x \frac{\partial v(\xi,t)}{\partial \xi} d\xi + R \int_0^x i(\xi,t) d\xi = 0 \tag{9}$$

$$C \int_0^x \frac{\partial v(\xi,t)}{\partial t} d\xi + \int_0^x \frac{\partial i(\xi,t)}{\partial \xi} d\xi + G \int_0^x v(\xi,t) d\xi = 0 \tag{10}$$

Since the relations between the current and voltage functions of the main and auxiliary problems are defined as

$$i(x,t) = \frac{\partial I(x,t)}{\partial x}, \tag{11}$$

$$v(x,t) = \frac{\partial V(x,t)}{\partial x}. \tag{12}$$

Then the set of equations of auxiliary problem can be written as below;

$$L \frac{\partial I(x,t)}{\partial t} + v(x,t) - v(0,t) + RI(x,t) = 0, \tag{13}$$

$$C \frac{\partial V(x,t)}{\partial t} + i(x,t) - i(0,t) + GV(x,t) = 0, \tag{14}$$

$$I(x,0) = I_0(x), \tag{15}$$

$$V(x,0) = V_0(x). \tag{16}$$

Here $I_0(x)$ and $V_0(x)$ are any continuously differentiable functions satisfying following relations:

$$\frac{dI_0(x)}{dx} = i_0(x), \tag{17}$$

$$\frac{dV_0(x)}{dx} = v_0(x). \tag{18}$$

For just simplicity, $v(0,t), i(0,t)$ can be replaced with $v_0(t), i_0(t)$ respectively. The boundary conditions for the auxiliary problem will be the same as the boundary conditions of the main problem.

The auxiliary problem is superior to the main problem. Firstly $V(x,t)$ and $I(x,t)$ functions of the auxiliary problem are used in the equations without their space derivatives. This will obviously lead to higher accuracy in the solution as derivative operations always tend to be more sensitive to the numerical values. Secondly, the differentiability property of these new functions of the auxiliary problem is one degree higher than the differentiability property of the main problem so that even discontinuous functions can be employed as the initial profiles in the model. Additionally the auxiliary problem allows us to employ well known numerical methods which are implemented easily by means of accurate and fast working algorithms.

3 Finite Difference Equations for the Auxiliary Problem

The solution of the telegraphers equation via the auxiliary problem can be obtained by using a finite difference scheme. For this purpose, the transmission line problem is divided into the grid of

$$\omega_{h,\tau} = \{(x_i, t_k) \mid x_i = ih, t_k = k\tau, i = 0, \pm 1, \pm 2, ... n-1; k = 0, 1, 2, ...; h > 0, \tau > 0\}.$$

where h and t are the space and time steps of the grid. The following common quadrature formula is used for calculation of the integral terms of the auxiliary problem equations.

$$\int_0^x \phi(z)dz \approx h\Sigma_{j=1}^{i}\phi(z_j). \tag{19}$$

Then, at any point (i,k) of the grid, the finite difference equations of the auxiliary problem can be obtained as

$$I_{i,k+1} = I_{i,k} - \frac{\tau}{hL}V_{i-1,k+1} - \frac{R\tau}{L}\sum_{j=1}^{i} I_{j,k} - \sum_{j=1}^{i-1}(I_{j,k+1} - I_{j,k}) + \frac{\tau}{hL}v_0(t_k), \tag{20}$$

$$V_{i,k+1} = V_{i,k} - \frac{\tau}{hC}I_{i,k+1} - \frac{G\tau}{C}\sum_{j=1}^{i} V_{j,k} - \sum_{j=1}^{i-1}(V_{j,k+1} - V_{j,k}) + \frac{\tau}{hC}i_0(t_k), \tag{21}$$

$$V_{i,0} = V_0(x_i). \tag{22}$$

$$I_{i,0} = I_0(x_i). \tag{23}$$

Here $V_{i,k}$ and $I_{i,k}$ represent the approximate values of $V(x,t)$ and $I(x,t)$ functions at the point (i,k) of the grid. After getting these $V_{i,k}$ and $I_{i,k}$ values then the main problem functions $v_{i,k}$ and $i_{i,k}$ can be found by just simply using (11),(12) equalities.

4 Results

The transient transmission line with a square pulse is selected as an example for validating the new method. This problem can be seen in many practical applications typically in computer networks. When a battery or a pulse generator being connected to a transmission line is switched on, it takes some time for the voltage and current to reach steady values. A lossy line with the length l=1 m and per unit line parameters R=7.5 Ohm/m, L=10 mH/m, G = 0.003 S/m, C = 0.4 F/m is considered. As depicted in figure 2, the line is connected to a dc battery of 2 volts.

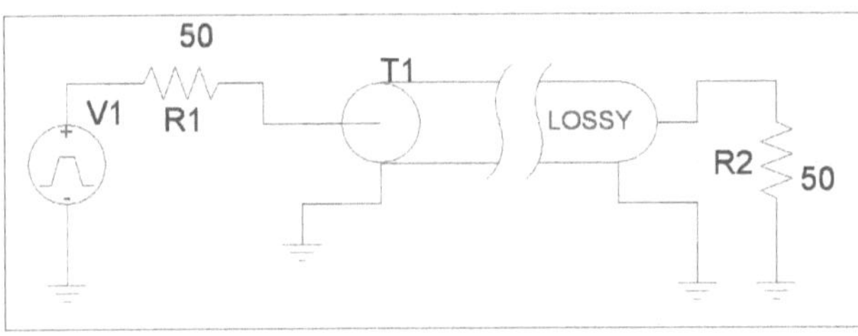

Fig. 2. Sample Lossy Transmission Line Representation

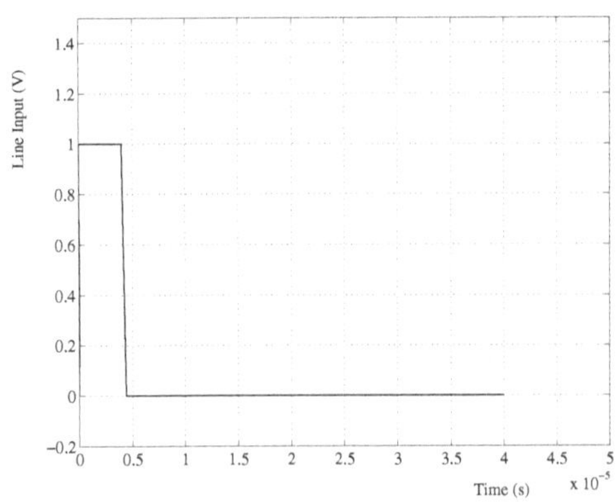

Fig. 3. Line Input Pulse with 0.004 ms width

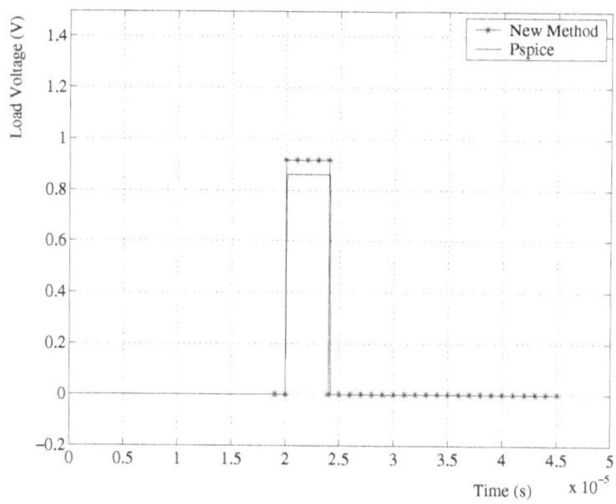

Fig. 4. Load Outputs (New Method versus PSPICE)

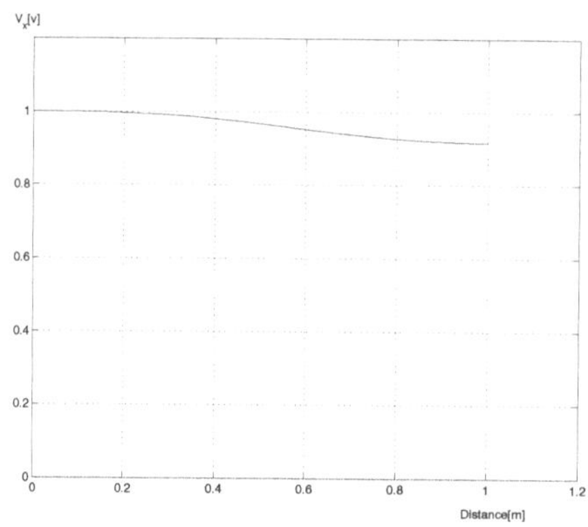

Fig. 5. Display of Line Loss from the New Method

The line ends with a purely resistive load $Z_l = 50$ Ohm. As the load and characteristic impedances are equal, the transmission line is said to be matched. This practically simplifies handling the load end boundary condition.

The result of the voltage at the load is presented in figure 4 for a square pulse input given in figure 3. The time delay between the input signal and the load end signal is $2\ 10^{-6}$ seconds. As seen from the figure that the result matches reasonably well with the result obtained from the PSPICE model. The voltage loss along the transmission line obtained from the new method is shown in figure 5. The loss in the line is about 10% of the input signal.

5 Conclusion

As the preliminary results show that the new technique is promising for the transient lossy transmission line analysis. Initially the method is tested with the simplest case. However the actual strength of the method can be observed in transmission line applications with nonlinear parameters and loads. This work may be extended to cover such complex cases. It should be noted that introduction of the auxiliary problem has several advantages over the main problem. Firstly, the well known numerical methods can easily be applied to the auxiliary problem. Further, the auxiliary problem lets higher ordered finite differences schemes be used. This leads to developing simple algorithms thus efficient and fast implemented model from computing point of view.

References

1. Hayt, W.H.: Engineering Electromagnetics. McGraw-Hill, New York (2001)
2. Sadiku, M.N.O.: Elements of Electromagnetics. Oxford University Press, Oxford (2001)
3. Brannin, F.H.: Transient Analysis of Lossless Transmission Lines. Proc. IEEE 55, 2012–2013 (1967)
4. Palusinski, O.A., Lee, A.: Analysis of Transients in Nonuniform and Uniform Multiconductor Transmission Lines. IEEE Trans. Microwave Theory Tech. 17, 127–138 (1989)
5. Koshylakov, N.S.: Partial Differential Equations of Mathematical Physics., Bishaya Skola, Moscow (1970)
6. Rasulov, M.A.: On a Method of Solving the Cauchy problem for a First Order Nonlinear Equation of Hyperbolic Type with a Smooth Initial Condition. Soviet Mathematics Doklady 43(1), 150–153 (1991)
7. Rasulov, M.A., Karaguler, T., Sinsoysal, B.: A Finite Difference Schemes for Solving System Equations of Gas Dynamic in a Class of Discontinuous Functions. Applied Mathematics and Computation 143, 145–164 (2004)
8. Rasulov, M.A., Karaguler, T., Sinsoysal, B.: Numerical Solution of Cauchy Problem for Second Order Nonlinear Wave Equation with Changeable Type in a Class of Discontinuous Functions. Applied Mathematics and Computation 147(2), 423–437 (2004)

On Superlinear PCG Methods for FDM Discretizations of Convection-Diffusion Equations

János Karátson and Tamás Kurics

Department of Applied Analysis and Computational Mathematics
ELTE University, H-1117 Budapest, Hungary
{karatson,fantom}@cs.elte.hu

Abstract. The numerical solution of linear convection-diffusion equations is considered. Finite difference discretization leads to an algebraic system solved by a suitable preconditioned CG method, where the preconditioning approach is based on equivalent operators. Our goal is to study the superlinear convergence of the preconditioned CG iteration and to find mesh independent behaviour on a model problem. This is an analogue of previous results where FEM discretization was used.

1 Introduction

The conjugate gradient method has become a widespread way of solving linear algebraic systems. One of its most important features is superlinear convergence, see [3] where an extensive summary is given on the convergence of the CG method. For discretized elliptic problems, the CG method is mostly used with suitable preconditioning, which sometimes relies on Hilbert space theory and then provides mesh independent convergence.

An important preconditioning tool which provides mesh independent convergence is the equivalent operator approach, in which one proposes the discretization of another suitable linear preconditioning operator as preconditioning matrix [6,7,8,10,11]. The cited papers give mesh independence results in the context of linear convergence, in particular, a rigorous framework is developed in [7]. More recently, mesh independent superlinear convergence has been demonstrated for finite element discretizations of nonsymmetric elliptic problems (convection-diffusion equations) in the authors' papers [4,5,9]. The Hilbert space setting of the FEM enables us to estimate the superlinear convergence factors in the discrete case from above by the analogous factors in the operator level, where the latter is based on the eigenvalues of the preconditioned operator.

Our goal is to study the same problem in the case of finite difference discretizations, i.e. to study the superlinear convergence of the preconditioned CG iteration under equivalent operator preconditioning and to verify mesh independent behaviour. Here an important difference arises between FEM and FDM discretizations, pointed out already in [7]. Namely, the FDM lacks the organized Hilbert space background that FEM is based on, hence a case-by-case study

S. Margenov, L.G. Vulkov, and J. Waśniewski (Eds.): NAA 2008, LNCS 5434, pp. 345–352, 2009.

of convergence is required when FDM discretizations are used with equivalent operator preconditioning. For linear convergence such a work has been started already in [6,8] and extended in [7,10,11]. The present paper aims to take a first step to verify mesh independence of superlinear convergence, and hence a model problem is considered with a uniform FD grid. The required mesh independent bound is proved for a certain class of coefficients, and numerical calculations show similar behaviour for other coefficients as well. Preliminaries are given in sections 2 and 3, the model problem and the convergence results are presented in sections 4 and 5, respectively.

2 Equivalent Operator Preconditioning

Let us consider an elliptic convection-diffusion equation

$$\begin{cases} Lu \equiv -\Delta u + \mathbf{b} \cdot \nabla u + \alpha u = g \\ u_{|\partial\Omega} = 0 \end{cases} \tag{1}$$

on a bounded domain $\Omega \subset \mathbf{R}^d$. We assume that $\mathbf{b} \in C^1(\overline{\Omega})^d$ and $\alpha \in L^\infty(\Omega)$; further, there holds the usual coercivity condition

$$\alpha - \frac{1}{2}\operatorname{div}\mathbf{b} \geq 0. \tag{2}$$

In this study we focus on regularly perturbed problems. The coercivity condition (2) implies that for all $g \in L^2(\Omega)$ problem (1) has a unique weak solution in $H_0^1(\Omega)$.

The FDM discretization of (1) on a given grid ω_h leads to a linear algebraic system

$$L_h c = g_h \,. \tag{3}$$

Our goal is to solve (1) by iteration, applying a suitable preconditioned conjugate gradient method. The proposed preconditioner is obtained via a symmetric preconditioning operator

$$Su := -\Delta u + \beta u \qquad \text{for } \; u_{|\partial\Omega} = 0, \tag{4}$$

where $\beta \in L^\infty(\Omega)$, $\beta \geq 0$: namely, the matrix S_h is defined as the FDM discretization of the operator S on the same grid ω_h. The preconditioned form of the discretized system is

$$S_h^{-1} L_h \, c = f_h \tag{5}$$

where $f_h := S_h^{-1} g_h$.

There exist several CG algorithms for such nonsymmetric systems (see e.g. [1,3]). We propose the generalized conjugate gradient–least square method (GCG-LS method) [2,3]. This algorithm uses least-square residual minimization and avoids the normal equations, hence one needs to solve only one auxiliary system in each step (in contrast to the widespread CGN method which requires stepwise two auxiliary systems). The next section gives a brief description of this method.

3 The Preconditioned Generalized Conjugate Gradient Method

The generalized conjugate gradient, least square (GCG-LS) method is defined in [2]. The full version of the GCG-LS method constructs a sequence of search directions d_k and simultaneously a sequence of approximate solutions u_k. Following the terminology of [2], the definition also involves an integer $s \in \mathbb{N}$, further, we let $s_k = \min\{k, s\}$ $(k \geq 0)$. The full version of the algorithm for the solution of the preconditioned system (5) is as follows:

$$\begin{cases} (1) \ \text{Let } c_0 \text{ be arbitrary, let } r_0 \text{ be the solution of } S_h r_0 = L_h c_0 - f_h; \\ \quad\ d_0 = -r_0; \text{ and } z_0 \text{ be the solution of } S_h z_0 = L_h d_0; \\ \qquad \text{for any } k \in \mathbb{N}: \ \text{when } c_k, d_k, r_k, z_k \text{ are obtained, let} \\ (2a) \ \text{the numbers } \alpha_{k-j}^{(k)} \quad (j = 0, \dots, k) \quad \text{be the solution of} \\ \quad \sum_{j=0}^{k} \alpha_{k-j}^{(k)} \langle S_h z_{k-j}, z_{k-l} \rangle = - \langle r_k, S_h z_{k-l} \rangle \qquad (0 \leq l \leq k); \\ (2b) \ c_{k+1} = c_k + \sum_{j=0}^{k} \alpha_{k-j}^{(k)} d_{k-j}; \\ (2c) \ r_{k+1} = r_k + \sum_{j=0}^{k} \alpha_{k-j}^{(k)} z_{k-j}; \\ (2d) \ \beta_{k-j}^{(k)} = \langle L_h r_{k+1}, z_{k-j} \rangle / \|z_{k-j}\|_{S_h}^2 \quad (j = 0, \dots, s_k); \\ (2e) \ d_{k+1} = -r_{k+1} + \sum_{j=0}^{s_k} \beta_{k-j}^{(k)} d_{k-j}; \\ (2f) \ z_{k+1} \text{ be the solution of } S_h z_{k+1} = L_h d_{k+1}. \end{cases} \tag{6}$$

We are interested in the superlinear convergence property of the GCG-LS method. Denoting by c_h the unique solution of (3), we study the error vector $e_k = c_k - c_h$ and use the 'energy' norm $\|c\|_{L_h}^2 = \langle L_h c, c \rangle$. The related results are formulated by considering the preconditioned matrix as a perturbation of the identity [3]. Let us decompose our operator as

$$L = S + Q,$$

that is, letting $\gamma = \alpha - \beta$,

$$Qu = \mathbf{b} \cdot \nabla u + \gamma u. \tag{7}$$

Further, let the matrix Q_h be defined as the FDM discretization of the operator Q on the same grid ω_h as for L in (3). Then $L_h = S_h + Q_h$, hence (5) can be written as

$$(I_h + S_h^{-1} Q_h)\, c = f_h \tag{8}$$

where I_h is the corresponding identity matrix. Let us define

$$\varrho := \frac{1}{\|L_h^{-1} S_h\|} \geq \min_{c \neq 0} \frac{\langle L_h c, c \rangle}{\langle S_h c, c \rangle}.$$

If $S_h^{-1} Q_h$ is normal then the following convergence result holds [3]:

Theorem 1. *The generalized conjugate gradient method applied for equation (5) yields*

$$\left(\frac{\|e_k\|_{L_h}}{\|e_0\|_{L_h}} \right)^{1/k} \le \frac{2}{\varrho} \left(\frac{1}{k} \sum_{i=1}^{k} \left| \lambda_i(S_h^{-1} Q_h) \right| \right) \tag{9}$$

($k = 1, 2, ...$) where $\lambda_i(S_h^{-1}Q_h)$ are the eigenvalues of the matrix $S_h^{-1}Q_h$.

This shows superlinear convergence if the eigenvalues $\lambda_i(S_h^{-1}Q_h)$ accumulate in zero. If $S_h^{-1}Q_h$ is not normal then, by [3], the number of iterations increases with the order of the largest Jordan block. The numerical experiments in [9] show that the superlinear behaviour remains similar in this case.

Remark 1. When symmetric part preconditioning is used, i.e., $S_h = \frac{1}{2}(L_h + L_h^T)$, then a simpler truncated algorithm called GCG-LS(0) is applicable where only the previous search direction d_k is used (see [2] for details). Further, due to the obvious identity $\langle L_h c, c\rangle = \langle S_h c, c\rangle$, we have $\varrho = 1$ in this case in (9).

In the case of FEM discretization, when L_h and Q_h are the stiffness matrices of L and Q in a FEM subspace, the analogue of the sequence (9) can be estimated in a mesh uniform superlinear way. Our goal is to demonstrate an analogous result for certain finite difference discretizations. For this we need to find a sequence

$$\varepsilon_k \to 0$$

independently of h such that for all $h > 0$ the eigenvalues $\lambda_i(S_h^{-1}Q_h)$ satisfy

$$\frac{1}{k} \sum_{i=1}^{k} \left| \lambda_i(S_h^{-1} Q_h) \right| \le \varepsilon_k . \tag{10}$$

4 A Model Problem and the Properties of the Eigenvalues

Let us consider a special case of (1) which has been analysed in [10] in the context of linear convergence. The convection-diffusion problem

$$\begin{cases} Lu \equiv -\Delta u + \mathbf{b} \cdot \nabla u + \alpha u = g \\ u_{|\partial\Omega} = 0 \end{cases} \tag{11}$$

is posed on the unit square $\Omega := [0,1]^2 \subset \mathbf{R}^2$ with constant coefficients $\mathbf{b} = (b_1, b_2) \in \mathbf{R}^2$ and $\alpha \in \mathbf{R}$. We assume $\alpha \ge 0$, then the coercivity condition (2) holds. Similarly, in the preconditioning operator

$$Su := -\Delta u + \beta u \qquad \text{for } u_{|\partial\Omega} = 0, \tag{12}$$

we set $\beta \in \mathbf{R}$, $\beta \ge 0$.

Let ω_h be a uniform grid on $[0,1]^2$ and let us define upwind or centered differencing for the first order and centered differencing for the second order derivatives, respectively. Denote by n the number of interior gridpoints in each direction, and by $h = 1/(n+1)$ the grid parameter. Let L_h, S_h and Q_h denote the matrices corresponding to the discretizations of L, S and $Q = L - S$, respectively. Then by [10], the eigenvalues

$$\mu_{jk} := \lambda_{jk}(S_h^{-1} Q_h) \tag{13}$$

of the preconditioned matrix $S_h^{-1} Q_h$ satisfy

$$-(4+\beta h^2)\mu_{jk} + \Big((\alpha-\beta)h^2 + (b_1+b_2)h\Big)$$
$$= -2\Big(\left(\mu_{jk}^2 - \mu_{jk} b_1 h\right)^{1/2} \cos k\pi h + \left(\mu_{jk}^2 - \mu_{jk} b_2 h\right)^{1/2} \cos j\pi h\Big) \tag{14}$$

(for $j, k = 1, \ldots, n$) in the case of backward differencing and

$$-(4+\beta h^2)\mu_{jk} + (\alpha-\beta)h^2$$
$$= -2\Big(\left(\mu_{jk}^2 - b_1^2 h^2/4\right)^{1/2} \cos k\pi h + \left(\mu_{jk}^2 - b_2^2 h^2/4\right)^{1/2} \cos j\pi h\Big) \tag{15}$$

(for $j, k = 1, \ldots, n$) in the case of centered differencing.

5 Some Mesh Independent Superlinear Convergence Results

Since the eigenvalues (13) are given with double indexing, in view of (10) we wish to find a mesh independent sequence $\varepsilon_n \to 0$ independently of h such that for all $h > 0$

$$\frac{1}{n^2} \sum_{j,k=1}^{n} |\mu_{jk}| \le \varepsilon_n .$$

In general, the relations (14)-(15) lead to fourth order algebraic equations whose roots cannot be handled in explicit form. In what follows, first we consider a special class of coefficients where μ_{jk} are obtained directly and an explicit expression can be derived for ε_n. Then some numerical calculations are given which show favourable convergence rates also for other types of coefficients.

Proposition 1. *Let us consider problem* (11) *with a convection term* $\mathbf{b} = (b, b)$, *where* $b \in \mathbf{R}$ *is arbitrary, and let* $\beta := \alpha$ *in* (12). *Then, using centered differencing, the eigenvalues* $\mu_{jk} := \lambda_{jk}(S_h^{-1} Q_h)$ *satisfy*

$$\frac{1}{n^2} \sum_{j,k=1}^{n} |\mu_{jk}| \le \varepsilon_n$$

where

$$\varepsilon_n := \frac{2\sqrt{2}b}{n^2} \sum_{j,k=1}^{[(n+1)/2]} \frac{1}{\sqrt{\beta + 4k^2 + 4j^2}} \to 0 \tag{16}$$

and ε_n *is independent of* h. *(Here* $[(n+1)/2]$ *denotes the integer part of* $(n+1)/2$.*)*

Proof. In the present case relation (15) turns into

$$-(4 + \beta h^2)\mu_{jk} = -2\left(\mu_{jk}^2 - b^2h^2/4\right)^{1/2}\left(\cos k\pi h + \cos j\pi h\right) \tag{17}$$

(for $j, k = 1, ..., n$), whose roots are purely imaginary and satisfy

$$|\mu_{jk}| = \frac{bh|\cos k\pi h + \cos j\pi h|}{\sqrt{\left(4 + \beta h^2\right)^2 - 4\left(\cos k\pi h + \cos j\pi h\right)^2}}. \tag{18}$$

The numerator is at most $2bh$, and in the denominator we can use the estimates $\left(4 + \beta h^2\right)^2 \geq 16 + 8\beta h^2$ and $\left(\cos k\pi h + \cos j\pi h\right)^2 \leq 2\left(\cos^2 k\pi h + \cos^2 j\pi h\right) = 4 - 2\left(\sin^2 k\pi h + \sin^2 j\pi h\right)$. Hence we obtain

$$|\mu_{jk}| \leq \frac{bh}{\sqrt{2\left(\beta h^2 + \sin^2 k\pi h + \sin^2 j\pi h\right)}}. \tag{19}$$

If $1 \leq j, k \leq (n + 1)/2 - 1/(2h)$, then we can use the estimate $\sin t \geq (2/\pi)t$, whence the expression under the root becomes $2h^2\left(\beta + 4k^2 + 4j^2\right)$ and we obtain

$$|\mu_{jk}| \leq \frac{b}{\sqrt{2\left(\beta + 4k^2 + 4j^2\right)}}. \tag{20}$$

If j or k is greater than $(n + 1)/2$ then the possible sin values in (19) lie in the same range as for $j, k \leq (n + 1)/2$. This symmetry reason yields

$$\frac{1}{n^2} \sum_{j,k=1}^{n} |\mu_{jk}| \leq \frac{4}{n^2} \sum_{j,k=1}^{[(n+1)/2]} |\mu_{jk}|,$$

which, together with (20), implies the required estimate. Since the r.h.s. of (20) tends to 0, and ε_n is constant times the arithmetic mean of this sequence, therefore $\varepsilon_n \to 0$ as well. Finally, ε_n is obviously independent of h. ∎

Remark 2. The eigenvalue bound (20) is almost the same as the one obtained in [4] for the FEM case, differing only in the constants. Hence we have the same rate as proved there if returning to simple indices in ε_n. Namely, let i_s, j_s $(s \in \mathbf{N}^+)$ denote the indices of the eigenvalues ordered as $|\mu_{i_1,j_1}| \geq |\mu_{i_2,j_2}| \geq \ldots$ Then there holds

$$\frac{1}{k} \sum_{s=1}^{k} |\mu_{i_s,j_s}| \leq \frac{c}{\sqrt{k}} \qquad (k = 1, 2, ...)$$

where $c > 0$ is independent of k.

Table 1. $\alpha = \beta = 20$, $b_1 = b_2 = 4$

	centered differencing					upwind differencing				
	1/h					1/h				
Itr.	8	16	32	64	128	8	16	32	64	128
1	0.4225	0.4413	0.4467	0.4482	0.4486	0.4497	0.4490	0.4488	0.4487	0.4487
2	0.4225	0.4413	0.4467	0.4482	0.4486	0.4497	0.4490	0.4488	0.4487	0.4487
3	0.3816	0.4037	0.4100	0.4117	0.4122	0.4169	0.4136	0.4127	0.4124	0.4124
4	0.3611	0.3849	0.3917	0.3935	0.3940	0.4005	0.3960	0.3946	0.3943	0.3942
5	0.3488	0.3736	0.3807	0.3826	0.3831	0.3907	0.3854	0.3838	0.3834	0.3833
6	0.3407	0.3661	0.3734	0.3753	0.3758	0.3842	0.3783	0.3766	0.3761	0.3760
7	0.3251	0.3523	0.3601	0.3622	0.3627	0.3725	0.3656	0.3636	0.3631	0.3629
15	0.2548	0.2903	0.3004	0.3031	0.3038	0.3220	0.3090	0.3053	0.3043	0.3041
16	0.2492	0.2856	0.2958	0.2986	0.2993	0.3182	0.3047	0.3009	0.2999	0.2996
17	0.2426	0.2798	0.2903	0.2931	0.2938	0.3139	0.2995	0.2955	0.2944	0.2941
63	0.1063	0.1710	0.1888	0.1936	0.1948	0.2446	0.2078	0.1984	0.1960	0.1954
64	0.1046	0.1698	0.1877	0.1924	0.1937	0.2441	0.2069	0.1974	0.1949	0.1943
65		0.1685	0.1866	0.1914	0.1926		0.2059	0.1963	0.1939	0.1932
255		0.0676	0.1020	0.1109	0.1132		0.1436	0.1210	0.1157	0.1144
256		0.0673	0.1018	0.1107	0.1130		0.1435	0.1208	0.1155	0.1142
257			0.1016	0.1106	0.1128			0.1206	0.1153	0.1140

Table 2. $\alpha = \beta = 4$, $b_1 = 0$, $b_2 = 4$

	centered differencing					upwind differencing				
	1/h					1/h				
Itr.	8	16	32	64	128	8	16	32	64	128
1	0.3850	0.4033	0.4086	0.4100	0.4104	0.4097	0.4103	0.4104	0.4105	0.4105
2	0.3850	0.4033	0.4086	0.4100	0.4104	0.4097	0.4103	0.4104	0.4105	0.4105
3	0.3398	0.3577	0.3630	0.3644	0.3648	0.3695	0.3662	0.3653	0.3650	0.3650
4	0.3172	0.3349	0.3402	0.3417	0.3420	0.3495	0.3442	0.3427	0.3423	0.3422
5	0.2981	0.3198	0.3262	0.3279	0.3283	0.3326	0.3296	0.3288	0.3286	0.3285
6	0.2853	0.3097	0.3168	0.3187	0.3192	0.3214	0.3198	0.3195	0.3194	0.3194
7	0.2694	0.2948	0.3024	0.3044	0.3049	0.3085	0.3056	0.3053	0.3052	0.3051
15	0.2024	0.2315	0.2410	0.2435	0.2442	0.2547	0.2459	0.2448	0.2445	0.2445
16	0.1970	0.2265	0.2364	0.2391	0.2398	0.2503	0.2413	0.2403	0.2401	0.2400
17	0.1916	0.2217	0.2313	0.2340	0.2346	0.2461	0.2368	0.2354	0.2350	0.2349
63	0.0914	0.1271	0.1416	0.1460	0.1472	0.1446	0.1542	0.1486	0.1478	0.1477
64	0.0902	0.1261	0.1407	0.1451	0.1463	0.1435	0.1534	0.1478	0.1469	0.1468
65		0.1251	0.1398	0.1442	0.1454		0.1526	0.1469	0.1460	0.1459
255		0.0544	0.0732	0.0805	0.0826		0.0820	0.0868	0.0839	0.0835
256		0.0542	0.0730	0.0803	0.0825		0.0819	0.0867	0.0837	0.0833
257			0.0728	0.0802	0.0823			0.0866	0.0836	0.0832

The superlinear and the mesh independent behaviour of ε_k is shown by the columns and rows of the presented tables, respectively. In the left part of Table 1, centered differencing is considered for a problem in the setting of Proposition 1.

The right part of Table 1 shows that the behaviour of ε_k is almost the same for upwind differencing. Table 2 shows similar results for $\beta_1 \neq \beta_2$. (We note that [10] suggests $\beta = \mathcal{O}(b_1^2 + b_2^2)$ as an efficient choice in S, which is in correlation with this table in the sense that a smaller β in Table 2 has produced similar numerical results for $b_1 = 0$ as a greater β in Table 1 for $b_1 > 0$.)

Acknowledgments

This research was supported by the Hungarian Research Foundation OTKA, grant no. T 043765 and K 67819.

References

1. Ashby, S.F., Manteuffel, T.A., Saylor, P.E.: A taxonomy for conjugate gradient methods. SIAM J. Numer. Anal. 27(6), 1542–1568 (1990)
2. Axelsson, O.: A generalized conjugate gradient least square method. Numer. Math. 51, 209–227 (1987)
3. Axelsson, O.: Iterative Solution Methods. Cambridge University Press, Cambridge (1994)
4. Axelsson, O., Karátson, J.: Superlinearly convergent CG methods via equivalent preconditioning for nonsymmetric elliptic operators. Numer. Math. 99(2), 197–223 (2004)
5. Axelsson, O., Karátson, J.: Mesh independent superlinear PCG rates via compact-equivalent operators. SIAM J. Numer. Anal. 45(4), 1495–1516 (2007)
6. D'yakonov, E.G.: On an iterative method for the solution of finite difference equations (in Russian). Dokl. Akad. Nauk SSSR 138, 522–525 (1961)
7. Faber, V., Manteuffel, T., Parter, S.V.: On the theory of equivalent operators and application to the numerical solution of uniformly elliptic partial differential equations. Adv. Appl. Math. 11, 109–163 (1990)
8. Gunn, J.E.: The numerical solution of $\nabla \cdot a\nabla u = f$ by a semi-explicit alternating direction iterative method. Numer. Math. 6, 181–184 (1964)
9. Karátson, J., Kurics, T.: Superlinearly convergent PCG algorithms for some nonsymmetric elliptic systems. J. Comp. Appl. Math. 212(2), 214–230 (2008)
10. Manteuffel, T., Otto, J.: Optimal equivalent preconditioners. SIAM J. Numer. Anal. 30, 790–812 (1993)
11. Manteuffel, T., Parter, S.V.: Preconditioning and boundary conditions. SIAM J. Numer. Anal. 27(3), 656–694 (1990)

Self-affine Fractals Generated by Nonlinear Systems

Ljubiša Kocić[1], Sonja Gegovska-Zajkova[2], and Elena Babače[2]

[1] University of Niš, Faculty of Electronic Engineering, P.O. Box 73, 18 000 Niš, Serbia
kocic@elfak.ni.ac.yu
[2] Ss Cyril and Methodius University, Faculty of Electrical Engineering and Information Technologies, P.O. Box 574, Skopje, R.Macedonia
{szajkova,elena.babace}@feit.ukim.edu.mk

Abstract. A system of ODE's with nonlinear terms exhibits a nonlinear dynamic behavior. Under some conditions these terms can be locally approximated by linear factors, which can be, after discretization transformed in the sequence of (hyperbolic) Iterated Function Systems (IFS) that generates a unique fractal attractor. This attractor reflects the dynamics in the vicinity of the approximated point of the nonlinear system. Here, the IFS is replaced with an associate AIFS (Affine invariant IFS), a kind of IFS that has affine invariance property and permits further manipulating of this fractal attractor.

1 Introduction

Let $S = [s_{ij}]_{i,j=1}^{m}$ be an $m \times m$ row-stochastic real matrix (its rows sum up to 1). We refer to the linear mapping $\mathcal{L} : \mathbb{R}^m \to \mathbb{R}^m$, such that $\mathcal{L}(\mathbf{x}) = S^T\mathbf{x}$ as the linear mapping *associated with* S. The mapping *projection*, $\text{proj}_\perp : \mathbb{R}^m \to \mathbb{R}^{m-1}$ is defined by $\text{proj}_\perp(x_1, x_2, \ldots, x_m) = (x_1, x_2, \ldots, x_{m-1})$. The mapping $w : \mathbb{R}^{m-1} \to \mathbb{R}^{m-1}$ is called *(orthogonal) projection* of $\mathcal{L}$ if $\text{proj}_\perp(\mathcal{L}(\mathbf{x})) = w(\text{proj}_\perp(\mathbf{x}))$.

Definition 1. Let $\mathbf{T} = \text{conv}\{\mathbf{T}_1, \mathbf{T}_2, \ldots, \mathbf{T}_m\}$ $(m > 1)$ be a non-degenerate simplex and let $\{S_i\}_{i=1}^{n}$ $(n > 1)$ be a set of real square nonsingular row-stochastic matrices of order m. The system $\Omega(\mathbf{T}) = \{\mathbf{T}; S_1, S_2, \ldots, S_n\}$ is called *hyperbolic affine invariant IFS (AIFS) in* $\mathbb{R}^m$, provided that the linear mappings associated with S_i are contractions in $(\mathbb{R}^m, d_E)$, where d_E is Euclidean metric ([3]).

The properties of the AIFS include easy control over continuity property, interpolation property, inclusive property, symmetry property, ability to produce smooth curve and convex hull property. But, what is the most important, AIFS has affine invariance property, which is highly desirable especially in processes of modeling.

Theorem 1. *For any affine mapping* w, $w(\mathbf{x}) = A\mathbf{x} + \mathbf{b}$, $\mathbf{x} \in \mathbb{R}^{m-1}$, *there is one and only one* $\mathcal{L}$ *such that* w *is the orthogonal projection of* $\mathcal{L}$. *More precisely, if* S *is* $m \times m$ *row-stochastic matrix associated with* $\mathcal{L}$, *than*

S. Margenov, L.G. Vulkov, and J. Waśniewski (Eds.): NAA 2008, LNCS 5434, pp. 353–360, 2009.

$$S^{\mathrm{T}}Q = \begin{bmatrix} A & \vdots & \mathbf{b} \\ \cdots\cdots & & \cdots \\ \mathbf{c}^T & \vdots & s_{mm} \end{bmatrix}, \tag{1}$$

where

$$\mathbf{c} = \begin{bmatrix} s_{1m} - s_{mm} \\ s_{2m} - s_{mm} \\ \vdots \\ s_{m-1,m} - s_{mm} \end{bmatrix}, \quad Q = \begin{bmatrix} I_{m-1} & \vdots & \mathbf{0} \\ \cdots & & \cdots\cdots \\ -\mathbf{1} & \vdots & 1 \end{bmatrix}, \quad I_{m-1} = [\delta_{ij}]_{i,j=1}^{m-1}.$$

2 Local Linear Attractor of a Simple Dynamical System

Let the functions F_1 and F_2 map $D \subset \mathbb{R}^2$ into $\mathbb{R}$, where D is some simple compact in $\mathbb{R}^2$ and $F_i \in C^r$, $r \geq 2$, $i = 1, 2$. Let the following system of differential equations

$$\begin{cases} \dfrac{dx_1}{dt} = F_1(x_1,\, x_2), \\[2ex] \dfrac{dx_2}{dt} = F_2(x_1,\, x_2), \end{cases} \tag{2}$$

having general solution $x_1 = \varphi(t, C_1, C_2),\ x_2 = \psi(t, C_1, C_2),\ t \in [a, b], C_1, C_2 \in \mathbb{R}$, characterizes a simple dynamical system.

Further, let $\alpha \in [a, b]$ and let the constants C_1 and C_2 be specified so that $x_{1,0} = \varphi(\alpha, C_1, C_2)$ and $x_{2,0} = \psi(\alpha, C_1, C_2)$. Obviously, $(x_{1,0}, x_{2,0}) \in D$. Now, let

$$\tau = \{a = t_0,\, t_1,\, \ldots,\, b = t_N\} \tag{3}$$

be an equidistant mesh on $[a,\, b]$ with a step h $(h > 0)$.

Applying Taylor expansion ([4]) on both F_1 and F_2 at the point $(x_{1,0}; x_{2,0}) \in D$, we obtain

$$\frac{dx_i}{dt} \approx F_i(x_{1,0};\, x_{2,0}) + \frac{\partial F_i(x_{1,0};\, x_{2,0})}{\partial x_1}(x_1 - x_{1,0}) + \frac{\partial F_i(x_{1,0};\, x_{2,0})}{\partial x_2}(x_2 - x_{2,0}), \tag{4}$$

where $i = 1, 2$.

Discretization of the last system yields:

$$\begin{aligned} \frac{x_{i,n+1} - x_{i,n}}{h} &\approx F_i(x_{1,0};\, x_{2,0}) + \frac{\partial F_i(x_{1,0};\, x_{2,0})}{\partial x_1}(x_{1,n} - x_{1,0}) + \\ &+ \frac{\partial F_2(x_{1,0};\, x_{2,0})}{\partial x_2}(x_{2,n} - x_{2,0}), \qquad i = 1, 2, \end{aligned} \tag{5}$$

or

$$
\begin{aligned}
x_{1,n+1} \approx & \left[1+\frac{\partial F_1(x_{1,0};\, x_{2,0})}{\partial x_1}h\right] x_{1,n} + \frac{\partial F_1(x_{1,0};\, x_{2,0})}{\partial x_2} h x_{2,n} + \\
& + h\left[F_1(x_{1,0};\, x_{2,0}) - \frac{\partial F_1(x_{1,0};\, x_{2,0})}{\partial x_1}x_{1,0} - \frac{\partial F_1(x_{1,0};\, x_{2,0})}{\partial x_2}x_{2,0}\right], \\
x_{2,n+1} \approx & \frac{\partial F_2(x_{1,0};\, x_{2,0})}{\partial x_1} h x_{1,n} + \left[1+\frac{\partial F_2(x_{1,0};\, x_{2,0})}{\partial x_2}h\right] x_{2,n} + \\
& + h\left[F_2(x_{1,0};\, x_{2,0}) - \frac{\partial F_2(x_{1,0};\, x_{2,0})}{\partial x_1}x_{1,0} - \frac{\partial F_2(x_{1,0};\, x_{2,0})}{\partial x_2}x_{2,0}\right].
\end{aligned}
\tag{6}
$$

For fixed $(x_{1,0};\, x_{2,0})$, i.e. for fixed α, we introduce the following notations

$$
\begin{aligned}
A_\alpha &= 1+\frac{\partial F_1(x_{1,0};\, x_{2,0})}{\partial x_1}h, \qquad B_\alpha = \frac{\partial F_1(x_{1,0};\, x_{2,0})}{\partial x_2}h, \\
E_\alpha &= h\left[F_1(x_{1,0};\, x_{2,0}) - \frac{\partial F_1(x_{1,0};\, x_{2,0})}{\partial x_1}x_{1,0} - \frac{\partial F_1(x_{1,0};\, x_{2,0})}{\partial x_2}x_{2,0}\right], \\
C_\alpha &= \frac{\partial F_2(x_{1,0};\, x_{2,0})}{\partial x_1}h, \qquad D_\alpha = 1+\frac{\partial F_2(x_{1,0};\, x_{2,0})}{\partial x_2}h, \\
F_\alpha &= h\left[F_2(x_{1,0};\, x_{2,0}) - \frac{\partial F_2(x_{1,0};\, x_{2,0})}{\partial x_1}x_{1,0} - \frac{\partial F_2(x_{1,0};\, x_{2,0})}{\partial x_2}x_{2,0}\right].
\end{aligned}
\tag{7}
$$

Then, the system (6) can be rewritten in the form:

$$
\begin{aligned}
x_{1,n+1} &= A_\alpha x_{1,n} + B_\alpha x_{2,n} + E_\alpha \\
x_{2,n+1} &= C_\alpha x_{1,n} + D_\alpha x_{2,n} + F_\alpha .
\end{aligned}
$$

For the square matrix M_α and the vector $\mathbf{n}_\alpha$, such that

$$
M_\alpha = \begin{bmatrix} A_\alpha & B_\alpha \\ C_\alpha & D_\alpha \end{bmatrix}, \quad \mathbf{n}_\alpha = \begin{bmatrix} E_\alpha \\ F_\alpha \end{bmatrix}, \tag{8}
$$

the following theorem holds ([4]).

Theorem 2. *If $h > 0$ is such that the matrix M_α, defined by (8), satisfies the inequality*

$$
||M_\alpha||_{sp} < 1 \tag{9}
$$

(where $||\cdot||_{sp}$ denotes spectral norm), then

$$
\left[\frac{\partial F_1}{\partial x_1}+\frac{\partial F_2}{\partial x_2}+\left(\frac{\partial F_1}{\partial x_1}\frac{\partial F_2}{\partial x_2}-\frac{\partial F_1}{\partial x_2}\frac{\partial F_2}{\partial x_1}\right)h\right]^2 > \left(\frac{\partial F_1}{\partial x_1}-\frac{\partial F_2}{\partial x_2}\right)^2+\left(\frac{\partial F_1}{\partial x_2}+\frac{\partial F_2}{\partial x_1}\right)^2 . \tag{10}
$$

Moreover, if (10) *and*

$$1 - 2\left(\frac{\partial F_1}{\partial x_1} + \frac{\partial F_2}{\partial x_2}\right)h > \left(\left(\frac{\partial F_1}{\partial x_1}\right)^2 + \left(\frac{\partial F_1}{\partial x_2}\right)^2 + \left(\frac{\partial F_2}{\partial x_1}\right)^2 + \left(\frac{\partial F_2}{\partial x_2}\right)^2\right)h^2, \tag{11}$$

are satisfied, then (9) *holds.*

It is known that the spectral norm of a given matrix is equal to the norm of the matrix induced by the Euclidean vector norm ([5]), i.e. if

$$||A||_2 = \max_{||\mathbf{x}|| \neq 0} \frac{||A\mathbf{x}||_2}{||\mathbf{x}||_2},$$

then

$$||A||_{sp} = ||A||_2. \tag{12}$$

So, if $||A||_{sp} < 1$, then w is contraction, since

$$d_E(w(\mathbf{x}), w(\mathbf{y})) = ||w(\mathbf{x}) - w(\mathbf{y})||_2 = ||A\mathbf{x} - A\mathbf{y}||_2 = ||A(\mathbf{x} - \mathbf{y})||_2$$
$$\leq ||A||||\mathbf{x} - \mathbf{y}||_2 = ||A||_{sp}||\mathbf{x} - \mathbf{y}||_2 < ||\mathbf{x} - \mathbf{y}||_2 = d_E(\mathbf{x}, \mathbf{y}).$$

Therefore, the following corollary is obvious.

Corollary 1. *If the functions system* (2) *for some* h *satisfies* (10) *and* (11), *then the mapping* $w_\alpha : \mathbb{R}^2 \to \mathbb{R}^2$ *given by* $\omega_\alpha(\mathbf{x}) = M_\alpha \mathbf{x} + \mathbf{n}_\alpha$, M_α *and* $\mathbf{n}_\alpha$ *are defined in* (8), *is a contraction.*

Let $\{\alpha_1, \alpha_2, \ldots, \alpha_n\}$ be a finite set of real numbers from the interval $[a, b]$. Then, using the same procedure described above, one can obtain a corresponding set of affine mappings $\{(M_{\alpha_1}, \mathbf{n}_{\alpha_1}), (M_{\alpha_2}, \mathbf{n}_{\alpha_2}), \ldots, (M_{\alpha_n}, \mathbf{n}_{\alpha_n})\}$. If all mappings from the last set are contractions, then they define a hyperbolic IFS. According to Banach's theorem, there exists a unique attractor of this IFS.

Also, if there are two or more (finite number) systems of differential equations, and corresponding contractions to each of the system, then again hyperbolic IFS can be obtained.

According to (1), the IFS obtained by this procedure can be easily transformed into AIFS.

Corollary 2. *The AIFS associated with* (2) *and a finite set of real numbers* $\{\alpha_1, \alpha_2, \ldots, \alpha_n\}$ *is given by* $\{\mathbf{T}; S_1, S_2, \ldots, S_n\}$, *where*

$$S_i = (Q^{-1})^{\mathrm{T}} \begin{bmatrix} M_{\alpha_i}^{\mathrm{T}} & \vdots & \mathbf{c}_i \\ \cdots & \cdots & \cdots \\ \mathbf{n}_{\alpha_i}^T & \vdots & s_{mm} \end{bmatrix}. \tag{13}$$

The real power of AIFS lies in ability of having more precise information of the location of the attractor. The next theorem is an example of this.

Theorem 3. *Let the system* (2) *and the set of real numbers* $\{\alpha_1, \alpha_2, \dots, \alpha_n\}$ *define hyperbolic AIFS* $S = \{\mathbf{T}; S_1, \dots, S_n\}$, *where* $\mathbf{T}$ *is a 2D symplex (triangle). Let the attractor of* S *be* A_S. *If*

$$\begin{aligned} &A_{\alpha_i} + E_{\alpha_i} > 0,\, B_{\alpha_i} + F_{\alpha_i} > 0,\, A_{\alpha_i} + B_{\alpha_i} + E_{\alpha_i} + F_{\alpha_i} \le 1, \\ &C_{\alpha_i} + E_{\alpha_i} > 0,\, D_{\alpha_i} + F_{\alpha_i} > 0, C_{\alpha_i} + D_{\alpha_i} + E_{\alpha_i} + F_{\alpha_i} \le 1, \end{aligned} \tag{14}$$

then $A_S \subset \mathbf{T}$.

Proof. By (13), having in mind that $\left(Q^{-1}\right)^{\mathrm{T}} = \begin{bmatrix} 1 & 0 & 1 \\ 0 & 1 & 1 \\ 0 & 0 & 1 \end{bmatrix}$, one gets

$$S_i = \begin{bmatrix} A_{\alpha_i} + E_{\alpha_i} & B_{\alpha_i} + F_{\alpha_i} & s_{13} \\ C_{\alpha_i} + E_{\alpha_i} & D_{\alpha_i} + F_{\alpha_i} & s_{23} \\ E_{\alpha_i} & F_{\alpha_i} & s_{33} \end{bmatrix},$$

where A_{α_i}, B_{α_i}, C_{α_i}, D_{α_i}, E_{α_i} and F_{α_i} are given by (7), and s_{13}, s_{23} and s_{33} have appropriate values making rows of S_i sum up to 1. It is known that rows of S_i represent *barycentric* (more correctly *areal*) coordinates with respect to 2D simplex $\mathbf{T} = \operatorname{conv}\{\mathbf{T_1}, \mathbf{T_2}, \mathbf{T_3}\}$. On the other hand, if $\mathbf{T}$ is given by the 2×3 matrix $T = [\mathbf{T_1} \quad \mathbf{T_2} \quad \mathbf{T_3}]$, where $\mathbf{T_i}$ are column-vectors of the vertices of $\mathbf{T}$, then $S_i T$ is again a 2×3 matrix representing a new simplex $\mathbf{T}^i$, obtained by $\mathbf{T}$ by the linear mapping defined by the matrix S_i. If (14) holds, then each vertex of $\mathbf{T}^i$ is a convex combination of vertices $\mathbf{T_1}$, $\mathbf{T_2}$ and $\mathbf{T_3}$. Thus $\mathbf{T}^i \subset \mathbf{T}$, for all i. The union of these image simplices, $W(\mathbf{T}) = \bigcup_{i=1}^{n} \mathbf{T}^i$ is Hutchinson operator associated with the AIFS S. It is clear that $\mathbf{T}^i \subset \mathbf{T} \Rightarrow W(\mathbf{T}) \subset \mathbf{T}$ so $A_s = \lim_{k\to\infty} W^{\circ k}(\mathbf{T}) \subset \mathbf{T}$. □

3 Numerical Experiments

In order to test this approach we made some experiments. Numerical calculations in all examples were performed by MATHEMATICA 6.0 package.

Example 1. Let us consider the system (known as Brusselator, dynamic model of tri-molecular chemical reaction)

$$\begin{aligned} \dot{x} &= a - (b+1)x + x^2 y \\ \dot{y} &= bx - x^2 y \end{aligned},$$

with $a = 1.3$ and $b = 2.9$. Note that $b > 1 + a^2$ which means that the system is in unstable zone. Using an equidistant mesh on the segment $[0, 50]$ with the step $h = 0.05$ and procedure described above, we construct two mappings: $\omega_\alpha(\mathbf{x}) = M_\alpha \mathbf{x} + \mathbf{n}_\alpha$, defined by the point $(1.2, 0.9)$, where

$$M_\alpha = \begin{bmatrix} 0.913 & 0.072 \\ 0.037 & 0.928 \end{bmatrix}, \quad \mathbf{n}_\alpha = \begin{bmatrix} -0.0646 \\ 0.1296 \end{bmatrix},$$

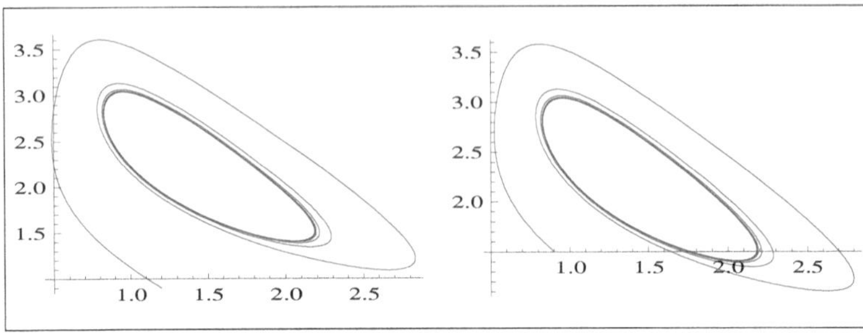

Fig. 1. Numerical solutions of the system through the points (1.2,0.9) and (0.9,1.5)

and $\omega_\beta(\mathbf{x}) = M_\beta\mathbf{x} + \mathbf{n}_\beta$, defined by the point $(0.9, 1.5)$, where

$$M_\beta = \begin{bmatrix} 0.94 & 0.0405 \\ 0.01 & 0.9595 \end{bmatrix}, \quad \mathbf{n}_\beta = \begin{bmatrix} -0.0565 \\ 0.1215 \end{bmatrix}.$$

The corresponding AIFS is represented by the matrices S_α and S_β:

$$S_\alpha = \begin{bmatrix} 0.8484 & 0.1666 & -0.015 \\ 0.0074 & 1.0576 & 0.065 \\ -0.0646 & 0.1296 & 0.935 \end{bmatrix}, \quad S_\beta = \begin{bmatrix} 0.8835 & 0.1315 & -0.015 \\ -0.016 & 1.081 & -0.065 \\ -0.0565 & 0.1215 & 0.935 \end{bmatrix}$$

In the Figure 1 numerical solutions of the system in two chosen points are presented. For the spectral norms of the matrices we obtain

$$\|M_\alpha\|_{sp} = 0.97568 \text{ and } \|M_\beta\|_{sp} = 0.976939.$$

The attractors of this IFS, as well as of the corresponding AIFS, are shown in Figure 2.

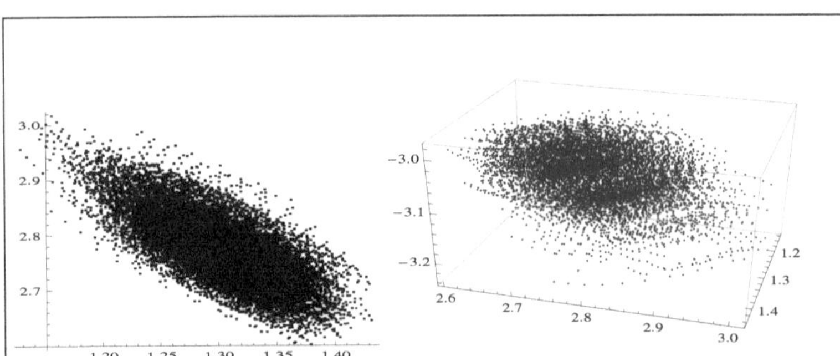

Fig. 2. The attractor of the IFS (left) and the corresponding AIFS (right)

Example 2. From the following system

$$\dot{x} = -y - x(2.5 + x^2 + y^2)$$
$$\dot{y} = x - y(2.5 + x^2 + y^2)$$

using an equidistant mesh on the segment $[0, 15]$ with a step $h = 0.05$ two mappings can be constructed:
$\omega_\alpha(\mathbf{x}) = M_\alpha \mathbf{x} + \mathbf{n}_\alpha$, defined by the point $(0, -0.05)$, where

$$M_\alpha = \begin{bmatrix} 0.874875 & -0.05 \\ 0.05 & 0.874625 \end{bmatrix}, \quad \mathbf{n}_\alpha = \begin{bmatrix} 0 \\ -0.0000125 \end{bmatrix},$$

and $\omega_\beta(\mathbf{x}) = M_\beta \mathbf{x} + \mathbf{n}_\beta$, defined by the point $(-1, 0.1)$, where

$$M_\beta = \begin{bmatrix} 0.7245 & -0.04 \\ 0.06 & 0.8235 \end{bmatrix}, \quad \mathbf{n}_\beta = \begin{bmatrix} -0.101 \\ 0.0101 \end{bmatrix}.$$

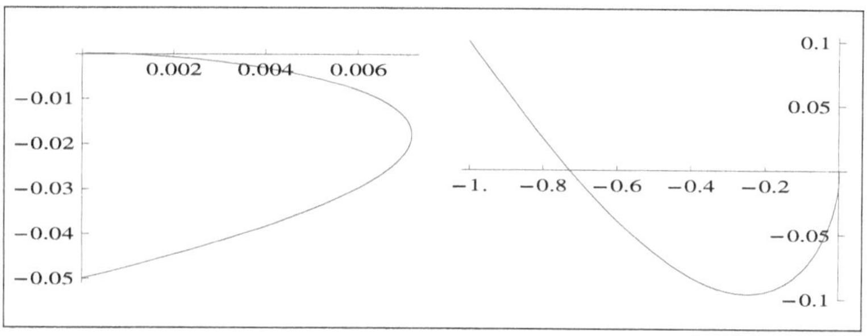

Fig. 3. Numerical solutions of the system through the points (0,-0.05) and (-1, 0.1)

The corresponding AIFS is represented by the matrices S_α and S_β,

$$S_\alpha = \begin{bmatrix} 0.874875 & 0.04999875 & 0.0751375 \\ -0.05 & 0.874613 & 0.175388 \\ 0 & -0.0000125 & 1.00001 \end{bmatrix},$$

$$S_\beta = \begin{bmatrix} 0.6235 & 0.0701 & 0.3064 \\ -0.141 & 0.8336 & 0.3074 \\ -0.101 & 0.0101 & 1.0909 \end{bmatrix}.$$

In the Figure 3 numerical solutions of the system in two chosen points are presented.

Since the spectral norms of the matrices

$$\|M_\alpha\|_{sp} = 0.876303 \text{ and } \|M_\beta\|_{sp} = 0.826113.$$

are less than 1, the mappings are contractions and they define a hyperbolic IFS, whose unique attractor is shown in Figure 4 (left). The attractor of the corresponding AIFS is also shown on the Figure 4 (right).

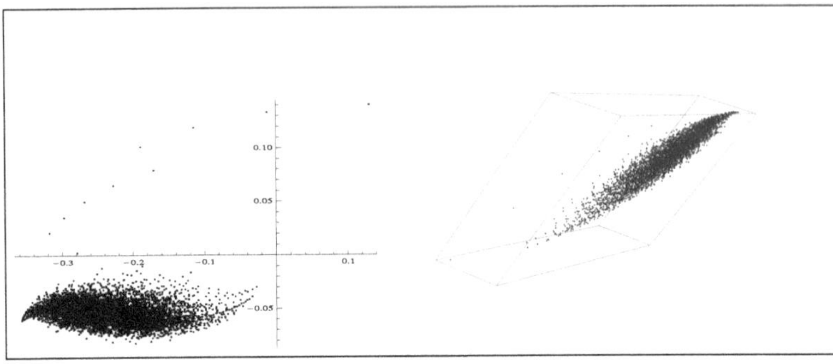

Fig. 4. The attractor of the IFS (left) and the corresponding AIFS (right)

4 Conclusion

Dynamical system may be considered as a system of nonlinear mappings. Under some conditions nonlinear terms in dynamical system can be locally approximated by linear factors. After discretization described above, these factors can be transformed in the sequence of hyperbolic Iterated Function Systems that generates a unique fractal attractor. Replacement of these IFS with associate affine invariant IFS enable us to use many adventages of AIFS over IFS, such as convex hull, continuity, interpolation, and especially affine invariant property, which is very important in the processes of modeling fractal sets.

As a confirmation of this approach, two examples are given. Both of them consider system of two differential equations.

Notes and Comments. All examples are made by using MATHEMATICA 6.0 package and original software written by authors. All programs are available upon request.

The authors are obliged to the referees for their valuable suggestions.

References

1. Clarke, F., Ekeland, I.: Nonlinear oscillations and boundary-value problems for Hamiltonian systems. Arch. Rat. Mech. Anal. 78, 315–333 (1982)
2. Barnsley, F.M.: Fractals Everywhere. Academic Press, San Diego (1993)
3. Kocić, L.M., Simoncelli, A.C.: Shape predictable IFS representations. In: Novak, M.M. (ed.) Emergent Nature, pp. 435–436. World Scientific, Singapore (2002)
4. Kocić, L.M., Gegovska-Zajkova, S., Babače, E.: Nonlinear systems and iterated function system, Differential geometry - Dynamical Systems. In: Balkan Society of geometers, vol. 10, pp. 197–205, M.S.C. 2000: 34A34, 28A80. Geometry Balkan Press (2008)
5. Meyer, C.D.: Matrix analysis and applied linear algebra. SIAM, Philadelphia (2000)
6. Sprott, J.C.: Chaos and time series analysis. Oxford University Press, Oxford (2003)

Numerical Modelling of Cellular Immune Response to Virus

Mikhail K. Kolev

Department of Mathematics and Computer Science,
University of Warmia and Mazury,
Olsztyn, Zolnierska 14, 10-561, Poland
kolev@matman.uwm.edu.pl

Abstract. We present a mathematical model of cellular immune response to viral infection. The model is a bilinear system of integro-differential equations of Boltzmann type. Results of numerical experiments are presented.

Keywords: Numerical modelling, kinetic model, integro-differential equations, nonlinear dynamics, virus, immune system.

1 Introduction

Various viruses cause diseases, some of which like AIDS, hepatitis etc. are quite dangerous for people. Over the past decades serious advance in understanding the mechanisms of the interactions between viruses and immune system has been achieved by the use of *in vitro* and *in vivo* experiments. Clinical and experimental observations have been successfully complemented by mathematical models. Numerical and mathematical modelling provide an essential tool for description and prediction of the complex and highly nonlinear dynamics of the virus - immune system interactions [2], [12], [14].

Viruses are intracellular pathogens. In order to reproduce, they have to enter susceptible cells and use the metabolic mechanisms of these host cells [17]. The defense system can apply innate and adaptive responses against the invading pathogens. The innate immune mechanisms are nonspecific and use various physical barriers and changes as well as immune cells possessing no memory. The adaptive (acquired) defense mechanisms are able to specifically recognize the physical structure of the foreign pathogens. This leads to the activation and expansion of populations of immune cells, which are able to fight the infection. The acquired immunity can be subdivided into two main categories: (i) humoral immunity, and (ii) cell-mediated (or cellular) immunity. The humoral responses are performed by antibodies, which are produced by B cells. The main immune cells involved in the cellular immunity are T lymphocytes. They include cytotoxic T lymphocytes (CTLs) and T helper (T_h) cells. The cytotoxic T lymphocytes are able to destroy infected cells. The main function of the T helper cells is to produce cytokines and signals inducing the proliferation and activation of

S. Margenov, L.G. Vulkov, and J. Waśniewski (Eds.): NAA 2008, LNCS 5434, pp. 361–368, 2009.

immune cells [11], [13]. A mathematical model of the interactions between the virus and the humoral immunity has been recently proposed and analysed in [9], [10]. Here, we propose a model describing the cellular immune response to viral infection.

The organization of the paper is as follows. In Section 2 we present the interacting populations and the mathematical model. Numerical approximations to the solutions of the model are constructed in Section 3. Results of our numerical experiments are presented in Section 4.

2 Mathematical Model

Following the idea of Wodarz et al. [18], [19] we consider the following four interacting populations, each denoted by the corresponding subscript i (see Table 1).

Table 1. Virus-cellular immune system dynamics variables

Variable i	Abbreviation	Population	Activation state $u \in [0, 1]$
1	Uninfected T_h	Uninfected helper T cells	0.5
2	Infected T_h	Infected helper T cells	virus replication, T_h destruction
3	Virus	Free virus particles	rate of infection of T_h
4	CTLs	Cytotoxic T lymphocytes	destruction of infected T_h

The interacting individuals (cells or particles) are characterized by a microscopic state variable $u \in [0, 1]$, which describes the specific biological function (activity) of each individual.

In our model, the state of activity of the infected helper T cells denotes the virus mediated killing rate of the infected cells as well as the rate of viral reproduction inside the host cell. We assume that the T helper cells infected by cytopathic viruses (i.e. viruses able to shorten the life-span of the host cells at a higher rate) possess higher activation states. Moreover, the infected cells with higher states of activity are assumed to produce larger amount of virus particles.

Here, the state of activity of free viruses denotes their ability to infect the susceptible T_h cells. The higher the ability of a virus to enter a cell, the higher the activation state of the virus.

In our model, we assume that the state of activity of the CTLs denotes their ability to destroy the infected T_h cells.

Here, the presence of internal degree of freedom of the population of the uninfected helper T cells is neglected. We assume that only some fixed state of activity of these cells (say $u = 0.5$) is possible.

The meaning of the states of activity of the interacting populations is presented in Table 1.

We denote by

$$f_i(t, u), \quad f_i : [0, \infty) \times [0, 1] \to R_+, \quad i = 1, \dots, 4,$$

the distribution density of the i-th population with activation state $u \in [0,1]$ at time $t \geq 0$. Moreover, let

$$n_i(t) = \int_0^1 f_i(t,u)du, \quad n_i : [0,\infty) \to R_+, \quad i = 1,\ldots,4, \tag{1}$$

be the concentration of the i-th individuals at time $t \geq 0$.

Further, we present a mathematical model describing the dynamics of the distribution densities of the interacting populations. Respective gain, loss and conservative terms corresponding to the most important processes of production, destruction and change of activity of the individuals are included in the system (2)-(5) of partial integro-differential equations. The modelling approach is similar to those used in nonequilibrium statistical mechanics and generalized kinetic theory. This is the reason for using the term "kinetic models". Their application in immunology has been introduced by Bellomo and Forni for modelling tumor growth [3] and has been later developed in a series of papers, cf. [6], [7] as well as the special issues [4], [5] for reference. The properties of the Generalized Kinetic Models have been extensively studied, see [1] for complete bibliography.

Our model of the interactions between the virus and the cellular immune system is given by the following system of partial integro-differential equations.

$$\tfrac{d}{dt}n_1(t) = S_1(t) - d_{11}n_1(t) - d_{13}n_1(t)\int_0^1 vf_3(t,v)dv, \tag{2}$$

$$\begin{aligned}\tfrac{\partial f_2}{\partial t}(t,u) = {} & p_{13}^{(2)}(1-u)n_1(t)\int_0^1 vf_3(t,v)dv - d_{24}f_2(t,u)\int_0^1 vf_4(t,v)dv \\ & -d_{22}uf_2(t,u) + c_{22}\left(2\int_0^u (u-v)f_2(t,v)dv - (1-u)^2 f_2(t,u)\right),\end{aligned} \tag{3}$$

$$\tfrac{\partial f_3}{\partial t}(t,u) = p_{22}^{(3)}\int_0^1 vf_2(t,v)dv - d_{33}f_3(t,u), \tag{4}$$

$$\tfrac{\partial f_4}{\partial t}(t,u) = p_{13}^{(4)}(1-u)n_1(t)\int_0^1 f_3(t,v)dv - d_{44}f_4(t,u), \tag{5}$$

with nonnegative initial conditions

$$n_1(0) = n_1^{(0)}, \quad f_i(0,u) = f_i^{(0)}(u), \quad i = 2,3,4.$$

All parameters denoted by $p_{ij}^{(k)}$, d_{ij} and c_{ij} are assumed to be nonnegative and $p_{13}^{(2)} = 2d_{13}$.

The function $S_1(t)$ denotes the rate of production of uninfected T helper cells. The parameter d_{11} characterizes the natural death of the uninfected cells, which become infected by the virus with a rate proportional to their concentration as well as to the activation state of the virus (see equation (2)).

The factor $(1-u)$ in the gain term of equation (3) describes our assumption that the activity of the newly infected T helper cells is low. This is connected with the experimental observations showing that the virus needs some time after entering the host cell in order to replicate. During this period the virus particle uncoats and the viral genome is exposed. Subsequently, the viral genome is replicated and viral proteins are made. New virus particles are produced after the association of the newly generated viral proteins with the viral genomes [17]. The rate of destruction of the infected cells by the virus is assumed to be higher for cells with higher states of activity. It is described by the loss term

$$d_{22} u f_2(t, u).$$

The parameter d_{24} characterizes the rate of destruction of infected cells by CTLs which is assumed to be proportional to the activation state of CTLs. The replication of the virus particles inside the infected cells leads to increase in the probability of the destruction of the infected cells by the virus. We describe this by the conservative term

$$c_{22}\left(2\int_0^u (u-v) f_2(t,v)dv - (1-u)^2 f_2(t,u)\right)$$

corresponding to raising the activation states of the infected cells (see equation (3)).

The parameter $p_{22}^{(3)}$ characterizes the rate of reproduction of the virus inside the host cells, which is assumed to be proportional to the activation state of the infected cells (see equation (4)). The parameter d_{33} characterizes the natural death of viruses.

There is experimental evidence that the newly produced CTLs need time for their development and activation [11]. The factor $(1-u)$ in the gain term of equation (5) describes our assumption that the activity of the newly generated CTLs is low. The rate of generation of the CTLs is assumed to be proportional to the concentrations of the uninfected helper T cells and of the virus, both of which stimulate the proliferation of cytotoxic T lymphocytes [13]. The parameter d_{44} characterizes the natural death of CTLs.

The simplifying assumptions in the model (2)-(5) were selected with the aim to make the model relatively simple as well as to describe the most important features of the interactions between the virus and the cellular immune system.

3 Approximate Solution of the Model

Here, we construct a numerical solution to the concentrations of individuals $n_i(t)$, $i = 1, \ldots, 4$, at any time variable $t > 0$. The concentrations $n_2(t)$, $n_3(t)$, and $n_4(t)$ can be computed from (1) by the use of the functions $f_2(t,u)$, $f_3(t,u)$, and $f_4(t,u)$. To compute numerical approximations to the functions $n_1(t)$, $f_2(t,u)$, $f_3(t,u)$, and $f_4(t,u)$, we perform a discretization of the system (2)-(5) with respect to the state of activity $u \in [0,1]$ by applying the uniform grid-points

$$u_i = i\Delta u, \quad i = 0, \ldots, N,$$

where N is a positive integer and $\Delta u = 1/N$. Then the values $f_2(t,u)$, $f_3(t,u)$, and $f_4(t,u)$ in (2)-(5) can be replaced by their approximations

$$f_j(t, u_i) \approx f_{j,i}(t), \quad j = 2, 3, 4, \tag{6}$$

at the state grid-points $u_i \in [0,1]$.

For every $t > 0$ and every $u_i \in [0, 1]$ with $i = 0, \ldots, N$, we apply the approximations (6) for quadrature formulae to approximate the integrals:

$$\begin{aligned} &\int_0^1 f_j(t,v)dv &&\approx Q_0^N\Big[f_j(t,v)\Big], \quad j=3, \\ &\int_0^1 v f_j(t,v)dv &&\approx Q_0^N\Big[v f_j(t,v)\Big], \quad j=2,3,4, \\ &\int_0^{u_i} (u_i - v) f_j(t,v)dv &&\approx Q_0^i\Big[(u_i - v) f_j(t,v)\Big], \quad j=2. \end{aligned} \tag{7}$$

The approximations in (7) represent arbitrary quadratures. For example, in Section 4, the values $Q_0^N\Big[f_j(t,v)\Big]$, $Q_0^N\Big[v f_j(t,v)\Big]$ and $Q_0^i\Big[(u_i - v) f_j(t,v)\Big]$ are computed by the use of the composite Simpson's rule [8], [16].

The approximations (6) and (7) applied to the partial integro-differential system (2)-(5) yield a system of ordinary differential equations.

This system is solved in Section 4. Its numerical solutions $f_{j,i}(t)$, with $j = 2, 3, 4$ and $i = 0, \ldots, N$, are then used to compute the approximations to the functions $n_2(t)$, $n_3(t)$, and $n_4(t)$. The approximations are computed from

$$n_j(t) \approx Q_0^N\Big[f_j(t,v)\Big], \quad j = 2, 3, 4. \tag{8}$$

4 Numerical Experiments and Discussion

The system of ordinary differential equations corresponding to the discretized model (2)-(5) was solved by using the code `ode15s` from the Matlab ODE suite [15] with $RelTol = 10^{-3}$ and $AbsTol = 10^{-4}$. With the choice of the parameter values used for our numerical experiments, the resulting ODE system is stiff. Therefore, we applied the `ode15s` solver, which is designed for stiff ODEs. In this case, the more commonly used higher-order `ode45` solver worked longer than ode15s.

The obtained approximate solutions for the functions $f_{2,i}(t)$, $f_{3,i}(t)$, and $f_{4,i}(t)$, with $i = 0, \ldots, N$, were applied to (8) to compute the concentrations $n_2(t)$, $n_3(t)$, and $n_4(t)$.

As initial conditions we assumed the presence of uninfected T helper cells, CTLs, free virus particles, and the absence of infected T helper cells, setting for every $i = 0, \ldots, N$:

$$n_1(0) = 1, \quad f_{2,i}(0) = 0, \quad f_{3,i}(0) = 0.1, \quad f_{4,i}(0) = 0.1.$$

The values of the parameters of the model were set as follows:

$$S_1(t) = 100, \quad t \geq 0,$$

$$d_{22} = d_{24} = 50, \quad c_{22} = 10,$$

$$d_{11} = d_{33} = d_{44} = p_{22}^{(3)} = p_{13}^{(4)} = 100.$$

Through our simulations we studied the role of the parameter d_{13} for the dynamics of the solutions to the system (2)-(5). This parameter describes the

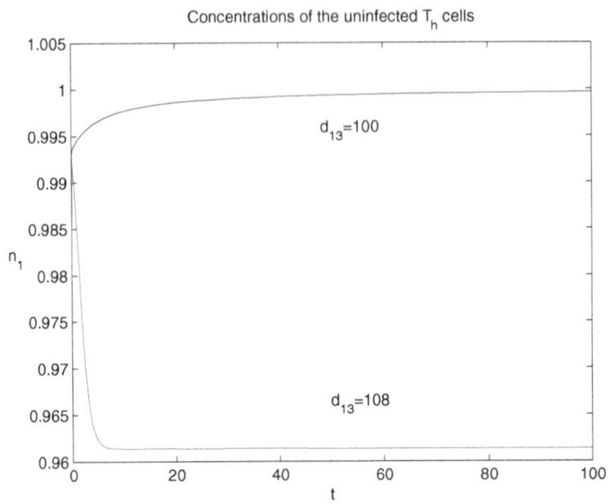

Fig. 1. Dynamics of the uninfected cells for $d_{13} = 100$ and $d_{13} = 108$

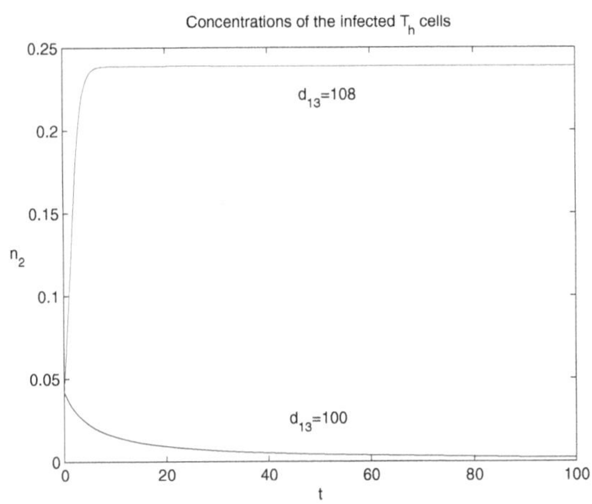

Fig. 2. Dynamics of the infected cells for $d_{13} = 100$ and $d_{13} = 108$

rate of viral infectivity. The infection rate is very important for the reproduction of the viruses. They have to use the metabolic machinery of the susceptible cells in order to replicate [17]. Consequently, the parameter d_{13} is connected with the possible development and expansion of the viral population. Therefore, the parameter d_{13} can play very significant role for the outcome of the interaction between the virus and the immune system.

The results of our simulations showed that for lower values of the parameter d_{13} an effective, sustained CTL response becomes established. In such cases virus load is contained at low levels. For higher values of the parameter describing the virus infectivity the viral load is at high levels and an effective, sustained CTL

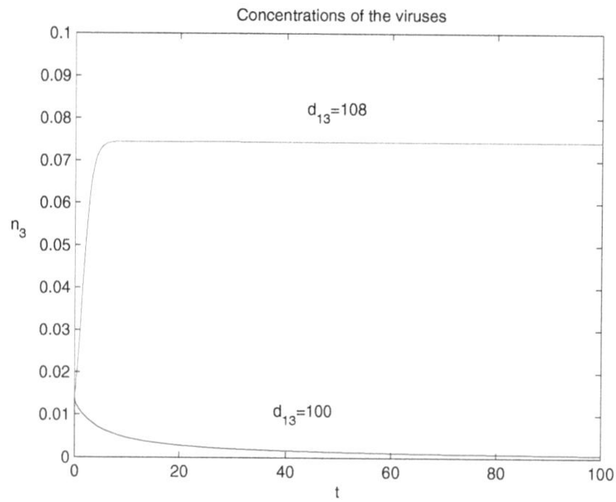

Fig. 3. Dynamics of the virus for $d_{13} = 100$ and $d_{13} = 108$

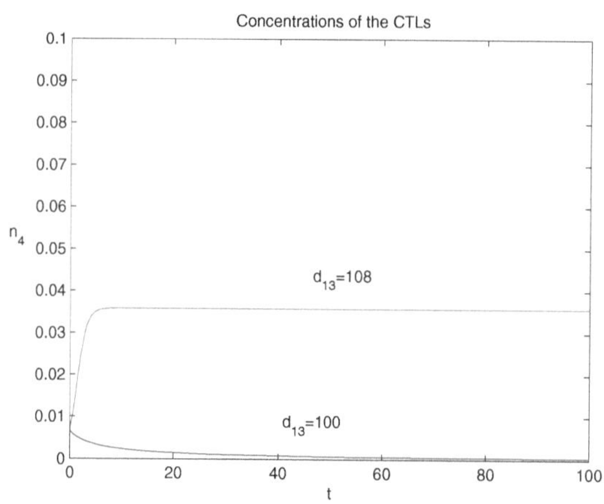

Fig. 4. Dynamics of the CTLs for $d_{13} = 100$ and $d_{13} = 108$

response is not established. The results for the concentrations of all populations for the values $d_{13} = 100$ and $d_{13} = 108$ are presented on Fig. 1 - Fig. 4.

Our future work will address analysis of the role of the other parameters of the model as well as an efficient algorithm for finding precise parameter values for the model equations.

Acknowledgement

The author wish to express his gratitude to Barbara Zubik-Kowal as well as to anonymous referees for their useful comments which led to the improvements in the presentation of the results.

References

1. Arlotti, L., Bellomo, N., De Angelis, E., Lachowicz, M.: Generalized Kinetic Models in Applied Sciences. World Sci., New Jersey (2003)
2. Bellomo, N., Bellouquid, A.: On the onset of nonlinearity for diffusion models of binary mixtures of biological materials by asymptotic analysis. Internat. J. Nonlinear Mech. 41(2), 281–293 (2006)
3. Bellomo, N., Forni, G.: Dynamics of tumor interaction with the host immune system. Math. Comput. Modelling 23, 11–29 (1996)
4. Bellomo, N., Maini, P.: Preface in: Cancer Modelling (II). Math. Models Methods Appl. Sci. 16(7b), iii–vii (2006) (special issue)
5. Bellomo, N., Sleeman, B.: Preface in: Multiscale Cancer Modelling. Comput. Math. Meth. Med. 20(2-3), 67–70 (2006) (special issue)
6. De Angelis, E., Lodz, B.: On the kinetic theory for active particles: A model for tumor-immune system competition. Math. Comput. Modelling 47(1-2), 196–209 (2008)
7. De Lillo, S., Salvatori, M.C., Bellomo, N.: Mathematical tools of the kinetic theory of active particles with some reasoning on the modelling progression and heterogeneity. Math. Comput. Modelling 45(5-6), 564–578 (2007)
8. Gautschi, W.: Numerical Analysis: An Introduction. Birkhäuser, Boston (1997)
9. Kolev, M.: Mathematical modelling of the interactions between antibodies and virus. In: Proc. of the IEEE Conf. on Human System Interactions, pp. 365–368. Krakow (2008)
10. Kolev, M.: Mathematical modelling of the humoral immune response to virus. In: Proc. of the Fourteenth National Conf. on Application of Mathematics in Biology and Medicine, Leszno, Poland, pp. 63–68 (2008)
11. Kuby, J.: Immunology. W.H. Freeman, New York (1997)
12. Lollini, P.L., Motta, S., Pappalardo, P.: Modeling tumor immunology. Math. Models Methods Appl. Sci 16(7b), 1091–1125 (2006)
13. Lydyard, P.M., Whelan, A., Fanger, M.W.: Instant Notes in Immunology. BIOS Sci. Publ. Ltd., Oxford (2000)
14. d'Onofrio, A.: Tumor–immune system interaction and immunotherapy: Modelling the tumor–stimulated proliferation of effectros. Math. Models Methods Appl. Sci. 16(8), 1375–1402 (2006)
15. Shampine, M.W., Reichelt, M.W.: The Matlab ODE suite. SIAM J. Sci. Comput. 18, 1–22 (2007)
16. Volkov, E.A.: Numerical Methods. Hemisphere/Mir, New York/Moscow (1990)
17. Wodarz, D.: Killer Cell Dynamics. Springer, Heidelberg (2006)
18. Wodarz, D., Bangham, C.R.: Evolutionary dynamics of HTLV-I. J. Mol. Evol. 50, 448–455 (2000)
19. Wodarz, D., Krakauer, D.C.: Defining CTL-induced pathology: implication for HIV. Virology 274, 94–104 (2000)

A Two-Grid Approximation of an Interface Problem for the Nonlinear Poisson-Boltzmann Equation

Miglena N. Koleva and Lubin G. Vulkov

Faculty of Natural Science and Education
University of Rousse, 8 Studentska str., Rousse 7017, Bulgaria
{mkoleva,lvalkov@ru.acad.bg}

Abstract. We present a robust and efficient numerical method for solution of an interface problem for a generalization of the Poisson-Boltzmann equation, arising in molecular biophysics. The differential problem is solved by FEM (finite element method) technique on two (coarse and fine) subspaces. The solution of the nonlinear system of algebraic equations on the fine mesh is reduced to the solution on two small (one linear and one nonlinear) systems on the coarse grid and a large linear one on the fine grid. It is shown, both theoretically and numerically, that the coarse space can be extremely coarse and still achieve asymptotically optimal approximation.

1 Introduction

In the region $\Omega = \Omega_1 \times \Omega_L \times \Omega_2$ we consider the following interface problem

$$-\left(p_1(x)u_{1x}\right)_x = f_1(x,u), \quad x \in \Omega_1 = (a_1, b_1), \tag{1}$$
$$-\left(p_L(x)u_{Lx}\right)_x + r(x)u_L = g_L(x), \quad x \in \Omega_L = (b_1, a_2), \tag{2}$$
$$-\left(p_2(x)u_{2x}\right)_x = f_2(x,u), \quad x \in \Omega_2 = (a_2, b_2), \tag{3}$$
$$u_1(b_1) = u_L(b_1), \quad u_L(a_2) = u_2(a_2), \tag{4}$$
$$p_1(b_1)u_{1x}(b_1) = p_L(b_1)u_{Lx}(b_1), \quad p_L(a_2)u_{Lx}(a_2) = p_2(a_2)u_{2x}(a_2), \tag{5}$$
$$u(a_1) = u^1, \quad u(b_2) = u^2. \tag{6}$$

The motivation for studying the problem (1)-(6) comes from some models, describing processes in the bimolecular electrostatics, see [2,4,5,7]. For example, in [4] the authors model a cell membrane by two charge electrolyte films Ω_i, $i = 1, 2$, separated by dielectric Ω_L. It is assumed that the electrolyte media in both regions, Ω_1 and Ω_2, consist of opposite charges $\pm q$ with farfield concentrations ρ_i^∞ and potentials ψ_i^∞, $i = 1, 2$. Then in (1)-(3):

$$\begin{gathered} p_i(x) = \varepsilon_i > 0, \quad f_i(x, u_i) = 2\rho_i^\infty q \sinh \frac{q}{kT}(u_i - \psi_i^\infty), \quad i = 1, 2, \\ p_L(x) = \varepsilon_L, \quad r(x) = \frac{q}{kT} \cosh \frac{q\psi_L^\infty}{kT}, \quad g_L(x) = -\sinh \frac{q\psi_L^\infty}{kT}, \end{gathered} \tag{7}$$

where u_i are potentials, ε_i are dielectric constants, k is Boltzmann constant and $T = const$ is the temperature.

S. Margenov, L.G. Vulkov, and J. Waśniewski (Eds.): NAA 2008, LNCS 5434, pp. 369–376, 2009.

2 Finite Element Approximation

Let $\mathcal{H}^p(\Omega)$ be the standard Sobolev space with norm $\|\cdot\|_p$. The weak formulation of problem (1)-(6) with $r(x) = 0$ for simplicity, reads as follows. Find $u = (u_1, u_L, u_2) \in \mathcal{H}^1(\Omega) \equiv \mathcal{H}^1(\Omega_1) \times \mathcal{H}^1(\Omega_L) \times \mathcal{H}^1(\Omega_2)$, which obeys the integral identity:

$$\begin{aligned} &\int_{a_1}^{b_1} p_1(x) u_x v_x dx + \int_{b_1}^{a_2} p_L(x) u_x v_x dx + \int_{a_2}^{b_2} p_2(x) u_x v_x dx \\ &= \int_{a_1}^{b_1} f_1(x,u) v dx + \int_{b_1}^{a_2} g_L(x) v dx + \int_{a_2}^{b_2} f_2(x,u) v dx, \quad \forall v \in \mathcal{H}^1(\Omega) \end{aligned} \tag{8}$$

and $v_1(b_1) = v_L(b_1) = v(b_1)$, $v_L(a_2) = v_2(a_2) = v(a_2)$.

Using the theory in [6], we can prove the next assertion.

Theorem 1. *Let the following assumptions hold:*

(A1) *There exist a constants ρ_{0i}, such that $0 < \rho_{0i} < \rho_i(x)$ and $\rho_i \in L_\infty(\Omega_i)$, $i = 1, L, 2$;*

(A2) *There exist a non-negative continuous functions c_i, μ_i, such that for some $\rho \in (1, \infty)$: $|f_i(x,\xi)| \le c_i(\xi)(\mu_i(x) + |\xi|^\rho)$, $\xi \in \mathcal{R}$, $x \in \Omega_i$, $i = 1, L, 2$. Then, there exists unique solution $u \in \mathcal{H}^1(\Omega)$ of the problem (1)-(6).*

We define the following uniform, on each interval Ω_1, Ω_L, Ω_2, mesh:

$$\begin{aligned} \bar{\omega}^h = \{ & x_i = (i-1)h_1, i = 1, \ldots, N_1, (N_1 - 1)h_1 = x_{N_1} = b_1, x_i = b_1 + (i-1)h_L, \\ & i = N_1 + 1, \ldots, N_2, b_1 + (N_2 - 1)h_L = x_{N_2} = a_2, x_i = a_2 + (i-1)h_2, \\ & i = N_2 + 1, \ldots N_2, a_2 + (N_2 - 1)h_2 = x_{N_2} = b_2 \}. \end{aligned}$$

We seek the FE solution u^h in the form: $u(x) \sim u^h(x) = \sum_{i=1}^{N} y_i \varphi_i(x)$, $\varphi_i(x)$ belongs to the piecewise finite element space, defined on $\bar{\omega}^h$. Also, for the non-linearities, we use product approximation formula $f(u) \simeq \sum_{i=1}^{N} f(y_i)\varphi_i(x)$, [9]. Finally from (8) we obtain for $j = 1, \ldots, N$:

$$\begin{aligned} &\sum_{i=1}^{N} y_i \int_{a_1}^{b_1} p_1 \varphi_i' \varphi_j' dx + \sum_{i=1}^{N} y_i \int_{b_1}^{a_2} p_L \varphi_i' \varphi_j' dx + \sum_{i=1}^{N} y_i \int_{a_2}^{b_2} p_2 \varphi_i' \varphi_j' dx \\ &= \sum_{i=1}^{N} f_1(y_i) \int_{a_1}^{b_1} \varphi_i \varphi_j dx + \sum_{i=1}^{N} \int_{b_1}^{a_2} g_L \varphi_j dx + \sum_{i=1}^{N} f_2(y_i) \int_{a_2}^{b_2} \varphi_i \varphi_j dx. \end{aligned} \tag{9}$$

In the case of a piecewise linear basis function, using the denotations

$$\overline{p}_{k_{i+1/2}} = \frac{1}{h_{ki}} \int_{x_i}^{x_{i+1}} p_k(x) dx, \qquad \overline{p}_{k_{i-1/2}} = \frac{1}{h_{ki}} \int_{x_{i-1}}^{x_i} p_k(x) dx,$$

$$A_k^i = \frac{\overline{p}_{k_{i-1/2}}}{h_k}, \quad B_k^i = \frac{\overline{p}_{k_{i+1/2}}}{h_k}, \quad C_{ks}^i = \frac{\overline{p}_{k_{i-1/2}}}{h_k} + \frac{\overline{p}_{s_{i+1/2}}}{h_s}, \quad k = 1, L, 2,$$

$$F^i_{ks} = \begin{cases} \frac{h_k}{6} f_k(y_{i-1}) + \frac{4h_k}{6} f_k(y_i) + \frac{h_k}{6} f_k(y_{i+1}) + h_k g_k(x_i), & k = s \neq L, \\ \frac{h_k}{6} f_k(y_{i-3+2k}) + \frac{h_k}{3} f_k(y_i) + \frac{h_k}{2} g_k(x_i) + \frac{h_s}{2} g_s(x_i), & k \neq s, \\ h_L g_L(x), & k = s = L. \end{cases}$$

from (9) we derive the scheme

$$y_i = u^1, \quad i = 1, \tag{10}$$
$$-A^i_1 y_{i-1} + C^i_{11} y_i - B^i_1 y_{i+1} = F^i_{11}, \quad i = 2, \ldots, N_1 - 1, \tag{11}$$
$$-A^i_1 y_{i-1} + C^i_{1L} y_i - B^i_L y_{i+1} = F^i_{1L}, \quad i = N_1, \tag{12}$$
$$-A^i_L y_{i-1} + C^i_{LL} y_i - B^i_L y_{i+1} = h_L g_L(x_i), \quad i = N_1 + 1, \ldots, N_2 - 1,$$
$$-A^i_L y_{i-1} + C^i_{L2} y_i - B^i_2 y_{i+1} = F^i_{2L}, \quad i = N_2, \tag{13}$$
$$-A^i_2 y_{i-1} + C^i_{22} y_i - B^i_2 y_{i+1} = F^i_{22}, \quad i = N_2 + 1, \ldots, N - 1, \tag{14}$$
$$y_i = u^2, \quad i = N. \tag{15}$$

Now, in (b_1, a_2) the system of linear equations is

$$A^i_L y_{i-1} - C^i_{LL} y_i + B^i_L y_{i+1} = -F^i_{LL}, \quad i = N_1 + 2, \ldots, N_2 - 2,$$

$$y_{N_1+1} = æ_1 y_{N_1+2} + \mu_1, \quad æ_1 = \frac{{B_L}^{N_1+1}}{C^{N_1+1}_{LL}},$$

$$y_{N_2-1} = æ_2 y_{N_2-2} + \mu_2, \quad æ_2 = \frac{{A_L}^{N_2-1}}{C^{N_2-1}_{LL}},$$

$\mu_1 = \frac{{A_L}^{N_1+1}}{C^{N_1+1}_{LL}} y_{N_1} + \frac{h_L g_L(x_{N_1+1})}{C^{N_1+1}_{LL}} = \widetilde{\mu}_1 y_{N_1} + \overline{\mu}_1$, $\mu_2 = \frac{{B_L}^{N_2-1}}{C^{N_2-1}_{LL}} y_{N_2} + \frac{h_L g_L(x_{N_2-1})}{C^{N_2-1}_{LL}} = \widetilde{\mu}_2 y_{N_2} + \overline{\mu}_2$. Using the left and right Thomas methods, we find

$$y_{N_1+1} = T_1 y_{N_1} + T_2 y_{N_2} + T_3, \tag{16}$$
$$y_{N_2-1} = K_1 y_{N_1} + K_2 y_{N_2} + K_3, \tag{17}$$

where

$$T_1 = \frac{\widetilde{\mu}_1}{\nu}, \quad T_2 = \frac{\widetilde{\mu}_2 æ_1}{\nu} \underbrace{\left(\frac{B^{N_1+2}_L}{\triangle^{N_1+2}_1} \cdot \frac{B^{N_1+3}_L}{\triangle^{N_1+3}_1} \cdot \ldots \cdot \frac{B^{N_2-2}_L}{\triangle^{N_2-2}_1} \right)}_{\delta_1}, \quad \nu = 1 - \beta_{N_1+2} æ_1,$$

$$T_3 = \frac{\overline{\mu}_1}{\nu} + \frac{æ_1}{\nu} \left(\overline{\mu}_2 \delta_1 + \frac{F^{N_1+2}_{LL}}{\triangle^{N_1+2}_1} + \sum_{k=N_2-2}^{N_1+2} \frac{F^k_{LL}}{\triangle^k_1} \cdot \frac{A^{k-1}_L}{\triangle^{k-1}_2} \cdot \ldots \cdot \frac{A^{N_1+2}_L}{\triangle^{N_1+2}_1} \right), \tag{18}$$

$$\triangle^i_1 = C^i_{LL} - \beta_{i+1} B^i_L, \quad \overset{(\leftarrow)}{\beta}_i = A^i_L / \triangle^i_1, \quad i = N_2 - 2, \ldots, N_1 + 2, \quad \beta_{N_2-1} = æ_2,$$

and

$$K_1 = \frac{\widetilde{\mu}_1 æ_2}{\kappa} \underbrace{\left(\frac{A^{N_2-2}_L}{\triangle^{N_2-2}_2} \cdot \frac{A^{N_2-3}_L}{\triangle^{N_2-3}_2} \cdot \ldots \cdot \frac{A^{N_1+2}_L}{\triangle^{N_1+2}_2} \right)}_{\delta_2}, \quad K_2 = \frac{\widetilde{\mu}_2}{\kappa}, \quad \kappa = 1 - \alpha_{N_2-1} æ_2,$$

$$K_3 = \frac{\overline{\mu}_2}{\kappa} + \frac{æ_2}{\kappa} \left(\overline{\mu}_1 \delta_2 + \frac{F^{N_2-2}_{LL}}{\triangle^{N_2-2}_2} + \sum_{k=N_1+2}^{N_2-2} \frac{F^k_{LL}}{\triangle^k_2} \cdot \frac{A^{k+1}_L}{\triangle^{k+1}_2} \cdot \ldots \cdot \frac{A^{N_2-2}_L}{\triangle^{N_2-2}_2} \right), \tag{19}$$

$$\triangle_2^i = C_{LL}^i - \alpha_i A_L^i, \quad \overset{(\rightarrow)}{\alpha}_{i+1} = B_L^i/\triangle_2^i, \; i = N_1+2,\dots,N_2-2, \quad \alpha_{N_1+2} = \text{æ}_1.$$

Substituting (16) and (17) in (12) and (13) respectively, we get

$$-A_1^i y_{i-1} + (C_{1L}^i - B_L^i T_1) y_i - B_L^i T_2 y_{N_2} = F_{1L}^i + B_L^i T_3, \quad i = N_1, \tag{20}$$

$$(C_{L2}^i - A_L^i K_2) y_i - B_2^i y_{i+1} - A_L^i K_1 y_{N_1} = F_{2L}^i + A_L^i K_3, \quad i = N_2. \tag{21}$$

The new scheme is (10), (11), (20), (21), (14) and (15). Thus, instead of solving the nonlinear-linear system of equations in $\bar{\omega}^h$, we compute the solution only in $[a_1, b_1]$ and $[a_2, b_2]$, and calculate the Thomas coefficients (18), (19) in the interval (b_1, a_2).

3 Two-Grig Algorithm

Up to now, we succeed to exclude the system of equation in region Ω_L. Next, using the two-grid linearization [1,8,10], we will reduce the number of nonlinear equations which have to solve, at the expense of linear equation and at the same time the convergence rate will be increased. For that purpose we involve another finite element space, defined on the **coarse mesh** $\bar{\omega}^H$, with step sizes H_1, H_L, H_2 and total grid nodes N^c, $a_1 = x_1$, $b_1 = x_{N_1^c}$, $a_2 = x_{N_2^c}$, $b_2 = x_{N^c}$. The previous defined mesh $\bar{\omega}^h$ we shall call **fine mesh** and for clarity, the total grid nodes of this mesh will be denoted with N^f, $a_1 = x_1$, $b_1 = x_{N_1^f}$, $a_2 = x_{N_2^f}$, $b_2 = x_{N^f}$, $H_i > h_i$, $i = 1, L, 2$. The two-grid algorithm is as follows:

Step 1. Solve the nonlinear system (10), (11), (20), (21), (14) and (15) on coarse mesh $\bar{\omega}^H$, using Newton method, [3]. The solution from this step is denoted by y^H, $u^H = \sum\limits_{i=1}^{N^c} y_i^H \Phi_i(x)$, $\Phi_i(x)$ belongs to the finite element space, defined on the coarse mesh.

Step 2. (***linearization***) We seek the correction term e^h on mesh $\bar{\omega}^h$, such that the fine mesh solution $y^h = e^h + y^H$ satisfy the scheme (10), (11), (20), (21), (14) and (15). In other words, we apply the Taylor expansion of the unknown fine mesh solution, y^h, around the already known (from step 1) coarse mesh solution, y^H. Thus, the resulting scheme is $\mathbb{A}^f e^h = \overline{F}$, namely

$$e_i^h = 0, \quad i = 1, \tag{22}$$

$$-[A_1^i + \frac{h_1}{6} f_1'(y_{i-1}^H)] e_{i-1}^h + [C_{11}^i - \frac{4h_1}{6} f_1'(y_i^H)] e_i^h - [B_1^i + \frac{h_1}{6} f_1'(y_{i+1}^H)] e_{i+1}^h$$
$$= \overline{F}_1^i, \quad i = 2, \dots, N_1^f - 1, \tag{23}$$

$$-[A_1^i + \frac{h_1}{6} f_1'(y_{i-1}^H)] e_{i-1}^h + (C_{1L}^i - B_L^i T_1 - \frac{h_1}{3} f_1'(y_i^H)) e_i^h - B_L^i T_2 e_{N_2}^h$$
$$= \overline{F}_{N_1}^i, \quad i = N_1^f, \tag{24}$$

$$[C_{L2}^i - A_L^i K_2 - \frac{h_2}{3} f_2'(y_i^H)] e_i^h - [B_2^i + \frac{h_1}{6} f_2(y_{i+1}^H)] e_{i+1}^h - A_L^i K_1 e_{N_1}^h$$
$$= \overline{F}_{N_2}^i, \quad i = N_2^f, \tag{25}$$

$$-[A_2^i + \frac{h_2}{6} f_2'(y_{i-1}^H)]e_{i-1}^h + [C_{22}^i - \frac{4h_2}{6} f_2'(y_i^H)]e_i^h - [B_2^i + \frac{h_1 2}{6} f_2'(y_{i+1}^H)]e_{i+1}^h$$
$$= \overline{F}_2^i, \quad i = N_2^f + 1, \ldots, N^f - 1, \tag{26}$$
$$e_i^h = 0, \quad i = N^f, \tag{27}$$

where

$$\overline{F}_k^i = \begin{cases} A_k^i y_{i-1}^H - C_{kk}^i y_i^H + B_k^i y_{i+1}^H + \frac{h_k}{6} f_k(y_{i-1}^H) + \frac{4h_k}{6} f_k(y_i^H) + \frac{h_k}{6} f_k(y_{i+1}^H) & \\ +h_k g_k(x_i), & k = 1,2, \\ A_1^i y_{i-1}^H - (C_{1L}^i - B_L^i T_1) y_i^H + B_L^i T_2 y_{N_2}^H + \frac{h_1}{6} f_1(y_{i-1}^H) + \frac{h_1}{3} f_1(y_i^H) & \\ +\frac{h_1}{2} g_1(x_i) + \frac{h_L}{2} g_L(x_i) + B_L^i T_3, & k = N_1, \\ -(C_{L2}^i - A_L^i K_2) y_i^H + B_2^i y_{i+1}^H + A_L^i K_1 y_{N_1}^H + \frac{h_2}{6} f_2(y_{i+1}^H) + \frac{h_2}{3} f_2(y_i^H) & \\ +\frac{h_2}{2} g_2(x_i) + \frac{h_L}{2} g_L(x_i) + A_L^i K_3, & k = N_2. \end{cases}$$

Step 3. Again we solve *linear* system of equations on coarse grid $\bar{\omega}^H$, in order to compute the correction term e^H, such that the solution $\widetilde{y}^h = y^H + e^h + e^H$ satisfy the scheme (10), (11), (20), (21), (14) and (15), written on $\bar{\omega}^H$, $\widetilde{u}^h = \sum_{i=1}^{N^c} \widetilde{y}_i^h \Phi_i(x)$. We will note, that obtaining the system of equations, we take into account the equations from step 2, (23)-(26) and cancel some of the addends, because of the fact that, if some relation is true on a fine mesh, it's remain true also on a coarse mesh. Thus, on $\bar{\omega}^H$ we solve the system $\mathbb{A}^c e^H = \widetilde{F}$, where

$$\widetilde{F}_k^i = \begin{cases} \frac{H_k}{6} f_k''(y_{i-1}^H)(e_{i-1}^h)^2 + \frac{4H_k}{6} f_k''(y_i^H)(e_i^h)^2 + \frac{H_k}{6} f_k''(y_{i+1}^H)(e_{i+1}^h)^2, & k = 1,2, \\ \frac{H_1}{6} f_1''(y_{i-1}^H)(e_{i-1}^h)^2 + \frac{H_1}{3} f_1''(y_i^H)(e_i^h)^2, & k = N_1, \\ \frac{H_2}{6} f_2''(y_{i+1}^H)(e_{i+1}^h)^2 + \frac{H_2}{3} f_2''(y_i^H)(e_i^h)^2, & k = N_2. \end{cases}$$

Step 4. Repeat step 2 on the fine mesh $\bar{\omega}^h$ and find $\widetilde{\widetilde{y}}^h$, $\widetilde{\widetilde{u}}^h = \sum_{i=1}^{N^f} \widetilde{\widetilde{y}}_i^h \varphi_i(x)$. In [10] is proven that a more dramatic results can be derived if one more Newton iteration is performed on the fine mesh after step 3.

Theorem 2. *Let u^h is the finite element solution of (8) on $\bar{\omega}^h$, $h = \max\{h_1, h_L, h_2\}$ and $H = \max\{H_1, H_L, H_2\}$. If $u \in \mathcal{W}_2^2(\Omega)$, then*

$$\|u^h - (u^H + e^h)\|_1 \lesssim H^4.$$

If $u \in \mathcal{W}_2^4(\Omega)$, then

$$\|u^h - \widetilde{u}^h\|_1 \lesssim H^5, \ \|u^h - \widetilde{u}^h\|_0 \lesssim H^6 \ \text{and} \ \|u - \widetilde{u}^h\|_1 \lesssim h + H^5, \ \|u - \widetilde{u}^h\|_0 \lesssim h^2 + H^6.$$

More over,

$$\|u^h - \widetilde{\widetilde{u}}^h\|_1 \lesssim H^{12} |\log h|^{\frac{1}{2}}.$$

The proof follows from arguments, given in [10].

4 Numerical Results

Example 1. The test example is problem (1)-(6), $p_1(x) = x^2+1$, $p_L(x) = 2x+3$, $p_2(x) = x^2+x$, $r(x) = 0$, $f_1(x,u) = -u^3+g_1(x)$ and $f_2(x,u) = -u^3-u+g_2(x)$. Functions $g_i(x)$, $i = 1, L, 2$ are chosen such that $u_1(x) = x\sin(x)$, $u_L(x) = ax^3+bx^2+cx+d$ and $u_2(x) = xe^{-x}$ is the exact solution of the model problem, coefficients a, b, c and d are determined from (4), (5). The mesh parameters are: $H_1 = H_L = H_2 = H$, $h_1 = h_L = h_2 = h$, $a_1 = 1$, $b_1 = 2$, $a_2 = 3$ and $b_2 = 4.5$. In Table 1, we give error in max norm (E^H_∞) and convergence rate (CR_∞) of the numerical solution after first and second step of the algorithm, $h = H^2$. Obviously, the convergence rate is $\mathcal{O}(h^2+H^4)$. In Table 2 we show error in max and L_2 discrete norms (E^H_∞, E^H_2), convergence rate (CR_∞, CR_2) of the numerical solution after third step of the algorithm, $h = H^3$. The results shows that the convergence rate in max norm (equivalent to $\|\cdot\|_1$ norm in 1D case) is $\mathcal{O}(h+H^5)$ and in L_2 norm (equivalent to $\|\cdot\|_0$ norm) is $\mathcal{O}(h^2+H^6)$. CPU time (in seconds) for both cases: CPU^f- computing on the whole region $[a_1, b_2]$ and CPU^r- computing only on $[a_1, b_1]$, $[a_2, b_2]$, using (18), (19) is also given in Table 2. The effect from excluding the domain (b_1, a_2) comes in the case, when this interval is large or if we use very fine mesh. The next Table 3 demonstrates the efficiency of step 4. We give CPU time and max errors of the fine and coarse mesh solution at each step of the two-grid algorithm, $H = 2^{-2}$, $h = H^5$. Note, that the difference between CPU^f and CPU^r increases after step 2 in CPU^r favor, because the coefficients (18), (19) are already calculated in step 1 and step 2, for coarse and fine mesh, respectively. The convergence rate is computed, using double mesh principle.

Table 1. Error and convergence rate in max norm, at step 1 and step 2, $h = H^2$

	step 1		step 2	
H	E^H_∞	CR_∞	E^H_∞	CR_∞
2^{-2}	7.075929e-2		4.877880e-3	
2^{-3}	1.759978e-2	*2.0074*	3.055009e-4	***4.0041***
2^{-3}	4.394422e-3	*2.0018*	1.904044e-5	***4.0031***
2^{-5}	1.098262e-3	*2.0005*	1.187477e-6	***4.0020***
2^{-6}	2.745441e-4	*2.0001*	7.411441e-8	***4.0020***
2^{-7}	6.863468e-5	*2.0000*	4.680361e-9	***3.9851***

Example 2. (***Poisson-Boltzmann equation***) From physical point of view, it is of interest to compute the solution of problem (1)-(5), (7), where $\psi_1^\infty = u_1(a_1) = 1$, $\psi_L^\infty = 0$, $\psi_2^\infty = u_2(b_2) = 2$, $\varepsilon_1 = 1$, $\varepsilon_L = 1$ $\varepsilon_2 = 2$, $q/kT = 1$ and $\rho_i^\infty = -1/q$. Having in mind the Taylor series of function $\sinh(\cdot)$, the most appropriate initial solution (for starting Newton iterations) is the solution of the linear problem $f_i = 2\rho_i^\infty q^2(u_i - \psi_i^\infty)/kT$, $i = 1, 2$. The coarse and fine mesh solution is plotted on Figure 1, for $a_1 = -2.5$, $b_1 = -1$, $a_2 = 1$, $b_2 = 2$, $H = 2^{-2}$ and $h = H^5$.

Table 2. Max and L_2 error, convergence rate and CPU time at step 3, $h = H^3$

H	E_∞^H	CR_∞	E_2^H	CR_2	CPU^f	CPU^r
2^{-2}	3.706981e-4		9.267452e-5		0.400	0.390
2^{-3}	7.765266e-6	*5.5771*	9.706582e-7	*6.5771*	0.625	0.584
2^{-4}	1.891633e-7	*5.3593*	1.182271e-8	*6.3593*	5.875	4.556
2^{-5}	6.136609e-9	*4.9460*	1.917690e-10	*5.9461*	335.875	289.096

Table 3. Error of fine and coarse mesh solution and CPU time at each step, $h = H^5$

step	E_∞^H	E_∞^h	E_2^H	E_2^h	CPU^f	CPU^r
step 1	7.075929e-2		1.768982e-2		0.032	0.030
step 2	6.339948e-4	6.672832e-4	1.584987e-4	1.668208e-4	6.078	5.470
step 3	1.049555e-4		2.623888e-5		6.359	5.722
step 4	9.233036e-6	9.595545e-6	9.016637e-9	9.370649e-9	9.875	7.205

Example 3. (***2D case***) Now, $\Omega = [(a_1,b_1)\times(0,d)]\times[(b_1,a_2)\times(0,d)]\times[(a_2,b_2)\times(0,d)]$, $d > 0$. The test example is

$$
\begin{aligned}
&-u_{1xx} - u_{1yy} + {u_1}^3 + u_1 = g_1(x,y), \quad (x,y)\in\Omega_1,\\
&-u_{Lxx} - u_{Lyy} + u_L = g_L(x,y), \quad (x,y)\in\Omega_L,\\
&-u_{2xx} - u_{2yy} + {u_2}^3 = g_2(x,y), \quad (x,y)\in\Omega_2,\\
&u_1(b_1,y) = u_L(b_1,y), \quad u_L(a_2,y) = u_2(a_2,y), \quad 0\le y\le d, \qquad (28)\\
&u_{1x}(b_1,y) = u_{Lx}(b_1,y), \quad u_{Lx}(a_2,y) = u_{2x}(a_2,y), \quad 0\le y\le d \qquad (29)
\end{aligned}
$$

with zero Dirihlet boundary conditions. Functions g_1, g_L and g_2 are chosen such that the exact solution is: $u_1(x,y) = x(x-a_1)\sin\frac{\pi y}{d}$, $u_L(x,y) = (Ax^3 + Bx^2 + Cx + D)\sin\frac{\pi y}{d}$, $u_2(x,y) = x(b_2 - x)\sin\frac{\pi y}{d}$, A, B, C, D are determined from conditions (28) and (29). The results from computation for $a_1 = 1$, $b_1 = 2$,

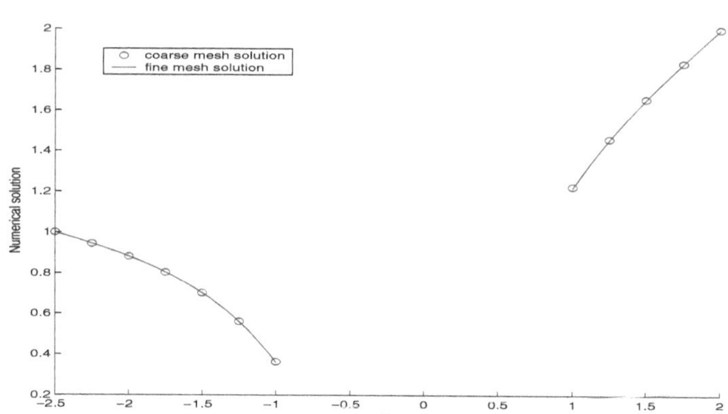

Fig. 1. Coarse and fine mesh solution of the Poisson-Boltzmann problem

Table 4. Error in max discrete norms (E_∞^H and E_∞^h), $H = 2^{-2}$, $h = H^2$

	step 1	step 2	step 3	step 4
E_∞^H	5.454353e-1	4.294958e-2	3.448029e-2	1.375294e-2
E_∞^h		4.632204e-2		1.476173e-2

$a_2 = 3$, $b_2 = 4.5$ and $d = 1$ are given in Table 4. The mesh step size is uniform on both directions - x and y, $H = 2^{-2}$, $h = H^2$.

5 Conclusions

The effectiveness of the two-grid method is obvious: we reach a high accuracy, solving nonlinear equations on a coarse grid and linear equations on a fine mesh. Moreover, the coarse grid can be very coarse and the results are still with good precision. It turned out, that the idea to reduce the number of linear equations, which have to solve, by excluding the region (b_1, a_2), comes in handy in the case of a large computational interval (b_1, a_2) (big distance between b_1 and a_2 or very fine mesh), especially after second step of the algorithm. The error decrease dramatically at the second step, if $h = H^2$, at the third step, if $h = H^3$ and at the fourth step, if $h \geq H^5$.

Acknowledgement. This research is supported by the Bulgarian National Fund of Science under Project VU-MI-106/2005.

References

1. Axelsson, O.: On mesh independence and Newton methods. Appl. Math. 38(4-5), 249–265 (1993)
2. Buike, M., Buikis, A.: Modelling of three-dimensional transport process in anisotropic layered stratum by consrevative averaging method. WSEAS Transactions of Heat and Mass Transfer 1(4), 430–437 (2006)
3. Cimrák, I.: Numerical solution of degenerate convection-diffusion problem using broyden scheme. In: Proceedings of ALGORITMY. Conf. on Sci. Comput., pp. 14–22 (2002)
4. Cook, B., Kazakova, T., Lyu, S., Madrid, P., Neal, J., Pauletti, M., Zhao, R.: Cell-Foreign Particle Interaction. IMA Preprint Series 2133(3) (2006)
5. Givoli, D.: Finite Element Modeling of Thin Layers. CMES 5(6), 497–531 (2004)
6. Kufner, A., Fučic, S.: Nonlinear Differential Equations. Nauka, Moskow (1988)
7. Ladyzhenskaya, A., Solnnikov, A., Uraltseva, N.: Linear and Quasi-Linear Equations of Parabolic Type. In: Translations of Mathematical Monographs, vol. 23, American Mathematical Society, Providence (1968)
8. Mu, M., Xu, J.: A two-grid method for a mixed Stokes-Darcy model for coupling fluid flow with porous media flow. Math. of Comp. 71(256), 1617–1626 (2008)
9. Sanz-Serna, J., Abia, L.: Interpolation of the coefficients in nonlinear elliptic Galerkin procedures. SIAM J. Numer. Anal. 21(1), 77–83 (1984)
10. Xu, J.: A novel two-grid method for semilinear elliptic equations. SIAM J. Sci. Comput. 15(1), 231–237 (1994)

Numerical Study of Rayleigh-Benard Convection in a Rectangular Box

V.V. Kolmychkov, O.S. Mazhorova, Yu.P. Popov, and O.V. Shcheritsa

Keldysh Institute of Applied Mathematics RAS
Miusskaya Sq.4, 125047, Moscow, Russia

Abstract. Numerical study for convective instability of a liquid with linear vertical initial temperature profile in a closed box is presented. The paper provides stability analysis of two-dimensional rolls. Results are compared with R. Krishnamurti, F.H. Busse and coworkers data.

1 Introduction

Rayleigh-Benard convection in a horizontal layer heated from below has become the favored example for the study of the spontaneous formation of structures in hydrodynamic systems. In its idealized form Rayleigh-Benard convection involves fluid placed between flat horizontal plates which are infinite in extent and are perfect heat conductors. The fluid is driven by maintaining the lower plate at a temperature δT above the upper plate temperature. For small driving the fluid remains at rest and a linear temperature profile is set up interpolating between the upper and lower plate temperatures. This is the "conducting" or "uniform" solution. Due to the thermal expansion, however, the fluid near the lower plate is less dense, an intrinsically unstable solution in the gravitational field. Of course the fluid cannot rise as a whole since there would be no place for the fluid above it to go. Thus, due to a conservation law (mass in this case) an instability at a finite wavelength is observed — a fundamental precursor of patten formation. This instability occurs when the driving δT is strong enough to overcome the dissipative effects of thermal conduction and viscosity. The "control parameter" describing the instability, the Rayleigh number Ra, is dimensionless ratio of the destabilizing buoyancy force $\rho_0 \alpha g \delta T$ to the stabilizing dissipative force $\nu\kappa/d^3$, where ρ_0 is the average mass density, α is the thermal expansion coefficient, g is the acceleration of gravity, ν is the kinematic viscosity, κ is the thermal diffusivity and d the plate separation. In the order of increasing Ra, the first transitions occurs at well-known critical Rayleigh number Ra_{cr}, which is independent of Prandtl number $Pr = \nu/\kappa$. The instability occurs at the value $Ra_{cr} = 1708$. The wavenumber k_{cr} of the instability can easily be seen to be of order d, the plates separation, since this is the only length scale available in this ideal, static problem. Obtained picture of an instability towards a pattern in which the fluid rises in some regions and falls in others with a characteristic horizontal length scale d. The simplest manifestation of such a solution is the familiar convective roll pattern.

S. Margenov, L.G. Vulkov, and J. Waśniewski (Eds.): NAA 2008, LNCS 5434, pp. 377–383, 2009.

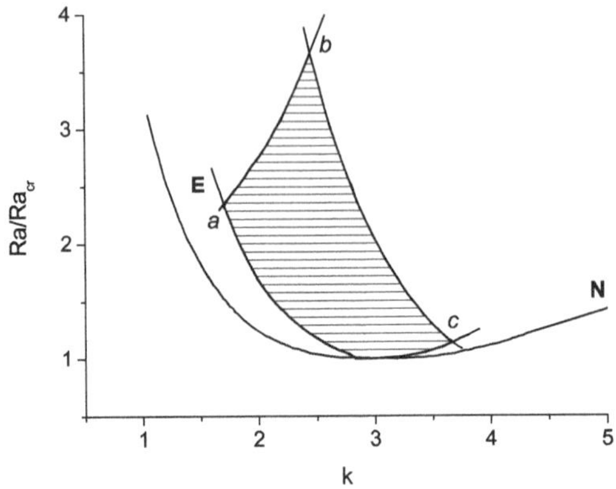

Fig. 1. "Busse balloon". $Pr = 0.71$. **N** — the marginal curve, **Eac** — the Eckhaus instability.

In linear approach the solution of the problem can be presented in the form $f(k, z, t, Ra)e^{i\mathbf{kx}}$, where $k = |\mathbf{k}|$, $\mathbf{x} = (x, y)$. The region in the plane (Ra, k) where $\partial_t f(k, z, t, Ra) < 0$ is stability region. Stability region boundary is called marginal curve. For the case of rigid boundaries the marginal curve is presented in fig.1. Minimal value of Ra on the marginal curve is critical Rayleigh number Ra_{cr}, corresponding k is critical wavenumber - $k_{cr} = 3.117$. For a given $Ra > Ra_{cr}$ there is a band of wavenumbers for which steady state is unstable. But linear analysis doesn't allow to predict the flow pattern.

The next step in the study of the problem is to find the flow for a given Ra and k and to investigate it's stability. The pioneer work in this field has been done by A.Shluter, D.Lortz and F.Busse [1]. In a series of papers, Busse and coworkers found the region in the (Ra, k) plane also known as the "Busse balloon" where steady straight rolls are predicted to be stable. The shape of the balloon depends on the Prandtl number. For $Pr = 0.71$, that corresponds to air, it's shown in fig.1. Inside the balloon the periodic roll structure is stable with respect to infinitesimal disturbances, but if we cross one of its boundaries, rolls are destabilized by a secondary instability. Curve a-c separates the stability region and a region of Eckhaus instability.

The results mentioned above have been obtained for infinite horizontal layer. Analysis of amplitude equations with usual boundary conditions shows that the local wavenumber in the bulk of the fluid is restricted to a much narrower band than the stability balloon suggests, no matter how far away the boundary [2].

This paper presents numerical study for the stability of steady rolls in a finite domain with rigid boundaries. The simulation has been carried out in 2D and 3D time-dependent approach.

2 Governing Equations

Rayleigh-Benard problem is governed by 3D time-dependent fluid flow and heat transport equations. Fluid motion is described by the Navier-Stokes equations in the Boussinesq-approximation. In Cartesian coordinates (x, y, z), the dimensionless equations take the form [3]:

$$\partial_t V + (V\nabla)V = -\nabla p + \Delta V + \sum_{i=1}^{2} GrTe_z \nabla V = 0 \tag{1}$$

$$\partial_t T + (V\nabla)T = \frac{1}{Pr}\Delta T \tag{2}$$

where t is time, $\partial_\xi \equiv \frac{\partial}{\partial \xi}$, $V = (V_x, V_y, V_z)$ – velocity, $\nabla = (\partial_x, \partial_y, \partial_z)$, p – pressure, $\Delta = \nabla^2 = \partial^2_{xx} + \partial^2_{yy} + \partial^2_{zz}$, T – temperature, $Gr = g\alpha d^3 \delta T/\nu^2$ – Grashoff number, e_z=$(0, 0, 1)$. The non-dimensional variables are introduced by scaling the length with the depth d, time is scaled with $t_\nu = d^2/\nu$.

All boundaries of the rectangular box are assumed to be rigid: V=0. Vertical boundaries are heat-insulated. Temperature on the bottom is $T|_{z=0} = 1$, on the top $T|_{z=1} = 0$.

At $t = 0$ the liquid is in the static state: $V = 0$, $T = 1 - z$.

3 Numerical Procedure

Governing equations (1)-(2) were solved by sequential procedure. The velocity is advanced using the temperature from the previous layer in the force term. The calculated velocity field is inserted into the heat transfer equation. The problem is approximated at staggered grid using control-volume procedure. Resulting finite-difference scheme conserves kinetic energy and heat. The scheme is implicit, has second order of accuracy in space and first in time.

The Navier-Stokes equations are calculated by the sort of projection method with pressure correction [4,5].

Prediction step. A predicted velocity field $\widetilde{V}$ is computed implicitly from equation (1) in which the pressure gradient at the current time step t_{n+1} is taken from previous one.

$$\frac{\widetilde{V} - V^n}{\tau} + (V^n\nabla)\widetilde{V} = -\nabla p^n + \Delta\widetilde{V} + F^n \tag{3}$$

On the boundary

$$\widetilde{V} = 0$$

The set of algebraic equations for each component of the momentum is solved in turn, treating the grid point values of dominant velocity component as a sole set of unknowns. To determine the velocity we use the conjugate gradient method [6].

Projection step. The velocity field $\widetilde{V}$ is corrected by taking into account the pressure gradient at the t_{n+1} to satisfy the incompressibility constraint:

$$V^{n+1} = \widetilde{V} - \tau\nabla(p^{n+1} - p^n) \tag{4}$$

This step is performed by computing the pressure correction $\delta p = p^{n+1} - \tilde{p}$ from Poisson problem:

$$\Delta(\delta p) = \frac{div\tilde{V}}{\tau} \qquad (5)$$

On the boundary:

$$\frac{\partial \delta p}{\partial n} = 0$$

The final velocity and pressure fields then read

$$V^{n+1} = \widetilde{V} - \tau\nabla\delta p \quad \text{with} \quad V^{n+1} = 0 \quad \text{on the boundary} p^{n+1} = \widetilde{p} + \delta p \qquad (6)$$

In 2D simulation velocity and temperature fields are also determined consequently. Conservative finite-difference scheme is used. The Navier-Stokes equations in stream-vorticity formulation are solved by direct matrix algorithm.

4 Numerical Results

The range of stable wavenumbers has been determined in scope of 2D aproach. For a fixed Rayleigh number initial linear vertical temperature profile was disturbed with a given wavenumber k_{dis}:

$$T(x, z = 1/3) = 2/3\cos(k_{dis}x). \qquad (7)$$

The initial perturbation evolves to a roll pattern with a wavenumber k. The set of resulting k forms the band of stable wavenumbers for the chosen Ra. The obtained region is crosshatched in fig.2. The boundaries of stable region

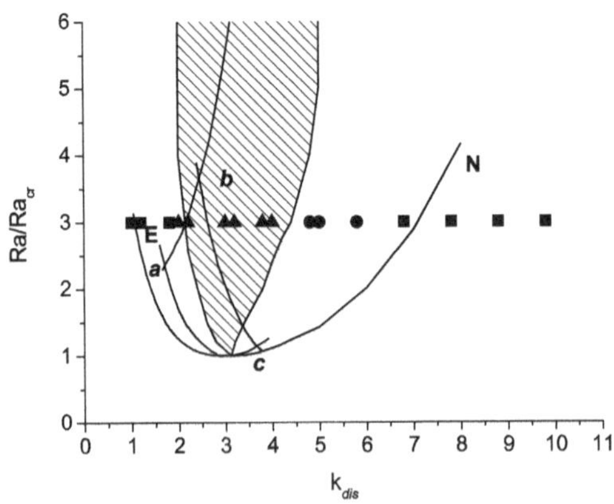

Fig. 2. Stability diagram. Crosshatched area is the stability region (2D simulation). **N** — the marginal curve, **Eac** — the Eckhaus instability.

are practically linear in the vicinity of the critical Rayleigh number. The width of the stable wavenumber band is essentially less then one predicted by linear analysis. These facts are in good agreement with the results found in the scope of amplitude equations.

Three different scenarios for onset of stable flow were registered in our calculations:

1. the initial perturbation with k_{dis} evolves to roll pattern with near the same wavenumber. It means that k_{dis} is stable and lies in the stability region (triangles in fig.2).
2. the initial perturbation grows slowly and cells with a wavenumber $k \neq k_{dis}$ appear near the vertical borders. The cells with new wavenumber gain kinetic energy much faster the than initial one and spread from the boundaries into the bulk easily as long as the whole box is filled up with the new pattern (fig.3a). Wavenumbers that evolved according to this scenario are marked by squares in fig.2.
3. the initial disturbance evolves into well developed roll pattern with $k = k_{dis}$ (fig.3b). Then several flow transitions occur until stable roll pattern is set on. Wavenumbers k_{dis} are marked by circle in fig.2.

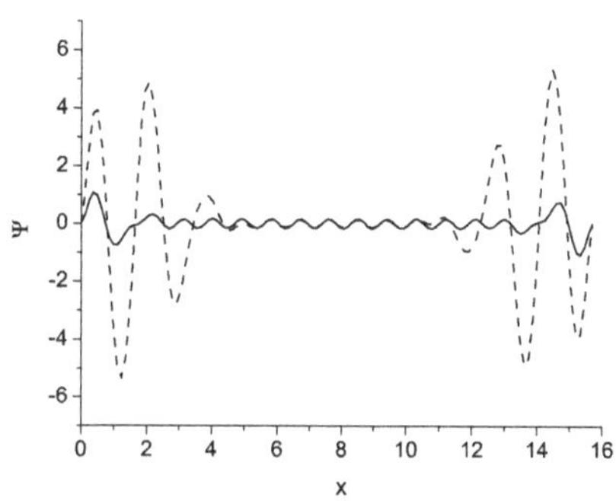

a) $k = 6.8$

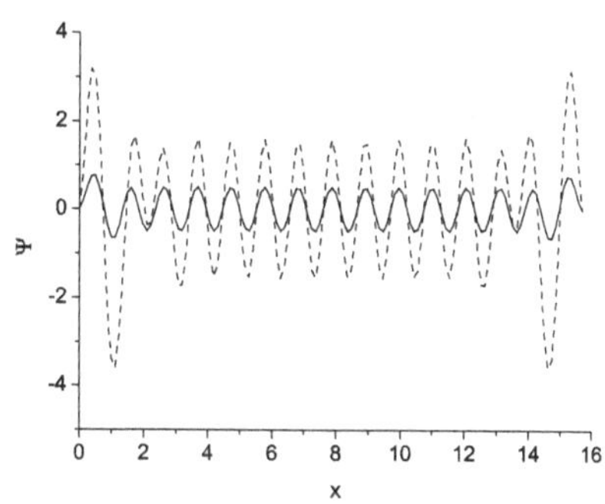

b) $k = 5.8$

Fig. 3. Stream function in the plane $z = 1/2$. Solid line corresponds to $t = 0.2t_\nu$, dash line — $t = 0.4t_\nu$.

To refine the results of 2D calculations the series of 3D simulations with similar initial data have been done. Three scenarios mentioned above have been registered in 3D case also. As expected, in 3D case stability area for roll flow pattern is narrower then in 2D case and close to Busse balloon. The typical example of flow evolution at low Rayleigh number is given in fig.4. The calculation has been done for $Pr = 1$, $Ra = 3500$, initial perturbation $k_{dis} = 5.8$ (fig.4a).

The rolls with new wavenumber $k \approx 3.5$ appear near the boundaries (fig.4b) and fill up the whole domain (fig.4c). $k = 3.5$ belongs to the stability region

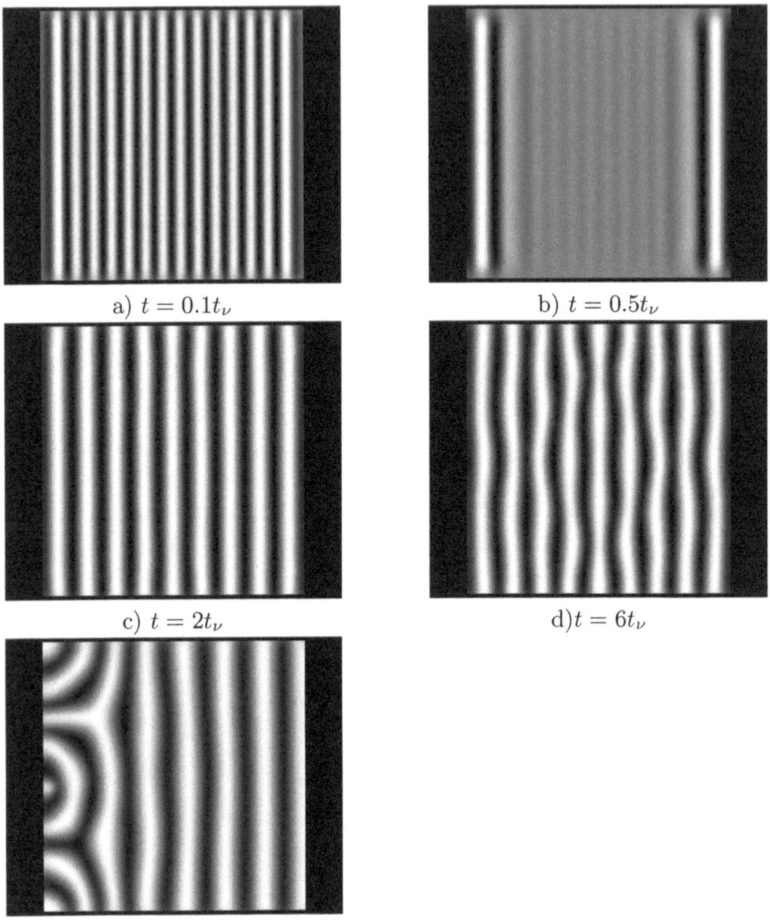

a) $t = 0.1t_\nu$ b) $t = 0.5t_\nu$ c) $t = 2t_\nu$ d)$t = 6t_\nu$ e) $t = 60t_\nu$

Fig. 4. Temperature in the plane $z = 0.5$ for $Ra = 3500, Pr = 1$ on the different moments. White areas are hot, black - cold.

obtained in 2D modelling but unstable to 3D disturbances (fig.4d). Up to the time $t = 60t_\nu$ the flow evolves to a roll-like pattern with $k = 2.7$ that lies in the Busse balloon (fig.4e).

5 Conclusions

Convective stability of two-dimensional rolls in a closed box has been studied by means of 2D and 3D computer simulation. In plane (Ra, k) the stability region has been obtained. The results are in good agreement with theoretical data. Three different scenarios for onset of stable flow were registered.

References

1. Schlüter, A., Lortz, D., Busse, F.: On the stability of steady finite amplitude convection. Journal of Fluid Mechanics 23, 129–144 (1965)
2. Hohenberg, P.C., Cross, M.C.: An introduction to pattern formation in nonequilibrium systems. In: Garido, L. (ed.) Fluctuations and Stochastic Phenomena in Condensed Matter Physics. Lecture Notes in Physics, vol. 268, p. 55. Springer, New York (1981)
3. Gershuni, G.Z., Zhukhovitskii, E.M.: Convective stability of incompressible fluids. Nauka, Moscow (1972) (in Russian)
4. Belotserkovskii, O.M.: Numerical modelling in mechanics of continua. Nauka, Moscow (1984) (in Russian)
5. Kolmychkov, V.V., Mazhorova, O.S., Popov, Y.P.: Computer Simulation for Subcritical Convection in Multi-Component Alloys. Journal of Mathematical Modelling and Analysis 11(1), 57–71 (2006)
6. Vinsome, P.K.: An orthomin, an iterative method for solving sparse set of simultaneous linear equations. In: Proc. 4 Symp. Reservoir simulation. Soc. Petroleum Eng. AIME, pp. 149–159 (1976)

On a Discrete Maximum Principle for Linear FE Solutions of Elliptic Problems with a Nondiagonal Coefficient Matrix

Sergey Korotov[1], Michal Křížek[2], and Jakub Šolc[2]

[1] Institute of Mathematics, Helsinki University of Technology, P.O. Box 1100, FI–02015 TKK, Finland
sergey.korotov@hut.fi

[2] Institute of Mathematics, Academy of Sciences, Žitná 25, CZ–115 67 Prague 1, Czech Republic
krizek@math.cas.cz, solc@math.cas.cz

Abstract. In this paper we give a sufficient condition for the validity of a discrete maximum principle (DMP) for a class of elliptic problems of the second order with a nondiagonal coefficient matrix, solved by means of linear finite elements (FEs). Numerical tests are presented.

1 Introduction

There are many works devoted to the validity of various DMPs for FE-type approximations. In general, all such papers can be split into two groups: those dealing with standard (linear) FE computational schemes, see e.g. [2], [5], [8], [9], [21], and papers, where certain nonlinear FE-type schemes are proposed, as in [4], [11], [15], [16].

There are two main reasons why nonlinear FE schemes are developed. First of all, linear FE schemes have been shown to produce approximations satisfying DMP for problems mostly with diagonal coefficient matrices. Second, such schemes often require a usage of FE triangulations with certain geometrical properties (e.g. the nonobtuseness of simplicial FE meshes, etc.). However, it is not always easy to construct such meshes, and further refine them preserving the desired geometrical conditions.

Note that the only obtuse triangle in a given triangulation may destroy the validity of DMP when solving the Poisson equation by standard linear finite elements, see [3]. The situation is even worse for anisotropic case.

In this paper we demonstrate that imposing slightly more severe condition on the triangulations used (forcing them be acute and not only nonobtuse as in [2], [5], [8], [9]), we can still produce DMP-adequate approximations for some class of elliptic problems with full diffusion tensors using only standard linear FE schemes, see [7] for treating similar situation in the parabolic case. We need to solve elliptic equations with nondiagonal coefficient matrices in many areas, e.g., for flow in porous media, transport of atmospheric gases, heat conduction in anisotropic media, financial mathematics [6], [13], [18], [19].

S. Margenov, L.G. Vulkov, and J. Waśniewski (Eds.): NAA 2008, LNCS 5434, pp. 384–391, 2009.

2 Model Problem and Maximum Principle

We consider the following elliptic problem: Find a function u such that

$$-\operatorname{div}(\mathcal{A}\nabla u) = f \quad \text{in} \quad \Omega, \tag{1}$$

$$u = 0 \quad \text{on} \quad \partial\Omega, \tag{2}$$

where $\Omega \subset \mathbf{R}^2$ is a bounded polygonal domain with Lipschitz boundary $\partial\Omega$, $f \in L^2(\Omega)$, $\mathcal{A}$ is a symmetric uniformly positive definite 2×2 matrix (often called a diffusion tensor) with smooth entries $\mathcal{A}_{km}$, $k, m = 1, 2$, defined on Ω.

The classical solution of (1)–(2), if it exists, is known to satisfy the following maximum principle [12]:

$$f \leq 0 \qquad \Longrightarrow \qquad u \leq 0 \quad \text{in} \quad \overline{\Omega}. \tag{3}$$

Remark 1. If the matrix $\mathcal{A}$ is constant then by the linear transformation $F(x) = \mathcal{A}^{-1/2}x$ equation (1) becomes the Poisson equation on the domain $F(\Omega)$ with zero boundary conditions on $\partial(F(\Omega))$.

3 FE Discretization

We shall use the standard Sobolev space notation. Assume that the coefficients $\mathcal{A}_{km} \in L^\infty(\Omega)$. Then the weak formulation of problem (1)–(2) reads: Find a function $u \in H_0^1(\Omega)$ such that

$$\int_\Omega \mathcal{A}\nabla u \cdot \nabla v \, dx = \int_\Omega f v \, dx \qquad \forall v \in H_0^1(\Omega). \tag{4}$$

Let $\mathcal{T}_h$ stand for a face-to-face triangulation of $\overline{\Omega}$ with a discretization parameter h and triangular elements denoted by symbol T (possibly with some subindices). Let $V_h \subset H_0^1(\Omega)$ be a FE space spanned by the standard Courant piecewise linear continuous basis functions $\varphi_1, \ldots, \varphi_N$ associated with the interior nodes $P_1, \ldots, P_N$, i.e., those vertices of triangles from $\mathcal{T}_h$ that do not belong to $\partial\Omega$.

The FE solution of problem (4) is a function $u_h \in V_h$ such that

$$\int_\Omega \mathcal{A}\nabla u_h \cdot \nabla v_h \, dx = \int_\Omega f v_h \, dx \qquad \forall v_h \in V_h. \tag{5}$$

Algorithmically, $u_h = \sum_{i=1}^{N} y_i \varphi_i$ with $\mathbf{y} = (y_1, \ldots, y_N)^\top$ being a vector-solution of the following system of linear equations

$$\mathbf{A}\mathbf{y} = \mathbf{F}, \tag{6}$$

where $\mathbf{A}$ is the finite element $N \times N$ matrix with entries $a_{ij} = \int_\Omega \mathcal{A}\nabla\varphi_j \cdot \nabla\varphi_i \, dx$, and the vector $\mathbf{F} = (f_1, \ldots, f_N)^\top$ has entries $f_i = \int_\Omega f \varphi_i \, dx$.

Consider acute triangulations, i.e., for any angle α of any element T from $\mathcal{T}_h$ we have

$$\alpha \leq \frac{\pi}{2} - \alpha_1, \tag{7}$$

where α_1 is a fixed positive constant. Obviously, this implies that

$$\alpha \geq 2\alpha_1 \tag{8}$$

for any angle α of any $T \in \mathcal{T}_h$.

Remark 2. For a given $\mathcal{T}_h$ there exists a constant $C > 0$ such that for any triangle $T \in \mathcal{T}_h$ and any of its altitude ℓ (which is always inside of T due to the acuteness property) we have

$$\ell \geq C h_T, \tag{9}$$

where h_T is the diameter of T.

4 DMP and Conditions for Its Validity

The following implication

$$f \leq 0 \qquad \Longrightarrow \qquad u_h \leq 0 \quad \text{in} \quad \overline{\Omega} \tag{10}$$

presents a natural discrete analogue (discrete maximum principle, or DMP in short) of the maximum principle (3). It is clear that (10) holds if $\mathbf{A}$ is a monotone matrix.

It is known that $\mathbf{A}$ is symmetric and positive definite. Hence, due to a well-known result on monotonicity of the Stieltjes matrices (cf. [20, p. 85]) we only need to provide the nonpositivity of the off-diagonal entries of $\mathbf{A}$ which, in turn, holds if

$$a_{ij}|_T = \int_T \mathcal{A} \nabla \varphi_j \cdot \nabla \varphi_i \, dx \leq 0 \quad (i \neq j) \tag{11}$$

for each triangular element $T \in \mathcal{T}_h$.

In what follows, we shall work with the following matrix decompositions of the diffusion tensor

$$\mathcal{A} = \mathcal{D} + \mathcal{B}, \tag{12}$$

where $\mathcal{D} = d\,I$, with I being the unit matrix and $d > 0$ being a positive constant. The entries of $\mathcal{B} = \mathcal{A} - \mathcal{D}$ are denoted by $\mathcal{B}_{km}$, $k, m = 1, 2$. Let

$$\bar{b} := \max_{k,m=1,2} \{\text{ess sup}_{x \in \Omega} |\mathcal{B}_{km}(x)|\}. \tag{13}$$

Remark 3. Note that there are infinitely many decompositions of the type (12).

Theorem 1. *Let*

$$\bar{b} \leq \frac{d\, C^2 \operatorname{ctg} 2\alpha_1}{4 \cos \alpha_1}, \tag{14}$$

where the constants α_1, C, and $\bar{b}$ are defined in (7), (9), and (13), respectively. Then the discrete maximum principle (10) holds.

Proof. We observe that, in view of (12),

$$a_{ij}|_T = \int_T \mathcal{A}\nabla\varphi_j \cdot \nabla\varphi_i \, dx = \int_T (\mathcal{D}+\mathcal{B})\nabla\varphi_j \cdot \nabla\varphi_i \, dx$$
$$= d\int_T \nabla\varphi_j \cdot \nabla\varphi_i \, dx + \int_T \mathcal{B}\nabla\varphi_j \cdot \nabla\varphi_i \, dx \le$$

$$\le d\int_T \nabla\varphi_j \cdot \nabla\varphi_i \, dx + 4\bar{b}\int_T \Big(\max_{i=1,\dots,N,k=1,2}\Big|\frac{\partial\varphi_i}{\partial x_k}\Big|\Big)^2 dx. \quad (15)$$

The following well-known formula

$$\int_T \nabla\varphi_i \cdot \nabla\varphi_j \, dx = -\frac{1}{2}\operatorname{ctg}\alpha_{ij}^T \quad (16)$$

is valid, where α_{ij}^T is the angle in T opposite to the edge P_iP_j.

Further, from (9) we have the following property for the basis functions

$$\Big|\frac{\partial\varphi_i}{\partial x_k}\Big|_T\Big| \le \frac{1}{Ch_T}, \quad (17)$$

where C is the constant from (9), $i = 1,\dots,N$, and $k = 1,2$. Therefore, using (15), (16), (17), (7), and the formula for the area of a triangle, we get

$$a_{ij}|_T \le -\frac{d}{2}\operatorname{ctg}\alpha_{ij}^T + \frac{2\bar{b}}{C^2}\sin(\frac{\pi}{2}-\alpha_1) \le -\frac{d}{2}\operatorname{ctg}2\alpha_1 + \frac{2\bar{b}}{C^2}\cos\alpha_1 \le 0,$$

provided (14) holds. ■

Remark 4. Consider a strongly regular family of triangulations $\mathcal{F} = \{\mathcal{T}_h\}_{h\to 0}$, see [1] for a definition. The same analysis can be now done again with another (uniform) constants.

5 Numerical Tests

Consider our problem with the diffusion tensor $\mathcal{A}$ defined by

$$\mathcal{A}(x) = \begin{bmatrix} r_2(x) & r_1(x) \\ -r_1(x) & r_2(x) \end{bmatrix} \begin{bmatrix} K & 0 \\ 0 & 1/K \end{bmatrix} \begin{bmatrix} r_2(x) & -r_1(x) \\ r_1(x) & r_2(x) \end{bmatrix}, \quad (18)$$

where $(r_1(x), r_2(x))^\top$ is the unit (normalized) radius-vector from the origin to the point x. The condition number of $\mathcal{A}(x)$ and the eccentricity of the associated ellipse is determined by a positive parameter K. The diffusion tensor field is radially symmetric, so it should behave in the same manner in every direction. Such a problem may describe, e.g., the temperature distribution in a wooden log or in a vulcanic basalt.

The shape of the domain $\Omega(s)$ is controlled by a slope parametr s. The rectangular domain $(-2.5, 2.5) \times (-1.5, 1.5)$ is discretized in the standard way

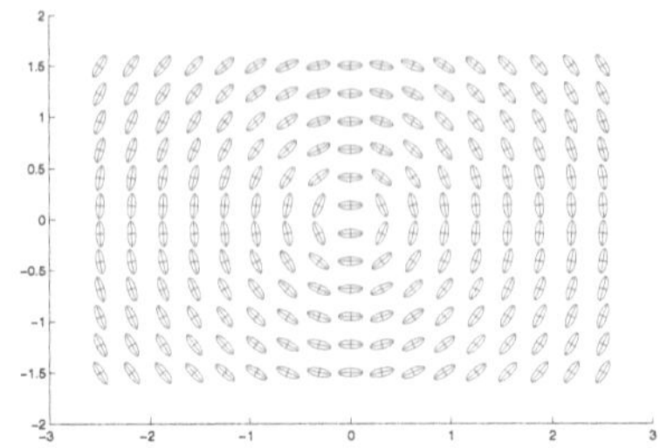

Fig. 1. Diffusion tensor field. The main axes of ellipsae indicate directions of the largest heat conductivities (for $K = 3$).

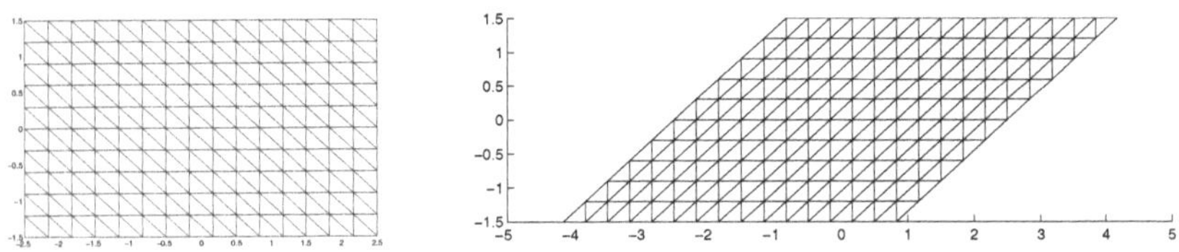

Fig. 2. The triangulation for $s = 0$ and $s = 1$

(see Fig. 2) and then it is deformed so that each subsequent row of elements is shifted against the preceding one by the distance sh_x, where h_x is width of an element in direction of the axis x. Thus,

$$\Omega(s) = \{(x, y) \in R^2 : y \in (-1.5, 1.5);\ -\frac{5}{2} + s\frac{10}{9}y \le x \le \frac{5}{2} + s\frac{10}{9}y\}.$$

If $s \in (0, 1)$, the triangulation has only acute angles. For $s = 0.5$ all elements are isosceles triangles. For $s = 0$ and $s = 1$ they have right angles and the triangulations have locally the same structure, but the shape of the domain is different.

We choose the parameter d from decomposition (12) as the mean value of diagonal entries of the middle matrix on the right-hand side of (18), i.e., $d = \frac{K^2+1}{2K}$.

If the DMP is not satisfied, the inverse of the stiffness matrix has to contain some negative entries (cf. Fig. 3 (left)).

We computed the stiffness matrices for many combinations of parameters K and s, and found the minimal values of entries of their inverses. The colour of each pixel in Fig. 3 (right) corresponds to the minimal value of the inverse stiffness matrix for a given pair of parameters. The vertical axis describes the dependence of the minimum on the slope parameter s. The range overlaps the interval $(0, 1)$, so it shows the area of non-acute triangulations, too.

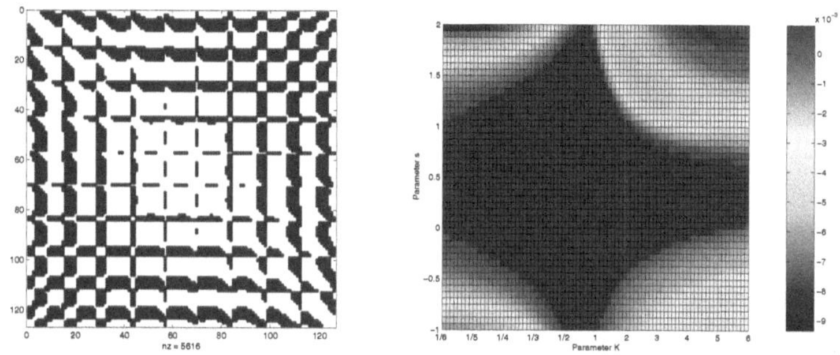

Fig. 3. Left: Negative entries in the inverse matrix for $K = 30$ and $s = 1.5$. Right: Dependence of the minimal entry of the inverse stiffness matrix on parameters s and K.

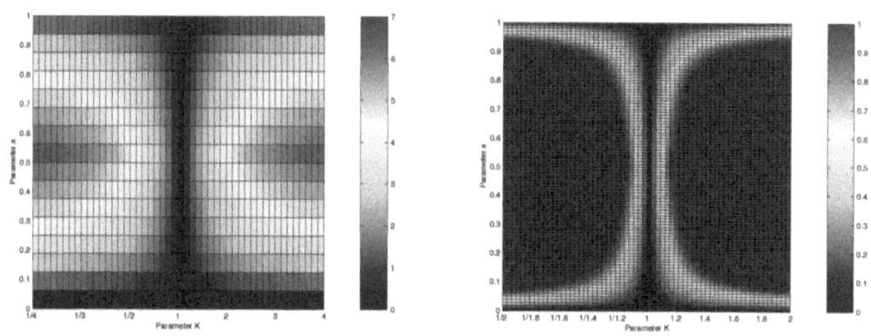

Fig. 4. Left: Behaviour of the left-side of condition (19). Right: Areas of the validity of (19). The values larger than 1 are indicated by a dark colour.

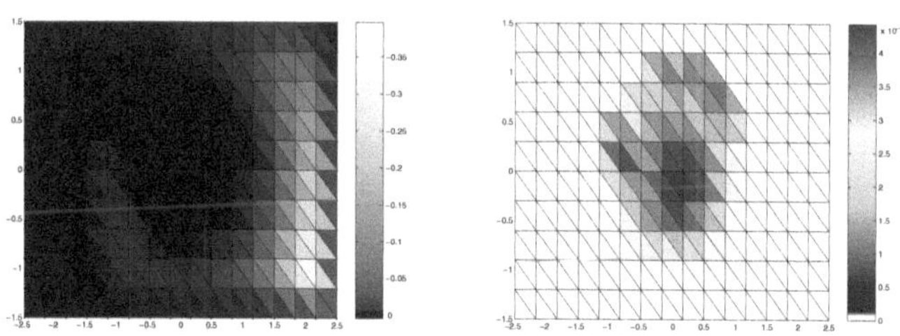

Fig. 5. Left: The discrete solution u_h for a certain F. Right: A domain, where DMP is violated.

Condition (14) can be modified to the form

$$\frac{4\,\bar{b}\,\cos\alpha_1}{d\,C^2\,\operatorname{ctg}2\alpha_1} \leq 1. \tag{19}$$

Values of the expression on the left-hand side of (19) are shown in Fig. 4. By Theorem 1, the DMP is satisfied for those values that are less or equal to one. Note that the functions in Figs. 3 and 4 need not be symmetric, because the shape of $\Omega(s)$ is different for $s = 0$ and $s = 1$.

Example: Fig. 5 shows that the area, where the DMP is violated, can be in the middle of the domain for $K = 8$ and $s = 0$.

6 Conclusions and Open Problems

The above analysis shows that DMP is valid on acute and regular triangulations for full diffusion tensors with a sufficiently strong diagonal dominance property.

Construction of acute triangulations is quite well studied, e.g., in works [14], [17]. The family of uniformly acute triangulations can be easily constructed using the standard $2d$ red refinement of triangles by midlines.

It is worth mentioning that the idea proposed here cannot be easily generalized for elliptic problems in dimension 3 and higher, since it is not clear how to construct and further refine acute simplicial meshes of a given polytopic domain (cf. [3], [10]).

Acknowledgement

This work is supported by the Academy Research Fellowship no. 208628 and project no. 124619 from the Academy of Finland, and by the Grant IAA 100190803 of the Grant Agency of the Academy of Sciences of the Czech Republic and the Institutional Research Plan AV0Z 10190503.

References

1. Brandts, J., Korotov, S., Křížek, M.: On the equivalence of regularity criteria for triangular and tetrahedral finite element partitions. Comput. Math. Appl. 55, 2227–2233 (2008)
2. Brandts, J., Korotov, S., Křížek, M.: The discrete maximum principle for linear simplicial finite element approximations of a reaction-diffusion problem. Linear Algebra Appl. 429, 2344–2357 (2008)
3. Brandts, J., Korotov, S., Křížek, M., Šolc, J.: On nonobtuse simplicial partitions. SIAM Rev. (accepted)
4. Burman, E., Ern, A.: Discrete maximum principle for Galerkin approximations of the Laplace operator on arbitrary meshes. C. R. Math. Acad. Sci. Paris 338, 641–646 (2004)
5. Ciarlet, P.G., Raviart, P.A.: Maximum principle and uniform convergence for the finite element method. Comput. Methods Appl. Mech. Engrg. 2, 17–31 (1973)
6. Fiebig-Wittmaack, M., Börsch-Supan, W.: Positivity preserving in difference schemes for the 2D diffusive transport of atmospheric gases. J. Comput. Phys. 115, 524–529 (1994)
7. Horváth, R.: Sufficient conditions of the discrete maximum-minimum principle for parabolic problems on rectangular meshes. Comput. Math. Appl. 55, 2306–2317 (2008)
8. Karátson, J., Korotov, S.: Discrete maximum principles for finite element solutions of nonlinear elliptic problems with mixed boundary conditions. Numer. Math. 99, 669–698 (2005)

9. Karátson, J., Korotov, S., Křížek, M.: On discrete maximum principles for nonlinear elliptic problems. Math. Comput. Simulation 76, 99–108 (2007)
10. Křížek, M.: There is no face-to-face partition of R^5 into acute simplices. Discrete Comput. Geom. 36, 381–390 (2006)
11. Kuzmin, D.: On the design of algebraic flux correction schemes for quadratic finite elements. J. Comput. Appl. Math. 218, 79–87 (2008)
12. Ladyzhenskaya, O.A., Ural'tseva, N.N.: Linear and quasilinear elliptic equations. Leon Ehrenpreis Academic Press, New York (1968)
13. Lamberton, D., Lapeyre, B.: Introduction to stochastic calculus applied to finance, 2nd edn. Chapman & Hall/CRC Financial Mathematics Series. Chapman & Hall/CRC, Boca Raton (2008)
14. Li, J.Y.S.: Nonobtuse meshes with guaranteed angle bounds. Master Thesis, Simon Fraser Univ. (2006)
15. Liska, R., Shashkov, M.: Enforcing the discrete maximum principle for linear finite element solutions of elliptic problems. Commun. Comput. Phys. 3, 852–877 (2008)
16. Lipnikov, K., Shashkov, M., Svyatskiy, D., Vassilevski, Y.: Monotone finite volume schemes for diffusion equations on unstructured triangular and shape-regular polygonal meshes. J. Comput. Phys. 227, 492–512 (2007)
17. Maehara, H.: Acute triangulations of polygons. European J. Combin. 23, 45–55 (2002)
18. Mlacnik, M.J., Durlofsky, L.J.: Unstructured grid optimization for improved monotonicity of discrete solutions of elliptic equations with highly anisotropic coefficients. J. Comput. Phys. 216, 337–361 (2006)
19. Le Potier, C.: Schéme volumes finis monotone pour des opérateurs de diffusion fortement anisotropes sur des maillages de triangles non structurés. C. R. Acad. Sci. Paris I(341), 787–792 (2005)
20. Varga, R.: Matrix Iterative Analysis. Prentice Hall, New Jersey (1962)
21. Vejchodský, T., Šolín, P.: Discrete maximum principle for higher-order finite elements in 1D. Math. Comp. 76, 1833–1846 (2007)

Comparative Analysis of High Performance Solvers for 3D Elasticity Problems

Ivan Lirkov[1], Yavor Vutov[1], Maria Ganzha[2], and Marcin Paprzycki[2]

[1] Institute for Parallel Processing, Bulgarian Academy of Sciences,
Acad. G. Bonchev, Bl. 25A, 1113 Sofia, Bulgaria
ivan@parallel.bas.bg, yavor@parallel.bas.bg
http://parallel.bas.bg/~ivan/, http://parallel.bas.bg/~yavor/
[2] Systems Research Institute, Polish Academy of Sciences
ul. Newelska 6, 01-447 Warsaw, Poland
paprzyck@ibspan.waw.pl, maria.ganzha@ibspan.waw.pl
http://mpaprzycki.swps.edu.pl, http://www.ganzha.euh-e.edu.pl

Abstract. We consider the numerical solution of 3D linear elasticity equations. The investigated problem is described by a coupled system of second order elliptic partial differential equations. This system is then discretized by conforming or nonconforming finite elements. After applying the Finite Element Method (FEM) based discretization, a system of linear algebraic equations has to be solved. In this system the stiffness matrix is large, sparse and symmetric positive definite. In the solution process we utilize a well-known fact that the preconditioned conjugate gradient method is the best tool for efficient solution of large-scale symmetric systems with sparse positive definite matrices. In this context, the displacement decomposition (DD) technique is applied at the first step to construct a preconditioner that is based on a decoupled block diagonal part of the original matrix. Then two preconditioners, namely the Modified Incomplete Cholesky factorization MIC(0) and the Circulant Block-Factorization (CBF) preconditioning, are used to precondition thus obtained block diagonal matrix.

As far as the parallel implementation of the proposed solution methods is concerned, we utilize the Message Passing Interface (MPI) communication libraries. The aim of our work is to compare the performance of the two proposed preconditioners: the DD MIC(0) and the DD CBF. The presented comparative analysis is based on the execution times of actual codes run on modern parallel computers. Performed numerical tests demonstrate the level of parallel efficiency and robustness of the proposed algorithms. Furthermore, we discuss the number of iterations resulting from utilization of both preconditioners.

1 Introduction

Our work concerns development and implementation of efficient parallel algorithms for solving elasticity problems arising in geosciences. Typical application problems include simulation of foundations of engineering constructions (that

S. Margenov, L.G. Vulkov, and J. Waśniewski (Eds.): NAA 2008, LNCS 5434, pp. 392–399, 2009.

transfer and distribute the total load into a bed of soil) and multilayer media with strongly varying material characteristics. Here, the spatial framework of the construction produces a complex stressed-strained state in the active interaction zones. The modern design of cost-efficient construction with a sufficient guaranteed reliability requires determining parameters of this stressed-strained state.

This type of engineering problems is described mathematically by a system of three-dimensional nonlinear partial differential equations. A finite element (or finite difference) discretization reduces the partial differential equation problem to a system of linear equations $K\mathbf{x} = \mathbf{f}$, where the stiffness matrix K is large, sparse and symmetric positive definite. It is a well known fact that Conjugate Gradient (CG) type methods are claimed to be the most cost-effective way to solve problems of this type, c.f. [1]. Furthermore, to accelerate convergence of the iterative process, a preconditioner M is combined with the CG algorithm.

To make a reliable prediction of the safety of the construction, which is sensitive to soil deformations, a very accurate model and thus a large system of sparse linear equations has to be solved. In real-life applications, such system can contain up to several millions of unknowns. Note that the numerical solution of linear systems is a fundamental operation in computer modeling of elasticity problems. Specifically, solving these linear systems is usually very time-consuming (requiring up to 90% of the total solution time). Hence, developing fast solvers for linear equations is essential. Furthermore, such algorithms should be expected to significantly speed up the simulation processes of real application problems. Due to the size of the system, an efficient iterative solver should not only have a fast convergence rate but also high parallel efficiency. Moreover, the resulting program has to be efficiently implementable on modern shared-memory, distributed memory, and shared-distributed memory parallel computers.

The remaining part of the paper is organized as follows. The considered elasticity problems are described in the next section. In section 3 we present the developed parallel solvers. The results of test experiments on parallel computer systems are provided in section 4.

2 Elasticity Problems

For simplicity, in this work we focus our attention on 3D linear elasticity problems following two basic assumptions: (1) displacements are small, and (2) material properties are isotropic. A precise mathematical formulation of the considered problem has been described in [5]. The 3D elasticity problem in the stressed-strained state can be described by a coupled system of three differential equations. Applying linearization, the nonlinear equations can be transformed into a system of three linear differential equations, which is often referred to as Lamé equations.

We restrict our considerations to the case when the computational domain Ω is a rectangular parallelepiped $\Omega = [0, x_1^{max}] \times [0, x_2^{max}] \times [0, x_3^{max}]$, where the boundary conditions on each face of Ω are of fixed type.

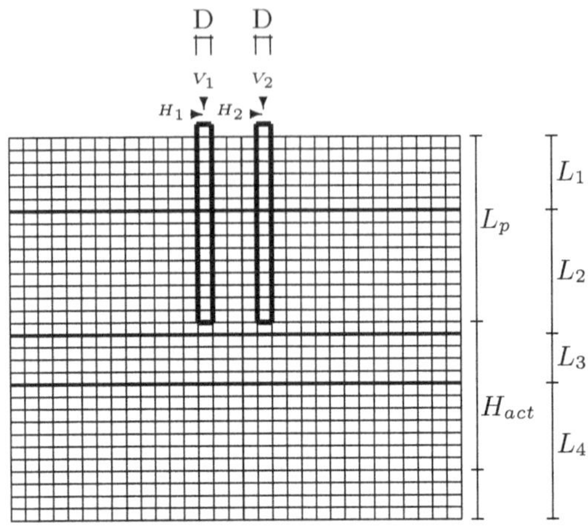

Fig. 1. Cross section of the computational domain Ω. $x_1^{max} = x_2^{max} = 37.2m$, $x_3^{max} = 31.0m$. $|H_1| = |H_2| = 150kN$, $|V_1| = 4000kN$, $|V_2| = 2000kN$, $E_{pile} = 31500MPa$, $\nu_{pile} = 0.2$, $E_{L_1} = 5.2MPa$, $\nu_{L_1} = 0.4$, $E_{L_2} = 9.4MPa$, $\nu_{L_2} = 0.35$, $E_{L_3} = 14.0MPa$, $\nu_{L_3} = 0.25$, $E_{L_4} = 21.4MPa$, $\nu_{L_4} = 0.2$.

A benchmark problem from [4] is used in numerical tests reported here. The engineering problem describes two piles in an inhomogeneous sandy clay soil (see Fig. 1). In the solution process, uniform grid is used with n grid points along each coordinate direction. In our experiments we used two kind of meshes: coarse mesh with step sizes $1.2 \times 1.2 \times 1$ and fine mesh with step sizes $0.6 \times 0.6 \times 0.5$.

3 Parallel Displacement Decomposition Solvers

There exists a substantial body of work dealing with preconditioning of iterative solution methods for elasticity systems discretized by using the Finite Element Method. For instance, in [2] Axelsson and Gustafson construct their preconditioners based on the point-ILU (Incomplete LU) factorization of the displacement decoupled block-diagonal part of the original matrix. This approach is known as the displacement decomposition (see, e.g., [3]).

Our first approach uses the DD MIC(0) (Modified Incomplete Cholesky) factorization, presented in [7]. It uses nonconforming Ranacher-Turek elements for the discretization. Modifying the block-diagonal displacement decomposed matrix, an auxiliary matrix B is obtained. This matrix has a block structure, with diagonal blocks being diagonal matrices. Then the MIC(0) factorization of the matrix B is used as a preconditioner.

The other parallel preconditioning technique used in this work is the circulant block-factorization used for preconditioning of the obtained block-diagonal matrix, c.f. [6]. Here, a displacement decomposition circulant block factorization preconditioner is constructed.

Suitable modification of the DD MIC(0) algorithm allows parallelization of the preconditioning, but results more communication steps in comparison with the DD CBF preconditioner. The estimate of the condition number of the DD CBF preconditioner shows that the convergence is asymptotically as fast as preconditioners based on the point-ILU factorization, c.f. [5,6]. Moreover the DD CBF solver has a good parallel efficiency (see, e.g., [5,6]).

4 Experimental Results

Before proceeding with describing results of performed benchmarking runs let us illustrate the nature of the solved problem. The obtained solution of the elasticity problem is used for computation of the vertical strain in the computational domain. Thus, in Fig. 2 vertical displacements and vertical strains are depicted in a cross section of the domain. An isoline connects points with equal values.

To efficiently solve the problem, portable parallel FEM codes were designed and implemented in C (the DD CBF code) and C++ (the DD MIC(0) code). In both cases, parallelization has been facilitated using the MPI library, c.f. [8,9]. The parallel code has been tested on cluster computers located in the National Energy Research Scientific Computing Center (NERSC). In our experiments, times have been collected using the MPI provided timer and for each problem size and number of processors we report the best results from multiple runs. For linear system with N unknowns we represent the number of iterations as N_{it}, the elapsed time as T_p (in seconds; obtained on p processors), the speed-up as $S_p = T_1/T_p$, and the parallel efficiency as $E_p = S_p/p$. Because of the NERSC-imposed limitations in available computational time and memory in some cases we were not able to establish single-processor performance for the largest problem. Therefore, for the largest problems we report parallel efficiency related to results collected on 2 processors.

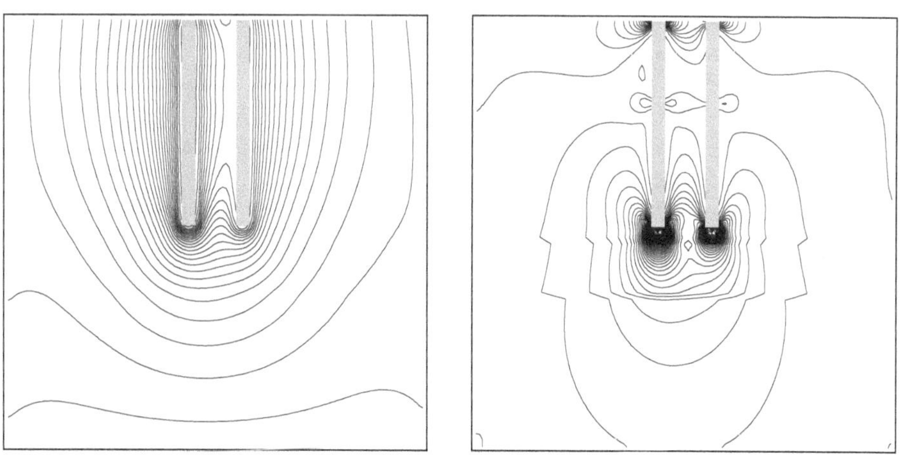

Fig. 2. Vertical displacements on the left, vertical strains on the right

Table 1. Experimental results on Bassi

p	N	N_{it}	T_p	S_p	E_p	N	N_{it}	T_p	S_p	E_p
DD MIC(0)										
1	2 179 548	1 533	1 514.3			17 298 000	2 840	23 101.0		
2		1 533	795.3	1.90	0.952		2 840	11 584.3	1.99	0.997
4		1 533	414.9	3.65	0.912		2 840	5 973.1	3.87	0.967
8		1 533	217.7	6.95	0.869		2 840	3 219.0	7.18	0.897
16		1 533	125.9	12.02	0.752		2 840	1 684.5	13.71	0.857
32		1 533	74.1	20.43	0.638		2 840	913.2	25.30	0.791
64							2 840	544.4	42.43	0.663
DD CBF										
1	786 432	1 297	1 912.6			6 291 456	2 641	41 075.5		
2		1 297	955.6	2.00	1.001		2 633	17 378.5	2.36	1.182
4		1 291	467.0	4.10	1.024		2 625	8 633.0	4.76	1.189
8		1 290	235.1	8.14	1.017		2 617	4 330.6	9.49	1.186
16		1 252	115.7	16.53	1.033		2 612	2 191.6	18.74	1.171
32		1 282	62.4	30.63	0.957		2 609	1 079.7	38.04	1.189
64		1 247	33.3	57.51	0.899		2 600	575.8	71.34	1.115

Table 1 summarizes results collected on the IBM p575 POWER 5 system, named Bassi [10]. Bassi is a distributed memory computer with 888 IBM POWER 5 processors (running at 1.9 GHz) distributed among 111 compute nodes with 8 processors per node. Each Bassi processor has a theoretical peak performance of 7.6 GFlop/s. Processors within each node have a shared memory pool of 32 GB. Bassi's network switch is the IBM "Federation" HPS switch which is connected to a two-link network adapter on each node. We have used IBM C and C++ compilers with options "-O3 -qstrict -qarch=auto -qtune=auto". According to the best of our knowledge, this and compiler switches used on other machines should result in maximal performance optimization.

Table 2 shows execution time on the NERSC Cray XT4 system, named Franklin [11]. Franklin is a massively parallel processing (MPP) system with 9 660 compute nodes, and the entire system has a total of 19 320 processor cores. Specifically, each of Franklin's compute nodes consists of a 2.6 GHz dual-core AMD Opteron processor with a theoretical peak performance of 5.2 GFlop/s. Each compute node has 4 GB of memory and is connected to a dedicated SeaStar2 router through the Hypertransport with a 3D torus topology. We have used the PGI C and C++ compilers with options "-fast -O3 -Minline".

The memory available on a single node of Franklin is not large enough to run our experiments for the fine mesh and we reported here the execution time starting from two processors on two different nodes.

In Table 3 we present results of experiments performed on the Jacquard [12]. It is a 712-CPU (356 dual-processor nodes running at 2.2 GHz) Opteron Linux cluster. Each processor has a theoretical peak performance of 4.4 GFlop/s. Processors within each node share 6 GB of memory and are interconnected through a high-speed InfiniBand network. We have used C/C++ compilers produced by PathScale with the ACML Optimized Math Library and compiled the code using

Table 2. Experimental results on Franklin

p	N	N_{it}	T_p	S_p	E_p	N	N_{it}	T_p	E_p
				DD MIC(0)					
1	2 179 548	1 959	3 787.7			17 298 000			
2		1 959	1 935.9	1.96	0.978		3 404	26 593.8	
4		1 959	1 001.6	3.78	0.945		3 404	13 712.8	0.970
8		1 959	526.9	7.19	0.899		3 404	6 921.7	0.961
16		1 959	286.9	13.20	0.825		3 404	3 573.8	0.930
32		1 959	167.2	22.65	0.708		3 404	1 937.2	0.858
64		1 959					3 404	1 115.7	0.745
				DD CBF					
1	786 432	1 298	1 392.6			6 291 456			
2		1 294	745.9	1.87	0.933		2 632	11 944.2	
4		1 291	380.9	3.66	0.914		2 630	6 443.3	0.927
8		1 292	184.5	7.55	0.943		2 621	3 243.1	0.921
16		1 251	88.8	15.69	0.980		2 612	1 600.8	0.933
32		1 286	50.4	27.63	0.864		2 613	800.1	0.933
64		1 281	33.8	41.16	0.643		2 608	420.8	0.887

Table 3. Experimental results on Jacquard

p	N	N_{it}	T_p	S_p	E_p	N	N_{it}	T_p	E_p
				DD MIC(0)					
1	2 179 548	1 959	1 083.6						
2		1 959	535.7	2.023	1.011				
4		1 959	301.4	3.594	0.899				
8		1 959	180.7	5.997	0.750				
16		1 959	117.0	9.264	0.579				
32		1 959	81.7	13.260	0.414				
				DD CBF					
1	786 432	1 260	1 277.8			6 291 456			
2		1 297	675.2	1.89	0.946		2636	12 465.1	
4		1 259	351.3	3.64	0.909		2629	6 693.9	0.931
8		1 290	185.1	6.90	0.863		2617	3 354.3	0.929
16		1 287	92.7	13.79	0.862		2610	1 664.9	0.936
32		1 286	57.7	22.16	0.692		2605	805.0	0.968
64		1 283	43.1	29.68	0.464		2602	556.0	0.701

"mpicc -Ofast $ACML" command. The "-Ofast" option is a generic option leading to a vendor suggested aggressive optimization.

Several jobs submitted on Jacquard are still waiting in the queue and this is the reason for the fine mesh to report only some results from DD CBF code.

First, let us note that the number of iterations for the DD CBF varies with the number of processors. This well known effect is caused by the different order of summations in the inner product computations involved in the PCG. The same effect is not observed in the DD MIC(0) code, because of a special precaution taken during the computation of the inner products. Here, the order of the additions is made independent of the number of processors.

An interesting phenomenon is observed concerning the number of iterations for the DD MIC(0). On Bassi they are notably smaller than those on both Franklin and Jacquard. For instance, for the smaller problem the number of iterations is 1533 on Bassi and 1959 on the other two clusters. For the larger problem these numbers are 2840 and 3404 respectively. This is probably caused by the difference in the processor architectures (Power 5 vs. x86).

The number of unknowns in the nonconforming discretization of the problem (used by the DD MIC(0) solver) is about three times greater than in the conforming one (used by the DD CBF solver). Furthermore, the number of iterations of the DD MIC(0) preconditioner is also greater than the number of iterations of the CBF one. Nevertheless, computing times of both solvers are comparable, with the DD MIC(0) solver being somewhat faster. A notable exception from this are runs on Franklin, where the DD MIC(0) code performs more than two times slower. We believe that the cause of this phenomenon is the PGI compiler and its inability to appropriately optimize the C++ code (vis-a-vis the C code).

As expected the parallel efficiency of the DD CBF solver is generally better than this of the DD MIC(0). On Bassi, the efficiency of 64% is obtained for the smaller problem on 32 processors for the MIC(0) preconditioner. For the larger problem, on 64 processors, the efficiency is 66% for the same solver. For the CBF preconditioner the lowest efficiencies are 90% and 94%, reached on 64 processors, for the smaller and larger problems respectively. There could be both software and hardware causes for the super-linear speed-up observed on Bassi. On the software side, when using more processors, the number of iterations needed for some convergence steps are smaller. On the hardware side, cache effect is usually

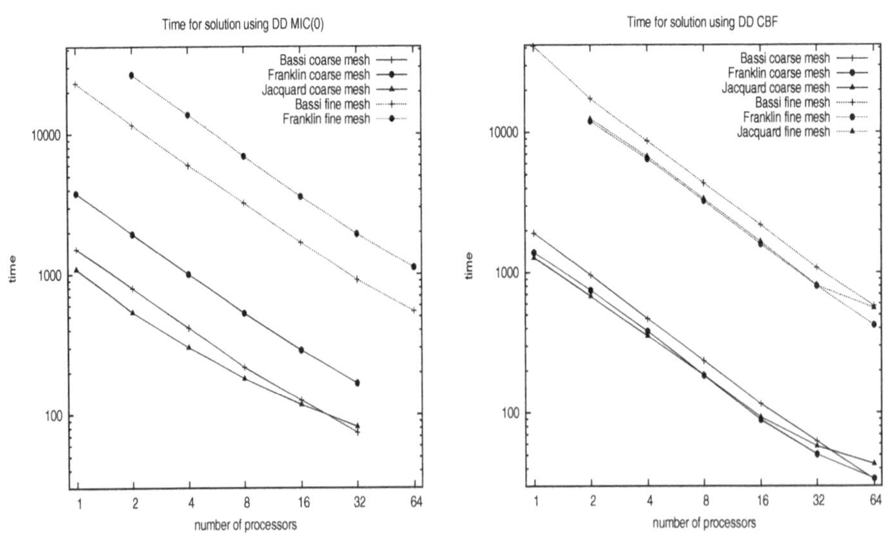

Fig. 3. Execution times for the coarse and fine mesh

the cause. When the sub-domains are smaller, the data each processor owns are more easily fit into cache.

The superiority of the parallel properties of the CBF solver can be tracked on Jacquard and Franklin as well, although they are not as pronounced.

To summarize, in Fig. 3 computing times on different clusters are shown for both algorithms. The left picture well illustrate the above mentioned phenomenon with slower execution of the DD MIC(0) solver on Franklin. Also, the theoretical peak performance of Bassi is the highest but with respect to the execution time on one processor Jacquard is the fastest machine.

Acknowledgments

Computer time grant from the National Energy Research Scientific Computing Center is kindly acknowledged. This research was partially supported by grant I-1402/2004 from the Bulgarian NSF. Work presented here is a part of the Poland-Bulgaria collaborative grant: "Parallel and distributed computing practices".

References

1. Axelsson, O.: Iterative solution methods. Cambridge Univ. Press, Cambridge (1994)
2. Axelsson, O., Gustafsson, I.: Iterative methods for the solution of the Navier equations of elasticity. Comp. Meth. Appl. Mech. Eng. 15, 241–258 (1978)
3. Blaheta, R.: Displacement decomposition-incomplete factorization preconditioning techniques for linear elasticity problems. Num. Lin. Alg. Appl. 1, 107–128 (1994)
4. Georgiev, A., Baltov, A., Margenov, S.: Hipergeos benchmark problems related to bridge engineering applications. REPORT HG CP 94–0820–MOST–4
5. Lirkov, I.: MPI solver for 3D elasticity problems. Math. and computers in simulation 61(3–6), 509–516 (2003)
6. Lirkov, I., Margenov, S.: MPI parallel implementation of CBF preconditioning for 3D elasticity problems. Math. and computers in simulation 50(1–4), 247–254 (1999)
7. Vutov, Y.: Parallel DD-MIC(0) Preconditioning of Nonconforming Rotated Trilinear FEM Elasticity Systems. In: Lirkov, I., Margenov, S., Waśniewski, J. (eds.) LSSC 2007. LNCS, vol. 4818, pp. 745–752. Springer, Heidelberg (2008)
8. Snir, M., Otto, S., Huss-Lederman, S., Walker, D., Dongara, J.: MPI: The Complete Reference. Scientific and engineering computation series. The MIT Press, Cambridge (1997) (second printing)
9. Walker, D., Dongara, J.: MPI: a standard Message Passing Interface. Supercomputer 63, 56–68 (1996)
10. Bassi — IBM POWER 5, `http://www.nersc.gov/nusers/systems/bassi/`
11. Franklin — Cray XT4, `http://www.nersc.gov/nusers/systems/franklin/`
12. Jacquard — Opteron Cluster, `http://www.nersc.gov/nusers/resources/jacquard/`

Damping Control Strategies for Vibration Isolation of Disturbed Structures

Daniela Marinova

Technical University-Sofia, Sofia 1756, Bulgaria
dmarinova@dir.bg

Abstract. This paper investigates thin layered structures as active systems and control strategies for their vibration suppressions and shape regulating. We focus on FEM modelled thin plates. Optimal voltages for shape control are obtained using genetic optimization. An optimal selection of actuators number and locations is considered. Data fusion for reduction of input data for simplifying the controlling process is studied. Numerical simulations are presented.

1 Introduction

The laminated composite structures are new generation materials with important applications in high technologies. Functionally graded materials (FGM) constitute a new class of materials with smooth variation of material properties across the thickness [1]. Finite element method (FEM) is a prospective approach for investigating of FGM structures.

For light-weight high-performance flexible structures the concept of structures with self-controlling and self-monitoring capabilities is very helpful. Longitudinally piezoelectric fibre reinforced composite (PFRC) materials are used as distributed actuators and sensors for these purposes and for active vibration control of the flexible structures [3]. The exact solutions for problems with more general boundary conditions and arbitrary loading can not be derived and the finite element method remains the most popular tool for solving such problems.

This paper presents a general dynamical FEM model for a FGM plate integrated with a piezoelectric PFRC actuator of the plate.

2 Preliminary

We consider a symmetric, simply supported rectangular composite plate with FGM core and PFRC layer acting as an actuator.

The extended Mindlin theory for thin plates is adopted in the FEM. The first order shear deformation theory for the displacements/thickness gives:

$$u = u_0(x,y) + z\theta_x(x,y) \quad v = v_0(x,y) + z\theta_y(x,y) \quad w = w_0(x,y) \qquad (1)$$

where u_0,v_0,w_0 are the displacements of the mid-plane and θ_x,θ_y are the rotational angles of the normal of the plate. The linear strain-displacement relations

S. Margenov, L.G. Vulkov, and J. Waśniewski (Eds.): NAA 2008, LNCS 5434, pp. 400–407, 2009.

are used. ε_z is assumed to be zero. The strain vector is presented with in-plane and out-plane parts:

$$\{\varepsilon\} = \{\varepsilon_b^T, \varepsilon_s^T\}^T \quad \{\varepsilon_b\} = \{\varepsilon_{bt}\} + z\{\varepsilon_{br}\} \quad \{\varepsilon_s\} = \{\varepsilon_{st}\} + z\{\varepsilon_{sr}\} \tag{2}$$

The FGM layer is isotropic with exponential of z Young modulus [4] and its electric potential equals zero. The PFRC thickness is small and its electric potential is linear in the **z** direction. Similarly, the stress vectors are split in two parts for in-plane and out-of-plane stresses. The constitutive equations for the FGM and PFRC layer stress vectors are:

$$\begin{aligned} \{\sigma_b^1\} &= [C_b^1]\{\varepsilon_b^1\} \quad \{\sigma_s^1\} = [C_s^1]\{\varepsilon_s^1\} \\ \{\sigma_b^2\} &= [C_b^2]\{\varepsilon_b^2\} - [e_b]\{E\} \quad \{\sigma_s^2\} = [C_s^2]\{\varepsilon_s^2\} - [e_s]\{E\} \end{aligned} \tag{3}$$

where C_b^1 is plane-stress matrix and C_s^1 is the transverse-shear elasticity matrix, $[e_b]$,$[e_s]$ are the dielectric constants, E is the electric field vector, C_b^2,C_s^2 are the transformed elastic coefficient matrices.

3 Finite Element Formulation

Eight node isoparametric elements is considered for the discretization. After applying the principle of total minimum potential energy $\delta T_p^e = 0$, the following system of equations for one finite element is derived:

$$[K_t^e]\{d_t^e\}+[K_{tr}^e]\{d_r^e\} = [F_{tp}^e]\{\Phi_0^e\}+\{F^e\} \quad [K_n^e]\{d_t^e\}+[K_r^e]\{d_r^e\} = [F_{rp}^e]\{\Phi_0^e\} \tag{4}$$

where $[K_t^e]$, $[K_{tr}^e]$, $[K_{rt}^e]$, $[K_r^e]$ are the stiffness matrices, $[d_t^e]$, $[d_r^e]$ are the displacement vectors, Φ_0^e is the electric potential. The assembling of these equations leads to the system of linear static equations:

$$[K]\{d\} = [F_{tp}]\{\Phi\} + \{F\} \tag{5}$$

where the in-plane displacements are expressed trough out-plane ones what is the advantage of the created FEM model.

3.1 Numerical Validation

The model of the FGM/PFRC plate is validated. The simulation results are compared with the results of other authors and sources.

In the first experiment of one FGM layered plate model with constant homogenous Young's modulus is considered. The result using FEMLAB with triangular elements is 0.0170. The results with the derived model for different FEM mesh discretization are in table 1 below.

The next simulation is done with a two layered plate. Sinusoidal distributions of the mechanical and electrical loads are used.

$$\begin{aligned} F(x,y) &= f_0 \sin(px).\sin(qy) \\ \Phi(x,y,z) &= V(z)\sin(px)\sin(qy) \end{aligned} \quad p = \frac{2\pi}{a}, \; q = \frac{2\pi}{b}, \quad f_0 = -40\mathrm{N} \,/\, \mathrm{m}^2$$

Table 1. Maximum Plate Displacement

Plate size	*Max displ.*	*Plate size*	*Max displ.*
10x10	0.01650915	20x20	0.01683623
12x24	0.01675899	24x24	0.01686863
12x32	0.01677569	32x32	0.01689886

The results are evaluated with and without applying the electrical potential distribution on the actuator surface for different values of length to thickness ratio s/h of the FGM plate.

The Table 2 below contains the comparison of the responses of a FGM/PFRC thin plate ($E_h/E_0 = 10,\quad s = 100$ with $(u_1^*, w^*, \sigma_{x1}^*, \sigma_{y1}^*, \sigma_{xy1})$ and without $(u_2^*, w^*, \sigma_{x2}^*, \sigma_{y2}^*, \sigma_{xy2})$ the applied sinusoidal voltage with the presented FEM model and the analytical solution results [5].

Table 2. Thin Plate $E_h/E_0 = 10$, $s = 100$

(V)	*Method*	u_1^*	u_2^*	w^*	σ_{x1}^*	σ_{x2}^*	σ_{y1}^*	σ_{y2}^*	σ_{xy1}^*	σ_{xy2}^*
100	*Numeric*	-0.0202	-0.0496	0.9341	0.0296	1.61260	-0.1143	0.173	0.023	-0.4817
	Analytic	-0.0202	-0.0497	0.9368	0.0291	1.61240	-0.1145	0.1751	0.0230	-0.4813
-100	*Numeric*	-0.0188	0.0681	-2.7629	0.145	-2.4429	0.2886	-1.005	-0.1173	0.9312
	Analytic	-0.0188	0.0681	2.7629	0.1457	-2.4446	0.2892	-1.0090	-0.1171	0.9298
0	*Numeric*	-0.0195	0.0093	-0.9144	0.0873	-0.4151	0.0872	-0.416	-0.0472	0.2247
	Analytic	-0.0195	0.0093	-0.9155	0.0874	-0.4161	0.0873	-0.4170	-0.0470	0.2243

3.2 Plate's Dynamics

The expression for the mass matrix of an element in FEM is expressed with an integral over the element's volume of the density, approximated with the shape functions N:

$$M^e = \int_V N^T \rho N dv \tag{6}$$

Coupling the element matrices we obtain the mass matrix $[M]$for the plate. The equation of motion will have the form

$$[M]\{\ddot{d}\} + [K]\{d\} = [F_{tr}]\{\Phi\} + \{F\} \tag{7}$$

where $\{\ddot{d}\}$ is the vector of the accelerations. The results obtained by FEMLAB and the proposed FEM dynamical model are compared. The first six eigenmodes of a plate with thickness 0.05 m. found by the proposed model are 918.6182, 2147.3176, 2147.3176, 3119.5268, 3119.5268, 3244.4108 and are closed to the results found by FEMLAB are 906.051, 2201.85, 2202.26, 3358.75, 4185.32, 4189.65. The additional fourth and fifth arisen eigenfrequencies could be explained with the used in-plane displacements. The corresponding eigenvectors are zeros in the **z** direction and have nonzero values in **x** and **y** directions. This phenomena happens for thicker plates. For control purposes (7) is rewritten in the form

$$\dot{X} = AX + B_f F + B_u U \tag{8}$$

where $X = \left[\{d\} \, \{\dot{d}\} \right]^T$ is the state vector, A is the system matrix with included damping, F is the generalized force vector, U is the generalized control vector and B_f,B_u are location matrices.

In designing smart structures with piezoelectric actuators the choice of actuators, their locations, and the charge amount are important factors. In this study the shape control problem is formulated and realized with several configurations of piezoelectric actuators based on a genetic algorithm [6]. A related problem concerning the damage identification using static data and genetic optimization is studied as well.

4 Shape Control

For shape control the static equilibrium of the whole discretized plate (8) is used. FGM with Young's modulus on the bottom equal to $E_0 = 70$GPa exponentially changes to the top. A uniform vertical load is applied on the upper surface of the plate.

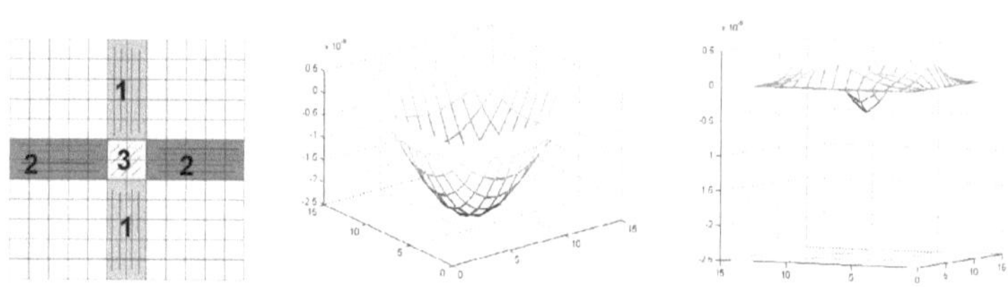

Fig. 1. a) Configuration, b) no control, c) 100V applied to patches 1,2,3

Only some of the elements are equipped with piezoelectric layers. Every actuator covers exactly the surface of one corresponding finite element. The first set of piezoelectric patches using three kinds of symmetrically placed patches with different orientations of the fibers and the results are shown in figure 1a.

5 Optimal Shaping Using Genetic Procedure

Dealing with smart structures two general problems are met: design and location of PZT and the amount of the applied voltage. The model defines a set of input–output data but the values of the parameters are unknown. Here the number and location of the piezoelectric patches are chosen heuristically and the optimal voltages are found minimizing an error function using Genetic Optimization Algorithms. The following optimization criterion is used

$$err(V^{el}) = \left\| X(V^{el}, F) - \overline{X}(F) \right\| \tag{9}$$

where $\| \ \|$ is L_2 norm in $\mathbf{R^n}$, $X(V^{el}, F) = [x_1 \ldots x_n]$ is the nodal displacement array, $\bar{X}(F) = [\bar{x}_1 \ldots x_n]$ is the desired value of the nodal displacement array, $V^{el} = [V_1 \ldots V_n]$ is the electric voltage array and F is the external mechanical loading acting on the plate.

Firstly, we solve the optimization problem for a plate with constant through thickness properties using GA MATLAB, default values with population size 100 and 67 generations. The maximum displacement of the loaded plate with static control is -3.2407e-007 m. that is approximately 23 times less than the maximum displacement -7.3553e-006 m. of the loaded plate without control.

fitness = -3.4338 (constant E, ν)

Patch num	1	2	3	4	5
Opt volt (V)	42.07	81.5	42.07	81.5	118.25

fitness = -5.3815 (variable E, ν)

1	2	3	4	5
61.5	142.7	60.2	88.4	193.75

In the case of loaded plate with variable stiffness and ratio $E^{top}/E^{bottom} = 10$ the best solution is achieved in 57 generations. The maximum displacement with applied control is 9.1319e-008, which is 27 times less then the maximum displacement when there is no control.

6 Eigenmodes Shaping

One helpful approach in the vibration suppression is the eigenmodes control. The purpose is to deform the plate as close as possible to eigenmode subjecting the piezoelectric patches with suitable charges. We have to choose the number, the position, the orientation and the amount of the charge. The first three parameters are defined and fixed. We investigate the best voltage.

The first eigenmode plays the most significant role in the deformation. To suppress it we have used the configuration of the piezo-electric patches shown in Figure 2a. The graphic of the difference between the eigenmode shape and the obtained deflections are depicted in Figure 2b.

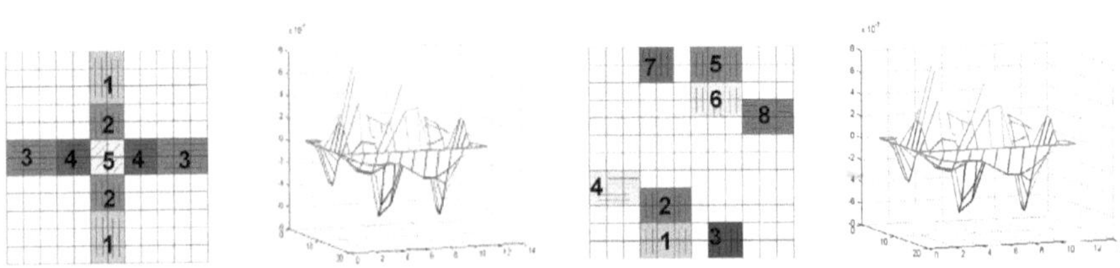

Fig. 2. a) Configuration, b) difference $< 0.25\%$, c) configuration, d) difference $< 20\%$

The following result for the input electricity is obtained:

Patch number	1	2	3	4	5	6
Optimal voltage (V)	65.70	97.75	65.95	114.80	246.00	257.05

To suppress the second eigenmode the configurations and orientations of the patches considered is shown in Figure 2c. The graphics of the difference between the eigenmode shape and the obtained deflections are depicted in the figure 2d. The optimal GA solutions are shown in the table below.

Patch number	1	2	3	4	5	6	7	8
Optimal voltage (V)	-9.38	92.855	10.85	27.23	1.12	-86.52	-15.26	-31.32

The material of a structure subjected to a dynamical load gets fatigue since cracks arise. That can cause serious problems.

A simplified damage modeling and identification is considered in this section. We assume that only one element is damaged and we search for its location and the degree of its crack. The influence of the damage is modelled with a reduction of the stiffness matrix of the corresponding element. The position of the cracked element is described with two variables: one for the horizontal position of the element and one for the vertical one. Results obtained using GA procedure are shown in the Table 3.

Table 3. Damage Identification – fixed element 80% stiffness

Popul. size	*Row Num*	*Column Num*	*Severity*	*Fitness value*
5	2	6	0.8171	7.037
10	2	8	0.8421	1.6039
20	2	8	0.8016	0.002476
50	2	8	0.7924	0.054899

7 Input Data Fusion

Sensor fusion addresses the problem of employing multiple data sources and combining their information to a more accurate and reliable representation of the structure's behavior. This paper discusses the application of stateless confidence-weighted averaging fusion of sensor data to structural control.

An algorithm for fusing data from replicated sensors based on weighted averages is suggested in [8]. The inputs to the fusion process are weighted according to the uncertainty associated to their sources. The measurement errors of all inputs are assumed to be independent and have a mean of zero. The fused value x_F is calculated as the weighted average of all measurements x_i. The weights minimize the expected variance of the fused value (9).

$$x_F = \sum_{i=1}^{n} w_i x_i \quad \sigma_F^2 = \sum_{i=1}^{n} w_i^2 \sigma_i^2 \tag{10}$$

Thus the optimal weights w_i are the solution to the minimization problem

$$\arg\min_{w_i} \sum_{i=1}^{n} w_i^2 \sigma_{i=1}^2 \quad w_i = \left(\sigma_i^2 \sum_{j=1}^{n} \sigma_j^{-2}\right)^{-1} \tag{11}$$

The sum of all weights is equal to 1 in order to obtain an undistorted result.

$$x_F = \sum_{i=1}^{n} x_i \sigma_i^{-2} \Big/ \sum_{i=1}^{n} \sigma_i^{-2} \quad \sigma_F^2 = \left(\sum_{i=1}^{n} \sigma_i^{-2}\right)^{-1} \tag{12}$$

Under the assumption of independence of errors between sensors and supposing the measurements have the expected deviation from the true value equal to 0, this method minimizes the expected variance of the fused value [7]. The transition between a faulty and an accurate measurement is smooth.

8 Active Optimal Control

For control design purposes of the structure (10) is used. The measurements are considered to be linear combinations of the system's states.$y = Cx$. The aim is to design a feedback control law proportional to the measurements $u = -Gx$. We choose the linear quadratic optimal control algorithm requesting the minimization of the following objective function

$$J = \int_0^\infty (y^T Q y + u^T R u) dt \tag{13}$$

The steady state case is considered and the optimization horizon is allowed to extend to infinity. In this case the control law is a linear time invariant function of the outputs of the system $G = R^{-1}B^T P$,where P is the positive definite solution of the Algebraic Riccati Equation. If the pair (A, B) is stabilizable, $R > 0$ and Q can be factored as $Q = C_q^T C_q$ such that the pair (C_q, A) is detectable, then the optimal controller asymptotically stabilizes the system (9). Minimizing the cost function J means to keep the control forces and the system response small. The matrices Q and R are the main design parameters.

9 Numerical Evaluation

To demonstrate the efficiency of the confidence weighted averaging, the data fusion method is evaluated on simulated data. For the generation of sensor data on which to analyze the performance of the proposed approaches, the smart structure is supposed to be equipped with distance sensors. The task is to regulate the shape of the beam undergone external excitations by employing sensor fusion measurements and the optimal control strategy.

The generated artificial data are based on the following assumptions. Data for the sources with different variances was generated with zero-mean noise following a normal distribution. Faults were then introduced into each dataset by randomly manipulating some of the nine sensors with equal probability of selection, generating random faults for these selected sensors following a uniform distribution. This data emulates low-quality sensor information in the sense that every single set of "concurrent" measurements contains faulty observations and no a-priori information about a sensor's probability of delivering faulty data outside of its usual range of fluctuation is available.

The performance of the averaging methods is with moderate improvement. With an increasing number of faults, the mean squared as well as the maximum error increases dramatically. Therefore, on extremely faulty data without any information about fault does not perform well.

Depending on the sensor configuration the uncertainty would vary since apparently the sensor signals interfered with each other.

The results show that ECWA generally performs better in terms of the mean squared error of the fusion result. In the case of CWA, the variance is significantly underestimated. The great discrepancy between estimated and true variance in the fusion result stems from error correlations within the data that are not considered in CWA.

10 Conclusion

Based on the extended Mindlin plate theory, an efficient and accurate FEM model of FGM/PFRC plates is developed to control the shape by internal stresses. Using genetic optimization procedures it is shown that only with a small number of actuators with their optimal placement and optimal voltages an effective shape control of the plate is achieved. The location and rotation of the patches are determined beforehand, by taking in account the desired shape. A related damage identification problem is considered.

Algorithms for sensor fusion - confidence weighted averaging is considered and tested for plate shape regulating. Unlike other methods CWA assumes independent error behavior, combines all observation by calculating a weighted average based on the uncertainty associated to each sensors. On simulated data, it is shown that the stateless fusion of data sensors is applicable in regulating of plate structure undergone external influences.

References

1. Vel, S., Batra, R.: Three-dimensional analysis of transient thermal stresses in functionally graded plates. Int. J. Solids and Structures 40, 7181–7196 (2003)
2. Dong, S., Tong, L.: Vibration control of plates using discretely distributed piezoelectric quasi-modal actuators/sensors. AIAA Journal 39, 1766–1772 (2001)
3. Sankar, B.: An elasticity solution for functionally graded beams. Composite Science and Technology 61, 689–696 (2001)
4. Kwon, Y., Bang, H.: The Finite Element Method Using Matlab. CRC Press, Boca Raton (2000)
5. Ray, M., Sachade, H.: Exact Solutions for the Functionally Graded Plates Integrated with a Layer of Piezoelectric Fibre-Reinforced Composite. ASME Journal of Applied Mechanics 73, 622–632 (2007)
6. Houck, C., Joines, J., Kay, M.: A genetic algorithm for function optimization: a Matlab implementation. NCSU-IE TR 95-09 (1995)
7. Elmenreich, W., Schorgendorfer, A.: Fusion of Continuous-valued Sensor Measurements using Confidence-weighted Averiging. JVC 13, 1303–1312 (2007)

Numerical Simulation of Shock Wave Diffraction on the Sphere in the Shock Tube

Sergey N. Martyushov and Yanina G. Martyushova

State Duma of Russian Federation, Moscow, Russia, Moskovsky Aviacionny Institute-Technical University, Moscow, Russia

Abstract. In non-stationary problems of flows, for example problem of shock wave diffraction on a flying body, the impulse of appearing drag force generates non-stationary forward and rotary movement of the body. The describing mathematical model is nonlinear not only with respect to coefficients of system of the equations, but also in the sense of dependence of the system of equations and boundary conditions from integral characteristics of the decision (drag force and its moment relative to the center of mass of the body). The purpose of this work is realization of computing aspects of numerical simulation of such problems by modern numerical algorithms on the example of model problem of shock wave diffraction on sphere in the shock tube. The problem was numerically simulated on the basis of the difference scheme of second order accuracy [1,2]. The 3-D structured grids construction was performed by algorithm based on the Poisson equations decision [5,6].

1 Difference Algorithm

System of equations for ideal non viscous gas for the moving finite volume V can be written in the form:

$$\frac{d}{dt}\oint_V \overrightarrow{U}dV + \oint_S \overrightarrow{n}\hat{F}dS = 0, \tag{1}$$

where conservative unknowns vector will be: $\overrightarrow{U} = (\rho, \overrightarrow{m}, e), \overrightarrow{m} = \rho(\overrightarrow{u} - \overrightarrow{v})$, vector of fluxes normal to boundary of control volume can be written in the form: $\overrightarrow{F_n} = \overrightarrow{n}\hat{F} = (\rho u', \overrightarrow{m}u' + P\overrightarrow{n}, eu' + Pu_n)$, $u' = \overrightarrow{n}(\overrightarrow{u} - \overrightarrow{v}), u_n = \overrightarrow{u}\,\overrightarrow{n}, v_n = \overrightarrow{v}\,\overrightarrow{n}, \overrightarrow{v}$ - velocity vector of the control volume boundary. Vector $\overrightarrow{F_n}$ is a nonlinear function of $\overrightarrow{U}$, then $d\overrightarrow{F_n} = Ad\overrightarrow{U}$. Let R will be matrix of right eigenvectors of matrix A, L – matrix of left eigenvectors of matrix A. Matrixes L, R not depend from $\overrightarrow{v}$. Eigenvalues of matrix A can be written in the form: $a_{2,3,4} = u', a_1 = u' + a, a_5 = u' - a$. Formulae for fluxes calculation and eigenvalues of matrix A depend on the velocity vector of boundary of control volume. Explicit step operator of difference algorithm for approximation of system (1) is splitted on symmetric sequence of step operators in directions:

$$\overrightarrow{U}^{n+2} = L(2\Delta t)\overrightarrow{U}^{n};\; L(2\Delta t) = L_i(\Delta t)L_j(\Delta t)L_k(\Delta t)L_k(\Delta t)L_j(\Delta t)L_i(\Delta t) \tag{2}$$

S. Margenov, L.G. Vulkov, and J. Waśniewski (Eds.): NAA 2008, LNCS 5434, pp. 408–414, 2009.

Operator in coordinate directions i , for example, can be written in the form:

$$L_i(\Delta t) = I - \frac{\Delta t}{V_i}[\hat{F}_{i+1/2}\overrightarrow{S}_{i+1/2} - \hat{F}_{i-1/2}\overrightarrow{S}_{i-1/2}],$$

where $\hat{F}_{i+1/2}\overrightarrow{S}_{i+1/2}$ is vector of fluxes throw the boundary $i+1/2$ of grid cell in normal to boundary direction and area of this boundary, consequently . Vector of fluxes in normal to the boundary direction, calculated on the basis of scheme [1], can be written in the form:

$$F_{j-1/2} = 1/2(F_j + F_{j+1} + \textstyle\sum_{l=1}^{m} R^l_{j+1/2}[1/2\Psi(a^l_{j+1/2})(g^l_j + g^l_{j+1})$$
$$-Q(a^l_{j+1/2} + \gamma_{j+1/2})\alpha^l_{j+1/2}]), \quad g^l_j = (1+\omega^l\theta_j)limiter(\alpha^l_{j+1/2}, \alpha^l_{j-1/2}), \quad (3)$$
$$\gamma^l_{j+1/2} = \begin{cases} 1/2\Psi(a^l_{j+1/2})(g^l_{j+1} - g^l_j)/\alpha^l_{j+1/2}, \alpha^l_{j+1/2} \neq 0 \\ 0, \ \alpha^l_{j+1/2} = 0. \end{cases}$$
$$\Psi(z) = Q(z) - \lambda z^2, \alpha^l_{j+1/2} = \overrightarrow{L}^l_{j+1/2}(\overrightarrow{U}_{j+1} - \overrightarrow{U}_j),$$

The operator of artificial compression was defined by function:

$$\theta^l_j = |\alpha^l_{j+1/2} - \alpha^l_{j-1/2}|/(|\alpha^l_{j+1/2}| + |\alpha^l_{j-1/2}|),$$

where $\omega^k \in [0,2]$ – the artificial compression parameters, which assume different values for every characteristic field; $\overrightarrow{\alpha}_{j+1/2}$ -vector of characteristic unknowns in delta form; $a^l_{j+1/2}, \overrightarrow{R}^l_{j+1/2}, \overrightarrow{L}^l_{j+1/2}$ — eigenvalues and eigenvectors of matrix $A = \partial\overrightarrow{F}_j/\partial\overrightarrow{U}$, calculated for middle point of boundary $j+1/2$ of the cell. As the function $Q(z)$ was chosen $Q(z) = |z|$. Two types of limiter operators where used: operator *minmod* introduced by Harten and operator *Superbee*, introduced by Roe:

$$limiter(x,y) = minmod(x,y) = sgn(x)\ max[0,\ min(|x|, y\ sign(x)],$$

$$limiter(x,y) = minmod[minmod(2x,y), minmod(x,2y)].$$

For gas dynamics parameters calculation on the boundary $j+1/2$ Roe's interpolation procedure [2] was used

$$\widetilde{\rho} = \sqrt{\rho_j\rho_{j+1}}, K = \sqrt{\rho_j}/(\sqrt{\rho_j} + \sqrt{\rho_{j+1}}), \quad (4)$$

$$\widetilde{u}, \widetilde{v}, \widetilde{h} = Ku, v, h_j + (1-K)u, v, h_{j+1}, h = \gamma P/(\gamma-1)\rho + \overrightarrow{V}^2/2, \overrightarrow{V} = (u,v),$$

where $\widetilde{\rho}, \widetilde{u}, \widetilde{v}, \widetilde{h}$,consequently, values of density, velocity vector components and enthalpy on $j+1/2$ boundary of the cell.

2 Harten Scheme Modification in Its Present Realization

1. As was shown in [3] Roe's procedure define value of sound speed by formulae:

$$c^2_{j+1/2} = (\gamma-1)/2\alpha(1-\alpha)(\overrightarrow{V}_{j+1} - \overrightarrow{V}_j) + \alpha c^2_j + (1-\alpha)c^2_{j+1}, 0 < \alpha < 1$$

This value can be outside of interval $[c_j, c_{j+1}]$. In this case eigen vectors calculation on the basis of (4) is wrong and eigen values can have sign different from the right one. In present realization of Harten scheme this possibility is verified and in case, when $c^2_{j+1/2}$ is outside the interval $[c_j, c_{j+1}]$ Roe interpolation for values on boundary between cells is changed by simply half sum of gas dynamics parameters in this cells.

2. For calculation of characteristic unknowns on $j+1/2$ boundary by Harten scheme geometric and gas dynamics parameters from five cells $j, j-1, j-2, j+1, j+2$ are used. For reducing the influence of geometric characteristics of far away cells instead of characteristic unknowns $\alpha_{-1/2} = L_{-1/2}\Delta U_{-1/2}, \alpha_{1/2} = L_{1/2}\Delta U_{1/2}, \alpha_{3/2} = L_{3/2}\Delta U_{3/2}$. In present realization pseudo- characteristic unknowns are used: $\alpha_{-1/2} = L_{1/2}\Delta U_{-1/2}, \alpha_{1/2} = L_{1/2}\Delta U_{1/2}, \alpha_{3/2} = L_{1/2}\Delta U_{3/2}$. Detailed investigation of algorithm (3)-(5) and its testing on the 1-D flows where made in [4].

3 The Grids Construction Algorithm

Calculation grids were generated by Thompson - type algorithm, based on decision of system of three Poisson equations [5,6]:

$$g_{11}(\vec{r}_{\xi\xi} + P\vec{r}_{\xi}) + g_{22}(\vec{r}_{\eta\eta} + Q\vec{r}_{\eta}) + g_{33}(\vec{r}_{\zeta\zeta}) + R\vec{r}_{\zeta}) \tag{5}$$

$$-2g_{12}\vec{r}_{\xi\eta} - 2g_{23}\vec{r}_{\eta\zeta} - 2g_{13}\vec{r}_{\xi\zeta} = 0,$$

where $\vec{r}(\xi,\eta,\zeta)$ - radius - vector of grid point in Cartesian coordinates, ξ,η,ζ-curvilinear coordinates, correspond to coordinate lines of grid, g_{ij} - components of metric tensor. Geometric adaptation of grid to specific features of calculation area is made by introducing of control functions P, Q, R. Those functions serve for compression or rarefaction of coordinate surfaces to fixed manifolds of coordinate surfaces, lines and points. In particular function P governs the ξ -coordinate surfaces behavior and has the form:

$$\begin{gathered} P = -\sum_{i=1}^{N} a_i sign(\xi-\xi_i) e^{-c_i|\xi-\xi_i|} - \sum_{k=1}^{M1} b_k sign(\xi-\xi_k) \\ e^{-d_k((\xi-\xi_k)^2+(\eta-\eta_k)^2)^{1/2}} - \sum_{k=1}^{M2} e_k sign(\xi-\xi_k) e^{-f_k((\xi-\xi_k)^2+(\zeta-\zeta_k)^2)^{1/2}} \\ -\sum_{j=1}^{L} g_j sign(\xi-\xi_j) e^{-h_j((\xi-\xi_j)^2+(\eta-\eta_j)^2+(\zeta-\zeta_j)^2)^{1/2}} \end{gathered} \tag{6}$$

The control functions R, Q can be received from P by cyclic inversion of independent unknowns. The similar numerical algorithm on the basis of system of Poisson equations with Beltramy operators in right hand side was used for 2-D grid construction on the boundary surfaces of the calculation region:

$$g_{22}(\vec{r}_{\xi\xi} + P\vec{r}_{\xi}) + g_{11}(\vec{r}_{\eta\eta} + Q\vec{r}_{\eta}) - 2g_{12}\vec{r}_{\xi\eta} = \vec{n}R, \tag{7}$$

Control functions P, Q in (7) are similar on sense to control functions (6) for the spatial case. In case when the surface on which the grid is under construction, is defined in the form of analytical function $z = f(x, y)$ or $F(x, y, z) = 0$ the right hand side of (7) will be:

$$\vec{n}R = \vec{n}G[(1+q^2)r - 2pqs + (1+p^2)t]/(1+p^2+q^2)^{3/2}, G = g_{11}g_{22} - g_{12}^2 \quad (8)$$

$$p = f_x, q = f_y, r = f_{xx}, s = f_{xy}, t = f_{yy}.$$

Difference approximation of system (5) (6) was made by using central differences. The numerical decision of appearing system of linear algebraic equations was performed with using iterative method, namely over relaxation method. Initial approximation for decision was calculated by transfinite interpolation.

4 Calculations on Block Structural Grids

For calculation of shock wave diffraction on the sphere in shock tube it is convenient to divide calculation area for two part, namely: upstream to the sphere and downstream to the sphere. The calculations in this case are performed on the block structural grids, consist of two blocks. For simultaneous calculation of flows in both regions it is necessary to establish rules of connection for numerical decisions in regions. It can be done in different ways. One way is to establish boundary condition, connecting numerical solutions in both regions. For difference scheme [1] it will be sufficient to calculate flows on the boundary by the formulae similar to difference scheme's one (3), where gas dynamics values on boundary between regions where calculated with Roe's procedure (4) from gas dynamics values in right and left regions.

4.1 Formulation of Problem

The flow, appears during the diffraction of shock wave around the sphere in the shock tube with quadratic or circle section is calculated. The plane shock wave initially situated on some distance from sphere and propagates along the shock tube. In connection with existing physical experiments data we consider two formulation of problem: a) the sphere start to move during the shock wave propagation; b) sphere is fixed in shock tube. In first case during shock wave propagating the sphere start to move and calculation grid moves with it. The velocity of grid points on every time step can be calculated by the next formulae:

$$S_{n+1} = S_n + \Delta t W_n, W_{n+1} = W_n + \Delta t a, a = F_z/m_{sphere}, \quad (9)$$

$$F_z = -\oint_{S_{sphere}} \vec{n}(P - P_0)dS\vec{I_z}, \vec{I_z} = (0,0,1),$$

where for system (1) $\vec{v} = (0, 0, W_n)$. From the value of velocity we can calculate transition way of sphere and compare it with experimental data.

4.2 2-D Calculations of Flow Near Fixed Sphere

For case of fixed sphere calculations where made for shock tube with circle cross section using 2-D axysymmetrical gas dynamic model. Calculation results where compared with experimental data [7,8]. Structures of main discontinuities of diffraction process for consequent time moments are shown by pictures of level lines of density on Fig.1 for $M_{shock} = 2.2$ (supersonic flow after diffraction). Comparing of physical experiments results [7,8] and present calculation results was made for $M_{shock} = 1.22$. In [7] results of experiments are shown for pressure. On Fig. 2 pressure profiles from present calculations are compared with pressure values (more precisely non dimensional value $(P - P_0)/(P_1 - P_0)$ where P_1 - pressure in frontal stagnation point , P_0 - pressure before the shock wave) from experiment [7] for time moments $t = 140ms, 208ms, 296ms, 380ms$ ($t = 0$ - beginning of the diffraction). Disagreements with experimental results on profiles C,D connected with existence of the strut near sphere model rear stagnation point in experiments and shock wave diffraction on it. In [8] profile of drag force from experiments are shown for the shock wave intensity $M_{shock} = 1.22$ and different values of radius of the sphere model. For comparing with calculation results instead of drag force non dimensional coefficient $C_D = 2f/(\rho_2 U_2^2 \pi R^2)$ was used. On Fig.3 profile of computed C_D (compared with experimental data, signed by little circles) are shown.

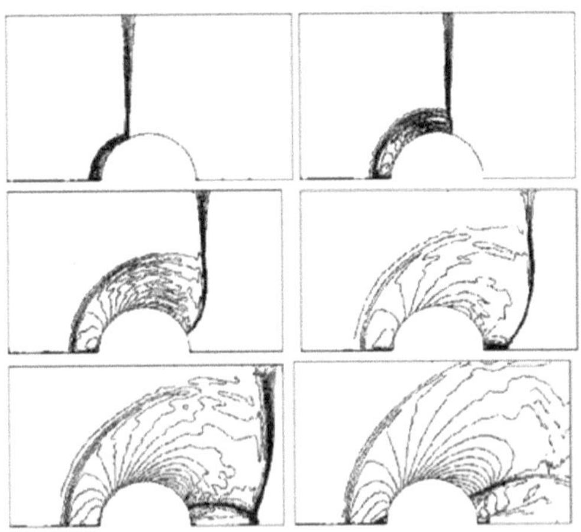

Fig. 1. Density level lines for supersonic flow after diffraction

Conclusion. Comparing experimental results [7,8] with numerical ones in 2-D formulation shows sufficient agreement and, so, reliability of algorithm using for 3-D formulation.

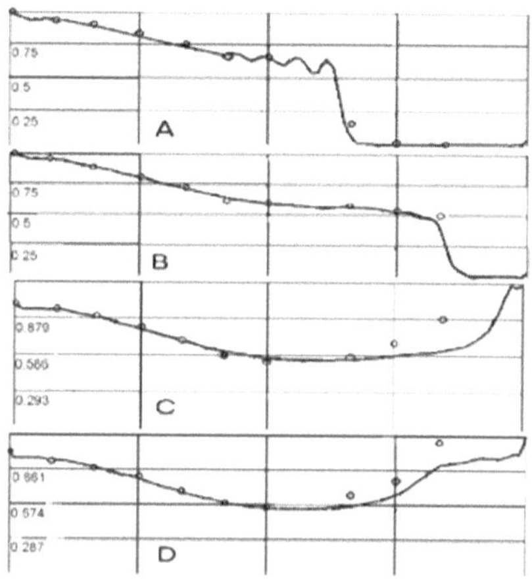

Fig. 2. Pressure contours for consequence time moments

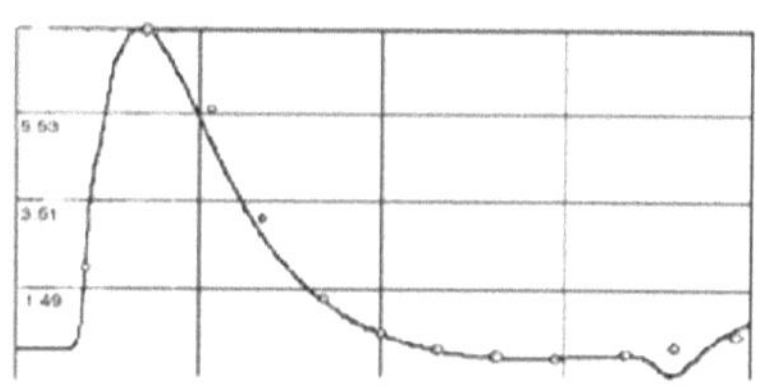

Fig. 3. Drag coefficient contour

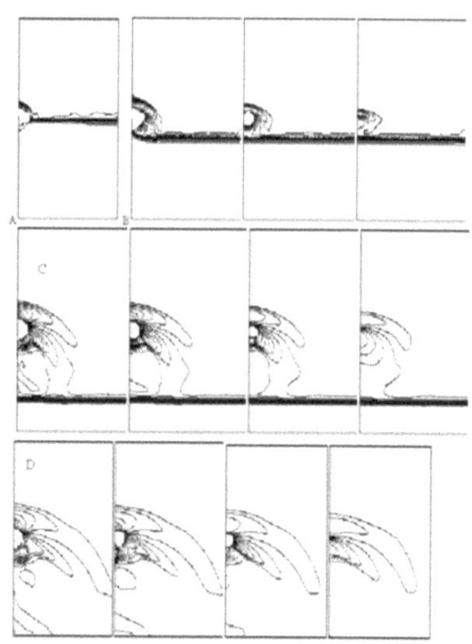

Fig. 4. Density level lines for supersonic 3-D flow after diffraction

4.3 Three-Dimensional Calculations

Calculation area consist of two part, namely: upstream to the sphere and downstream to the sphere. The size of calculation sub region along the tube was chosen equal to 6, radius of sphere is 0.6 size. For establishing the convergence of difference decision to exact one sequence of calculation grids was used: grid of size 61*61*50 points (with 10*10 points on the sphere), 65*65*50 points (with 20*20 points on the sphere) and grid of size 76*76*60 points (with 25*25 points on the sphere).

CALCULATION RESULTS. Visualization of calculation results was made by pictures of level lines of density. Those pictures are shown on Fig. 4 for consequent moments of time (denoted by letters A-D) and consequent coordinate surfaces ($I = 1, 10, 20, 30, 40$) from center of sphere to wall of shock tube (collected in one row of Fig. 4).

CONCLUSION. Calculations where made for shock wave intensity $M_{shock} = 1.8$. Number of time steps in calculations differs from 1000 (for the first grid) to 3000 (for the third grid). There exist convergence of calculation results for consequence of grids for value of way of sphere during diffraction.

References

1. Yee, H.C., Warming, R.E., Harten, A.: Implicit Total Variation Diminishing (TVD) Schemes for Steady-State Calculation. J. Comp. Phys. 57, 327–361 (1985)
2. Martyushov, S.N.: Calculation of two non stationary problems of diffraction by explicit algorithm of second order of accuracy. Comp. Technol., Novosibirsk. 1(4), 82–89 (1996)
3. Vinokur, M.: An Analysis of Finite-Difference and Finite-Volume Formulations of Conservation Lows. J. Comp. Phys. 81, 1–51 (1989)
4. Il'in, S.A., Timofeev, E.V.: Comparison of quasi monotony difference scheme. N 2 FTI Ioffe Institute publishing, St. Petersburg. 1550 (1991)
5. Thompson, J.F., Warsi, Z.U.A., Mastin, C.W.: Numerical Grid Generation, p. 306. North Holland, NY (1985)
6. Martyushov, S.N.: Numerical grid generation in computational field simulation. In: Proceedings of the 6-th International Conf. Greenwich Great Britain, p. 249 (1998)
7. Tanno, H., Itoh, K., Saito, T., Abe, A., Tokayama, K.: Interaction of a shock wave with a sphere suspended in vertical shock tube. Shock Waves 13, 249 (2003)
8. Sun, M., Saito, T., Tokayama, K., Tanno, H.: Unsteady drag on a sphere by shock wave loading. Shock Waves 14, 3 (2005)

Numerical Solution of a Class of Boundary Value Problems Arising in the Physics of Josephson Junctions

Hristo T. Melemov and Todor L. Boyadjiev

Plovdiv University (brunch Smolyan),
Sofia University & JINR, Dubna, Russia
{hristo_melemov,todorlb}@fmi.uni-sofia.bg

Abstract. In this paper we propose a method of numerical solution of non-linear boundary value problems for systems of ODE's given on the embedded intervals. The algorithm is based on the continuous analog of Newton method coupled with spline-collocation scheme of fourth order of accuracy. Demonstrative examples of similar problems take place in physics of stacked Josephson junctions with different layers lengths. As a concrete example we consider the problem for calculation the possible distributions of magnetic flux in a system of two magnetically coupled long Josephson junctions. The influence of length's ratio on the main physical properties of basic bound states is investigated numerically. The existence of bifurcations by change the lengths of the layers for some couples of solutions has been proved.

1 Problem Statement

Nonlinear systems of ordinary differential equations given on embedded intervals occur in many physical problems and especially in the theory of stacked Josephson junctions (JJ) [1]-[4].

We consider a model of two-layers JJ [5] with different layer lengths $2L$ and $2l$ correspondingly, where $L \geq l$. We suppose that the short layer is situated symmetrically to the long one. The coordinate origin is in the middle of the stack i.e. for long (first) layer we have $x \in [-L, L]$ and for short (second) one - $x \in [-l, l]$. We denote by $s \in (-1, 0]$ the coupling coefficient between the layers [1].

In order to obtain the model equations for stacked JJ we consider the full energy functional $F[\varphi]$ which can be represented as a sum

$$F = F_1[\varphi_1] + F_2[\varphi_2] + F_{12}[\varphi]. \tag{1}$$

Here $\varphi(x) = [\varphi_1(x), \varphi_2(x)]^T$ is the vector of magnetic fluxes in the layers (the superscript T means transposition), $F_i[\varphi_i]$, $(i = 1, 2)$, are the partial energies of the uncoupled layers $(s = 0)$. The functional $F_{12}[\varphi]$ represents the coupling

S. Margenov, L.G. Vulkov, and J. Waśniewski (Eds.): NAA 2008, LNCS 5434, pp. 415–422, 2009.

energy between the layers. In the symmetric overlap case [5] the corresponding expressions can be represented in the form

$$F_1[\varphi_1] = \int\limits_{-L}^{L} \left(\frac{1}{2}\,\varphi_{1,x}^2 + 1 - \cos\varphi_1 - \gamma\varphi_1\right) dx - h_e \Delta\varphi_1, \tag{2a}$$

$$F_2[\varphi_2] = \int\limits_{-l}^{l} \left[\frac{1}{2}\,\varphi_{2,x}^2 + \rho\,(1 - \cos\varphi_2) - \gamma\varphi_2\right] dx - h_e \Delta\varphi_2, \tag{2b}$$

$$F_{12}[\varphi_1, \varphi_2] = \frac{s}{1-s^2} \int\limits_{-l}^{l} \left[\frac{s}{2}\left(\varphi_{1,x}^2 + \varphi_{2,x}^2\right) - \varphi_{1,x}\varphi_{2,x}\right] dx + \frac{s}{1+s}\, h_e \left[\varphi_1(l) - \varphi_1(-l) + \Delta\varphi_2\right], \tag{2c}$$

where h_e — the external magnetic field, γ — the external current and vector $\Gamma = \gamma\,(1,1)^T$. The full magnetic fluxes across the layers are defined by

$$\Delta\varphi_1 = \varphi_1(L) - \varphi_1(-L), \quad \Delta\varphi_2 = \varphi_2(l) - \varphi_2(-l). \tag{3}$$

From the necessary extremum conditions [12] of the functional (1) we obtain the following non-linear boundary value problem (BVP)

$$\varphi_{1,x}(-L) = h_e\,, \tag{4a}$$
$$-\varphi_{1,xx} + \sin\varphi_1 - \gamma = 0, \quad x \in (-L, -l)\,, \tag{4b}$$
$$\varphi_{2,x}(-l+0) - s\varphi_{1,x}(-l+0) = (1-s)h_e\,, \tag{4c}$$
$$-A\varphi_{xx} + J_z(\varphi) + \Gamma = 0\,, \quad x \in (-l, l)\,, \tag{4d}$$
$$\varphi_{2,x}(l-0) - s\varphi_{1,x}(l-0) = (1-s)h_e\,, \tag{4e}$$
$$-\varphi_{1,xx} + \sin\varphi_1 - \gamma = 0\,, \quad x \in (l, L)\,, \tag{4f}$$
$$\varphi_{1,x}(L) = h_e\,. \tag{4g}$$

The square interaction matrix A depends only on coupling coefficient [3]

$$A(s) = \frac{1}{1-s^2}\begin{pmatrix} 1 & -s \\ -s & 1 \end{pmatrix}.$$

We denote by $J_z = (\sin\varphi_1, \rho\sin\varphi_2)^T$ the Josephson currents vector. Physically the parameter $\rho = L/l \geq 1$ represents the amplitude of Josephson current in the short layer. All the quantities are in dimensionless form (see for example, [6]).

The equations (4a) and (4g) are the corresponding boundary conditions for $\varphi_1(x)$ at the boundaries $x = \pm L$, and (4c) and (4e) — the boundary conditions for $\varphi_2(x)$ at $x = \pm l$. On the boundaries $x = \pm l$ the standard smoothing conditions for $\varphi_1(x)$ are fulfilled.

When $L = l$ ($\rho = 1$) the traditional BVP [7] for two-layers JJ follows from (4).

2 Solution Algorithm

The solution of non-linear BVP (4) is based on the Continuous analog of Newton method [8]. At each iteration we solve the following linear BVP

$$w_{1,x}(-L) = -\varphi_{1,x}(-L) + h_e\,, \tag{5a}$$

$$-w_{1,xx} + \cos\varphi_1 w_1 = \varphi_{1,xx} - \sin\varphi_1 + \gamma\,, \tag{5b}$$

$$w_{2,x}\,(-l+0) - s\,w_{1,x}\,(-l+0) = -\varphi_{2,x}\,(-l+0) + s\varphi_{1,x}\,(-l-0) + (1-s)\,h_e\,, \tag{5c}$$

$$-A(s)w_{xx} + Q(x)w = A(s)\varphi_{xx} - J_z(\varphi) - \Gamma\,, \tag{5d}$$

$$w_{2,x}(l-0) - s\,w_{1,x}(l-0) = -\varphi_{2,x}(l-0) + s\varphi_{1,x}(l-0) + (1-s)\,h_e\,, \tag{5e}$$

$$-w_{1,xx} + \cos\varphi_1 w_1 = \varphi_{1,xx} - \sin\varphi_1 + \gamma, \tag{5f}$$

$$w_{1,x}(L) = -\varphi_{1,x}(L) + h_e, \tag{5g}$$

where 2-matrix $Q(x)$ is defined by $Q(x) = \mathrm{diag}(\cos\varphi_1(x), \rho\cos\varphi_2(x))$.

For numerical solution of the problem (5) formally we can extend smoothly the "short" function $\varphi_2(x)$ on the interval $[-L, L]$ and then apply usual discretization technique to this "extended" BVP. But in this way there will be unduly null in the matrix of the algebraic linear system which increases its dimension.

In order to solve the linear problem (5) in this paper the spline-collocation scheme [9] is applied.

Let in interval $[-L, L]$ an irregular grid is given

$$\{x_i, i = 1, 2, \ldots, n, x_{i+1} = x_i + h_i, x_1 = -L, x_k = -l, x_r = l, x_n = L\}$$

with n nodes and steps h_i, $1 \le k \le r \le n$. Note, that boundary points $x = \pm l$ are included as nodes in the grid. The case $k = 1$ and $r = n$ corresponds to usual JJ with equal layers.

In every subinterval $[x_i, x_{i+1}]$, the solution is searched as a cube hermitian spline [10]

$$S_1(x) = \Phi(t)\,u_i + \Psi(t)\,h_i\,m_i + \bar{\Phi}(t)\,u_{i+1} + \bar{\Psi}(t)\,h_i\,m_{i+1}\,, \tag{6}$$

where $t = (x - x_i)/h_i$, $t \in [0, 1]$ is local coordinate, $\{u_i, m_i\}$ — value of the spline $S_1(x)$ and their derivative $m(x) \equiv S_{1,x}(x)$ in nodes $i = 1, 2, \ldots, k-1$ and $i = r+1, \ldots, n$ of the grid. Basic functions $\Phi(t) = (1-t)^2(1+2t)$ and $\Psi(t) = t\,(1-t)^2$ satisfying conditions $\Phi(0) = 1$ and $\dot{\Psi}(0) = 1$ (super-dote indicates differentiation with respect to local variable t). Remaining values of basic functions and their derivatives in nodes are equal to zero. About functions $\bar{\Phi}(t)$ and $\bar{\Psi}(t)$ we have $\bar{\Phi}(t) = \Phi(1-t)$, $\bar{\Psi}(t) = -\Psi(1-t)$.

Similarly, for subintervals in $[-l, l]$ an approximate solution is searched in the form

$$S(x) = \begin{pmatrix} S_1(x) \\ S_2(x) \end{pmatrix} = \Phi(t)\begin{pmatrix} u_{1,i} \\ u_{2,i} \end{pmatrix} + \Psi(t)h_i\begin{pmatrix} m_{1,i} \\ m_{2,i} \end{pmatrix} + \bar{\Phi}(t)\begin{pmatrix} u_{1,i+1} \\ u_{2,i+1} \end{pmatrix} + \bar{\Psi}(t)h_i\begin{pmatrix} m_{1,i+1} \\ m_{2,i+1} \end{pmatrix}.$$

We choose the collocation dots to be Gaussian nodes $t_j = (1 \pm \sqrt{3}/3)/2$, $j = 1, 2$ in $[0, 1]$. Simultaneously with smoothing conditions for the unknown functions and accounting all the boundary conditions, we obtain the following block-diagonal system of algebraic equations

$$WU = P,$$

Here U — vector nodes variables and matrix W have a structure

$$W = \begin{pmatrix}
e_0 & 0 & 0 & \cdot & \cdot & \cdot & \cdot & \cdot & \cdot & \cdot \\
A_1 & B_1 & 0 & \cdot & \cdot & \cdot & \cdot & \cdot & \cdot & \cdot \\
\cdot & \cdot & \cdot & \cdot & \cdot & \cdot & \cdot & \cdot & \cdot & \cdot \\
\cdot & \cdot & A_{k-1} & B_{k-1} & 0 & 0 & \cdot & \cdot & \cdot & \cdot \\
\cdot & \cdot & 0 & e_s & e_0 & 0 & \cdot & \cdot & \cdot & \cdot \\
\cdot & \cdot & \cdot & \cdot & \cdot & \cdot & \cdot & \cdot & \cdot & \cdot \\
\cdot & \cdot & \cdot & \cdot & A_s & B_s & \cdot & \cdot & \cdot & \cdot \\
\cdot & \cdot & \cdot & \cdot & \cdot & \cdot & \cdot & \cdot & \cdot & \cdot \\
\cdot & \cdot & \cdot & \cdot & \cdot & e_s & e_0 & 0 & 0 & \cdot \\
\cdot & \cdot & \cdot & \cdot & \cdot & A_r & 0 & B_r & 0 & \cdot \\
\cdot & \cdot & \cdot & \cdot & \cdot & \cdot & \cdot & \cdot & \cdot & \cdot \\
\cdot & \cdot & \cdot & \cdot & \cdot & \cdot & 0 & 0 & A_{n-1} & B_{n-1} \\
\cdot & \cdot & \cdot & \cdot & \cdot & \cdot & 0 & 0 & 0 & e_0
\end{pmatrix}$$

where 2-vectors $e_0 = (0, 1)$, $e_s = (0, -s) = -se_0$, and remaining elements of W are 2-matrices with elements

$$A_{i,j1} = -\frac{1}{h_i^2}\ddot{\Phi}_j + \Phi_j c_{ij}, \quad A_{i,j2} = -\frac{1}{h_i}\ddot{\Psi}_j + \Psi_j h_i c_{ij},$$
$$B_{i,j1} = -\frac{1}{h_i^2}\ddot{\bar{\Phi}}_j + \bar{\Phi}_j c_{ij}, \quad B_{i,j2} = -\frac{1}{h_i}\ddot{\bar{\Psi}}_j a_{ij} + \bar{\Psi}_j h_i c_{ij},$$

for $i = 1, \ldots, k-1$ and $i = r, \ldots, n-1$, and $c_{ij} = \cos\varphi_1(x_{ij})$. Elements W in $i = k, \ldots, r-1$, $j = 1, 2$ are 4-matrixes with elements

$$[A_{i,j1}]_{mn} = -\frac{1}{h_i^2}\ddot{\Phi}_j a_{mn} + \Phi_j q_{mn,ij}, \quad [A_{i,j2}]_{mn} = -\frac{1}{h_i}\ddot{\Psi}_j a_{mn} + \Psi_j h_i q_{mn,ij}$$
$$[B_{i,j1}]_{mn} = -\frac{1}{h_i^2}\ddot{\bar{\Phi}}_j a_{mn} + \bar{\Phi}_j q_{mn,ij}, \quad [B_{i,j2}]_{mn} = -\frac{1}{h_i}\ddot{\bar{\Psi}}_j a_{mn} + \bar{\Psi}_j h_i q_{mn,ij}$$

Values $\{a_{mn}\}$, $m, n = 1, 2$ — elements matrix A and $\{q_{mn,ij}\}$ elements $Q(x)$ in the relating Gauss nodes.

The number of the blocks of matrix W is equal to the number $n-1$ of grid's subintervals. The number of the columns in every block is fixed (8 in the case under consideration), the number of rows in blocks depends on the number of the blocks. In the external intervals $x \in [-L, -l]$ and $x \in [l, L]$ every block has two rows except the first and last blocks. This two blocks contain additional rows, which take into account the boundary conditions (4a) and (4g) in long subjunction. The internal blocks in the intervals $x \in [-l, l]$ have four rows. The blocks corresponding to boundaries $x = \pm l$ have three rows: two rows from discretization (4b) and (4f), and one row from boundary conditions (4c) and (4e) for function $\varphi_2(x)$. The number of the node's variables is $2(k-1)+4(r-k+1)+2(n-r)$.

In order to solve this system of algebraic equations we use specialized subprogram CWIDTH, which is described in detail in [10]. This program realizes Gauss method modified for block-diagonal systems of algebraic equations.

Based on the represented algorithm a program for investigation of static distributions of magnetic flux in two-layers JJs with different layer lengths is made.

3 Numerical Results

Further we will discuss some numerical results obtained by means of discussed above algorithm.

The solutions of the boundary problem (4) depend on the coordinate x, as well as on the set of parameters $p \equiv (L, l, s, h_e, \gamma)$, i.e. $\varphi_i = \varphi_i(x, p)$, $i = 1, 2$. Further the dependence on p is denoted only if it is necessary.

The basic numerical characteristics of every solution of non-linear BVP (4) are full (1), partial (2a), (2b) and coupling energies (2c), the full magnetic fluxes through the layers (3), as well as the average magnetic fluxes [8]:

$$N_1(p) = N[\varphi_1] = \frac{1}{2L\pi}\int_{-L}^{L} \varphi_1(x,p)dx, \; N_2(p) = N[\varphi_2] = \frac{1}{2l\pi}\int_{-l}^{l} \varphi_2(x,p)dx. \quad (7)$$

The calculation of the solutions of BVP (4) and possible their bifurcations at change the parameters is an important but difficult problem. In this paper we investigate the influence of the length of short contact $2l$ on some typical bound states in the stack. All numerical results are received for "long enough" contact ($2L = 10$) and fixed coupling coefficient $s = -0.3$.

The fluxon (vortex) distributions of magnetic flux play an important role in the theory and application of JJs. It is well known that in the "infinity" JJ, which is described by unperturbed sine-Gordon equation, there exists countable set of solutions [11].

For physical reasons it is convenient to discriminate unipolar fluxon solutions, composed from equally oriented vortices of magnetic field and heteropolar solutions, whose internal magnetic field is result of nonlinear interaction between heteropolar vortices. Further we shall consider only simple unipolar solutions of kind $\Phi^{\pm n}$, where $n = 1, 2, \ldots$

In case of finite length JJs the possible solutions become deformed as a result of interactions with the boundaries as well as with the applied external magnetic field h_e and external current γ [8]. But thus the values of the average magnetic fields remain constants

$$N\left[\Phi^{\pm n}\right] = N\left[\Phi_\infty^{\pm n}\right] = \pm n.$$

Here $\Phi_\infty^{\pm n}$ refers to corresponding solutions in the"infinity" contact.

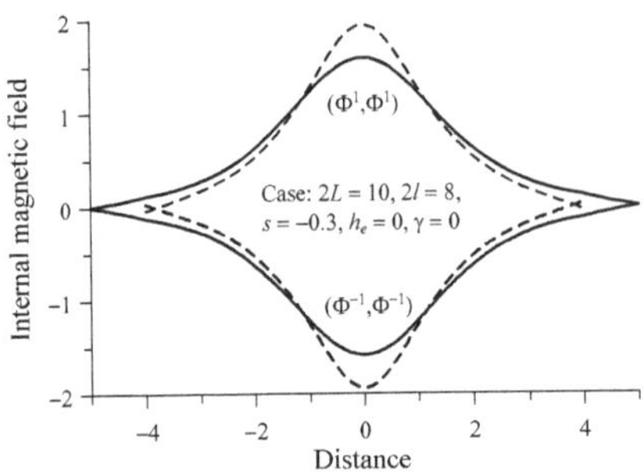

Fig. 1. $(\Phi^{\pm 1}, \Phi^{\pm 1})$-distributions of the internal magnetic field

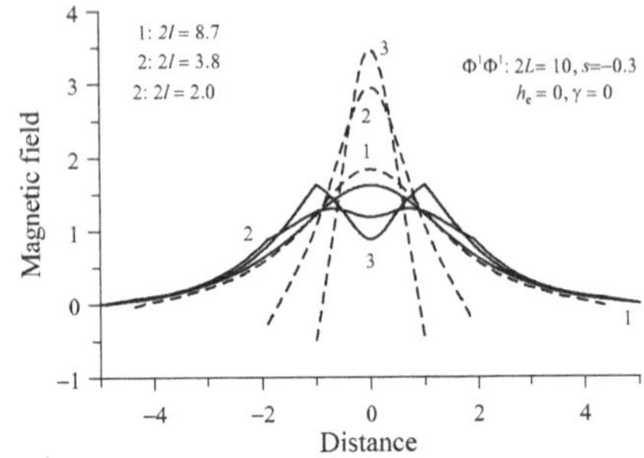

Fig. 2. (Φ^1, Φ^1)-states for $h_e = 0$, $\gamma = 0$, and different l

A special case in two-layer JJs are vortex solutions of type $(\Phi^{\pm 1}, \Phi^{\pm 1})$, composed of two unipolar fluxons, on one in each layer. Concrete examples are presented in Fig. 1 where $2L = 10$, $2l = 8$, $h_e = 0$ and $\gamma = 0$ (solid and dashed lines correspond to the states in long and short layers). It is visible, that the length decreasing is compensated by rise of the amplitude of internal magnetic field $\varphi_{2,x}(x)$ in short layer.

The influence of half-length l of the short layer on internal magnetic field for solutions of type (Φ^1, Φ^1) is presented in Fig. 1 and Fig. 2. The decrease of l increases the amplitude of the internal magnetic field $\varphi_{2,x}(x)$ in the short layer. In turn the interaction between the layers leads to $\varphi_{2,x}(\pm l) \neq 0$, so the smoothness of the field $\varphi_{1,x}(x)$ in points $\pm l$ worsens and the inductive current $\varphi_{1,xx}(\pm l)$ become broken. According to (4c) and (4e) it follows that the corresponding jump depends mainly on coupling coefficient s. This effect increases when the parameter l decrease (see Fig. 2). For large enough values of l the graphic of $\varphi_{1,x}(x)$ has an extremum in the middle of contact (curve 1). The amplitude of this extremum decreases when half-length l decreases as well. For $2l \sim 5 \div 6$ the graphic of the magnetic field $\varphi_{1,x}(x)$ has a plateau (curve 2). At further decrease of l graphics $\varphi_{1,x}(x)$ gets a minimum in the middle of JJ (curve 3).

Specified above feature well explains the behaviour of curves $F_{12}(l)$, which are shown on Fig. 3 and Fig. 4. Really, for (Φ^1, Φ^1)-state and in general, for every unipolar couple of vortices, the integrand in (2c) is always negative. But because of a negative factor $s/(1-s^2)$, the integral remains always positive (see the dashed line in Fig. 3). Note that when $2l < 1.53$ for $h_e = 0$ and $\gamma = 0$ the couple (Φ^1, Φ^1) does not exist. Hence the value $2l_B \approx 1.53$ is a bifurcation point for this solution at change of l. Physically this bifurcation means that l_B is the minimal length providing existence of the solution (Φ^1, Φ^1). In case of single JJ this fact is noticed in [13],[14].

For heteropolar states the integrand in (2c) can vary — the first term is negative, but the second one is positive. Especially for (Φ^1, Φ^{-1}) the integral remains always negative as at decrease of l(dashed line in Fig. 4).

A comparison of the dependences of partial energies $F_i(l)$, $i = 1, 2$, for solutions (Φ^1, Φ^1) and (Φ^1, Φ^{-1}) is made on Fig. 5. One can see that the change of partial energies $F_1(l)$ of long layers in all range of l does not exceed several

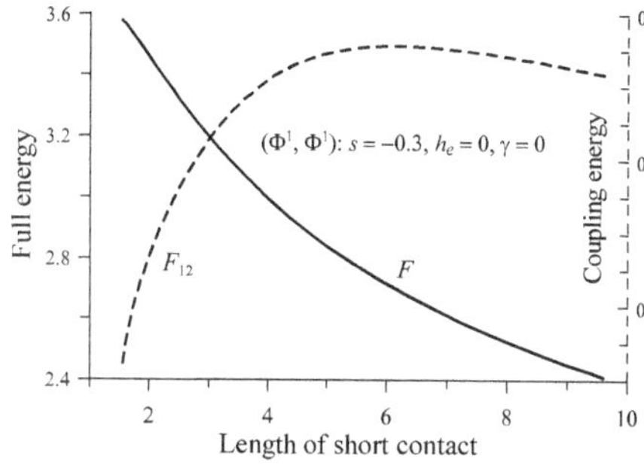

Fig. 3. Full energy of (Φ^1, Φ^1)-state

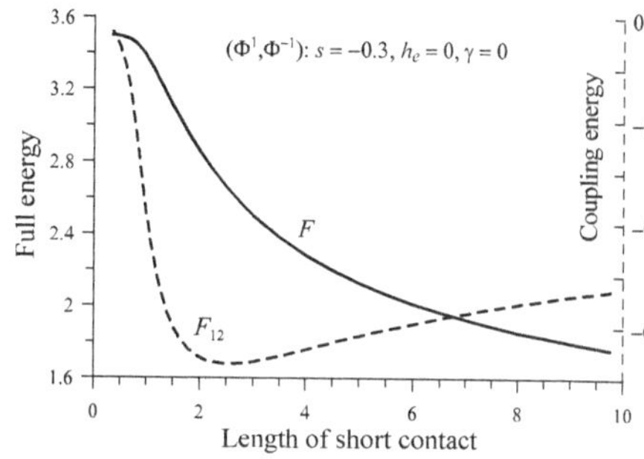

Fig. 4. Full energy of (Φ^1, Φ^{-1})-state

percents. On the other hand, the graphs of partial energies of short layers $F_2(l)$ practically coincide till l_B. Hence, the full energy of (Φ^1, Φ^{-1}) bound state is less then the full energy of (Φ^1, Φ^1) — the couple (Φ^1, Φ^{-1}) is more stable than (Φ^1, Φ^1). This means that in the experiment the probability of detection of (Φ^1, Φ^1)-state is less than the probability of detection of (Φ^1, Φ^{-1})-state.

In addition we shall note that the values of average magnetic fluxes (7) are $N\left[\Phi^{\pm 1}\right] = \pm 1$ in all admissible range of l, so the derivatives $\partial N_i(l)/\partial l = 0$, $i = 1, 2$.

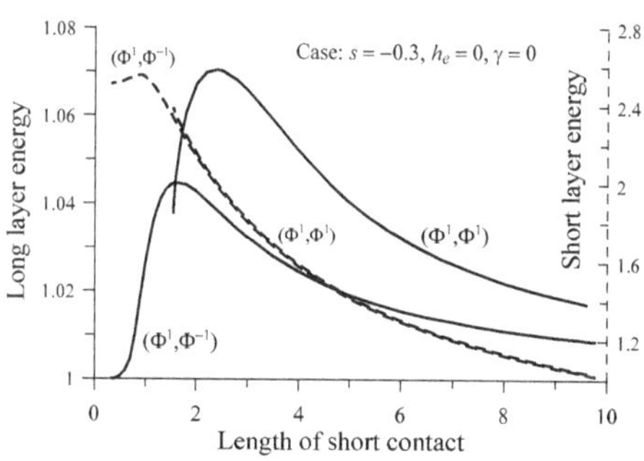

Fig. 5. Comparison the partial energies for (Φ^1, Φ^1) and (Φ^1, Φ^{-1}) states

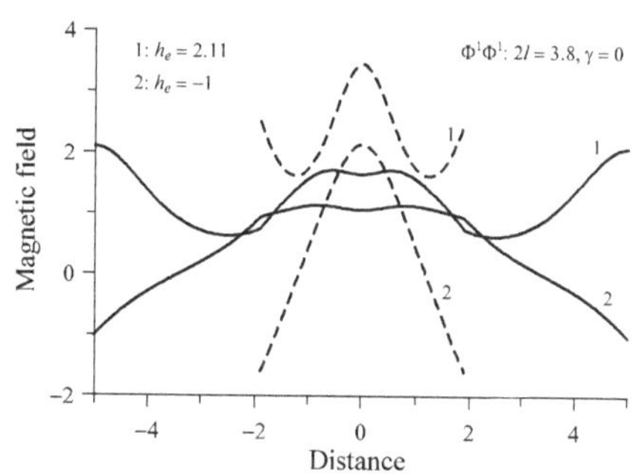

Fig. 6. Graphics of (Φ^1, Φ^1) in critical values of h_e

Every magnetic flux distribution $\varphi(x)$ in JJ has a region of existence by change of h_e for fixed values of other parameters. This region is limited by lower h_{min} and upper h_{max} values of the external magnetic field. If $\gamma = 0$ the values h_{min} and h_{max} are called critical fields for solution under consideration [6]. For unipolar solutions the points $(h_{min}, 0)$ and $(h_{max}, 0)$ on the plane (h_e, γ) are bifurcation points at change of h_e, where the transitions from Josephson to resistive regimes become.

In Fig. 6 the distributions of magnetic fields for (Φ^1, Φ^1)-type solution of (4) for $2l = 3.8$, $h_{max} \approx 2.11$ (curves 1) and $h_{min} \approx -1$ (curves 2) are demonstrated. It can be seen, that the external magnetic field h_e changes the long layer distribution $\varphi_{1,x}(x)$ mainly in the neighborhood of the boundaries $\pm L$. At the same time the deformation of the internal magnetic field $\varphi_{2,x}(x)$ in short layer is considerable along the whole length.

Conclusions. A spline-collocation scheme for numerical solution of non-linear BVP for systems of ODE's given on the embedded intervals is worked out. The scheme realization leads to a system of the block-diagonal system of algebraic equations. Such scheme can be easily extended to problems with discontinuous coefficients without violation of structure of algebraic system.

The developed technique gives a possibilities for detailed investigation of many multiparametric physical problems. Especially we analyze the existence and stability of some types magnetic flux distributions in magnetically coupled 2-layer JJ's.

Acknowledgments. Authors thank Prof. I.V. Puzynin and Prof. Yu.M. Shukrinov (JINR, Dubna) for useful remarks.

This work is supported by Sofia University Scientific foundation under Grant No 135/2008 and Plovdiv University Scientific foundation under Grant No IC-M-4/2008.

References

1. Volkov, A.F.: Solitons in Josephson superlaticess. JETP Lett. 45(6), 299–301 (1987)
2. Boyadjiev, T.L., Pavlov, D.V., Puzynin, I.V.: Computation of Bifurcations of Stable States in Two-Layer Inhomogeneous Josephson Junctions. Comm. JINR. Dubna. P5-89-173 (1989)
3. Sakai, S., Bodin, P., Pedersen, N.F.: Fluxons in thin-film superconductor-insulator superlattices. J. Appl. Phys. 73(5), 2411–2418 (1993)
4. Machida, M., Sakai, S.: Unifed theory for magnetic and electric field coupling in multistacked Josephson junctions. PRB. 70, 144520 (2004)
5. Melemov, H.T., Boyadjiev, T.L.: Numerical solution of system of ODE's on embedded intervals. Comm. JINR. Dubna. P11-2008-31 (2008)
6. Licharev, K.K.: Dynamics of Josephson Junctions and Circuits, vol. 634. Gordon and Breach, New York (1986)
7. Goldobin, E., Ustinov, A.V.: Current locking in magnetically coupled long Josephson junctions. Phys. Rev. B. 59(17), 11532–11538 (1999)
8. Puzynin, I.V., et al.: Methods of computational physics for investigation of models of complex physical systems. Physics of Particles and Nuclei. 38(1), 70–116 (2007)
9. Boyadjiev, T.L.: Spline-collocation scheme of higer order of accuracy. Comm. JINR. Dubna. P2-2002-101 (2002)
10. de Boor, C.: A Practical Guide to Splines. Springer, Heidelberg (1978)
11. Iliev, I.D., Khristov, E.K., Kirchev, K.P.: Spectral methods in soliton equations. Longman Sci. & Techn., Wiley (1994)
12. Gelfand, I.M., Fomin, S.V.: Calculus of Variations. Prentice-Hall, Englewood Cliffs (1963)
13. Boyadjiev, T., Todorov, M.: Numerical Investigation of a Bifurcation Problem with free Boundaries Arising from the Physics of Josephson Junctions. Mathematical modeling 12(4), 61–72 (2000)
14. Boyadjiev, T., Todorov, M.: Minimal Length of Josephson Junctions with Stable Fluxon Bound States. Superconducting Science and Technology 14, 1–7 (2002)

How to Choose Basis Functions in Meshless Methods?

Vratislava Mošová

Institute of Exact Science, Moravian University
Jeremenkova 42, 772 00 Olomouc, Czech Republic
vratislava.mosova@mvso.cz

Abstract. Technical progress in construction of computers, their higher speed, and larger memory gave possibility to develop new numerical methods for solution of boundary value problems. A lot of meshless methods have been developed in last years. Some of them are in a way identical with the Galerkin method, where the trial space is formed by especially constructed functions. The choice of the proper trial space is important, because the error in the Galerkin method is determined by the fact how well the exact solution can be approximated by the elements from this finite dimensional space.

In this contribution different trial spaces are considered. We will construct spaces generated by means of B-splines and shape functions received by means of the RKP technique. Examples of using trial spaces mentioned for solution of some boundary value problems will be given. We focus our attention on error estimates in these cases too.

Keywords: Meshless methods, Galerkin method, B-spline basis, RKP shape function basis.

1 Introduction

The meshless methods began to be used for solving boundary value problems from the seventies of the 20th century. The Smooth particle hydrodynamic method (SPHM, see [7]) was the first of these methods that was used to solve problems from astrophysics. Soon a lot of new different meshless methods were developed. We can name for example Element free Galerkin method (EFGM, see [2]), Reproducing kernel particle method (RKPM see [3]), FEM with B-splines (see [5]) or Generalized finite element method (GFEM, see [1]). Name "meshless" was chosen for these methods because all of them have one important property - they need no explicitly given mesh. It is the reason why the meshless methods are used when it is necessary to change the mesh in the course of computation. It is for example in case of large deformation, crack propagation or moving boundary. The second feature of the meshless methods is that they construct new shape functions that do not have to be only of polynomial type.

In this contribution we focus on two of the meshless methods - Galerkin method where the space of basis functions is generated by means of weighted B-splines and then Galerkin method with basis space generated by means of RKP

S. Margenov, L.G. Vulkov, and J. Waśniewski (Eds.): NAA 2008, LNCS 5434, pp. 423–430, 2009.

shape functions. The paper consists of 4 sections. We summarize the construction and some properties of the B-splines in the 2nd section. We deal with the RKP shape functions in the 3rd section. Also the numerical solutions received by means of these methods are studied in these sections. For the sake of simplicity we work with 1D case in this paper, however the techniques presented have a direct 2D or 3D counterpart.

2 B-Splines

It is known that success of the Galerkin method depends on the way how the basis functions are chosen. A proper candidate on such basis can be cardinal B-splines we will describe now.

Definition 1. *Let $b^0(x)$ be the characteristic function of the interval* $[0,1]$ *and*

$$b^n(x) = \int_{x-1}^{x} b^{n-1}(\xi)\mathrm{d}\xi, \; n = 1, 2, \ldots. \tag{1}$$

For $k \in \mathbb{Z}$, $h > 0$ denote

$$b^n_{k,h}(x) = b^n(x/h - k). \tag{2}$$

The linear combination

$$\sum_{k\in\mathbb{Z}} c_k b^n_{k,h}(x) \tag{3}$$

is called the cardinal B-spline of degree n.

Note 1. The B-spline $b^n_{k,h}$ is a polynomial of order n. It is nonzero only in the interval $(kh, (k+n+1)h)$.

We use the cardinal B-splines to solve boundary value problems now.

Example 1. Consider the 1D Neumann boundary value problem

$$u''(x) + 16^2 u(x) = x, \; x \in (0,1), \tag{4}$$

$$u'(0) = u'(1) = 0. \tag{5}$$

Find an approximation of the weak solution using B-splines defined above.

Solution 1. We find $u \in W^{1,2}(0,1)$ such that

$$-\int_0^1 u'v' \,\mathrm{d}x + 16^2 \int_0^1 uv \,\mathrm{d}x = \int_0^1 xv \,\mathrm{d}x, \; \forall v \in W^{1,2}(0,1). \tag{6}$$

We approximate the unknown function $u(x)$ by $\tilde{u}(x) = \sum_{k=1}^{N} b^3_{k-2,h}(x)u_k$ over uniformly distributed nodes $x_1, x_2, \ldots, x_N \in [0,1]$ that form a grid with width $h = \frac{1}{N-1}$. This approximation in conjunction with the Galerkin method provides a mesh-free computational formulation of the BVP. Results for different number of nodes are given in Figure 1 and Table 1. The full line represents the exact solution and the dashed one is the approximate solution.

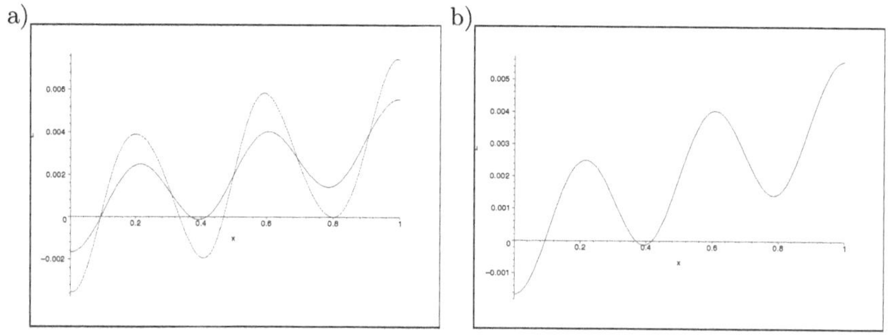

Fig. 1. The exact solution and its approximation for a) $N = 8$ and b) $N = 11$

Table 1. Dependence of the error on the number of nodes

N	8	11	14
$\max\lvert u - \tilde{u}\rvert$	1.8×10^{-3}	8×10^{-5}	14×10^{-6}

In case of Dirichlet boundary value problem with homogeneous boundary conditions, the linear combination of functions $b^n_{k,h}$ must vanish at boundary points. But generally the linear combination $\sum_{k\in\mathbb{Z}} c_k b^n_{k,h}(x) = 0$ only if all $c_k = 0$. Another way to achieve the zero value at boundary points can be multiplying B-splines by a proper weight function (i.e., a nonnegative continuous function that vanishes at boundary points of the considered interval J.) For example, the function

$$w(x) = c\,\mathrm{dist}\,(x, \partial J),\;\; x \in J, \tag{7}$$

represents the standard weight function.

Example 2. Solve the following 1D Dirichlet boundary value problem

$$u''(x) + 16^2 u(x) = x,\;\; x \in (0,1), \tag{8}$$

$$u(0) = u(1) = 0, \tag{9}$$

using the weighted B-splines.

Solution 2. We find $u \in W_0^{1,2}(0,1)$ such that

$$-\int_0^1 u'v'\,\mathrm{d}x + 16^2 \int_0^1 uv\,\mathrm{d}x = \int_0^1 xv\,\mathrm{d}x,\;\; \forall v \in W_0^{1,2}(0,1). \tag{10}$$

We suppose that nodes $x_1, x_2, \ldots, x_N \in [0,1]$ are uniformly distributed. We find the Galerkin approximation in the form $\tilde{u}(x) = \sum_{k=1}^N w(x) b^3_{k-2,h}(x) u_k$, where the weight function $w(x) = \min\{1, \frac{x}{s}, \frac{1-x}{s}\}$, $s = 0.2$. Note that the choice of weight function (its form and the parameter s that define the width of the strip inside [0,1] where $w(x) \neq 1$) affects the quality of approximate solution. The results for $N = 11$ are given in Figure 2. The error estimate for different N is in Table 2.

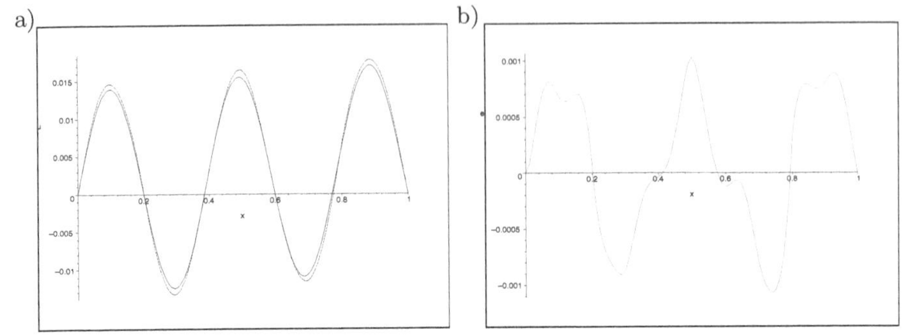

Fig. 2. a) The approximate and the exact solutions. b) The error of the approximation.

Table 2. Dependence the error on the number N of nodes

N	8	11	14
$\max\|u - \tilde{u}\|$	1.5×10^{-2}	10^{-3}	4×10^{-4}

In the 1D case it is possible to prove the following error estimate that is the analogy of the Jackson inequality given in [5].

Theorem 1. *Let w be a standard weight function and $\tilde{u}(x) = \sum_{k=1}^{N} w(x) b^n_{k-\frac{n+1}{2},h}(x) u_k$ be an approximation of u on a bounded interval J. Then*

$$\|u - \tilde{u}\|_l \leq c(J, w, n) h^{k-l} \|u\|_k, \; l < k \leq n+1 \tag{11}$$

for all $u \in W_0^{k,2}(J)$.

3 RKP Shape Functions

The RKP shape functions (reproducing kernel particle shape functions) were introduced (see [3]) in the context of approximate solution of partial differential equations in the 1990's. The main idea of construction is to develop the shape functions that are k consistent, i.e., the shape functions approximate a polynomial of degree $\leq k$ exactly.

In this paper, we describe the construction of RKP shape functions for uniformly distributed particles $x_1, \ldots, x_N \in [a, b]$, where $x_{I+1} = x_I + h$, $h = \frac{b-a}{N-1}$. Shape functions will be developed from the polynomial basis of order m, $m \in \mathbb{N}$, that has the form $p(x) = (1, x, \ldots, x^m)$, and from a window function[1] v of one variable.

The technique of construction of the RKP shape functions is given in the following definition.

[1] The window function has to be a nonnegative smooth function with compact support. Usually cubic splines, cone functions or exponential functions are used.

Definition 2. *Let a polynomial basis p of degree m, particles $x_1, \ldots, x_N \in [a,b]$, a window function v and a number $\rho > 0$ be given. We define the RKP shape function of order α ($\alpha = 0, 1, \ldots, m$) associated with the particle x_I by the formula*

$$\Phi_I^\alpha(x) = \alpha!\, p\left(\frac{x_I - x}{\rho}\right) b^\alpha(x) \frac{1}{\rho} v\left(\frac{x_I - x}{\rho}\right). \tag{12}$$

Here the $(m+1)$-dimensional vector $b^\alpha(x)$ is a solution of the system

$$M(x) b^\alpha(x) = p^\alpha(0), \tag{13}$$

$$M(x) = (m_{ij}(x))_{i,j=1}^{m+1}, \quad m_{ij}(x) = \sum_{I=1}^{N} \left(\frac{x_I - x}{\rho}\right)^{i+j-2} \frac{1}{\rho} v\left(\frac{x_I - x}{\rho}\right) h,$$

$$p^\alpha(0) = (\underbrace{0, \ldots, 0, 1}_{\alpha}, 0, \ldots, 0)^T.$$

Note 2. The smoothness of the shape function Φ_I^α is the same as the smoothness of the window function v. The size of support of the shape function Φ_I^α depends on the size of support of the window function v and on the parameter ρ.

Example 3. Solve the 1D Neumann boundary value problem (4), (5) using the RKP shape functions.

Solution 3. We take for the construction of shape functions the uniformly distributed nodes $x_1, x_2, \ldots, x_N \in [0,1]$, the polynomial basis $p(x) = (1, x)^T$, the window function $v(x) = (1 - x^2)^2$ for $|x| \leq 1$, $v(x) = 0$ for $|x| > 1$ and the parameter $\rho = 0.3$. We assume the Galerkin approximation in the form $\overline{u}(x) = \sum_{I=1}^{N} \Phi_I^0(x) u_I^0 h$ first[2]. The approximation $\overline{u}$ and the exact solution u for $N = 11$ and $N = 14$ is drawn in Figure 3. The errors for different N are in Table 3.

a)

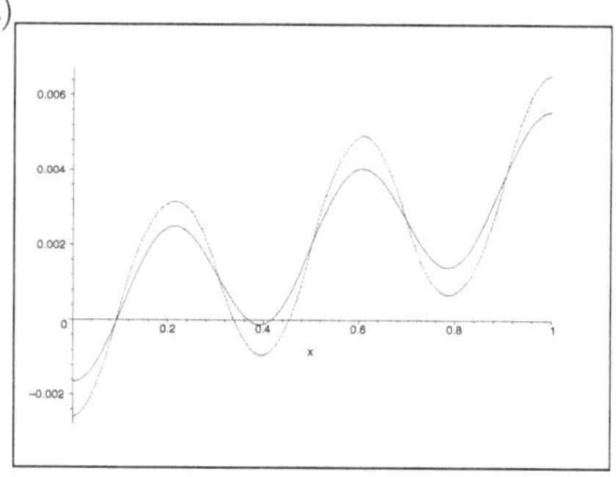

b)

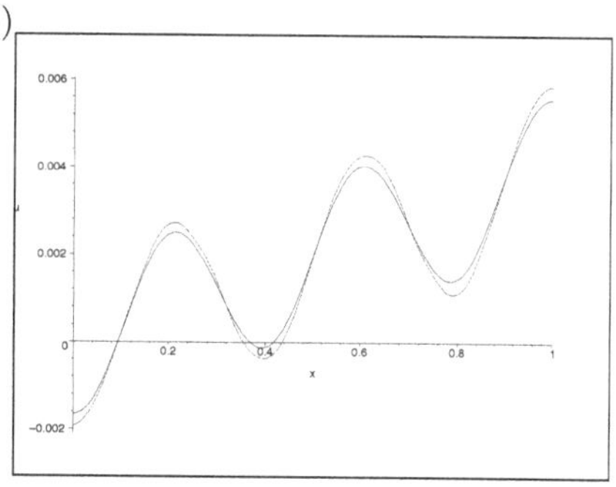

Fig. 3. RKPM - the approximation and the exact solution for a) $N = 11$ and b) $N = 14$

Table 3. Dependence of the error on the number of nodes

N	8	11	14
$\max\|u - \tilde{u}\|$	2.4×10^{-3}	9.2×10^{-4}	3×10^{-4}

To accelerate the convergence process we can apply the RKP shape functions of the higher order. It is necessary to be careful. While the RKP shape functions of order 0 are linearly independent, some functions from the set of the higher order functions can be redundant (see [6]). We will assume[3] that $\overline{u}(x) = \sum_{I=1}^{N} \Phi_I^0(x) u_I^0 h + \sum_{I=2}^{N-1} \Phi_I^1(x) u_I^1 h$. We can see the results for $N = 11$ in Figure 4. The error for different N is in Table 4.

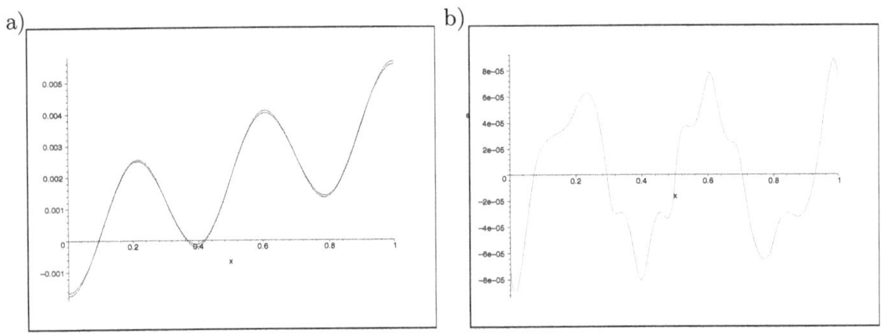

Fig. 4. a) The approximation and the exact solution. b) The course of the error $u - \overline{u}$.

Table 4. Dependence of the error on the number of nodes

N	8	11	14
$\max\|u - \overline{u}\|$	2.3×10^{-3}	9.6×10^{-5}	5.6×10^{-5}

Similar as in Section 2, the RKP shape functions do not satisfy the Dirichlet boundary conditions. There are many ways how to fix it. One possibility is to multiply the shape functions by such a weight function, which has zero values on boundary.

Example 4. Solve the 1D Dirichlet boundary value problem (8), (9).

Solution 4. We consider $\rho = 0.3$, the polynomial basis $p(x) = (1, x)^T$, the window function $v(x) = (1 - x^2)^2$ for $|x| \leq 1$, $v(x) = 0$ for $|x| > 1$, and the weight function $w = x(1 - x)$. We find the solution in the form $\overline{u}(x) = \sum_{I=1}^{N} w(x) \Phi_I^0(x) u_I^0 h + \sum_{I=2}^{N-1} w(x) \Phi_I^1(x) u_I^1 h$. The results for $N = 11$ are in Figure 5. The error for different number of particles is given in Table 5.

---

[2] In this case $\alpha = 0$ and the vector b^0 is the solution of the system $M(x) b^0(x) = (1, 0)^T$.

[3] In this case $\alpha = 0, 1$ and the vectors b^0, b^1 are the solution of the system $M(x) b^0(x) = (1, 0)^T$, $M(x) b^1(x) = (0, 1)^T$.

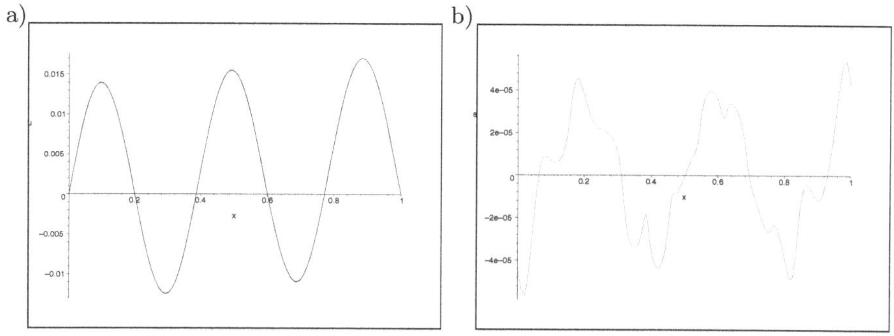

Fig. 5. a) The approximate and the exact solutions. b) The course of the error $u - \overline{u}$.

Table 5. Dependence of the error on the number of nodes

N	6	8	11	14
$\max\lvert u - \overline{u}\rvert$	4×10^{-3}	6.8×10^{-5}	6.4×10^{-5}	6×10^{-5}

In the 1D case we can prove the following error estimate (compare with Theorem 3.2 in [6])

Theorem 2. *Let J be a bounded interval, $v \in W^{m+1,2}(J) \cap C_0^m(J)$, $m \geq 1$ and $\overline{u}(x) = \sum_{\alpha=0}^{m} \sum_{I=1}^{N} \Phi_I^\alpha(x) u_I^\alpha h$ be an RKP approximation of the function $u \in W^{m+1,2}(J) \cap C^0(J)$. Then*

$$\|u - \tilde{u}\|_l \leq c\rho^{m+1-l} \|u\|_{m+1}, \quad 0 \leq l \leq m. \tag{14}$$

4 Conclusion

In this contribution we presented solutions of boundary value problems by means of two meshless methods - the Galerkin method with weighted B-spline basis and the Galerkin method with RKP shape functions basis. To realize these methods, no explicitly given mesh was required. The size of support and smoothness of shape functions can be chosen at the beginning of the computation. We do it by means of the choice of parameters n and h (see Note 1) in the case of B-splines. And, in the RKPM, it depends on the choice of the window function v and the dilatation parameter ρ (see Note 2).

The weighted B-splines are a simple and comfortable tool from a computational point of view (recursive formulas give possibility to compute the derivative or the scalar products of B-splines in a simple way - see [5]). The error of any approximation depends not only on the number of particles but it depends on other factors, too. We saw, in the Example 2 above, that the error of the approximation obtained can become smaller if a proper weight function is added to a Dirichlet problem.

The construction of RKP shape functions is rather complicated. But the pains taken pay, because only small number of particles is necessary to receive a good approximation of the solution. We also saw that higher order of RKP shape functions can speed up the convergence and that the quality of the approximate solution depends on the choice of the weight function. (Approximate properties of the RKP shape function system in dependence on the weight function selection are also studied in [1].)

The influence of the weight function choice (its form, order or width of strips) on approximative solution can be the subject of next study.

References

1. Babuška, I., Banerjee, U., Osborn, J.E.: Survey of meshless and generalized finite element mehods: An unified approach. Acta Numer., 1–125 (2003)
2. Belytschko, T., Lu, Y., Gu, I.: Element-Free Galerkin Methods, Internat. J. Numer. Methods Engrg. 37, 229–256 (1994)
3. Chen, J.S., Pan, C., Wu, C.T., Liu, W.K.: Reproducing kernel particle methods for large deformation analysis of non-linear structures. Comput. Methods Appl. Mech. Engrg. 139, 195–227 (1996)
4. Chen, J.S., Pan, C., Wu, C.T.: Large deformation analysiss of rubber based on a reproducing kernel particle methods. Comput. Mech. 19, 211–227 (1997)
5. Hollig, W.: Finite Element Methods with B-Splines. SIAM, Philadelphia (2003)
6. Li, S., Liu, W.K.: Reproducing kernel hierarchical partition of unity. Internat. J. Numer. Methods Engrg. 45, 251–317 (1999)
7. Monaghan, J.J.: Why Particle Methods Work. Sci. Stat. Comput. 3(4), 422–433 (1982)

Question of Existence and Uniqueness of Solution for Navier-Stokes Equation with Linear "Do-Nothing" Type Boundary Condition on the Outflow

Tomáš Neustupa

Faculty of Mechanical Engineering, Czech Technical University Prague,
Karlovo nám 13, 121 35 Prague, Czech Republic,
tneu@centrum.cz

Abstract. The paper deals with the mathematical model of a flow of a viscous incompressible fluid through a 2D cascade of profiles. We consider a splited "do–nothing" type boundary condition on the outflow. The existence of a weak solution of a corresponding steady boundary value problem is known, see [2] and [3]. We recall the weak formulation, the theorem on existence and we study the uniqueness of the weak solution in this paper.

1 Introduction

The paper is concerned with the theoretical analysis of the model of incompressible, viscous, stationary flow through a plane cascade of profiles. The model of cascade of profiles is defined in a domain which represents the exterior to an infinite row of profiles, periodically spaced in one direction. The problem is formulated in a bounded domain of the form of one space period and completed by the Dirichlet boundary condition on the inlet and the profile, a suitable natural boundary condition on the outlet and periodic boundary conditions on artificial cuts. We will study the question of uniqueness of the weak solution of this problem for linear separated "do nothing" type boundary condition (which we derive).

2 Geometry of the Problem

We study the flow through a plane cascade of profiles. This domain represents the 2D model of the flow through a bladed machine (i.e. turbine or compressor). The model can be obtained by doing an artificial cut in an fixed distance from the axis of rotation. We will get a domain which is infinite, but periodic in one direction and which contain an infinite number of profiles.

Let us suppose that C_0 is a simple closed curve in $\mathbb{R}^2$ which is piecewise of the class C^2 and whose interior and exterior are domains with a Lipschitz–continuous boundary. For $k \in \mathbb{Z}$ we put $C_k = \{ (x_1, x_2 + k\tau);\ (x_1, x_2) \in C_0 \}$,

S. Margenov, L.G. Vulkov, and J. Waśniewski (Eds.): NAA 2008, LNCS 5434, pp. 431–438, 2009.

where τ is a positive constant. We assume that τ is so large that the curves C_k are mutually disjoint. By $\mathrm{Int}\, C_k$ we denote the interior of the curve C_k. The set $M = \bigcup_{k=-\infty}^{+\infty} \overline{\mathrm{Int}\, C_k}$ is called a *cascade of profiles* and each of its components $\overline{\mathrm{Int}\, C_k}$ is called a *profile*. Number τ is called the *period* of the cascade. The resulting domain is on the FIG. 1.

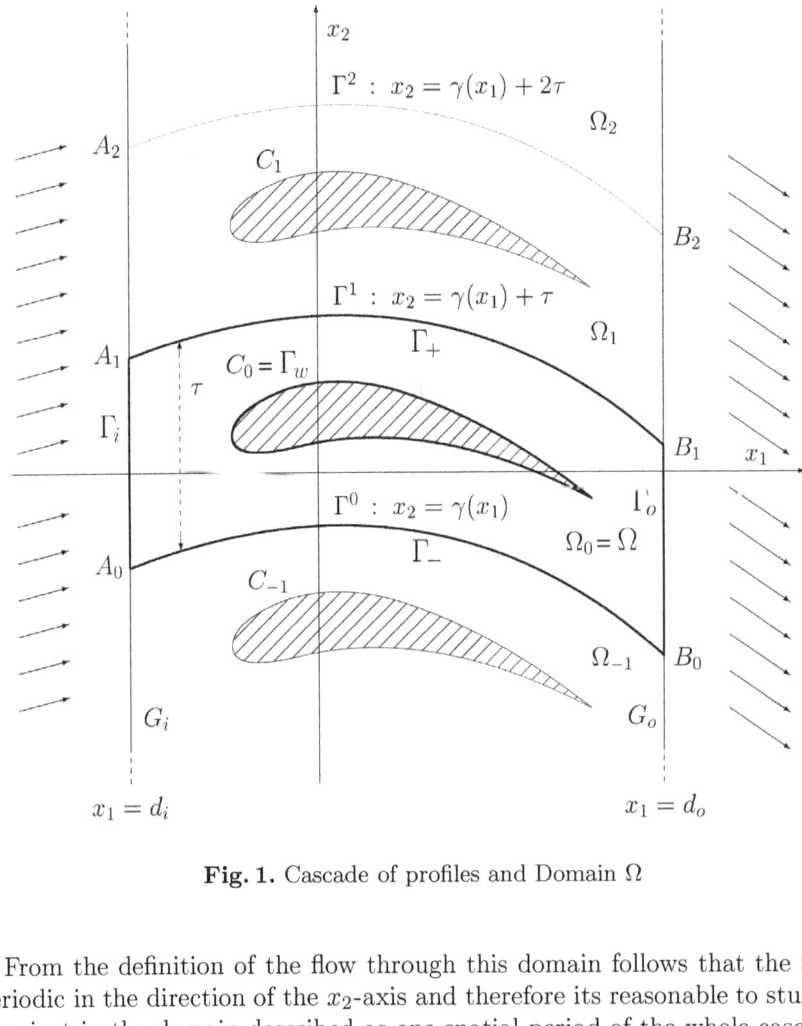

Fig. 1. Cascade of profiles and Domain Ω

From the definition of the flow through this domain follows that the flow is periodic in the direction of the x_2-axis and therefore its reasonable to study the flow just in the domain described as one spatial period of the whole cascade of profiles. Let as denote this one period Ω and its boundary $\partial\Omega = \Gamma_i + \Gamma_o + \Gamma_+ + \Gamma_- + \Gamma_w$. The shape of the domain Ω is on FIG. 1, in thicker line.

It is possible to prove that the restriction to the one spatial period is allowed (i. e. if we will find a solution on the one period, then we can prolong it periodically and get the solution on the whole cascade of profiles) for the proof see [3].

3 Definitions and Auxiliary Result

In the process of the deriving of the weak solution we will need these tools.

3.1 Definition of Function Spaces

We shall work with the following function spaces.

By $H^1(\Omega)$ we denote the usual Sobolev space and by $H^1(\Omega)^2 := \left[H^1(\Omega)\right]^2$ we denote functions with two components, both in $H^1(\Omega)$. Space X is a space of test functions constructed for the deriving of the weak solution.

$$X = \{\boldsymbol{v} \in H^1(\Omega)^2;\ \boldsymbol{v} = \boldsymbol{0} \ \text{ a.e. in } \Gamma_i \cup \Gamma_w,\ \boldsymbol{v}(x_1, x_2 + \tau) = \boldsymbol{v}(x_1, x_2) \\ \text{for a.a. } (x_1, x_2) \in \Gamma_-\} \ .$$

The boundary conditions on the curves Γ_i, Γ_w and Γ_- are interpreted in the sense of traces. Let

$$V = \left\{\boldsymbol{v} \in X;\ \operatorname{div} \boldsymbol{v} = 0 \ \text{ a.e. in } \Omega\right\} .$$

The norm in X, defined in this way

$$|||\boldsymbol{v}||| = \left(\int_\Omega \sum_{i,j=1}^{2} \left(\frac{\partial v_i}{\partial x_j}\right)^2 \mathrm{d}\boldsymbol{x}\right)^{1/2}$$

is an equivalent norm in $H^1(\Omega)^2$.

3.2 Function $\boldsymbol{g}^*$ – The Realization of the Boundary Condition on Γ_i

Lemma 1. *Let $s \in (\frac{1}{2}, 1]$ and let function $\boldsymbol{g}$ belongs to the Sobolev–Slobodetskii space $H^s(\Gamma_i)^2$. Then there exists a constant $c_g > 0$ independent of $\boldsymbol{g}$ and a divergence-free extension $\boldsymbol{g}^* \in H^1(\Omega)^2$ of function $\boldsymbol{g}$ from Γ_i onto Ω such that $\boldsymbol{g}^* = \boldsymbol{0}$ on Γ_w, $\boldsymbol{g}^*$ satisfies the condition of periodicity*

$$\boldsymbol{g}^*(x_1, x_2 + \tau) = \boldsymbol{g}^*(x_1, x_2) \qquad \text{for } (x_1, x_2) \in \Gamma_-$$

and the estimate

$$\|\boldsymbol{g}^*\|_1 \leq c_g \, \|\boldsymbol{g}\|_{s;\,\Gamma_i} \ .$$

We will seek the weak solution $\boldsymbol{u}$ in the form $\boldsymbol{u} = \boldsymbol{g}^* + \boldsymbol{z}$ where $\boldsymbol{z} \in V$ will be a new unknown function. This form of $\boldsymbol{u}$ guarantees that $\boldsymbol{u}$ will satisfy the prescribed velocity profile on the boundary Γ_i.

4 The Problem with a Linear Boundary Condition on the Outflow

We will consider the flow with linear boundary condition on the outlet of the domain Ω.

4.1 Governing Equations

Here we denote $\boldsymbol{u} = (u_1, u_2)$ the velocity, p the kinematic pressure and $\boldsymbol{n}$ denotes the outer normal to the boundary. We study the flow described by 2D steady Navier–Stokes equation in the form

$$\omega(\boldsymbol{u})\,\boldsymbol{u}^{\perp} \;=\; -\nabla q + \nu\,(-\partial_2, \partial_1)\,\omega(\boldsymbol{u}) + \boldsymbol{f} \tag{1}$$

where $\omega(\boldsymbol{u}) = \partial_1 u_2 - \partial_2 u_1$, $\boldsymbol{u}^{\perp} = (-u_2, u_1)$ and $q := p + \frac{|\boldsymbol{u}|^2}{2}$. $\omega(\boldsymbol{u})$ denotes the vorticity of the flow and q is the Bernoulli pressure. The condition of incompressibility says that

$$\operatorname{div}\,\boldsymbol{u} = 0\ . \tag{2}$$

4.2 Boundary Conditions

We prescribe the no slip Dirichlet boundary condition on the inlet

$$\boldsymbol{u}\,|_{\Gamma_i} = \boldsymbol{g}\ , \tag{3}$$

further we assume that there is a noslip boundary condition on the profile

$$\boldsymbol{u}\,|_{\Gamma_w} = \boldsymbol{0}\ . \tag{4}$$

We suppose that the periodicity boundary conditions are fullfiled on the artificial periodic boundaries Γ_+ and Γ_-

$$\boldsymbol{u}(x_1, x_2 + \tau) = \boldsymbol{u}(x_1, x_2) \qquad \text{for } (x_1, x_2) \in \Gamma_-\ , \tag{5}$$

$$\frac{\partial \boldsymbol{u}}{\partial \boldsymbol{n}}(x_1, x_2 + \tau) = -\frac{\partial \boldsymbol{u}}{\partial \boldsymbol{n}}(x_1, x_2) \qquad \text{for } (x_1, x_2) \in \Gamma_-\ . \tag{6}$$

The Bernoulli pressure q is naturally supposed to be τ–periodic in the x_2–direction too, i.e.

$$q(x_1, x_2 + \tau) \;=\; q(x_1, x_2) \qquad \text{for } (x_1, x_2) \in \Gamma_-\ . \tag{7}$$

The boundary condition used on the outflow Γ_o is

$$q = h_1, \qquad -\nu\,\omega(\boldsymbol{u}) = h_2 \tag{8}$$

where $\boldsymbol{h} = (h_1, h_2)$ is a given function on Γ_o. This condition will naturally arise (as a boundary condition of the "do nothing" type) from an appropriate weak formulation.

Remark 1. Considering the nonlinear term in the Navier–Stokes equation in the form $\omega(\boldsymbol{u})\,\boldsymbol{u}^{\perp}$ has the advantage that if we formally multiply (1) by $\boldsymbol{u}$, the product $\omega(\boldsymbol{u})\,\boldsymbol{u}^{\perp} \cdot \boldsymbol{u}$ equals zero point–wise a.e. in Ω and it is therefore not necessary to integrate by parts in order to remove or transform this term. Thus, we avoid problems on the part Γ_o of the boundary, caused by possible backward flows. On the other hand, this approach implies that we must deal with the Bernoulli pressure $q = p + \frac{1}{2}|\boldsymbol{u}|^2$ instead of the physical pressure p on the right hand side of (1) and consequently, also in the first of the boundary conditions (8).

4.3 Weak Formulation of the Problem

In order to arrive formally at the weak formulation of the problem (1)–(8), we multiply (1) by an arbitrary test function $\boldsymbol{v} = (v_1, v_2) \in V$, integrate over Ω, apply Green's theorem and use the condition of incompressibility (2), and boundary conditions and conditions of periodicity (5)–(7). We obtain the equation

$$\nu \int_\Omega \omega(\boldsymbol{u}) \cdot \omega(\boldsymbol{v})\, \mathrm{d}\boldsymbol{x} + \int_\Omega \omega(\boldsymbol{u})\, \boldsymbol{u}^\perp \cdot \boldsymbol{v}\, \mathrm{d}\boldsymbol{x} - \nu \int_{\Gamma_o} \omega(\boldsymbol{u})\, (v_2 n_1 - v_1 n_2)\, \mathrm{d}S$$
$$+ \int_{\Gamma_o} q\, \boldsymbol{v} \cdot \boldsymbol{n}\, \mathrm{d}\boldsymbol{x} = \int_\Omega \boldsymbol{f} \cdot \boldsymbol{v}\, \mathrm{d}\boldsymbol{x}\ .$$

Using the identities $n_1 = 1$ and $n_2 = 0$ on Γ_o we have

$$\nu \int_\Omega \omega(\boldsymbol{u}) \cdot \omega(\boldsymbol{v}) \mathrm{d}\boldsymbol{x} + \int_\Omega \omega(\boldsymbol{u}) \boldsymbol{u}^\perp \cdot \boldsymbol{v} \mathrm{d}\boldsymbol{x} - \int_{\Gamma_o} [\nu\, \omega(\boldsymbol{u}) v_2 - q v_1]\, \mathrm{d}S = \int_\Omega \boldsymbol{f} \cdot \boldsymbol{v} \mathrm{d}\boldsymbol{x}.$$

Substituting here for the terms in the integrand on Γ_o from (8), we obtain

$$\nu \int_\Omega \omega(\boldsymbol{u}) \cdot \omega(\boldsymbol{v})\, \mathrm{d}\boldsymbol{x} + \int_\Omega \omega(\boldsymbol{u})\, \boldsymbol{u}^\perp \cdot \boldsymbol{v}\, \mathrm{d}\boldsymbol{x} + \int_{\Gamma_o} [h_2\, v_2 + h_1\, v_1]\, \mathrm{d}S = \int_\Omega \boldsymbol{f} \cdot \boldsymbol{v}\, \mathrm{d}\boldsymbol{x}.$$

This integral equation can be written in the form

$$a(\boldsymbol{u}, \boldsymbol{v}) = (\boldsymbol{f}, \boldsymbol{v}) + b(\boldsymbol{h}, \boldsymbol{v}) \tag{9}$$

where $\boldsymbol{h}$ is given by (8) and the forms a and b are defined below:

$$a(\boldsymbol{u}, \boldsymbol{v}) = a_1(\boldsymbol{u}, \boldsymbol{v}) + a_2(\boldsymbol{u}, \boldsymbol{u}, \boldsymbol{v})\ , \quad a_1(\boldsymbol{u}, \boldsymbol{v}) = \nu \left(\omega(\boldsymbol{u}), \omega(\boldsymbol{v})\right)_{L^2(\Omega)}\ ,$$

$$a_2(\boldsymbol{u}, \boldsymbol{v}, \boldsymbol{w}) = \int_\Omega \omega(\boldsymbol{u})\, \boldsymbol{v}^\perp \cdot \boldsymbol{w}\ \mathrm{d}\boldsymbol{x}\ , \quad b(\boldsymbol{h}, \boldsymbol{v}) = -\int_{\Gamma_o} \boldsymbol{h} \cdot \boldsymbol{v}\ \mathrm{d}S\ .$$

The weak problem now reads as follows:

Definition 1. *Let function $\boldsymbol{g} \in H^s(\Gamma_i)^2$ (for some $s \in (\frac{1}{2}, 1]$) satisfy the condition $\boldsymbol{g}(A_1) = \boldsymbol{g}(A_0)$ (where A_0 and A_1 are the end points of Γ_i). Let $\boldsymbol{f} \in L^2(\Omega)^2$ and $\boldsymbol{h} \in L^2(\Gamma_o)^2$. The* **weak solution** *of the problem (1)–(8) is a vector function $\boldsymbol{u} \in H^1(\Omega)^2$ which satisfies the condition of incompressibility (2) a.e. in Ω, the identity (9) for all test functions $\boldsymbol{v} \in V$, the boundary conditions (3), (4) in the sense of traces on Γ_i and Γ_w and the condition of periodicity (5) in the sense of traces on Γ_- and Γ_+.*

Remark 2. The deriving of the "do-nothing" type boundary condition is based on the selection of the behavior on the boundary Γ_o in such a way that the integral over Γ_o will be zero ($\boldsymbol{h} = \boldsymbol{0}$) or equal to some chosen function ($\boldsymbol{h}$) and on the construction of the function spaces in such a way that we do not restrict the behavior on this part of boundary (see definition of the space X and V).

Let us further denote by $\boldsymbol{g}^*$ the extension of the function $\boldsymbol{g}$ from the line segment Γ_i into Ω, constructed in SUBSEC. (3.2). We can further seek for the weak solution $\boldsymbol{u}$ in the form $\boldsymbol{u} = \boldsymbol{g}^* + \boldsymbol{z}$ where $\boldsymbol{z} \in V$ is a new unknown function.

This form of $\boldsymbol{u}$ guarantees that $\boldsymbol{u}$ satisfies the condition (3) and the boundary and periodicity conditions (4)–(6). Substituting the sum $\boldsymbol{g}^* + \boldsymbol{z}$ for $\boldsymbol{u}$ into the (1), we arrive at the following problem: *Find a function $\boldsymbol{z} \in V$ such that it satisfies the equation*

$$a(\boldsymbol{g}^* + \boldsymbol{z}, \boldsymbol{v}) = (\boldsymbol{f}, \boldsymbol{v}) + b(\boldsymbol{h}, \boldsymbol{v}) \tag{10}$$

for all $\boldsymbol{v} \in V$.

Let us formulate the problem of existence of the weak solution as the following theorem:

Theorem 1. *There exists $\varepsilon > 0$ such that if $\|\boldsymbol{g}\|_{H^s(\Gamma_i)^2} < \varepsilon$ then there exists a solution $\boldsymbol{u} = \boldsymbol{g}^* + \boldsymbol{z}$ of the problem defined in DEFINITION 1. Moreover $\boldsymbol{z}$ satisfies the estimate*

$$|||\boldsymbol{z}||| \leq R_1 \ .$$

Consequently, the weak problem (9) has a solution $\boldsymbol{u}$ $(= \boldsymbol{z} + \boldsymbol{g}^*)$ that satisfies

$$\|\nabla \boldsymbol{u}\|_{L^2(\Omega)^2} \leq R_1 + \|\nabla \boldsymbol{g}^*\|_{L^2(\Omega)^2} \leq R_1 + c\,\|\boldsymbol{g}\|_{H^s(\Gamma_i)^2} := R_2 \ .$$

Here R_1 and c are constants based on the construction of the weak solution in the proof of the theorem.

Proof. The proof of this theorem is carried out by using the method of Galerkin approximations. We need to prove the coercivity of the form a and construct the weak solution. The value of the constant ε comes from the proof to ensure coercivity. The proof can be found in [2] and in [3].

Remark 3. The special Navier–Stokes formulation (1) naturally leads to the linear "do-nothing" type boundary condition (8) which, on the contrary to the basic "do-nothing" boundary condition, enables us to prove existence of the weak solution.

4.4 On Uniqueness of a Weak Solution

Suppose that $\boldsymbol{u}_1$ and $\boldsymbol{u}_2$ are two solutions of the weak problem (9). We shall prove that the solutions coincide if at least one of them is "sufficiently small" in the norm $|||\,.\,|||$. The special structure of the form a_2 implies

$$a_2(\boldsymbol{u}, \boldsymbol{v}, \boldsymbol{v}) = \int_\Omega \omega(\boldsymbol{u})\, \boldsymbol{v}^\perp \cdot \boldsymbol{v} \,\mathrm{d}\boldsymbol{x} = 0$$

for all $\boldsymbol{u}$ and $\boldsymbol{v}$ from $H^1(\Omega)^2$.

Theorem 2. *There exists $R > 0$ such that if $\boldsymbol{u}_1$ and $\boldsymbol{u}_2$ are two solutions of the problem (9) such that $\|\nabla \boldsymbol{u}_1\|_{L^2(\Omega)^4} \leq R$ then $\boldsymbol{u}_1 = \boldsymbol{u}_2$.*

Proof. The solutions $\boldsymbol{u}_1$ and $\boldsymbol{u}_2$ fulfil

$$a(\boldsymbol{u}_1, \boldsymbol{v}) = (\boldsymbol{f}, \boldsymbol{v}) + b(\boldsymbol{h}, \boldsymbol{v}) \ , \qquad a(\boldsymbol{u}_2, \boldsymbol{v}) = (\boldsymbol{f}, \boldsymbol{v}) + b(\boldsymbol{h}, \boldsymbol{v}) \tag{11}$$

for all $\boldsymbol{v} \in V$. Subtracting these equations, we get

$$a(\boldsymbol{u}_1, \boldsymbol{v}) - a(\boldsymbol{u}_2, \boldsymbol{v}) = 0 \ .$$

Substituting here the form of a from (9) and choosing $\boldsymbol{v} = \boldsymbol{u}_1 - \boldsymbol{u}_2$, we get

$$a_1(\boldsymbol{u}_1 - \boldsymbol{u}_2, \boldsymbol{u}_1 - \boldsymbol{u}_2) + a_2(\boldsymbol{u}_1, \boldsymbol{u}_1, \boldsymbol{u}_1 - \boldsymbol{u}_2) - a_2(\boldsymbol{u}_2, \boldsymbol{u}_2, \boldsymbol{u}_1 - \boldsymbol{u}_2) = 0 \ . \tag{12}$$

The difference $a_2(\boldsymbol{u}_1, \boldsymbol{u}_1, \boldsymbol{u}_1 - \boldsymbol{u}_2) - a_2(\boldsymbol{u}_2, \boldsymbol{u}_2, \boldsymbol{u}_1 - \boldsymbol{u}_2)$ can be estimated as

$$\begin{aligned}
&\left|a_2(\boldsymbol{u}_1, \boldsymbol{u}_1, \boldsymbol{u}_1 - \boldsymbol{u}_2) - a_2(\boldsymbol{u}_2, \boldsymbol{u}_2, \boldsymbol{u}_1 - \boldsymbol{u}_2)\right| \\
&= \left|a_2(\boldsymbol{u}_1 - \boldsymbol{u}_2, \boldsymbol{u}_1, \boldsymbol{u}_1 - \boldsymbol{u}_2) - a_2(\boldsymbol{u}_2, \boldsymbol{u}_1 - \boldsymbol{u}_2, \boldsymbol{u}_1 - \boldsymbol{u}_2)\right| \\
&= \left|a_2(\boldsymbol{u}_1 - \boldsymbol{u}_2, \boldsymbol{u}_1, \boldsymbol{u}_1 - \boldsymbol{u}_2)\right| = \left|\int_\Omega \omega(\boldsymbol{u}_1 - \boldsymbol{u}_2)\, \boldsymbol{u}_1^\perp \cdot (\boldsymbol{u}_1 - \boldsymbol{u}_2)\, \mathrm{d}\boldsymbol{x}\right| \\
&\le \|\omega(\boldsymbol{u}_1 - \boldsymbol{u}_2)\|_{L^2(\Omega)} \|\boldsymbol{u}_1\|_{L^4(\Omega)^2} \|\boldsymbol{u}_1 - \boldsymbol{u}_2\|_{L^4(\Omega)^2} \\
&\le c\, \|\nabla(\boldsymbol{u}_1 - \boldsymbol{u}_2)\|_{L^2(\Omega)} \|\boldsymbol{u}_1\|_{L^4(\Omega)^2} \|\boldsymbol{u}_1 - \boldsymbol{u}_2\|_{L^4(\Omega)^2} \\
&\le c\, \|\nabla(\boldsymbol{u}_1 - \boldsymbol{u}_2)\|_{L^2(\Omega)}\, c_1^2\, \|\nabla \boldsymbol{u}_1\|_{L^2(\Omega)^2} \|\nabla(\boldsymbol{u}_1 - \boldsymbol{u}_2)\|_{L^2(\Omega)^2} \\
&\le c\, c_1^2\, R\, |||\boldsymbol{u}_1 - \boldsymbol{u}_2|||^2
\end{aligned} \tag{13}$$

where the constant c_1 is from the inequality $\|\boldsymbol{u}\|_{L^4(\Omega)^2} \le c_1 \|\nabla \boldsymbol{u}\|_{L^2(\Omega)^4}$. The proof of this inequality can be found in [3]. Substituting (13) to (12), we get

$$\nu\, |||\boldsymbol{u}_1 - \boldsymbol{u}_2|||^2 \le c\, c_1^2\, R\, |||\boldsymbol{u}_1 - \boldsymbol{u}_2|||^2 \ .$$

This inequality easily implies the statement of the theorem. □

Corollary 1. *If the given functions* $\boldsymbol{g}$, $\boldsymbol{f}$ *and* $\boldsymbol{h}$ *are so small in comparison with* ν *that number* R_1 *(given by THEOREM 1) is "small enough" (the question of uniqueness depends on the viscosity and the velocity of the fluid. I.e. "small enough" mean that the velocity is not too high according to the viscosity of the fluid) then the weak solution* $\boldsymbol{u}$ *is unique in the class of all solutions of the problem (9).*

Remark 4. Concerning the situation on the outflow, the flow through a cascade has similar features as a flow through a channel. The choose of the boundary condition on the outflow is studied extensively. J. Heywood, R. Rannacher and S. Turek [4] explicitly did not involve any boundary condition on the outflow into the weak formulation and by means of a backward integration by parts have shown that this induces the so called "do nothing" boundary condition $-\nu \frac{\partial \boldsymbol{u}}{\partial \boldsymbol{n}} + p\boldsymbol{n} = \boldsymbol{0}$. This boundary condition do not enable us to control the possible backward flow into the domain through the outflow part of boundary, which mean we cant prove the existence of the weak solution. In numerical computation we can often get the non-convergence of the weak solution.

In paper [5], we studied the nonlinear modification of the "do-nothing" boundary condition proposed by C. H. Bruneau, F. Fabrie in [1]

$$-\nu \frac{\partial \boldsymbol{u}}{\partial \boldsymbol{n}} + p\boldsymbol{n} - \frac{1}{2}(\boldsymbol{u}\cdot\boldsymbol{n})^{-}\boldsymbol{u} = \boldsymbol{h}$$

where the superscript $-$ denotes the negative part. Here its possible to obtain coercivity and thus to prove the existence of the weak solution. However the proof is rather complicated due to the nonlinearity.

We presented the linear type of "do-nothing" boundary condition which enables us to prove existence of the weak solution. Moreover in the case of uniqueness, we are able to prove stronger result than with the nonlinear boundary condition. I.e. we only need one of the two solution ($\boldsymbol{u}_1$ and $\boldsymbol{u}_2$) to be bounded to prove uniqueness in nonlinear case this can be done only if we restrict both solutions. The behavior of this condition in numerical simulation is yet to be studied.

Conclusion. We showed a construction of the linear "do-nothing" type boundary condition, which allows to prove the existence of a weak solution. We showed uniqueness result for this problem as well. Our goal in the future is to study this condition and its behavior in a numerical computation. The weak formulation can be (after recovering pressure) used as the velocity-pressure formulation and considered as the basis for the finite element discretization.

Acknowledgement. The research was supported by the research plan of the Ministry of Education of the Czech Republic No. MSM 6840770010.

References

1. Bruneau, C.H., Fabrie, P.: New efficient boundary conditions for incompressible Navier–Stokes equations: A well–posedness result. Mathematical Modelling and Numerical Analysis 30(7), 815–840 (1996)
2. Neustupa, T.: Modelling of a Steady Flow in a Cascade with Separate Boundary Conditions for Vorticity and Bernoulli's Pressure on the Outflow. WSEAS Transactions on Mathematics 3(5), 274–279 (2006)
3. Neustupa, T.: Mathematical Modelling of Viscous Incompressible Flow through a Cascade of Profiles. Dissertation Thesis. Faculty of Mathematics and Physics, Charles University Prague (2007)
4. Heywood, J.G., Rannacher, R., Turek, S.: Artificial boundaries and flux and pressure conditions for the incompressible Navier-Stokes equations. Int. J. for Numerical Methods in Fluids 22, 325–352 (1996)
5. Feistauer, M., Neustupa, T.: On some aspects of analysis of incompressible flow through cascades of profiles. In: Operator Theory, Advances and Applications, vol. 147, pp. 257–276. Birkhäuser, Basel (2004)

Geometrical Analysis of Model Predictive Control: A Parameterized Polyhedra Approach

S. Olaru[1], I. Dumitrache[2], and D. Dumur[1]

[1] Automatic Control Dpt., SUPELEC, France
{sorin.olaru,didier.dumur}@supelec.fr
[2] University Polytehnica Bucharest,Romania
idumitrache@ics.pub.ro

Abstract. The paper deals with the receding horizon optimal control schemes. The presence of input and state constraints is shown to lead at the implementation stage to the resolution of a parametric optimization problem. A geometrical analysis of the set of constraints can be done using the concept of parameterized polyhedra.

1 Introduction

The philosophy behind Model-based Predictive Control (MPC) is to exploit in a "receding horizon" manner the simplicity of the Euler-Lagrange approach for the optimal control. To be more specific, the control action u_t at state x_t is obtained from the control sequence $\mathbf{k}_u^* = [u_t^T, \ldots, u_{t+N-1}^T]^T$ obtained as a result of the optimization problem:

$$\begin{array}{ll} \min\limits_{\mathbf{k}_u} & \varphi(x_{t+N}) + \sum\limits_{k=0}^{N-1} l(x_{t+k}, u_{t+k}) \\ \text{subj. to}: & x_{t+1} = f(x_t) + g(x_t)u_t; \\ & h(x_t, \mathbf{k}_u) \leq 0 \end{array} \tag{1}$$

constructed for a finite prediction horizon N, cost per stage $l(.)$, terminal weight $\varphi(.)$, the system dynamics described by $f(.), g(.)$ and the constraints written in a compact form using elementwise inequalities on functions linking the states and the control actions, $h(.)$.

For the optimization problem (1), the current state serves as an initial condition and influences both the objective function and the feasible domain:

$$\begin{array}{rl} \mathbf{k}_u^*(x_t) = \min\limits_{\mathbf{k}_u} & F(x_t, \mathbf{k}_u) \\ \text{subj. to}: & \left\{ \begin{array}{l} C_{in}(\mathbf{k}_u, x_t) \leq 0 \\ C_{eq}(\mathbf{k}_u, x_t) = 0 \end{array} \right. \end{array} \tag{2}$$

The system state can be interpreted as a vector of parameters, and the problems to be solved are part of the multiparametric optimization programming family. From the cost function point of view, the parametrization is somehow easier to deal with and eventually can be entirely translated towards the set of constraints.

S. Margenov, L.G. Vulkov, and J. Waśniewski (Eds.): NAA 2008, LNCS 5434, pp. 439–446, 2009.

Unfortunately, similar observation cannot be made about the feasible domain and its adjustment with respect to the parameters evolution.

In the class of convex sets, the use of polyhedral domains is not hazardous since they offer important advantages, like the closeness over the intersection or the fact that the polyhedral invariant sets (largely used for enforcing stability) are less conservative than the ellipsoidal ones for example. More than that, any convex and compact set is known to be approximated within a ϵ approximation by a finitely generated polytope.

In the current paper, these polyhedral feasible domains will be analyzed with a focus on the parametrization leading to parameterized polyhedra [4]:

$$\begin{aligned} &\mathbf{k}_u^*(x_t) = \min_{\mathbf{k}_u} F(x_t, \mathbf{k}_u) \\ &\text{subj. to :} \quad \begin{cases} A_{in}\mathbf{k}_u \leq b_{in} + B_{in}x_t \\ A_{eq}\mathbf{k}_u = b_{eq} + B_{eq}x_t \end{cases} \end{aligned} \tag{3}$$

where the objective function $F(x_t, \mathbf{k}_u)$ can be either linear, quadratic or other nonlinear type.

2 From Polyhedral Domains to Parameterized Polyhedra

A mixed system of linear equalities and inequalities defines a polyhedron [7]. It can be represented in the dual (Minkowski) formulation:

$$\mathcal{P} = \{\mathbf{k}_u \in \mathbb{R}^p \,|\, A_{eq}\,\mathbf{k}_u = b_{eq}; A_{in}\mathbf{k}_u \leq b_{in}\} \Longleftrightarrow \mathcal{P} = \underbrace{conv.hull\mathbf{V} + cone\mathbf{R} + lin.space\mathbf{L}}_{generators} \tag{4}$$

where $conv.hull\mathbf{V}$ denotes the set of convex combinations of vertices $\mathbf{V} = \{\mathbf{v}_1, \ldots, \mathbf{v}_\vartheta\}$, $cone\mathbf{R}$ denotes nonnegative combinations of unidirectional rays in $\mathbf{R} = \{\mathbf{r}_1, \ldots, \mathbf{r}_\rho\}$ and $lin.space\mathbf{L} = \{\mathbf{l}_1, \ldots, \mathbf{l}_\lambda\}$ represents a linear combination of bidirectional rays (with ϑ, ρ and λ the cardinals of the related sets). This dual representation in terms of generators can be rewritten as:

$$\mathcal{P} = \left\{\mathbf{k}_u = \sum_{i=1}^{\vartheta} \alpha_i \mathbf{v}_i + \sum_{i=1}^{\rho} \beta_i \mathbf{r}_i + \sum_{i=1}^{\lambda} \gamma_i \mathbf{l}_i; 0 \leq \alpha_i \leq 1, \; \sum_{i=1}^{\vartheta} \alpha_i = 1 \;, \beta_i \geq 0 \;, \forall \gamma_i \right\} \tag{5}$$

with α_i, β_i, γ_i the coefficients describing the convex, non-negative and linear combinations in (4). Numerical methods like the Chernikova algorithm are implemented for constructing the double description, either starting from constraints (4) either from the generators (5) representation.

A *parameterized polyhedron* is defined in the implicit form by a finite number of inequalities and equalities with the note that the affine part depends linearly on a vector of parameters $x \in \mathbb{R}^n$ for both equalities and inequalities:

$$\begin{aligned} \mathcal{P}(x) &= \{\mathbf{k}_u(x) \in \mathbb{R}^p \,|\, A_{eq}\,\mathbf{k}_u = B_{eq}x + b_{eq}; A_{in}\mathbf{k}_u \leq B_{in}x + b_{in}\} \\ &= \left\{\mathbf{k}_u(x) |\; \mathbf{k}_u(x) = \sum_{i=1}^{\vartheta} \alpha_i(x)\mathbf{v}_i(x) + \sum_{i=1}^{\rho} \beta_i \mathbf{r}_i + \sum_{i=1}^{\lambda} \gamma_i \mathbf{l}_i \right\} \\ &0 \leq \alpha_i(x) \leq 1, \; \sum_{i=1}^{\vartheta} \alpha_i(x) = 1 \;, \beta_i \geq 0 \;, \forall \gamma_i . \end{aligned} \tag{6}$$

This dual representation of the parameterized polyhedral domain reveals the fact that only the vertices are concerned by the parametrization (resulting the so-called *parameterized vertices* - $\mathbf{v}_i(x)$), whereas the rays and the lines do not change with the parameters' variation. The basic idea is to identify the parameterized polyhedron with a non-parameterized one in an augmented space:

$$\tilde{\mathcal{P}} = \left\{ \begin{bmatrix} \mathbf{k}_u \\ x \end{bmatrix} \in \mathbb{R}^{p+n} \middle| \, [A_{eq}| - B_{eq}] \begin{bmatrix} \mathbf{k}_u \\ x \end{bmatrix} = b_{eq}; \, [A_{in}| - B_{in}] \begin{bmatrix} \mathbf{k}_u \\ x \end{bmatrix} \leq b_{in} \right\} \quad (7)$$

The original polyhedron in (6) can be found for any particular value of the parameters vector x through $P(x) = \text{Proj}_{\mathbf{k}_u}\left(\tilde{P} \cap H(x)\right)$, for any given hyperplane $H(x_0) = \left\{ \begin{pmatrix} \mathbf{k}_u \\ x \end{pmatrix} \in \mathbb{R}^{p+n} \middle| \, x = x_0 \right\}$ and using $\text{Proj}_{\mathbf{k}_u}(.)$ as the projection from $\mathbb{R}^{p+n}$ to the first p coordinates $\mathbb{R}^p$.

Within the polyhedral domains $\tilde{\mathcal{P}}$, the correspondent of the parameterized vertices in (6) can be found among the faces of dimension n. After enumerating these n-faces: $\left\{F_1^n(\tilde{\mathcal{P}}), \dots F_j^n(\tilde{\mathcal{P}}), \dots, F_\varsigma^n(\tilde{\mathcal{P}})\right\}$, one can write: $\forall i, \exists j \in \{1, \dots, \varsigma\}$ *s.t.* $\left[\mathbf{v}_i(x)^T \; x^T\right]^T \in F_j^n(\tilde{\mathcal{P}})$ or equivalently:

$$\mathbf{v}_i(x) = \text{Proj}_{\mathbf{k}_u}\left(F_j^n(\tilde{P}) \cap H(x)\right) \quad (8)$$

From this relation it can be seen that not all the n-faces correspond to parameterized vertices. However it is still easy to identify those which can be ignored in the process of construction of parameterized vertices based on the relation: $\text{Proj}_x\left(F_j^n(\tilde{P})\right) < n$ with $\text{Proj}_x(.)$ the projection from $\mathbb{R}^{p+n}$ to the last n coordinates $\mathbb{R}^n$ (corresponding to the parameters' space). Indeed the projections are to be computed for all the n-faces, those which are degenerated are to be discarded and all the others are stored as validity domains - $D_{\mathbf{v}_i} \in \mathbb{R}^n$, for the parameterized vertices that they are identifying:

$$D_{\mathbf{v}_i} = \text{Proj}_n\left(F_j^n(\tilde{P})\right) \quad (9)$$

Once the parameterized vertices identified and their validity domain stored, the dependence on the parameters vector can be found using the supporting hyperplanes for each n-face:

$$\mathbf{v}_i(x) = \begin{bmatrix} A_{eq} \\ \bar{A}_{in_j} \end{bmatrix}^{-1} \begin{bmatrix} B_{eq} \\ \bar{B}_{in_j} \end{bmatrix} x + \begin{bmatrix} b_{eq} \\ \bar{b}_{in_j} \end{bmatrix} \quad (10)$$

where $\bar{A}_{in_j}$, $\bar{B}_{in_j}$, $\bar{b}_{in_j}$ represent the subset of the inequalities, satisfied by saturation for $F_j^n(\tilde{P})$. The inversion is well defined as long as the faces with degenerate projections are discarded.

The double representation of the parameterized polyhedra offers a complete description of the feasible domain for the predictive control law as long as this is based on a multiparametric optimization with linear constraints. Using the generators representation, with simple difference operations on convex sets one can compute the region of the parameters space where no parameterized vertex is defined:

$$\aleph = \mathbb{R}^n \setminus \{\cup D_{\mathbf{v}_i};\ i = 1 \ldots \vartheta\} \tag{11}$$

representing from the MPC point of view, the set of infeasible states.

2.1 Regular Feasible Domains and the Link with Constraints Redundancy

The problem of redundancy can influence the performance of the multiparametric optimization routines. When the feasible domains are described as parameterized polyhedra, the redundancy has to be characterized with respect to the parameters space. Indeed one can distinguish constraints which are redundant globally (for all possible parameters) in the sense that by extracting them from (6) the domain $\mathcal{P}(x)$ suffers no modification. In contrast with global redundancy one can observe also a local redundancy that can be interpreted from the dual description point of view as in the following definitions.

Definition 1. *A parameterized polyhedron (6) is defined locally over $\mathcal{D} \subset \mathbb{R}^n$ by a regular set of constraints if the set of non-redundant constraints remains constant for all $x \in \mathcal{D}$.*

Definition 2. *The polyhedral domain is said to have a* regular shape *locally over $\mathcal{D} \subset \mathbb{R}^n$ if the set of parameterized vertices remains constant for $x \in \mathcal{D}$.*

Even if these two definitions are theoretically equivalent, it is practically difficult to deal with the constraints redundancy as long as there is no a priori knowledge about the regions in the parameters space where they become redundant. On the contrary as showed in the relation (9), each parameterized vertex $\mathbf{v}_i$ is defined over a certain validity domain $D_{\mathbf{v}_i}$. Thus, by their juxtaposition, a cutting of the parameters space in zones with regular shapes $R_1, \ldots, R_{n_r} \in \mathbb{R}^n$ can be performed [5]. Knowing that for the feasible parameters it corresponds at least a parameterized vertex, one can write the simple relation between the validity domains and the regions with regular shape:

$$\cup D_{\mathbf{v}_i} = \cup R_j;\ i = 1 \ldots \vartheta;\ j = 1 \ldots n_r \tag{12}$$

The multiparametric optimization performances can take advantage of this partition of the parameters space in zones with regular shape. One can restate the problem in terms of a look-up table of problems with a non-redundant set of constraints (Table 1). The non-redundant sets of inequalities are uniquely defined by the pairs $(A_{in_i}; B_{in_i}; b_{in_i})$ for each convex region R_i based on the valid parameterized vertices for the same region.

A singularity of the parameterized polyhedra approach is the design freedom offered for the granularity of the parameter's space partitioning. This is an important advantage and it can be seen as a compromise between the implementation schemes based exclusively on on-line optimization routines on one

Table 1. Equivalence between the original optimization and the one based on piecewise redundancy-free sets of constraints

$\min_{\mathbf{k}_u}$	$F(x_t, \mathbf{k}_u)$
subj. to :	$\begin{cases} A_{in}\mathbf{k}_u \le b_{in} + B_{in}x_t \\ A_{eq}\mathbf{k}_u = b_{eq} + B_{eq}x_t \end{cases}$

$\Longleftrightarrow$

if $x_t \in R_1$	$\min_{\mathbf{k}_u} F(x_t, \mathbf{k}_u)$ subj.to : $\begin{cases} A_{in_1}\mathbf{k}_u \le b_{in_1} + B_{in_1}x_t \\ A_{eq}\mathbf{k}_u = b_{eq} + B_{eq}x_t \end{cases}$
...	...
if $x_t \in R_\rho$	$\min_{\mathbf{k}_u} F(x_t, \mathbf{k}_u)$ subj.to : $\begin{cases} A_{in_\rho}\mathbf{k}_u \le b_{in_\rho} + B_{in_\rho}x_t \\ A_{eq}\mathbf{k}_u = b_{eq} + B_{eq}x_t \end{cases}$

hand and the evaluation of the explicit solution found off-line on the other hand. Sometimes a mixture of these two techniques might improve the computational time. The scheme in table 1, with cuttings of the parameters space corresponding to simpler optimization problems can offer the freedom in choosing the right balance between positioning mechanisms and on-line solvers.

3 Towards Explicit Solutions

In the case of sufficiently large memory resources, construction of the explicit solution for the multiparametric optimization problem can be an interesting alternative to the iterative optimization routines. In this direction recent results where presented at least for the case of linear and quadratic cost functions (see [8],[2],[1],[11]). In the following it will be shown that a geometrical approach based on the parameterized polyhedra can bring a useful insight as well.

3.1 Linear Cost Function

The linear cost functions are extensively used in connection with model based predictive control and especially for robust case [3]. In a compact form, the multiparametric optimization problem is:

$$\begin{aligned} &\mathbf{k}_u{}^*(x_t) = \min_{\mathbf{k}_u} f^T \mathbf{k}_u \\ &\text{subject to } A_{in}\mathbf{k}_u \le B_{in}x_t + b_{in} \end{aligned} \tag{13}$$

The problem deals with a polyhedral feasible domain which can be described as previously in a double representation. Further the explicit solution can be constructed based on the relation between the parameterized vertices and the linear cost function. The next result resumes this idea.

Proposition: The solution for a multiparametric linear problem is characterized as follows:

a) For the subdomain $\aleph \in \mathbb{R}^n$ where the associated parameterized polyhedron has no valid parameterized vertex the problem is infeasible;

b) If there exists a bidirectional ray $\mathbf{l}$ such that $f^T\mathbf{l} \neq 0$ or a unidirectional ray $\mathbf{r}$ such that $f^T\mathbf{r} \le 0$, then the minimum is unbounded;

c) If all bidirectional rays $\mathbf{l}$ are such that $f^T\mathbf{l} = 0$ and all unidirectional rays $\mathbf{r}$ are such that $f^T\mathbf{r} \ge 0$ then there exists a cutting of the parameters in zones where the parameterized polyhedron has a regular shape $\bigcup_{j=1\ldots\rho} R_j = \mathbb{R}^n - \aleph$. For each region R_j the minimum is computed with respect to the given linear cost function and for all the valid parameterized vertices:

$$\underline{m}(x) = \min\left\{f^T\mathbf{v}_i(x)|\mathbf{v}_i(x) \text{ vertex of } \mathcal{P}(x)\right\} \tag{14}$$

The minimum $\underline{m}(x)$ is attained by constant subsets of parameterized vertices of $\mathcal{P}(x)$ over a finite number of polyhedral zones in the parameters space R_{ij} ($\cup R_{ij} = R_j$). The complete optimal solution of the multiparametric optimization is given for each R_{ij} by:

$$\begin{aligned} S_{R_{ij}}(x) = conv.hull\left\{\mathbf{v}_1^*(x), \ldots, \mathbf{v}_s^*(x)\right\} + \\ + cone\left\{\mathbf{r}_1^*, \ldots, \mathbf{r}_r^*\right\} + lin.spaceP(\mathbf{p}) \end{aligned} \tag{15}$$

where $\mathbf{v}_i^*$ are the vertices corresponding to the minimum $\underline{m}(x)$ over R_{ij} and $\mathbf{r}_i^*$ are such that $f^T\mathbf{r}_i^* = 0$□

This result provides *the entire family of solutions* for the linear multiparametric optimization, even for the cases where this family is not finite (for example there are several vertices attaining the minimum).

Remark: For the regions of the parameters space characterized by the case *(a)*, the set of constraints cannot be fulfilled and the feasible domain is empty.

Remark: If the solution of the optimization problem is characterized by the case *(b)*, then the control law based on such an optimization is not well-posed as the optimal control action needs an infinite energy in order to be effectively applied.

Remark: Due to the fact that the parameterized vertices have a linear dependence on the parameter vector, the explicit solution will be piecewise linear. However, the solution is not unique as it can be seen from the case *(c)* and equation (15) and thus for the practical control purposes a continuous piecewise candidate is preferred, eventually by minimizing the number of partitions in the parameters space.

3.2 Quadratic Cost Function

The case of a quadratic const function is one of the most popular at least for the linear MPC. The explicit solution based on the exploration of the parameters space ([1], [11]) is extensively studied lately. Alternative methods based on geometrical arguments or dynamical programming ([2], [8]) improved also the awareness of the explicit MPC formulations. The parameterized polyhedra can serve as a base in the construction of such explicit solution [4], for a quadratic multiparametric problem:

$$\begin{aligned} &\mathbf{k}_u{}^*(x_t) = \arg\min_{\mathbf{k}_u} \mathbf{k}_u^T H\mathbf{k}_u + 2\mathbf{k}_u{}^T F x_t \\ &\text{subject to } A_{in}\mathbf{k}_u \leq B_{in}x_t + b_{in} \end{aligned} \tag{16}$$

In this case the main idea is to consider the unconstrained optimum:

$$\mathbf{k}_u^{sc}(x_t) = H^{-1}Fx_t$$

and its position with respect to the feasible domain given by a parameterized polyhedron as in (6).

If a simple transformation is performed:

$$\tilde{\mathbf{k}}_u = H^{1/2}\mathbf{k}_u$$

then the isocost curves of the quadratic function are transformed from ellipsoid into circles centered in $\tilde{\mathbf{k}}_u^{sc}(x_t) = H^{-1/2}Fx_t$. Further one can use the Euclidean projection in order to retrieve the multiparametric quadratic explicit solution.

Indeed if the unconstrained optimum $\tilde{\mathbf{k}}_u^{sc}(x_t)$ is contained in the feasible domain $\tilde{\mathcal{P}}(x_t)$ then it is also the solution of the constrained case, otherwise existence and uniqueness are assured as follows:

Proposition: For any exterior point $\tilde{\mathbf{k}}_u(x_t) \notin \tilde{\mathcal{P}}(x_t)$, there exists an unique point characterized by a minimal distance with respect to $\tilde{\mathbf{k}}_u^{sc}(x_t)$. This point satisfies:

$$(\tilde{\mathbf{k}}_u^{sc}(x_t) - \tilde{\mathbf{k}}_u^*(x_t))^T(\tilde{\mathbf{k}}_u - \tilde{\mathbf{k}}_u^*(x_t)) \leqslant 0, \forall \tilde{\mathbf{k}}_u \in \tilde{\mathcal{P}}(x_t) \Box$$

The construction mechanism uses the parameterized vertices in order to split the regions neighboring the feasible domain in zones characterized by the same type of projection.

Remark: The use of these geometrical arguments makes the construction of explicit solution to deal in a natural manner with the so-called *degeneracy*. This phenomenon is identified by the parameters' values where the feasible domain changes its shape (the set of parameterized vertices is modified).

4 Generalizations and Open Problems

If the feasible domain is described by a nonlinear set of constraints then a general procedure for the construction of an exact explicit solution is not available. However, one can still use the parameterized polyhedra concepts by following the ideas [9]: i) if the feasible domain is non convex an approximation in terms of a finite union of parameterized convex sets can be constructed; ii) for each convex subset find a parameterized vertices interpretation; iii) build partial explicit solutions by reporting the unconstrained optimum to each parameterized polyhedron and finally compare them in order to obtain a global solution.

A crucial feature that any control system must have, especially in critical applications for modern society (e.g., aerospace, transport, power generation and distribution, etc.), is fault tolerant capabilities. By fault tolerant control systems, it is understood systems which continue to perform satisfactorily in the presence of deviations from a pre-specified mode of behaviour, special attention being given to actuators and sensors which are typical components prone to malfunctioning. It was shown that fault tolerant capabilities can be guaranteed by the construction of the invariant sets for the healthy/faulty behavior [15], [14]. The use of MPC and parameterized polyhedra in this framework can provide the maximal range of references which can be tracked despite faults by accordingly switching between sensors information.

An interesting research direction is provided by the class of hybrid system. The MPC leads in this case to multiparametric mixed integer optimisation problems. The parameterized polyhedra approach can be combined with the Voronoi diagrams to obtain analytic solutions and thus provide explicit control laws in terms of both continuous and switching control actions [13]. In the case of control actions in a finite alphabet, an interesting problem for systems subject to constraints is to preserve the system within a feasible region characterized by a rich family of control alternatives. These problems are related to the volume of lattice polyhedra and can use results such as Erhart polynomials.

References

1. Borelli, F.: Constrained Optimal Control of Linear and Hybrid Systems. Springer, Heidelberg (2003)
2. Goodwin, G.C., Seron, M.M., De Dona, J.A.: Constrained Control and Estimation. Springer, London (2004)
3. Kerrigan, E.C., Maciejowski, J.M.: Feedback min-max model predictive control using a single linear program: robust stability and the explicit solution. International Journal of Robust and Nonlinear Control 14(4), 395–413 (2004)
4. Olaru, S., Dumur, D.: A Parameterized Polyhedra Approach for Explicit Constrained Predictive Control. In: 43rd IEEE CDC, pp. 1580–1585 (2004)
5. Olaru, S., Dumur, D.: Avoiding Constraints Redundancy in Predictive Control Optimization Routines. IEEE Transactions on Automatic Control 50(9), 1459–1466 (2005)
6. Olaru, S., Dumur, D.: Compact explicit MPC with guarantee of feasibility for tracking. In: 44th IEEE Conference on Decision and Control, pp. 969–974 (2005)
7. Motzkin, T.S., Raiffa, H., Thompson, G.L., Thrall, R.M.: The Double Description Method (1953); republished in Motzkin, T.S.: Selected Papers. Birkhauser, Boston, 51–73 (1983)
8. Seron, M.M., Goodwin, G.C., De Dona, J.A.: Characterisation of Receding Horizon Control for Constrained Linear Systems. Asian J. of Control 5-2, 271–286 (2003)
9. Olaru, S., Dumur, D., Dobre, S.: On the geometry of predictive control with nonlinear constraints. In: Filipe, J., Ferrier, J.-L., Cetto, J.A., Carvalho, M. (eds.) Informatics in Control, Automation and Robotics IV, pp. 301–314. Springer, Heidelberg (2008)
10. Mayne, D.Q., Rawlings, J.B., Rao, C.V., Scockaert, P.O.M.: Constrained model predictive control: Stability and optimality. Automatica 36, 789–814 (2000)
11. Tondel, P., Johansen, T., Bemporad, A.: An algorithm for multi-parametric quadratic programming and explicit MPC solutions. In: Proc. 40th IEEE CDC, pp. 1199–1204 (2001)
12. Wilde, D.K.: A library for doing polyhedral operations. Technical report IRISA 785 (1993)
13. Olaru, S., Dumur, D., Thomas, J., Zainea, M.: Predictive control for hybrid systems. Implications of polyhedral pre-computations. Nonlinear Analysis: Hybrid Systems and Applications Elsevier 2(2), 510–531 (2008)
14. Olaru, S., De Dona, J.A., Seron, M.M.: Positive invariant sets for fault tolerant multisensor control schemes. In: Proceedings of the IFAC World Congress, Seoul, Korea, pp. 1224–1229 (2008)
15. Seron, M.M., Zhuo, X.W., De Doná, J.A., Martínez, J.J.: MultiSensor Switching Control Strategy with Fault Tolerance Guarantees. Automatica (2008)

Parallel Performance and Scalability Experiments with the Danish Eulerian Model on the EPCC Supercomputers

Tzvetan Ostromsky[1], Ivan Dimov[1], and Zahari Zlatev[2]

[1] Institute for Parallel Processing, Bulgarian Academy of Sciences,
Acad. G. Bonchev str., bl. 25-A, 1113 Sofia, Bulgaria
ceco@parallel.bas.bg, ivdimov@bas.bg
http://www.bas.bg/clpp/

[2] National Environmental Research Institute, Department of Atmospheric Environment, Frederiksborgvej 399 P. O. Box 358, DK-4000 Roskilde, Denmark
zz@dmu.dk
http://www.dmu.dk/AtmosphericEnvironment

Abstract. The Danish Eulerian Model (DEM) is a powerful air pollution model, designed to calculate the concentrations of various dangerous species over a large geographical region (e.g. Europe). It takes into account the main physical and chemical processes between these species, the actual meteorological conditions, emissions, etc. . This is a huge computational task and requires significant resources of storage and CPU time. Parallel computing is essential for the efficient practical use of the model. Some efficient parallel versions of the model were created over the past several years.

A suitable parallel version of DEM by using the Message Passing Interface library (MPI) was implemented on two powerful supercomputers of the EPCC - Edinburgh, available via the HPC-Europa programme for transnational access to research infrastructures in EC: a Sun Fire E15K and an IBM HPCx cluster. Although the implementation is in principal, the same for both supercomputers, few modifications had to be done for successful porting of the code on the IBM HPCx cluster. Performance analysis and parallel optimization was done next. Results from benchmarking experiments will be presented in this paper.

Another set of experiments was carried out in order to investigate the sensitivity of the model to variation of some chemical rate constants in the chemical submodel. Certain modifications of the code were necessary to be done in accordance with this task. The obtained results will be used for further sensitivity analysis studies by using Monte Carlo simulation.

1 Introduction

The problem for air pollution modelling has been studied for years [6,13]. An air pollution model is generally described by a system of partial differential equations for calculating the concentrations of a number of chemical species (pollutants and other components of the air that interact with the pollutants)

S. Margenov, L.G. Vulkov, and J. Waśniewski (Eds.): NAA 2008, LNCS 5434, pp. 447–453, 2009.

in a large 3-D domain (part of the atmosphere above the studied geographical region). The main physical and chemical processes (horizontal and vertical wind, diffusion, chemical reactions, emissions and deposition) should be adequately represented in the system.

The Danish Eulerian Model (DEM) [1,8,12,13,14] is mathematically represented by the following system of partial differential equations:

$$\begin{aligned}\frac{\partial c_s}{\partial t} &= -\frac{\partial(uc_s)}{\partial x} - \frac{\partial(vc_s)}{\partial y} - \frac{\partial(wc_s)}{\partial z} + \\ &+ \frac{\partial}{\partial x}\left(K_x\frac{\partial c_s}{\partial x}\right) + \frac{\partial}{\partial y}\left(K_y\frac{\partial c_s}{\partial y}\right) + \frac{\partial}{\partial z}\left(K_z\frac{\partial c_s}{\partial z}\right) + \\ &+ E_s + Q_s(c_1, c_2, \ldots c_q) - (k_{1s} + k_{2s})c_s, \quad s = 1, 2, \ldots q\,.\end{aligned} \tag{1}$$

where

- c_s – the concentrations of the chemical species;
- u, v, w – the wind components along the coordinate axes;
- K_x, K_y, K_z – diffusion coefficients;
- E_s – the emissions;
- k_{1s}, k_{2s} – dry / wet deposition coefficients;
- $Q_s(c_1, c_2, \ldots c_q)$ – non-linear functions describing the chemical reactions between species under consideration ([3]).

2 Splitting into Submodels

The above rather complex system (1) is split into 3 subsystems / submodels, according to the major physical / chemical processes and the numerical methods applied in their solution.

$$\frac{\partial c_s^{(1)}}{\partial t} = -\frac{\partial(uc_s^{(1)})}{\partial x} - \frac{\partial(vc_s^{(1)})}{\partial y} + \frac{\partial}{\partial x}\left(K_x\frac{\partial c_s^{(1)}}{\partial x}\right) + \frac{\partial}{\partial y}\left(K_y\frac{\partial c_s^{(1)}}{\partial y}\right) = A_1 c_s^{(1)}(t)$$

horizontal advection & diffusion

$$\frac{\partial c_s^{(2)}}{\partial t} = E_s + Q_s(c_1^{(2)}, c_2^{(2)}, \ldots c_q^{(2)}) - (k_{1s} + k_{2s})c_s^{(4)} = A_2 c_s^{(2)}(t)$$

chemistry, emissions & deposition

$$\frac{\partial c_s^{(3)}}{\partial t} = -\frac{\partial(wc_s^{(3)})}{\partial z} + \frac{\partial}{\partial z}\left(K_z\frac{\partial c_s^{(3)}}{\partial z}\right) = A_3 c_s^{(3)}(t)$$

vertical transport

Various splitting schemes have been proposed and analysed in [6,7]. The sequential splitting scheme is used in this version of DEM.

3 Parallelization Strategy and Numerical Methods, Used in the Solution of the Submodels

The splitting is an important step in the efficient numerical treatment of the model, but suitable well-balanced parallelization should be used for each of the submodels. After the discretization of the large computational domain each submodel becomes itself a huge computational task. Special attention should be payed to the chemical part. The dynamics of the chemical and photo-chemical processes requires using of small time-step to ensure stability of the computations.

A distributed memory parallelization strategy via MPI is used in the proposed implementation of the Danish Eulerian Model. It is based on decomposition of the spatial domain. Some previous work on this topic, very important from the performance viewpoint, can be found in [2,8,9,10,11]. The Message Passing Interface library (MPI) is described in more detail in [4]. For maximum portability only standard MPI routines are used in the proposed implementation of DEM.

Parallelization is based on domain decomposition of the horizontal grid, which implies certain restrictions on the number of MPI tasks and requires communication on each time step. Improving the data locality for more efficient cache utilization is achieved by using *chunks* to group properly the small tasks in the chemistry-deposition and vertical exchange stages. The value of the corresponding parameter `CHUNKSIZE` needs some tuning for the target parallel system.

Additional **pre-processing** and **post-processing** stages are needed for scattering the input data and gathering the results. These are cheap, but their relative weight grows up with increasing the number of MPI tasks, affecting the total *speed-up*[1] and scalability.

The numerical methods, used in the solution of the submodels, are given below.

- **Advection-diffusion part:** Finite elements, followed by predictor-corrector schemes with several different correctors.
- **Chemistry-deposition part:** An improved version of the QSSA (Quazi Steady-State Approximation) - [5].
- **Vertical transport:** Finite elements, followed by theta-methods.

4 Scalability Experiments

In this section we present some tables with time and speed-up results, showing the scalability of the MPI implementation of the Danish Eulerian Model on the EPCC supercomputers. The time and the **speed-up (Sp)** of the main computational stages of the model, as well as in total, are given in separate columns. The total time includes also the times for pre-processing and post-processing

[1] The ratio between the execution times of an algorithm in parallel (on n processors) and in sequential (on one processor) of the same machine, is called *speed-up* on n processors for the corresponding algorithm and machine and is denoted by $Sp(n)$ throughout this paper.

stages and some data transfer procedures, which are either inherently sequential or cannot be efficiently parallelized. In addition, the total *efficiency* E (in percent) is given in the last column, where $E = Sp(n)/n\,.100\%$, where n is the number of processors (given in the first column).

The Sunfire E15000 system consists of 52 Ultrasparc III processors (900 MHz, 2 level cache). 4 of them are "front-end" processors for interactive use, the rest 48 are "back-end" processors, with 48 GBytes shared memory in total. The results of one-year experiments with the 3-D version with medium resolution spatial grid (96 x 96), are presented in Table 1 below.

The IBM high-performance cluster "HPCx" is a huge system, based on the IBM Power5 processor (64-bit RISC processor, frequency 1.5 GHz). The cluster is built of 160 IBM P5 eServer 575 LPAR nodes (each with 16 IBM Power5 processors, 32 GByte RAM, 3-level cache). They communicate via IBM's High Performance

Table 1. Time in seconds and the **speed-up** (given in brackets) of the 3-D MPI version of DEM for running one-year experiments on the SUN cluster ("Lomond") at EPCC.

3-D MPI version of DEM on a Sunfire E15000 ($96 \times 96 \times 10$) grid, 35 species, CHUNKSIZE=32									
# proc.	Advection – diffusion		Chemistry and emissions		Vertical transport		TOTAL		
	Time [s]	**(Sp)**	Time [s]	**(Sp)**	Time [s]	**(Sp)**	Time [s]	**(Sp)**	E [%]
1	4180	—	15685	—	1500	—	21590	—	—
2	2155	**(1.9)**	7882	**(2.0)**	744	**(2.0)**	11058	**(2.0)**	98%
4	1037	**(4.0)**	3924	**(4.0)**	339	**(4.4)**	5453	**(4.0)**	99%
8	509	**(8.2)**	1976	**(7.9)**	165	**(9.1)**	2728	**(7.9)**	99%
16	265	**(15.8)**	975	**(16.1)**	86	**(17.4)**	1377	**(15.7)**	98%
24	187	**(22.4)**	648	**(24.2)**	58	**(25.9)**	920	**(23.5)**	98%
32	139	**(30.1)**	476	**(31.7)**	47	**(31.0)**	702	**(30.8)**	96%
48	127	**(32.9)**	317	**(47.5)**	32	**(46.9)**	690	**(31.3)**	65%

Table 2. Time in seconds and the **speed-up** (given in brackets) of the 3-D MPI version of DEM for running one-year experiments on the IBM high-performance cluster "HPCx"

3-D MPI version of DEM on the IBM "HPCx" ($96 \times 96 \times 10$) grid, 35 species, CHUNKSIZE=32									
# proc.	Advection – diffusion		Chemistry and emissions		Vertical transport		TOTAL		
	Time [s]	**(Sp)**	Time [s]	**(Sp)**	Time [s]	**(Sp)**	Time [s]	**(Sp)**	E [%]
1	958	—	8216	—	385	—	9598	—	—
4	246	**(3.9)**	2064	**(4.0)**	97	**(4.0)**	2438	**(3.9)**	98%
8	136	**(7.0)**	1035	**(7.9)**	48	**(8.0)**	1244	**(7.7)**	96%
16	76	**(12.6)**	518	**(15.9)**	25	**(15.8)**	637	**(15.1)**	94%
24	50	**(19.2)**	345	**(23.8)**	16	**(24.0)**	441	**(21.8)**	91%
32	43	**(22.3)**	259	**(31.7)**	12	**(32.0)**	339	**(28.3)**	88%
48	34	**(28.2)**	173	**(47.5)**	8	**(48.1)**	242	**(39.7)**	83%

Switch. This machine, located in Daresbury (England), is one of the most powerful supercomputers (second in Europe and 12-th in the world) with a total of 2560 processors, peak performance 15.36 TFlop/s, 5.12 TByte RAM and 72 TByte disk storage. The results of one-year experiments with the 3-D version with medium resolution spatial grid (96 x 96), are presented in Table 2 below.

5 Sensitivity Analysis Experiments

Chemistry is the most difficult and computationally intensive part of the model, as can be seen in Tables 1,2 in the previous section. It is described as a system of non-linear equations between the unknown concentrations of the chemical species, some other parameters (e.g. temperature, light, etc.) and constants. The condensed Carbon Bond-IV Mechanism (CBM IV) is used in DEM [13]. Each chemical rate constant determines the intensity of certain chemical process. For the purpose of the sensitivity analysis it can be considered as a random variable. The first stage of such study should determine the most important chemical rate constants with respect to a given criterion. Ozone is known to be one of the most dangerous pollutants in the air. That is why the mean monthly ozone concentration was chosen as a primary criterion. By extensive experiments with perturbation of a large number of coefficients of the suspicious chemical reactions, we were trying to determine the most critical of them. The results of these experiments revealed the chemical reaction 22 ($NO + OH \rightarrow HNO_2$) as one of the most critical. The sensitivity of five important pollutants to its rate constant perturbation is shown in Fig. 1 .

The results are yet to be analysed and furtherly used in sensitivity analysis study of DEM by using Monte Carlo methods.

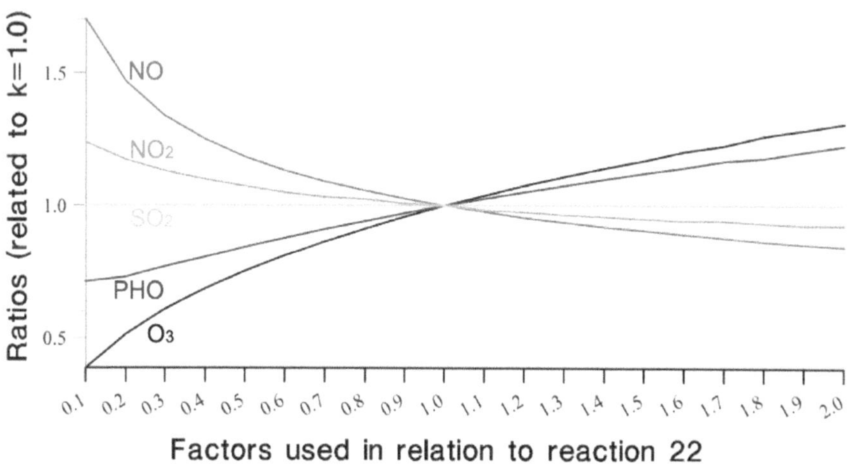

Fig. 1. Sensitivity of the mean monthly concentrations of 5 chemical species (nitrogen oxide (NO), nitrogen dioxide (NO_2), sulphur dioxide (SO_2), peroxy radical (PHO), ozone (O_3)) to the rate of the chemical reaction 22 (July 1998)

6 Conclusions and Plans for Future Work

- The parallel MPI implementation of DEM, based on horizontal partitioning of the domain, is well balanced and proved its efficiency on some of the most powerful parallel supercomputers in Europe.
- When using the full capacity of the SUN cluster "Lomond" (all its 48 back-end processors), a significant breakdown in the efficiency (to about 65% from more than 95% for up to 32 processors) is observed. The bottleneck appeared to be the communication network, which shows certain slowdown in its work on full load.
- The efficiency and speed-up is relatively higher in the computationally-intensive stages. In particular, the chemistry stage has almost linear speed-up (and it takes up to 85% of the total time on the "HPCx").
- The results of the sensitivity analysis experiments with perturbation of the reaction coefficients will be processed by using Monte Carlo methods. This would possibly lead to improvements in the chemical scheme of the model as well as to its more efficient use in some time-critical applications.

Acknowledgments

This research was supported in part by the Bulgarian IST Centre of Competence in 21 Century – BIS-21++ (contract # INCO-CT-2005-016639), by the NATO project "Monte Carlo Sensitivity Studies of Environmental Security" (PDD(TC)-ESP.EAP.CLG 982641), and by the HPC-Europa project (RII3-CT-2003-506079) with the support of the European Community - Research Infrastructure Action under the FP6 "Structuring the European Research Area" Programme.

References

1. Alexandrov, V., Sameh, A., Siddique, Y., Zlatev, Z.: Numerical integration of chemical ODE problems arising in air pollution models. Env. Modeling and Assessment 2, 365–377 (1997)
2. Dimov, I., Georgiev, K., Ostromsky, T., Zlatev, Z.: Computational challenges in the numerical treatment of large air pollution models. Ecological Modelling 179, 187–203 (2004)
3. Gery, M.W., Whitten, G.Z., Killus, J.P., Dodge, M.C.: A photochemical kinetics mechanism for urban and regional modeling. J. Geophys. Res. 94, 12925–12956 (1989)
4. Gropp, W., Lusk, E., Skjellum, A.: Using MPI: Portable programming with the message passing interface. MIT Press, Cambridge (1994)
5. Hesstvedt, E., Hovand, Ø., Isaksen, I.A.: Quasi-steady-state approximations in air pollution modeling: comparison of two numerical schemes for oxidant prediction. Int. Journal of Chemical Kinetics 10, 971–994 (1978)
6. Marchuk, G.I.: Mathematical modeling for the problem of the environment. Studies in Mathematics and Applications, vol. 16. North-Holland, Amsterdam (1985)
7. McRae, G.J., Goodin, W.R., Seinfeld, J.H.: Numerical solution of the atmospheric diffusion equations for chemically reacting flows. J. Comp. Physics 45, 1–42 (1984)

8. Ostromsky, T., Owczarz, W., Zlatev, Z.: Computational Challenges in Large-scale Air Pollution Modelling. In: Proc. 2001 International Conference on Supercomputing in Sorrento, pp. 407–418. ACM Press, New York (2001)
9. Ostromsky, T., Zlatev, Z.: Parallel Implementation of a Large-scale 3-D Air Pollution Model. In: Margenov, S., Waśniewski, J., Yalamov, P. (eds.) LSSC 2001. LNCS, vol. 2179, pp. 309–316. Springer, Heidelberg (2001)
10. Ostromsky, T., Zlatev, Z.: Flexible two-level parallel implementations of a large air pollution model. In: Dimov, I.T., Lirkov, I., Margenov, S., Zlatev, Z. (eds.) NMA 2002. LNCS, vol. 2542, pp. 545–554. Springer, Heidelberg (2003)
11. Ostromsky, T., Zlatev, Z.: Parallel and GRID implementation of a large scale air pollution model. In: Boyanov, T., Dimova, S., Georgiev, K., Nikolov, G. (eds.) NMA 2006. LNCS, vol. 4310, pp. 475–482. Springer, Heidelberg (2007)
12. WEB-site of the Danish Eulerian Model, `http://www.dmu.dk/AtmosphericEnvironment/DEM`
13. Zlatev, Z.: Computer treatment of large air pollution models. Kluwer, Dordrecht (1995)
14. Zlatev, Z., Dimov, I., Georgiev, K.: Three-dimensional version of the Danish Eulerian Model. Zeitschrift für Angewandte Mathematik und Mechanik 76, 473–476 (1996)

LAPACK-Based Condition Estimates for the Discrete-Time LQG Design

P.Hr. Petkov[1], M.M. Konstantinov[2], and N.D. Christov[3]

[1] Technical University of Sofia, 1000 Sofia, Bulgaria
[2] University of Architecture, Civil Engineering and Geodesy, 1046 Sofia, Bulgaria
[3] Université des Sciences et Technologies de Lille, 59655 Villeneuve d'Ascq, France

Abstract. The Linear-Quadratic Gaussian (LQG) design is the most efficient and widely used design approach in the field of linear stochastic control systems. From theoretical point of view this approach is reduced to the synthesis of a LQ state regulator and of a Kalman filter for the controlled system. From computational point of view the LQG design consists of solving a pair of matrix Riccati equations: one for the LQ regulator design and a second one (dual to the first Riccati equation) for the Kalman filter design. In this paper we present reliable algorithms for estimation of condition numbers of the discrete Riccati equations in the discrete-time LQG design. Efficient LAPACK-based condition estimators are proposed involving the solution of triangular Lyapunov equations along with one-norm computation.

1 Introduction

Consider the discrete-time linear stochastic system

$$\begin{aligned} x_{i+1} &= Ax_i + Bu_i + v_i,\ i \geq 0 \\ y_i &= Cx_i + w_i \end{aligned} \tag{1}$$

where $x_i \in \mathcal{R}^n$, $u_i \in \mathcal{R}^m$, $y_i \in \mathcal{R}^r$ are the system state, input and output vectors, respectively, $A \in \mathcal{R}^{n\times n}$, $B \in \mathcal{R}^{n\times m}$, $C \in \mathcal{R}^{r\times n}$ are known constant matrices, and $\{v_i\}$ and $\{w_i\}$ are independent zero-mean Gaussian white-noise sequences with variance matrices $V \geq 0$ and $W > 0$, respectively. It is supposed that the pairs (A, B), $(A, V^{1/2})$ are stabilizable and the pair (A, C) is detectable.

The discrete-time LQG design problem consists to find for system (1) a control u that minimizes the quadratic performance index

$$J(u) = \lim_{N\to\infty} \mathcal{E}\left\{ \frac{1}{N}\sum_{i=0}^{N} (x_i^T Q x_i + u_i^T R u_i) \right\} \tag{2}$$

having partial knowledge of the system state x_i via the output vector y_i. It is assumed that $Q \geq 0$, $R > 0$ and the pair $(A, Q^{1/2})$ is detectable.

S. Margenov, L.G. Vulkov, and J. Waśniewski (Eds.): NAA 2008, LNCS 5434, pp. 454–460, 2009.

As it is well known the LQG design problems split into two parts: an optimal estimation of the system state and a linear quadratic regulator problem [1]. For the discrete-time LQG problem (1), (2), the optimal control u^* is given by

$$u_i^* = -(R + B^T X B)^{-1} B^T X A \hat{x}_i \tag{3}$$

where $\hat{x}_i$ is the optimal estimate of x_i obtained using the Kalman filter

$$\hat{x}_{i+1} = A\hat{x}_i + Bu_i + AYC^T(W + CYC^T)^{-1}(y_i - C\hat{x}_i). \tag{4}$$

Here X and Y are the unique non-negative definite solutions of the dual discrete matrix Riccati equations

$$A^T X A - X + Q - A^T X B(R + B^T X B)^{-1} B^T X A = 0 \tag{5}$$

and

$$AYA^T - Y + V - AYC^T(W + CYC^T)^{-1} CYA^T = 0. \tag{6}$$

The numerical solution of the Riccati equations (5), (6) may face some difficulties. First, the corresponding equation may be ill conditioned, i.e. small perturbations in the coefficient matrices may lead to large variations in the solution. The second difficulty is connected with the stability of the numerical method and the reliability of its implementation. It is well known [2] that the methods for solving the Riccati equations are generally unstable. Therefore, it is very useful to have easily computable estimates of the equation conditioning and error bounds for the solutions.

This paper deals with the efficient and reliable condition estimation of the discrete matrix Riccati equations (5), (6). The estimates implemented involve the solution of triangular Lyapunov equations along with usage of the LAPACK norm estimator.

In what follows we shall consider the conditioning estimation for the discrete matrix Riccati equation (5). The corresponding results for the Riccati equation (6) can be obtained using the duality of (5) and (6). In the sequel we shall write equation (5) as

$$X = Q + A^T X(I_n + SX)^{-1} A \tag{7}$$

where $S = BR^{-1}B^T$.

2 Conditioning of Discrete Riccati Equations

Suppose that the matrices A, Q, S in (7) are subject to perturbations ΔA, ΔQ, ΔS, so that instead of the initial data we have the matrices

$$\tilde{A} = A + \Delta A, \quad \tilde{Q} = Q + \Delta Q, \quad \tilde{S} = S + \Delta S.$$

It is supposed that the perturbations preserve the symmetric structure of the equation, i.e. the perturbations ΔQ and ΔS are symmetric.

Let $\tilde{X} = X + \Delta X$ be a solution obtained by replacing A, Q, S in (7) with the matrices $\tilde{A}$, $\tilde{Q}$, $\tilde{S}$. If $\|\Delta A\|$, $\|\Delta Q\|$ and $\|\Delta S\|$ are sufficiently small, then the perturbed solution $\tilde{X} = X + \Delta X$ is well defined.

The perturbation analysis of (7) is aimed at studying the variation ΔX in the solution $\tilde{X} = X + \Delta X$ due to the perturbations ΔA, ΔQ, ΔS. If small perturbations in the data lead to small variations in the solution, the corresponding equation is said to be well-conditioned and if these perturbations lead to large variations in the solution this equation is ill-conditioned.

The condition number of the Riccati equation (7) is defined as

$$K_R = \lim_{\delta \to 0} \sup \left\{ \frac{\|\Delta X\|}{\delta \|X\|} : \|\Delta A\| \leq \delta \|A\|,\ \|\Delta Q\| \leq \delta \|Q\|,\ \|\Delta S\| \leq \delta \|S\| \right\}.$$

For sufficiently small δ we have (within first order terms)

$$\frac{\|\Delta X\|}{\|X\|} \leq K_R \delta.$$

Denote by $\bar{X}$ the solution of the Riccati equation computed by a numerical method in finite arithmetic with relative precision ε. If the method is backward stable, the relative error in the solution can be estimated by

$$\frac{\|\bar{X} - X\|}{\|X\|} \leq p(n) K_R \varepsilon$$

where $p(n)$ is some low-order polynomial of n. This shows the importance of the condition numbers in the numerical solution of Riccati equations.

The computation of the exact value of K_R requires the construction and manipulation of $n^2 \times n^2$ matrices which is not practical for large n. That is why it is useful to derive approximations of K_R that can be obtained inexpensively.

In first order approximation ΔX can be represented as

$$\Delta X = -\Omega^{-1}(\Delta Q) - \Theta(\Delta A) + \Pi(\Delta S) \tag{8}$$

where

$$\begin{aligned} \Omega(Z) &= A_c^T Z A_c - Z \\ \Theta(Z) &= \Omega^{-1}(Z^T X A_c + A_c^T X Z) \\ \Pi(Z) &= \Omega^{-1}(A_c^T X Z X A_c) \end{aligned} \tag{9}$$

are linear operators in the space of $n \times n$ matrices, which determine the sensitivity of X with respect to the perturbations in Q, A, S, respectively, and $A_c = (I_n + SX)^{-1} A$.

Based on (8), an approximate condition number can be defined as [3]

$$K_B := \frac{\|\Omega^{-1}\| \|Q\| + \|\Theta\| \|A\| + \|\Pi\| \|S\|}{\|X\|} \tag{10}$$

where

$$\|\Omega^{-1}\| = \max_{Z\neq 0} \frac{\|\Omega^{-1}(Z)\|}{\|Z\|}$$

$$\|\Theta\| = \max_{Z\neq 0} \frac{\|\Theta(Z)\|}{\|Z\|}$$

$$\|\Pi\| = \max_{Z\neq 0} \frac{\|\Pi(Z)\|}{\|Z\|}$$

are the corresponding induced operator norms.

It follows that the Riccati equation (7) is ill-conditioned if $\|\Omega^{-1}\|$ is large, that is, if the matrix A_c is almost unstable. Note that Θ and Π depend on X.

3 Condition Estimates

An important practical issue is how to inexpensively estimate the quantities $\|\Omega^{-1}\|$, $\|\Theta\|$ and $\|\Pi\|$ in the condition number (10).

The spectral norm of the inverse operator Ω^{-1} can be written as

$$\|\Omega^{-1}\| = \max_{Z\neq 0} \frac{\|Z\|}{\|A_c^T Z A_c - Z\|} = \frac{1}{\mathrm{sep}(A_c^T, -A_c)}$$

where

$$\mathrm{sep}(A_c^T, -A_c) := \min_{Z\neq 0} \frac{\|A_c^T Z A_c - Z\|_F}{\|Z\|} = \sigma_{\min}(A_c^T \otimes A_c - I_{n^2})$$

is the separation between the matrices A_c^T and $-A_c$. The evaluation of $\|\Omega^{-1}\|$ by computing the minimal singular value $\sigma_{\min}(A_c^T \otimes A_c - I_{n^2})$ is not practical for medium-size or large-scale problems, due to the excessive computational effort and memory requirements.

In practice, an estimate of the one-norm of Ω^{-1} is efficiently obtained using the one-norm estimator [4], which estimates the norm $\|T\|_1$ of a linear operator T by evaluating this operator and its dual on suitably selected values in the operator domain. Formally, one has to compute Tv and $T^T w$, where v and w are chosen by a special optimization algorithm. This estimator is implemented in the LAPACK subroutine `xLACON` [5] which is called via a reverse communication interface providing the products Tv and $T^T w$.

The use of xLACON to estimate $\|\Omega^{-1}\|_1$ means to solve the linear equations

$$Py = v, \ P^T z = v$$

with $P = A_c^T \otimes A_c - I_{n^2}$ and v being determined by `xLACON`. This is equivalent to the solution of the discrete Lyapunov equations

$$\begin{aligned} A_c^T Y A_c - Y &= V \\ A_c Z A_c^T - Z &= V \end{aligned} \tag{11}$$

where $\text{vec}(V) = v$, $\text{vec}(Y) = y$, $\text{vec}(Z) = z$. The solution of these equations can be obtained in a numerically reliable way using the discrete counterpart of the Bartels-Stewart algorithm, proposed by Barraud [6,2].

An estimate of $\|\Theta\|_1$ can be obtained in a similar way by solving the Lyapunov equations

$$\begin{aligned} A_c^T Y A_c - Y &= V^T X A_c + A_c^T X V \\ A_c Z A_c^T - Z &= V^T X A_c + A_c^T X V \end{aligned} \tag{12}$$

To estimate $\|\Pi\|_1$ by xLACON, it is necessary to solve the equations

$$\begin{aligned} A_c^T Y A_c - Y &= A_c^T X V X A_c \\ A_c Z A_c^T - Z &= A_c^T X V X A_c. \end{aligned} \tag{13}$$

The accuracy of the estimates that we obtain via this approach depends on the ability of `xLACON` to find a right-hand side vector v which maximizes the ratios

$$\frac{\|y\|}{\|v\|}, \frac{\|z\|}{\|v\|}$$

when solving the equations $Py = v$, $P^T z = v$. As in the case of other condition estimators it is always possible to construct special examples where the value produced by `xLACON` underestimates the true value of the corresponding norm by an arbitrary factor. Note, however, that this may happens in rare circumstances.

The software implementation of the proposed Riccati equation condition estimator for is based entirely on LAPACK and BLAS [7,8] subroutines.

4 Numerical Example

In this section we present an example that demonstrates the performance of the estimates implemented in the solution of families of discrete Riccati equations whose conditioning vary very much. All computations were carried out on a Sun workstation with relative machine precision $\varepsilon = 2.22 \times 10^{-16}$.

In order to have a closed form solution, the matrices in the Riccati equations are chosen as

$$A = TA_0T^{-1}, \quad Q = T^{-T}Q_0T^{-1}, \quad S = TS_0T^T$$

where A_0, Q_0, S_0 are diagonal matrices and T is a nonsingular transformation matrix. The solution is then given by $X = T^{-T}X_0T^{-1}$, where X_0 is a diagonal matrix whose elements are determined simply from the elements of A_0, Q_0, S_0. To avoid large rounding errors in constructing and inverting T, this matrix is chosen as $T = H_2DH_1$, where H_1 and H_2 are elementary reflectors and D is a diagonal matrix,

$$\begin{aligned} &H_1 = I_n - 2ee^T/n, e = [1, 1, ..., 1]^T \\ &H_2 = I_n - 2ff^T/n, f = [1, -1, 1, ..., (-1)^{n-1}]^T \\ &D = \text{diag}(1, s, s^2, ..., s^{n-1}), s > 1. \end{aligned}$$

Using different values of the scalar s, it is possible to change the condition number of the matrix T with respect to inversion, $\mathrm{cond}_2(T) = s^{n-1}$.

Consider a family of discrete Riccati equations whose matrices A_0, Q_0, S_0 are chosen as

$$A_0 = \mathrm{diag}(A_1, A_1), \quad Q_0 = \mathrm{diag}(Q_1, Q_1), \quad S_0 = \mathrm{diag}(S_1, S_1),$$

where

$$A_1 = \mathrm{diag}(0,\ 1,\ 2)$$
$$Q_1 = \mathrm{diag}(10^k,\ 1,\ 10^{-k})$$
$$S_1 = \mathrm{diag}(10^{-k},\ 10^{-2k},\ 10^{-k}).$$

The conditioning of these equations deteriorates with the increase of k and s.

In Figure 1 we show the ratio of the condition number estimate and the exact condition number as functions of k and s.

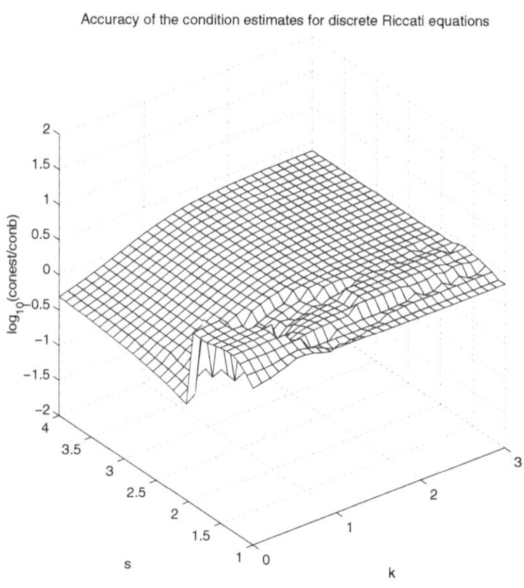

Fig. 1. Accuracy of the condition number estimate for a family of discrete Riccati equations

5 Conclusions

Efficient and reliable condition estimates are presented for the discrete matrix Riccati equations arising in the discrete-time LQG design. The estimates implemented involve the solution of discrete Lyapunov equations along with usage of the LAPACK norm estimator. The numerical experiments show that the condition estimates are always of the same order as the true condition numbers.

References

1. Anderson, B.D.O., Moore, J.B.: Optimal Control: Linear Quadratic Methods. Prentice-Hall International, London (1989)
2. Petkov, P.H., Christov, N.D., Konstantinov, M.M.: Computational Methods for Linear Control Systems. Prentice-Hall, N.Y (1991)
3. Byers, R.: Numerical condition of the algebraic Riccati equation. Contemp. Math. 47, 35–49 (1985)
4. Higham, N.J.: FORTRAN codes for estimating the one-norm of a real or complex matrix, with applications to condition estimation (Algorithm 674). ACM Trans. Math. Software 14, 381–396 (1988)
5. Anderson, E., Bai, Z., Bischof, C., Demmel, J., Dongarra, J., Croz, J.D., Greenbaum, A., Hammarling, S., McKenney, A., Ostrouchov, S., Sorensen, D.: LAPACK Users' Guide, 2nd edn. SIAM, Philadelphia (1995)
6. Barraud, A.Y.: A numerical algorithm to solve $A^TXA - X = Q$. IEEE Trans. Autom. Control 22, 883–885 (1977)
7. Lawson, C.L., Hanson, R.J., Kincaid, D.R., Krogh, F.T.: Basic Linear Algebra Subprograms for FORTRAN usage. ACM Trans. Math. Software 5, 308–323 (1979)
8. Dongarra, J.J., Du Croz, J., Duff, I., Hammarling, S.: A set of Level 3 Basic Linear Algebra Subprograms. ACM Trans. Math. Software 16, 1–17 (1990)

Smooth and Nonsmooth Solutions of Several Equations of Mathematical Physics and Their Cellular Neural Network Realization

P. Popivanov and A. Slavova

Institute of Mathematics
Bulgarian Academy of Sciences
Sofia 1113, Bulgaria
{popivano,slavova}@math.bas.bg

Abstract. This paper deals with travelling wave solutions of several equations of mathematical physics (Camassa-Holm and its generalizations, etc.) Under several conditions we propose a full classification of the corresponding solutions which can be smooth or nonsmooth at isolated points. This way we obtain peakons, compactons, cuspons, solitons, periodic waves. At the end of the paper Cellular Neural Networks realization is given.

1 Introduction and Formulation of the Main Results

1. The well known Camassa-Holm equation

$$u_t - u_{xxt} + 3uu_x = 2u_x u_{xx} + uu_{xxx}, t > 0, x \in \mathbf{R} \tag{1}$$

models the propagation of water waves in the shallow water regime when the wave length is considerably larger than the average water depth [1]. Recently the following Camassa-Holm - γ (CH- γ) equation was studied by many authors [2],[3]:

$$u_t + c_0 u_x + 3uu_x - \alpha^2(u_{xxt} + uu_{xxx} + 2u_x u_{xx}) + \gamma u_{xxx} = 0, \tag{2}$$

where α^2, c_0 and γ are real-valued parameters.

Rosenau has considered several generalizations of the Korteweg-de Vries (KdV) equation [4],[5] and has shown that they possess a new form of solitary wave with compact support. This is for example the equation

$$u_t + (u^m)_x + (u^n)_{xx} = 0, \tag{3}$$

where m, n are parameters.

A different generalization of the KdV equation was studied in [6], namely:

$$u_t = u_x u^{l-2} + \alpha[2u_{xxx}u^p + 4pu^{p-1}u_x u_{xx} + p(p-1)u^{p-1}(u_x)^3], \tag{4}$$

where p, l, α are constants.

S. Margenov, L.G. Vulkov, and J. Waśniewski (Eds.): NAA 2008, LNCS 5434, pp. 461–469, 2009.

In our paper [8] we investigated the following generalization of Camassa-Holm equation:

$$u_t + K(u^m)_x - (u^n)_{xxt} = [\frac{((u^n)_x)^2}{2} + u^n(u^n)_{xx}]_x, \tag{5}$$

where $K = const > 0$ and $m, n \in \mathbf{N}$.

As we show in [8] smooth compact travelling wave solutions exist for the equation describing the vibrations of a chain of particles interconnected by springs:

$$u_{tt} + \Phi'(u) = (T(u_x))_x, \tag{6}$$

where $T(p) = p + p^\alpha$, $\alpha > 1$.

Recently the cusped solutions of the CH equation, the cuspon solitary waves and the antipeakon limits were considered in [7]. Therefore, there are a lot of papers devoted to the existence of travelling wave solutions of several third order nonlinear PDE of mathematical physics. They can be smooth or they can develop singularities of the following two types: peaks, cusps.

To be more precise, the travelling wave solutions $u = \varphi(x - ct)$, $c = const > 0$ of the above mentioned equations satisfy ODE with separate variables and of special type.

In the case of CH- γ equation (2) the corresponding ODE is:

$$(\varphi')^2(\gamma + c\alpha^2 - \alpha^2\varphi) = -\varphi^3 - (c_0 - c)\varphi^2 - C_1\varphi - C_2, \tag{7}$$

where C_1, C_2 are arbitrary constants; $\varphi - \varphi(\xi)$, $\xi = x - ct$. As it concerns (4), we have the ODE:

$$\alpha\varphi^p(\varphi')^2 + \frac{\varphi^l}{l(l-1)} - \frac{c}{2}\varphi^2 + C_1\varphi + C_2 = 0, \tag{8}$$

where $l \neq 0, 1$ and C_1, C_2 are arbitrary constants.

The generalized CH equation (5) is reduced to the following ODE:

$$(\varphi')^2(c - \varphi^n) = c_1\varphi^{3-n} - K_1\varphi^{m+2-n} + C_1\varphi^{2-n} + C_2\varphi^{2-2n}, \tag{9}$$

where $c_1 = \frac{2c}{n^2+n}$, $K_1 = \frac{2K}{m(m+n)}$ and C_1, C_2 are arbitrary constants.

All these equations can be generalized in a simple way. Thus, consider the nonlinear third order PDE of 2 independent variables:

$$F(u, u_t, u_x, u_{tt}, u_{tx}, \ldots, u_{xxx}) = 0, u = u(t, x). \tag{10}$$

Putting $u = \varphi(x - ct)$, $c \neq 0$, $\varphi = \varphi(\xi)$, $\xi = x - ct$ we get:

$$A(\varphi) = F(\varphi, -c\varphi', \varphi', c^2\varphi'', -c\varphi'', \ldots, \varphi''') = 0.$$

We shall assume further on that for some $c \neq 0$

$$A(\varphi) = \frac{d}{d\xi}G_c(\varphi, \varphi', \varphi''), \tag{11}$$

where G_c is a smooth function depending on c.

Thus, $G_c(\varphi,\varphi^{'},\varphi^{''}) = C_1 = const$. Multiplying G_c by $\varphi^{'}$ we get:

$$\varphi^{'} G_c(\varphi,\varphi^{'},\varphi^{''}) = C_1\varphi^{'}. \tag{12}$$

Our main assumption is the following one:

$$\varphi^{'} G_c(\varphi,\varphi^{'},\varphi^{''}) = \frac{d}{d\xi}(P_1(\varphi)(\varphi^{'})^2 + P_2(\varphi)), \tag{13}$$

where $P_1(\varphi)$, $P_1(\varphi)$ are polynomials of φ.

So we obtain that

$$P_1(\varphi)(\varphi^{'})^2 + P_2(\varphi) = C_1\varphi + C_2, \tag{14}$$

where C_1, C_2 are arbitrary constants.

A glance at (7)-(9) shows that in the case l, p, m, n - integers these equations are of the form (14) for appropriate P_1 and P_2.

2. This way we come to the necessity of studying of the following Cauchy problem:

$$\begin{cases} Q(y)(y^{'})^2 = P(y), y = y(x) \\ y(x_0) = y_0, \end{cases} \tag{15}$$

where $P(y)$, $Q(y)$ are polynomials of y, $Q(y)P(y) > 0$, $\forall y \in (\alpha,\beta)$ and without loss of generality $P(y) > 0$, $Q(y) > 0$ for each $y \in (\alpha,\beta)$, $\alpha < \beta$. Eventually $P(\alpha)Q(\alpha) = 0$, $P(\beta)Q(\beta) = 0$.

We suppose that:

$$\left| \begin{array}{l} P(y) = c_1(y-\alpha)^{m_1}(1+o(1)), m_1 \geq 0, c_1 > 0, y > \alpha, y \to \alpha \\ Q(y) = c_2(y-\alpha)^{m_2}(1+o(1)), m_2 \geq 0, c_2 > 0, y > \alpha, y \to \alpha \end{array} \right. \tag{16}$$

$$\left| \begin{array}{l} P(y) = c_3(\beta-y)^{n_1}(1+o(1)), n_1 \geq 0, c_3 > 0, y < \beta, y \to \beta \\ Q(y) = c_4(\beta-y)^{n_2}(1+o(1)), n_2 \geq 0, c_4 > 0, y < \beta, y \to \beta \end{array} \right. \tag{17}$$

We will concentrate on the case:

$$\left| \begin{array}{l} y^{'} = \sqrt{\frac{P(y)}{Q(y)}} \\ y(x_0) = y_0 \in (\alpha,\beta), \text{i.e. } y \in (\alpha,\beta) \end{array} \right. \tag{18}$$

and consequently, $y(x)$ is strictly monotonically increasing solution of (15). Certainly, we will use in many cases even continuation of the solution and when it is possible - periodic continuation on the real line $\mathbf{R}^1_x$. The continuations satisfy (15). They are continuous but non smooth at the peak and cusp points.

Denote now

$$F(y) = \int_{y_0}^{y} \sqrt{\frac{Q(\lambda)}{P(\lambda)}} d\lambda, \alpha < y < \beta.$$

Evidently $F^{'}(y) > 0$, $F(y_0) = 0$, $F(y) < 0$ for $\alpha < y < y_0$ and $F(y) > 0$ for $\beta > y > y_0$. Certainly, $x - x_0 = F(y)$, $y = y(x)$.

There are 4 different cases concerning the behavior of $F(y)$ for $y \to \alpha$ and $y \to \beta$.

Thus,

$$(A) \quad 1)\ lim_{y\to\alpha}F(y) = \tilde{c} < 0 \iff \frac{m_2 - m_1}{2} + 1 > 0$$

$$(A) \quad 2)\ lim_{y\to\alpha}F(y) = -\infty \iff \frac{m_2 - m_1}{2} + 1 \leq 0$$

$$(A) \quad 3)\ lim_{y\to\beta}F(y) = d > 0 \iff \frac{n_2 - n_1}{2} + 1 > 0$$

$$(A) \quad 4)\ lim_{y\to\beta}F(y) = \infty \iff \frac{n_2 - n_1}{2} + 1 \leq 0.$$

In a natural way we come to the following subcases of the cases (A) 1)- (A) 4):

$$1a)\ m_2 - m_1 > 0 \Rightarrow F^{'}(\alpha) = 0$$

$$1b)\ m_2 - m_1 = 0 \Rightarrow F^{'}(\alpha) > 0$$

$$1c)\ -2 < m_2 - m_1 < 0 \Rightarrow F^{'}(\alpha) = +\infty$$

$$2)\ m_2 - m_1 \leq -2 \Rightarrow F^{'}(\alpha) = +\infty$$

$$3d)\ n_2 - n_1 > 0 \Rightarrow F^{'}(\beta) = 0$$

$$3e)\ n_2 = n_1 \Rightarrow F^{'}(\beta) > 0$$

$$3f)\ -2 < n_2 - n_1 < 0 \Rightarrow F^{'}(\beta) = +\infty$$

$$4)\ n_2 - n_1 \leq -2 \Rightarrow F^{'}(\beta) = +\infty.$$

This is the main Theorem 1 of our paper.

Theorem 1

(i) Suppose that (A) 1), (A) 3) hold and denote by $\frac{T}{2} = \int_\alpha^\beta \sqrt{\frac{Q(\lambda)}{P(\lambda)}}d\lambda > 0$. *Then (15) possesses a strictly monotonically increasing solution for* $x \in [0, \frac{T}{2}]$. *Moreover, in the case (A) 1)* $y \approx \alpha + [(\frac{m_2-m_1}{2}+1)\sqrt{\frac{c_1}{c_2}}x]^{\frac{1}{1+\frac{m_2-m_1}{2}}}$, $x \approx 0$, $x > 0$, *while in the case (A) 3)* $y \approx \beta - [(\frac{T}{2}-x)\sqrt{\frac{c_3}{c_4}}(\frac{n_2-n_1}{2}+1)]^{\frac{1}{1+\frac{n_2-n_1}{2}}}$, $x \approx \frac{T}{2}$, $x < \frac{T}{2}$.

(ii) Suppose that (A) 2) and (A) 3) hold.

Then (15) possesses a strictly monotonically increasing solution on the half line $(-\infty, 0]$. *Moreover, the solution has a horizontal asymptote at* $y = \alpha$, *while* $y \approx \beta - [-(1+\frac{n_2-n_1}{2})\sqrt{\frac{c_3}{c_4}}x]^{\frac{1}{1+\frac{n_2-n_1}{2}}}$, $x \approx 0$, $x < 0$.

(iii) Assume that (A) 2), (A) 4) hold. Then (15) has a kink type solution, i.e. strictly monotonically increasing solution with horizontal asymptotes at $y = \alpha$ *and* $y = \beta$.

(iv) Assume that (A) 1), (A) 4) hold. Then (15) has a monotonically increasing solution in the interval $[0, +\infty)$*. Moreover, its behavior for* $x \to 0$ *is given in (i), while it has a horizontal asymptote at* $y = \beta$*.*

Geometrical interpretation of several cases of Theorem 1 is given on Fig. 1 and Fig.2.

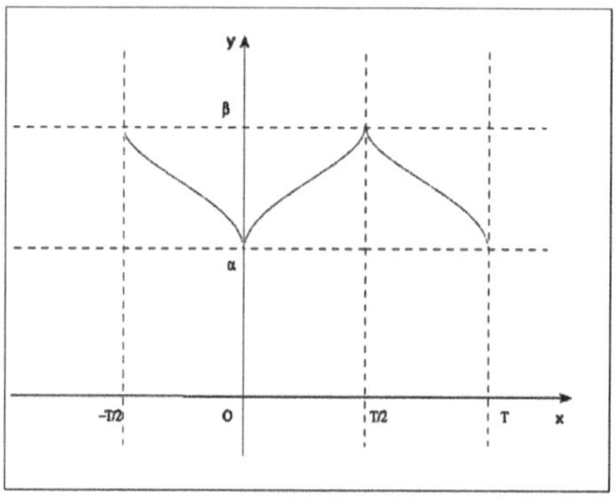

1a), 3d)

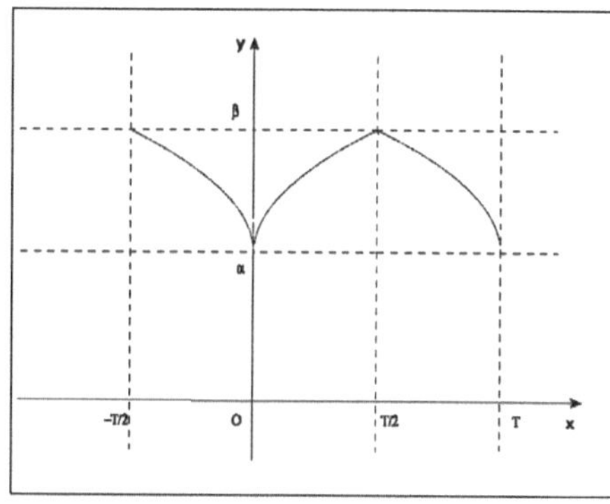

1a), 3e)

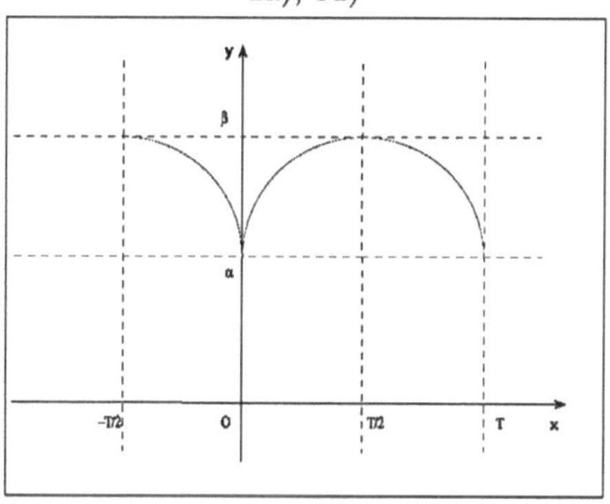

1a), 3f)

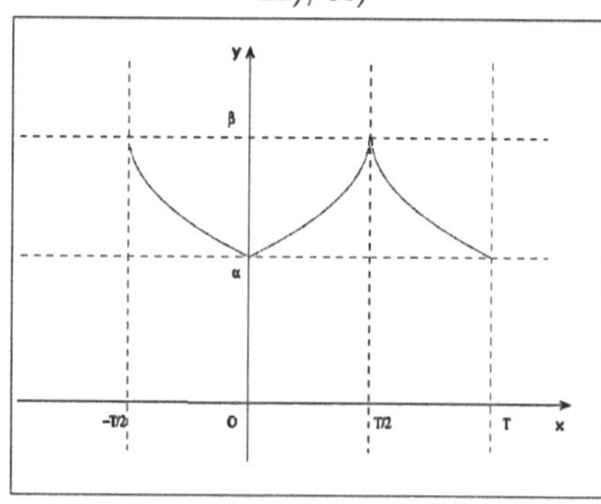

1b), 3d)

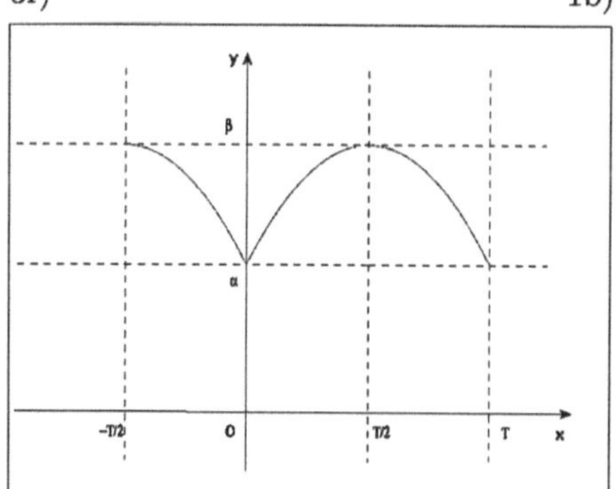

1b), 3f)

Fig. 1. Peaks and cusps configurations

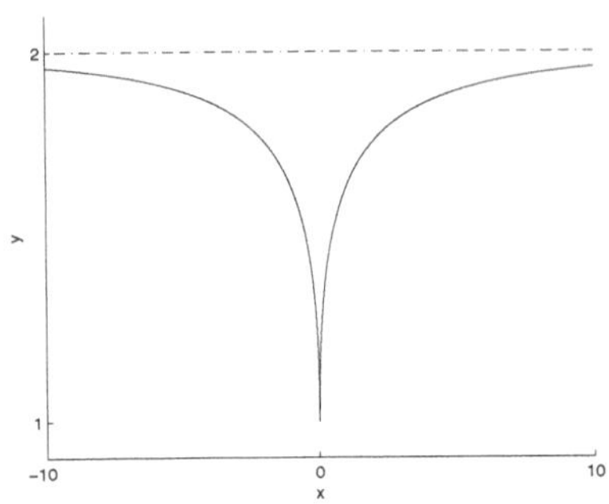

1a), 4)

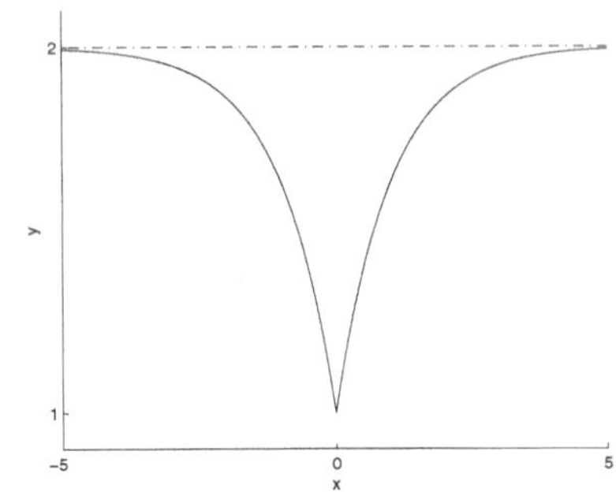

1b), 4)

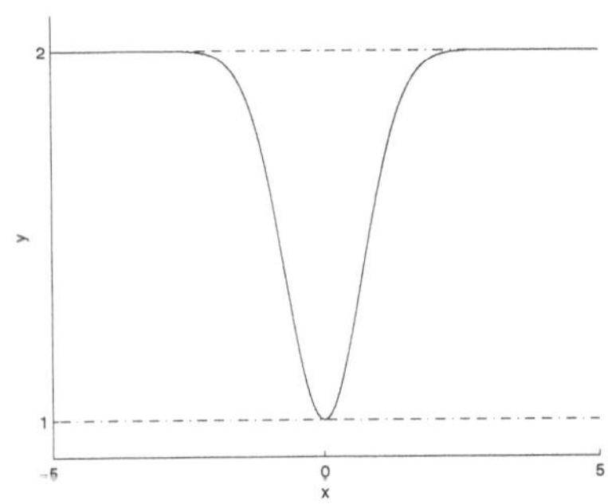

1c), 4)

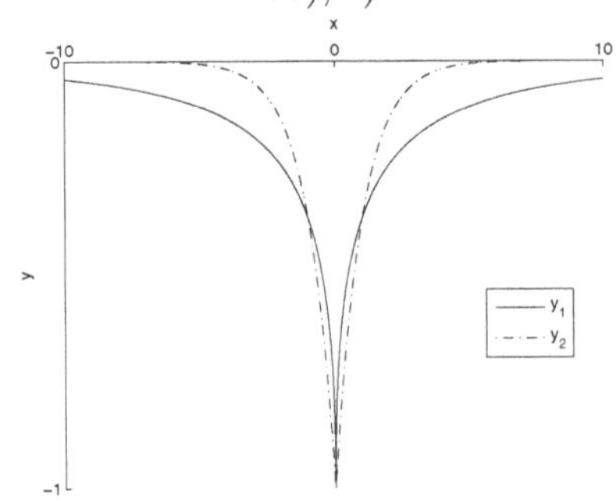

Peakon y_2 and cuspon y_1 with the same vertex

Fig. 2. Peakon-soliton, cuspon-soliton waves

2 Cellular Neural Networks Realization

Cellular Neural Networks (CNNs) are complex nonlinear dynamical systems, and therefore one can expect interesting phenomena like bifurcations and chaos to occur in such nets. It was shown that as the cell self-feedback coefficients are changed to a critical value, a CNN with opposite-sign template may change from stable to unstable. Namely speaking, this phenomenon arises as the loss of stability and the birth of a limit cycles.

CNN model for the generalization of the Camassa-Holm equation (5) will be the following:

$$\frac{du_{ij}}{dt} + KA_1 * u_{ij}^m - \frac{d}{dt}(A_2 * u_{ij}^n) = A_1 * [\frac{(A_1 * u_{ij}^n)^2}{2} + u_{ij}^n A_2 * u_{ij}^n], 1 \leq i, j \leq M, \tag{19}$$

where $A_1 = (1, -2, 1)$, is one-dimensional discretized Laplacian CNN template, $A_2 = \begin{pmatrix} 0 & 1 & 0 \\ 1 & -4 & 1 \\ 0 & 1 & 0 \end{pmatrix}$ is two-dimensional discretized Laplacian CNN template, $*$ is the convolution CNN operator [9].

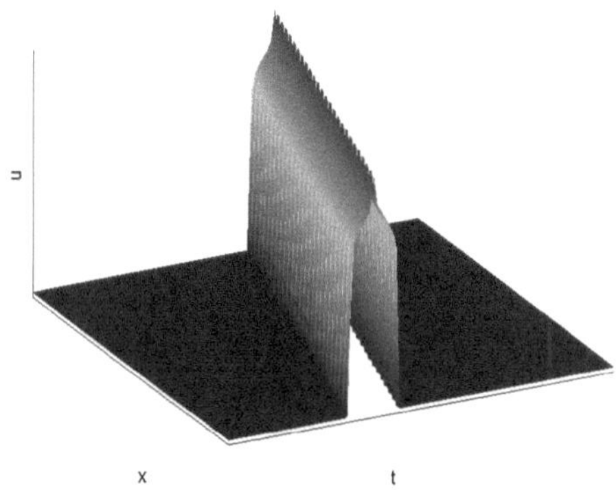

Fig. 3. The wave solution of the type compacton-peakon

We study here the structure of the travelling wave solutions of the CNN model (19) of (5) having the form:

$$u_{ij} = \Phi(icos\Theta + jsin\Theta - ct), \tag{20}$$

Φ being a continuous function. Let us substitute (20) in (19). Therefore we consider solution $\Phi(s;c)$ of:

$$-c\Phi^{'}(s;c) + G_1(\Phi(s;c)) = 0, \tag{21}$$

where $G_1(\Phi) \in \mathbf{R}^1$ and $\Phi(s;c)$ is satisfying:

$$lim_{s\to\pm\infty}\Phi(s;c) = 0 \tag{22}$$

for some $c > 0$.

Below we propose the following result.

Theorem 2. *Suppose that $u_{ij}(t) = \Phi(icos\Theta + jsin\Theta - ct)$ is a travelling wave solution of the CNN model of (5) and $\Omega = \frac{2\pi k}{L}$, $0 \le k \le L-1$. Then there exists $c > 0$ and $s_0 > 0$ such that*

(i) for $s < s_0$ the solution $\Phi(s;c)$ of (21) satisfying (22)is increasing;
(ii) for $s > s_0$ the solution $\Phi(s;c)$ of (21) satisfying (22)is decreasind;
(iii) for $s = s_0$ the solution $\Phi(s;c)$ of (21) has maximum of cusp type or maximum of angle type with positive opening.

Moreover, the solution $\Phi(s;c)$ is either nonvanishing everywhere or compactly supported, i.e. $\Phi(s;c) = 0$ for $|s - s_0| \ge d$, d being an appropriate positive constant.

We do not give the proofs of Theorem 1,2 in this short communication. We shall give the simulation of the CNN model (19) on Figure 3.

References

1. Camassa, R., Holm, D.: An integrable shallow water equation with peaked solution. Phys. Rev. Lett. 71, 1661–1664 (1993)
2. Liu, Z., Qian, T.: Peakons of the Camassa-Holm equation. Appl. Math. Modeling 26, 473–480 (2002)
3. Tang, M., Zang, W.: Four types of bounded solutions of CH - γ equation. Science in China Series A: Mathematics 50(1), 132–152 (2007)
4. Rossenau, P.: What is.. a compacton? Notices of the AMS 52, 738–739 (2005)
5. Rossenau, P.: On a model equation of travelling and stationary compactons. Phys. Lett. A 356, 44–50 (2006)
6. Shen, J., Xu, W.: Travelling wave solutions in a class of generalized Korteweg-deVries equation. Chaos, solitons and fractals 34, 1299–1306 (2007)
7. Parker, A.: Cusped solutions of the Camassa-Holm equation. I.Cuspon solitary wave and antipeakon limit. Chaos, solitons and fractals 34, 730–739 (2007)
8. Popivanov, P., Slavova, A., Zecca, P.: Compact travelling waves and peakon types solutions of several equations of mathematical physics and their Cellular Neural Networks realization, Nonlinear Analysis: Real World Applications (accepted) doi:10.1016/j.nonrwa.2008.01.020
9. Chua, L.O., Yang, L.: Cellular neural networks: Theory. IEEE Trans. Circuit Syst. 35, 1257–1271 (1988)

Finite Difference Method for Two-Dimensional Equations of Gas Dynamics Using Artificial Viscosity

Igor V. Popov and Igor V. Fryazinov

Institute for Mathematical Modeling, Russ. Ac. Sci.
4A Miusskaya square, 125047, Moscow, Russia
{popov,friazinov}@imamod.ru

Abstract. A new finite difference method is proposed for gas dynamics equations. It is a homogeneous, monotonic scheme of second order of accuracy on time and space outside domains of discontinuity and shock waves. A new way to introduce artificial viscosity is proposed for two-dimensional schemes. Test simulations of discontinues and shock waves are presented.

1 Introduction

We propose a new two-dimensional difference scheme for gas dynamics equations. It is homogeneous, practically monotonic, uses only a 9-points space stencil, well keeps weak contact discontinuities and shock waves. An approximation accuracy of the proposed scheme is $O\left(\tau^2 + h_x^2 + h_y^2\right)$ for domains with smooth boundaries and outside of compression waves. Here τ is time step, h_x and h_y are space steps. In this scheme along with the Lax-Wendroff correction we introduce a new monotonic artificial viscosity μ.

There is a lot of papers on methods for solving gas dynamics equations. Let us mention only some of them: [1] (Lagrange variables), [2,3] (Euler variables), kinetic self-consistent schemes [4]. A review of modern "monotonic" schemes with the flux correction techniques (FCT, TVD, ENO, WENO-methods) can be found in [5,6]. In [7] the results of 8 tests using 10 differences schemes are presented. In the present paper the numerical results [7] are obtained using the following method. A comparison with the methods of [7] is given.

2 Problem Formulation

Let us consider the 2-D Euler equations.

$$\frac{\partial}{\partial t}\begin{pmatrix}\rho\\ I_x\\ I_y\\ E\end{pmatrix} + \frac{\partial}{\partial x}\begin{pmatrix}\rho u\\ I_x u + p\\ I_y u\\ (E+p)u\end{pmatrix} + \frac{\partial}{\partial y}\begin{pmatrix}\rho v\\ I_x v\\ I_y v + p\\ (E+p)v\end{pmatrix} = 0.$$

S. Margenov, L.G. Vulkov, and J. Waśniewski (Eds.): NAA 2008, LNCS 5434, pp. 469–475, 2009.

Here ρ is density, $\boldsymbol{v} = (u, v)$ is velocity, $\boldsymbol{I} = \rho\boldsymbol{v}$ is impulse, $E = e + \rho\dfrac{u^2+v^2}{2}$ is total energy of gas, $e = \rho\varepsilon$ is internal energy of gas. These equations are solved in domain $0 < x < l_x$, $0 < y < l_y$, $t > 0$. We use the ideal gas equation $p = (\gamma - 1)\, e$.

At the boundary and at the first moment we prescribe ρ, $\boldsymbol{v}$ and E (or e). At the solid walls we set zero density flux, impulse and one of two velocity components.

3 Approximation

In domain $0 < x < l_x$, $0 < y < l_y$ we introduce a grid with the steps $h_x = \dfrac{l_x}{N_x - 1}$, $h_y = \dfrac{l_y}{N_y - 1}$, where N_x, N_y are numbers of points in x and y directions. The time step τ is introduced a little later.

Now we use the expansion

$$\rho^{n+1} = \rho^n + \tau\frac{\partial \rho^n}{\partial t} + +\frac{\tau^2}{2}\frac{\partial^2 \rho^n}{\partial t^2} + \dots. \tag{1}$$

We take the first time derivative from the continuous equation. Let's replace $\dfrac{\partial \rho^n}{\partial t}$ with its expression

$$-\left(\frac{\partial}{\partial x}(\rho u)^n + \frac{\partial}{\partial y}(\rho v)^n\right).$$

To find the second time derivative, we differentiate the continuous equation on time and use the momentum equation. Then for the third term of (1) we have

$$\frac{\tau}{2}\left(\frac{\partial^2}{\partial x^2}(\rho u^2 + p)^n + \frac{\partial^2}{\partial y^2}(\rho v^2 + p)^n + 2\frac{\partial^2}{\partial x \partial y}(\rho u v)^n\right).$$

So we have the following equation

$$\frac{\rho^{n+1} - \rho^n}{\tau} + \frac{\partial}{\partial x}(\rho u)^n + \frac{\partial}{\partial y}(\rho v)^n -$$

$$-\frac{\tau}{2}\left(\frac{\partial^2}{\partial x^2}(\rho u^2 + p)^n + \frac{\partial^2}{\partial y^2}(\rho v^2 + p)^n + 2\frac{\partial^2}{\partial x \partial y}(\rho u v)^n\right) = 0.$$

Here the last term is the Lax-Wendroff correction. In the flux form it is

$$\frac{\rho^{n+1} - \rho^n}{\tau} + \frac{\partial}{\partial x}(W_\rho^x)^n + \frac{\partial}{\partial y}(W_\rho^y)^n = 0.$$

Here

$$W_\rho^x = \rho u - \frac{\tau}{2}\left(\frac{\partial}{\partial x}(\rho u^2 + p) + \frac{\partial}{\partial y}(\rho u v)\right),$$

$$W_\rho^y = \rho v - \frac{\tau}{2}\left(\frac{\partial}{\partial y}\left(\rho v^2 + p\right) + \frac{\partial}{\partial x}\left(\rho u v\right)\right).$$

In the finite-difference form these equations are

$$\frac{\rho_i^{n+1} - \rho_i^n}{\tau} = \frac{\left(W_\rho^x\right)_{i+1/2,k} - \left(W_\rho^x\right)_{i-1/2,k}}{h_x} + \\ + \frac{\left(W_\rho^y\right)_{i,k+1/2} - \left(W_\rho^y\right)_{i,k-1/2}}{h_y},$$

where

$$\left(W_\rho^x\right)_{i+1/2,k} = \frac{u_{i+1,k}^n + u_{i,k}^n}{2}\frac{\rho_{i+1,k}^n + \rho_{i,k}^n}{2} - \left(LW_\rho^x\right)_{i+1/2,k},$$

$$\left(LW_\rho^x\right)_{i+1/2,k} = \frac{\tau}{2}\left[\frac{\left(\rho u^2 + p\right)_{i,k+1}^n - \left(\rho u^2 + p\right)_{i,k}^n}{h_y}\right] + \\ + \frac{\tau}{8}\left[\frac{(\rho u v)_{i+1,k+1}^n - (\rho u v)_{i+1,k-1}^n}{h_y} + \frac{(\rho u v)_{i,k+1}^n - (\rho u v)_{i,k-1}^n}{h_y}\right],$$

$$\left(W_\rho^y\right)_{i,k+1/2} = \frac{v_{i,k+1}^n + v_{i,k}^n}{2}\frac{\rho_{i,k+1}^n + \rho_{i,k}^n}{2} - \left(LW_\rho^y\right)_{i,k+1/2},$$

$$\left(LW_\rho^y\right)_{i,k+1/2} = \frac{\tau}{2}\left[\frac{\left(\rho v^2 + p\right)_{i,k+1}^n - \left(\rho v^2 + p\right)_{i,k}^n}{h_y}\right] + \\ + \frac{\tau}{8}\left[\frac{(\rho u v)_{i+1,k+1}^n - (\rho u v)_{i-1,k+1}^n}{h_x} + \frac{(\rho u v)_{i+1,k}^n - (\rho u v)_{i-1,k}^n}{h_x}\right].$$

Here LW_ρ^x, LW_ρ^y are the Lax-Wendroff corrections.

Analogously, using the Lax-Wendroff corrections let us the write the momentum and energy equations in flux form.

$$\frac{I_i^{n+1} - I_i^n}{\tau} = \frac{\left(W_I^x\right)_{i+1/2,k} - \left(W_I^x\right)_{i-1/2,k}}{h_x} + \\ + \frac{\left(W_I^y\right)_{i,k+1/2} - \left(W_I^y\right)_{i,k-1/2}}{h_y},$$

$$\frac{E_i^{n+1} - E_i^n}{\tau} = \frac{\left(W_E^x\right)_{i+1/2,k} - \left(W_E^x\right)_{i-1/2,k}}{h_x} + \\ + \frac{\left(W_E^y\right)_{i,k+1/2} - \left(W_E^y\right)_{i,k-1/2}}{h_y}.$$

They have the approximation accuracy $O\left(\tau^2 + h_x^2 + h_y^2\right)$. These corrections are not enough to provide a "monotonicity" of the scheme and we need to introduce some additional viscosity

$$\frac{\rho^{n+1} - \rho^n}{\tau} + \frac{\partial\left(W_\rho^x\right)^n}{\partial x} + \frac{\partial\left(W_\rho^y\right)^n}{\partial y} - \frac{\partial\left(W_{\mu\rho}^x\right)^n}{\partial x} - \frac{\partial\left(W_{\mu\rho}^y\right)^n}{\partial y} = 0.$$

We can do that in a way that the scheme accuracy becomes near the one of the scheme WENO5.

4 Numerical Results

The first problem was about reflection of oblique shock wave from hard wall. Initial values were $\rho = 1$, $u_0 = 4$, $v_0 = 0$, $e = 1$ and $\gamma = 2$. Boundary conditions were the following:

1) $\rho = 1.5$, $u = 11/3$, $v = -\sqrt{3}/3$ and $e = 14/9$ at AF;
2) $\rho = 2.1$, $u = 10/3$, $v = 0$ and $e = 20/9$ at FE.

At boundary BD the condition of hard wall were used. At the boundary ED the condition of free exit was set. The sizes of domain were $|AB| = 1$, $|AE| = 5\sqrt{3}/2$ and $|AF| = 7\sqrt{3}/4$.

The Gugonio conditions were fulfilled for gas dynamics parameters at the boundary.

The calculations were fulfilled at orthogonal grid with steps $h_x = 2.5\sqrt{3}/(N_x - 1)$, $h_y = 1/(N_y - 1)$ and $Ku = 0.66$.

The results are shown at Fig. 2. Parameters were taken from [9].

The second test was about channel with edge for supersonic flux. The channel length is equal 3, and width is equal 1. Edge was placed in 0.6 distance from

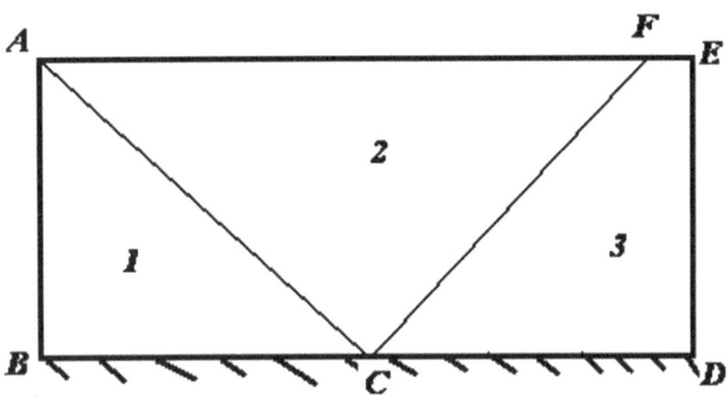

Fig. 1. Calculation domain for oblique shock wave

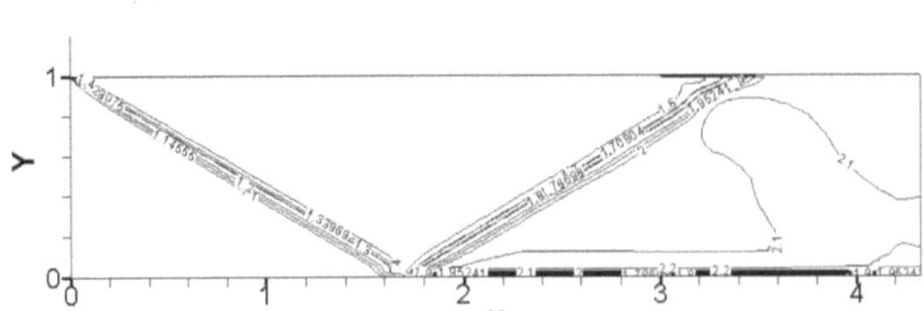

Fig. 2. Density distribution for oblique shock wave on excite

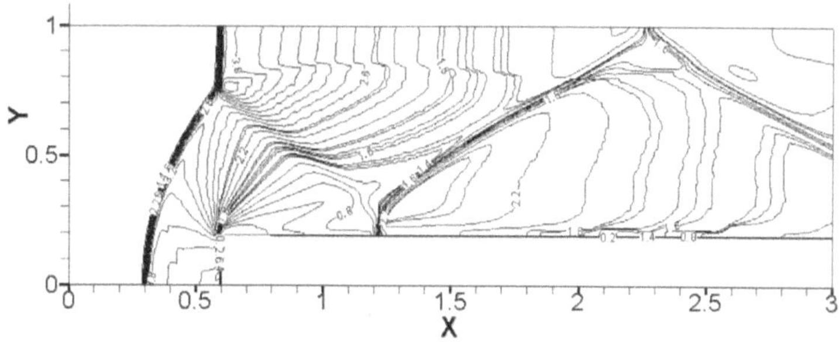

Fig. 3. Density distribution

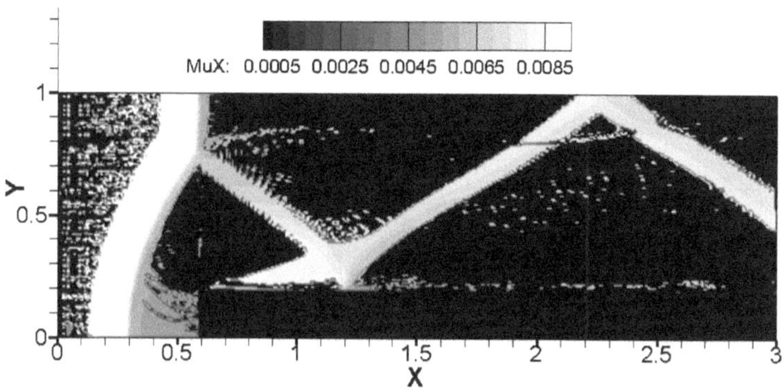

Fig. 4. Isolines of artificial viscosity for X-component

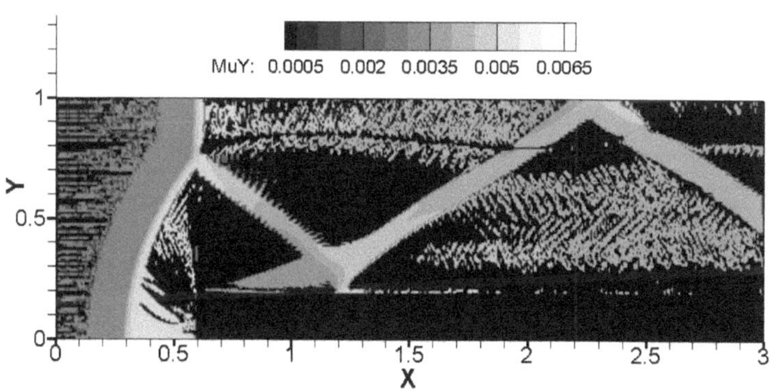

Fig. 5. Isolines of artificial viscosity for Y-component

enter and has high 0.2. Gas was not viscous and heat conductive. Gas parameters are $\gamma = 1.4$, $\rho = 1$, $p = 1/\gamma$, $u = 3$ and $v = 0$. At exit free boundary conditions are posed. That leads to heavy shock waves. The results are given at Fig. 3. For the same problem with artificial viscosity see Fig. 4-5.

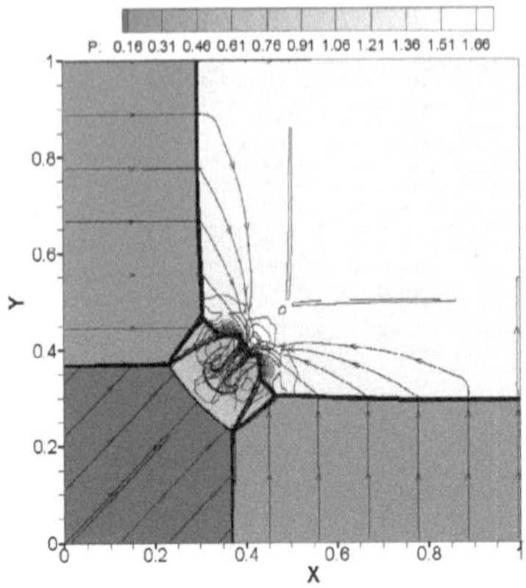

Fig. 6. Test 3

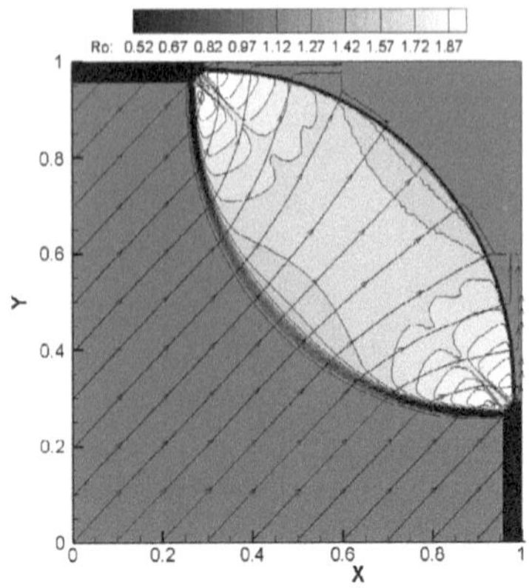

Fig. 7. Test 4

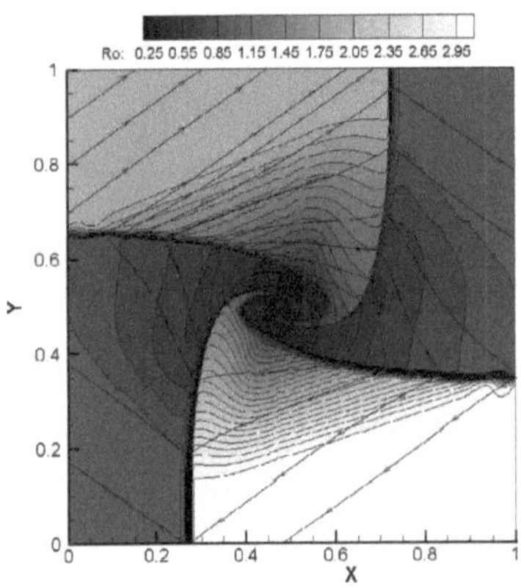

Fig. 8. Test 6

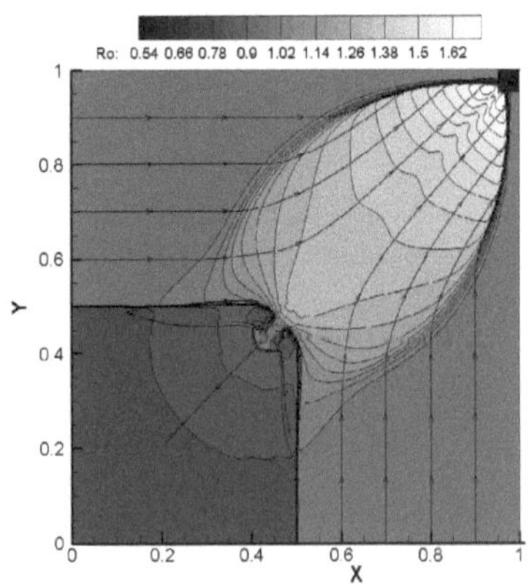

Fig. 9. Test 12

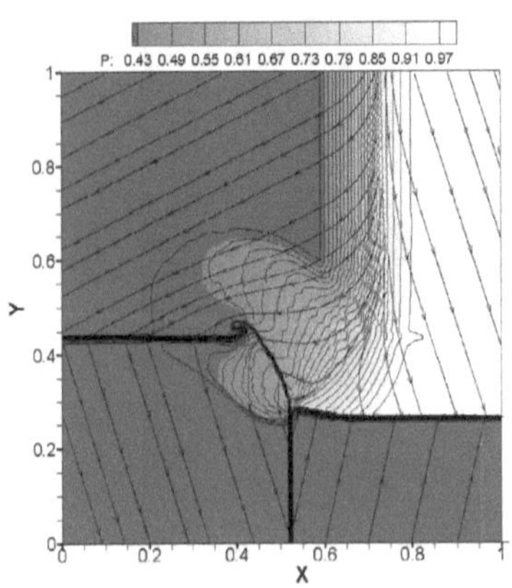

Fig. 10. Test 15

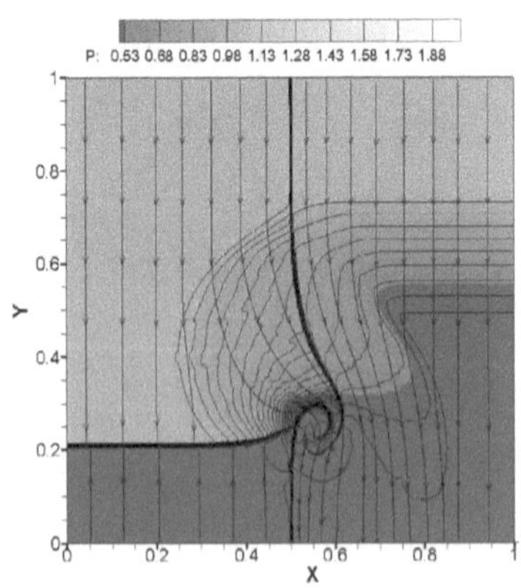

Fig. 11. Test 17

5 Gas Dynamics Test Cases

We took several test cases from [7] for CFLF, CFLFh, WAFT, WAFC, CLAW, PPM, WENO5, CWENO3, LL, JT , schemes. Tests 3, 4 are the interaction of four shock waves. Test 6 is the problem of 4 contact discontinuous. Test 12 is two discontinuous plus two shock waves. Number 15-17 are the problem of calculation of one discontinuous, two shock and one relaxation waves. All the test problems are simulated on the grids of 401x401 nodes.

6 Conclusions

The paper introduces a new explicit difference scheme to simulate gas dynamics equations. The order of approximation accuracy is $O(\tau^2 + h_x^2 + h_y^2)$ on smooth solutions. The scheme is homogeneous and provides an automatic shock capturing. It exploits a 9-points stencil for 2D problems and may be used for 3D problems. The numerical results show that our method (AUB2D) may be competitive with the modern gas dynamics finite-difference methods.

References

1. Samarskii, A.A., Popov, Y.P.: Finite Difference Methods for Problems in Gas Dynamics. Nauka, Moscow (1992) (in Russian)
2. Rozhdestvensky, B.L., Yanenko, N.N.: Systems of quasi-linear equations. Nauka, Moscow (1978) (in Russian)
3. Godunov, K.S., Zabrodin, V.A., Ivanov, Y.A., et al.: Numerical solution of multidimensional problems in the gas dynamics. Nauka, Moscow (1976) (in Russian)
4. Chetverushkin, B.N.: Kinetic schemes in gas dynamic. Moscow State University (2004)
5. Kulikovsky, A.G., Pogorelov, N.V., Semenov, A.Y.: Mathematical Problems of the Numerical Solution of Hyperbolic Systems of Equations. Phizmathlit, Moscow (2001) (in Russian)
6. Bondarenko, Y., Bashurov, V.V., Yanilkin, Y.V.: A mathematical model and numerical methods. for solving nonmstationary gas dynamic problems. Survey of foreign publications. RFNC-VNIIFF 88-2003 (in Russian) (Preprint)
7. Richard, L., Burton, W.: Comparison of several difference schemes on 1D and 2D test problems for the Euler equations. SIAM J. Sci. Comput. 259(30), 995–1017 (2003)
8. Breslavsky, P.V., Mazhukin, V.I.: Simulation of integrating discontinuous solutions on dynamically adaptive grids. Computational methods in applied mathematics 7(2), 103–107 (2007)
9. Vasilevskii, V.F., Vyaznikov, K.V., Tishkin, V.F., Favorskii, A.P.: Monotonous difference schemes of high order of accuracy for nonregular grid. Preprint No. 124 (1990); IAM name Keldysh (in Russian)

A Second Order Central Scheme for Hamilton-Jacobi Equations on Triangular Grids

Peter Popov[1] and Bojan Popov[2]

[1] Institute for Scientific Computation, Texas A&M University,
College Station, TX-77843, USA
ppopov@tamu.edu

[2] Department of Mathematics, Texas A&M University,
College Station, TX-77843, USA
popov@tamu.edu

Abstract. In this paper, we describe a Godunov-type fully discrete scheme for Hamilton-Jacobi equations on triangular meshes. This scheme is an extension of the Lin-Tadmor and Kurganov-Tadmor fully discrete nonoscillatory central schemes to unstructured triangular meshes. In this new construction, we propose a new, "genuinely multidimensional", nonoscillatory reconstruction. The construction is simple, universal and deviates from the existing high-order extensions of the central and central-upwind schemes for Hamilton-Jacobi equations.

1 Introduction

Nonoscillatory central schemes are a class of Godunov-type numerical methods for solving first order nonlinear partial differential equations. A special case of such equations are Cauchy problems for Hamilton-Jacobi equations of the form

$$\begin{aligned} \frac{\partial u}{\partial t}(\mathbf{x},t) + H(\mathbf{x},\nabla u) &= 0, \qquad (\mathbf{x},t) \in \mathbb{R}^2 \times [0,T], \\ u(0,\mathbf{x}) &= \tilde{u}(\mathbf{x}), \qquad \mathbf{x} \in \mathbb{R}^2. \end{aligned} \tag{1}$$

These equations are used in a variety of applications such as optimal control problems, geometical optics and calculus of variations. It is well known that for nonlinear Hamiltonians solutions of such equations with smooth initial data are continuous but may develop discontinuities in their gradient in finite time. Hence, it is necessary to consider such equations in some weak/generalized sense. The theory of the so-called viscosity solition answers the question of well-posedness of the Cauchy problem for a large class of Hamiltonians. This celebrated theory was first developed for convex Hamiltonians [1] and later extended to a wide class of Hamiltonians [2,3,4,5,6].

Given the importance of problems, there has been an enourmous amount of activities based on this theory and its applications. Converging first order approximations were introduced by Souganidis [7] and high-order upwind methods were introduced by Osher and Sethian [8] and Osher and Shu [9]. Various extensions of these methods were considered in [10,11,12] and many other authors.

S. Margenov, L.G. Vulkov, and J. Waśniewski (Eds.): NAA 2008, LNCS 5434, pp. 476–485, 2009.

A different class of Godunov-type schemes, first introduced for hyperbolic conservation laws, the so-called central schemes, was introduced in the context of Hamilton-Jacobi equations by Lin and Tadmor [13]. The main advantage of central schemes is their simplicity and computational efficiency. A variety of central schemes was later developed, see for example [14,13,15,16,17,18,19]. In all of the above mentioned references, the central idea was exploited to derive a semidiscrete central or central-upwind formulation (see [18]) and use it to evolve the nodal values of the approximate solution. Even though this approach can be used in the context of unstructured meshes [14], it is best suited for uniform mesh: the formulation is very simple and the local equations are the same. Unfortunately, the high-order extensions of all numerical schemes are based on special nonlinear weights/procedures and the convergence theory for such schemes, especially in the multidimensional case, is not developed at this time.

We will deviate from the mainstream central schemes because we are interested in approximating solutions of the Cauchy problem (1) on a given unstructured conforming triangulation of the domain $\Omega \subset \mathbb{R}^2$. We follow the original central approach of Kurganov and Tadmor [15] and derive a fully discrete scheme. Namely, the evolution stage is carried out at points that are located away from the discontinuities. That is, the equation is solved in regions where the solution remains smooth. Then, a reconstruction is used to recover a global numerical solution at the next time step. Our main contribuiont is to propose a new simple and robust piece-wise parabolic reconstruction in the context of an arbitrary triangular grid. This makes the scheme suitable for large problems with complicated geometry and easy to adjust in case when adaptive refinement is needed. The new scheme is very similar in spirit to the Christov-Popov central scheme [20] in the context of nonlinear conservation laws.

The paper is organized as follows. In Section 2 we describe our scheme and in Section 3 we present different numerical tests for both linear and convex Hamiltonians.

2 A Fully-Discrete Godunov-Type Scheme for the Hamilton-Jacobi Equations

We are interested in computing numerical solutions approximating the Cauchy problem for the Hamilton-Jacobi equation (1). The numerical method we consider is a Godunov-type central scheme, see e.g. [15], in which the approximate solution u_h, a C^0 function, is known at the beginning of a time step. Equation (1) is then solved in the time slab $[t_n, t_{n+1}]$:

$$\begin{aligned} \frac{\partial u^n}{\partial t}(\mathbf{x}, t) + H(\mathbf{x}, \nabla u^n) &= 0 && \text{for } \forall(\mathbf{x}, t) \in \Omega \times [t_n, t_{n+1}] \\ u^n(t_n, \mathbf{x}) &= u_h(\mathbf{x}, t_n) && \text{for } \forall \mathbf{x} \in \Omega \end{aligned} \quad (2)$$

and $u^n(\mathbf{x}, t_{n+1})$, which is not necessarily a C^0 function anymore is then projected on the original grid to obtain a conforming C^0 approximation $u_h(\mathbf{x}, t_{n+1})$.

The scheme proposed here uses a piecewise quadratic, conforming approximation of $u(:,t)$ on triangles. We assume that the domain Ω is partitioned by a regular, conforming, triangular mesh $\mathcal{T}^h$:

$$\mathcal{T}^h = \bigcup_{i=1}^{N_e} e_i. \tag{3}$$

Here, the mesh triangles are denoted by e_i, $i = 1,\dots,N_e$, where N_e is the number of triangles in the mesh, they are assumed to have a minimum angle requirement and the overall mesh parameter is denoted by h. Given some discrete time instance t_n, the numerical solution to (2) is defined on $\mathcal{T}^h$ as a piece-wise quadratic polynomial, conforming over the triangulation edges:

$$u_h(\mathbf{x},t_n) \in C^0(\Omega), \qquad u_h(\mathbf{x},t_n)|_{e_i} \in P_2(e_i), \qquad i = 1,\dots,N_e \tag{4}$$

In general this type of approximation of u implies that the spatial derivatives of u_h may be discontinuous across the edges of the triangles. As a result the numerical propagation speeds will be discontinuous at element boundaries and it is not clear how to evolve u_h over a time step. For that reason, we also consider a set $\tilde{\mathcal{T}}^h$ of *interior triangles*. Each $\tilde{e} \in \tilde{\mathcal{T}}^h$ is a triangle similar to the corresponding triangle $e \in \mathcal{T}^h$ in the original mesh. Furthermore, $\tilde{e}$ is a homothetic copy of e and is entirely contained in e, that is $\tilde{e} \cong e$ and $\tilde{e} \subseteq e$. For the later reason, the triangle $\tilde{e}$ are called interior.

The idea behind the introduction of $\mathcal{T}^h$ is to select each $\tilde{e}_i$, $i = 1,\dots,N_e$ in such a way that any discontinuity at the edges of e_i will not be able to propagate into $\tilde{e}_i$ during one time step $dt = t_{n+1} - t_n$. In such an interior triangle the solution is locally smooth (a quadratic polynomial) and it can be evolved by any reasonable time-stepping method. This suggests the following time-stepping procedure. First, starting with $u_h(\mathbf{x},t_n)$ at the beginning of the time step, one takes the restriction:

$$\tilde{u}_h|_{\tilde{e}} = u_h|_{\tilde{e}}, \qquad \text{for each } \tilde{e} \in \tilde{\mathcal{T}}^h. \tag{5}$$

The restriction $\tilde{u}_h$, defined on the disjoint set of triangles $\tilde{\mathcal{T}}^h$ is then evolved for a time interval dt locally in each interior element $\tilde{e}$, thus obtaining $\tilde{u}_h(\mathbf{x},t_{n+1})$. Finally a reconstruction procedure is used to recover $\tilde{u}_h(\mathbf{x},t_{n+1})$ from $u_h(\mathbf{x},t_{n+1})$. The procedure is summarized below:

- For each mesh triangle e, determine an interior triangle $\tilde{e}$ in such a way that any discontinuity of ∇u_h at the edges of e_i will not be able to propagate inside $\tilde{e}_i$ during the time-step
- Restrict $u_h(\mathbf{x},t_n)$ to $\tilde{\mathcal{T}}^h$ and evolve locally each $u_h|_{\tilde{e}}$, $\tilde{e} \in \tilde{\mathcal{T}}^h$ using a suitable time-integrator, e.g. a Strong-Stability Preserving (SSP) method [21].
- Reconstruct $\tilde{u}_h(\mathbf{x},t_{n+1})$ on the original grid $\mathcal{T}^h$, thus obtaining $u_h(\mathbf{x},t_{n+1})$.

The construction of the interior triangles $\tilde{e}$ and the local evolution of $\tilde{u}_h$ is described in the next Section 2.2. The reconstruction of the evolved restricted solution $\tilde{u}_h(\mathbf{x},t_{n+1})$ on the original grid is describe in Section 2.3

2.1 Element-Wise Restriction to Interior Triangles

At the beginning of time step t_n we have a continuous, piecewise quadratic solution $u_h(\cdot, t_n)$. The solution cannot be evolved directly, since at element interfaces the first derivatives of the solution are not continuous. The objective of this step is to select interior triangles such that the restriction of $u_h(\cdot, t_n)$ to these interior triangles remains smooth over the entire time-step $[t_n, t_{n+1}]$. The so restricted solution will then be evolved locally as described in Section 2.2. In this section we are deling with the solution only at time t_n, therefore, for the sake of simplicity, we write u_h.

The selection of interior triangles is performed as follows. Consider the set $\mathcal{V}_2$ all of all vertices and midpoints of the triangles of $\mathcal{T}^h$, in other words the nodes defining the standard Lagrange quadratic basis functions over $\mathcal{T}^h$. Further, for any node $\mathbf{v} \in \mathcal{V}_2$, denote by $\mathcal{E}(\mathbf{v})$ the set of all elements of $\mathcal{T}^h$ which have $\mathbf{v}$ as one of its six nodes. Note that $\mathcal{E}(v)$ is the support of the (quadratic) basis function $\phi_{\mathbf{v}}$ which corresponds to the node $\mathbf{v}$. If $\mathbf{v}$ is a midpoint, then $\mathcal{E}(\mathbf{v})$ consists of two elements, while if $\mathbf{v}$ is a triangle vertex, $\mathcal{E}(\mathbf{v})$ contains all the triengles that share this vertex.

Recall, that we use a standard quadratic Lagrangian representation of the discrete solution u_h. Therefore, $u_h \in C^0(\Omega)$, and in particular, $u_h \in C^0(\mathcal{E}(\mathbf{v}))$. However, its gradient ∇u_h is piece-wise linear function over each element but discontinuous over element edges. In particular, at triangle edge midpoints, ∇u_h is a two-valued function and at triangle vertices ∇u_h is a multi-valued function. Thus, given a triangle e, we would like to estimate the maximum extent to which the discontinuities in ∇u_h at the border of the triangle ∂e can propagate inside e for an interval of time dt. This in term will define the interior triangle $\tilde{e}$ in which u_h stays smooth over a time-step dt.

So, let $e \in \mathcal{T}^h$ be a triangle. Let $\mathbf{v}_i$, $i = 1, \ldots, 6$, be the six nodes (the three triangle vertices and the three edge midpoints of e) which define the six Lagrange basis function $\{\phi_{\mathbf{v}_i}\}_{i=1}^6$ over e. Consider the restriction $u_h|_A$, where $A = \{e | e \in \mathcal{E}(\mathbf{v}_i)\}$, in other words, all the piece-wise quadratic polynomials defined on the elements that form the support of the basis function $\phi_{\mathbf{v}_i}$. Next, let $\{\mathbf{q}_e = \nabla u_h|_e(\mathbf{v}_i, t_n)\}_{e \in \mathcal{E}(\mathbf{v}_i)}$ be the set of all possible gradients of u_h at that vertex. In the case of *convex* Hamiltonian, the maximum speed $a(\mathbf{v}_i)$ at that vertex is determined by:

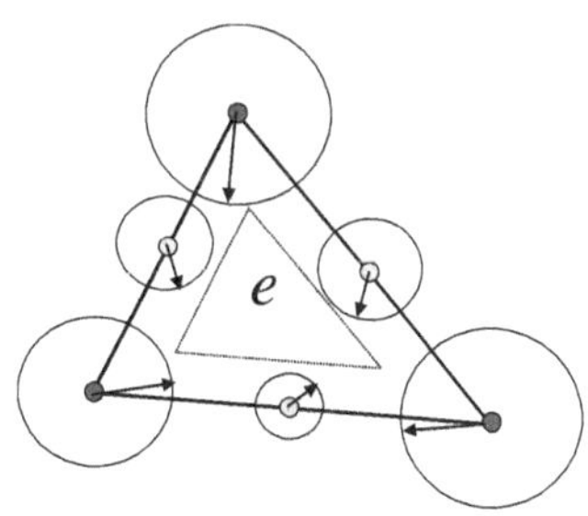

Fig. 1. Interior triangle

$$a(\mathbf{v}_i) = \max_{e \in \mathcal{E}(\mathbf{v}_i)} ||\nabla H(\mathbf{q}_e)||$$

Further, define the balls B_i, $i = 1, \ldots, 6$ to be the six balls centered at v_i, respectively, and having radius $a(\mathbf{v}_i)\,dt$, respectively (Figure 1). The selection of an interior triangle $\tilde{e}$ is then performed as follows: we take e and shrink it homothetically with respect to its center of gravity. We take the largest such $\tilde{e}$ which does not intersect with any of the six balls $B_1, \ldots, B_6$ (Figure 1).

2.2 Local Evolution

Once an interior triangle $\tilde{e}$ is chosen, the solution can be evolved locally in that triangle without regard for the neighbours. That is, the restriction $\tilde{u}_h|_{\tilde{e}}$ (c.f. equation (5)) remains smooth over an entire time step. Therefore, we can use a second order ODE method to solve:

$$\frac{d\tilde{u}_h(\mathbf{x},t)}{dt} = -H(\mathbf{x}, \nabla\tilde{u}_h) \quad \text{where } \mathbf{x} \in \tilde{e} \text{ and } t \in [t_n, t_{n+1}]. \tag{6}$$

Indeed, $\tilde{u}_h|_{\tilde{e}}$ is a quadratic polynomial, therefore uniquely defined by its values at the vertices of $\tilde{e}$ and at the midpoints of its sides. Following standard notation from finite element methods, we write

$$\tilde{u}_h(\mathbf{x},t) = \sum_1^6 \tilde{U}_i(t)\phi_i(\mathbf{x})$$

where $\phi_i(\mathbf{x})$, $i = 1, ..., 6$ are the standard six nodal basis function for a lagrangian finite element defined over the triangle $\tilde{e}$. Therefore, equation (6) can be rewritten as:

$$\frac{d\tilde{\mathbf{U}}(\mathbf{t})}{dt} = -H(\mathbf{x}, \tilde{\mathbf{U}}(\mathbf{t}) \cdot \nabla\boldsymbol{\Psi}(\mathbf{x})) \quad \text{where } \mathbf{x} \in \tilde{e} \text{ and } t \in [t_n, t_{n+1}]. \tag{7}$$

where the $\tilde{\mathbf{U}}(\mathbf{t}) = \left(\tilde{U}_1(t), ..., \tilde{U}_6(t)\right)^T$ is the column vector of the point values of $\tilde{u}_h$ at the six nodes and $\boldsymbol{\Psi}(\mathbf{x}) = (\psi_1(\mathbf{x}), ..., \psi_6(\mathbf{x}))^T$ is the vector function composed of the six nodal basis functions. The above equation is solved by a Modified Euler method [22,23] which provides second order accuracy in time. This is also an SSP method [21]. One could also use a different higher order SSP scheme, such as the Runge-Kutta schemes proposed in [21].

2.3 Reconstruction on Triangular Grids

At the end of the local evolution (7), one has an piecewise quadratic $\tilde{u}_h(\cdot, t_{n+1})$ approximation defined over $\tilde{\mathcal{T}}^h$. Note that all quantities in this section are defined at time t_{n+1}, hence we will skip the time step from the notation. So, let us extend $\tilde{u}_h$ to the entire mesh $\mathcal{T}^h$ by simply extending the domain of each piecewise quadratic polynomial $\tilde{u}_h|_{\tilde{e}}$ from $\tilde{e}$ to e. That is, the extension $E\tilde{u}_h$ is defined by:

$$E\tilde{u}_h : \mathcal{T}^h \to \mathbb{R}, \quad (E\tilde{u}_h)|_e \in P_2 \text{ and } \quad (E\tilde{u}_h)|_{\tilde{e}} = \tilde{u}_h|_{\tilde{e}}.$$

Since there is no chance of confusion, we will just write $\tilde{u}_h$, instead of $E\tilde{u}_h$. This function is now piecewise quadratic over each element of $\mathcal{T}^h$, but it is discontinuous over element boundaries. The goal of the reconstruction is to generate a piecewise continuous function u_h from $\tilde{u}_h$.

Consider an element e and its three neighbors e_1, e_2 and e_3 (Figure 2). Further, let $\tilde{u}_h|_e$, $\tilde{u}_h|_{e_1}$, $\tilde{u}_h|_{e_2}$, $\tilde{u}_h|_{e_3}$ be the restrictions of $\tilde{u}_h$ on e, e_1, e_2 and e_3, respectively. Given the element e, we consider five quadratic polynomials: p, $p_\star$,

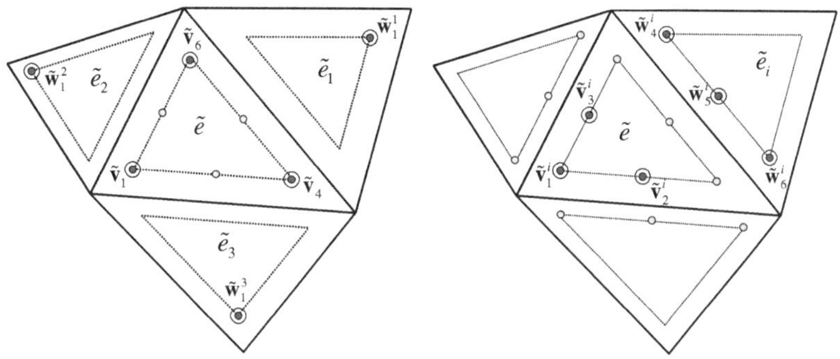

(a) 4-Patch Interpolant (b) One of the three side interpolant

Fig. 2. Patterns use for the reconstruction of the solution

p_1, p_2, p_3, defined as follows. First, we take $p = \tilde{u}_h|_e$, that is, p is just the restriction of $\tilde{u}_h$ on the element e itself, which, of course is a quadratic polynomial. Next, the polynomial $p_\star$ interpolates the six vertices $\tilde{\mathbf{v}}_1$, $\tilde{\mathbf{v}}_4$, $\tilde{\mathbf{v}}_6$, $\tilde{\mathbf{w}}_1^1$, $\tilde{\mathbf{w}}_1^2$, $\tilde{\mathbf{w}}_1^3$, shown on Figure 2(a). They are, the three vertices $\tilde{\mathbf{v}}_1$, $\tilde{\mathbf{v}}_4$, $\tilde{\mathbf{v}}_6$ of the interior triangle $\tilde{e}$, and $\tilde{\mathbf{w}}_1^i$, where $i = 1, 2, 3$, is the vertex of the the interior neighbor $\tilde{e}_i$, opposite to the shared edge between e and e_i (see Figure 2(a)).

Next, each of the three quadratic interpolants p_i, $i = 1, 2, 3$ is defined using information from $\tilde{e}$ and each of its three neighbours. For each i, we choose six vertices in order to define p_i. First, we take the vertex $\tilde{\mathbf{v}}_1^i$ of $\tilde{e}$ which is opposite to the side, that e and e_i share. Next, we also add the two midpoints $\tilde{\mathbf{v}}_2^i$ and $\tilde{\mathbf{v}}_3^i$ of the two sides of $\tilde{e}$ which contain $\tilde{\mathbf{v}}_1^i$ (Figure 2(b)). To these three point, we also add the points $\tilde{\mathbf{w}}_4^i$, $\tilde{\mathbf{w}}_5^i$ and $\tilde{\mathbf{w}}_6^i$. These are the two vertices ($\tilde{\mathbf{w}}_4^i$ and $\tilde{\mathbf{w}}_6^i$) and the midpoint ($\tilde{\mathbf{w}}_5^i$) of the side of $\tilde{e}_i$ which is parallel to the shared side of e and e_i (if is also parallel to the line connecting the two midpoints $\tilde{\mathbf{v}}_2^i$ and $\tilde{\mathbf{v}}_3^i$). Now, p_i is the quadratic polynomial which interpolates the six vertices $\tilde{\mathbf{v}}_1^i$, $\tilde{\mathbf{v}}_2^i$, $\tilde{\mathbf{v}}_3^i$, $\tilde{\mathbf{w}}_4^i$, $\tilde{\mathbf{w}}_5^i$ and $\tilde{\mathbf{w}}_6^i$.

Let us also define a measure of the convexity of a given quadratic polynomial $q = a_{11}x^2 + a_{12}xy + a_{22}y^2 + b_1 x + b_2 y + c$:

$$conv(q) := \max\left(|\lambda_1|, |\lambda_2|\right), \tag{8}$$

where λ_1 and λ_2 are the roots of the quadratic equation $\lambda^2 - 2(a_{11} + a_{22})\lambda + (4a_{11}a_{22} - a_{12}^2) = 0$.

With this definition of p, $p_\star$, p_1, p_2 and p_3 the reconstruction proceeds as follows. We first traverse all the elements of $\widetilde{\mathcal{T}}^h$. For each $\tilde{e} \in \widetilde{\mathcal{T}}^h$ we construct the five quadratic polynomials p, $p_\star$, p_1, p_2, p_3 and we select the one which has minimal convexity. Let this quadratic polynomial be denoted by $p_{\tilde{e}}$:

$$\begin{aligned} &p_{\tilde{e}} \in \{p, p_\star, p_1, p_2, p_3\}, \\ &conv\left(p_{\tilde{e}}\right) = \min\left(conv(p), conv(p_\star), conv(p_1), conv(p_2), conv(p_3)\right). \end{aligned}$$

Next, we define a piecewise quadratic, continuous function u_h over the original triangulation $\mathcal{T}^h$ as follows. Over each element $e \in \mathcal{T}^h$ of the original triangulation

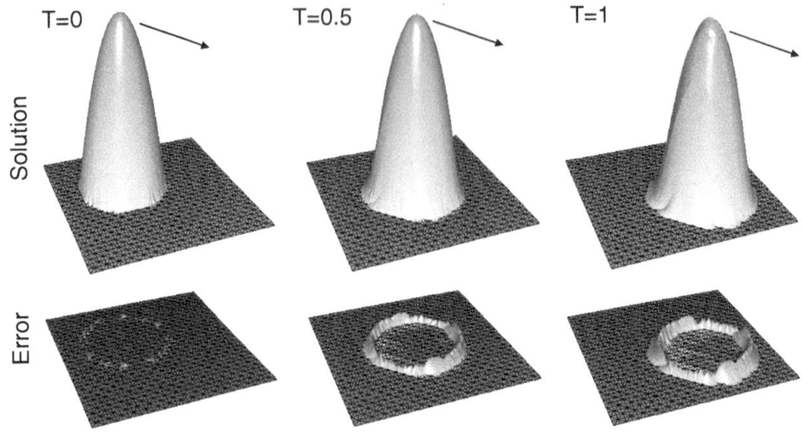

Fig. 3. Linear transport of a quadratic function

u_h is quadratic and uniquely defined by its values at the vertices $\mathbf{v}_1$, $\mathbf{v}_4$ and $\mathbf{v}_6$ and midpoints $\mathbf{v}_2$, $\mathbf{v}_3$, and $\mathbf{v}_5$ of e. These values are defined as the averaged of the values of all relevant interpolants $\tilde{p}_i$:

$$u_h(\mathbf{v}_i, t_n + 1) = \frac{1}{\dim\left(\mathcal{E}\left(\mathbf{v}_i\right)\right)} \sum_{\tilde{e} \in \mathcal{E}(\mathbf{v}_i)} p_{\tilde{e}}\left(\mathbf{v}_i\right)$$

3 Numerical Examples

3.1 Linear Transport

In order to investigate the behavior of the proposed scheme we first solved a simple problem with linear Hamiltonian. The Hamiltonian, $H(u_x, u_y) = u_x + u_y$, implies transport of the initial condition along the vector $(1, 1)$. We first consider a quadratic function with compact support for the initial condition:

$$u(x, 0) = \max(0, 1 - x^2 - y^2) = \begin{cases} (1 - x^2 - y^2) & \text{for} \quad x^2 + y^2 \leq 1 \\ 0 & \text{for} \quad x^2 + y^2 > 1 \end{cases}$$

In the interior of its support, the initial condition is interpolated exactly by the scheme. Since the mesh is not aligned with the boundary of its support, the error is initially generated along the unit circle (Figure 3). As the solution is transported with time, the error is sill concentrated at the edge of its support. While it grows in magnitude, elsewhere (both in the wake and inside the solution's support) it is eventually damped out by the convexity minimizing scheme.

In order to assess the convergence of the scheme we performed a series of numerical runs with slightly different initial condition:

$$u(x, 0) = \max(0, 1 - 16x^2 - 16y^2)^4 = \begin{cases} (1 - 16x^2 - 16y^2)^4 & \text{for} \quad x^2 + y^2 \leq \frac{1}{4} \\ 0 & \text{for} \quad x^2 + y^2 > \frac{1}{4} \end{cases}$$

Table 1. Convergence results for transport of a compactly supported initial condition

Level	$(dx = 0.1, dt = 0.0007)$ L_2 Error	Order	$(dx = 0.1, dt = 0.0014)$ L_2 Error	Order	$(dx = 0.1, dt = 0.0056)$ L_2 Error	Order
1 (dx, dt)	5.717×10^{-2}		5.127×10^{-2}		3.425×10^{-2}	
2 $(dx/2, dt/2)$	1.636×10^{-2}	1.80	1.515×10^{-2}	1.76	6.581×10^{-3}	2.37
3 $(dx/4, dt/4)$	2.396×10^{-3}	2.77	2.133×10^{-3}	2.83	1.829×10^{-3}	1.85
4 $(dx/8, dt/8)$	5.682×10^{-4}	2.08	5.222×10^{-4}	2.03	7.651×10^{-4}	1.26
5 $(dx/16, dt/16)$	1.385×10^{-4}	2.04	1.401×10^{-4}	1.89	3.564×10^{-4}	1.1
6 $(dx/32, dt/32)$	3.882×10^{-5}	1.83	4.783×10^{-5}	1.54	1.836×10^{-4}	0.95

We again considered a linear hamiltonian $H(u_x, u_y) = u_x + u_y$. We further selected a unit domain centered at the origin. The problem was solved in the time interval $t \in [0, 0.14]$ for a series of regular meshes with parameter h and corresponding timesteps dt. The results are reported in Table 1. At each level, both h and dt are refined simultaneously by a factor of 2. Each column reports the results for a particular fixed CFL ratio, e.g. h/dt is held constant. For all CFL considered, initially the scheme displays second order. As the mesh is refined, the order of the larger CFL ratios deteriorates. However, for sufficiently small CFL ratio, the scheme retains second order accuracy.

3.2 Nonlinear Examples with Convex Hamiltonian

Radial Burger's equation. In the first example with nonlinear Hamiltonian (Figure 4) we use $H(u_x, u_y) = u_x^2 + u_y^2$ and an the initial guess:

$$u(x, 0) = \max(0, 1 - x^2 - y^2) = \begin{cases} 1 - x^2 - y^2 & \text{for} \quad x^2 + y^2 \leq 1 \\ 0 & \text{for} \quad x^2 + y^2 > 1 \end{cases}$$

The exact solution to this problem can be easily obtained in analytical form by reformulating the problem in cylindrical coordinates and reducing it to a 1D Burgers equation. The solution maintains its support in the unit circle at all times. While initially smooth inside its support, it develops a sharp corner at the origin at $t = 0.25$. This is followed by continual decrease in its amplitude.

The problem is solved on a general triangular mesh. The triangulation is obtained by meshing a circle centered at the origin and with radius 0.95 inside a square domain of dimensions 4×4. The numerical solution as well as he error is shown in Figure 4. It should be emphasized that the mesh is neither aligned with the boundary of the solution's support (the unit circle), nor with the origin, where a sharp corner is developed. Due to the initial condition being interpolated exactly, the error is first generated along the circle $x^2 + y^2 = 1$. For $t > 0$, it can be seen that the error propagates form the boundary of the support $(x^2 + y^2 = 1)$ to the interior of the domain. No discernible oscillations are present at any time, even after the sharp point is developed after $t > 0.25$.

2D Burger's equation. In this second example with nonlinear Hamiltonian, we solve a problem with $H(u_x, u_y) = \frac{1}{2}(u_x + u_y + 1)^2)$ on a 30×30 regular triangular grid of a 2 domain. The problem is equivalent to a 2D Burger's equation. We use periodic boundary conditions with an initial condition:

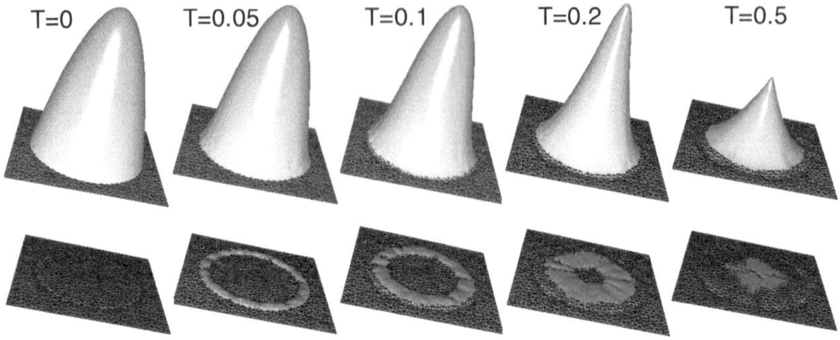

Fig. 4. Solution and error for $H(u_x, u_y) = u_x^2 + u_y^2$ and $u(x,0) = \max(0, 1 - x^2 - y^2)$

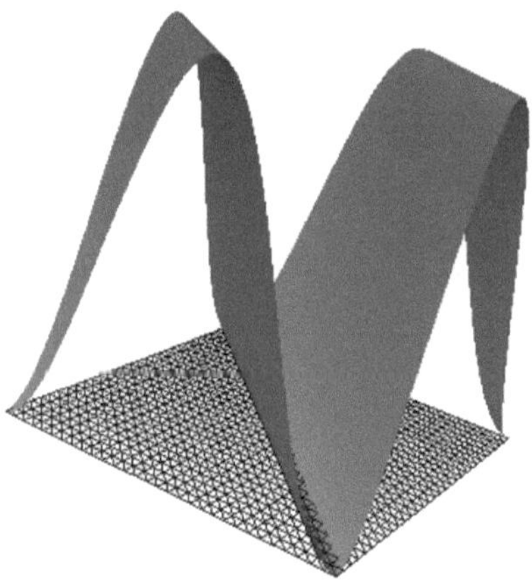

(a) IC: $u(\mathbf{x}) = -\frac{1}{2}\cos(\pi(x+y))$

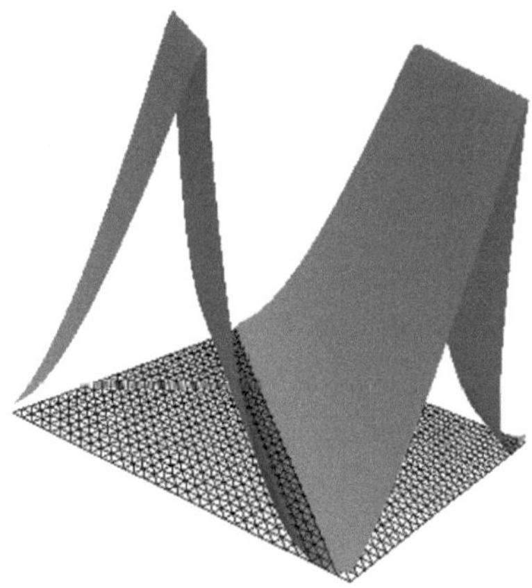

(b) Solution at $t = 1.5/\pi$

Fig. 5. Initial condition (a) and solution (b) for $H(u_x, u_y) = \frac{1}{2}(u_x + u_y + 1)^2$

$$u(\mathbf{x}, 0) = -\frac{1}{2}\cos(\pi(x+y))$$

The solution, initially smooth, develops discontinous derivatives across two edges, as seen from Figure 5. As in the previous examples, there are no discernible oscillations. The two sharp edges are captured well by the scheme, even though they are not aligned with element edges.

References

1. Kružkov, S.N.: Generalized solutions of Hamilton-Jacobi equations of eikonal type. I. Statement of the problems; existence, uniqueness and stability theorems; certain properties of the solutions. Mat. Sb. 98(140), 3(11), 450–493, 496 (1975)
2. Barles, G.: Solutions de viscosité des équations de Hamilton-Jacobi. Mathématiques & Applications, vol. 17. Springer, Heidelberg (1994)

3. Crandall, M.G., Ishii, H., Lions, P.L.: User's guide to viscosity solutions of second order partial differential equations. Bull. Amer. Math. Soc. 27(1), 1–67 (1992)
4. Crandall, M.G., Lions, P.L.: Viscosity solutions of Hamilton-Jacobi equations. Trans. Amer. Math. Soc. 277(1), 1–42 (1983)
5. Crandall, M.G., Evans, L.C., Lions, P.L.: Some properties of viscosity solutions of Hamilton-Jacobi equations. Trans. Amer. Math. Soc. 282(2), 487–502 (1984)
6. Lions, P.L.: Generalized solutions of Hamilton-Jacobi equations. Research Notes in Mathematics, vol. 69. Pitman, Boston (1982)
7. Souganidis, P.E.: Approximation schemes for viscosity solutions of Hamilton-Jacobi equations. J. Differential Equations 59(1), 1–43 (1985)
8. Osher, S., Sethian, J.A.: Fronts propagating with curvature-dependent speed: algorithms based on Hamilton-Jacobi formulations. J. Comput. Phys. 79(1), 12–49 (1988)
9. Osher, S., Shu, C.W.: High-order essentially nonoscillatory schemes for Hamilton-Jacobi equations. SIAM J. Numer. Anal. 28(4), 907–922 (1991)
10. Abgrall, R.: Numerical discretization of the first-order Hamilton-Jacobi equation on triangular meshes. Comm. Pure Appl. Math. 49(12), 1339–1373 (1996)
11. Jiang, G.S., Peng, D.: Weighted ENO schemes for Hamilton-Jacobi equations. SIAM J. Sci. Comput. 21(6), 2126–2143 (2000)
12. Zhang, Y.T., Shu, C.W.: High-order WENO schemes for Hamilton-Jacobi equations on triangular meshes. SIAM J. Sci. Comput. 24(3), 1005–1030 (2002)
13. Lin, C.T., Tadmor, E.: High-resolution nonoscillatory central schemes for Hamilton-Jacobi equations. SIAM J. Sci. Comput. 21(6), 2163–2186 (2000)
14. Levy, D., Nayak, S., Shu, C.W., Zhang, Y.T.: Central WENO schemes for Hamilton-Jacobi equations on triangular meshes. SIAM J. Sci. Comput. 28(6), 2229–2247 (2006)
15. Kurganov, A., Tadmor, E.: New high-resolution semi-discrete central schemes for Hamilton-Jacobi equations. J. Comput. Phys. 160(2), 720–742 (2000)
16. Bryson, S., Levy, D.: High-order semi-discrete central-upwind schemes for multidimensional Hamilton-Jacobi equations. J. Comput. Phys. 189(1), 63–87 (2003)
17. Bryson, S., Levy, D.: High-order central WENO schemes for multidimensional Hamilton-Jacobi equations. SIAM J. Numer. Anal. 41(4), 1339–1369 (2003)
18. Kurganov, A., Noelle, S., Petrova, G.: Semidiscrete central-upwind schemes for hyperbolic conservation laws and Hamilton-Jacobi equations. SIAM J. Sci. Comput. 23(3), 707–740 (2001)
19. Bryson, S., Kurganov, A., Levy, D., Petrova, G.: Semi-discrete central-upwind schemes with reduced dissipation for Hamilton-Jacobi equations. IMA J. Numer. Anal. 25(1), 113–138 (2005)
20. Christov, I., Popov, B.: New non-oscillatory central schemes on unstructured triangulations for hyperbolic systems of conservation laws. J. Comput. Phys. 227(11), 5736–5757 (2008)
21. Gottlieb, S., Shu, C.W., Tadmor, E.: Strong stability-preserving high-order time discretization methods. SIAM Rev. 43(1), 89–112 (2001)
22. Shu, C.W.: Total-variation-diminishing time discretizations. SIAM J. Sci. Statist. Comput. 9(6), 1073–1084 (1988)
23. Shu, C.W., Osher, S.: Efficient implementation of essentially nonoscillatory shock-capturing schemes. J. Comput. Phys. 77(2), 439–471 (1988)

Grid Method for Solving the Flow of Traffic Problem on the Highway in a Class of Discontinuous Functions

Mahir Rasulov[1] and Kenan Gocer[2]

[1] Beykent University, Department of Mathematics and Computing
[2] Beykent University, Department of Engineering and Architecture
Sisli-Ayazaga Campus, 34396 Istanbul, Turkey
{mresulov,kgocer}@beykent.edu.tr

Abstract. In this paper a new method for obtaining an exact and numerical solution of initial value problem for a first order nonlinear partial equation which describes the traffic flow on highway in a class of discontinuous functions is developed. For this goal, at first, the special auxiliary problem having some advantages over the main problem has been suggested and using the solution of the auxiliary problem an original method for finding the weak solution of the main problem has been suggested. Using the advantages of the auxiliary problem the new numerical method for obtaining a solution which expresses the all physical properties accurately is purposed, too.

1 Introduction

It is known that many problems of sciences and technology are reduced to find the solution of the initial or initial-boundary value problems of the first order nonlinear partial differential equations [1], [7].

The traffic flow problem on highways is one of the above mentioned important problem. Using the kinematic theory of wave this problem has been presented for the first time in [1], [6].

We denote the density of vehicles per unit length of highway, and the flux function of posing x section of the highway at per unit of time by $\rho(x,t)$ and $q(x,t)$, respectively. If we assume that there will be no vehicle joining to or leaving from the highway, the following balance equation

$$\frac{d}{dt}\int_{x_1}^{x_2} \rho(x,t)dx + q(x_1,t) - q(x_2,t) = 0 \qquad (1)$$

holds.

As the first approach, in traffic flow problem the flux function $q(x,t)$ is expressed by the local density $\rho(x,t)$ as $q = Q(\rho) = \rho V(\rho) = v_{max}\rho(1-\frac{\rho}{\rho_{max}})$, here v_{max} and ρ_{max} are the maximum values of the speed of vehicles and density, respectively.

S. Margenov, L.G. Vulkov, and J. Waśniewski (Eds.): NAA 2008, LNCS 5434, pp. 486–493, 2009.

If the functions $\rho(x,t)$ and $Q(\rho)$ are continuously differentiable then, the equation (1) is equivalent to

$$\frac{\partial \rho(x,t)}{\partial t} + \frac{\partial Q(\rho(x,t))}{\partial x} = 0. \tag{2}$$

In practical problems it is important to know the functional relation between Q and ρ. But, in many problems of the theory of kinematic waves it would be a better approximation to assume that q is dependent not only on ρ as well as on ρ_x, [5], [7]. A simple all together assumption is to take $q = Q(\rho) - \nu\rho_x$, here ν is certain constant. In traffic flow, for example, it can be claimed that drivers will reduce their speed to account for an increasing density of vehicles ahead, and conversely. This leads to taking account of diffusion of wave, and in this case

$$v = V(\rho) - \frac{\nu}{\rho}\rho_x = V(\rho) - \nu\frac{\partial \ln \nu}{\partial x}. \tag{3}$$

The second effect is the time lag in the response of the driver and of his car to any changes in the flow conditions. One way to introduce this effect is to consider the expression for v in (3) as a desired velocity towards in which the driver accelerates, therefore the equation $\frac{\partial v}{\partial t} + v\frac{\partial v}{\partial x} = -\frac{1}{\tau}(v - V(\rho) + \frac{\nu}{\rho}\rho_x)$ may be introduced for the acceleration.

When ν and τ are both small in a suitable non dimensional measure, v is approximated by $V(\rho)$, and we get the simpler theory. With the higher order terms included in last equation, we expect shocks to appear as smooth steps and so on.

In this paper we will consider the equation (2) with following

$$\rho(x,0) = \rho_0(x) \tag{4}$$

initial condition. Here, $\rho_0(x)$ is an initial distribution of vehicles on highway. The problem (2),(4) is called as the main problem.

From a physical point of view the function $Q(\rho(x,t))$ must be a concave function, that is, $Q''(\rho) \leq 0$ and the function $\rho_0(x)$ may be as a piecewise continuous or as a continuous function with compact support having a negative and a positive slops.

In [2],[3],[7] proved that there is a such number T_0 that the derivatives of the solution of problem (2),(4) with respect to x and t become infinite for any $t \geq T_0$. Hence, the solution of the problem (2),(4) in a classical meaning does not exist. Therefore, it is necessary to find the weak solution of this problem. On the order hand the solution of the problem (2),(4) obtained by the method of characteristics is a closed form $\rho(x,t) = \rho_0(x - Q'(\rho)t)$, but it is often impossible to obtain an explicit expression for the unknown function. The obtained functional relation is called as the alternative form of the main problem.

2 Exact Solution of the Main Problem

In this section we will obtain the exact solution of the problem (2),(4).

2.1 The Cauchy Problem with Continuous Initial Profile

In this section we will solve the eq.(2) when the function $\rho_0(x)$ is continuous and has a compact support, that is, supp $\rho_0(x) = \{x \mid \rho_0(x) \neq 0\}$ is bounded and has a negative and a positive slopes. As the initial distribution of initial density of vehicles we will take the function

$$\rho_0(x) = \begin{cases} -\frac{\rho_{max}-\rho_{min}}{l^2}x^2 + \rho_{max}, & \mid x \mid \leq l \\ \rho_{min}, & \mid x \mid > l. \end{cases} \tag{5}$$

Here, ρ_{max} and ρ_{min} are the maximum and minimum values of the density of vehicles on a highway with lengths of l.

As it is known that, there is the number T_0 such that for $t > T_0$ in the solution of the problem (2),(4) a jump whose location is not known beforehand occurs. The weak solution of the problem is defined as

Definition 1. *Nonnegative and satisfying the initial condition (4) the function $\rho(x,t)$ is called a weak solution of the problem (2),(4) if the following integral relation*

$$\int_{D_T} \{\rho\varphi_t + Q(\rho)\varphi_x\}\, dxdt + \int_{-\infty}^{\infty} \varphi(x,0)\rho_0(x)dx = 0 \tag{6}$$

holds for every test functions $\varphi(x,t)$ defined and twice differentiable in the upper half plane and vanishes for $\mid x \mid \to l$ and $\varphi(x,T) = 0$.

The weak solution in the meaning of (6) of the problem (2),(4) consists of two parts, the continuous and shock parts. The continuous differentiable part satisfies the equation (2) and the discontinuous part satisfies the Rankino- Hugoniot condition.

In order to find the weak solution, according to [3], [4], the auxiliary problem

$$\frac{\partial v(x,t)}{\partial t} + Q(\rho(x,t)) = 0, \tag{7}$$

$$v(x,0) = v_0(x) \tag{8}$$

is introduced. Here, $v_0(x)$ is any absolutely continuous function satisfying the following equation $\frac{dv_0(x)}{dx} = u_0(x)$.

Theorem 1. *If $v(x,t)$ is a soft solution of the auxiliary problem (7),(8), then the function $\rho(x,t)$ defined by*

$$\rho(x,t) = \frac{\partial v(x,t)}{\partial x} \tag{9}$$

is the solution of the main problem (2),(4).

Theorem 2. *If $v(x,t)$ is the solution of the auxiliary problem (7),(8), then*
1^0. *the function $\rho(x,t)$ defined by (9) is the weak solution of the main problem;*
2^0. *$v(x,t)$ is an absolutely continuously function.*

The auxiliary problem has the following advantages:

- The function $v(x,t)$ is smoother than $\rho(x,t)$;
- $\rho(x,t)$ can be determined without using the derivatives $\frac{\partial\rho}{\partial x}$ and $\frac{\partial\rho}{\partial t}$ which are not defined at the neighborhoods of the points of discontinuities.

The function $v_0(x)$ is defined as

$$v_0(x) = \begin{cases} v_\ell, & x < -l, \\ -\frac{\rho_{max}-\rho_{min}}{3l^2}x^3 + \rho_{max}x, & \mid x \mid < l, \\ v_r, & x > l, \end{cases} \tag{10}$$

here, $v_\ell = \frac{\rho_{max}-\rho_{min}}{3}l - \rho_{max}l$ and $v_r = -\frac{\rho_{max}-\rho_{min}}{3}l + \rho_{max}l$.

The solution of the problem (7),(8) obtained by characteristic method is

$$v(x,t) = \left[-Q\left(\frac{dv_0(\xi)}{d\xi}\right) + \left(\frac{dv_0}{d\xi}\right)Q'\left(\frac{dv_0}{d\xi}\right)\right]t + v_0(\xi), \tag{11}$$

here $\xi = x - Q'(\rho)t$, [3]. Taking the relation $Q(\rho) = v_{max}(1-\frac{\rho}{\rho_{max}})\rho$ into account we have

$$v(x,t) = -\frac{v_m}{\rho_m}\rho^2(x,t)t + v_0(\xi). \tag{12}$$

Since the equation (2) describes the conservation of mass, the value of the integral $E(t) = \int_{-\infty}^{\infty}\rho(x,t)dx$ is constant for any t. The number $E(0) = \int_{-\infty}^{\infty}\rho_0(x,0)dx = \frac{2l}{3}(2\rho_{max}+\rho_{min})$ is called the critical value of the function $v(x,t)$.

Definition 2. *The function defined by*

$$v_{ext}(x,t) = \begin{cases} v(x,t), \ v < E(0) \\ E(0), \ v \geq E(0) \end{cases} \tag{13}$$

is called the extended solution of the problem (7),(8).

From Theorem 1, for the weak solution of the main problem (2),(4), we have

$$\rho_{ext}(x,t) = \frac{\partial v_{ext}(x,t)}{\partial x}. \tag{14}$$

2.2 The Cauchy Problem with Discontinuous Initial Profile

In this section we investigate the following problem. We consider a stream of vehicles stopped by a red signal at point $x = 0$. The road is jammed initially behind the signal and there is no traffic ahead of the signal. As soon as the signal turns green, the stream of vehicles starts moving across $x = 0$. The initial state of vehicles of traffic is given by

$$\rho(x,0) = \begin{cases} \rho_l, \ x < 0 \\ \rho_r, \ x > 0 \end{cases} \tag{15}$$

and their graphs are illustrated in Figure 1. Here, ρ_l and ρ_r are known constants. The problem (2),(15) is called the Riemann problem. We will investigate the following two cases: (i) $\rho_l > \rho_r$, (ii) $\rho_l < \rho_r$.

Since $Q(\rho)$ is a concave function, according to the general theory when $\rho_l < \rho_r$ the solution of the problem (2),(15) has the points of discontinuity of which the location is not known beforehand and for this solution the entropy condition is satisfied too. However, when $\rho_l > \rho_r$, in the solution the shock does not occur, but the first derivatives have first type discontinuities [2]. It is clear that the classical solution for both cases does not exist. By simple calculations for the function $\rho(x,t)$ we have the following

$$\rho(x,t) = \begin{cases} \rho_l, & \frac{x}{t} < v_m(1 - \frac{2}{\rho_m})\rho_l, \\ -\frac{\rho_m}{v_m}\frac{x}{2t} + \frac{\rho_m}{2}, & v_m(1 - \frac{2}{\rho_m})\rho_l < \frac{x}{t} < v_m(1 - \frac{2}{\rho_m})\rho_r, \\ \rho_r, & \frac{x}{t} > v_m(1 - \frac{2}{\rho_m})\rho_r \end{cases} \qquad (16)$$

expression.

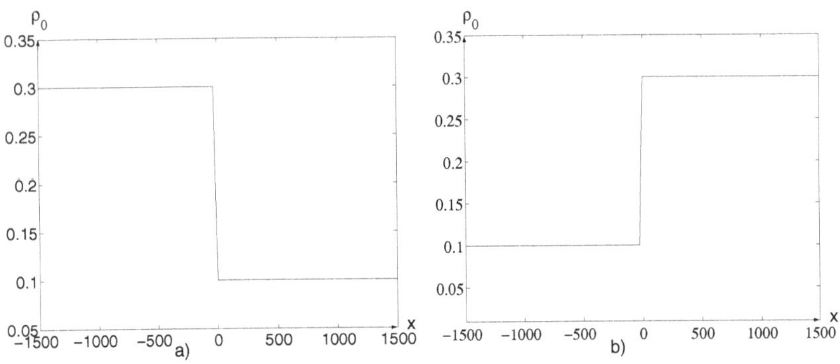

Fig. 1. Initial profiles a) $\rho_l > \rho_r$; b) $\rho_l < \rho_r$

Definition 3. *Nonnegative and satisfying the initial condition (15) the function $\rho(x,t)$ is called a weak solution of the problem (2),(15) if the following integral relation*

$$\int_{D_T} \{\rho\varphi_t + Q(\rho)\varphi_x\}\, dxdt + \int_{-\infty}^{0} \varphi(x,0)\rho_\ell dx + \int_{0}^{\infty} \varphi(x,0)\rho_r dx = 0 \qquad (17)$$

holds for every test functions $\varphi(x,t)$ defined and twice differentiable in the upper half plane and vanishes for $\mid x \mid \rightarrow l$ and $\varphi(x,T) = 0$.

In this case the solution of the auxiliary problem (7),(8) has the following

$$v(x,t) = \begin{cases} v_-, & \xi < 0 \\ v_+, & \xi > 0 \end{cases} \qquad (18)$$

form. Here, $v_- = -\frac{v_{max}}{\rho_{max}}\rho_l^2 t + \rho_l\left[x - v_{max}\left(1 - \frac{2\rho_l}{\rho_{max}}\right)\right]t$ and $v_+ = -\frac{v_{max}}{\rho_{max}}\rho_r^2 t + \rho_r\left[x - v_{max}\left(1 - \frac{2\rho_r}{\rho_{max}}\right)\right]t$.

It is clear that the shock of $v(x,t)$ from v_- to v_+ must be found through $v_- = v_+$. From this equation we have $\frac{x}{t} = v_{max} - \frac{v_{max}}{\rho_{max}}(\rho_r + \rho_l) \equiv S$, that this expression is the solution of $\frac{dx}{dt} = \frac{Q(\rho_r)-Q(\rho_l)}{\rho_r-\rho_l}$ which is known as the Rankine-Hugoniot condition.

Taking Theorem 1 into account and (18) for the weak solution of the problem (2), (15) we get the following

$$\rho(x,t) = \begin{cases} \rho_l, \ \frac{x}{t} < S \\ \rho_r, \ \frac{x}{t} > S \end{cases} \tag{19}$$

expression. The graphs of the solutions (16) and (19) are shown in Figure 2.

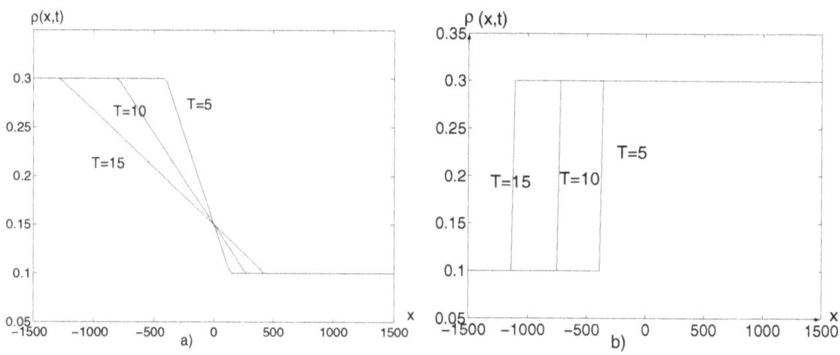

Fig. 2. Dynamical Distributions a) $\rho_l > \rho_r$; b) $\rho_l < \rho_r$

3 Finite Differences Schemes in a Class of Discontinuous Functions

In this section, we develop a sensitive numerical method for the solution of the problem (2), (4), and investigate some properties of it. As shown in section 1, the solution of this problem has discontinuous points, whose location are unknown beforehand. These properties in the exact solution do not permit the application of classical numerical methods to this problem directly. By using the advantages of the suggested auxiliary problem, a new numerical algorithm is proposed. In [3], [4] the suggested numerical method was studied for only a nonlinear scalar equation.

3.1 The Finite Differences Scheme for Cauchy Problem

In order to construct the method, the domain of definition of the problem is covered by the following grid,

$$\omega_{h,\tau} = \{(x_i, t_k) \mid x_i = ih, t_k = k\tau, i = 0, \pm1, \pm2, ..., k = 0, 1, 2, ...; h > 0, \tau > 0\}$$

where, h and τ are steps of the grid for x and t variables, respectively.

Firstly, the problem of (7),(8) is approximated by the finite difference scheme at any point (i,k) of the grid $\omega_{h,\tau}$ as follows

$$V_{i,k+1} = V_{i,k} - \tau Q\left(\wp_{i,k}\right), \tag{20}$$

$$V_{i,0} = v_0(x_i). \tag{21}$$

A function $v_0(x_i)$ is any solution of the following finite difference equation

$$(V_0)_{\bar{x}} = \rho_0(x_i). \tag{22}$$

It is easy to prove that

$$\wp_{i,k+1} = \frac{V_{i,k+1} - V_{i-1,k+1}}{h}. \tag{23}$$

Here, the grid functions $\wp_{i,k}$ and $V_{i,k}$ represent approximate values of the functions $\rho(x,t)$ and $v(x,t)$ at point (i,k) respectively.

In order to prove (23), it is sufficient first to write the equation (20) at a point $(i-1,k)$, then subtract it from (20) and divide by 2. By taking (23) into consideration, it is seen that $\wp_{i,k}$ satisfies the following nonlinear system of algebraic equations

$$\wp_{i,k+1} = \wp_{i,k} - \frac{\tau}{h}\left(Q\left(\wp_{i,k}\right) - Q\left(\wp_{i-1,k}\right)\right). \tag{24}$$

Theorem 3. *The expression $E_1(t_k) = h\sum_i \wp_{i,k}$, is independent of time.*

Proof. Firstly, considering $E_1(t_{k+1})$, that is

$$E_1(t_{k+1}) = h\sum_{i=-\infty}^{\infty} \wp_{i,k+1} = h\sum_{i=-\infty}^{\infty} \frac{V_{i,k+1} - V_{i-1,k+1}}{h} =$$

$$h\sum_{i=-\infty}^{\infty}\left\{\frac{V_{i,k} - V_{i-1,k}}{h} - \frac{\tau}{h}\left[Q(\wp_{i,k}) - Q(\wp_{i-1,k})\right]\right\} = \sum_{i=-\infty}^{\infty} \wp_{i,k}.$$

This means that $E_1(t_k)$ is independent of k. This completes the proof.

Definition 4. *The quantities $E_1(0)$ defined by $\bar{E}_1(0) = h\sum_i \wp_{i,0}$, is called the critical value for the grid functions $V_{i,k}$.*

Definition 5. *The mesh function defined by*

$$V_{i,k}^{ext} = \begin{cases} V_{i,k}, & V_{i,k} < E_1(0) \\ E_1(0), & V_{i,k} \geq E_1(0), \end{cases}$$

is called the extended solutions of the problem (20),(21).

From Theorem 4, we have $\wp_{i,k}^{ext} = \left(V_{i,k}^{ext}\right)_{\bar{x}}$, and this expression is called the extended numerical solution of the main problem.

As it can be seen from (20)-(21), the suggested algorithms are very effective and economic from a computational point of view.

Additionally, considering (9), we rewrite (7) as $\frac{\partial v(x,t)}{\partial t} + Q(\rho(x,t)) = 0$.

Then, by applying, for example, the Runge-Kutta method to the equations above, we can write a higher order finite- difference scheme for the main problem with respect to τ.

References

1. Lighthill, M.J., Whitham, G.B.: On Kinematic waves. II. A Theory of Traffic Flow on Long Crowded Roads. Proc. Roy. Soc. London ser. A 229, 317–345 (1955)
2. Oleinik, O.A.: Discontinuous Solitions of Nonlinear Differential Equations. Usp.Math. Nauk 12 (1957)
3. Rasulov, M.A.: On a Method of Solving the Cauchy Problem for a First Order Nonlinear Equation of Hyperbolic Type with a Smooth Initial Condition. Soviet Math. Dok. 43(1) (1991)
4. Rasulov, M.A., Ragimova, T.A.: A Numerical Method of the Solution of ne Nonlinear Equation of a Hyperbolic Type of the First Order Differential Equations. Minsk 28(7), 2056–2063 (1992)
5. Zhang, P., Lin, R.-X., Wong S.C.: High-Resolution Numerical Approximation of Traffic Flow Problems with Variable Lanes and Free-Flow velocities. Physical Review, E-71, 056704 (2005)
6. Richards, P.I.: Shock Waves on the Highway. Oper. Res. 4(42) (1956)
7. Whitham, G.B.: Linear and Nonlinear Waves. Wiley Int., New York (1974)

The Study of Filtration of Two Phase Fluid in a Porous Medium in a Class of Discontinuous Functions

Mahir Rasulov[1] and R. Haluk Kul[2]

[1] Beykent University, Department of Mathematics and Computing
[2] Beykent University, Department of Management Information Systems
Sisli-Ayazaga Campus, 34396 Istanbul, Turkey
{mresulov,hkul}@beykent.edu.tr

Abstract. In this study a new numerical method for obtaining the solution of the problem for 2-D system of equations which describes the filtration of the multi-phase fluid in porous medium with suitable initial and boundary conditions is proposed. For this aim, the special auxiliary problem having some advantages over the main problem is introduced. An efficient and accurate numerical algorithm based on the auxiliary problem is derived. Furthermore, some results of numerical experiments from the related subjects of physics are presented.

1 Introduction

It is known that the filtration of multi-phase fluids in porous medium is described by the Muskat-Mires's system of differential equations given below, [1]

$$m\frac{\partial}{\partial t}\left[\frac{\sigma_w}{\beta_w}\right] + div\,(w_w) + q_w = 0, \tag{1}$$

$$m\frac{\partial}{\partial t}\left[\frac{\sigma_o}{\beta_o}\right] + div\,(w_o) + q_o = 0, \tag{2}$$

$$w_\ell = \frac{kk_\ell(\sigma)}{\mu_\ell \beta_\ell}\nabla p, \quad (\ell = w, o), \tag{3}$$

$$\sigma_o + \sigma_w = 1. \tag{4}$$

Here, m is porosity; σ_w and σ_o are water and oil saturations, respectively; $k_0(\sigma)$ and $k_w(\sigma)$ are permeability of oil and water, respectively; k is the absolute permeability of rock; p is the pressure of the petrol reservoir; μ_0 and μ_w are the oil and water phase viscosities; F_o and F_w are fractional flow of oil and water; q_0 is the oil production rate; q_w is the water production rate.

For the sake of simplicity, we assume that $\beta_0 = \beta_w = 1$ and $\sigma = \sigma_w(x,t)$. It is known that the system of equations (1)- (4) can be reduced to the form

S. Margenov, L.G. Vulkov, and J. Waśniewski (Eds.): NAA 2008, LNCS 5434, pp. 494–500, 2009.

$$div\left[k\left(\frac{k_w(\sigma)}{\mu_w}+\frac{k_o(\sigma)}{\mu_o}\right)grad\ p\right]=-q, \tag{5}$$

$$m\frac{\partial\sigma}{\partial t}+div\,(w(t)F_w(\sigma))=-q_w. \tag{6}$$

Indeed, taking (4) into consideration and summing (1) and (2) we have equation

$$div\,(w_w+w_o)=-q, \tag{7}$$

where $q=q_w+q_o$.

Introducing the following notation $w(t) = w_w + w_o$, and considering the equation (3) we get

$$w=-k\left[\left(\frac{k_o(\sigma)}{\mu_o}+\frac{k_w(\sigma)}{\mu_w}\right)\nabla p\right]. \tag{8}$$

From this relation we have

$$\nabla p=-\frac{w(t)}{k\left[\frac{k_w(\sigma)}{\mu_w}+\frac{k_o}{\mu_o}\right]}. \tag{9}$$

Substituting (9) into (6), we obtain

$$m\frac{\partial\sigma}{\partial t}+div\left[-k\frac{k_w\,(\sigma)}{\mu_w}\frac{w(t)}{-k\left(\frac{k_w(\sigma)}{\mu_w}+\frac{k_o(\sigma)}{\mu_o}\right)}\right]+q_w=0. \tag{10}$$

Let us denote the following notations $F_w(\sigma)$ and $F_o(\sigma)$ which describe fractional flow of oil and water in total value of fluid, respectively,

$$F_w(\sigma)=\frac{\frac{k_w(\sigma)}{\mu_w}}{\frac{k_w(\sigma)}{\mu_w}+\frac{k_o(\sigma)}{\mu_o}},\quad F_o(\sigma_o)=\frac{\frac{k_o(\sigma)}{\mu_o}}{\frac{k_w(\sigma)}{\mu_w}+\frac{k_o(\sigma)}{\mu_o}}. \tag{11}$$

The functions $F_w(\sigma)$ and $F_o(\sigma)$ are called Leverett's functions. It is seen from (11) that $F_w+F_o=1$. It should be noted that for the function F_w the following relations are fulfilled $q_w = qF_w$ and $q_o=(1-F_w)q$. Then the equation (10) can be rewritten as

$$m\frac{\partial\sigma}{\partial t}+div\,(wF_w(\sigma))=q_w.$$

Hence, the system of equations (1)-(4) has been reduced to the following form

$$div\left[k\left(\frac{k_w}{\mu_w}+\frac{k_0}{\mu_0}\right)grad\ p\right]=-q, \tag{12}$$

$$m\frac{\partial\sigma}{\partial t}+div\,(wF_w(\sigma))=q_0. \tag{13}$$

The volume of water and oil are expressed by the given formulas

$$q_w = \sum_{\nu=1}^{N_1} Q_w^{(\nu)} \delta_{i,j}, \quad q_o = \sum_{\mu=1}^{N_2} Q_o^{(\mu)} \delta_{i,j}.$$

Here, N_1 and N_2 are the numbers of water and oil wells, and $\delta_{i,j}$ is the Kronecker's function given by $\delta_{i,j} = \begin{cases} 1, & x = x_i^{(\ell)}, y = y_j^{(\ell)} \\ 0, & x \neq x_i^{(\ell)}, y \neq y_j^{(\ell)} \end{cases}, \ (\ell = \nu, \mu)$.

Now, we formulate the initial and boundary conditions for the equations (12) and (13). Let us assume that D denotes the contour of the oil reservoir and ∂D denotes the boundary of the domain D. Here, the symbols appear in the domain indicate the oil and water wells. The points of locations of oil and water wells are shown as $\Gamma_\nu^{(o)} = \left({}^o x^{(\nu)}, {}^o y^{(\nu)}\right)$, $(\nu = 1, 2, \ldots, N_1)$, and $\Gamma_\mu^{(w)} = \left({}^w x^{(\mu)}, {}^w y^{(\mu)}\right)$, $(\mu = 1, 2, \ldots, N_2)$, respectively.

The initial and boundary conditions for the equation (12) are

$$p(x, y, 0) = p_0(x, y), \tag{14}$$

$$\left.\frac{\partial p}{\partial \vec{n}}\right|_{\partial D} = 0. \tag{15}$$

If we assume that the connected water in the porous medium is $\sigma_w^{(0)}$, for the equation (13), the initial and boundary conditions can be written as follows

$$\sigma_w\,(x, y, 0) = \sigma_w^{(0)}, \tag{16}$$

$$\sigma_w(x, y, t)|_{\Gamma_\nu^{(0)}} = 1 - \sigma_w^{(0)}, (\nu = 1, 2, \ldots, N_1). \tag{17}$$

It is known that the solution of such equation (13) even in one dimensional case with very smooth initial data has the points of discontinuities whose locations are unknown beforehand, [1], [3], [6]. As a matter of fact, in two dimensional case the number of discontinuities are so many as that it is impossible to define contact of oil water accurately. Existence of the points of discontinuities in the solution whose locations are unknown beforehand causes many problems when we apply a well known classical method to obtain solution of the investigated problem. In [2], [7] for the numerical solution of the investigated problem, a classical solution method has been applied. In [4], the front-tracking method for the one dimensional system of which describes the motion of three phase fluid is suggested.

Therefore, in order to solve the system of equations (1)-(3) (or (12)-(13)) with initial and boundary conditions (14)-(17) it is necessary to develop the more effective and sensitive numerical methods. In the literature, similar problems are attempted [2], [3].

Within the limit of this study, the finite differences method for finding the numerical solutions of the investigated problem is suggested. For this goal, the special auxiliary problem having some advantages over main problem is introduced, [1], [6]. At first, we substitute the following transformation.

$$\sigma = u - \int_0^t q_w(\tau)d\tau. \tag{18}$$

For the new function the equation (13) can be written as given below

$$m\frac{\partial u}{\partial t} + div\,(\boldsymbol{w}F_w) = 0. \tag{19}$$

2 The Auxiliary Problem

The following auxiliary problem for obtaining the water saturation is introduced

$$m\frac{\partial v(x,y,t)}{\partial t} + \boldsymbol{w}F_w(u) = 0, \tag{20}$$

$$v(x,y,0) = v_0(x,y), \tag{21}$$

$$u|_{\Gamma_\nu} = (1-\sigma_w) + \int_0^t q_w(\tau)d\tau, (\nu = 1, 2, ..., N_1) \tag{22}$$

$$\sigma_w(x,y,0) = \sigma_w^{(0)}. \tag{23}$$

Here, the function $v_0(x,y)$ is any continuous differentiable solution of the equation

$$\frac{\partial v_0(x,y)}{\partial x} + \frac{\partial v_0(x,y)}{\partial y} = \sigma_0(x,y). \tag{24}$$

2.1 Finite Differences Method

In order to construct the numerical method, the domain where x, y and t change continuously is covered by the following grid

$$\omega_{h_x h_y \tau} = \{x_i = ih_x, y_j = jh_y, t_n = n\tau, i = \overline{0, N_x}, j = \overline{0, N_y}, n = 0, 1, ...\}$$

where, h_x and h_y and τ are steps of the grid for x, y and t variables, respectively.

At first, we will approximate the problem (20)-(23) as

$$V_{i,j}^{n+1} = V_{i,j}^n - \tau \mathbf{W}_{i,j}^n F_w(U_{i,j}^n), \tag{25}$$

$$V_{i,j}^0 = V_0(x,y), \tag{26}$$

$$V_{i,j}^n\,|_{\Gamma_\nu} = \sigma_0\,|_{\Gamma_\nu} + \tau\sum_{\nu=1}^{n} q_w \tag{27}$$

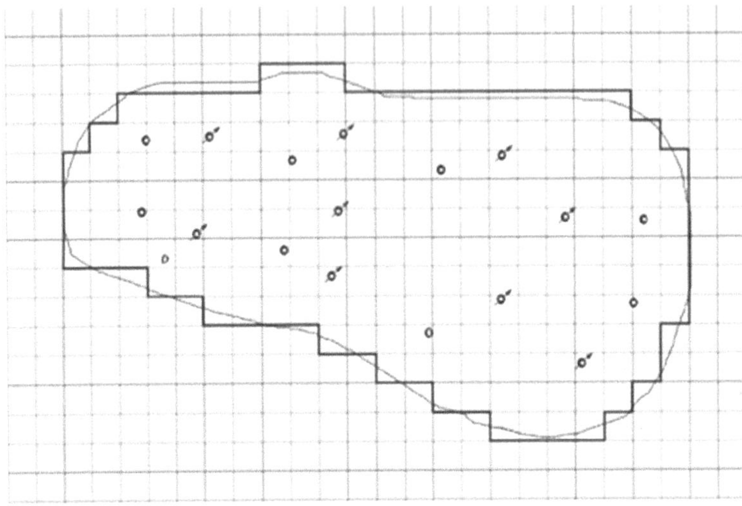

Fig. 1. Domain of the Solution with Oil and Water Wells

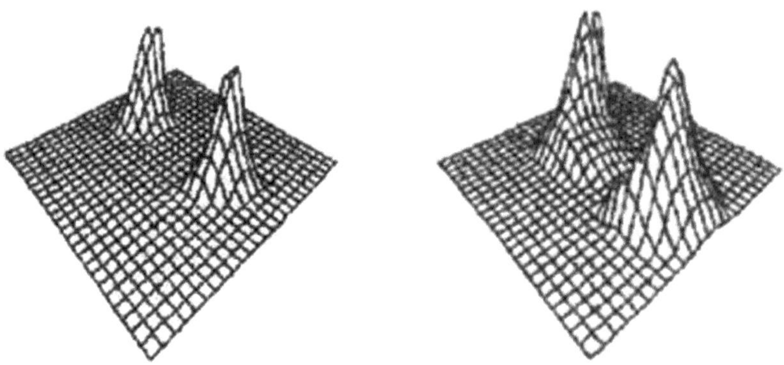

Fig. 2. a) $t = 600$ days; b) $t = 1800$ days

Here, $\boldsymbol{W}^n_{i,j} = \{{}_xW^n_{i,j}, {}_yW^n_{i,j}\}$, $V^n_{i,j}$ are the value of speed vector $\boldsymbol{w}$ and the function $v(x, y, t)$ in the point of (x_i, y_i, t) of the grid.

It can be easily shown that

$$U^n_{i,j} = \frac{{}_xV^n_{i,j} - {}_xV^n_{i-1,j}}{h_x} + \frac{{}_yV^n_{i,j} - {}_yV^n_{i,j-1}}{h_y}. \tag{28}$$

Now, we approximate the equation (12) with conditions (14) and (15) by the following finite differences scheme

$$\Phi^n_{i+1/2,j}(\wp^{n+1}_{i+1,j} - \wp^{n+1}_{i,j}) - \Phi^n_{i-1/2,j}(\wp^{n+1}_{i,j} - \wp^{n+1}_{i-1,j}) +$$

$$\Phi^n_{i,j+1/2}(\wp^{n+1}_{i,j+1} - \wp^{n+1}_{i,j}) - \Phi^n_{i,j-1/2}(\wp^{n+1}_{i,j} - \wp^{n+1}_{i,j}) = \frac{q}{bk}, \tag{29}$$

$$(i = \overline{0, N_x}), (j = \overline{0, N_y}).$$

Here, $\Phi(\sigma, p) = \frac{k_w(\sigma)}{\mu_w} + \frac{k_0(\sigma)}{\mu_0}$, $\Phi^n_{i \pm 1/2, j} = \frac{2\wp^n_{i,j} \cdot 2\wp^n_{i\pm1,j}}{\wp^n_{i,j} + \wp^n_{i\pm1,j}}$, $\Phi^n_{i, j\pm 1/2} = \frac{2\wp^n_{i,j} \cdot 2\wp^n_{i,j\pm1}}{\wp^n_{i,j} + \wp^n_{i,j\pm1}}$, b- is the height of the porous medium, N_x and N_y are number of points with respect to x and y respectively, and $\wp^n_{i,j}$ - is approximate value of the pressure $p(x, y, t)$at point (x_i, y_i, t_n).

The initial and boundary conditions for the equation (28) are $\wp^n_{i,j} = p_0(x_i, y_i)$,

$$\wp_{-1,j} = \wp_{1,j}, \wp_{N_x-1,j} = \wp_{N_x+1,j}, (j = \overline{0, N_y});$$

$$\wp_{i,-1} = \wp_{i,1}, \wp_{i,N_y-1} = \wp_{i,N_y+1}, (i = \overline{0, N_x}).$$

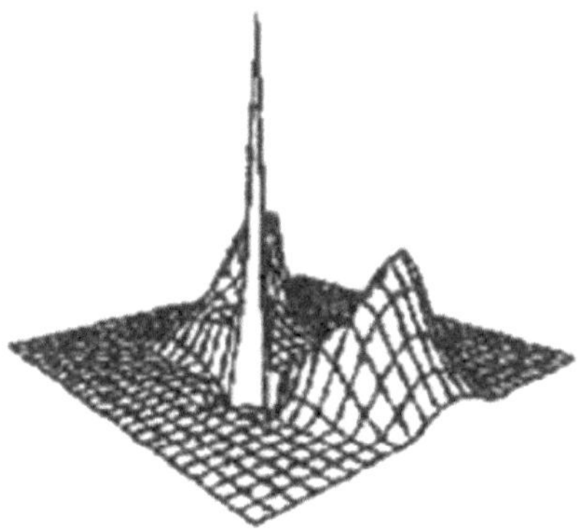

Fig. 3. $t = 2200$ days

3 Application

Using the suggested algorithms and data given below, some numerical experiments are carried out.

$$\mu_w = 0.00075pa.s,\ \mu_o = 0.00075pa.s,\ m = 0.2,\ k = 0.50mkm^2,\ \sigma_0 = 0.1,$$

$$\sigma_1 = 0.7,\ \ \sigma_* = 0.8,\ \ p_0 = 194.10^5pa,\ \ q_w = 200\ m^3/day,$$

$$\ell = 600, b = 600\ m, N_1 = 2,$$

$$k_w(\sigma) = \begin{cases} \left(\frac{\sigma - 0.1}{\sigma_* - \sigma_0}\right)^2, & \sigma_0 \le \sigma \le \sigma_1, \\ 0.8\left(\frac{\sigma - \sigma_0}{\sigma_* - \sigma_0}\right)^{1/2}, & \sigma_1 \le \sigma \le \sigma_*, \end{cases}$$

$$k_0(\sigma) = \left(\frac{\sigma_* - \sigma}{\sigma_* - \sigma_0}\right)^2, \sigma_0 \le \sigma \le \sigma_*.$$

In figures 2-3, the dynamical distributions of water saturation which is pumped from two wells at different value of time are shown. In Figure 4, The pressure distribution is shown with isobar contours.

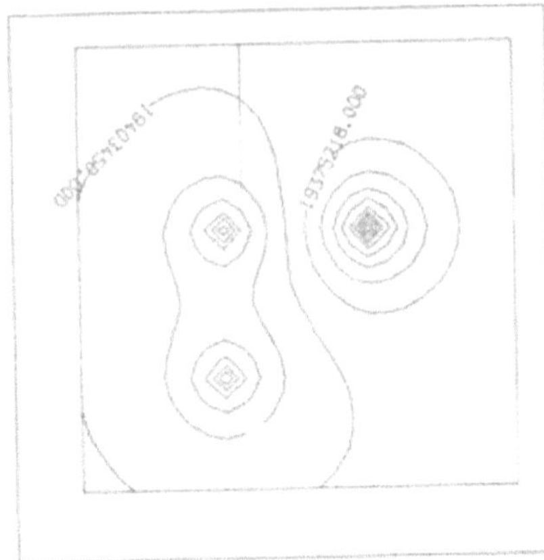

Fig. 4. $t = 2200$ days

4 Conclusions

The results obtained in this study are itemized below:

- The system of the partial differential equations which describes the motion of two-phase compressible fluid in porous medium is described.
- A special auxiliary problem is introduced which can solve cases that discontinuous functions appear in the solution domain..
- Using the advantages of the auxiliary problem the efficient numerical algorithms are suggested in a class of discontinuous functions. The obtained solutions express that the mathematical model works adequately with physical models and measurements.

References

1. Abasov, M.T., Rasulov, M.A., Ragimova, A.T.: Identification of the Saturation Jump in the Process of Oil Displacement by Water in a Two-Dimensional Domain. Soviet Math. Dokl. USSR 319(4), 943–947 (1992)
2. Aziz, K., Settari, A.: Petroleum Reservoir Simulation. Elsevier Applied Science Pub., London (1979)
3. Collins, R.E.: Flow of Fluids through Porous Materials. Penn-Well Books, Tulsa (1976)
4. Lie, K.A., Juanes, R.: A Front-Tracking Method for the Simulation of Three-Phase Flow in Porous Media. Computational Geosciences 9(1), 29–59 (2005)
5. Muskat, M.: The Flow of Homogeneous Fluids through Porous Medium. McGraw-Hill, New York (1946)
6. Rasulov, M.A.: On the Method of Calculation of the First Phase Saturation during the Process of Displacement of Oil by Water from Porous Medium. Appl. Math. Comput. 85(1), 1–16 (1997)
7. Rasulov, M.A., Abasov, M.T.: To the Theory Filtration of Three-Phase Mixes in View of Them Compressible. Soviet Math. Dokl. USSR 325(1), 130–133 (1992)

Grid Approximation of a Singularly Perturbed Parabolic Reaction-Diffusion Equation on a Ball

Lidia Shishkina and Grigory Shishkin

Institute of Mathematics and Mechanics,
Ural Branch of Russian Academy of Sciences, Ekaterinburg 620219, Russia
lida@convex.ru, shishkin@.imm.uran.ru

Abstract. We consider a Dirichlet problem on a ball for a singularly perturbed parabolic reaction-diffusion equation. The Laplacian in the equation is multiplied by a perturbation parameter ε^2, where $\varepsilon \in (0, 1]$. The solution of such a problem exhibits the parabolic boundary layer in a neighbourhood of the ball boundary as $\varepsilon \to 0$. Using the integro-interpolational method and the condensing mesh method, we construct conservative finite difference schemes whose solutions converge ε-uniformly.

1 Introduction

For singularly perturbed boundary value problems for parabolic reaction-diffusion equations, methods of the construction of ε-uniformly convergent finite difference schemes are well developed; see, e.g., [1,2,3,4] and the bibliography therein. When constructing the finite difference schemes, an approach based on the condensing mesh method is applied. Piecewise-uniform meshes are used that condense in a neighbourhood of boundary layers. The finite difference schemes are constructed by a term-by-term (termwise) grid approximation of differential equations. Quite often in mathematical modeling, differential equations are written in *divergent form* (see, e.g., [5]). Such a form of differential equations allows us to construct conservative finite difference schemes for which conservation laws are fulfilled; see, e.g., [6,7]. In [6] for regular parabolic equations in divergent form, conservative finite difference schemes are constructed whose solutions converge in the maximum norm at the rate $\mathcal{O}\big(N^{-2} + N_0^{-1}\big)$, where $N + 1$ and $N_0 + 1$ are the number of nodes in the meshes in space and temporal variables. In the same book, conservative difference schemes are constructed for the heat conduction problem on a circle and a ball in the case of cylindrical and spherical symmetries; to construct the difference schemes, the integro-interpolational method is applied.

In the present paper, a boundary value problem on a ball for a singularly perturbed parabolic reaction-diffusion equation in *divergent form* with third-kind boundary conditions is considered in the case of *spherical symmetry.* When the perturbation parameter ε tends to zero (ε takes arbitrary values in the open-closed interval $(0, 1]$), the solution of the problem exhibits the parabolic boundary layer in a neighbourhood of the ball boundary.

S. Margenov, L.G. Vulkov, and J. Waśniewski (Eds.): NAA 2008, LNCS 5434, pp. 501–508, 2009.

Using the integro-interpolational method, conservative finite difference schemes are constructed whose solutions converge ε-uniformly at the rate $\mathcal{O}(N^{-2}\ln^2 N + N_0^{-1})$. The finite difference schemes are considered on traditional grids and "flow grids" [6] that are condensed in a neighbourhood of the boundary layer.

2 Problem Formulation

2.1. In the domain $\overline{G}$, where

$$\overline{G} = G \bigcup S, \quad G = D \times (0, T], \quad D = (0,\, d), \tag{1}$$

we consider the boundary value problem for the parabolic equation[1]

$$L_{(2)}\, u(r,t) = f(r,t), \quad (r,t) \in G, \quad u(r,t) = \varphi(r,t), \quad (r,t) \in S^p, \tag{2}$$

$$\frac{\partial}{\partial r}\, u(r,t) = 0, \quad (r,t) \in S^0.$$

Here
$$L_{(2)} = L_2 + L_1,$$

$$L_2 \equiv \varepsilon^2 r^{-2} \frac{\partial}{\partial r}\left(r^2 \frac{\partial}{\partial r}\right) = \varepsilon^2 \left(\frac{\partial^2}{\partial r^2} + 2\, r^{-1} \frac{\partial}{\partial r}\right), \quad L_1 \equiv -c(r,t) - p(r,t)\,\frac{\partial}{\partial t}.$$

The coefficients $c(r,t)$, $p(r,t)$ and the right-hand side $f(r,t)$ are assumed to be sufficiently smooth on the set $\overline{G}$, moreover,

$$c(r,t) \geq 0, \quad 0 < p_0 \leq p(r,t) \leq p^0, \quad (r,t) \in \overline{G}, \tag{3}$$

the boundary function $\varphi(r,t)$ is assumed to be sufficiently smooth on the sets S_0 and $\overline{S}^d$ and continuous on S^p, where $S^p = S_0 \bigcup S^d$ is the parabolic part of the boundary, and $S_0 = \overline{D} \times \{t = 0\}$, $S^0 = \{r = 0\} \times (0, T]$, $S^d = \{r = d\} \times (0, T]$. The perturbation parameter ε takes arbitrary values in the interval $(0, 1]$.

By a solution of the boundary value problem, we mean a function $u \in C(\overline{G}) \bigcap C^{2,1}(G \bigcup S^0)$ that satisfies the differential equation on G and the initial-boundary condition on S.

The problem (2), (1) is a boundary value problem on the ball $\widetilde{D}$ with the radius d for a singularly perturbed parabolic reaction-diffusion equation in the case of spherical symmetry

$$\widetilde{L}\,\widetilde{u}(x,t) = \widetilde{f}(x,t), \quad (x,t) \in \widetilde{G}, \qquad \widetilde{u}(x,t) = \widetilde{\varphi}(x,t), \quad (x,t) \in \widetilde{S}. \tag{4}$$

Here

$$\overline{\widetilde{G}} = \widetilde{G} \bigcup \widetilde{S}, \quad \widetilde{G} = \widetilde{D} \times (0, T], \quad \widetilde{S} = \widetilde{S}^L \bigcup \widetilde{S}_0; \tag{5}$$

$$\widetilde{L} \equiv \varepsilon^2 \sum_{s=1}^{3} \frac{\partial^2}{\partial x_s^2} - \widetilde{c}(x,t) - \widetilde{p}(x,t)\,\frac{\partial}{\partial t},$$

[1] The notation $L_{(j.k)}$ ($\overline{G}_{(j.k)}$, $M_{(j.k)}$) means that these operators (domains, constants) are introduced in formula $(j.k)$.

the functions $c(r,t)$, …, $\varphi(r,t)$, $u(r,t)$ and $\widetilde{c}(x,t)$, …, $\widetilde{\varphi}(x,t)$, $\widetilde{u}(x,t)$ are related by the relation $v(r,t) = \widetilde{v}(x,t)$, $r = r(x) = (x_1^2 + x_2^2 + x_3^2)^{1/2}$.

2.2. We discuss the behaviour of the solution to problem (2), (1). When the parameter ε tends to zero, the solution of the boundary value problem exhibits the parabolic boundary layer in a neighbourhood of the boundary S^d.

Errors of solutions to classical finite difference schemes that approximate the boundary value problem (2), (1) depend on the parameter ε and can be commensurable with the exact solutions for small values of the parameter ε. It is known that in the case of singularly perturbed problems in the presence of parabolic boundary layers, the fitted operator method is not applicable for constructing ε-uniformly convergent finite difference schemes [1,8,9].

Our aim for the boundary value problem (2), (1) is to construct a conservative finite difference scheme that converges ε-uniformly.

3 A Priori Estimates

We give *a priori* estimates of the solution and its derivatives. To derive the estimates we use the technique of [1,10,11,12].

Using the comparison theorem, we find

$$|u(r,t)| \le M\left[\max_{\overline{G}}|f(r,t)| + \max_{S^p}|\varphi(r,t)|\right], \quad (r,t) \in \overline{G}. \tag{1}$$

Assume that the data of the problem (4), (5) satisfy the condition

$$\widetilde{c},\ \widetilde{p},\ \widetilde{f} \in C^{l+\alpha,(l+\alpha)/2}(\overline{\widetilde{G}}), \tag{2a}$$

$$\widetilde{\varphi} \in C^{l+2+\alpha}(\widetilde{S}_0)\bigcap C^{l+2+\alpha,(l+2+\alpha)/2}(\widetilde{S}^L)\bigcap C(\widetilde{S}), \quad l \ge 0, \quad \alpha > 0, \tag{2b}$$

moreover, on the set $\widetilde{S}^c = \overline{\widetilde{S}}^L \bigcap \widetilde{S}_0$, compatibility conditions are fulfilled [11] that guarantee the inclusion

$$\widetilde{u} \in C^{l+2+\alpha,(l+2+\alpha)/2}(\overline{\widetilde{G}}) \tag{2c}$$

for fixed values of the parameter ε; some additional conditions are given below.

3.1. We find some estimates of the problem (2), (1) using *a priori* estimates up to the boundary [11,12]. The boundary value problem (4), (5) in the variables ξ, t, where $\xi = (\xi_1, \xi_2, \xi_3)$, $\xi_s = \xi_s(x) = \varepsilon^{-1} x_s$, $s = 1,2,3$, transforms into the following problem

$$\widehat{L}\,\widehat{u}(\xi,t) = \widehat{f}(\xi,t), \quad (\xi,t) \in \widehat{G}, \tag{3a}$$

$$\widehat{u}(\xi,t) = \widehat{\varphi}(\xi,t), \quad (\xi,t) \in \widehat{S}. \tag{3b}$$

Here $\widehat{v}(\xi,t) = \widetilde{v}(x(\xi),t)$, $\widetilde{v}(x,t)$ is one of the functions $\widetilde{u}(x,t)$, …, $\widetilde{\varphi}(x,t)$, and $\widehat{G}^0 = \{(\xi,t) : \xi = \xi(x),\ (x,t) \in \widetilde{G}^0\}$, where $\widetilde{G}^0$ is either $\widetilde{G}$ or $\widetilde{S}$. The differential equation (3a) in $\widehat{G}$ and the boundary condition (3b) on $\widehat{S}$ are regular in ε. Using *a priori* estimates up to the boundary, we find

$$\left| \frac{\partial^{k+k_0}}{\partial \xi_1^{k_1} \partial \xi_2^{k_2} \partial \xi_3^{k_3} \partial t^{k_0}} \widehat{u}(\xi, t) \right| \le M, \quad (\xi, t) \in \overline{\widehat{G}}.$$

In the variables x, t we obtain the estimate

$$\left| \frac{\partial^{k+k_0}}{\partial x_1^{k_1} \partial x_2^{k_2} \partial x_3^{k_3} \partial t^{k_0}} \widehat{u}(\xi, t) \right| \le M\, \varepsilon^{-k}, \quad (x, t) \in \overline{\widehat{G}},\ k + 2k_0 \le K, \tag{4a}$$

where $k = k_1 + k_2 + k_3$, $K = l + 2$, $l = l_{(2)}$.

Thus, the following estimate holds

$$\left| \frac{\partial^{k+k_0}}{\partial r^k \partial t^{k_0}} u(r, t) \right| \le M\, \varepsilon^{-k}, \quad (r, t) \in \overline{G},\ k + 2k_0 \le K,\ K = K_{(4a)}. \tag{4b}$$

3.2. When derive estimates based on asymptotic constructs, we assume that the following condition holds:

$$\widetilde{c},\ \widetilde{p},\ \widetilde{f} \in C^{l+2+\alpha,(l+2+\alpha)/2}(\overline{\widetilde{G}}), \tag{5}$$

and also the conditions (2b) and (2c) are fulfilled.

We find estimates for the regular and singular components of the solution to the boundary value problem (2), (1).

Write the solution of the problem (4), (5) as the sum of the functions

$$\widetilde{u}(x, t) = \widetilde{U}(x, t) + \widetilde{V}(x, t), \quad (x, t) \in \overline{\widetilde{G}}, \tag{6}$$

where $\widetilde{U}(x, t)$ and $\widetilde{V}(x, t)$ are the regular and singular parts of the solution. The decomposition of the solution to the boundary value problem (2), (1)

$$u(r, t) = U(r, t) + V(r, t), \quad (r, t) \in \overline{G}, \tag{7}$$

corresponds to the decomposition (6) for the problem (4), (5).

The function $\widetilde{U}(x, t)$ is the restriction to $\overline{\widetilde{G}}$ of the function $\widetilde{U}^e(x, t)$, $(x, t) \in \overline{\widetilde{G}}^e$, where $\widetilde{U}^e(x, t)$ is the solution of the problem

$$\widetilde{L}^e_{(4)}\, \widetilde{U}^e(x, t) = \widetilde{f}^e(x, t), \quad (x, t) \in \widetilde{G}^e, \qquad \widetilde{U}^e(x, t) = \widetilde{\varphi}^e(x, t), \quad (x, t) \in \widetilde{S}^e.$$

Here $\widetilde{S}^e = \widetilde{S}(\widetilde{G}^e)$; the domain $\widetilde{G}^e$ the extension of $\widetilde{G}$ beyond the set $\widetilde{S}^L$, the set $\widetilde{G}^e$ includes the domain $\widetilde{G}$ along with its m_2-neighbourhood; the coefficients of the operator $\widetilde{L}^e_{(4)}$ and the right-hand part $\widetilde{f}^e(x, t)$ are smooth continuations of the corresponding data of the problem (4), (5); the function $\widetilde{\varphi}^e(x, t)$ is a smooth one, moreover, $\widetilde{\varphi}^e(x, t) = \widetilde{\varphi}(x, t)$, $(x, t) \in \widetilde{S}_0$. The functions $\widetilde{f}^e(x, t)$ and $\widetilde{\varphi}^e(x, t)$

are assumed to be equal zero outside an $(2^{-1}m_2)$-neighbourhood of the set $\overline{\widetilde{G}}$. The function $\widetilde{V}(x,t)$ is the solution of the problem

$$\widetilde{L}_{(4)}\,\widetilde{V}(x,t) = 0, \quad (x,t) \in \widetilde{G}, \qquad \widetilde{V}(x,t) = \widetilde{\varphi}(x,t) - \widetilde{U}(x,t), \quad (x,t) \in \widetilde{S}.$$

For the components $\widetilde{U}_0^e(x,t)$ and $\widetilde{v}_0^e(x,t)$ in the representation

$$\widetilde{U}^e(x,t) = \widetilde{U}_0^e(x,t) + \widetilde{v}_0^e(x,t), \quad (x,t) \in \overline{\widetilde{G}}^e,$$

where the functions $\widetilde{U}_0^e(x,t)$ and $\widetilde{v}_0^e(x,t)$ are solutions of the problems

$$\widetilde{L}^e_{(8)}\,\widetilde{U}_0^e(x,t) \equiv \left\{-\widetilde{c}^e(x,t) - \widetilde{p}^e(x,t)\,\frac{\partial}{\partial t}\right\}\widetilde{U}_0^e(x,t) = \widetilde{f}^e(x,t), \quad (x,t) \in \overline{\widetilde{G}}^e \setminus \widetilde{S}_0^e,$$

$$\widetilde{U}_0^e(x,t) = \widetilde{\varphi}^e(x,t), \quad (x,t) \in \widetilde{S}_0^e; \tag{8}$$

$$\widetilde{L}^e_{(4)}\,\widetilde{v}^e(x,t) = -\varepsilon^2 \sum_{s=1}^{3} \frac{\partial^2}{\partial x_s^2}\,\widetilde{U}_0^e(x,t), \quad (x,t) \in \widetilde{G}^e,$$

$$\widetilde{v}^e(x,t) = \widetilde{\varphi}^e(x,t) - \widetilde{U}_0^e(x,t), \quad (x,t) \in \widetilde{S}^e,$$

one has the estimates

$$\left|\frac{\partial^{k+k_0}}{\partial x_1^{k_1}\partial x_2^{k_2}\partial x_3^{k_3}\partial t^{k_0}}\,\widetilde{U}_0^e(x,t)\right| \le M,$$

$$\left|\frac{\partial^{k+k_0}}{\partial x_1^{k_1}\partial x_2^{k_2}\partial x_3^{k_3}\partial t^{k_0}}\,\widetilde{v}^e(x,t)\right| \le M\,\varepsilon^{2-k}, \quad (x,t) \in \overline{\widetilde{G}}^e, \ k + 2k_0 \le K.$$

For the function $\widetilde{U}(x,t)$ we obtain the estimate

$$\left|\frac{\partial^{k+k_0}}{\partial \xi_1^{k_1}\partial \xi_2^{k_2}\partial \xi_3^{k_3}\partial t^{k_0}}\,\widetilde{U}(x,t)\right| \le M\,[1 + \varepsilon^{2-k}], \quad (x,t) \in \overline{\widetilde{G}}, \quad k + 2k_0 \le K,$$
$$k = k_1 + k_2 + k_3.$$

For the functions $U(r,t)$ and $V(r,t)$ in (7) we have the estimates

$$\left|\frac{\partial^{k+k_0}}{\partial r^k\partial t^{k_0}}\,U(r,t)\right| \le M\,[1 + \varepsilon^{2-k}], \tag{9}$$

$$\left|\frac{\partial^{k+k_0}}{\partial r^k\partial t^{k_0}}\,V(r,t)\right| \le M\,\varepsilon^{-k}\exp\left(-\,m\,\varepsilon^{-1}(d - r)\right), \quad (r,t) \in \overline{G}, \ k + 2k_0 \le K,$$

where $K = l + 2$ with $l = l_{(5)}$, and m is an arbitrary number.

4 Grid Approximations of the Boundary Value Problem

4.1. We now construct a finite difference scheme based on a classical finite difference approximation of the boundary value problem (2), (1) on rectangular grids. On the set $\overline{G}$ we introduce the grid

$$\overline{G}_h = \overline{D}_h \times \overline{\omega}_0 = \overline{\omega} \times \overline{\omega}_0, \tag{1}$$

where $\overline{\omega}$ and $\overline{\omega}_0$ are, in general, arbitrary nonuniform meshes on the intervals $[0, d]$ and $[0, T]$, respectively. Set $r^0 = 0$, $h^i = r^{i+1} - r^i$, r^i, $r^{i+1} \in \overline{\omega}$, $h = \max_i h^i$, $h_t^j = t^{j+1} - t^j$, t^j, $t^{j+1} \in \overline{\omega}_0$, $h_t = \max_j h_t^j$. We denote by $N+1$ and N_0+1 the numbers of nodes in the meshes $\overline{\omega}$ and $\overline{\omega}_0$, assume that the conditions $h \le M N^{-1}$, $h_t \le M N_0^{-1}$ hold.

When constructing a difference scheme, the integro-interpolational method is used [6]. To solve the problem (2), (1), we use the finite difference scheme

$$\begin{aligned} &\Lambda_{(2)}\, z(r,t) = f(r,t), \quad (r,t) \in G_h, \quad z(r,t) = \varphi(r,t), \quad (r,t) \in S_h^p, \\ &\lambda^0\, z(r,t) = 0, \quad (r,t) \in S_h^0. \end{aligned} \tag{2}$$

Here $G_h = G \bigcap \overline{G}_h$, $S_h^p = S^p \bigcap \overline{G}_h$, $S_h^0 = S^0 \bigcap \overline{G}_h$,

$$\Lambda_{(2)}\, z(r,t) \equiv \varepsilon^2 r^{-2}\, \delta_{\widehat{r}} \left(r_{i-1/2}^2\, \delta_{\overline{r}}\, z(r,t) \right) - c(r,t)\, z(r,t) - p(r,t)\, \delta_{\overline{t}}\, z(r,t),$$

$(r,t) \in G_h$, $r = r_i$; $\lambda^0\, z(r,t) \equiv \delta_r\, z(r,t)$; $\delta_{\overline{r}}\, z(r,t)$, $\delta_r\, z(r,t)$, $\delta_{\widehat{r}} \left(r_{i-1/2}^2\, \delta_{\overline{r}}\, z(r,t) \right)$ and $\delta_{\overline{t}}\, z(r,t)$ are the difference derivatives (see [6]):

$$\delta_r\, z(r,t) = \left(h^i\right)^{-1} \left(z(r^{i+1},t) - z(r,t) \right), \quad \delta_{\overline{r}}\, z(r,t) = \left(h^{i-1}\right)^{-1} \left(z(r,t) - z(r^{i-1},t) \right),$$

$$\delta_{\widehat{r}} \left(r_{i-1/2}^2\, \delta_{\overline{r}}\, z(r,t) \right) = 2\,(h^i + h^{i-1})^{-1} \left\{ r_{i+1/2}^2\, \delta_r\, z(r,t) - r_{i-1/2}^2\, \delta_{\overline{r}}\, z(r,t) \right\},$$

$$\delta_{\overline{t}}\, z(r,t) = \left(h^{j-1}\right)^{-1} \left(z(r,t) - z(r,t^{j-1}) \right), \quad (r,t) = (r^i, t^j),$$

$$r_{i-1/2} = 2^{-1}\,(r^{i-1} + r^i), \quad r_{i+1/2} = 2^{-1}\,(r^i + r^{i+1}).$$

Note that the grid function

$$w_h(r_{i+1/2}, t) = -\varepsilon^2 r^2\, \delta_r z(r,t), \quad r_{i+1/2} = 2^{-1}(r^i + r^{i+1}), \quad r = r^i, \quad (r,t) \in G_h$$

corresponds to the function $w(r,t) = -\varepsilon^2 r^2 \frac{\partial}{\partial r} u(r,t)$, $(r,t) \in G$ which is a "substance flow" (up to the constant-multiplier $4\,\pi$) through the sphere with the radius $r = r_{i+1/2}$. The difference scheme (2), (1) belongs to conservative schemes [6,15].

The difference operator $\Lambda_{(2)}$ is ε-uniformly monotone [6].

For the difference scheme (2), (1), the maximum principle holds.

Using the comparison theorem (see, e.g., [6]), the ε-uniform boundedness of the solution to problem (2), (1) is established: $|z(r,t)| \le M$, $(r,t) \in \overline{G}_h$.

Taking into account the estimate (4b) for $K = 4$, for the solution of the difference scheme (2), (1), we obtain the estimate

$$|u(r,t) - z(r,t)| \leq M\,\left[(\varepsilon^2 + N^{-1})^{-1} N^{-1} + N_0^{-1}\right], \quad (r,t) \in \overline{G}_h. \tag{3}$$

We now estimate an error generated by the grid approximation of the boundary condition on the set S^0. We use the following function as a majorant

$$w(r,t) = \delta(r+\delta)^{-1} \quad (r,t) \in \overline{G}_h, \quad \textit{where} \;\; \delta \approx h^0.$$

On the uniform grid

$$\overline{G}_h = \overline{G}{}^u_h = \overline{\omega}^u \times \overline{\omega}{}^u_0, \tag{4}$$

where $\overline{\omega}^u$ and $\overline{\omega}{}^u_0$ are uniform meshes, we have the estimate

$$|u(r,t) - z(r,t)| \leq M\,\left[(\varepsilon + N^{-1})^{-2} N^{-2} + N_0^{-1}\right], \quad (r,t) \in \overline{G}_h. \tag{5}$$

The estimate (5) is unimprovable with respect to the values N, N_0 and ε.

The difference scheme (2), (4) converges under the condition

$$N^{-1} = o(\varepsilon), \quad N_0^{-1} = o(1). \tag{6}$$

4.2. We now construct a grid condensing in a neighbourhood of the boundary layer on which the solution of the difference scheme converges ε-uniformly. On the set $\overline{G}$ we introduce the grid

$$\overline{G}_h = \overline{G}{}^s_h = \overline{\omega}^s \times \overline{\omega}_0, \tag{7}$$

where $\overline{\omega}_0 = \overline{\omega}_{0(4)}$, and $\overline{\omega}^s = \overline{\omega}^s(\sigma)$ is a piecewise-uniform mesh on $[0,d]$ that condenses in a neighbourhood of $r = d$; the parameter σ depends on N and ε. The mesh sizes in $\overline{\omega}^s$ (see, e.g., [1,13,14]) are constant on the intervals $[d-\sigma, d]$ and $[0, d-\sigma]$ and equal to $h^{(1)} = 2\,\sigma\,N^{-1}$ and $h^{(2)} = 2\,(d-\sigma)\,N^{-1}$, respectively. The value σ is specified by

$$\sigma = \sigma(\varepsilon,\, N) = \min\left[2^{-1}\, d,\;\, M\,\varepsilon\,\ln N\right],$$

where M is an arbitrary number.

$\mathcal{E}$-uniform convergence of the difference scheme (2), (7) to the solution of the boundary value problem (2), (1) follows from the estimates (9) for $K = 4$. The convergence rate of the solution of the difference scheme is estimated by using a technique in [1,13,14]. For the solution of the difference scheme we obtain the ε-uniform estimate

$$|u(r,t) - z(r,t)| \leq M\,\left[N^{-2}\,\ln^2 N + N_0^{-1}\right], \quad (r,t) \in \overline{G}_h, \tag{8a}$$

and also the ε-dependent estimate

$$|u(r,t) - z(r,t)| \leq M\,\left[\min[\varepsilon^{-2},\, \ln^2 N]\, N^{-2} + N_0^{-1}\right], \quad (r,t) \in \overline{G}_h. \tag{8b}$$

Theorem 1. *Let the components in the decomposition (7) of the solution of the boundary value problem (2), (1) satisfy the estimates (9) for $K = 4$. Then the solution of the conservative finite difference scheme (2), (7) converges to the solution of the boundary value problem ε-uniformly. The discrete solutions satisfy the estimate (8).*

Acknowledgments

This research was supported by the Russian Foundation for Basic Research under grants Nos. 07–01–00729, the Boole Centre for Research in Informatics at the National University of Ireland, Cork, and by the Mathematics Applications Consortium for Science and Industry in Ireland (MACSI) under the Science Foundation Ireland (SFI) Mathematics Initiative.

References

1. Shishkin, G.I.: Discrete Approximations of Singularly Perturbed Elliptic and Parabolic Equations. Russian Academy of Sciences, Ural Section, Ekaterinburg (1992) (in Russian)
2. Hemker, P.W., Shishkin, G.I., Shishkina, L.P.: ε-uniform schemes with high-order time-accuracy for parabolic singular perturbation problems. IMA J. Numer. Anal. 20(1), 99–121 (2000)
3. Shishkin, G.I.: A finite difference scheme on a priori adapted meshes for a singularly perturbed parabolic convection-diffusion equation. Numer. Math. Theory Methods Appl. 1(2), 214–234 (2008)
4. Shishkin, G.I.: Grid approximation of singularly perturbed parabolic equations with piecewise continuous initial–boundary conditions. In: Proc. Steklov Inst. Math., vol. 2, pp. 213–230 (2007)
5. Grigoriev, V.A., Zorin, V.M. (eds.): Teplotechnicheskii experiment: Spravochnik. Energoizdat, Moscow (1982) (in Russian)
6. Samarskii, A.A.: The Theory of Difference Schemes. Marcel Dekker, Inc., New York (2001)
7. Kalitkin, N.N.: Numerical methods. Nauka, Moscow (1978) (in Russian)
8. Shishkin, G.I.: Approximation of solutions of singularly perturbed boundary value problems with a parabolic boundary layer. USSR Comput. Maths. Math. Phys. 29(4), 1–10 (1989)
9. Miller, J.J.H., O'Riordan, E., Shishkin, G.I.: Fitted Numerical Methods for Singular Perturbation Problems. In: Error Estimates in the Maximum Norm for Linear Problems in One and Two Dimensions. World Scientific, Singapore (1996)
10. Il'in, A.M., Kalashnikov, A.S., Oleinik, O.A.: Second-order linear equations of parabolic type. Uspehi Mat. Nauk 17(3), 3–146 (1962) (in Russian)
11. Ladyzhenskaya, O.A., Solonnikov, V.A., Ural'tseva, N.N.: Linear and Quasilinear Equations of Parabolic Type. Translations of Mathematical Monographs, vol. 23. American Mathematical Society, Providence (1967)
12. Friedman, A.: Partial Differential Equations of Parabolic Type. Prentice-Hall, Inc., Englewood Cliffs (1964)
13. Shishkin, G.I.: Grid approximation of singularly parturbed boundary value problem for quasi-linear parabolic equations in case of complete degeneracy in spatial variables. Soviet J. Numer. Anal. Math. Modelling 6(3), 243–261 (1991)
14. Shishkin, G.I.: Grid approximation of a singularly perturbed boundary value problem for a quasilinear elliptic equation in the case of complete degeneration. Comput. Maths. Math. Phys. 31(12), 33–46 (1991)
15. Samarskii, A.A.: Introduction to the Theory of Difference Schemes. Nauka, Moscow (1971) (in Russian)

Properties of Generalized Polynomial Spaces in Three Variables

Dana Simian

University "Lucian Blaga" of Sibiu, Faculty of Sciences
5-7 dr. I. Raţiu str, 550012 Sibiu, Romania
dana.simian@ulbsibiu.ro

Abstract. Multivariate interpolation is a topic which often appears in practical modeling problems. Different type of spaces of functions are used for solving interpolation problems. When the interpolation conditions are of different kind, by example, spacial and temporal, one possibility for modeling the problem is to use a generalize degree, in which the monomials exponents are weighted with a weight vector with integer components. In order to use such a generalize polynomial space as interpolation space, it is necessary to know the dimension and a basis of it. The aim of this article is to study and prove many properties of the generalize polynomial spaces in three variables.

1 Introduction

Many modeling problems impose the existence of weighted interpolation conditions as well as the utilization of a generalized degree which also weighted the solution. The notion of degree is very important in algebraic problems regarding H-bases and Gröbner bases of an ideal. On the other hand, ideal interpolation schemes are connected with the reduction process with respect to a H-basis or a Gröbner basis of the ideal $ker(\Lambda)$, where Λ is the set of interpolation conditions.

The grading decomposition of the polynomial spaces gives different generalized degrees. Let consider $(\Gamma, +)$ an orderer monoid, with respect to the total ordering $\prec$, which is compatible with the addition operation, that is

$$\alpha \prec \beta \Rightarrow \gamma + \alpha \prec \gamma + \beta, \; \forall\, \alpha, \beta, \gamma \in \Gamma.$$

A grading induced by $(\Gamma, +)$ on the space of polynomials is represented by a direct sum $\Pi = \bigoplus_{\gamma \in \Gamma} \mathcal{P}_\gamma^{(\Gamma)}$ satisfying

$$f \in \mathcal{P}_\alpha^{(\Gamma)}, \; g \in \mathcal{P}_\beta^{(\Gamma)} \Rightarrow f \cdot g \in \mathcal{P}_{\alpha+\beta}^{(\Gamma)}, \; \forall\, \alpha, \beta \in \Gamma.$$

Every polynomial has an unique representation in terms of this grading,

$$f = \sum_{i=1}^{s} f_{\gamma_i}, \; f_{\gamma_i} \in \mathcal{P}_{\gamma_i}^{(\Gamma)}; \quad f_{\gamma_i} \neq 0.$$

S. Margenov, L.G. Vulkov, and J. Waśniewski (Eds.): NAA 2008, LNCS 5434, pp. 509–516, 2009.

The components f_{γ_i}, $i \in \{1,\dots,s\}$ represent the Γ- homogeneous terms of degree γ_i.

In [2] is introduced a generalized weighted degree, named w-degree.

Definition 1. *(T. Sauer [2]) The w-degree of the monomial x^α is*

$$\delta_w(x^\alpha) = w \cdot \alpha = \sum_{i=1}^{d} w_i \cdot \alpha_i, \ \forall\ \alpha \in N^d,\ w \in N^d,\ x \in R^d. \tag{1}$$

We studied in [6] and [7] some properties of w-degree in the space of bivariate polynomials. In practical problems it is necessary to handle polynomials with more than 2 variables. The aim of this article is to extend our previous results regarding the w- degree, for the space of polynomials in three variables and to obtain a frame capable to offer us the possibility to extend the results for d variables.

We will next use the following notations:

$$A^{0,d}_{n,w} = \{\alpha \in N^d \mid w \cdot \alpha = n\},\ w \in (N^*)^d,\ n \in N \tag{2}$$

$$r^{(d)}_{n,w} = \#(A^{0,d}_{n,w}) \tag{3}$$

2 Properties for w-Degree Spaces of Polynomials in Three Variables

The aim of this section is to generalize the results given in [6] for polynomial in two variables for the case of d variables, $d \geq 3$. In order to use the spaces of w-degree as polynomial interpolation spaces it is necessary to know the dimensions of w-degree homogeneous polynomial subspaces, $r^{(d)}_{n,w}$ and the exponents of the monomials from these subspaces, $A^{0,d}_{n,w}$. We want to obtain a recursive relation for computing $r^{(d)}_{n,w}$ and $A^{0,d}_{n,w}$ starting from $r^{(2)}_{n,w}$ and $A^{0,2}_{n,w}$.

We will take first $d = 3$. A monomial $x_1^{\alpha_1} x_2^{\alpha_2} x_3^{\alpha_3}$ is in $A^{0,3}_{n,w}$ if $\sum_{i=1}^{3} w_i \cdot \alpha_i = n$. Hence, the exponents are linear dependent. We will use next the general form of a monomial: $X_{ijk} = x_i^{\alpha_i} x_j^{\alpha_j} x_k^{\alpha_k}$, $i \neq j \neq k \in \{1,2,3\}$. If we consider α_i given, then $\alpha_j \cdot w_j + \alpha_k \cdot w_k = m = n - \alpha_i \cdot w_i$, that is $(\alpha_j, \alpha_k) \in A^{0,2}_{m,w}$ and we reduced the initial problem in three variable to many problems in two variables. The properties of the dimensions of bivariate w-homogeneous spaces depends of the relative divisibility properties of components of the weight w (see [6]). That is why, from a computational point of view is very important which exponent α_i will be "bind" in our approach.

First, we will reformulate in an appropriate way the results from [6]. We will give only the proves wich are different from [6].

Lemma 1. *Let be $(w_i, w_j) \in (N^*)^2$, $i \neq j \in \{1,2,3\}$ and*

$$r_{ijk} = (kw_j)\ mod\ w_i,\ k \in \{0,\dots,w_i - 1\} \tag{4}$$

Then $r_{ijk} \neq r_{ijl}$, $\forall\ k \neq l$.

Lemma 2. *Let be*

$$m_{ij} = k_i w_i + k_j w_j, \tag{5}$$

$m_{ij} \not\vdots w_i$, $m_{ij} \not\vdots w_j$, $i \neq j$, $i,j \in \{1,2,3\}$, $k_i, k_j, w_i, w_j \in Z_+$ *and* r_{ijk} *given in (4). Then*

1. *The function* $q_{ij} : \{0, \ldots, w_i - 1\} \rightarrow \{0, \ldots, w_i - 1\}$, $q_{ij}(k) = r_{ijk}$ *is one to one.*
2. *For all* m_{ij} *there is an unique* $k \in \{0, \ldots, w_i - 1\}$, *such that*

$$m_{ij} \ mod \ w_i = r_{ijk}.$$

3. *If* m_{ij} *mod* $w_i = r_{ijl_{ij}}$, *then*

$$k_j = p_{ij} w_i + l_{ij}, \tag{6}$$

with $p_{ij} \in M_{ij} = \{[0, [a_{ij}]] \bigcap N\}$ *for* $[a_{ij}] \geq 0$ *and there is no* $k_j \in N$ *which satisfy (6) if* $[a_{ij}] < 0$, *where* $a_{ij} = \dfrac{m_{ij} - l_{ij} w_j}{w_i \cdot w_j}$ *and* $[\cdot]$ *is the integer part function.*

4. *With the notation from item 3, the following equality holds:*

$$M_{ij} = M_{ji} \tag{7}$$

Proof .
1. Obviously results from lemma 1.
2. The unique value k is given by $k = q_{ij}^{-1}(s_{ij})$, where $s_{ij} = m_{ij} \ mod \ w_i$.
3. From the proof of item 2 we obtain $l_{ij} = q_{ij}^{-1}(s_{ij})$. From the definition of q_{ij} we obtain $k_j = p_{ij} \cdot w_i + l_{ij}$, $p_{ij} \in Z_+$, $p_{ij} < w_i$. From (5) we obtain

$$k_i = \frac{m_{ij} - w_j(p_{ij} w_i + l_{ij})}{w_i} \tag{8}$$

We also have, $m_{ij} \ mod \ w_i = (l_{ij} w_j) \ mod \ w_i$. Therefore, we have $k_i \in Z_+$ if the numerator from (8) is grater or equal 0.
4. The equality (7) results from the equality $\left[\dfrac{m_{ij} - l_{ij} w_j}{w_i w_j}\right] = \left[\dfrac{m_{ij} - l_{ji} w_i}{w_i w_j}\right]$.

With the previous notations from this section, the following theorem holds:

Theorem 1. *Let be* $w = (w_i, w_j) \in N^2$, $i \neq j$, $i,j \in \{1,2,3\}$.

1. *If* $n \vdots w_1$, *that is* $n = c_i w_i$, *then* $r_{n,w}^{(2)} = \left[\dfrac{c_i}{w_j}\right] + 1$, *and* $(\alpha_i, \alpha_j) \in A_{n,w}^{0,2}$ *are given by* $\alpha_j = l_j w_i$, *with* $0 \leq l_j \leq \left[\dfrac{c_i}{w_j}\right]$; $\alpha_i = \dfrac{n - w_j \alpha_j}{w_i}$
2. *For any* n, $0 < n < \min(w_i, w_j)$, $r_{n,w}^{(2)} = 0$.

3. If $n \,\vdots\!\!/ w_i$ *and* $n \,\vdots\!\!/ w_j$, *with* $n \geq \min(w_i, w_j)$, *then*

$$r^{(2)}_{n,w} = \#(M_{ij}), \text{ with } M_{ij} = \{[0, a_{ij}] \cap N\}, \text{ if } a_{ij} \geq 0, \; a_{ij} = \left[\frac{n - l_{ij} w_j}{w_i w_j}\right]$$

where $[\cdot]$ *is the integer part function,* $l_{ij} = q_{ij}^{-1}(s_{ij})$, *with* $s_{ij} = n \bmod w_i$. $(\alpha_i, \alpha_j) \in A^{0,2}_{n,w}$ *are given by*

$$\alpha_i = \frac{n - (p_{ij} w_i + l_{ij}) w_j}{w_i}, \text{ with } p_{ij} \in M_{ij} \text{ and } \alpha_j = \frac{n - w_i \alpha_i}{w_j}$$

Corollary 1. $r^{(2)}_{0,w} = 1$ *and* $(0,0) \in A^{0,2}_{0,w}$, $\forall\, w \in (N^*)^2$.

The dimensions of w-degree polynomial spaces have an interesting property of "periodicity" in the sense defined in [6]. Let be $r_w : Z_+ \to Z_+$, $r_w(n) = r^{(2)}_{n,w}$

Theorem 2. *If* $n \geq w_i w_j$, $i \neq j$, $i, j \in \{1, 2, 3\}$ *and* $(w_i, w_j) = 1$, *then*

$$r_{n,w} = r_{n \bmod w_i w_j, w} + (n \text{ div } w_i w_j), \; \forall n \in N. \tag{9}$$

with $n \text{ div } w_i w_j = \left[\frac{n}{w_i w_j}\right]$ *and* $[\cdot]$ *the integer part function. If* $(w_i, w_j) = c > 1$ *then* $r_w(n) = 0$, $\forall\; n \in A = \{j \in N |\; j \,\vdots\!\!/ d\}$ *and equality (9) holds for every* $n \in N \setminus A$.

It can be seen that the case in which $(w_i, w_j) = c \neq 1$ is more unfavorable that the case in which $(w_i, w_j) = 1$, be cause there are many values of n such that the homogeneous polynomial space of w-degree n has the dimension equal 0. In the case $(w_i, w_j) = 1$ only the homogeneous spaces of w-degree n, with $n < min(w_1, w_2)$ are missing.

Theorem 1 can be used with any combination of indices (i, j). This is very useful when we want to "pass" from $d - 1$ variables to d variables. It is easier to generate the exponents from the monomials if we bind the first or the last exponent (α_1 or α_d). If one of these two exponents is 1, we will bind it. If not, we will choose that exponent which leads us to a more favorable situation in $d - 1$ variables taking into account theorem 2. By example if $w = (4, 2, 3)$ it is more convenable to bind the first exponent be cause $(2, 3) = 1$. Of course we can solve the problem binding another exponent, not necessary the first or the last, but in these cases we will have more complicated problems in writing and displaying the exponents.

Having a polynomial space in d variables, we will use a recursive scheme starting from the case in 2 variables which is solved in theorems 1 and 2. We can extend, for several variables, the recursive relation obtained for the case of three variables.

Let consider the space of polynomials in three variables. We calculate de dimension of w-homogeneous spaces in three variables, reducing the problem to one in the space in two variables. Next we will use the notation $r^{(2,[i,j])}_{n,w}$ for the dimension of a w- homogeneous space of polynomials of degree n, in two variables: x_i and x_j, $i \neq j$, $i, j \in \{1, 2, 3\}$. Usually we chose $i = 1$, $j = 2$ or $i = 2$, $j = 3$.

In the following theorems we suppose, for the simplicity of notations, that we bind α_1 and use the results from the spaces of w-degree polynomial spaces in 2 variables, x_2 and x_3.

Theorem 3. *If $w = (1, w_2, w_3) \in N^3$ and $(w_2, w_3) = 1$ than the dimensions of the w-degree homogeneous polynomial spaces satisfy the relations:*

$$r_{w,0}^{(3)} = 1 \text{ and } (0,0,0) \in A_{0,w}^{0,3} \tag{10}$$

$$r_{w,k}^{(3)} = r_{w,k-1}^{(3)} + r_{w,k}^{(2),[2,3]}, \ \forall k \in Z^+ \tag{11}$$

$$r_{w,n}^{(3)} \neq 0, \ \forall n \in Z_+ \tag{12}$$

The monomials from the homogeneous space of w- degree n are $(\alpha_1, \alpha_2, \alpha_3)$ with $\alpha_1 \in \{0, \ldots, n\}$ and $(\alpha_2, \alpha_3) \in A_{n-\alpha_1,w}^{0,2}$, with $A_{n-\alpha_1,w}^{0,2}$ given in theorem 1 in which we choose $i = 2$ and $j = 3$.

Proof. The equality (10) obviously results from corollary 1.

For proving (11) we use theorem 2 and induction on k and on α_1. The induction process leads us to a up triangular table. On the line i of the table, we will have $\alpha_1 = i$, $i \in \{0, 1, \ldots, n\}$ and, starting from the column i, we will have the values $r_{w,i}^{(2),[2,3]}$. Therefore

$$r_{w,k}^{(3)} = \sum_{s=0}^{k} r_{w,k-s}^{(2),[2,3]} \tag{13}$$

Hence

$$r_{w,k}^{(3)} = r_{w,k-1}^{(3)} + r_{w,k}^{(2),[2,3]}, \ \forall k > 0. \tag{14}$$

Taking into account (10) we obtain

$$r_{w,n}^{(3)} = r_{w,0}^{(3)} + \sum_{s=1}^{n} r_{w,n-s}^{(2),[2,3]} \neq 0, \ \forall \ n \in N.$$

The recursive relation (11) is illustrated in Table 1.

Table 1. Calculus of $r_{k,w}^{(3)}$ for $(w_1, w_2, w_3) = (1, 2, 3)$, using recursive relation (11)

k	0	1	2	3	4	5	6	7	8	9	10	11	12	...
$r_{k,w}^{(3)}$	*1*	1	2	3	4	5	7	8	10	12	14	16	19	...
		0+1	*1+1*	*2+1*	*3+1*	*4+1*	*5+2*	*7+1*	*8+2*	*10+2*	*12+2*	*14+2*	*16+3*	...
$r_{k,w}^{(2)[2,3]}$	1	0	1	1	1	1	2	1	2	2	2	2	3	...

For illustrating the proof of theorem 3 we build the triangular table of dimensions for $(w_1, w_2, w_3) = (1, 2, 3)$:

Table 2. Calculus of $r^{(3)}_{k,w}$ for $(w_1, w_2, w_3) = (1, 2, 3)$, using relation (13)

k	0	1	2	3	4	5	6	7	8	9	10	11	12	...
$\alpha_1 = 0$	1	0	1	1	1	1	2	1	2	2	2	2	3	...
$\alpha_1 = 1$	-	1	0	1	1	1	1	2	1	2	2	2	2	...
$\alpha_1 = 2$	-	-	1	0	1	1	1	1	2	1	2	2	2	...
$\alpha_1 = 3$	-	-	-	1	0	1	1	1	1	2	1	2	2	...
$\alpha_1 = 4$	-	-	-	-	1	0	1	1	1	1	2	1	2	...
$\alpha_1 = 5$	-	-	-	-	-	1	0	1	1	1	1	2	1	...
$\alpha_1 = 6$	-	-	-	-	-	-	1	0	1	1	1	1	2	...
$\alpha_1 = 7$	-	-	-	-	-	-	-	1	0	1	1	1	1	...
...	...	...	...	...	...	...	...	...	...	...	...	...	...	...
$r^{(3)}_{k,w}$	1	1	2	3	4	5	7	8	10	12	14	16	19	...

The next theorem gives a more general result:

Theorem 4. *Let be $w = (w_1, w_2, w_3) \in N^3$. The dimensions of the w-degree homogeneous polynomial spaces in 3 variables satisfy the relations:*

$$r^{(3)}_{w,0} = 1 \text{ and } (0,0,0) \in A^{0,3}_{0,w} \tag{15}$$

$$r^{(3)}_{w,q} = r^{(2),[2,3]}_{w,q}, \; q \in \{0, 1, \ldots, w_1 - 1\},$$

$$r^{(3)}_{w,k\cdot w_1+q} = r^{(3)}_{w,(k-1)\cdot w_1+q} + r^{(2),[2,3]}_{w,k\cdot w_1+q}, \; \forall k \in Z, \; k > 0, \; 0 \le q \le w_1 - 1 \tag{16}$$

The monomials from the homogeneous space of w- degree n are $(\alpha_1, \alpha_2, \alpha_3)$ with $\alpha_1 \in \left\{0, \ldots, \left[\frac{n}{w_1}\right]\right\}$ and $(\alpha_2, \alpha_3) \in A^{0,2}_{n-\alpha_1\cdot w_1}$, with $A^{0,2}_{n-\alpha_1\cdot w_1}$ given in theorem 1 in which we choose $i = 2$ and $j = 3$.

Proof. The proof is similar with the proof of theorem 3. The difference is, that in the table of dimensions, in every line we advance with w_1 columns to the right. The induction process leads us to a up triangular table. On the line i of the table, we will have $\alpha_1 = i$, $i \in \{0, 1, \ldots, n\}$ and, starting from the column $w_1 + i$, we will have the values $r^{(2),[2,3]}_{w,i}$. Hence

$$r^{(3)}_{w,k\cdot w_1+q} = \sum_{s=0}^{k} r^{(2),[2,3]}_{w,(k-s)\cdot w_1+q}, \; \forall k > 0. \tag{17}$$

Theorem 4 is illustrated in Table 3 and Table 4, for $(w_1, w_2, w_3) = (2, 2, 3)$:

The recursive relation (16) can be generalized, using induction on d, for d variables.

Theorem 5. *Let be $w \in N^d$. The dimensions of the w-degree homogeneous polynomial spaces in d variables, satisfy the relations:*

$$r^{(d)}_{w,0} = 1 \text{ and } (0, \ldots, 0) \in A^{0,d}_{0,w} \tag{18}$$

$$r^{(d)}_{w,q} = r^{(d-1),[2,\ldots,d]}_{w,q}, \forall q \in \{0, \ldots, w_1 - 1\} \tag{19}$$

$$r^{(d)}_{w,k\cdot w_1+q} = r^{(d),[2,\ldots,d]}_{w,(k-1)\cdot w_1+q} + r^{(d-1),[2,\ldots,d]}_{w,k\cdot w_1+q}, \forall k \in Z^+, \; q \in \{0, \ldots, w_1 - 1\} \tag{20}$$

Table 3. Calculus of $r^{(3)}_{k,w}$ for $(w_1, w_2, w_3) = (2, 2, 3)$, using relation (17)

k	0	1	2	3	4	5	6	7	8	9	10	11	12	...
$\alpha_1 = 0$	1	0	1	1	1	1	2	1	2	2	2	2	3	...
$\alpha_1 = 1$	-	-	1	0	1	1	1	1	2	1	2	2	2	...
$\alpha_1 = 2$	-	-	-	-	1	0	1	1	1	1	2	1	2	...
$\alpha_1 = 3$	-	-	-	-	-	-	1	0	1	1	1	1	2	...
$\alpha_1 = 4$	-	-	-	-	-	-	-	-	1	0	1	1	1	...
$\alpha_1 = 5$	-	-	-	-	-	-	-	-	-	-	1	0	1	...
$\alpha_1 = 6$	-	-	-	-	-	-	-	-	-	-	-	-	1	...
...	...	...	...	...	...	...	...	...	...	...	...	...	...	...
$r^{(3)}_{k,w}$	1	0	2	1	3	2	5	3	7	5	9	7	12	...

Table 4. Calculus of $r^{(3)}_{k,w}$ for $(w_1, w_2, w_3) = (2, 2, 3)$, using recursive relation (15)

k	0	1	2	3	4	5	6	7	8	9	10	11	12	...
$r^{(3)}_{k,w}$	*1*	0	2	1	3	2	5	3	7	5	9	7	12	...
			1+1	*0+1*	*2+1*	*1+1*	*3+2*	*2+1*	*5+2*	*3+2*	*7+2*	*5+2*	*9+3*	...
$r^{(2)[2,3]}_{k,w}$	1	0	*1*	1	1	1	2	1	2	2	2	2	3	...

3 Conclusions and Further Work

This article presents original results regarding the w-degree polynomial spaces in three variables. Theorems 3 and 4 allow us to find the dimension of homogeneous space of w-degree n and the monomial basis for these spaces. Having these elements, we can use w-degree polynomial spaces in three variables as interpolation spaces for different interpolation conditions, like in [7]. The method we used for generalization the results from bivariate polynomial spaces to spaces of polynomials in three variables can be used for obtaining similar results in case of d- variables. The recurrence relation between the dimension of polynomial spaces of w-degree n in d variables and some dimensions of polynomial spaces of w-degree k in $d-1$ dimensions can be easily implement. Our further work focus on implementing the theoretical results from this article and discuss various computational aspects of this implementation. We want to study the influence of the choice of exponent we bind to every step on the complexity of our algorithms.

Acknowledgment

This work was completed with the support of the research grant of the Romanian Ministry of Education and Research, CNCSIS 33/2007.

References

1. Möller, M., Michael, H., Sauer, T.: H – bases for polynomial interpolation and system solving. Advances in Comp. Math. 12, 335–362 (2000)
2. Sauer, T.: Gröbner basis, H-basis and interpolation. Transactions of the American Mathematical Society (1994)
3. Sauer, T.: Polynomial interpolation of minimal degree. Numer. Math. 78(1), 9–85 (1997)
4. Sauer, T.: Polynomial interpolation of minimal degree and Gröbner bases, Groebner Bases and Applications. In: Buchberger, B., Winkler, F. (eds.) Proc. of the Conf. 33 Years of Groebner Bases. London Math. Soc. Lecture Notes, vol. 251, pp. 483–494. Cambridge University Press, Cambridge (1998)
5. Sauer, T.: Gröbner bases, H – bases and interpolation. Trans. Amer. Math. Soc. 353, 2293–2308 (2001)
6. Simian, D., Simian, C., Moiceanu, A.: Computational aspects in spaces of bivariate polynomial of w-degree n. In: Li, Z., Vulkov, L.G., Waśniewski, J. (eds.) NAA 2004. LNCS, vol. 3401, pp. 486–494. Springer, Heidelberg (2005)
7. Simian, D., Simian, C.: On some interpolation problems in polynomial spaces with generalized degree. NAUN Int. J. of Math. Models and Meth. in Appl. Sci. 1(1), 147–150 (2007)

A Discrete Model for a Network Having Broken Packages

Dana Simian[1], Vladislav Georgiev[2], and Corina Simian[1]

[1] University "Lucian Blaga" of Sibiu, Romania
[2] St. Kliment Ohridski University of Sofia, Bulgaria

Abstract. Broken packages are packages with not enough information regarding the destination address and they have a negative impact on a computer network. Broken packages stay in the router too much time and overload it. This is a real problem which appears in the computer network of Sofia University. We know only the time when a broken package arrives in the router. The problem is to find the computer which send broken packages. The aim of this article is to propose a model for defining the data transmission stability and correctness in a computer network. Our theoretical model is temporal dependent and captures the transmission properties of a computer network.

1 Introduction and Motivation

The study made in this article is starting from a real problem which Sofia University Computer Network (SU) is confronted with.

One of the research area in Internet management and applications is to deliver real time applications with minimum network resources. A physical network is a group of interconnected computers, each with its distinct IP address. An IP address is an unique address that certain electronic devices use in order to identify and communicate with each other on a computer network utilizing the Internet Protocol standard. Usually it is used Ipv4. In the beginning the IP addresses were divided in 5 classes, from A to E. The dividing depends on the binary configuration of the first octet of the address. The SU is using 16 networks Class C, divided in 30 subnetworks. Sofia University has got many campuses in different places in the city. This campuses are connected by MAN, which uses optical fiber. When you receive or send data (e-mail, web pages) using Internet Protocol, the message is divided in small parts named packages. Every package contains the address of the persons who send the message and the address of the person who should receive it. Every packet is sent first to a "Gateway Computer", which is reading the address to where the packet should arrive and send the packet to another "Gateway". This happens until the packet arrives to a "Gateway" neighbor with the computer where the packet should be received. According to how much information concerning the destination address a packet contains, we have 3 possibilities:

- the package contain complete information concerning the destination address, it passes trough the router and arrives to the destination;

S. Margenov, L.G. Vulkov, and J. Waśniewski (Eds.): NAA 2008, LNCS 5434, pp. 517–524, 2009.

- there is not enough information about destination address, very little is missing and the router can manage with these packages and send them out
- too much information about the destination address is missing and these packages remain in the router, which has not enough information for layer 3 and doesn't know what to do with them. These kind of packages are named broken packages.

In SU is used a special software for routing, Zebra. Zebra is a routing software package that provides TCP/IP based routing services with routing protocols support such as RIP, OSPF and BGP. Zebra also supports special BGP Route Reflector and Route Server behavior. In addition to traditional IPv4 routing protocols, Zebra also supports IPv6 routing protocols. This software is working with packages, read the information inside them and send the packages to the destination.

Some problems that appear in SU are given by the broken packages. Broken packages provoke problems when they arrive in the router at the third level, because they don't contain enough information. The router tries to read the information that the package contains and tries to calculate what is necessary to be done with these packages and where to send them. The processor calculates too much for the broken packages and because of this appear delays for the other packages. The system load average is getting too big and the operating system can crush. We need to find where these packages come from, but we can not make this directly, because we don't have enough information. A measurement of the traffic through the router gives us the time when this happens. An unjustifiable growing of the traffic thought the router is almost always made by broking packages. The packages start to come in the morning around 9 and finish around 17 and 30 and we suppose the broken packages come from a work station. The SU has many little networks and we can locate the network where the broken packet is coming from, using the information from vlan, but we can not locate the host.

The aim of this article is to give a model for defining the transmission properties in a computer network.

The paper is organized as follows: in section 2 are presented some existing models for computer networks and flow traffic. In section 3 is presented our model and in section 4 are presented conclusions and further directions of study.

2 Computer Networks Models

In [4] a computer network is represented by a connected, undirected graph $G = (V, E)$, with the nodes as computers and the edges as the (physical) communication links. Communications between any arbitrary pair of nodes takes place through a path in G. This graph model is also used in representation of logical, or virtual network. Such examples of graphs are enumerated in [4]. We can have an e-mail graph, in which a node represents an user and the edges emanating from this node go to all individual addresses in the e-mail address book of this user. We can use also a web-graph in which a node is represented by a web page and edges by hyperlinks. The Internet can be modeled using a Microscopic Internet Graph or a Macroscopic Internet Graph. In the Microscopic Internet Graph nods are represented by routers and hosts, while edges correspond to communication links. In the macroscopic model, each node incorporates a number of routers. Empirical studies demonstrated that the graphs associated to real world-networks are scale - free (power-low) random graph, characterized at least by the degree distribution, the clustering coefficient and average distance.

In [1] is considered a model to a multi-server integrated network and a traffic model within this scenario.

In [5] are discussed many possibilities of worm modeling, but these alternatives mimic the ones of network modeling in general. The authors define the scalability and the fidelity of a model, in order to compare different models and propose packet-logic simulation, as a logical alternative for testing the models.

In [9] a cellular automaton is used for modeling a large network. Behavior of each cell captures some important details related to network protocols. The model is used for studying what is happening when many network connections are active simultaneously.

In [6] is given a discrete computer network model for a simple computer network with one server. This model allows a discussion about the stability of the network in time.

We did not find models for the correctness of data transmission.

3 A Model for Stability and Transmission Correctness of a Network

The subnetwork from SU where the broken packages come from can be identified using the information from vlan. First, we suppose, for simplicity that, in a finite time interval, not very long, only a subnetwork is responsible for transmission of broken packages. Therefore, we will give a discrete model for a network with one server and with possible errors in information transmission.

Our graph model contains n nodes, represented by the computers from the network. In the same moment a node can send and receive data. We take into account only the case of finite memory, that is every node and the server have a maximum allowable storage. We work with discrete time, $t \in \{t_0, t_1, \ldots\} \subset Z_+$. The time intervals $[t_k, t_{k+1}]$ are assumed equal. Let $x_i(t)$, $i \in \{1, \ldots n\}$ be the difference between the input and the output amount of data, of node i, $i \in \{1, \ldots, n\}$, at the moment t. The storage of the node i at the time t is $\mid x_i(t) \mid$. The storage of the whole network, or of the server at time t is

$$|\Delta x(t)| = \left| \sum_{i=1}^{n} x_i(t) \right| \quad (1)$$

We can analyze the network by the point of view of storage and by the point of view of data transmission.

An analysis of a simple network, by the point of view of storage, can be found in [6]. If at the moment t the amount of data stored on every computer is less than or equal to its maximum admissible storage and the amount of data stored on the server is not greater than the maximum admissible storage on the server, than the network is running well at this moment. If the network is running well at all moments, than it is called *stable* by the point of view of storage.

Definition 1. *(Shi Y., Chen G. [6]) A network is said to be stable if there exists a positive constant* $r_0 < M_0$ *such that for all initial point* $x_0 \in R^n$, $x(t_0) = (x_1(t_0), \ldots, x_n(t_0))^T$, *satisfying* $\sum_{i=1}^{n} |x_i(t_0)| \leq r_0$, *one has* $|x_i(t)| \leq S_i, \forall\, i \in \{1, \ldots n\}$ *and* $|\Delta x(t)| \leq M_0$, *for all* $t \in Z_+$. *Otherwise, it is said to be unstable. In particular, the network is called a devil network if it is unstable and, further, for any small positive constant* $r \leq M_0$ *there exist an initial point* $x(t_0)$ *with* $\sum_{i=1}^{n} |x_i(t_0)| \leq r$, *such that* $|x_i(t)| \leq S_i$, $i \in \{1, \ldots, n\}$, *for all* $t \in Z_+$ *and* $|\Delta x(t_k)| \leq \alpha M_0$ *for infinitely many times* t_k, *but* $|\Delta x(t'_k)| > M_0$ *for infinitely many times* $t'_k > t_k$.

In order to define our model, we introduce the notions of *correct transmission, incomplete transmission, transmission with damages* and *incorrect transmission* of data.

If, all the packages sent by the node i arrive to their destination, $\forall i \in \{1, \ldots, n\}$ then we say that the network is with *correct transmission.* If some information from the package is incorrect or incomplete, there are two situations: the packet returns to its source or the package is lost. In the first case we will say that the transmission in the network is *incomplete* and in second case we will say that the transmission is *with damages.* In both cases the transmission is *incorrect.* The lost packages overload the router.

We introduce the following notations.

$is_i(t)$ is the amount of data sent by the node i at the time t.
$ir_i(t)$ is the amount of data received by the node i at the time t.
The amount of data will be measured in number of packages. Obviously $x_i(t) = ir_i(t) - is_i(t)$.

Ignoring nonlinear factors, simply assume that $x(t_k) = (x_1(t_k), \ldots, x_n(t_k))^T$, $ir(t_k) = (ir_1(t_k), \ldots, ir_n(t_k))^T$, $is(t_k) = (is_1(t_k), \ldots, is_n(t_k))^T$ satisfy the following discrete linear systems:

$$ir(t_{k+1}) = R(t_k) ir(t_k) + q(t_k) \tag{2}$$
$$is(t_{k+1}) = S(t_k) is(t_k) \tag{3}$$
$$x(t_{k+1}) = B(t_k) x(t_k) + u(t_k) \tag{4}$$

R, S are coupling matrices with elements 0 or 1, with dimension $n \times n$. q, u are column vectors with n lines which express the properties of the transmission process of data. A correct process of data transmission correspond to zero matrices q, u at any time.

The elements of the matrices R, S, A are evaluated as follows:

- $r_{i,j}(t) = 1$ if the node i received data from the node j at the moment t and is zero otherwise. We consider that $r_{i,i}(t) = 0$.
- $s_{i,j}(t) = 1$ if the node i send data to the node j at the moment t and is zero otherwise.

We define the matrix $A(t) = R(t) - S(t)$. A is also a coupling matrix. After a simple computation we obtain $B(t) = (A(t) - S(t))x(t)$ and $u(t) = q(t) - is(t) \cdot A(t)$.

There are four possible cases for the transmission of data realized by a computer, from the network.

Case 1: Correct transmission
A computer i has a correct transmission of data at the moment t if all the packages sent by it arrive to the destination. Let denote

$$d_{i,j}(t) = s_{i,j}(t) \cdot is_i(t) - r_{j,i}(t) \cdot ir_j(t), \tag{5}$$

the difference between the number of the packages sent from the computer i to the computer j and the number of the packages received by the computer j from the computer i.

Definition 2. *The computer i, from a network, is with correct transmission at the moment t if,* $\forall j \in \{1, \ldots, n\}$, $i \neq j$, *we have*

$$d_{i,j}(t) = 0 \text{ and } q_i(t) = 0 \tag{6}$$

From 6 results that, for a correct transmission of computer i, we have $a_{i,j}(t) = -a_{j,i}(t)$, $\forall j \in \{1, \ldots, n\}$.

Definition 3. *A network is with correct transmission at moment* $t \in I$ *if every computer from the network is with correct transmission at that moment.*

In this case, the matrix A is a antisymmetrical T-matrix.

Case 2: Incomplete transmission
A computer is with incomplete transmission at the moment t if it send some packages with missing information in the destination address and these packages returned to the source.

Definition 4. *A computer i is with incomplete transmission at the moment t if*

$$q_i(t) > 0 \text{ and } \sum_{\substack{j=1 \\ j \neq i}}^{n} d_{i,j}(t) = q_i(t) \tag{7}$$

Case 3: Damaged transmission
A computer is with damaged transmission at the moment t if it send at the moment t some packages which can not be delivered to the destination and can not be returned either to the source. We suppose that, for node i, we have not both damaged transmission and incomplete transmission at the moment t.

Definition 5. *A computer i is with damaged transmission at the moment t, if,*

$$q_i(t) = 0 \text{ and } \sum_{\substack{j=1 \\ j \neq i}}^{n} d_{i,j}(t) > 0 \tag{8}$$

Case 4: Incomplete and damaged transmission

A computer has an incomplete and damaged transmission at the moment t if it send at the moment t, both packages which return to the source and packages which are lost.

Definition 6. *A computer i is with incomplete and damaged transmission at the moment t if*

$$q_i(t) > 0 \text{ and } \sum_{\substack{j=1 \\ j \neq i}}^{n} d_{i,j}(t) > q_i(t) \tag{9}$$

For the case of a network we give the following definition

Definition 7. *A network is with incomplete, or damaged transmission at the moment t if there exists at least one computer with incomplete or damaged transmission at the moment t. If there is one computer with incorrect or damaged transmission at the moment t, the network is with incorrect transmission at this moment.*

A network is with incomplete transmission or damaged transmission in a temporal interval I, if there exists a moment $t \in I$, such that, at this moment the network is with incomplete or damaged transmission.

The case presented in [6] is obtained, for correct transmission, replacing in matrix S the values 1 with -1 and imposing that bidirectional transmission of data are not allowed. In this case we obtain $x(t_{k+1}) = A(t_k)x(t_k)$, and A is a symmetrical matrix with elements in $\{-1, 0, 1\}$.

Given the initial vectors $ir(t_0)$, $is(t_0)$, the solution of the systems (2)-(4) can be written as:

$$ir(t_n) = \sum_{i=1}^{n-1}\prod_{j=1}^{i-1} R(t_{n-j}) \cdot q(t_{n-i}) + R(t_{n-1}) \cdots R(t_1)(q(t_0) + R(t_0)ir(t_0)) \tag{10}$$

$$is(t_n) = S(t_{n-1}) \cdots S(t_0)is(t_0) \tag{11}$$

$$x(t_n) = \sum_{i=1}^{n-1}\prod_{j=1}^{i-1} B(t_{n-j}) \cdots u(t_{n-i}) + B(t_{n-1}) \cdot B(t_1)(u(t_0) + B(t_0)x(t_0)) \tag{12}$$

Equation 12 captures both the storage and the transmission aspects in a computer network.

In order to mathematical define the stability and the transmission correctness of a network, let denote by MR, the maximum storage of the router.

For an interval $I_k = [t_k, \ldots, t]$, we define the number of lost packages, which can be considered as a measure of the overload of router in the interval I.

$$V_I(t) = \sum_{\tau \in I_k} \sum_{i=1}^{n} \left(\sum_{\substack{j=1 \\ j \neq i}}^{n} d_{i,j}(\tau) - q_i(\tau) \right) \tag{13}$$

Definition 8. *A network is said with good transmission in interval $I = [t_0, t]$ if there are not any nodes with incomplete or damaged transmission, in the network in this interval of time, that is $\sum_{\substack{j=1 \\ j \neq i}}^{n} d_{i,j}(\tau) = 0$, $\forall i \in \{1, \ldots, n\}$ and $\forall \tau \in I$.*

There are many causes which can make a computer to send broken packages: a broken Eternet adaptor, cables too long, with an inadequate voltages, corrupted signals due to the electricity lines which interact with the computer cables. In some cases, not very often, it is possible that a computer to send broken packages at a moment t_k, but this problem to not appear any more at a moment $t_j > t_k$. If in an interval there are accidentally broken packages, but they do not overload the router we can consider that we have an admissible transmission.

Definition 9. *A network is said with α - admissible transmission in interval $I = [t_0, t]$, if there exists a positive constant $0 \leq p_0 \leq \alpha \cdot MR$, $\alpha \in [0, 1]$ such that $\sum_{\substack{j=1 \\ j \neq i}}^{n} d_{i,j}(t_0) \leq p_0$, for every $i \in \{1, \ldots, n\}$ and $V_I(t) \leq \alpha \cdot MR$.*

In definition 9, α is a transmission parameter. As α is smaller as the transmission is better. This parameter allows us to extend our transmission analysis to a big network with m small subnetworks.

Definition 10. *A network with m subnetworks is with β admissible transmission in interval $I = [t_0, t]$, if there exists $\beta \in (0, 1)$ and $\alpha_1, \ldots, \alpha_m$, with $\alpha \in [0, 1]$, $i \in \{1, \ldots, m\}$ such that the subnetwork i is with α_i admissible transmission and $\sum_{i=1}^{m} (V_I)_i(\tau) \leq \beta \cdot MR$ for every $\tau \leq t$.*

Obviously, a network is running well in interval I, if it is stable and is with good transmission or at least with α - admissible transmission in interval I.

The most simple case happens when only one computer has at moment t_0 damaged transmission (without incomplete transmission). Let suppose that computer "a" sends broken packages, in the interval $I = [t_0, t_k]$. Sufficient conditions for α - admissible transmission are:

$$\sum_{j=1}^{n} (s_{a,j}(t_0) \cdot is_a(t_0) - r_{j,a}(t_0) \cdot ir_j(t_0)) \leq p_0$$

$$\sup_{l} \sum_{j=1}^{n} \sum_{m=1}^{n} (s_{a,j}(t_l) W_{a,m}(t_{l-1}) is_m(t_0) - r_{j,a}(t_l) V_{j,m}(t_{l-1}) ir_m(t_0)) \leq \frac{\alpha \cdot MR}{n}$$

$$l \in \{1, \ldots, k\}$$

with $V(t_k) = R(t_k) \cdots R(t_0)$, $W(t_k) = S(t_k) \cdots S(t_0)$.

4 Conclusions and Further Work

In this paper we built a discrete model for a network, taking into account not only the stability of the network in the sense given in [6] but also the transmission properties of the network. Our model consider a network with a constant number of computers be cause the analysis of the transmission properties will be done in practice for a temporal interval not greater than 24 hours. The model can be extended for a network with variable number of computers as in [6].

Our further direction of study consist in validation our model using a network simulation software. Using this model we want to obtain an algorithm for optimize the process of identification the node with bad transmission in a network and implement it.

Our model gives us the possibility to theoretically discuss various scenarios of data transmission in a network, but because of incomplete data obtained for measurements in practice, we might use techniques from approximation theory for solving the systems (2) - (4).

Acknowledgment

This work was completed with the support of the research grant of the Romanian Ministry of Education and Research, CNCSIS 33/2007.

References

1. Ahmed, N.U., Wang, Q., Barbosa, L.O.: System Approach to Modeling the Token Bucket Algorithm in Computer Networks. Mathematical Problems in Engineering 8(3), 265–279 (2002)
2. Ahmed, N.U., Dabbous, T.E., Lee, Y.E.: Dynamic routing for computer queuing networks. Int. J. Systems SCI 19(6) (1988)
3. Faloutsos, M., Faloutsos, P., Faloutsos, C.: On powerlow relationships of the Internet topology. In: Proceeding SIGCOMM Cannes (1999)
4. Nikoloski, Z., Deo, N., Pucera, L.: Correlation Model of Worm Propagation on Scale - Free Networks. Complexus Network Modelling 3, 169–182 (2006)
5. Perumalla, K.S., Sundaragopalan, S.: High-Fidelity Modeling of Computer Network Worms. Technical Report GIT-CERCS-04-23, Center of Experimental Research in Computer Science, Georgia Institut of Technology (2004)
6. Shi, Y., Chen, G.: A discrete computer network model with expanding dimensions, Nonlinear Sciences, Chaotic Dynamics, 14 p. (2007) arXiv:0705.0855v1 [nlin.CD]
7. Wahida, C., Ahmed, N.U.: Congestion control using dynamic routing and flow control. Stochastic analysis and applications 10(2), 123–142 (1992)
8. Yook, S.H., Jeong, H., Barabasi, A.I.: Modeling the Internet's large scale topology. Proc. Nat. Acad. Sci. USA 99, 13382–13386 (2003)
9. Yuan, J., Mills, K.: Exploring Collective Dynamics in Communication Networks. Journal of Research of national Institute of Standards and Technology 107, 179–191 (2002)

Applications of the Connection between Approximation Theory and Algebra

Dana Simian and Corina Simian

University “Lucian Blaga” of Sibiu, Faculty of Sciences
5-7 dr. I. Raţiu str, 550012 Sibiu, Romania

Abstract. The aim of this paper is to illustrate a possibility of obtaining various theoretical results using the connection between multivariate interpolation and reduction process with respect to a H-basis of an ideal. Using this connection we can switch between interpolation theory and the theory of ideals. As a application of this connection, we found and proved an interesting identity, which is satisfied for all polynomials in d variables from an interpolation polynomial subspace.

1 Introduction

Multivariate interpolation is an useful tool for modeling processes in which it is necessary to approximate an unknown function, preserving some known information about it. This information is represented by the values of some functionals applied to the unknown function. Usually we know the values of this function on a set of points from R^d, or the values of different types of derivatives of the function on a set of points, or different integrals. There are situations in which are known the values of the function on a set of points given implicitly, for example points given as common zeros of some orthogonal polynomials. When we choose to use polynomials for interpolation of the unknown function we obtain a polynomial interpolation. Let be $\mathcal{F} \supset \Pi$ the space of functions to which the unknown function belongs and Λ the set of interpolation conditions. The multivariate interpolation problem is to find a polynomial subspace $\mathcal{P}$ such that, for an arbitrary $f \in \mathcal{F}$ there exist an unique polynomial $p \in \mathcal{P}$ satisfying the interpolation conditions: $\lambda(f) = \lambda(p), \forall \lambda \in \Lambda$. In this case, $\mathcal{P}$ is named the interpolation space for the conditions Λ and the pair $(\Lambda, \mathcal{P})$ defines an interpolation scheme, or is called a correct pair.

There are many approaches to generalize the univariate interpolation case to several variables and there are many interpolation schemes with interesting and useful properties. A survey of these methods and of most important interpolation schemes can be found in [4], [5], [8].

From computational point of view, minimal interpolation schemes are very important. We say that $\mathcal{P}$ is a minimal interpolation space for the conditions Λ if there are not other interpolation polynomial subspaces with total degree less than $\mathcal{P}$ for the conditions Λ. Generally, there are many minimal interpolation spaces for a given set of interpolation conditions. If there exist $n \in N$ such that, Π_n^d the

S. Margenov, L.G. Vulkov, and J. Waśniewski (Eds.): NAA 2008, LNCS 5434, pp. 525–531, 2009.

subspace of polynomials of total degree n in d variables is a minimal interpolation space for the conditions Λ, then Π_n^d is the unique minimal interpolation space for these conditions.

If $ker(\Lambda)$ is an ideal of polynomials we obtain an ideal interpolation scheme. Ideal interpolation schemes are generalizations of Hermite-Birchoff schemes for the multivariate case, in the sense given in the following theorem:

Theorem 1. *(M. Gasca, T.Sauer, [4]) A finite set of functionals $\Lambda \subset \Pi'$, defines an ideal interpolation scheme if and only if there exist points x_j and D-invariant polynomial subspaces Q_j, $j = 1, \ldots, m$, such that*

$$\Lambda = \{(q_{j,k}(D)f)(x_j),\ q_{j,k} \in Q_j,\ j = 1, \ldots, m;\ k = 0, \ldots, l_j - 1\}. \tag{1}$$

On the other hand, if $ker(\Lambda)$ is a polynomial ideal, we can apply the known results concerning the theory of ideals, related to the ideal $ker(\Lambda)$. It is proved in [11] that in the case of ideal interpolation schemes, we can switch between interpolation and reduction process with respect to a H-basis of the ideal $ker(\Lambda)$. In order to make the reduction process the choice of an inner product is necessary. Various inner products give us different ideal interpolation schemes.

The aim of this article is to formulate and prove some theoretical results using the connection between multivariate interpolation and the theory of ideals.

The paper is organized as follows: in the section 2 we present the least interpolation scheme and some results related to it, which are necessary for our purpose. In section 3 we introduce the main elements regarding the reducing process with respect to a H-basis of an ideal and present the theorem which makes the connection between multivariate interpolation and the theory of ideals. The main results can be found in section 4. Conclusions and further directions of work are formulated in section 5.

2 Least Interpolation Scheme

For our purpose the least interpolation scheme introduced by C. de Boor and A. Ron in [1], [2] has a great importance. We will denote next, the least interpolation scheme by $(\Lambda, H_\Lambda\downarrow)$. $H_\Lambda\downarrow$ is the vector space generated by the least terms of the power series representation of $\lambda \in \Lambda$, that is

$$H_\Lambda\downarrow = span\{g\downarrow; g \in H_\Lambda\}, \quad \text{with} \tag{2}$$

$$H_\Lambda = span\{\lambda^\nu;\ \lambda \in \Lambda\}; \tag{3}$$

where λ^ν is the generating function of the functional $\lambda \in \Lambda$.

Theorem 2. *(C. de Boor [2]) $H_\Lambda\downarrow$ is a minimal interpolation space for Λ.*

In order to obtain a formula for the interpolation operator it is introduced the pair between an analytic function and a polynomial

$$< f, p >= (p(D)f)(0) = \sum_{\alpha \in N^d} \frac{D^\alpha p(0) D^\alpha f(0)}{\alpha!} \tag{4}$$

$p(D)$ is the differential operator with constant coefficients associated to the polynomial p. This pair is a veritable inner product on polynomials' spaces.

The Lagrange case is obtained for the evaluation functionals. This case is studied in details in [1]. Here it is used the notation Π_Θ, for the least interpolation space in the Lagrange case.

$$\Pi_\Theta = H_\Lambda\downarrow;\ \Lambda = \{\delta_\theta;\ \theta \in \Theta \subset R^d\} \tag{5}$$

Obviously, Lagrange conditions yield to an ideal interpolation scheme.

3 H-bases and Interpolation

Definition 1. *For given $f,\ f_1,\dots,f_m \in \Pi^d$, we say that f reduces to $\tilde{f}$ with respect to polynomials vector $\mathcal{F} = \{f_1,\dots,f_m\}$ if*

$$\tilde{f} = f - \sum_{i=1}^{m} g_i{\cdot}f_i \text{ and } deg(\tilde{f}) < deg(f)$$

holds with polynomials g_i satisfying $deg(g_i) \le deg(f) - deg(f_i),\ i \in \{1,\dots,m\}$. In this case we write $f \to_{\mathcal{F}} \tilde{f}$.

Definition 2. *A finite set $\mathcal{H} \subset \Pi^d$ is an H-basis for an ideal I if and only if for all $p \in I$, $p \neq 0$, has a unique representation*

$$p = \sum_{i=1}^{s} h_i g_i,\ g_i \in \Pi^d \text{ and } deg(h_i) + deg(g_i) \le deg(p). \tag{6}$$

$\mathcal{H}$ is a H-basis for $I \Longleftrightarrow \{p\uparrow \mid p \in I\} =< p\uparrow \mid p \in \mathcal{H} >$.

The process of reduction can be generalized for all homogeneous terms of f. To do this let consider a m-vector of polynomials $(p_1,\dots,p_m)$ and define the following vector space of homogeneous polynomials:

$$V_n(p_1,\dots,p_m) = \left\{\sum_{i=1}^{m} q_i \cdot p_i\uparrow \mid q_i \in \Pi^0_{n-deg(p_i)};\ i = 1,\dots,m\right\} \subset \Pi^0_n, \tag{7}$$

$\Pi^0_k = \{0\}$, if $k < 0$.
Let any inner product defined on Π^d be given. This inner product induces a notion of orthogonality and we obtain the next decomposition into successive orthogonal complements:

$$W_n(p_1) = V_n(p_1)$$
$$\dots$$
$$W_n(p_1,\dots,p_j) = V_n(p_1,\dots,p_j) \ominus V_n(p_1,\dots,p_{j-1}),\ j = 2,\dots,m,$$

$$V_n(p_1,\dots,p_m) = \bigoplus_{j=1}^{m} W_n(p_1,\dots,p_j)$$

Definition 3. *A polynomial $f \in \Pi^d$ is called reduced with respect to the vector of polynomials $\mathcal{P} = (p_1, \dots, p_m)$, if each homogeneous term of f is reduced to zero. In other words, if we write $f = \sum_{j=0}^{deg(f)} f_j;\ f_j \in \Pi_j^0,\ j = 0, \dots, deg(f)$ then f is reduced if and only if $f_j \perp V_j(p_1, \dots, p_m),\ j = 0, \dots, deg(f)$.*

It is proved in [7] that if $\mathcal{H}$ is a H-basis, then the reduced polynomial of a polynomial f, modulo $\mathcal{H}$, is independent of the elements in $\mathcal{H}$. For an ideal I and a given inner product on Π^d we define the decomposition

$$\Pi_n^0 = V_n(I) \oplus W_n(I), \text{ with } V_n(I) = \{f\uparrow \mid f \in I;\ deg(f) = n\} \subset \Pi_n^0. \tag{8}$$

The inner product used in reduction process modulo an H-basis is essential. Different inner products will usually give different classes of reduced polynomials.

Proposition 1. *(T.Sauer, [7]) A polynomial p is reduced modulo an H-basis, $\mathcal{H} \subset \Pi$, of the ideal $I =< \mathcal{H} >$, with respect to the inner product $< f, p >= (p(D)f)(0)$, if and only if*

$$q \in \bigcap_{p \in \mathcal{H}} ker\ p\uparrow (D) = \bigcap_{p \in <\mathcal{H}>} ker\ p\uparrow (D) \tag{9}$$

The connection between ideal interpolation and reduction process modulo a H-basis of the ideal $ker(\Lambda)$ is given in the following theorem:

Theorem 3. *(T.Sauer, [7]) Let $\Lambda \subset \Pi'$ a set of functional with $I = ker\ \Lambda$ an ideal of polynomials and $\mathcal{H}$ a H-basis for I. Then the space $\mathcal{P}_{\mathcal{H}} = \Pi_{\to\mathcal{H}}$, of reduced polynomials modulo $\mathcal{H}$, is a minimal degree interpolation space with respect to Λ, and the interpolation operator is the reduction operator modulo $\mathcal{H}$, that is*

$$L_{\mathcal{P}_{\mathcal{H}}}(q) = q \to_{\mathcal{H}};\ q \in \Pi. \tag{10}$$

Given any Λ, there is usually a multitude of minimal degree interpolation spaces. However, any H-basis defines a minimal degree interpolation space.

Theorem 4. *(T.Sauer, [7]) If Λ define an ideal interpolation scheme and $\mathcal{H}$ is a H-basis of the ideal $I = ker\ \Lambda$, with respect to the inner product given in (4) then*

$$H_\Lambda \downarrow = \Pi_{\to\mathcal{H}} \tag{11}$$

4 Main Results

The main result of this paper is given in theorem 5 which offers a characterization of the polynomials from the interpolation space Π_Θ by means of a differential equation. The proof of this theorem uses the connection between ideal interpolation and the theory of ideals, presented in section 3.

Theorem 5. *Let be $\Theta \subset R^d$ a set of N points situated on the unit sphere, $S = \{x \in R^d;\ x_1^2 + \ldots x_d^2 = 1\}$, $d > 1$ with*

$$N = \binom{2m+d}{d},\ m \in Z_+. \tag{12}$$

Then, $\forall u \in \Pi_\Theta$, the following identity holds

$$\sum_{k_1=0}^{m} \sum_{k_2=0}^{k_1} \cdots \sum_{k_{d-1}=0}^{k_{d-2}} C_m^{k_1} C_{k_1}^{k_2} \ldots C_{k_{d-2}}^{k_{d-1}} \cdot D^{(2(m-k_1),2(k_1-k_2),\ldots,2(k_{d-2}-k_{d-1}),2k_{d-1})}(u) = 0 \tag{13}$$

with Π_Θ, the least interpolation space defined in (5).

Proof: We observe that $N = dim(\Pi_{2m}^d)$, but Π_{2m}^d is not an interpolation space for points Θ, be cause the interpolation points are situated on the unit sphere, that is there exist the polynomial

$$p(x_1, \ldots, x_d) = (x_1^2 + \ldots x_d^2)^m - 1 \tag{14}$$

having the property $p \in \Pi_{2m}^d \bigcap ker(\Theta)$. On the other hand, from theorem 2 we know that the pair (Θ, Π_Θ) is minimal correct. More, $\Pi_\Theta \neq \Pi_{2m}^d$ and $\Pi_\Theta \subset \Pi_n$, with $n > 2m$.

Let be $u \in \Pi_\Theta$ and H a H-basis of the ideal $ker(\Theta)$. Taking into account the theorem 3 we obtain that $u \in \Pi_{\to\mathcal{H}}$. Hence, from proposition 1 results

$$u \in \bigcap_{q\in H} ker(q\uparrow(D)) = \bigcap_{q\in ker(\Theta)} ker(q\uparrow(D)) \Rightarrow u \in ker(q\uparrow(D)),\ \forall q \in ker(\Theta)$$

$$\Rightarrow (q\uparrow(D))(u) = 0,\ \forall q \in ker(\Theta),\ \forall u \in \Pi_\Theta.$$

Therefore

$$(p\uparrow(D))(u) = 0, \tag{15}$$

with p given in (14).

We calculate the homogeneous polynomial $p\uparrow$.

$$p\uparrow(x_1, \ldots, x_d) = \sum_{k_1=0}^{m} C_m^{k_1} x_1^{2(m-k_1)} (x_2^2 + \ldots + x_d^2)^{k_1} = \ldots =$$

$$= \sum_{k_1=0}^{m} \sum_{k_2=0}^{k_1} \cdots \sum_{k_{d-1}=0}^{k_{d-2}} C_m^{k_1} C_{k_1}^{k_2} \ldots C_{k_{d-2}}^{k_{d-1}} x_1^{2(m-k_1)} x_2^{2(k_1-k_2)} \ldots x_{d-1}^{2(k_{d-2}-k_{d-1})} x_d^{2k_{d-1}}$$

Using (15) we obtain (14).

Example 1. Bivariate case

We choose $d = 2$. In bivariate case, the points from Θ are situated on the unit circle. The cardinality of Θ is $N = (m+1)(2m+1)$ and (14) becomes

$$\sum_{k_1=0}^{m} C_m^{k_1} D^{(2(m-k_1),2k_1)}(u) = 0,\ \forall u \in \Pi_\Theta \tag{16}$$

The case $m = 2q + 1,\ q \in Z_+$ was obtained in [13].

If we choose $m = 1$ we obtain a set of 6 points situated on the unit circle and (14) becomes

$$D^{(2,0)}(u) + D^{(0,2)}(u) = 0, \forall u \in \Pi_\Theta \tag{17}$$

that is, in this case the interpolation space is a space of harmonic polynomials.

In [3] is discussed the case of interpolation on a set of the vertices of a regular hexagon, $\theta_k = (cos(2\pi k/6), sin(2\pi k/6)), k = 1, \ldots, 6$. It is prove that the interpolation space is

$$\Pi_\Theta = \Pi_1 + (\Pi_2^0 - span(p_2)) + span(p_3), \tag{18}$$

with $p_2(x_1, x_2) = 1 - x_1^2 - x_2^2$, Π_2^0 the space of homogeneous second-degree polynomials and p_3 is a particular homogeneous cubic polynomial. To obtain the interpolation polynomial we can use as additional condition the differential equation given by (17).

For the interpolation conditions $f(\theta_k) = (-1)^j$, in [3] is obtained the interpolation polynomial

$$l(x_1, x_2) = ((1 + 2x_1 + 2x_1^2 + x_1^3) - x_2^2(2 + 3x_1))/6$$

It is easy to verify that l satisfy (17).

Example 2. Univariate case

The case $d = 1$ is not included in theorem 5. We want to find the equivalent of theorem 5 for $d = 1$. The unit sphere in R is the set $\{-1, 1\}$. The nontrivial case is obtained for $\Theta = \{-1, 1\}$. We know that in the univariate case, the cardinality of the set of interpolation points defines in unique way the interpolation space. For two interpolation conditions, the interpolation space is given by the set of polynomial of degree equal to 1, $\Pi_\Theta = \Pi_1^1$. The equivalent of the differential equation (13) is

$$D^2(u) = 0,\ \forall u \in \Pi_1^1 \tag{19}$$

5 Conclusions and Further Work

The connection between ideal interpolation and reduction process with respect to the ideal $ker(\Theta)$ is an important tool which allows us to find, in this paper, a differential equation which is verified by the all polynomials from the interpolation space Π_Θ, for a particular choice of Θ. This equation can be used as an additional condition and can help us in finding interpolation space and interpolation polynomial.

Further we want to obtain a generalization of the theorem 5 for a general ideal interpolation scheme and to make a connection between least interpolation and differential equations.

Acknowledgment

This work was completed with the support of the research grant of the Romanian Ministry of Education and Research, CNCSIS 33/2007.

References

1. de Boor, C., Ron, A.: On multivariate polynomial interpolation. Constr. Approx. 6, 287–302 (1990)
2. de Boor, C.: Polynomial interpolation in several variables. Math. Z. 210, 347–378 (1992)
3. de Boor, C.: On the error in multivariate polynomial interpolation. Math. Z. 220, 221–230 (1992)
4. Gasca, M., Sauer, T.: Polynomial interpolation in several variables. Advances in Computational Mathematics (1999)
5. Lorentz, R.A.: Multivariate Hermite interpolation by algebraic polynomials: a survey. J. Assoc. Comput. Mach. 122(1-2), 167–201 (2000)
6. Möller, H.M., Sauer, T.: H – bases for polynomial interpolation and system solving. Advances in Comp. Math. 12, 335–362 (2000)
7. Sauer, T.: Gröbner basis, H-basis and interpolation. Transactions of the American Mathematical Society (1994)
8. Phillips, G.M.: Interpolation and Approximation by Polynomials Series: CMS Books in Mathematics. XIV, 312 p. (2003)
9. Sauer, T.: Polynomial interpolation of minimal degree. Numer. Math. 78(1), 59–85 (1997)
10. Sauer, T.: Polynomial interpolation of minimal degree and Gröbner bases. In: Buchberger, B., Winkler, F. (eds.) Groebner Bases and Applications (Proc. of the Conf. 33 Years of Groebner Bases). London Math. Soc. Lecture Notes, vol. 251, pp. 483–494. Cambridge University Press, Cambridge (1998)
11. Sauer, T.: Gröbner bases, H – bases and interpolation. Trans. Amer. Math. Soc. 353, 2293–2308 (2001)
12. Sauer, T.: Ideal bases for graded polynomial rings and applications to interpolation. In: Gasca, M. (ed.) Multivariate Approximation and Interpolation with Applications, Academia de Ciencias, Zaragoza. Monograph. Acad. Cienc. Zaragoza, vol. 20, pp. 97–110 (2002)
13. Simian, D., Simian, C.: H-bases and Interpolation. General Mathematics 11(1-2), 63–76 (2003)

Efficient Numerical Method of the 1D Motion of Two-Phase Fluid through Porous Medium in a Class of Discontinuous Functions

Bahaddin Sinsoysal and Mahir Rasulov

Beykent University, Department of Mathematics and Computing,
Sisli-Ayazaga Campus, 34396 Istanbul, Turkey
{bsinsoysal,mresulov}@beykent.edu.tr

Abstract. In this paper a new numerical method for nonlinear system of partial differential equations which describes motion of a two-phase compressible fluid in microscopic porous medium in a class of discontinuous functions is suggested. For this aim, at first, the system of equations is split in two equations according to physical parameters. First one with respect to water saturation, and the other one with respect to pressure. Then, the special auxiliary problem having some advantages over the main problem is introduced. This auxiliary problem permits to develop the efficient numerical algorithm for obtaining the solution. Using the solution of the auxiliary problem, the solution of the main problem, which expresses all the physical properties is found.

1 Introduction

In order to exploit a petrol reservoir efficiently, the adequate mathematical models which accurately describe the filtration of multiphase fluids in porous medium and their efficient solutions are required.

Using this mathematical model and its solution the dynamical distributions of oil, water saturations, pressure of reservoir, and evaluation of the technological exploitation can be defined.

It is known that the flow of two-phase fluid in porous medium is described by the Muskat-Mires system of equations [4] as

$$\frac{m}{k}\frac{\partial}{\partial t}\left(\frac{\sigma_\ell}{\beta_\ell}\right) = div\left(\frac{k_\ell}{\mu_\ell \beta_\ell} gradp_\ell\right) + q_\ell, \quad (\ell = o, w). \tag{1}$$

Here, $q_\ell = \sum_{\nu=1}^{N_\ell} Q_\ell^{(\nu)}(t)\delta(x-x_\nu)\delta(y-y_\nu)$, $(\ell = o, w)$ are debits of oil and water, respectively; $\delta(x)$ is the Dirac function; β_0 and β_w are volume coefficients of oil and water phases; k_o and k_w are permeability of oil and water; μ_o and μ_w are oil and water phases viscosity; N_o and N_w are the number of oil and water wells; $Q_o^{(\nu)}$, $(\nu = \overline{1, N_o})$ and $Q_w^{(\mu)}$, $(\mu = \overline{1, N_w})$ are volume debits; $p_o(x,t)$, $p_w(x,t)$, $\sigma_o(x,t)$ and $\sigma_w(x,t)$ are oil and water phases pressure, oil and water saturations at the points (x,t), respectively.

S. Margenov, L.G. Vulkov, and J. Waśniewski (Eds.): NAA 2008, LNCS 5434, pp. 532–539, 2009.

In general, the system of equations which describes the motion of compressible multiphase fluid with regard to capillary pressure is called as mixed type. In other words, one of equations with respect to water (or oil) saturation is hyperbolic type, and the other equation regarding the pressure is elliptic type. These properties demand to develop the sensitive method for solving this system. For this aim, an original finite differences method is proposed in this paper. The obtained solution expresses all the physical properties of the problem.

Since the number of unknowns in the equation (1) is four we add the following two equations

$$p_o - p_w = p_c(\sigma_w), \quad \sigma_o + \sigma_w = 1. \tag{2}$$

Here, $p_c(\sigma_w)$ is capillary pressure.

According to the strategies of exploitation of a petrol reservoir to the system of equations (1),(2) the suitable initial and boundary conditions are added.

As it is known that, the solutions of the equations (1),(2) have the points of discontinuities whose locations are unknown beforehand. Existence of the points of discontinuities in the solutions involves difficulties in applying the classical numerical methods to the equations (1),(2) [1], [2], [3], [5].

In the macroscopical case, the effect of the capillary pressure can be ignored, that is $p_c(\sigma_w) = 0$. When $p_c(\sigma_w) = 0$, the system of the equations (1),(2), is investigated in [2]. In this reference, the system of equations (1),(2) is reduced to one first order nonlinear equation for the water (or oil) saturation. In the case $p_c(\sigma_w) \neq 0$, the system of equations (1),(2) is split in two equations with respect to water saturation and pressure, respectively

$$\frac{m}{k}\frac{\partial \sigma_w}{\partial t} = div\left(-\omega F_w(\sigma_w) + F_o(\sigma_w)p_c^{'}(\sigma_w)grad\sigma_w\right), \tag{3}$$

$$grad p_w = -\left(\omega(t)F_w(\sigma_w)\frac{\mu_w}{k_w} + F_o(\sigma_w)grad p_c(\sigma_w)\right). \tag{4}$$

Here, the functions F_w and F_o are called the Buckley-Leverett's functions.

In this paper the system of equations (1),(2) is split in two equations regarding the physical parameters such that the first equation with respect to water saturation, and the second equation with respect to pressure. For the sake of simplicity, at first, we investigate the one dimensional equations.

2 Splitting System of Equations

In this section we consider one dimensional flow of compressible fluids. Assume that, the initial pressure is equal to $p(x,0) = p_o$ and water is injected by point $x = 0$ with the debit $q_w(t)\ m^3/day$. The initial water saturation in porous medium is $\sigma_w^{(0)}$. In this case the system of equations (1),(2) can be written as

$$\frac{m}{k}\frac{\partial}{\partial t}\left(\frac{\sigma_\ell}{\beta_\ell}\right) = \frac{\partial}{\partial x}\left(\frac{k_\ell}{\mu_\ell\beta_\ell}\frac{\partial p_\ell}{\partial x}\right), \ (\ell = o, w). \tag{5}$$

The initial and boundary conditions for the equations (9),(10) are

$$\sigma_w(x,0)=\sigma_w^{(0)},\quad \sigma_o(x,0)=1-\sigma_w^{(0)}, \tag{6}$$

$$\sigma_w(0,t)=\sigma_w^{(1)}(t),\quad \sigma_o(0,t)=\sigma_o^{(1)}(t), \tag{7}$$

$$\frac{k_w}{\mu_w\beta_w}\frac{\partial p_w(0,t)}{\partial x}=-q_w(t),\quad \frac{k_o}{\mu_o\beta_o}\frac{\partial p_o(0,t)}{\partial x}=0. \tag{8}$$

As it is noted that the solutions of the equations (1),(2) are discontinuous functions, and therefore, the equations (1),(2) have not the classical solutions. The weak solutions are defined as follows.

Definition 1. *The functions $\sigma_\ell(x,t),(\ell=o,w)$ satisfying the initial conditions are called the weak solutions of the problem (1),(2), if the following integral relations*

$$\int_{R_+^2}\left\{\frac{m}{k}\frac{\sigma_\ell}{\beta_\ell}f_t+\frac{k_\ell}{\mu_\ell\beta_\ell}\frac{\partial p_\ell}{\partial x}f_x\right\}dxdt+\int_{-\infty}^{\infty}\frac{m}{k}\frac{\sigma_\ell(x,0)}{\beta_\ell(p(x,0))}f(x,0)dx=0,\ (\ell=o,w) \tag{9}$$

hold for every test function from $\overset{o}{C}{}^{2}_{1,1}\left(R_+^2\right)$ and $f(x,T)=0$. Here, $R_+^2=\{(x,t)\mid 0\le x<\infty, t\ge 0\}$.

As it is seen from equations (9) the functions $\sigma_o(x,t)$ and $\sigma_w(x,t)$ may be discontinuous functions too, and from a physical point of view the functions $\frac{k_o}{\mu_o\beta_o}\frac{\partial p_o}{\partial x}$ and $\frac{k_w}{\mu_w\beta_w}\frac{\partial p_w}{\partial x}$ must be continuous.

According to [5], [6], [7] we introduce the following system of auxiliary equations

$$\frac{m}{k}\frac{\partial\vartheta_\ell(x,t)}{\partial t}=\frac{k_\ell}{\mu_\ell\beta_\ell}\frac{\partial p_\ell(x,t)}{\partial x},\ (\ell=o,w) \tag{10}$$

here, $\vartheta_\ell(x,t)=\int_0^x\frac{\sigma_\ell(\xi,t)}{\beta_\ell}d\xi,\ (\ell=o,w)$. Taking into account (2), from (5) we get

$$\frac{\partial p_w}{\partial x}=\frac{\mu_w\beta_w}{k_w}F_w(\sigma_w)\left[\frac{m}{k}\frac{\partial(\vartheta_o+\vartheta_w)}{\partial t}-q_w\right]-F_o(\sigma_w)\frac{\partial p_c(\sigma_w)}{\partial x}. \tag{11}$$

Here, the functions $F_\ell(\sigma_\ell)=\frac{\frac{k_\ell}{\mu_\ell\beta_\ell}}{\frac{k_o}{\mu_o\beta_o}+\frac{k_w}{\mu_w\beta_w}},\ (\ell=o,w)$ are called Buckley-Leverett's functions. Substituting (11) into (10) (for $\ell=w$) we have

$$\frac{m}{k}\frac{\partial}{\partial t}\int_0^x\frac{\sigma_w(\xi,t)}{\beta_w}d\xi=F_w(\sigma_w)\left[-q_w(t)+\frac{m}{k}\frac{\partial}{\partial t}\int_0^x\left(\frac{1}{\beta_w}-\frac{1}{\beta_o}\right)\times\right.$$

$$\left.\times\,\sigma_w(\xi,t)d\xi+\frac{m}{k}\frac{\partial}{\partial t}\int_0^x\frac{d\xi}{\beta_o}\right]-\frac{k_w}{\mu_w\beta_w}F_o(\sigma_w)\frac{\partial p_c(\sigma_w)}{\partial x}+q_w. \tag{12}$$

As mentioned above, the system of equations (5) is split in two equations; the first one (equation (12)) with respect to $\sigma_w(x,t)$, and the second equation (11) with respect to pressure $p(x,t)$.

If $p_c(\sigma_w)=0$ then the equations (11), (12) take the following forms

$$\frac{\partial p_w}{\partial x}=\frac{\mu_w\beta_w}{k_w}F_w(\sigma_w)\left[\frac{m}{k}\frac{\partial(\vartheta_o+\vartheta_w)}{\partial t}-q_w\right], \tag{13}$$

$$\frac{m}{k}\frac{\partial}{\partial t}\int_0^x\frac{\sigma_w(\xi,t)}{\beta_w}d\xi=F_w(\sigma_w)\Bigg[-q_w(t)+\frac{m}{k}\frac{\partial}{\partial t}\int_0^x\frac{d\xi}{\beta_o}+$$

$$+\frac{m}{k}\frac{\partial}{\partial t}\int_0^x\left(\frac{1}{\beta_w}-\frac{1}{\beta_o}\right)\sigma_w(\xi,t)\Bigg]+q_w. \tag{14}$$

Thus, the system of equations (5) is reduced to the auxiliary equations (13) and (14). The initial conditions for the equations (13) and (14) are given as in (6). As it is seen from (14) this equation is integro-differential equation with respect to unknown function $\sigma_w(x,t)$.

3 Numerical Solution in a Class of Discontinuous Functions

In this section the efficient numerical method for the system of equations (13) and (14) is developed. In order to establish a numerical algorithm for the problem (13) and (14), at first, we cover the region R_+^2 by the grid as

$$\Omega_{h,\tau}=\left\{(x_i,t_k)\mid x_i=ih;t_k=k\tau,i=0,\pm1,\pm2,...;k=0,1,2,...;h>0,\tau>0\right\}.$$

Here, h and τ are the steps of the grid $\Omega_{h,\tau}$ with respect to x and t variables, respectively.

At first, the case when $p_c(\sigma_w)=0$ is considered. The integral entering under the derivative with respect to t in equations (13) and (14) is approximated by the following formula

$$\int_0^x\frac{\sigma_w(\xi,t)}{\beta_w(p(\xi,t))}d\xi\cong h\sum_{j=1}^{i}\frac{\Sigma_{w_j}}{\text{ß}_{w_j}}. \tag{15}$$

For the equation (14) the two differences schemes can be written.

3.1 Explicit Scheme

In this section the equation (14) at the point (x_i,t_k) of the grid $\Omega_{h,\tau}$ is approximated as follows

$$\hat{\Sigma}_{w_i} = \left[\frac{1}{\hat{ß}_{w_i}} - F_w(\Sigma_{w_i})\left(\frac{1}{\hat{ß}_{w_i}} - \frac{1}{\hat{ß}_{o_i}}\right)\right]^{-1} \left\{\frac{\Sigma_{w_i}}{ß_{w_i}} + \frac{\tau}{h}\frac{k}{m} q_w(t_k) F_o(\Sigma_{w_i}) + \right.$$

$$F_w(\Sigma_{w_i}) \left[\sum_{j=1}^{i}\left(\frac{1}{\hat{ß}_{o_j}} - \frac{1}{ß_{o_j}}\right) + \sum_{j=1}^{i-1}\left(\frac{\hat{\Sigma}_{w_j}}{\hat{ß}_{w_j}} - \frac{\hat{\Sigma}_{w_j}}{\hat{ß}_{o_j}}\right) - \sum_{j=1}^{i}\left(\frac{\Sigma_{w_j}}{ß_{w_j}} - \frac{\Sigma_{w_j}}{ß_{o_j}}\right)\right] +$$

$$\left. - \sum_{j=1}^{i-1}\left(\frac{\hat{\Sigma}_{w_j}}{\hat{ß}_{w_j}} - \frac{\Sigma_{w_j}}{ß_{w_j}}\right)\right\}. \tag{16}$$

Here, Σ_{w_i}, $ß_{w_i}$ and $ß_{o_i}$ are approximation values of the functions $\sigma_w(x,t)$, β_w and β_o at the point (x_i, t_k), respectively. The difference scheme requires the CFL (Courant-Friedrichs-Levy) condition on the steps of the grid $\Omega_{h,\tau}$.

3.2 Implicit Scheme

In case the right side of the equation (14) at the $t = t_{k+1}$ time layer is approximated as

$$\hat{\Sigma}_{w_i} = \left[\frac{1}{\hat{ß}_{w_i}} - F_w(\Sigma_{w_i})\left(\frac{1}{\hat{ß}_{w_i}} - \frac{1}{\hat{ß}_{o_i}}\right)\right]^{-1} \left\{\frac{\Sigma_{w_i}}{ß_{w_i}} + \frac{\tau}{h}\frac{k}{m} q_w(t_{k+1}) F_o\left(\hat{\Sigma}_{w_i}\right) + \right.$$

$$F_w\left(\hat{\Sigma}_{w_i}\right) \left[\sum_{j=1}^{i}\left(\frac{1}{\hat{ß}_{o_j}} - \frac{1}{ß_{o_j}}\right) + \sum_{j=1}^{i-1}\left(\frac{\hat{\Sigma}_{w_j}}{\hat{ß}_{w_j}} - \frac{\hat{\Sigma}_{w_j}}{\hat{ß}_{o_j}}\right) - \sum_{j=1}^{i}\left(\frac{\Sigma_{w_j}}{ß_{w_j}} - \frac{\Sigma_{w_j}}{ß_{o_j}}\right)\right]$$

$$\left. - \sum_{j=1}^{i-1}\left(\frac{\hat{\Sigma}_{w_j}}{\hat{ß}_{w_j}} - \frac{\Sigma_{w_j}}{ß_{w_j}}\right)\right\}. \tag{17}$$

The (17) is nonlinear algebraical equations for unknown $\hat{\Sigma}_{w_i}$ and in order to obtain their solutions the Newton iteration method is applied.

Now, we will develop the algorithm for the pressure $p(x_i, t_k)$. For this purpose, the equation (13) can be written as follows

$$\frac{\partial p_w}{\partial x} = \tilde{F}_w(\sigma_w) \times$$

$$\left\{\frac{m}{k}\int_0^x [C'(p) + B'(p)\sigma_w(\xi,t)]\frac{\partial p}{\partial t} d\xi + \frac{m}{k}\int_0^x B(p)\frac{\partial \sigma_w(\xi,t)}{\partial t} d\xi - q_w\right\}. \tag{18}$$

Here, $\tilde{F}_w(\sigma_w) = \left(\frac{\mu_w \beta_w}{k_w}\right) F_w(\sigma_w)$, $B(p) = \frac{1}{\beta_w(p)} - \frac{1}{\beta_o(p)}$, $C(p) = \frac{1}{\beta_o(p)}$.

The equation (18) is approximated with the finite differences scheme given below

$$\hat{\wp}_i = \wp_i + \left\{\frac{m}{k}\frac{h}{\tau}\tilde{F}_w(\Omega_i^{(w)})\left[\bar{C}'(\wp_i) + \bar{B}'(\wp_i)\Omega_i^{(w)}\right]\right\}^{-1} \times$$

$$\left\{\frac{\wp_{i+1} - \wp_{i-1}}{2h} - \frac{m}{k}\frac{h}{\tau}\tilde{F}_w(\Omega_i^{(w)})\sum_{j=1}^{i-1}\left[\bar{C}'(\wp_j) + \bar{B}'(\wp_j)\Omega_j^{(w)}\right](\hat{\wp}_j - \wp_j)\right.$$

$$-\frac{m}{k}\frac{h}{\tau}\tilde{F}_w(\Omega_i^{(w)})\sum_{j=1}^{i}\bar{B}(\wp_i)\left[\tilde{\Omega}_j^{(w)}-\Omega_j^{(w)}\right]+\tilde{F}_w(\Omega_i^{(w)})q_w(t_k)\Bigg\}. \quad (19)$$

The initial condition for equation (19) is $\wp_{i,0}=p_o(x_i)$. Now, we will investigate the case when $p_c(\sigma_w)\neq 0$.

4 The Case $p_c(\sigma_w)\neq 0$

In this section we consider the effect of the capillary pressure. For this goal, the equations (11) and (12) are approximated by the differences schemes given below

$$\hat{\Sigma}_{w_i}=\left[\frac{1}{\hat{ß}_{w_i}}-F_w\left(\Sigma_{w_i}\right)\left(\frac{1}{\hat{ß}_{w_i}}-\frac{1}{\hat{ß}_{o_i}}\right)\right]^{-1}\Bigg\{\frac{\Sigma_{w_i}}{ß_{w_i}}+\frac{\tau}{h}\frac{k}{m}q_w(t_k)F_o\left(\Sigma_{w_i}\right)+$$

$$F_w\left(\Sigma_{w_i}\right)\left[\sum_{j=1}^{i}\left(\frac{1}{\hat{ß}_{o_j}}-\frac{1}{ß_{o_j}}\right)+\sum_{j=1}^{i-1}\left(\frac{\hat{\Sigma}_{w_j}}{\hat{ß}_{w_j}}-\frac{\hat{\Sigma}_{w_j}}{\hat{ß}_{o_j}}\right)-\sum_{j=1}^{i}\left(\frac{\Sigma_{w_j}}{ß_{w_j}}-\frac{\Sigma_{w_j}}{ß_{o_j}}\right)\right]$$

$$-\sum_{j=1}^{i-1}\left(\frac{\hat{\Sigma}_{w_j}}{\hat{ß}_{w_j}}-\frac{\Sigma_{w_j}}{ß_{w_j}}\right)-\frac{k}{m}\frac{\tau}{2h^2}\frac{k_w}{\mu_w\beta_w}F_o\left(\Sigma_{w_i}\right)\left[\wp_c\left(\Sigma_{w_{i+1}}\right)-\wp_c\left(\Sigma_{w_{i-1}}\right)\right]\Bigg\} \quad (20)$$

$$\hat{\wp}_i=\wp_i+\left\{\frac{m}{k}\frac{h}{\tau}\tilde{F}_w(\Omega_i^{(w)})\left[\bar{C}'(\wp_i)+\bar{B}'(\wp_i)\Omega_i^{(w)}\right]\right\}^{-1}\times$$

$$\Bigg\{\frac{\wp_{i+1}-\wp_{i-1}}{2h}-\frac{m}{k}\frac{h}{\tau}\tilde{F}_w(\Omega_i^{(w)})\sum_{j=1}^{i-1}\left[\bar{C}'(\wp_j)+\bar{B}'(\wp_j)\Omega_j^{(w)}\right](\hat{\wp}_j-\wp_j)$$

$$-\frac{m}{k}\frac{h}{\tau}\tilde{F}_w(\Omega_i^{(w)})\sum_{j=1}^{i}\bar{B}(\wp_i)\left[\tilde{\Omega}_j^{(w)}-\Omega_j^{(w)}\right]+\tilde{F}_w(\Omega_i^{(w)})q_w(t_k)$$

$$+\frac{1}{2h}F_o(\Omega_i^{(w)})\left[\wp_c\left(\Omega_{i+1}^{(w)}\right)-\wp_c\left(\Omega_{i+1}^{(w)}\right)\right]\Bigg\}. \quad (21)$$

5 Numerical Experiments

In order to illustrate the efficiency of the suggested method, first we adapt this algorithm for solving to the problem (5) with the following data

$$k_o(\sigma_w)=\begin{cases}\left[\frac{0.85-\sigma_w}{0.8}\right]^{2.8}\left[1+(2.4)\sigma_w\right], & 0\leq\sigma_w\leq 0.85,\\ 0, & 0.85\leq\sigma_w\leq 1,\end{cases}$$

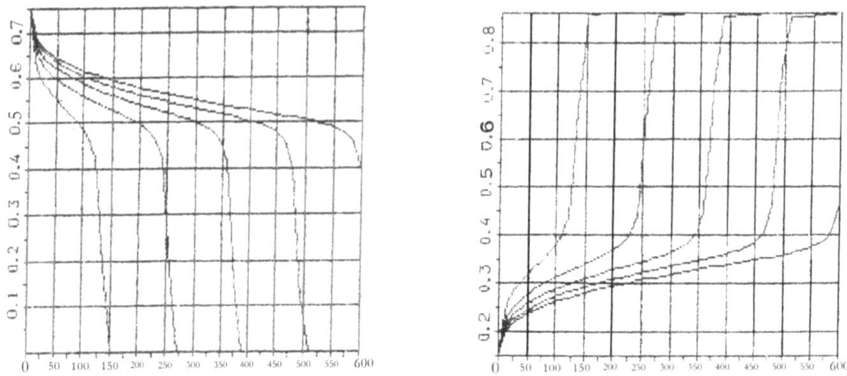

Fig. 1. Distributions of (a) water, (b) oil saturations at various values of time

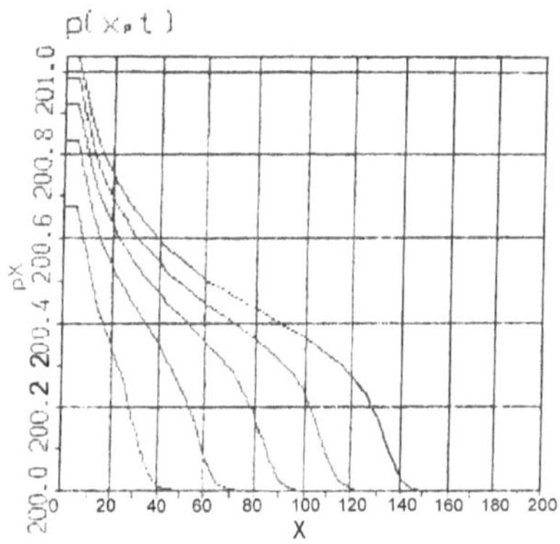

Fig. 2. Distribution of oil saturation at various values of time

$$k_w(\sigma_w) = \begin{cases} \left(\frac{\sigma_w - 0.2}{0.8}\right)^{3.5}, & 0.2 \leq \sigma_w \leq 1, \\ 0, & 0 \leq \sigma_w \leq 0.2, \end{cases}$$

$$\frac{\beta_o^0}{\beta_o} = exp\left[c_f^o(p_o - p_o^0)\right], \quad \frac{\beta_w^0}{\beta_o} = exp\left[c_f^w(p_w - p_w^0)\right], \quad c_f^o = c_f^w = 10^{-5},$$

$$\mu_o = 810^{-3}Pa.s, \ \mu_w = 10^{-3}Pa.s, \ H = 10m, \ L{=}600m, \ k{=}0.24m^2, \ H{=}1.5m.$$

The initial distributions of oil and water saturations with respect to efficient height of a porous medium are $\sigma_o^o(x) = 0.8;\ \sigma_w^o(x) = 0.2$. The initial distribution of the pressure is $p_0 = 175\frac{kgs}{sm^2}$.

As it is shown from Fig. 1 that, the solutions obtained by suggested algorithms express the physical properties of the problem, accurately. It should be noted that, it is possible to use the solutions $\sigma_o(x,t)$ and $\sigma_w(x,t)$ obtained by algorithms (16) for projection of development and exploitation of porous medium.

In addition, by using these solutions some technological indices of development may be computed.

The difference scheme of (16) is the first order with respect to τ, however, the order of it can be increased by a higher value by applying, for example, the Runge-Kutta method.

6 Conclusion

The results obtained in this paper can be listed as follows:

- The system of the partial differential equations which describes the motion of two-phase compressible fluid in microscopic porous medium is separated regarding the physical parameters.
- The special auxiliary problem of which the differentiable properties of the solution one order higher than the differentiable properties of the main problem is introduced.
- Using the advantages of the auxiliary problem the efficient numerical algorithms are suggested in a class of discontinuous functions. The obtained solutions express the all physical properties accurately.

References

1. Aziz, K., Settari, A.: Petroleum Reservoir Simulation. Elsevier Applied Science Pub., London (1979)
2. Buckley, S.E., Leverett, M.C.: Mechanism of Fluid Displacement in Sands. Trans. AIME 146, 107–116 (1942)
3. Collins, R.E.: Flow of Fluids through Porous Materials. Penn-Well Books, Tulsa (1976)
4. Muskat, M.: The Flow of Homogeneous Fluids through Porous Medium. McGraw-Hill, New York (1946)
5. Rasulov, M.A.: Numerical Solution of One Dimensional Filtration of Three Phase Compressible Fluid Through Medium in a Class of Discontinuous Functions. Appl. Math. Comput. 167(2), 1249–1266 (2005)
6. Rasulov, M.A.: Finite Difference Scheme for Solving of Some Nonlinear Problems of Mathematical Physics in a Class of Discontinuous Functions. Baku (1996) (in Russian)
7. Rasulov, M.A., Abasov, M.T.: To the Theory Filtration of Three-Phase Mixes in View of Them Compressible. Soviet Math. Dokl. USSR 325(1), 130–133 (1992)

The Neural Networks Approach to Identification of Local Damages in Elastic Structures

Arkady N. Soloviev, Polina S. Kourbatova, Nikolai I. Saprounov, and Sergey N. Shevtsov

Southern Federal University,
Milchakova st., 8-a, 344091. Rostov-on-Don, Russia
soloviev@math.rsu.ru

Abstract. With aid of neural network technology a reconstruction of crack and holes in elastic bodies was performed. On basis of multilayered perceptron (MLP) the artificial neural networks (ANN) were developed and programmatically realized. These ANN were used for identification of damaged state and for defects geometry reconstruction. At resolving of inverse problems of defects reconstruction an eigenfrequencies spectra or displacement field on free surface of body at stationary oscillation was used as additional information. The finite element soft package ACELAN acting in batch mode for learning sets generation was utilized. The examples of feedforward network training by backpropagation algorithm and also examples of damages like surface flaws and holes were considered. The numerical experiment results were shown a good reliability of developed method and performance of its program implementation.

Keywords: Perceptron, identification, defect's reconstruction, learning sets, finite element.

1 Introduction

One application of artificial neural networks (ANN) is the resolving of coefficient geometric problems of elasticity applying to important practical area – nondestructive evaluation (NDE) and defectoscopy. An efficiency of ANN in problems of determination and classification of different defects in structural materials is convincingly presented in [1]. Ref. [2] contain a surrey of artificial intelligent methods, using the ANN with applications to ultrasound defectoscopy. ANN technology was implemented in means of program supporting some equipment for defectoscopy and damages diagnostics [3,4].

All NDE methods it is possible conditionally to divide onto two groups. First group include the methods utilized the monitoring of non mechanical values, such as vortex electric current, a magnetic flux, X-ray imaging, etc. The second group uses the measurements of mechanical quantities: an elastic wave speed, a frequency response and time dependent response of displacements or accelerations of points located on free sections of body surfaces accessible to measuring. The information necessary for defects identification usually derive in the discrete

S. Margenov, L.G. Vulkov, and J. Waśniewski (Eds.): NAA 2008, LNCS 5434, pp. 540–547, 2009.

form. This is a spectrum of eigenfrequencies, a density of natural vibrations modes energy (a spectral probe), or oscillations amplitudes of finite set points on the body surface (a positional probe).

For identification of damaged state, for reconstruction of defect's kind and geometry the multilayered perceptron (MLP) architecture more often is used. Such MLPs were learned by error back propagation (BP) method [5]. Creation of learning vectors require a considerable set of experimental data obtained at different condition or the other way it is require an efficient numerical solving method of forward dynamic problems for structures with previously well-known imperfections. For simplest structures, e.g. homogenous beams, plates, etc. it is possible to utilize the analytical representations. However, for practice applications the main means is finite element method (FEM). But some unconformity between FEM model behavior and measured vibration data can cause to mistaken or not responsible conclusions. These considerations have fairly indicated in [6] where the statistical approach for elimination of this deficiency was proposed.

In presented work the method of above mentioned problems resolving is proposed. This method based on a linked FEM and ANN allows determining the geometrical properties of the boundary flaw, of cavities and holes in carrier structural elements. We suppose below that the flaw-sides not interact and that inners boundaries of cavities are stresses - free. Thus the reconstruction of flaws was performed on the basis of spectral probing, and definition of geometry of holes (coordinates of center and reference size) - on the basis of positional probing. A learning of the developed networks was performed by finite-element software ACELAN working in batch mode and interacting with Neural Network Toolbox MATLAB 7.1 trough xml – interface.

2 Problems Statement

1. Problems statement. In forward problem the stationary or natural oscillation of elastic bodies (see Fig.1) were considered. These problems were described by boundary value problem

$$\sigma_{ij,j} = -\rho\omega^2 u_i, \quad \sigma_{ij} = c_{ijkl} u_{k,l}, \quad i = 1, 2, 3 \tag{1}$$

$$u_i|_{S_u} = u_i^{(0)}, \quad \sigma_{ij} n_j|_{S_t} = p_i, \quad \sigma_{ij} n_j|_{S_d} = q_i \ , \tag{2}$$

where: u_i - quest components of displacement vector, $u_i^{(0)}$, p_i, and q_i - known components of displacement vector and surface loading, σ_{ij} and c_{ijkl} - components of stress tensor and elastic constants respectively; ρ - density; ω - oscillations angular frequency, S_d - the inner surface of flaw (Fig.1, left) or hole (Fig.1, right).

A defects identification require to determine a flaw configuration (coordinates of its outflow on body surface, length and inclination angle) or geometry of hole (coordinates of center, the reference size – radius). The surfaces S_d are unknown that attributes considered problems to inverse geometrical problems of elasticity.

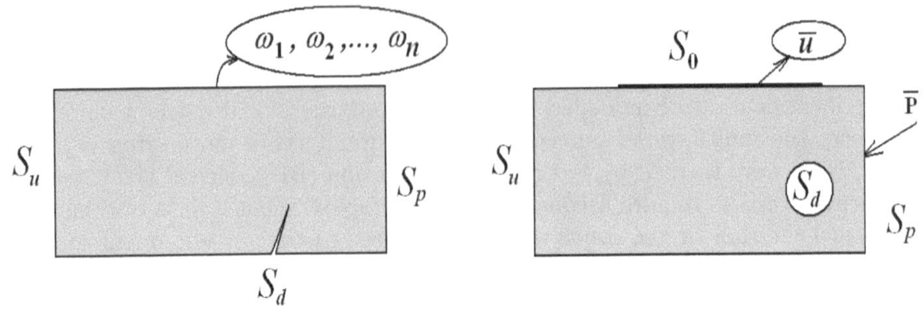

Fig. 1. Modeled bodies with flaw and hole

For resolving of inverse problems of surface S_d reconstruction it is necessary to complement the boundary conditions (2) by the additional information. Such additional information the displacement field $\overline{U}$ measured on a free surface S_0 can serve

$$u_i|_{S_0} = U_i(x,\, \omega), \quad x \in S_0 \quad \omega \in [\omega_b,\, \omega_e]. \tag{3}$$

On basis of $\overline{U}$ the discretized information is derived: Ω - set of eigenfrequencies (spectral probe)

$$\Omega = \{\omega_{r1},\, \omega_{r2},\, ...,\, \omega_{rN}\}, \tag{4}$$

or set of displacement amplitudes Ψ at stationary oscillations of some points x_k (positional scan) at selected frequencies ω_m (frequency scanning)

$$\Psi = \{U_i(x_k,\, \omega_m)|\, k = 1, 2, ..., K,\; m = 1, 2, ..., M\}\; x_k \in S_0\; \omega_m \in [\omega_b,\, \omega_e] \tag{5}$$

Thus Ω or Ψ represent some discrete set of the input information$X$.

3 Resolving of Inverse Problem

Resolving of inverse problem assume a reconstruction of the surfaceS_d, which must be parameterized. And at some prior information the number of these parameters can be made finite. In cases of flat or rectilinear flaws such parameters are coordinates of its outflow on body surface, length and inclination angle. At elliptic holes parameters reconstruction these parameters are coordinates of center, length and direction of semi axes. Besides, for a practical estimation of the structural damage are important for answering a question: whether there is an injured state and, if those is, what parameters of this damage is.

This reason allow to reduce the problem damaged state reconstruction to identification problem of finite number of parameters Y. The operator $F : X \rightarrow Y$ is implemented by ANN with MLP architecture defined by numbers of layers and numbers of neurons in each layer, by kind of transfer function, by the weight matrix W, and by the biases.

A perceptron learning was executed by BP – algorithm. The learning sets $\{X_\alpha,\, Y_\alpha\}$ were created on basis of resolving of forward problems (1)-(2) with known dimensions of defects (the surface S_d well determined) at conditions (4) or (5). The learning vectors were generated by finite-element software ACELAN [7] working in batch mode

During training the weights and biases of the network are iteratively adjusted to minimize the network performance function

$$E(W) = \sum_{\alpha=1}^{P} ||F(X_\alpha,\ W) - Y_\alpha|| \tag{6}$$

It is necessary to mark that the success of multiparameter optimization problems essentially depends on successful selection of initial guess values. In the considered case the initial values of weights were chosen as uniformly distributed with mean equal to null, and range of their variation was determined by number of neurons (m) in current hidden layer such manner that a dispersion $\sigma_w = m^{-1/2}$. Besides all input variable were previously treated so that the mean value on all training set was close to zero. Otherwise their comparison with standard deviation was hindered. And also at a partition on classes binary target values [0,1] during training of a network changed on small value ε, so that $[\varepsilon,\ 1-\varepsilon]$, to eliminate of weights saturation. If as a transfer function has used a hyperbolic tangent output set shifted so that its values placed on a linear segment of a transfer function.

4 Numerical Experiments

4.1 Problem of Flaws Classification

For the flaw that outflow perpendicular to the body surface the problem of flaw dimension determination was solved in the framework of 2D elasticity. Complementary information was derived from a spectral scan investigation. The statement of inverse problem include the relations (1), (2), (4) and represent a minimization of the functional (6) on the sample $\{X_\alpha,\ Y_\alpha\}$ $\alpha = 1, 2, ..., P$, which generated in ACELAN at solving of forward problem – modal analysis for rectangular area (steel). The left side of modeled subdomain was fixed. The flaw went out on an underside of area in a point apart 2/5 its lengths from an anchored edge (see Fig.2). Due the stress singularity near vertex of the flaw the finite element mesh was essentially condensed. The learning sequence represented a set of 50 vectors which elements were the eigenfrequencies. To each vector there corresponded a flaw depth in range [0, 0.5] varying with a step 0.01 (hereinafter all sizes in cm). Dimensionality of vectors (quantity of considered eigenfrequencies) and a domain of a spectrum selected after the analysis of their relation from the sizes or other parameters of classified object. All results shown below correspond to the first ten eigenfrequencies though the satisfactory results were reached already at use first two (see Tab. 1).

A learned MLP divided input space into four classes. To these classes there corresponded depth of a flaw nether than 0.125, 0.25, 0.375, 0.5 respectively. MLP generates output vectors, which are the normalized orthogonal basis vectors

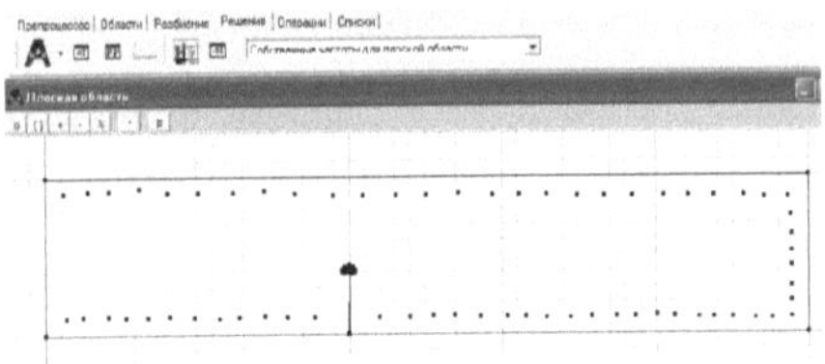

(a)

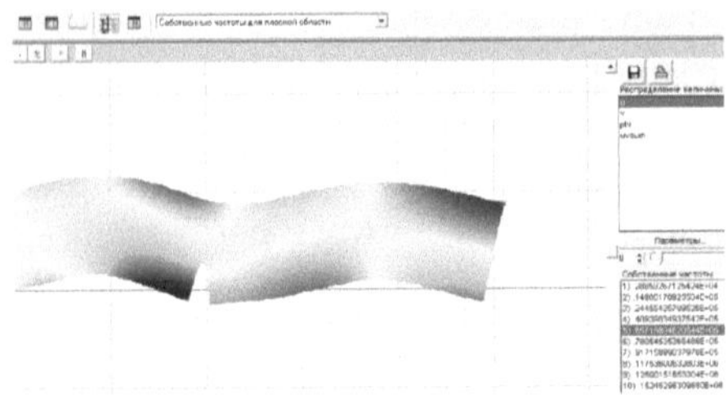

(b)

Fig. 2. Resolving of forward problem in ACELAN

(goals) of a 4- dimensions space. On Fig.2 (a) the errors are plotted with respect to training epochs. As one can see after 71 training epochs of MLP-classifier with 12 layers and alternating transfer function (linear and sigmoidal) the error not reached 0.0066.

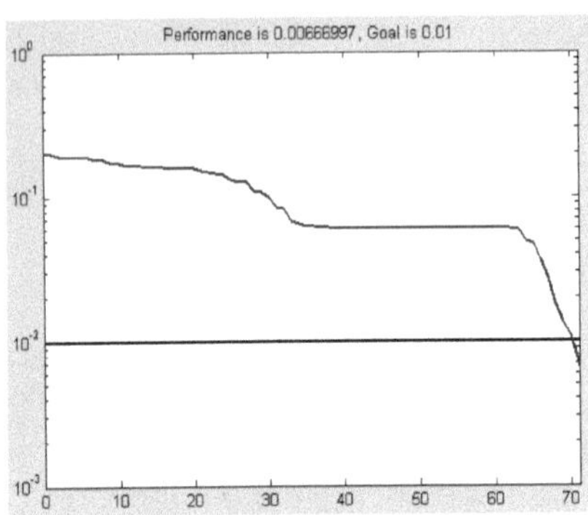

(a) convergence at training process

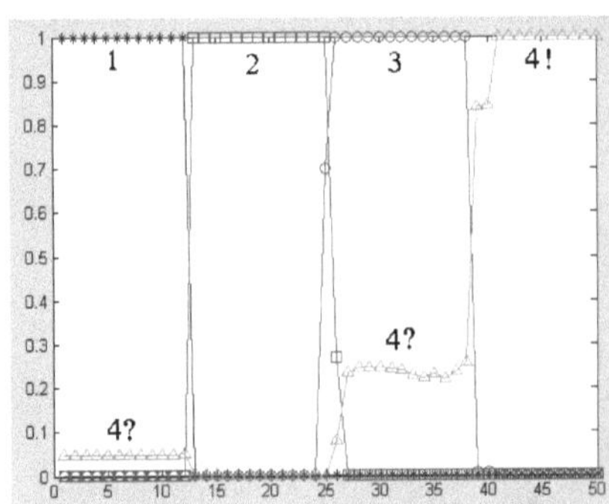

(b) results of classification

Fig. 3. Artificial neural networks

Some classification results were shown on a Fig.3(a). Each class was marked by a specific sign. The first, second and third classes are precisely classified with output values equal to unity; whereas at classification of the fourth class marked by triangles there is some reference probability both in first, and in the third classes (the mistaken classification has noted by question-mark). The Table 1 demonstrates high efficiency of the created ANN for a flaw depth classification.

For classification of flaws by position on the body surface and on flaw's length the training sets were enlarged up to 300 vectors and each vector contained 5 eigenfrequencies. Classification process was executed by two ANN. First ANN classified presence of an imperfection and attributed it to one of three classes (that corresponded to localization of a flaw on the body's basis). Second ANN subdivided flaws on depth into five classes, At training with learning rate 0.01

Table 1. Results of MLP model validation

Eigenfrequncies $\times 10^4$ Hz		Flaw depth, cm	Network's output				Class
1.7758	3.3117	0.05	1.00	0.00	0.00	0.00	1!
1.6641	3.1594	0.17	0.00	1.00	0.00	0.00	2!
1.6181	3.3090	0.3	0.00	0.00	1.00	0.00	3!
1.2257	2.7733	0.48	0.01	0.00	0.00	0.99	4!

first MLP reached an error E=0.0121 after 1000 epoch, and second MLP - error 0.0521 after 3000 epoch.

4.2 Hole Parameters Reconstruction

In second group of numerical experiments the problem of hole's dimension identification was solved. The flat rectangular domain (steel, 10×4 cm) made stationary oscillation beyond from resonances with frequency 12860 Hz (see Fig.4, b). Left side of rectangular was fixed, and left one was loaded by distributed normal force linearly increasing from bottom to top. The amplitude values of displacements in 9 points equidistant from each other were used as learning vector. An inverse problem described by equations (1), (2), by additional information obtained after positional probe (5), and by minimization of functional (6). The learning set was generated at variation of centered hole radius from 0 to 1 cm, step 0.05 cm.

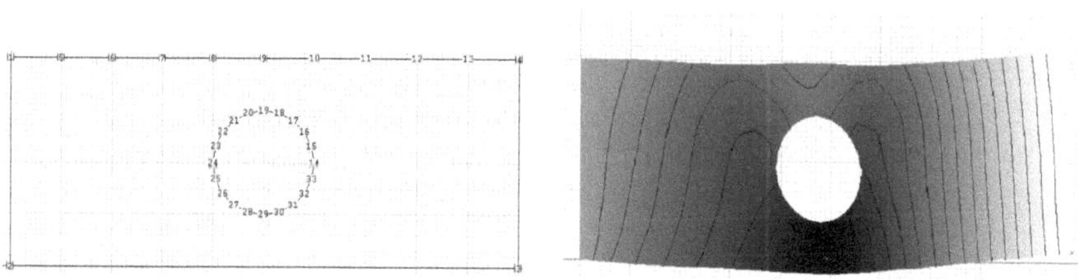

(a) the points of displacements on a top boundary

(b) distribution of the vertical displacement

Fig. 4. The body with hole

For determination of unknown radius the calculations at different values of a mean-square error were performed. A dramatic growth of MLP training duration at decreasing of a permissible error was observed. So for error tolerance 0.1 the learning rate by gradient based algorithm has made 1 sec, for 0.01 - 2 sec, 0.001 - 60 sec, 0.0001 - 1800 sec respectively.

The reconstruction of a hole radius was performed by trained MLP receiving input data which were generated at the solution of the forward problems in ACELAN. The values of reconstruction errors are shown in a Fig.5.

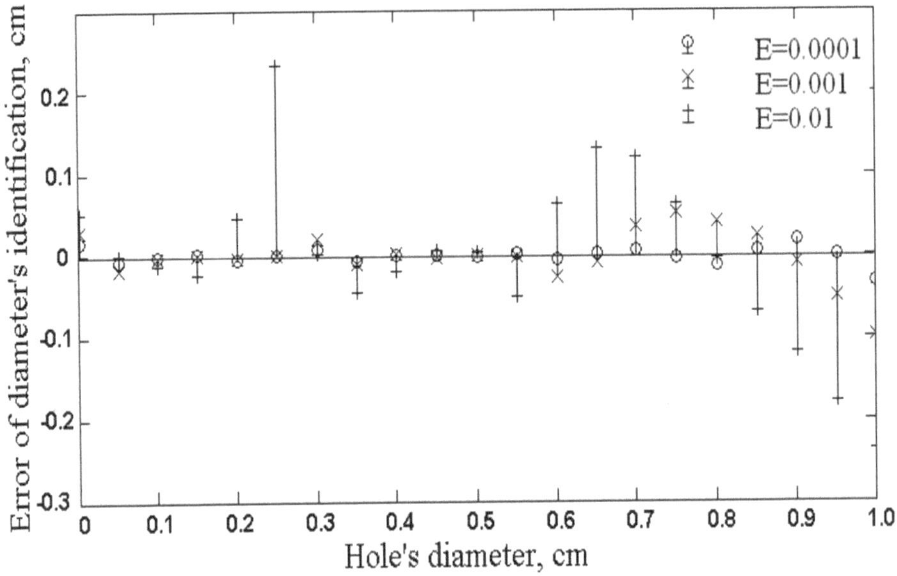

Fig. 5. Error of hole's radius identification by MLP trained with different error tolerance E

For determination of defect's center location the stationary oscillations of rectangular domain with circular hole (with invariable radius 1 cm) at frequency 11100 Hz were considered. At generation of learning set coordinates of hole center varied: on a horizontal from 3 cm to 7 cm with step 0.4 cm, and on a vertical from 1.5 cm to 2.5 cm with step 0.1 cm. MLP with 3 hidden layers learned at performance goal (error tolerance) $E=10^{-3}$ was determined the coordinates of hole's center with inaccuracy not more than 1.5 mm at probability 95%.

5 Conclusion

The developed methods and carried numerical experiments confirmed an efficiency of neural network approach complemented by finite-element method for identification of structures damaged state. For this purpose the mobile devices equipped with software similar designed can be used. The practical embodying of such verification methods requires also searching of an optimum sensors placement on a tested structure. This represents a separate problem for each structure with appropriate anisotropy of material and expected imperfections type.

The authors wish to acknowledge the partial financial support of Russian Foundation for Basic Research (Grants 06-01-08041, 07-08-13589, 07-08-12193).

References

1. Achenbach, J.D.: Quantitative nondestructive evaluation. International Journal of Solids and Structures 37, 13–27 (2000)
2. McNab, A., Dunlop, I.: A review of artificial intelligence applied to ultrasonic defect evaluation. Insight 37(1), 11–16 (1995)
3. Wong, B.K., Bodnovich, T.A, Selvi, Y.: A bibliography of neural network business applications research: 1988 - September 1994. Expert Systems 1995, 253–262 (1988)
4. Wong, B.K., Vincent, S.L., Lam, J.: A bibliography of neural network business applications research: 1994-1998. Computers & Operations Research 27, 1045–1076 (2000)
5. Haykin, S.: Neural Networks. University McMaster Ed., Hamilton, Ontario, Canada
6. Bakhary, N., Hong, H., Deeks, A.J.: Damage detection using artificial neural network with consideration of uncertainties. Engineering Structures (2007)
7. Petoushkov, A.L., Soloviev, A.N.: On realization of batch mode for finite-element package ACELAN, Modern problems of continuous media mechanics. In: Proc. of VIII Int., Conf., Rostov-on-Don, 2nd New Book edn., Rostov-on-Don, Russia, October 14-18, 2002, pp. 149–153 (2003)

Numerical Approximation of a Free Boundary Problem for a Predator-Prey Model

Răzvan Ştefănescu and Gabriel Dimitriu

University of Medicine and Pharmacy "Gr. T. Popa",
Department of Mathematics and Informatics,
700115 Iasi, Romania
rastefanescu@yahoo.co.uk, dimitriu.gabriel@gmail.com

Abstract. This paper is concerned with the numerical approximation of a free boundary problem associated with a predator-prey ecological model. Taking into account the local dynamic of the system, a stable finite difference scheme is used, and numerical results are presented.

1 Introduction

In recent years, the two-species predator-prey ecological models have received increasing research attention. Various forms of the systems have been proposed, especially on the following coupled systems of two reaction-diffusion equations:

$$\begin{cases} P_t - d_1 \varDelta P = P(a_1 - b_{11}P + c_{12}Q), & x \in \Omega, t > 0, \\ Q_t - d_2 \varDelta Q = Q(a_2 - b_{21}P - c_{22}Q), & x \in \Omega, t > 0, \end{cases} \tag{1}$$

where d_i, a_i, b_{ij}, c_{ij} are positive constants. In biological terms, P and Q represent, respectively, the spatial densities of predator and prey species that are interacting and migrating in the habitat Ω, d_i denotes its respective diffusion rate, and the real number a_i describes its net birth rate. b_{11} and c_{22} are the coefficients of intra-specific competitions, and b_{21} and c_{12} are the coefficients of inter-specific competitions. In the case that c_{12} is replaced by $-c_{12}$, (1) is the well-known Lotka-Volterra competition model.

Movement plays a role in structuring the interactions between individuals, their environment and their species. Next to the first passage time concept (in the context of animal movement, first passage time is the time taken for an animal to reach a specified site for the first time), the free boundary predator-prey models represent a new modality of understanding the effect of the landscape on animal movement and search time.

The remainder of the article is organized as follows. In the next section we briefly describe the free boundary predator-prey model. Section 3 is devoted to the description of the numerical approach. The numerical results are presented and discussed in Section 4. Finally, Section 5 is dedicated to the presentation of some conclusions and objectives for future work.

S. Margenov, L.G. Vulkov, and J. Waśniewski (Eds.): NAA 2008, LNCS 5434, pp. 548–555, 2009.

2 Free Boundary Problem

The asymptotic behavior of species have been known in the literature for domains with fixed boundary. In what follows, we consider that the predator species are initially limited to a specific part of the domain. To be more specific, let us consider the one-dimensional case. Assume that the prey species migrates in the habitat $(0, l)$ and the predator disperses through random diffusion only in a part of the habitat $(0, l)$, namely $0 < x < h(t)$, then there is no predator in the remaining part.

Let the diffusivity of the predator be d_1. Then the number of predator populations flowing across the boundary $x = h(t)$ from time t to time $t + \Delta t$ is $J\Delta t = -d_1(\partial P/\partial x)\Delta t$. These predators disperse from $x = h(t)$ to $x = h(t+\Delta t)$ during the time interval $[t, t + \Delta t]$ and the size of the population decides the length $h(t + \Delta t) - h(t)$. Supposing

$$-d_1 \frac{\partial P}{\partial x} \Delta t = f[h(t + \Delta t) - h(t)],$$

we know that the function f is increasing and $f(0) = 0$. Ecologically, this means that the size is increasing with respect to the moving length. Using the Taylor expansion of the function f one obtains

$$f[h(t+\Delta t) - h(t)] = 0 + f^{'}(0)[h(t+\Delta t) - h(t)] + \frac{1}{2} f^{''}(0)[h(t + \Delta t) - h(t)]^2 + ...$$

and therefore

$$-d_1 \frac{\partial P}{\partial x} = f^{'}(0) \frac{[h(t + \Delta t) - h(t)]}{\Delta t} + \frac{1}{2} f^{''}(0) \frac{[h(t + \Delta t) - h(t)]^2}{\Delta t} + ...$$

Now, letting $\Delta t \to 0$, we arrive at

$$-d_1 \frac{\partial P}{\partial x} = f^{'}(0) h^{'}(t).$$

Here, $f^{'}(0)$ is a positive constant since f is increasing and depends on the diffusivity of the predator in the part where no predator exists. If $f^{'}(0)$ is big enough, then the predator can disperse easily in the new area.

Denoting $\mu = d_1/f^{'}(0)$, then the conditions on the interface (free boundary) are

$$P = 0, \qquad -\mu \frac{\partial P}{\partial x} = h^{'}(t).$$

If all populations do not attempt to emigrate from inside, then there is no flux crossing the fixed boundary, that is, the homogeneous Neumann boundary conditions hold

$$\frac{\partial P}{\partial x}(0, t) = \frac{\partial Q}{\partial x}(0, t) = \frac{\partial Q}{\partial x}(0, t) = 0.$$

In such case, we have the problem for $P(x,t)$ and $Q(x,t)$ with a free boundary $x = h(t)$ such that

$$(P)\quad \begin{cases} P_t - d_1 P_{xx} = P(a_1 - b_{11}P + c_{12}Q), & 0 < x < h(t),\ t > 0, \\ Q_t - d_2 Q_{xx} = Q(a_2 - b_{21}P - c_{22}Q), & 0 < x < l,\ t > 0, \\ P(x,t) = 0, & h(t) < x < l,\ t > 0, \\ P = 0,\ h'(t) = -\mu \frac{\partial P}{\partial x}, & x = h(t), \quad t \geq 0, \\ \frac{\partial P}{\partial x}(0,t) = \frac{\partial Q}{\partial x}(0,t) = \frac{\partial Q}{\partial x}(l,t) = 0, & t > 0, \\ h(0) = b, \quad (0 < b < l), & \\ P(x,0) = P_0(x) \geq 0, & 0 \leq x \leq b, \\ Q(x,0) = Q_0(x) \geq 0, & 0 \leq x \leq l, \end{cases}$$

where the initial values P_0, Q_0 are nonnegative and satisfy $P_0(x) \in C^2[0,b]$, $P_0(x) > 0$, for $x \in [0,b)$, $P_0^{'} < 0$, $Q_0(x) \in C^2[0,l]$ and the consistency conditions $P_0^{'}(0) = Q_0^{'}(0) = Q_0^{'}(l) = 0$.

In the absence of Q, the problem is reduced to the one-phase Stefan problem, which accounts for phase transitions between solid and fluid states such as the melting of ice in contact with water. The existence, uniqueness and asymptotic behavior of the solution for (1) are known ([8]).

The results for free boundary problems have been applied to many areas, for example, the decrease of oxygen in a muscle in the vicinity of a clotted bloodvessel, the etching problem, the combustion process, the American option pricing problem ([6]), chemical vapour deposition in a hot wall reactor, image processing ([1]), wound healing and tumour growth ([4], [5] and [7]), the temperature distribution for polythermal ice sheets ([3]).

3 Numerical Approximation

The discretization is carried out by finite differences. The grids with equidistant nodes are denoted by:

$$0 = x_1 < x_2 < ... < x_{2n+1} = l,\ x_{n+1} = b; \qquad 0 = t_1 < t_2 < ... < t_{m+1} = T.$$

Furthermore, we choose : $x_j = (j-1)h,\ \ j = 1,2,...,2n+1$ with $h = \frac{1}{n}$, and $t_i = (i-1)k,\ \ i = 1,2....,m+1$, with $k = \frac{T}{m}$.

To obtain the numerical approximation of problem (P), we use the standard implicit scheme which is unconditionally stable.

Let $P_j^{(i)}$ and $Q_j^{(i)}$ be the approximations of $P(t_i,x_j)$ and $Q(t_i,x_j)$. The initial conditions yield:

$$P_j^{(1)} = P_0(x_j),\ \ j = \overline{1,n}; \qquad Q_j^{(1)} = Q_0(x_j),\ \ j = \overline{1,2n+1}.$$

Our next goal is to pass from some level i ($t = t_i$) to the next level i+1 ($t = t_{i+1}$), for $i = \overline{1,m}$. First, we need to determine the free boundary $h(t_{i+1})$, before calculating the solution of the system (P) at time level t_{i+1}. Thus, we search for an interval $[x_p, x_{p+1}]$ such that $h(t_i) \in [x_p, x_{p+1}]$. Then, in order to match the points from the free boundary with the grid points, we evaluate and compare the distances between $h(t_i)$ and x_p and $h(t_i)$ and x_{p+1}, respectively. The lowest value of this distance gives us the point x_f which represents the corresponding grid boundary point at time level t_i.

From the free boundary conditions we get:

$$h(t_{i+1}) = -\mu\frac{k}{h}[P(x_f,t_i) - P(x_{f-1},t_i)] + h(t_i).$$

Using a Taylor's series expansion we get the following discretization for the system equations:

$$-d_1kP_{j-1}^{(i+1)} + (h^2 + 2d_1k - a_1kh^2)P_j^{(i+1)} - kh^2c_{12}P_{j+1}^{(i+1)}Q_{j+1}^{(i+1)}$$

$$+b_{11}kh^2(P_j^{(i+1)})^2 - d_1kP_{j+1} - h^2P_j^{(i)} = 0, \qquad j = \overline{2, f-1},$$

for the first equation, and

$$-d_2kQ_{j-1}^{(i+1)} + (h^2 + 2d_2k - a_2kh^2)Q_j^{(i+1)} + b_{21}kh^2Q_{j+1}^{(i+1)}P_{j+1}^{(i+1)}$$

$$+c_{22}kh^2(Q_j^{(i+1)})^2 - d_2kQ_{j+1}^{(i+1)} - h^2Q_j^{(i)} = 0, \qquad j = \overline{2, 2n},$$

for the second one.

The discretization of equation corresponding to the boundary condition leads to

$$Q_2^{(i+1)} = Q_1^{(i+1)}; \qquad Q_{2n+1}^{(i+1)} = Q_{2n}^{(i+1)}; \qquad P_2^{(i+1)} = P_1^{(i+1)};$$

Moreover, we have $P_j^{(i+1)} = 0,\ j = \overline{f, 2n}$.

If we denote by $\alpha_1 = h^2 + d_2k - a_2kh^2$, $\alpha_2 = h^2 + 2d_2k - a_2kh^2$, $\beta_1 = h^2 + d_1k - a_1kh^2$ and $\beta_2 = h^2 + 2d_1k - a_1kh^2$, we obtain the discrete problem $(\mathit{\Gamma}_h)$:

$$\begin{cases} \alpha_1Q_2^{(i+1)} + b_{21}kh^2Q_2^{(i+1)}P_2^{(i+1)} + c_{22}kh^2(Q_2^{(i+1)})^2 - d_2kQ_3^{(i+1)} - h^2Q_2^{(i)} = 0, \\ -d_2kQ_{j-1}^{(i+1)} + \alpha_2Q_j^{(i+1)} + b_{21}kh^2Q_{j+1}^{(i+1)}P_{j+1}^{(i+1)} + c_{22}kh^2(Q_j^{(i+1)})^2 \\ -d_2kQ_{j+1}^{(i+1)} - h^2Q_j^{(i)} = 0, \qquad j = \overline{3, 2n-1}, \\ -d_2kQ_{2n-1}^{(i+1)} + \alpha_1Q_{2n}^{(i+1)} + b_{21}kh^2Q_{2n}^{(i+1)}P_{2n}^{(i+1)} + c_{22}kh^2(Q_{2n}^{(i+1)})^2 - h^2Q_{2n}^{(i)} = 0, \\ \beta_1P_2^{(i+1)} - c_{12}kh^2P_2^{(i+1)}Q_2^{(i+1)} + b_{11}kh^2(P_2^{(i+1)})^2 - d_1kP_3^{(i+1)} - h^2P_2^{(i)} = 0, \\ -d_1kP_{j-1}^{(i+1)} + \beta_2P_j^{(i+1)} - c_{12}kh^2P_j^{(i+1)}Q_j^{(i+1)} + b_{11}kh^2(P_j^{(i+1)})^2 \\ -d_1kP_{j+1}^{(i+1)} - h^2P_j^{(i)} = 0, \qquad j = \overline{3, f-2}, \\ -d_1kP_{f-2}^{(i+1)} + \beta_2P_{f-1}^{(i+1)} - c_{12}kh^2P_{f-1}^{(i+1)}Q_{f-1}^{(i+1)} + b_{11}kh^2(P_{f-1}^{(i+1)})^2 - h^2P_{f-1}^{(i)} = 0, \\ P_j^{i+1} = 0, \qquad j = \overline{f, 2n}, \end{cases}$$

which is a nonlinear algebraic system with $4n-2$ equations and $4n-2$ unknowns.

Remark 1. We use $P_j^{(i+1)}$ for $j = \overline{f, 2n}$ as unknowns, even if these are equal to zero, in order to get an algebraic system which has the number of equations equal to the number of unknowns.

To solve this nonlinear system, we use the Newton-Raphson method. We introduce the unknowns x_i according to

$$x_i = Q_{i+1},\ i = \overline{1, 2n-1}; \qquad x_j = P_{j-(2n-2)},\ j = \overline{2n, 4n-2}$$

Next, the problem (P_h) can be rewritten as:

$$f_l(x_1, x_2, ..., x_{4n-2}) = 0, \qquad l = \overline{1, 4n-2}, \tag{2}$$

where f_l represents the equation (1) of the discretized system.

If we denote by $X = (x_1, x_2, ..., x_{4n-2})$, then in the neighborhood of X, each of the functions f_l can be expanded in Taylor series as

$$f_l(X + \delta X) = f_l(X) + \sum_{v=1}^{4n-2} \frac{\partial f_l}{\partial x_v} \delta x_v + O(\delta X^2).$$

By neglecting the terms of order δX^2 and higher, we obtain a set of linear equations for the correction δX, that move each function closer to zero simultaneously:

$$\sum_{v=1}^{4n-2} a_{lv} \delta x_v = \beta_l, \qquad l = \overline{1, 4n-2},$$

where

$$a_{lv} = \frac{\partial f_l}{\partial x_v};\ \beta_l = -f_l(X).$$

This linear system can be solved by a direct method or an iterative one. The corrections are then added to solution vector:

$$x_i^{new} = x_i^{old} + \delta x_i.$$

To start this algorithm, we have to choose an initial estimate of the system's solution(at every time level i), and also a stopping criterion for ending the iteration process, which in our case was defined by

$$||\delta X||_{max} \leq 10^{-3}.$$

In the sequel, we present the Jacobian matrix corresponding to the system (2).

$$J = \begin{pmatrix} A & B \\ C & D \end{pmatrix},$$

where:

$$A = \begin{pmatrix} q_2 & q_1 & 0 & 0 & \cdots & \cdots & 0 \\ q_1 & q_3 & q_1 & 0 & \cdots & \cdots & 0 \\ 0 & q_1 & q_4 & q_1 & 0 & \cdots & 0 \\ 0 & 0 & q_1 & q_5 & q_1 & 0 & \cdots \\ \cdots & \cdots & \cdots & \ddots & \ddots & \ddots & \cdots \\ 0 & 0 & \cdots & \cdots & q_1 & q_{2n-1} & q_1 \\ 0 & 0 & \cdots & \cdots & 0 & q_1 & q_{2n} \end{pmatrix}, \quad B = \begin{pmatrix} p_2 & 0 & \cdots & \cdots & 0 \\ 0 & p_3 & 0 & \cdots & 0 \\ \cdots & \cdots & \ddots & \cdots & \cdots \\ 0 & \cdots & \cdots & p_{2n-1} & 0 \\ 0 & 0 & \cdots & 0 & p_{2n} \end{pmatrix},$$

$$C=\begin{pmatrix} s_2 & 0 & \cdots & \cdots & \cdots & \cdots\cdots & 0 \\ 0 & s_3 & 0 & \cdots & \cdots & \cdots\cdots & 0 \\ \cdots\cdots & \ddots & \cdots & \cdots & \cdots & \cdots\cdots & \\ 0 & \cdots\cdots & s_{f-2} & 0 & & \cdots\cdots & 0 \\ 0 & 0 & \cdots & 0 & s_{f-1} & 0 \cdots & 0 \\ 0 & \cdots\cdots & \cdots & \cdots & & 0 & \cdots\cdots \\ 0 & \cdots\cdots & \cdots & \cdots & & \cdots\cdots & 0 \end{pmatrix}, D=\begin{pmatrix} r_2 & r_1 & 0 & 0 & \cdots & \cdots\cdots\cdots & 0 \\ r_1 & r_3 & r_1 & 0 & \cdots & \cdots\cdots\cdots & 0 \\ 0 & r_1 & r_4 & r_1 & 0 & \cdots\cdots\cdots & 0 \\ \cdots\cdots & \ddots & \ddots & \ddots & & \cdots\cdots\cdots & \\ 0 & 0 & \cdots & r_1 & r_{f-1} & 0 \cdots\cdots & 0 \\ 0 & 0 & \cdots\cdots & 0 & 1 & 0 \cdots & 0 \\ 0 & \cdots\cdots & \cdots & \cdots & & 1 \cdots & 0 \\ \cdots\cdots & \cdots & \cdots & \cdots & & \cdots\cdots & \\ 0 & \cdots\cdots & \cdots & \cdots & & \cdots\cdots & 1 \end{pmatrix},$$

with

$$\begin{cases} q_1 = -d_2k, \\ q_2 = h^2 + d_2k - a_2kh^2 + b_{21}kh^2P_2 + 2c_{22}kh^2Q_2, \\ q_l = h^2 + 2d_2k - a_2kh^2 + b_{21}kh^2P_l + 2c_{22}kh^2Q_l, & l = \overline{3, 2n-1}, \\ q_{2n} = h^2 + d_2k - a_2kh^2 + b_{21}kh^2P_{2n} + 2c_{22}kh^2Q_{2n}, \\ p_l = b_{21}kh^2Q_l, \quad l = \overline{2, 2n}, \\ s_l = -kh^2c_{12}P_l, \quad l = \overline{2, f-1}, \\ r_1 = -d_1k, \\ r_2 = h^2 + d_1k - a_1kh^2 - kh^2c_{12}Q_2 + 2b_{11}kh^2P_2, \\ r_l = h^2 + 2d_1k - a_1kh^2 - kh^2c_{12}Q_l + 2b_{11}kh^2P_l, & l = \overline{3, f-1}. \end{cases}$$

Remark 2

a) Matrices C and D have a special form, owing to the free boundary $h(t)$.
b) The overall error satisfies :

$$(i). \ ||Q_j^{(i)} - Q(x_j, t_i)||_{max} = O(h^2) + O(k), \qquad j = \overline{1, 2n+1}, i = \overline{2, n+1},$$
$$(ii). \ ||P_j^{(i)} - P(x_j, t_i)||_{max} = O(h^2) + O(k), \qquad j = \overline{1, 2n+1}, i = \overline{2, n+1}.$$

These error estimates are deduced from the Taylor series expansions of $Q(x_j, t_i)$ and $P(x_j, t_i)$.

4 Numerical Results

The numerical tests have been performed with the following values of the parameters: $l = 2$, $T = 1$, $d_1 = 0.31$, $d_2 = 0.7$, $a_1 = 0.5$, $a_2 = 1.5$, $b_{11} = 0.9$, $c_{22} = 1.3$, $b_{21} = 0.4$, $c_{21} = 1$, $maxit = 100$, $eps = 0.001$, $\mu = 0.35$, $k_1 = 1$, $k_2 = 0.5$, $b = 1$, $P_0(b) = 0.3$, $n = m = 31$.

Here $maxit$ is the maximum number of iterations for Newton-Raphson method, eps the stopping criterion also for Newton-Raphson method, k_1 and k_2 the initial values (at time step $t = 1$) for Q and P.

The free boundary corresponding to the prey and predator is shown in Figure 1.

We used the Newton Raphson method for nonlinear systems to get these results. As we told before, this algorithm needs, at every time step, some initial

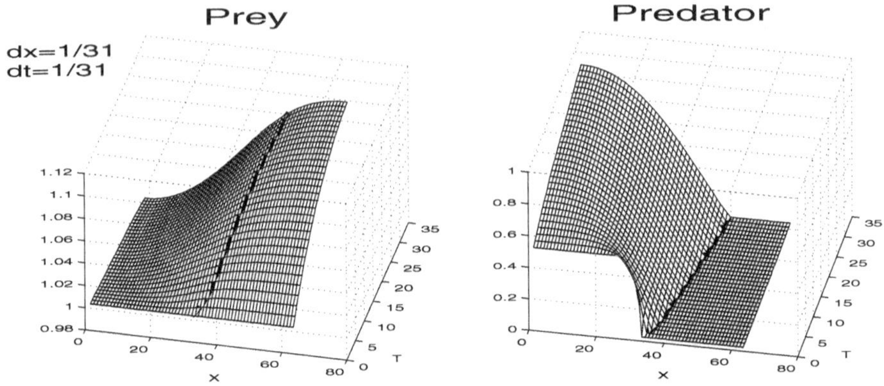

Fig. 1. The profiles of the state variables Q (left side) and P (right side), respectively

values for the solutions Q and P, in order to start it. In our case, at time level i+1, we choose as initial estimations to be the values of Q and P obtained at time step i.

Next, if we analyze the results in a different way (for a similar interpretation of the inverse Stefan problem see [2], pp. 132-137), separating the values of P respectively Q depending on the initial values, we see that a new free boundary arises. The behavior of this free boundary is shown in Figure 2. The meaning of the symbols in these figures is specified as

I : a grid node (t_i, x_j) such that $Q(t_i, x_j) = k_1,\ P(t_i, x_j) = k_2$;
$+$: a grid node (t_i, x_j) such that $Q(t_i, x_j) > k_1,\ P(t_i, x_j) > k_2$;
$-$: a grid node (t_i, x_j) such that $Q(t_i, x_j) < k_1,\ P(t_i, x_j) < k_2$;
0 : a grid node (t_i, x_j) such that $P(t_i, x_j) = 0$;

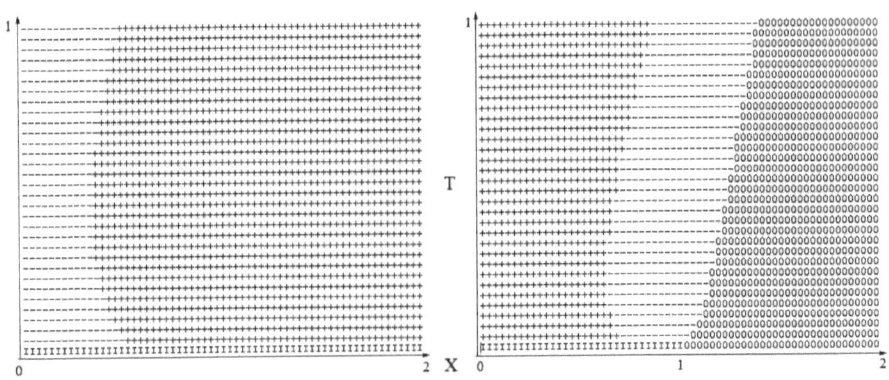

Fig. 2. The projection on the horizontal plane (x, t) of the variation profiles for Q (left side) and P (right side), respectively. The free boundary delimits the areas between negative and positive values for state variables.

5 Conclusions

In this study we focused on the numerical approximation of a free boundary problem for a predator-prey model. Using an implicit scheme, we obtained a non-linear system of algebraic equations, which was solved with Newton-Raphson method. The numerical solution was determined by using Matlab software.

For the model considered here, it is assumed that the number of predators flowing across the free boundary is increasing with respect to the moving length. Similar free boundary condition can be found in [4] for modeling the corneal stimulus cell density in the healed region. From ecological point of view, one more reasonable assumption is possibly that the motion of predators relies on the prey. In other words, the higher the density of the prey at the free boundary, the larger the flux of the predators should be.

Compared with the existing models such as tumor growth, to the best of our knowledge, there are very few results from the free boundary problems describing ecological models. As a future task, we intend to discretize this problem using spectral methods both in 1D and 2D cases, and also, in order to manipulate the free boundary, we want to place the problem into an optimal control framework.

Acknowledgments

The research is partially supported by CNCSIS-Romania under grant TD-201 no.569/ 1.10.2007.

References

1. Amadori, A.L., Vazquez, J.L.: Singular free boundary problem from image processing. Math. Models Metthods in Appl. Sci. 15, 689–715 (2005)
2. Arnăutu, V., Neittaanmaki, P.: Optimal Control from Theory to Computer Programs. Kluwer Academic Publishers, Dordrecht (2003)
3. Calvo, N., Durany, J., Vazquez, C.: Numerical approach of temperature distribution in a free boundary model for polythermal ice sheets. Numerische Mathematik 83(4), 557–580 (1999)
4. Chen, X.F., Friedman, A.: A free boundary problem arising in a model of wound healing. SIAM J. Math. Anal. 32, 778–800 (2000)
5. Chen, X.F., Friedman, A.: A free boundary problem for an elliptic-hyperbolic system: an application to tumor growth. SIAM J. Math. Anal. 35, 974–986 (2003)
6. Goodman, J., Ostrov, D.N.: On the early exercise boundary of the American put option. SIAM J. Appl. Math. 62, 1823–1835 (2002)
7. Tao, Y.S., Chen, M.J.: An elliptic-hyperbolic free boudary problem modelling cancer therapy. Nonlinearity 19, 419–440 (2006)
8. Lin, Z.: A free boundary problem for a predator-prey model. School of Mathematical Science, Yangzhou University, Yangzhou (2007)

A Second Order Accurate Difference Scheme for the Hyperbolic Problem with Concentrated Data

Zhi-zhong Sun

Department of Mathematics, Southeast University,
Nanjing 210096, Jiangsu province, P.R. China
zzsun@seu.edu.cn

Abstract. A numerical solution to the one-dimensional hyperbolic equation with concentrated data is considered. A second-order accurate difference scheme is derived by the method of the reduction of order on non-uniform meshes. The solvability, stability and second order L_∞ convergence are proved. A numerical example demonstrates the theoretical results.

1 Introduction

The study of properties of numerical schemes for discretizing of hyperbolic equations is of great interest in applied mathematics. Numerous works have been concerned with classical schemes for these equations in homogeneous whole space (especially studies of stability and dispersion relations). Other works have studied the approximations of waves at plane boundaries or interfaces. The problem of hyperbolic equations in the cases when the coefficients are discontinuous or sharply and their numerical solutions have been considered in [1]-[7]. Jovanović and Vulkov [4] presented two difference schemes approximating the one dimensional boundary value problem for the hyperbolic equations with concentrated data and proved that the first one is convergent with the convergence order of $O(\tau^2 + h^2)$ in the mesh-dependent L^2 norm and the second one is convergent with the convergence order of $O(\tau + h^2)$ in the mesh-dependent H^1 norm. In this article, we give a difference scheme for the hyperbolic equations with concentrated data and prove that the difference scheme is second order convergent in space and in time in L_∞ norm. The construction of the difference scheme proceeds along the same lines as in the paper [8] for the one-dimensional heat equation, but there are a great of difference in the analysis of the difference scheme because of the existence of the derivative $\partial^2 u/\partial t^2$ here instead of the derivative of $\partial u/\partial t$ there.

Let us consider the first initial-boundary value problem for the equation of vibrating string with concentrated mass at the interior point $x = \xi$ (see [4,10]):

$$[c(x) + K\delta(x-\xi)]\frac{\partial^2 u}{\partial t^2} - \frac{\partial}{\partial x}\left(a(x)\frac{\partial u}{\partial x}\right) = f(x,t), \qquad 0 < x < 1, \quad 0 < t \le T, \tag{1}$$

S. Margenov, L.G. Vulkov, and J. Waśniewski (Eds.): NAA 2008, LNCS 5434, pp. 556–563, 2009.

$$u(0,t)=0, \qquad u(1,t)=0, \qquad 0\le t\le T, \tag{2}$$

$$u(x,0)=\phi(x), \qquad \frac{\partial u(x,0)}{\partial t}=\psi(x), \qquad 0<x<1 \tag{3}$$

where $K>0, 0<c_1\le a(x)\le c_2, 0<c_3\le c(x)\le c_4$, and $\delta(x)$ is the Dirac distribution [11]. For the sake of conciseness, we assume

$$\begin{aligned} &a(x)\in C^3\left([0,\xi)\cup(\xi,1]\right),\ c(x)\in C[0,1],\ \phi(x)\in C^4\left([0,\xi)\cup(\xi,1]\right),\\ &\psi(x)\in C^0\left([0,\xi)\cup(\xi,1]\right),\ \phi(0)=\phi(1)=\psi(0)=\psi(1)=0, \end{aligned} \tag{4}$$

where

$$C^l\left([0,\xi)\cup(\xi,1]\right)\equiv\{g\mid g(\cdot)|_{[0,\xi]}\in C^l[0,\xi], g(\cdot)|_{[\xi,1]}\in C^l[\xi,1]\}.$$

It follows from (1) that the solution for this problem satisfies, for $Q_1=(0,\xi)\times(0,T]$ and $Q_2=(\xi,1)\times(0,T]$, the differential equation

$$c(x)\frac{\partial^2 u}{\partial t^2}-\frac{\partial}{\partial x}(a(x)\frac{\partial u}{\partial x})=f(x,t)$$

and for $x=\xi$ the conditions of conjugation

$$[u]_{x=\xi}\equiv u(\xi+0,t)-u(\xi-0,t)=0, \qquad \left[a\frac{\partial u}{\partial x}\right]_{x=\xi}=K\frac{\partial^2 u(\xi,t)}{\partial t^2}.$$

Similarly to those in [4], we assume that the problem (1)-(3) has solution $u(x,t)$ satisfying

$$u\in C^4(\{[0,\xi)\times[0,T]\}\cup\{(\xi,1]\times[0,T]\}) \quad \text{and} \quad u(\xi,\cdot)\in C^4[0,T]. \tag{5}$$

2 The Difference Scheme

We divide the intervals $[0,\xi]$ and $[\xi,1]$ into m and $M-m$ subintervals respectively with $0=x_0<x_1<\cdots<x_m=\xi<x_{m+1}<\cdots<x_M=1$ and set $\omega_1=\{x_0,x_1,\cdots,x_m\}$ and $\omega_2=\{x_m,x_{m+1},\cdots,x_M\}$. Let $h_i=x_i-x_{i-1}, 1\le i\le M$ and $h=\max_{1\le i\le M}\{h_i\}$. Usually, we let h_i be ξ/m for $1\le i\le m$ and $(1-\xi)/(M-m)$ for $m+1\le i\le M$. The interval $[0,T]$ is partitioned into N equal parts of width $\tau=T/N$ and set $\omega_\tau=\{t_k\,|\,t_k=k\tau, k=0,1,\cdots,N\}$. In addition, we denote $x_{i-\frac12}=\frac12(x_i+x_{i-1})$, $a_{i-\frac12}=a(x_{i-\frac12})$, $c_{i-\frac12}=c(x_{i-\frac12})$, $f^k_{i-\frac12}=f(x_{i-\frac12},t_k)$.

Let $u=\{u_i^k\,|\,0\le i\le M, 0\le k\le N\}$ be a grid function on $\{\omega_1\times\omega_\tau\}\cup\{\omega_2\times\omega_\tau\}$. Introduce the following notations:

$$u_i^{\bar k}=\frac12(u_i^{k+1}+u_i^{k-1}), \qquad \Delta_{\hat t}u_i^k=\frac{1}{2\tau}\left(u_i^{k+1}-u_i^{k-1}\right),$$

$$\Delta_t u_i^k=\frac{1}{\tau}(u_i^{k+1}-u_i^k), \qquad \Delta_t^2 u_i^k=\frac{1}{\tau}\left(\Delta_t u_i^k-\Delta_t u_i^{k-1}\right),$$

$$u^k_{i-\frac12}=\frac12(u_i^k+u_{i-1}^k), \qquad \delta_x u^k_{i-\frac12}=\frac{1}{h_i}(u_i^k-u_{i-1}^k),$$

$$\delta_x(a\delta_x u)_i^k=\frac{2}{h_{i+1}+h_i}\left(a_{i+\frac12}\delta_x u^k_{i+\frac12}-a_{i-\frac12}\delta_x u^k_{i-\frac12}\right),$$

where $u_i^{\bar{k}}$ is an average of u at the points (x_i, t_{k+1}) and (x_i, t_{k-1}) and $\Delta_{\hat{t}} u_i^k$ is the difference quotient of u based on these two points, $\Delta_t u_i^k$ is the difference quotient of u based on the points (x_i, t_{k+1}) and (x_i, t_k); $u_{i-\frac{1}{2}}^k$ is an average of u at the points (x_i, t_k) and (x_{i-1}, t_k) and $\delta_x u_{i-\frac{1}{2}}^k$ the difference quotient of u based on these two points; $\delta_x(a\delta_x u)_i^k$ and $\Delta_t^2 u_i^k$ are the second order difference quotients. It is clear that

$$\Delta_{\hat{t}} u_i^k = \frac{1}{2}\left(\Delta_t u_i^k + \Delta_t u_i^{k-1}\right).$$

We define

$$\|g\|_\infty = \max_{0\le i\le M} |g_i|, \quad \|g\| = \sqrt{\sum_{i=1}^M h_i \left(\frac{g_i + g_{i-1}}{2}\right)^2},$$

$$\|\delta_x g\| = \sqrt{\sum_{i=1}^M h_i \left(\frac{g_i - g_{i-1}}{h_i}\right)^2},$$

where $g = (g_0, g_1, \cdots, g_M)$ is a grid function on $\omega_1 \cup \omega_2$. In addition, if $g_0 = 0$ and $g_M = 0$, we have $\|g\|_\infty \le \|\delta_x g\|/2$.

The difference scheme we construct for (1)-(3) is as follows:

$$\alpha_i c_{i-\frac{1}{2}} \Delta_t^2 u_{i-\frac{1}{2}}^k + \beta_i c_{i+\frac{1}{2}} \Delta_t^2 u_{i+\frac{1}{2}}^k - \delta_x(a\delta_x u)_i^{\bar{k}} = \alpha_i f_{i-\frac{1}{2}}^k + \beta_i f_{i+\frac{1}{2}}^k,$$
$$1 \le i \le m-1, m+1 \le i \le M-1, 1 \le k \le N-1, \quad (6)$$

$$\alpha_m c_{m-\frac{1}{2}} \Delta_t^2 u_{m-\frac{1}{2}}^k + \beta_m c_{m+\frac{1}{2}} \Delta_t^2 u_{m+\frac{1}{2}}^k + K\frac{2}{h_m + h_{m+1}} \Delta_t^2 u_m^k$$
$$-\delta_x(a\delta_x u)_m^{\bar{k}} = \alpha_m f_{m-\frac{1}{2}}^k + \beta_m f_{m+\frac{1}{2}}^k, \quad 1 \le k \le N-1, \quad (7)$$

$$u_0^k = 0, \quad u_M^k = 0, \quad 0 \le k \le N, \quad (8)$$

$$u_i^0 = \phi(x_i), \quad 1 \le i \le M-1, \quad (9)$$

$$u_i^1 = \phi(x_i) + \tau\psi(x_i) + \frac{1}{2}\tau^2 \Psi_i, \quad 1 \le i \le M-1, \quad (10)$$

where $\alpha_i = h_i/(h_i + h_{i+1}), \beta_i = 1 - \alpha_i$ and

$$\Psi_i = \begin{cases} \frac{1}{c(x_i)}\left[\delta_x(a\delta_x\phi)_i + f(x_i, 0)\right], & \text{when } 1 \le i \le M-1 \text{ and } i \ne m, \\ \frac{1}{K}\left(a(x_m + 0)\delta_x\phi_{m+\frac{1}{2}} - a(x_m - 0)\delta_x\phi_{m-\frac{1}{2}}\right), & \text{when } i = m. \end{cases}$$

At each time level, (6)-(10) is a tridiagonal system of linear algebraic equations, which can be solved by the Thomas method.

The outline of the derivation of the difference scheme is as follows.

Let $v = a(x)\frac{\partial u}{\partial x}$, then (1)-(3) is equivalent to the following system of equations

$$c(x)\frac{\partial^2 u}{\partial t^2} - \frac{\partial v}{\partial x} = f(x,t), \quad x \in (0,\xi) \cup (\xi, 1), \quad 0 < t \le T, \quad (11)$$

$$\frac{1}{a(x)}v - \frac{\partial u}{\partial x} = 0, \quad x \in (0,\xi) \cup (\xi,1), \quad 0 < t \le T, \tag{12}$$

$$[u]_{x=\xi} = 0, \qquad 0 < t \le T, \tag{13}$$

$$[v]_{x=\xi} = K\frac{\partial^2 u(\xi,t)}{\partial t^2}, \qquad 0 < t \le T, \tag{14}$$

$$u(0,t) = 0, \qquad u(1,t) = 0, \qquad 0 \le t \le T, \tag{15}$$

$$u(x,0) = \phi(x), \qquad \frac{\partial u(x,0)}{\partial t} = \psi(x), \qquad 0 < x < 1 \tag{16}$$

We construct the difference scheme for (11)-(16) as follows:

$$c_{i-\frac{1}{2}}\Delta_t^2 u_{i-\frac{1}{2}}^k - \delta_x(v_1)_{i-\frac{1}{2}}^{\bar{k}} = f_{i-\frac{1}{2}}^k, \quad 1 \le i \le m,\ 1 \le k \le N-1, \tag{17}$$

$$\frac{1}{a_{i-\frac{1}{2}}}(v_1)_{i-\frac{1}{2}}^k - \delta_x u_{i-\frac{1}{2}}^k = 0, \quad 1 \le i \le m,\ 0 \le k \le N, \tag{18}$$

$$c_{i-\frac{1}{2}}\Delta_t^2 u_{i-\frac{1}{2}}^k - \delta_x(v_2)_{i-\frac{1}{2}}^{\bar{k}} = f_{i-\frac{1}{2}}^k, \qquad m+1 \le i \le M,\ 1 \le k \le N-1, \tag{19}$$

$$\frac{1}{a_{i-\frac{1}{2}}}(v_2)_{i-\frac{1}{2}}^k - \delta_x u_{i-\frac{1}{2}}^k = 0, \quad m+1 \le i \le M,\ 0 \le k \le N, \tag{20}$$

$$(v_2)_m^{\bar{k}} - (v_1)_m^{\bar{k}} = K\Delta_t^2 u_m^k, \quad 1 \le k \le N-1, \tag{21}$$

$$u_0^k = 0, \quad u_M^k = 0, \quad 0 \le k \le N, \tag{22}$$

$$u_i^0 = \phi(x_i), \quad 1 \le i \le M-1, \tag{23}$$

$$u_i^1 = \phi(x_i) + \tau\psi(x_i) + \frac{1}{2}\tau^2\Psi_i, \quad 1 \le i \le M-1 \tag{24}$$

and have the following equivalence theorem.

Theorem 1. *The difference scheme (17)-(24) is equivalent to (6)-(10) and*

$$\frac{1}{a_{i-\frac{1}{2}}}(v_1)_{i-\frac{1}{2}}^k - \delta_x u_{i-\frac{1}{2}}^k = 0, \quad 1 \le i \le m,\ k = 0,1, \tag{25}$$

$$\frac{1}{a_{i-\frac{1}{2}}}(v_2)_{i-\frac{1}{2}}^k - \delta_x u_{i-\frac{1}{2}}^k = 0, \quad m+1 \le i \le M,\ k = 0,1, \tag{26}$$

$$(v_1)_i^{\bar{k}} = a_{i+\frac{1}{2}}\delta_x u_{i+\frac{1}{2}}^{\bar{k}} - \frac{1}{2}h_{i+1}(c_{i+\frac{1}{2}}\Delta_t^2 u_{i+\frac{1}{2}}^k - f_{i+\frac{1}{2}}^k), \qquad 0 \le i \le m-1,\ 1 \le k \le N-1, \tag{27}$$

$$(v_1)_m^{\bar{k}} = a_{m-\frac{1}{2}}\delta_x u_{m-\frac{1}{2}}^{\bar{k}} + \frac{1}{2}h_m(c_{m-\frac{1}{2}}\Delta_t^2 u_{m-\frac{1}{2}}^k - f_{m-\frac{1}{2}}^k), \qquad 1 \le k \le N-1, \tag{28}$$

$$(v_2)_m^{\bar{k}} = a_{m+\frac{1}{2}}\delta_x u_{m+\frac{1}{2}}^{\bar{k}} - \frac{1}{2}h_{m+1}(c_{m+\frac{1}{2}}\Delta_t^2 u_{m+\frac{1}{2}}^k - f_{m+\frac{1}{2}}^k), \qquad 1 \le k \le N-1, \tag{29}$$

$$(v_2)_i^{\bar{k}} = a_{i-\frac{1}{2}}\delta_x u_{i-\frac{1}{2}}^{\bar{k}} + \frac{1}{2}h_i(c_{i-\frac{1}{2}}\Delta_t^2 u_{i-\frac{1}{2}}^k - f_{i-\frac{1}{2}}^k), \qquad m+1 \le i \le M,\ 1 \le k \le N-1. \tag{30}$$

At the $(k+1)-$th time level when $k \geq 1$, we regard (17)-(24) as a system of linear algebraic equations with respect to $\{u_i^{k+1}, 0 \leq i \leq M\} \cup \{(v_1)_i^{\bar{k}}, 0 \leq i \leq m\} \cup \{(v_2)_i^{\bar{k}}, m \leq i \leq M\}$. If $\{u_i^{k+1}, 0 \leq i \leq M\}$ has been determined, then $\{(v_1)_i^{\bar{k}}, 0 \leq i \leq m\} \cup \{(v_2)_i^{\bar{k}}, m \leq i \leq M\}$ can be obtained directly from (27)-(30).

The following three theorems can be proved.

Theorem 2. *The difference scheme (6)-(10) is uniquely solvable.*

Theorem 3. *Let the assumptions (4) and (5) hold. Then the solution $\{u_i^k\}$ of the difference scheme (6)-(10) is convergent to the solution $u(x,t)$ of the problem (1)-(3) with the convergence rate of $O(\tau^2 + h^2)$ in the L_∞ norm. In more precisely, there exists a constant c independent of h and τ such that*

$$\max_{0 \leq k \leq N} \max_{0 \leq i \leq M} \left| u(x_i, t_k) - u_i^k \right| \leq c(\tau^2 + h^2).$$

Theorem 4. *Let $\{u_i^k\}$ be the solution of the difference scheme (6)-(10), then there exists a constant c independent of h and τ such that*

$$\|\delta_x u^k\|^2 \leq c\left[\|\delta_x u^0\|^2 + \|\delta_x u^1\|^2 + \|\Delta_t u^0\|^2 + |\Delta u_m^0|^2 + \tau \sum_{l=1}^{k-1} \sum_{i=1}^{M} h_i (f_{i-\frac{1}{2}}^l)^2\right], \quad 1 \leq k \leq N.$$

3 A Numerical Example

Consider the following problem

$$\left[1 + 2\delta(x - \frac{2}{3})\right] \frac{\partial^2 u}{\partial t^2} - \frac{\partial^2 u}{\partial x^2} = \left[u_0(x) - u_0''(x)\right] e^{-t}, \quad 0 < x < 1, \ 0 < t \leq 1, \tag{31}$$

$$u(0,t) = 0, \quad u(1,t) = 0, \quad 0 < t \leq 1, \tag{32}$$

$$u(x,0) = u_0(x), \quad \frac{\partial u}{\partial t}(x,0) = -u_0(x), \quad 0 \leq x \leq 1. \tag{33}$$

where $u_0(x) = x^2(1-x)(|x - \frac{2}{3}| + 1)$. Its exact solution is $u(x,t) = u_0(x)e^{-t}$.

Take two positive integers M (divisible by 3) and N. Denote $h = 1/M$ and $\tau = 1/N$. Let $\{u_i^k \mid 0 \leq i \leq M, 0 \leq k \leq N\}$ is the solution of the difference scheme (6)-(10). Define the grid function

$$\tilde{u}_i^k = U_i^k - u_i^k, \quad 0 \leq i \leq M, \ 0 \leq k \leq N.$$

We compute numerical solutions of this problem by using difference scheme (6)-(10). The maximum norm errors of the numerical solutions with different mesh sizes are given in the third column at Table 1. From this table, we may see the errors of the difference scheme (6)-(10) decrease by a factor 4 as the mesh size is reduced by a factor of 2. This coincides with our theoretical results.

We compute the problem (31)-(33) by difference scheme (3.4)-(3.6) in [4] (denoted by JV1) and by difference scheme (3.20), (3.5) and (3.21) in [4] (denoted by JV2). The maximum norm errors of the numerical solutions with different mesh sizes are also given in the fourth column and fifth column at Table 1. We see that errors of the both difference schemes decrease by a factor 2 as the mesh size is reduced by a factor of 2 for this numerical example. We replace the first level approximation (3.6) in the difference scheme JV1 and (3.21) in the difference scheme JV2 by our first level approximation (10). These modified difference schemes are denoted by MJV1 and MJV2 respectively. Compute the same problem by the difference schemes MJV1 and MJV2 again. The maximum norm errors of the numerical solutions with different mesh sizes are also given in the last two columns at Table 1. We find that the errors of the MJV1 and MJV2 decrease by a factor 4 as the mesh size is reduced by a factor of 2 and they are approximately two times that of the difference scheme (6)-(10).

The maximum norm errors of the first level approximations of the difference scheme (6)-(10), JV1 and JV2 are presented at Table 2. We see that our approximation (10) has high accuracy, which satisfies

$$\|\delta_x \tilde{u}^0\| = 0, \quad \|\delta_x \tilde{u}^1\| \le c_5(\tau^2 + h^2),$$
$$\|\Delta_t \tilde{u}^0\| \le c_5(\tau^2 + h^2), \quad |\Delta_t \tilde{u}^0_m| \le c_5(\tau^2 + h^2).$$

Table 1. The maximum norm errors $\|\tilde{u}(h,\tau)\|_\infty$ of different schemes

h	τ	(6)-(10)	JV1	JV2	MJV1	MJV2
1/30	1/30	0.27431D-03	0.17052D-02	0.17297D-02	0.59653D-03	0.58144D-03
1/60	1/60	0.66135D-04	0.79014D-03	0.79630D-03	0.14823D-03	0.14438D-03
1/120	1/120	0.16230D-04	0.37897D-03	0.38055D-03	0.36947D-04	0.35978D-04
1/240	1/240	0.40216D-05	0.18542D-03	0.18582D-03	0.92237D-05	0.89805D-05
1/480	1/480	0.10009D-05	0.91689D-04	0.91789D-04	0.23043D-05	0.22434D-05
1/960	1/960	0.24967D-06	0.45588D-04	0.45613D-04	0.57589D-06	0.56063D-06
1/1920	1/1920	0.62315D-07	0.22730D-04	0.22736D-04	0.14395D-06	0.14015D-06

Table 2. The maximum norm errors $\|\tilde{u}^1\|_\infty$ of the first level approximations

h	τ	(10)	(3.6) in JV1	(3.21) in JV2
1/30	1/30	0.18229D-04	0.58144D-03	0.57540D-03
1/60	1/60	0.22396D-05	0.14438D-03	0.14362D-03
1/120	1/120	0.27750D-06	0.35978D-04	0.35884D-04
1/240	1/240	0.34535D-07	0.89805D-05	0.89686D-05
1/480	1/480	0.43074D-08	0.22434D-05	0.22419D-05
1/960	1/960	0.53783D-09	0.56063D-06	0.56044D-06
1/1920	1/1920	0.67191D-10	0.14015D-06	0.14012D-06

Now, suppose $\tau = h$, and denote

$$\|\tilde{u}(h)\|_\infty = \|\tilde{u}(h,h)\|_\infty.$$

If

$$\|\tilde{u}(h)\|_\infty \approx ch^p,$$

then we have

$$\ln \|\tilde{u}(h)\|_\infty \approx \ln c + p \ln h,$$

or,

$$-\ln \|\tilde{u}(h)\|_\infty \approx -\ln c + p(-\ln h).$$

Thus, the slope of the curve $(-\ln \|\tilde{u}(h)\|_\infty)$ against $(-\ln h)$ represents the convergence rate.

Using the data in Table 1 and with the help of MATLAB, we obtain the following linear least-squares fit to the data.

- difference scheme (6)-(10): $-\ln \|\tilde{u}(h)\|_\infty = 1.3670 + 2.0153(-\ln h)$
- difference scheme JV1: $-\ln \|\tilde{u}(h)\|_\infty = 2.8979 + 1.0345(-\ln h)$
- difference scheme JV2: $-\ln \|\tilde{u}(h)\|_\infty = 2.8768 + 1.0376(-\ln h)$
- difference scheme MJV1: $-\ln \|\tilde{u}(h)\|_\infty = 0.6170 + 2.0025(-\ln h)$
- difference scheme MJV2: $-\ln \|\tilde{u}(h)\|_\infty = 0.6422 + 2.0027(-\ln h)$

4 Conclusion

In this article, we derived a difference scheme (6)-(10) on non-uniform meshes for the problem (1)-(3) by the method of the reduction of order and proved the solvability, unconditional stability and L_∞ convergence of the difference scheme. A slight modification is provided for the difference schemes given by Jovanović and Vulkov.

The method is applicable to the following general problem

$$[c(x) + \sum_{j=1}^{l} K_j \delta(x-\xi_j)] \frac{\partial^2 u}{\partial t^2} - \left[\frac{\partial}{\partial x}\left(a(x)\frac{\partial u}{\partial x}\right) + b(x)\frac{\partial u}{\partial x} + d(x)u\right] = f(x,t),$$
$$0 < x < 1, \quad 0 < t \le T,$$
$$\alpha_0 \frac{\partial u}{\partial x}(0,t) + \alpha_1 u(0,t) = \alpha_2(t), \qquad \beta_0 \frac{\partial u}{\partial x}(1,t) + \beta_1 u(1,t) = \beta_2, \qquad 0 \le t \le T,$$
$$u(x,0) = \phi(x), \qquad \frac{\partial u(x,0)}{\partial t} = \psi(x), \qquad 0 < x < 1$$

where $K_j > 0, 1 \le j \le l, 0 < \xi_1 < \xi_2 < \cdots < \xi_l < 1; 0 < c_1 \le a(x) \le c_2, 0 < c_3 \le c(x) \le c_4$, $|\alpha_0| + |\alpha_1| \ne 0, |\beta_0| + |\beta_1| \ne 0$. Similar results may be obtained.

Acknowledgment

This research is supported by Natural Science Foundation of China (Contract grant number 10871044).

References

1. Jovanović, B.S.: Finite difference scheme method for boundary value problems with weak solutions. Technical report, Posebna izdanja Mat, Instituta 16 Belgrade,Yugoslavia (1993)
2. Jovanović, B.S.: Convergence of a finite-difference scheme for hyperbolic equations with variable coefficients. Z. Angew. Math. Mech. 72, 493–496 (1972)
3. Jovanović, B.S., Ivanović, L.D., Süli, E.E.: Convergence of a finite difference scheme for second-order hyperbolic equations with variable coeffients. IMA J. Numer. Anal. 7, 39–45 (1987)
4. Jovanović, B.S., Vulkov, L.G.: On the convergence schemes for hyperbolic problems with concentrated data. SIAM J. Numer. Anal. 41, 516–538 (2003)
5. Jovanović, B.S., Vulkov, L.G.: On the convergence schemes for heat equation with concentrated capacity. Numer. Math. 89, 715–734 (2001)
6. Lazarov, R.D., Makarov, V.L., Samarskii, A.A.: Applications of exact difference scheme for construction and studies of difference schemes on generalized solutions. Math. Sbornic 117, 469–480 (1982)
7. Samarskiĭ, A.A., Lazarov, R.D., Makarov V.L.: Difference Schemes for Differential Equations with Generalized Solutions. Vyshaya Shkola, Moscow (1987) (in Russian)
8. Sun, Z.Z., Zhu, Y.L.: A second order accurate difference scheme for the heat equation with concentrated capacity. Numer. Math. 97, 379–395 (2004)
9. Sun, Z.Z.: A second order accurate linearized difference scheme for the two dimensional Cahn-Hilliard equation. Math. Comp. 64, 1463–1471 (1995)
10. Tikhonov, A.N., Samarskiĭ, A.A.: Equations of Mathematical Physics. GITL, Moscow (1953) (in Russian)
11. Wloka, J.: Partial Differential Equations. Cambridge University Press, Cambridge (1987)

On the Discretization Time-Step in the Finite Element Theta-Method of the Discrete Heat Equation

Tamás Szabó

Eötvös Loránd University, Institute of Mathematics
1117 Budapest, Pázmány P. S. 1/c, Hungary

Abstract. In this paper the numerical solution of the one dimensional heat conduction equation is investigated, by applying Dirichlet boundary condition at the left hand side and Neumann boundary condition was applied at the right hand side. To the discretization in space, we apply the linear finite element method and for the time discretization the well-known theta-method. The aim of the work is to derive an adequate numerical solution for the homogenous initial condition by this approach. We theoretically analyze the possible choice of the time-discretization step-size and establish the interval where the discrete model is reliable to the original physical phenomenon.

As the discrete model, we arrive at the task of the one-step iterative method. We point out that there is a need to obtain both lower and upper bounds of the time-step size to preserve the qualitative properties of the real physical solution. The main results of the work is to determine the interval for the time-step size to be used in this special finite element method and analyze the main qualitative characterstics of the model.

1 Preliminaries

Minimum time step sizes for different diffusion problems have been analyzed by many researchers [7]. Thomas and Zhou [4] have constructed an approach to develop the minimum time step size, that can be used in the finite element method of diffusion problems. However, these approach is rigorous. We point out its imperfections and extend the analysis to the theta method as well, and develop an upper limit for the maximum time step size. In this paper, for the analysis of the one-dimensional classical diffusion problem, the heat conduction equation is considered. Heat conduction or, in other terminology, the thermal conduction is the self-generated transfer of thermal energy through the space, from a place of higher temperature to a place of lower temperature, and thus is at work to even out the temperature gradients. From mathematical point of view this equation is the prototypical parabolic partial differential equation.

S. Margenov, L.G. Vulkov, and J. Waśniewski (Eds.): NAA 2008, LNCS 5434, pp. 564–571, 2009.

The general form of this equation is

$$\begin{aligned} c\frac{\partial T}{\partial t} &= k\frac{\partial^2 T}{\partial x^2},\ x \in (0,1],\ t>0, \\ T(0,t) = \tau\,;\ \frac{\partial T}{\partial x}&(1,t) = 0,\ t \geq 0, \\ T(x,0) &= u_0(x),\ x \in (0,1], \end{aligned} \tag{1}$$

where c represents the specific heat capacity, that is the measure of the thermal energy required to increase the temperature of a matter by a certain temperature level, T is the temperature of the analyzed domain, t and x denotes the time and space variables, respectively, k is the coefficient of the thermal conductivity, that is the property of a material that indicates its ability to conduct thermal energy. Moreover, τ is the temperature at $x = 0$, a non-negative real number. The left-hand side of this equation expresses the rate of the temperature change at a point in space over time and the right-hand side indicates the spatial thermal conduction in direction x. During the analysis of the problem the space was divided into $n-1$ elements. The heat capacity and the coefficient of thermal conductivity are assumed to be constants. The boundary conditions are so-called mixed boundary conditions. The physical meaning of this type of boundary condition is that, at the end of the body the heat flux is zero, in other words the thermal energy can not leave the system. The weak form of the problem (1) is

$$\int_0^1 c\frac{\partial T}{\partial t}v(x)dx + kv(0)\frac{\partial T}{\partial x}(0,t) + \int_0^1 k\frac{\partial T}{\partial x}\frac{dv}{dx}dx = 0 \tag{2}$$

for all $v \in H_0^1(0,1)$, where $H_0^1(0,1)$ denotes the sub-space of Sobolev space $H^1(0,1)$ with $v(0) = 0$. Hence, we seek such a function $T(x,t)$, which belongs to $H^1(0,1)$ for all fixed t, morover, there exists $\frac{\partial T}{\partial t}$, and it satisfies (2) for all $v \in H_0^1(0,1)$.

We seek the spatially discretized temperature T_d in the form:

$$T_d(x,t) = \sum_{i=0}^{n} \phi_i(t)N_i(x), \tag{3}$$

where $N_i(x)$ are given shape functions, (Fig. 1) and ϕ_i are unknown, and n is the ordinal number of nodes. The unknown temperature index starts from 1, because, due to the boundary condition at the first node the temperature is known, namely, $\phi_0(t) = \tau$.

Substituting (3) into (2), we get the weak semidiscretized equation

$$\begin{aligned} &\sum_{i=0}^{n} \phi_i'(t)\int_0^1 cN_i(x)N_j(x)dx+ \\ &\qquad\qquad + \sum_{i=0}^{n} \phi_i(t)\int_0^1 kN_i'(x)N_j'(x)dx = 0,\ j = 1,2\ldots n. \end{aligned} \tag{4}$$

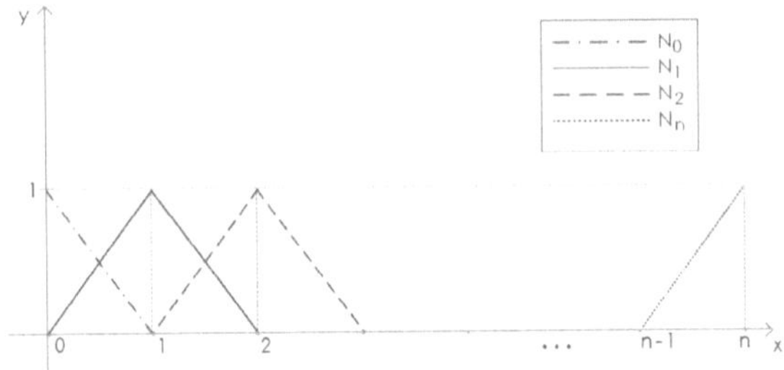

Fig. 1. Linear shape functions

Let $\underline{K}, \underline{M} \in \mathbf{R}^{(n+1)\times n}$ denote the so-called mass and stiffness matrices, respectively, defined by:

$$(\underline{K})_{ij} = \int_0^1 kN_i^{'}(x)N_j^{'}(x)dx, \tag{5}$$

$$(\underline{M})_{ij} = \int_0^1 cN_i(x)N_j(x)dx. \tag{6}$$

Then (4) can be expressed as:

$$\underline{M}\,\underline{\Phi}^{'} + \underline{K}\,\underline{\Phi} = 0, \tag{7}$$

where $\underline{\Phi} \in \mathbf{R}^{n+1}$ is a vector function with the components ϕ_i. For the time discretization of the system of ODE (7) we apply the well-known theta-method, which results in the equation

$$\underline{M}\frac{\Phi^{m+1} - \Phi^m}{\Delta t} + \underline{K}\left(\Theta\Phi^{m+1} + (1-\Theta)\Phi^m\right) = 0. \tag{8}$$

Clearly, this is a system of linear algebraic equations w.r.t. the unknown vector Φ^{m+1} being the approximation of the temperature at the new time-level. Here the parameter Θ is related to the applied numerical method and it is an arbitrary parameter on the interval $[0, 1]$. It is worth to emphasize that for $\Theta = 0.5$ the method yields the Crank-Nicolson implicit method which has higher accuracy for the time discretization [6].

In order to preserve the qualitative characteristics of the solution, the connections between the equations and the real problem must be analyzed. To obtain a lower bound for the time-step size, equation (6)-(8) should be analyzed. As it is well known, the temperature (in Kelvin) is a non-negative function in physics. In this article the following sufficient condition will be shown for the time-step size of the finite element theta-method to retain the physical characteristics of the solution:

$$\frac{h^2c}{6\Theta k} < \Delta t \leq \frac{h^2c}{3(1-\Theta)k}, \tag{9}$$

where h is the length of the spatial approximation. This sufficient condition is well known for problems with pure Dirichlet boundary conditions but not for the problems with mixed boundary conditions (Newton and Dirichlet), see e.g., [5] [3].

2 Analysis of FEM Equation

After performing the integral in (5) and (6) for the linear shape functions, the mass and the stiffness matrices have the following form

$$\underline{K} = k\frac{1}{h}\begin{bmatrix} -1 & 2 & -1 \dots & 0 \\ \vdots & \ddots & \ddots & \ddots & \vdots \\ 0 & \dots & -1 & 2 & -1 \\ 0 & \dots & 0 & -1 & 1 \end{bmatrix}, \quad \underline{M} = c\frac{h}{6}\begin{bmatrix} 1 & 4 & 1 & \dots & 0 \\ \vdots & \ddots & \ddots & \ddots & \vdots \\ 0 & \dots & 1 & 4 & 1 \\ 0 & \dots & 0 & 1 & 2 \end{bmatrix} \tag{10}$$

respectively. Using (10), the system (8) can be rewritten as:

$$a\Phi_0^{m+1} + b\Phi_1^{m+1} + a\Phi_2^{m+1} + e\Phi_0^m + f\Phi_1^m + e\Phi_2^m = 0 \tag{2.2(1)}$$

$$a\Phi_1^{m+1} + b\Phi_2^{m+1} + a\Phi_3^{m+1} + e\Phi_1^m + f\Phi_2^m + e\Phi_3^m = 0 \tag{2.2(2)}$$

$$\dots$$

$$a\Phi_{n-2}^{m+1} + b\Phi_{n-1}^{m+1} + a\Phi_n^{m+1} + e\Phi_{n-2}^m + f\Phi_{n-1}^m + e\Phi_n^m = 0 \tag{2.2(n-1)}$$

$$a\Phi_{n-1}^{m+1} + \frac{b}{2}\Phi_n^{m+1} + e\Phi_{n-1}^m + \frac{f}{2}\Phi_n^m = 0 \tag{2.2(n)}$$

where

$$a = \frac{hc}{6\Delta t} - \frac{\Theta k}{h}, \quad b = 2\left(\frac{hc}{3\Delta t} + \frac{\Theta k}{h}\right), \tag{11}$$

$$e = -\frac{hc}{6\Delta t} - \frac{(1-\Theta)\,k}{h}, \quad f = 2\left(\frac{(1-\Theta)\,k}{h} - \frac{hc}{3\Delta t}\right). \tag{12}$$

Clearly $b > 0$.

First we analyze the case when homogenous initial condition is given, i.e., $u_0(x) = 0$. Then $\Phi_i^0 = 0$, $(i = 1, 2, ..., n)$. Since $\tau > 0$, therefore, it is worth to emphasizing that, if τ is greater than zero, there is a discontinuity in the initial conditions at the point $(0,0)$. We investigate the condition under which the first iteration, denoted by $\Phi{=}\Phi^1$, results in non-negative approximation. The equations (2.2(1))-(2.2(n)) can be rewritten as

$$a\Phi_0 + b\Phi_1 + a\Phi_2 = 0 \tag{2.5(1)}$$

$$a\Phi_1 + b\Phi_2 + a\Phi_3 = 0 \tag{2.5(2)}$$

$$\dots$$

$$a\Phi_{n-2} + b\Phi_{n-1} + a\Phi_n = 0 \tag{2.5(n-1)}$$

$$a\Phi_{n-1} + \frac{b}{2}\Phi_n = 0 \tag{2.5(n)}$$

When $a = 0$, then the solution of this equation system is equal to zero. This means that the numerical scheme doesn't change the initial state which contradicts to the physical process. Therefore, in the sequel we assume that $a \neq 0$.

We seek the solution in the following form

$$\Phi_i = Z_i \Phi_0, \ i = 0, 1, ..., n. \tag{13}$$

Obviously, $Z_0 = 1$. Using (2.5(n)), Z_n can be expressed as

$$Z_n = -\frac{2a}{b} Z_{n-1} = X_{n-1} Z_{n-1}, \tag{14}$$

where

$$X_{n-1} = -\frac{2a}{b}. \tag{15}$$

In the next step, Z_{n-1} can be expressed from (2.5(n-1)). applying (7):

$$Z_{n-1} = -\frac{1}{\frac{b}{a} + X_{n-1}} Z_{n-2} = X_{n-2} Z_{n-2}, \tag{16}$$

where

$$X_{n-2} = -\frac{1}{\frac{b}{a} + X_{n-1}}. \tag{17}$$

For the i-th equation the following relation holds:

$$Z_i = -\frac{1}{\frac{b}{a} + X_i} Z_{i-1} = X_{i-1} Z_{i-1}, \ i = 1, 2, ..., n-1, \tag{18}$$

where

$$X_{i-1} = -\frac{1}{\frac{b}{a} + X_i}, \ i = n-1, n-2, ..., 1. \tag{19}$$

Hence we obtained the following statement.

Theorem 1. *The solution of the system of linear algebraic equations (2.5) can be defined by the following algorithm.*

1. *We put* $Z_0 = 1$;
2. *We define* $X_{n-1}, X_{n-2}, ..., X_0$ *by the formulas* (8) *and* (12), *respectively;*
3. *We define* $Z_1, Z_2, ..., Z_n$ *by the formulas* (7) *and* (11), *respectively;*
4. *By the formula* (6) *we define the values of* Φ_i.

The relation $\Phi_i \geq 0$, holds only under the condition $Z_i \geq 0$. From (11) we can see that it is equivalent to the non-negativity of X_i for all $i = 0, 1, ..., n-1$.

Therefore, based on (12), we have the condition $a < 0$ since $b > 0$.

For the analysis of the non-negativity of the numerical solution, produced by the above algorithm, we will use the following trivial statement.

Lemma 2. *Assume that $c > 2$ and let us define the recursion as follows $a_{i+1} = \frac{1}{c - a_i}$. When $a_1 \in (0,1)$ then $a_i \in (0,1)$ for any indices.*

This lemma implies that under the condition $-b/a > 2$ each element X_i is non-negative, because the condition $X_{n-1} = -2a/b \in (0,1)$ is automatically satisfied.

The non-negativity of a yields the condtion

$$a = \frac{hc}{6\Delta t} - \frac{\Theta k}{h} < 0. \tag{20}$$

that is, we got the condition

$$\frac{h^2 c}{6\Theta k} < \Delta t. \tag{21}$$

Hence, the following statement is proven.

Theorem 3. *Let us assume that the condition (14) holds. Then for the problem (1) with homogenous initial condition the linear finite element method results in a non-negative solution on the first time level.*

Naturally we are interested in the non-negativity preservation property not only at the first time level but on each ones. This means that rewritting the system (2.2(1))-(2.2(n)) in the matrix-vector form

$$A\Phi^{m+1} = f^m, \tag{22}$$

we must garantee the inverse-positivity of the matrix A and the non-negativity of right hand side f^m. Since under the condition (14) the realtions $b > 2|a|$ and $a < 0$ are valid, therefore, A is a strictly diagonally dominant M-matrix and hence its inverse is non-negative matrix [2]. The second condition can be guaranteed for arbitrary Φ^m if and only if e and f are non-positives. Obviously the condition $e < 0$ is always true, therefore the only condition, which should be satisfied, is the requirement $f \leq 0$, i.e.,

$$\Delta t \leq \frac{h^2 c}{3(1-\Theta)k}. \tag{23}$$

Theorem 4. *Let us assume that the time discretization parameter Δt satisfies the condition (9). Then for the problem (1) with arbitrary non-negative initial condition the linear finite element method results in a non-negative solution on any time level.*

3 Numerical Experiments

In the numerical experiments for the boundary condition at left hand side of the space domain we put the value $\tau = 273$. For the numerical experiments, a

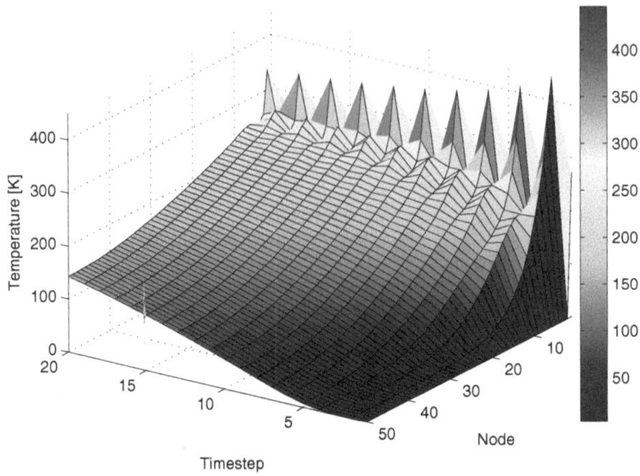

Fig. 2. The solution applying to high time step

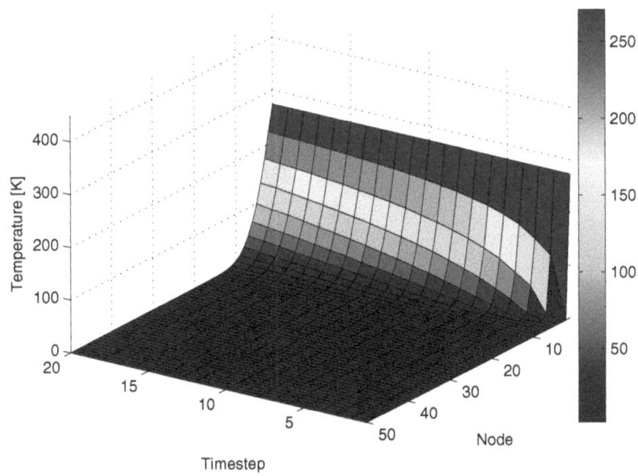

Fig. 3. The solution applying time step from the interval (16)

special type of Gauss elimination was used for the inversion of the sparse tri-diagonal matrices [1]. The following figures are in three dimensions, the first dimension is the length of a node, the second one is the temperature at the nodes, and the third one is the estimated time since the model start. First, we apply relatively high time-step, that causes the positivity of F. In (Fig. 2) one can see the numerical method is quite unstable, hence there is an oscillation with decreasing tendency in the results.

When we apply smaller time steps than, in (16), close to the first node, there will be small negative peaks, that is an unrealisctic solution, since the absolute temperature should be non-negative.

For the sake of completeness In Fig. 3 we applied the time-step size from the interval (16), and it can be seen that the oscillation disappears and we have got more stable numerical method. It is easy to see, that, by use of appropriate time steps, the solution becomes much smoother than in the Fig. 2.

4 Conclusions and Further Works

In this article the sufficient condition was given for the time-step size of the finite element theta-method to preserve the physical characteristics of the solution. For the homogenous initial condition we have shown that there exists only the lower bound for the time-step size of the finite element theta method., in order to preserve the non-negativity at the first time level. When we were interested in the non-negativity preservation property not only at the first time level but on the whole discretized time domain, then, by applying arbitrary initial condition, we shown the existence the bounds from both directions, i.e., there are upper and lower bounds for the time-step, as well.

Finally, we note that all results can be extended to the higher dimensional parabolic heat equation. In this case in (1.8) the mass ($\underline{M}$) and stiffnes ($\underline{K}$) matrices are block tridiagonal matrices, in the equations (2.2(1))-(2.2(n)) the corresponding coefficients are matrices and the unkown are vectros in each row. Therefore the conditions (2.14) and (2.16) can computed analogically. Detailed analysis of this problem will be down in the future.

References

1. Samarskiy, A.A.: Theory of different schemes. Moscow, Nauka (1977) (in Russian)
2. Berman, A., Plemmons, R.J.: Nonnegative matrices in the mathematical sciences. Computer Science and Applied Mathematics, 316 p. Academic Press (Harcourt Brace Jovanovich, Publishers), New York-London (1979)
3. Farkas, H., Faragó, I., Simon, P.: Qualitative properties of conductive heat transfer. In: Sienuitycz, S., De Voseds, A. (eds.) Thermodynamics of Energy Conversion and Transport, pp. 199–239 (2000)
4. Thomas, H.R., Zhou, Z.: An analysis of factors that govern the minimum time step size to be used in finite element analysis of diffusion problems. Commun. Numer. Meth. Engng. 14, 809–819 (1998)
5. Farago, I.: Non-negativity of the difference schemes. Pour Math. Appl. 6, 147–159 (1996)
6. Crank, J., Nicolson, P.: A practical method for numerical evaluation of solutions of partial differential equations of the heat conduction type. Proceedings of the Cambridge Philosophical Society 43, 50–64 (1947)
7. Murti, V., Valliappan, S., Khalili-Naghadeh, N.: Time step constraints in finite element analysis of the Poisson type equation. Comput. Struct. 31, 269–273 (1989)

On an Adaptive Semirefinement Multigrid Algorithm for Convection-Diffusion Problems

Daniela Vasileva

Institute of Mathematics and Informatics, Bulgarian Academy of Sciences,
Acad. G. Bonchev str., bl. 8, BG-1113 Sofia, Bulgaria
vasileva@math.bas.bg

Abstract. A multigrid algorithm with adaptive semirefinement is presented for the solution of convection-diffusion problems. The method is based on the discontinuous Galerkin discretisation, which can conveniently handle grid adaptation. The algorithm is presented here for 2D problems, but it can be generalized for 3D. Rectangular finite elements are used and during the process of adaptation they may be refined in the x, y or in both (x and y) directions.

The adaptation criterion is based on the comparison of the discrete solution on the finest grid and its restrictions to the next (in the x and y directions) grids. It refines in the x or/and y direction those cells, where the corresponding difference is too large.

The numerical experiments show that the algorithm may be successfully used for resolution of boundary and interior layers. The comparison with a similar adaptive refinement multigrid algorithm shows that significantly less computer resources may be used for layers, almost parallel to the x or y axis.

Keywords: Adaptive refinement, semirefinement, multigrid, discontinuous Galerkin method, convection-diffusion problems.

1 Introduction

The efficient resolution of boundary or/and interior layers is of great importance in mathematical modeling of many physical, engineering and other processes. Adaptive grid refinement may be successfully used in such cases. Often the layers are almost parallel to an axis and then an anisotropic type of refinement will significantly reduce the memory and CPU time, used in computations.

In the present paper we propose a combination of an adaptive semirefinement approach with a multigrid solver, which allows to achieve optimal efficiency. Section 2 concerns the governing equation and its discretisation. The third section describes the multigrid algorithm and the local semirefinement–multigrid algorithm is presented in the fourth section. An adaptive criterion is proposed in Section 5 and the last section contains results from numerical experiments.

S. Margenov, L.G. Vulkov, and J. Waśniewski (Eds.): NAA 2008, LNCS 5434, pp. 572–579, 2009.

2 Discontinuous Galerkin (DG) Discretisation

We consider the linear boundary value problem:

$$-\Delta u+\nabla\cdot(\mathbf{b}u)+cu \;=\; f \;\text{ in }\; \Omega\subset\mathbb{R}^2,\quad u(x,y) \;=\; u_0(x,y) \;\text{ on }\; \Gamma=\partial\Omega, \quad (1)$$

where the coefficients $\mathbf{b}(x,y) = (b_1(x,y), b_2(x,y)) \in (C^1(\Omega))^2$, $c(x,y) \geq 0$, $c(x,y) \in L_\infty(\Omega)$ and the right-hand side $f(x,y) \in L_2(\Omega)$. We assume that Ω allows a regular partitioning $\Omega_h = \{\Omega_e \mid \cup_e \Omega_e = \Omega,\ \Omega_i \cap \Omega_j = \emptyset,\ i \neq j\}$, into equally sized rectangular cells Ω_e. As weak form for (1) we use Baumann-Oden's [1,2] DG formulation and the space $S_h = \{\sum_e \phi_e,\ \phi_e \in \mathcal{P}_3(\Omega_e),\ \Omega_e \in \Omega_h\}$ of piecewise cubic polynomials on the partitioning Ω_h is used in the discretisation (for more details see [3,4,5,6]). The discretisation yields a linear system $L_h u_h = f_h$, where the matrix L_h has a diagonal block structure with blocks of size 16×16. We order the basic functions *point-wise* (for details see [3,4,5]) and the obtained linear system is iterated using the block Jacobi method with underrelaxation ($\omega = 0.5$).

3 Multigrid (MG) Algorithm

A detailed description of a "standard" MG approach and the corresponding smoothing analysis in the case of DG methods with constant coefficients may be found in [3,4,5]. Here we combine the DG discretisation with a *semicoarsening MG technique* (see [7] and references therein). The Sawtooth MG and full approximation storage (FAS) algorithms [8,9] are implemented. Sawtooth MG iteration ensures path-independent restrictions (see [7] for details) and FAS provides a way for local refinement (LR). The grid structure consists of the finest grid Ω_{mn} and all coarser (with $k \leq m$ and $l \leq n$) semicoarsened grids Ω_{kl} (see Fig.1). The number of cells on the finest grid is $2^m \times 2^n$, and the total number of cells on all grids is $(2^{m+1}-1)\times(2^{n+1}-1) \approx 4\times 2^m \times 2^n$, i.e., 3 times more than in the standard MG.

A description of one semicoarsening MG iteration for the discrete system $L_{mn}u_{mn} = f_{mn}$ on the grid Ω_{mn} with $level = m+n$ is given below:

1. $r_{mn} = f_{mn} - L_{mn}u_{mn}$, $\tilde{f}_{mn} = f_{mn}$;
2. `Repeat`
   `for all` $k+l = level$
       `if` $k>0$ `then` $r_{k-1,l} = \bar{R}_{kl}^{k-1,l} r_{kl}$, $u_{k-1,l} = R_{kl}^{k-1,l} u_{kl}$,
           $\bar{u}_{k-1,l} = u_{k-1,l}$, $\tilde{f}_{k-1,l} = L_{k-1,l}\bar{u}_{k-1,l} + r_{k-1,l}$;
       `if` $l>0$ `then` $r_{k,l-1} = \bar{R}_{kl}^{k,l-1} r_{kl}$, $u_{k,l-1} = R_{kl}^{k,l-1} u_{kl}$,
           $\bar{u}_{k,l-1} = u_{k,l-1}$, $\tilde{f}_{k,l-1} = L_{k,l-1}\bar{u}_{k,l-1} + r_{k,l-1}$;
   $level--$;
   `until` $level = 0$;
3. `Solve` $L_{00}u_{00} = \tilde{f}_{00}$; $c_{00} = u_{00} - \bar{u}_{00}$;
4. `Repeat`
   $level++$;

```
for all k+l = level
   if k > 0 then  c_kl = P_kl^{k-1,l} c_{k-1,l};   else  c_kl = 0;
   if l > 0 then  c_kl += P_kl^{k,l-1} c_{k,l-1};
   if k,l > 0 then  c_kl -= P_kl^{k-1,l-1} c_{k-1,l-1};
   u_kl += c_kl;
   Perform ν relaxations on  L_kl u_kl = f̃_kl;
   c_kl = u_kl - ū_kl;
until level = m+n;
```

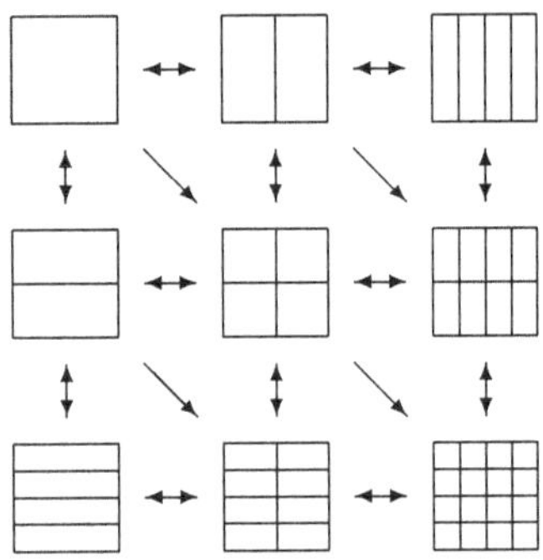

Fig. 1. Grid structure

Restrictions and prolongations are defined according to [3,4,5]: Let S_{kl} be the space of piecewise cubic polynomials on Ω_{kl}. Then the natural prolongation $P^{kl}_{k+1,l} : S_{kl} \rightarrow S_{k+1,l}$ on $\Omega_{k+1,l}$ preserves the coarse grid functions on the fine grid. The restriction operator *for the residues*, $\bar{R}^{kl}_{k+1,l} : S_{k+1,l} \rightarrow S_{kl}$ is the adjoint of $P^{kl}_{k+1,l}$ and the Galerkin relation exists between the discretisation on the coarse and fine grid $L_{kl} = \bar{R}^{kl}_{k+1,l}\; L_{k+1,l}\; P^{kl}_{k+1,l}$. Another restriction $R^{kl}_{k+1,l}$ is used *for the solution*, which preserves the function values and the derivatives at the coarse cell vertexes. It is a left-inverse of the prolongation, i.e., $R^{kl}_{k+1,l}\; P^{kl}_{k+1,l} = I_{kl}$.

4 Local Semirefinement (LSR) Multigrid Algorithm

The LSR-MG algorithm is a generalization of a local refinement (LR) MG algorithm [10,11,12].

On the first stage the equation is discretised and solved on the global coarsest grid Ω_{00}. On later stages, some unrefined cells on existing grids are selected for x, y or xy-refinement and divided into smaller cells. Two different types of grid structures – *full* and *reduced* may be created (see Fig.2 for an example). In the first variant of the algorithm (denoted as LSR-MG$_f$), if a new cell has a brother (in the x or y direction), but the common x or y father cell does not exist, we add it to the grid structure. The process is repeated recursively for the new father cells. In the second variant of the algorithm (denoted as LSR-MG$_r$), common fathers are added only on the existing grids (where some cells are already created

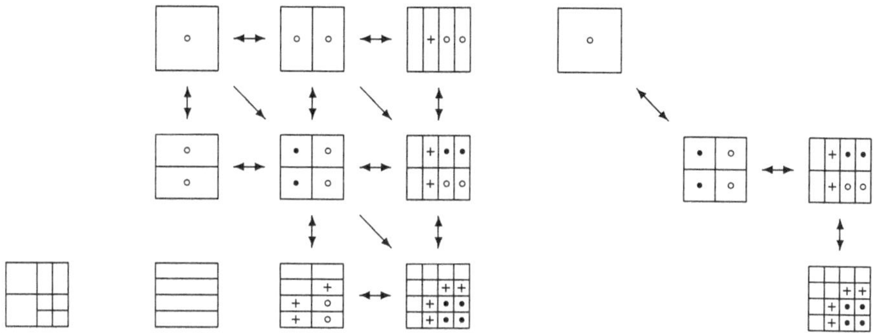

Fig. 2. A semirefined grid, the full and reduced LSR-MG structures

during the refinement process) or on the grids providing the only path to the origin. Thus a cell may not have x and/or y father, but only xy grandfather.

The solution in the new cells is interpolated from the previous grids and several relaxation sweeps (MG Sawtooth cycles) are performed. As we created only fathers with two children (in the x and/or y direction), the restrictions there may be always performed. The interpolation is defined as

$$u_{kl} = \sigma_x P_{kl}^{k-1,l}\ u_{k-1,l} + \sigma_y P_{kl}^{k,l-1}\ u_{k,l-1} + \sigma_{xy} P_{kl}^{k-1,l-1}\ u_{k-1,l-1}, \qquad (2)$$

where $(\sigma_x, \sigma_y, \sigma_{xy}) = (1, 1, -1), (1, 0, 0), (0, 1, 0), (0, 0, 1)$ or $(0.5, 0.5, 0)$ (depending on the existence of x, y fathers and xy grandfather).

Outside the refined region on $\Omega_{k+1,l}$ we may define a *virtual solution*

$$u_{k+1,l} = P_{k+1,l}^{kl}\ u_{kl},$$

then

$$r_{kl} = \bar{R}_{k+1,l}^{kl} \left(f_{k+1,l} - L_{k+1,l}\ P_{k+1,l}^{kl}\ u_{kl}\right) = \bar{R}_{k+1,l}^{kl}\ f_{k+1,l} - L_{kl}\ u_{kl},$$

$$\tilde{f}_{kl} = L_{kl}\ u_{kl} + r_{kl} = \bar{R}_{k+1,l}^{kl}\ f_{k+1,l} \approx f_{kl},$$

i.e., the right-hand side $\tilde{f}_{kl}$ on the coarser grid Ω_{kl} may be defined as in the standard LR-MG algorithm. Note, these considerations are not strict, as the prolongation for the virtual solution may not be as in (2).

The internal boundaries. A cell on $\Omega_{k+1,l}$ may have no neighbours on the same grid at some of its faces, although these faces are not on the boundary $\partial\Omega$. Thus for the discretisation on $\Omega_{k+1,l}$ *auxiliary cells* are used at the internal boundaries. The solution values are derived by interpolation (see (2)) from coarser grids. But, unlike the standard LR-MG, some coarse cells, used in this interpolation, may have children on another finer grid, i.e., the solution there may be further improved. And in order to decrease the length and the influence of the internal boundaries, we add the fathers of new cells to the grid structure (according to the rules in the LSR-MG$_f$ or LSR-MG$_r$ algorithm).

An example of a locally semirefined grid and the corresponding full and reduced grid structures for MG semicoarsening is given in Fig.2. The finest cells are marked with "•", the fathers, used in MG sweeps, are marked with "o", and the auxiliary cells are marked with "+". Note, only cells marked with "•" or "o" really belong to the grid structure.

5 Grid Adaptation

We use an adaptation criterion, based on the comparison of the discrete solution on the finest grid and its restrictions to the previous (in the x and y directions) grids. Details about a similar adaptation criterion in the LR-MG case may be found in [6].

Let $u(x,y)$ be the exact solution of our problem and $R_{kl}u$ be its restriction to Ω_{kl}, where $R_{kl} : H^1(\Omega) \to S_{kl}$ preserves the function values and the derivatives at the vertexes of each $\Omega_e \in \Omega_{kl}$. Then, if $u \in C^M(\Omega_e)$ and M=1,...,4

$$\|R_{k+1,l}u - R_{kl}u\|_{C(\Omega_e)} < C\|u\|_{C^M(\Omega_e)} 2^{-kM} ,$$
$$\|R_{k,l+1}u - R_{kl}u\|_{C(\Omega_e)} < C\|u\|_{C^M(\Omega_e)} 2^{-lM} .$$

From this we may suppose that for piecewise C^M-functions, $M \leq 4$, and *asymptotically* for large k, l, m and n, $m \geq k$, $n \geq l$,

$$\|R_{m+1,n}u - R_{mn}u\|_{C(\Omega_e)} \lessapprox q^{m+1-k} \|R_{kl}u - R_{k-1,l}u\|_{C(F^x(\Omega_e))} ,$$
$$\|R_{m,n+1}u - R_{mn}u\|_{C(\Omega_e)} \lessapprox q^{n+1-l} \|R_{kl}u - R_{k,l-1}u\|_{C(F^y(\Omega_e))} ,$$

with $q = 2^{-M}$. Here the cell $\Omega_e \in \Omega_{kl}$, $F^x(\Omega_e)$ and $F^y(\Omega_e)$ are its x and y fathers. Then we may estimate $\|u - R_{kl}u\|_{C(\Omega_e)}$ only by using $\|R_{kl}u - R_{k-1,l}u\|_{C(F^x(\Omega_e))}$ and $\|R_{kl}u - R_{k,l-1}u\|_{C(F^y(\Omega_e))}$. But as we do not know $R_{kl}u$, we use the approximate solution u_{kl}. Let

$$r^x_{kl} := \|u_{kl} - R^{k-1,l}_{kl}\; u_{kl}\|_{C(F^x(\Omega_e))}, \quad r^y_{kl} := \|u_{kl} - R^{k,l-1}_{kl}\; u_{kl}\|_{C(F^y(\Omega_e))}.$$

Then

$$\|u - R_{kl}u\|_{C(\Omega_e)} = \lim_{m\to\infty} \|R_{mm}u - R_{kl}u\|_{C(\Omega_e)} \leq$$
$$\lim_{m\to\infty} \left(\sum_{n=l+1}^{m} \|R_{mn}u - R_{m,n-1}u\|_{C(\Omega_e)} + \sum_{n=k+1}^{m} \|R_{nl}u - R_{n-1,l}u\|_{C(\Omega_e)} \right) \lessapprox$$
$$\lim_{m\to\infty} \left[(q + \ldots + q^{m-l}) r^y_{kl} + (q + \ldots + q^{m-k}) r^x_{kl} \right] = \frac{q}{1-q} (r^x_{kl} + r^y_{kl}) .$$

Thus the grid adaptation is performed as:

1. All cells of Ω_{00} are refined in the x and y direction at least once;
2. Let $\Omega_e \in \Omega_{kl}$, $k + l > 1$, be an unrefined cell, $Q = \dfrac{2q}{1-q}$ and T be a desired tolerance. If $r^x_{kl}\ Q > \mathsf{T}$, the cell is refined in the x direction. Correspondingly, if $r^y_{kl}\ Q > \mathsf{T}$, the cell is refined in the y direction.

Notice that we can estimate the local smoothness of the solution by estimating q similarly to [6], and we can use different factors q_x and q_y in the x and y direction. However, this will be a subject of further research.

6 Numerical Experiments

In the examples $\Omega = [0, 1]^2$ and the initial grid has 1 cell. The factor $q = 2^{-4}$ is used in the adaptation criterion and the prescribed tolerance is $\mathsf{T} = 0.01$. Ten MG sweeps with 2 post-smoothing iterations are performed on each stage. Three algorithms - LR-MG, LSR-MG$_f$ and LSR-MG$_r$ are compared.

Example 1. The following problem is considered

$$\Delta u - xu_x - yu_y - 2u = f(x, y),$$

where $f(x, y)$ and the Dirichlet boundary conditions correspond to the exact solution

$$u(x, y) = -xy(1 - \exp(50(x - 1)).$$

The solution and the finest grids in the cases of LR and LSR are plotted in Fig.3. The C-norm of the error is less than T in all cases and the grid is properly refined in the boundary layer. The number of cells N on the finest grid, as well as the numbers N_f and N_r of all cells in the full and reduced grid structures are significantly less for the LSR-MG cases (especially for LSR-MG$_r$). From the other side, our observations show that the solution converges faster (i.e., less MG sweeps may be performed) in the LSR-MG$_f$ case. As it is seen on Fig.3, in the LSR-MG cases the grid is refined 2 times in the y direction, although this is not necessary. But on the first stages the approximate solution is very rough (nonmonotone in y) and thus the adaptive criterion should use additional information in order to avoid this refinement.

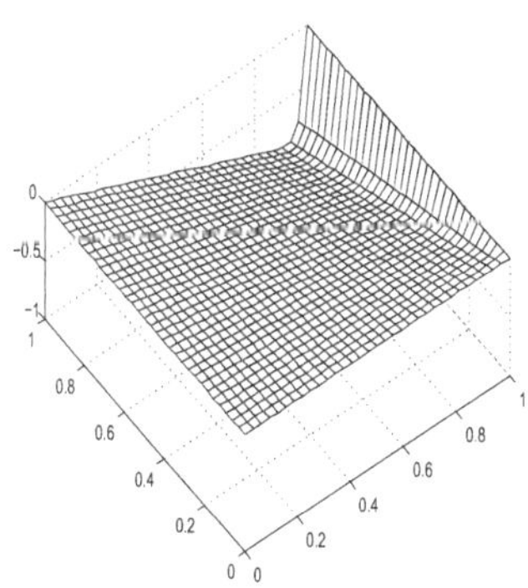

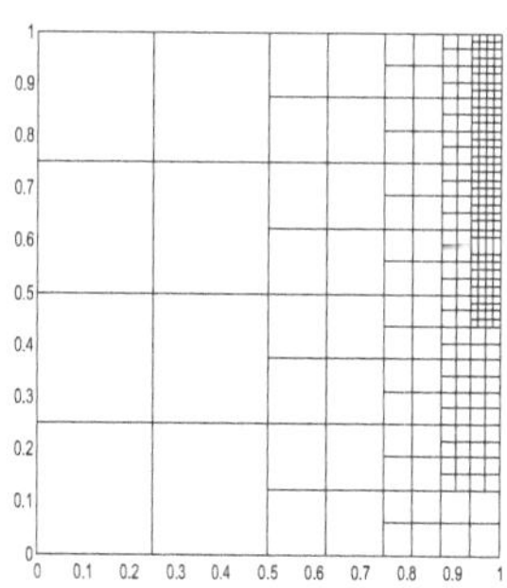

$N = 280$,
$N_f = N_r = 373$

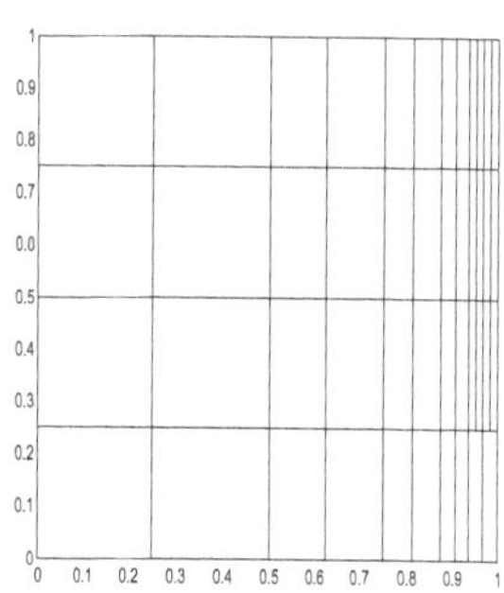

$N = 46$,
$N_f = 149$, $N_r = 81$

Fig. 3. The solution, LR and LSR for Example 1

Example 2. We consider the equation

$$\Delta u + (x - 0.5)u_x + (y - 0.5)u_y \;=\; f(x, y),$$

with $f(x, y)$ and Dirichlet boundary conditions, corresponding to the exact solution

$$u(x, y) = \arctan \frac{400x + 50y - 225}{8}.$$

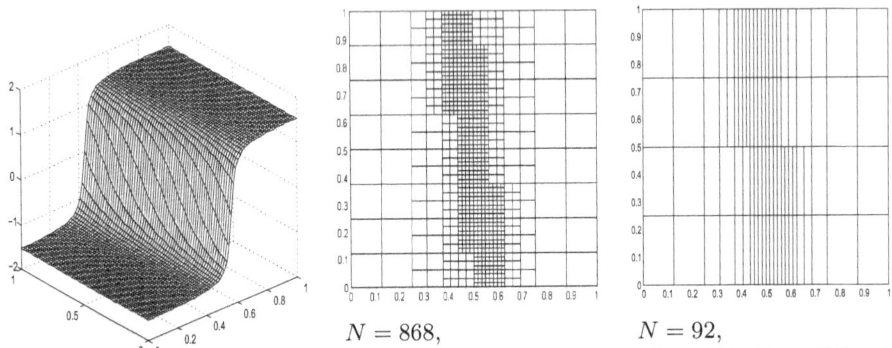

$N = 868$, $N_f = N_r = 1125$

$N = 92$, $N_f = 281$, $N_r = 173$

Fig. 4. The solution, standard LR and LSR for Example 2

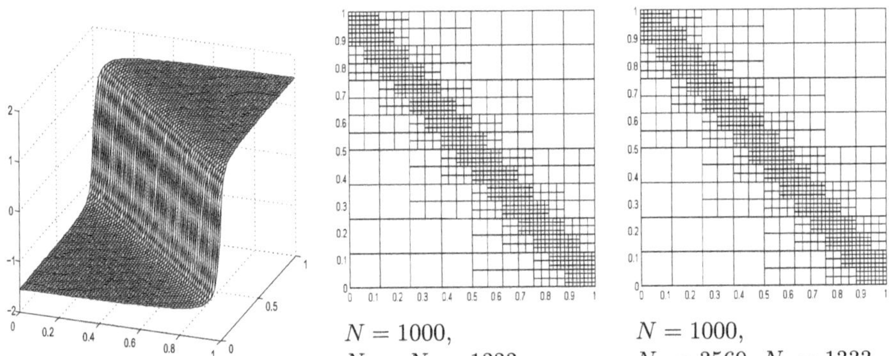

$N = 1000$, $N_f = N_r = 1333$

$N = 1000$, $N_f = 3569$, $N_r = 1333$

Fig. 5. The solution, LR and LSR for Example 3

The solution and the finest grids in the cases of LR and of LSR are plotted in Fig.4. The C-norm of the error is less than T in all cases, the grid is properly refined in the interior layer and the computing resources (memory and CPU time) are much less in the LSR-MG cases.

Example 3. The third example is for the same equation, as in Example 2

$$\Delta u + (x - 0.5)u_x + (y - 0.5)u_y \;=\; f(x, y),$$

but with $f(x, y)$ and Dirichlet boundary conditions, corresponding to an exact solution with a diagonal to the axes interior layer

$$u(x, y) = \arctan(50x + 50y - 50).$$

The solution and the finest grids in the cases of standard LR and of LSR are presented in Fig.5. The C-norm of the error is less than T and the finest grids are the same in all cases, but the number of cells in the grid structure for LSR-MG$_f$ is 2.68 times greater than for LR-MG and LSR-MG$_r$. In the LSR-MG$_r$ case the grid structure and the performance of the algorithm are the same as for LR-MG.

7 Concluding Remarks

LSR-MG may be successfully used for resolution of boundary and interior layers. The comparison with standard LR-MG shows that significantly less resources may be used for layers, (almost) parallel to the x or y axis. For layers diagonal to the axes or problems without layers almost three times more computations may be performed with LSR-MG$_f$. This could be avoided using LSR-MG$_r$ – then the grid structure and the amount of computations may be almost the same as in LR-MG. Other modifications of the grid structure and improvements of the adaptive criterion will be a subject of further investigations.

Acknowledgments. D. Vasileva thanks Prof. P.W. Hemker (CWI) for proposing the idea for LSR development and for the many fruitful discussions about MG for DG.

References

1. Baumann, C.E., Oden, J.T.: A discontinuous hp finite element method for convection-diffusion problems. Comput. Meth. Appl. Mech. Engrg. 175, 311–341 (1999)
2. Baumann, C.E.: An hp-adaptive discontinuous finite element method for computational fluid dynamics. PhD thesis, University of Texas at Austin (1997)
3. Hemker, P.W., Hoffmann, W., van Raalte, M.H.: Two-level Fourier analysis of a multigrid approach for discontinuous Galerkin discretisation. SIAM J. Sci. Comp. 25, 1018–1041 (2004)
4. Hemker, P.W., van Raalte, M.H.: Fourier two-level analysis for higher dimensional discontinuous Galerkin discretisation. Comp. Vis. Sci. 7, 159–172 (2004)
5. Van Raalte, M.H., Hemker, P.W.: Two-level multigrid analysis for the convection-diffusion equation discretized by a discontinuous Galerkin method. Num. Lin. Alg. Appl. 12, 563–584 (2005)
6. Vasileva, D., Kuut, A., Hemker, P.W.: An adaptive multigrid strategy for convection-diffusion problems. In: Lirkov, I., Margenov, S., Waśniewski, J. (eds.) LSSC 2005. LNCS, vol. 3743, pp. 138–145. Springer, Heidelberg (2006)
7. De Zeeuw, P.M.: Development of semi-coarsening techniques. Appl. Numer. Math. 19, 433–465 (1996)
8. Hackbusch, W.: Multi-Grid Methods and Applications. Springer, Berlin (1985)
9. Wesseling, P.: An Introduction to Multigrid Methods. Wiley, New York (1991)
10. Brandt, A.: Multi-level adaptive solutions to boundary value problems. Math. Comp. 31, 333–390 (1977)
11. Hemker, P.W.: On the structure of an adaptive multi-level algorithm. BIT 20, 289–301 (1980)
12. McCormick, S.F.: Multilevel Adaptive Methods for Partial Differential Equations. SIAM, Philadelphia (1989)

A Two-Grid Algorithm for Solution of the Difference Equations of a System of Singularly Perturbed Semilinear Equations

L.G. Vulkov[1] and A.I. Zadorin[2]

[1] University of Rousse, Department of Mathematics, Bulgaria
vulkov@ami.ru.acad.bg
[2] Institute of Mathematics, RAS, Siberian branch, Omsk 644099, Russia
zadorin@ofim.oscsbras.ru

Abstract. We propose a two-grid algorithm for implementation of a generalized A.M.Il'in's scheme to a system of semilinear diffusion convection-dominated equations. To solve the nonlinear algebraic system of difference equations we use Newton method. We derive the difference scheme on a coarse mesh and, then using uniform interpolation, taking into account the boundary layers, we obtain the initial guess for an iterative method on a fine mesh. Estimates of the accuracy and the computational work are obtained. The main advantage of the proposed algorithm is the computational cost.

1 Introduction

In the case of nonlinear boundary value problems, the order of ε uniform convergence rate of the two class special methods does not exceed 2 for elliptic reaction-diffusion equations and is not higher than 1 for convection-diffusion equations (e.g. see, [5]) and the bibliography therein. There are a few papers where for 1D nonlinear reaction-diffusion problems high-order difference schemes are constructed [5]. In contrast to the reaction-diffusion case, [5], there are no publications on construction of high-order finite difference or finite element approximations to nonlinear singularly perturbed convection-diffusion problems.

It is known, that classical difference schemes lose the property of convergence under applications to problems with boundary layers. In order to avoid these difficulties, exponentially-fitting techniques are frequently used [3,5,8]. The next difficulty concerning the nonlinear problems is in the efficient method to compute the solution of the nonlinear systems of difference equations. In this paper we shall employ the two-grid finite element method that was originally proposed by Axellson [2] and Xu [7], independently of each other. The idea in [2], [7] is basically to use a coarse grid to produce a rough approximation of the solution, and then use it as the initial guess for one Newton iteration on the fine grid.

Two-grid algorithms for singularly perturbed problems on adaptive meshes are proposed in [1], while the two-grid method was implemented to solve the Il'in's scheme for singularly perturbed convection-diffusion equations in [6].

S. Margenov, L.G. Vulkov, and J. Waśniewski (Eds.): NAA 2008, LNCS 5434, pp. 580–587, 2009.

Consider the boundary value problem:

$$-\varepsilon \mathbf{u}'' - a(x)\mathbf{u}' + \mathbf{F}(x, \mathbf{u}(x)) = \mathbf{0}, \quad x \in I = [0,1], \tag{1}$$

$$\mathbf{u}(0) = \mathbf{A}, \mathbf{u}(1) = \mathbf{B}, \tag{2}$$

where $a(x)$ is diagonal square matrix of order K with diagonal elements $a_i(x), i = 1, 2, \dots, K$, A and B are K-component vectors; ε is numerical parameter ($\varepsilon > 0$) and may be $\varepsilon \ll 1$; $\mathbf{F}$ is a known continuous vector function; $\mathbf{u}(x)$ is the vector of solution function . Suppose , that for all i $a_i \in C^2[0,1]$, $F_i \in C^2([0,1] \times l^K)$, where

$$l = \alpha_0^{-1} \|\mathbf{F}(x,0)\|_K + \max\{\|\mathbf{A}\|_\infty, \|\mathbf{B}\|_\infty\}$$

and

$$a_i(x) \geq \alpha_i > 0, \alpha_0 = \min_i \alpha_i. \tag{3}$$

Let $J(x,u)$ be the Jacobi matrix of the vector-function $\mathbf{F}(x, \mathbf{u})$:

$$J_{i,j} = \frac{\partial F_i}{\partial u_j}, \quad i,j = 1, 2, \dots, K.$$

Suppose, that J is M-matrix:

$$\sum_{j=1}^{K} J_{i,j}(x, \mathbf{w}) \geq 0, \; J_{i,j}(x, \mathbf{w}) \leq 0, \; j \neq i, \; x \in I, |w_i| \leq l, \quad 1 \leq i \leq K. \tag{4}$$

The problem (1)-(4) can be considered as a model for a pollution transfer with chemical reactions.

Notation. Through the paper C and C_i denote generic positive constants independent of ε and mesh size. Write for the norms: $\|q\| = \max |q(x)|, x \in I$ for a scalar function $q(x)$,

$$\|\mathbf{q}\|_K = \max_{1 \leq i \leq K} \max_{x \in I} |q_i(x)| \text{for an K-component vector function } \mathbf{q}(x);$$

$$\|\mathbf{q}\|_\infty = \max_i |q_i|, \text{ for vector } \mathbf{q};$$

$$\|\mathbf{q}^h\|_{K,\Omega} = \max_{1 \leq i \leq K} \max_{x \in \Omega} |q_i^h(x)| \text{for a vector-function } \mathbf{q}^h \text{on a grid } \Omega;$$

$$\|G\| = \max_i \sum_{j=1}^{K} |G_{i,j}| \text{for a square matrix } G \text{ of order } K.$$

In the next Section, we establish some a priori estimates of the solution and its derivatives. In Section 3 we prove uniform convergence of a generalized Ilin's scheme on a classical mesh. In Section 4 we first analyze the Newton method for solving of the nonlinear systems of the difference equations. Then we describe a two-grid algorithm for solving the difference schemes on two grid (coarse and fine). Numerical experiments for the scalar case are discussed in [1,6].

2 Analysis of the Differential Problem

Consider the auxiliary linear boundary value problem:

$$L\mathbf{w} = -\varepsilon \mathbf{w}'' - a(x)\mathbf{w}' + G(x)\mathbf{w} = \mathbf{P}(x),\ 0 < x < 1;\quad \mathbf{w}(0) = \mathbf{A},\ \mathbf{w}(1) = \mathbf{B},$$

where $a(x)$ is defined as above, $G(x)$ is an K-order square matrix, $G_{i,j}$, $P_i(x) \in C[0,1]$. Suppose, that

$$\sum_{j=1}^{K} G_{i,j}(x) \geq 0,\ G_{i,j}(x) \leq 0,\ j \neq i,\ 1 \leq i \leq K,\ x \in I. \tag{5}$$

According to [8] for the operator L the maximum principle holds and if the vector function $\mathbf{\Psi}(x)$ with components from $C^2[0,1]$ is such that

$$\mathbf{\Psi}(0) \geq 0, \mathbf{\Psi}(1) \geq 0, L\mathbf{\Psi}(x) \geq 0, x \in I, \tag{6}$$

then $\mathbf{\Psi}(x) \geq 0,\ x \in I$.

Lemma 1. *Let conditions (5) are fulfilled. Then*

$$\|\mathbf{w}\|_K \leq \alpha_0^{-1}\|\mathbf{P}\|_K + \max\{\|\mathbf{A}\|_\infty, \|\mathbf{B}\|_\infty\}. \tag{7}$$

Proof. Introduce the vector function $\mathbf{\Psi}(x)$ with components:

$$\Psi_i(x) = \alpha_0^{-1}\|\mathbf{P}\|_K(1-x) + \max\{\|\mathbf{A}\|_\infty, \|\mathbf{B}\|_\infty\} \pm w_i(x),\quad 1 \leq i \leq K.$$

Then conditions (6) are satisfied. This proves the lemma. □

Lemma 2. *Let* $\mathbf{u}(x)$ *is the solution of problem (1)-(4). Then*

$$\|\mathbf{u}\|_K \leq l. \tag{8}$$

Proof. Rewrite problem (1) in the form:

$$-\varepsilon \mathbf{u}'' - a(x)\mathbf{u}' + \mathbf{J}(x, \mathbf{s}(x))\mathbf{u} = -\mathbf{F}(x, \mathbf{0}),\quad x \in I = [0,1],$$

$$\mathbf{u}(0) = \mathbf{A}, \mathbf{u}(1) = \mathbf{B},$$

where $\mathbf{s}(x) \in [\mathbf{0}, \mathbf{u}(x)]$. Now we apply Lemma 1 to prove this lemma. □

Make decomposition of $\mathbf{u}(x)$

$$\mathbf{u}(x) = \mathbf{V}(x) + \mathbf{p}(x), \tag{9}$$

where

$$V_i(x) = \rho_i \exp(-a_i(0)\varepsilon^{-1}x),\ \rho_i = -\varepsilon u_i'(0)/a_i(0),\ 1 \leq i \leq K.$$

Lemma 3. *For some* C_1 *and any* i

$$|p_i'(x)| \leq C_1. \tag{10}$$

Proof. Using decomposition (9) in (1), we get:

$$\varepsilon p_i''(x) + a_i(x)p_i'(x) = Q_i(x),\ 1 \le i \le K, \tag{11}$$

where

$$Q_i(x) = F_i(x, \mathbf{u}(x)) + \frac{a_i(0)}{\varepsilon}(a_i(x) - a_i(0)) \exp(-a_I(0)\varepsilon^{-1}x).$$

Taking into account (8), we have for some C_2

$$|Q_i(x)| \le C_2. \tag{12}$$

Using (11), we find:

$$p_i'(x) = \frac{1}{\varepsilon}\int_0^x Q_i(s) \exp\Big(-\int_s^x \varepsilon^{-1}a_i(r)\,dr\Big)\,ds. \tag{13}$$

Now the estimates (3), (12) and an integration imply (10). □

The function $F(x, \mathbf{u})$ is continuous in domain the $I \times [-l, l]^K$, therefore $F(x, \mathbf{u})$ is bounded in the considered domain. It follows, that as in the case of a linear equation [2], the estimates on the derivatives hold:

$$|u_i^{(j)}(x)| \le C_3\Big[\frac{1}{\varepsilon^j}\exp(-\alpha_i\varepsilon^{-1}x) + 1\Big],\ 1 \le j \le 4. \tag{14}$$

Therefore the solution of problem (1)-(4) on each component has a boundary layer near the boundary $x = 0$. We must to take into account this fact at the difference scheme construction.

3 Difference Scheme

Introduce the uniform mesh on the interval $[0, 1]$:

$$\Omega_h = \{x_k = kh, k = 0, 1, \ldots, N,\ x_0 = 0, x_N = 1\}.$$

We generalize the Il'in difference scheme [3] for the system (1)-(2):

$$L_k^h\mathbf{y}^h = -\varepsilon_k^h\lambda_{xx}^h\mathbf{y}_k^h - \mathbf{a}(x_k)\lambda_x^h\mathbf{y}_k^h + \mathbf{F}(x_k, \mathbf{y}_k^h) = 0,\ 1 < k < N;\ \mathbf{y}_0^h = \mathbf{A},\ \mathbf{y}_N^h = \mathbf{B}, \tag{15}$$

where

$$\varepsilon_k^h = \frac{\mathbf{a}(x_k)h}{2}\mathrm{cth}\frac{\mathbf{a}(x_k)h}{2\varepsilon},,\ \lambda_x^h\mathbf{y}_k^h = \frac{\mathbf{y}_{k+1}^h - \mathbf{y}_{k-1}^h}{2h},\ \lambda_{xx}^h\mathbf{y}_k^h = \frac{\mathbf{y}_{k+1}^h - 2\mathbf{y}_k^h + \mathbf{y}_{k-1}^h}{h^2}.$$

Theorem 1. *There exists a constant, C_0 , such that*

$$||\mathbf{y}^h - [\mathbf{u}]_{\Omega_h}||_{K\Omega_h} \le C_0 h. \tag{16}$$

Proof. For any node x_k :

$$L_k^h \mathbf{y}^h - L_k^h[\mathbf{u}]_{\Omega_h} = \mathbf{Q}_k,\ \mathbf{Q}_k = \varepsilon_k^h \lambda_{xx}^h \mathbf{u}_k - \varepsilon \mathbf{u}_k'' + \mathbf{a}(x_k)(\lambda_x^h \mathbf{u}_k - \mathbf{u}_k').$$

Let $\mathbf{z}^h = \mathbf{y}^h - [\mathbf{u}]_{\Omega_h}$. Then for all $1 \leq k < N$ there exists a vector $\mathbf{s}_k^h$: , such that

$$T_k^h \mathbf{z}^h = -\varepsilon_k^h \lambda_{xx}^h \mathbf{z}_k^h - \mathbf{a}(x_k)\lambda_x^h \mathbf{z}_k^h + J(x_k, \mathbf{s}_k^h)\mathbf{z}_k^h = \mathbf{Q}_k,$$

where T^h is a linear operator. We use the estimates (14) on derivatives and the approach [4], where Il'in scheme for a linear problem was considered, to obtain the estimate.

$$|Q_k^j| \leq C_4 h \Big[\frac{1}{h+\varepsilon} \exp\Big(- \alpha_j x_{k-1}/(2\varepsilon)\Big) + 1\Big],\ \ 0 < k < N,\ 1 \leq j \leq K.$$

Let introduce the barrier grid vector function $\mathbf{\Psi}^h$ with components:

$$\Psi_k^j = C_5 h \Big[\exp\Big(- \alpha_0 x_{k-1}/(2\varepsilon)\Big) + 1 - x_k\Big] \pm z_k^j \ \ 1 \leq j \leq K.$$

We can choose the constant C_5, such that

$$\mathbf{\Psi}_0^h \geq \mathbf{0},\ \mathbf{\Psi}_N^h \geq \mathbf{0},\ T_k^h \mathbf{\Psi}^h \geq \mathbf{0},\ k = 1, 2, \ldots, N-1.$$

It follows from the maximum principle, that $\mathbf{\Psi}_k^h \geq \mathbf{0},\ \ k = 0, 1, \ldots, N$, which implies (16) . □

4 Newton Method

Newton method on a fine mesh. To compute the solution of the scheme (15) we apply Newton method:

$$-\varepsilon_k^h \lambda_{xx}^h \mathbf{y}_k^{m+1} - \mathbf{a}(x_k)\lambda_x^h \mathbf{y}_k^{m+1} + \mathbf{f}(x_k, \mathbf{y}_k^m) + J(x_k, \mathbf{y}_k^m)(\mathbf{y}_k^{m+1} - \mathbf{y}_k^m) = 0,$$

$$0 < k < N,\ \ \mathbf{y}_0^{m+1} = \mathbf{A},\ \mathbf{y}_N^{m+1} = \mathbf{B}. \tag{17}$$

Let $\mathbf{y}^0$ be the initial guess and $||\mathbf{y}^0 - \mathbf{y}^h||_{K,\Omega_h} \leq \delta$.

We will investigate the convergence of method (17). Let $\mathbf{z}^m = \mathbf{y}^m - \mathbf{y}^h$. Then:

$$-\varepsilon_k^h \lambda_{xx}^h \mathbf{z}_k^{m+1} - \mathbf{a}(x_k)\lambda_x^h \mathbf{z}_k^{m+1} + J(x_k, \mathbf{y}_k^m)\mathbf{y}_k^{m+1} = \mathbf{Q}_k^m,\ 1 < k < N,$$

$$\mathbf{z}_0^{m+1} = \mathbf{0},\ \mathbf{z}_N^{m+1} = \mathbf{0}, \tag{18}$$

where

$$\mathbf{Q}_k^m = [J(x_k, \mathbf{y}_k^h) - J(x_k, \mathbf{s}_k^m)]\mathbf{z}_k^m,\ \ \mathbf{s}^m \in [\mathbf{y}^m, \mathbf{y}^h]. \tag{19}$$

Now we use the introduced in the notation matrix norm to get:

$$||\mathbf{Q}_k^m||_\infty \leq \max_j \sum_{k=1}^{K} \sum_{m=1}^{K} \max |F_{u_k u_m}^j(x, \mathbf{w})| \times ||\mathbf{z}_k^m||_\infty^2,\ \ x \in I, \mathbf{w} \in [\mathbf{y}^m, \mathbf{y}^h].$$

Let

$$|F^j_{u_k u_m}(x,\mathbf{w})| \le D,\ x \in I, \mathbf{w} \in [\mathbf{y}^0, \mathbf{y}^h].$$

Then

$$||\mathbf{Q}^m_k||_\infty \le DK^2||\mathbf{z}^m_k||^2_\infty.$$

An application of the maximum principle to the linear problem (18) implies:

$$||\mathbf{y}^{m+1} - \mathbf{y}^h||_{K,\Omega_h} \le \alpha_0^{-1} DK^2 ||\mathbf{y}^m - \mathbf{y}^h||^2_{K,\Omega_h}. \tag{20}$$

Hence, the Newton method will converge, if

$$\alpha_0^{-1} DK^2 \delta < 1. \tag{21}$$

Using (20), one can show , that

$$||\mathbf{y}^m - \mathbf{y}^h||_{K,\Omega_h} \le \alpha_0 D^{-1} K^{-2} (\alpha_0^{-1} DK^2 \delta)^{2^m},\quad m \ge 0. \tag{22}$$

According to Theorem 1, the scheme (15) has accuracy $O(h)$. So, we have to do a number of iterations (17) to achieve the accuracy:

$$||\mathbf{y}^m - \mathbf{y}^h||_{K,\Omega_h} \le h.$$

Let m_h be this necessary number of iterations. Taking into account (22), we find:

$$m_h \ge \log_2 \frac{\ln(\alpha_0^{-1} DK^2 h)}{\ln(\alpha_0^{-1} DK^2 \delta)} \ge \log_2 \frac{\ln(h)}{\ln(\delta)}. \tag{23}$$

Let us estimate the number of the arithmetical operations. Suppose that on each iteration of method (17) we perform about $d\,N$ operations.Then for m_h iterations we need approximately

$$N_h \approx d\,N \log_2 \frac{\ln(\alpha_0^{-1} DK^2 h)}{\ln(\alpha_0^{-1} DK^2 \delta)} \approx d\,N \log_2 \frac{\ln(h)}{\ln(\alpha_0^{-1} DK^2 \delta)}.$$

operations. To reduce the number of operations let consider the two-grid Newton method.

Two-grid Newton algorithm. We introduce the coarse grid

$$\Omega_H = \{X_k = kH, k = 0, 1, \ldots, n,\ X_0 = 0, X_n = 1\}.$$

and write the Il'in scheme on Ω_H :

$$-\varepsilon^H_k \lambda^H_{xx} \mathbf{y}^H_k - \mathbf{a}(x_k) \lambda^H_x \mathbf{y}^H_k + \mathbf{F}(x_k, \mathbf{y}^H_k) = 0,\ 1 < k < n;\ \mathbf{y}^H_0 = \mathbf{A},\ \mathbf{y}^H_n = \mathbf{B}. \tag{24}$$

According to Theorem 1 we have

$$||\mathbf{y}^H - [\mathbf{u}]_{\Omega_H}||_{K\Omega_H} \le C_0 H.$$

To compute the solution of (24), we use Newton method:

$$-\varepsilon_k^H \lambda_{xx}^H \mathbf{y}_k^{m+1} - \mathbf{a}(x_k)\lambda_x^H \mathbf{y}_k^{m+1} + \mathbf{f}(x_k, \mathbf{y}_k^m) + J(x_k, \mathbf{y}_k^m)(\mathbf{y}_k^{m+1} - \mathbf{y}_k^m) = 0,$$

$$0 < i < n, \quad \mathbf{y}_0^{m+1} = \mathbf{A}, \ \mathbf{y}_n^{m+1} = \mathbf{B}. \tag{25}$$

We have to do a number of iterations (25) to achieve the accuracy:

$$||\mathbf{y}^m - \mathbf{y}^H||_{K,\Omega_H} \le H.$$

If m_H is the necessary number of iterations, in analogous way as (23) we obtain the estimate

$$m_H \ge \log_2 \frac{\ln(\alpha_0^{-1} D K^2 H)}{\ln(\alpha_0^{-1} D K^2 \delta)} \ge \log_2 \frac{\ln(H)}{\ln(\delta)}.$$

Let $\mathbf{y}^{m_H}$ be an approximate solution of scheme (24). Now we investigate, how to interpolate the solution $\mathbf{y}^{m_H}$ from nodes of coarse grid Ω_H to nodes of fine grid Ω_h.

We discuss this question of interpolation on the solution $\mathbf{u}(x)$ of problem (1)-(4). It was shown in [9] for a function with a boundary layer component, that the method of linear interpolation on uniform mesh leads to significant errors. In [9] is used the method of exponential interpolation for the solution of a scalar problem with exponential boundary layer. We use this approach for each component $u_i(x)$ of the vector function $\mathbf{u}(x)$:

$$Int([u_i]_{\Omega_H}, x) = (u_i(X_k) - u_i(X_{k-1})) \frac{\exp(-a_0\varepsilon^{-1}x) - \exp(-a_0\varepsilon^{-1}X_k)}{\exp(-a_0\varepsilon^{-1}X_k) - \exp(-a_0\varepsilon^{-1}X_{k-1})} + u_i(X_k) \tag{26}$$

for $x \in [X_{k-1}, X_k], \ \ X_k \in \Omega_H$, where $[u_i]_{\Omega_H}$ is the projection of $u_i(x)$ on grid Ω_H.

According to [9], if for the function $p_i'(x)$ the estimate (10) holds, then

$$|u_i(x) - Int([u_i]_{\Omega_H}, x)| \le 2C_1 H, \ \ x \in I.$$

Formula (26) is stable to perturbation of $[u_i]_{\Omega_H}$,

$$||y_i^{m_H} - [u_i]_{\Omega_H}||_\infty \le (C_0 + 1)H,$$

hence there exists a constant C_4 , such that

$$|u_i(x) - Int(y_i^{m_H}, x)| \le C_4 H, \quad x \in I.$$

Let

$$\mathbf{u}_H^I = \{Int(y_i^{m_H}, x), \ i = 1, 2, \ldots, K\}, \quad \mathbf{y}_H^I = [\mathbf{u}_H^I]_{\Omega_h}.$$

We estimate $||\mathbf{y}_H^I - \mathbf{y}^h||_{K,\Omega_h}$ as follows

$$||\mathbf{y}_H^I - \mathbf{y}^h||_{K,\Omega_h} \le ||\mathbf{y}_H^I - [\mathbf{u}]_{\Omega_h}||_{K,\Omega_h} + ||[\mathbf{u}]_{\Omega_h} - \mathbf{y}^h||_{K,\Omega_h} \le C_4 H + C_0 h.$$

So, using iterations on the coarse grid and exponential interpolation for each solution component, we got an initial guess $\mathbf{y}_H^I$ for the Newton method (17) on a fine grid with accuracy $O(H)$. Then we perform iterations (17) to find $\mathbf{y}^h$ with accuracy $O(h)$.

Let us account the number of arithmetical operations for the two-grid method:

$$N_{hH} \approx d\,n \log_2 \frac{\ln(\alpha_0^{-1} DK^2 H)}{\ln(\alpha_0^{-1} DK^2 \delta)} + d\,N \log_2 \frac{\ln(\alpha_0^{-1} DK^2 h)}{\ln(\alpha_0^{-1} DK^2 H)} + I_H,$$

where I_H is number of operations for an interpolation. Suppose, that $H \ll \alpha_0^{-1} DK^2$. Then

$$N_h \approx d\,N \log_2 \frac{\ln(h)}{\ln(\alpha_0^{-1} DK^2 \delta)}, \quad N_{hH} \approx d\,n \log_2 \frac{\ln(H)}{\ln(\alpha_0^{-1} DK^2 \delta)} + d\,N \log_2 \frac{\ln(h)}{\ln(H)} + I_H.$$

and the account economy of operations is as follows:

$$N_h - N_{hH} \approx d(N-n) \log_2 \frac{\ln(H)}{\ln(\alpha_0^{-1} DK^2 \delta)} - I_H.$$

Consider the case $h = H^2$. According to (20) we have to do only one iteration (17) on a fine grid to find the solution of scheme (15) with accuracy $O(h)$. In the case $h = H^4$ we need two iterations (17) to find $\mathbf{y}^h$ with accuracy $O(h)$.

Acknowledgement

Supported by National Fund of Bulgaria under project HS-MI-106/2005 and Russian Foundation for Basic Research under Grant 07-01-00729.

References

1. Angelova, I.T., Vulkov, L.G.: Comparison of the two-grid method for singularly perturbed reaction-diffusion problems on different meshes. Amer. Inst. of Phys. CP (in press)
2. Axelsson, O.: On mesh independence and Newton methods. Appl. of Mathematics 4-5(38), 249–265 (1993)
3. Il'in, A.M.: A difference scheme for a differential eqution with a small parameter affecting the highest derivative. Mat. Zametki. 6, 237–248 (1969) (in Russian)
4. Kellogg, R.B., Tsan, A.: Analysis of some difference approximations for a singular perturbation problems without turning pointd. Math. Comput. 32(144), 1025–1039 (1978)
5. Roos, H.-G., Stynes, M., Tobiska, L.: Numerical Methods for Singularly Perturbed Differential Equations. Convection-Diffusion and Flow problems. Springer, Berlin (2008)
6. Vulkov, L.G., Zadorin, A.I.: Two-grid interpolation algorithms for difference schemes of exponential type for semilinear diffusion convection-dominated equations. Amer. Inst. of Phys. CP (in press)
7. Xu, J.: A novel two-grid method for semilinear elliptic equations. SIAM J. Sci. Comput. 15(1), 231–237 (1994)
8. Zadorin, A.I.: Numerical solution of a boundary value problem for a set of equations with a small parameter. Comp. Math. and Math. Phys. 38(8), 1255–1265 (1998)
9. Zadorin, A.I.: Method of interpolation for a boundary layer problem. Sib. J. of Numer. Math. 10(3), 267–275 (2007)

Numerical Quadrature for Bessel Transformations with High Oscillations

Shuhuang Xiang

Central South University, Changsha, Hunan 410083, China
xiangsh@mail.csu.edu.cn

Abstract. We explore higher order numerical quadrature for the integration of systems containing Bessel functions such as $\int_a^b f(x)J_\nu(rx)dx$ and $\int_a^b f(x)\cos(r_1x)J_\nu(rx)dx$. The decay of the error of the these methods drastically improves as frequency grows.

1 Introduction

The integration of systems containing Bessel functions (Bessel transformations)

$$I(f) = \int_a^b f(x)J_\nu(rx)dx, \qquad I(f) = \int_a^b f(x)\cos(r_1x)J_\nu(rx)dx \tag{1}$$

is a central point in many practical problems in physics, chemistry and engineering, where $J_\nu(rx)$ is the Bessel function of first kind and of order ν, ν is a real number and r is a positive number. In most of the cases, these transformations cannot be done analytically and one has to rely on numerical methods. For large r_1 or r, the integrand becomes highly oscillatory and thereby presents serious difficulties in obtaining numerical convergence of the integration. Levin [10] presented a collocation method.

Levin's collocation method [10]: Given a general class of highly oscillatory integrals of the form

$$I(F) = \int_a^b \sum_{k=1}^m f_k(x)w_k(r,x)dx \equiv \int_a^b F(x)\cdot W(r,x)dx, \tag{2}$$

where $F(x) = (f_1(x), f_2(x), \cdots, f_m(x))^T$ is an m-vector of non-highly oscillatory functions, $W(r,x) = (w_1(r,x), w_2(r,x), \cdots, w_m(r,x))^T$ is an m-vector of linearly independent highly oscillatory functions also depending on r and "$\cdot$" denotes the inner product, assume that

$$W'(r,x) = A(r,x)W(r,x), \tag{3}$$

where $A(r,x)$ is an $m \times m$ matrix of non-highly oscillatory functions. Select linearly independent basis m-vector functions $\{U_k(x)\}_{k=1}^v$ and let $P(x) = \sum_{k=1}^v a_kU_k(x)$ satisfy

$$P'(c_j) + A^T(r,c_j)P(c_j) = F(c_j), \quad j = 1, 2, \cdots, v, \tag{4}$$

S. Margenov, L.G. Vulkov, and J. Waśniewski (Eds.): NAA 2008, LNCS 5434, pp. 588–595, 2009.

at nodes $c_1, c_2, \cdots, c_v$ and calculate the approximation

$$Q^L(F) = \int_a^b (P'(x) + A^T(r,x)P(x)) \cdot W(r,x)dx = P(b) \cdot W(r,b) - P(a) \cdot W(r,a). \quad (5)$$

Here $A^T(r,x)$ denotes the transpose of $A(r,x)$. The collocation method is applicable to a wide class oscillatory integrals with weight functions $W(r,x)$ satisfying certain differential conditions. It is efficient for computing integrals of the form

$$\int_a^b f(x)J_\nu(rx)dx, \qquad \int_a^b f(x)\cos(r_1 x)J_\nu(rx)dx, \quad (6)$$

by choosing respectively

$$F(x) = (0, f(x))^T, \qquad W(r,x) = (J_{\nu-1}(rx), J_\nu(rx))^T,$$

and

$$F(x) = (0, f(x))^T, \qquad W(r,x) = (e^{ir_1 x}J_{\nu-1}(rx), e^{ir_1 x}J_\nu(rx))^T,$$

where $A(r, r_1, x)$ is defined sa follows

$$A(r, r_1, x) = \begin{pmatrix} ir_1 + \frac{m-1}{x} & -r \\ r & ir_1 - \frac{m}{x} \end{pmatrix}.$$

Generalized quadrature rule [2,5,6]: Product integration rules are one of the most widely used quadrature methods for dealing with integrals with singular or oscillatory factors. More often than not a formula of the form

$$I(f) = \int_a^b f(x)w(x)dx = \sum_{k=1}^{v} w_k f(c_k) + E(f) = Q(f) + E(f)$$

is sought in which the error $E(f)$ if forced to be zero for the functions $1, x, x^2, \ldots, x^{v-1}$, where $Q(f) = \sum_{k=1}^{v} w_k f(c_k)$.

Generalized quadrature methods, developed by Evans and Webster [5], Chung, Evans and Webster [2] and Evans and Chung [6], generate quadrature rules for more general irregular oscillatory weight functions $w(x)$, based on Lagrange's identity

$$z\mathcal{L}w - w\mathcal{M}z = Z'(w,z) \quad (7)$$

by choosing the weight function $w(x)$ to satisfy $\mathcal{L}w = 0$, where $\mathcal{L}$ is a differential operator and $\mathcal{M}$ is the adjoint operator of $\mathcal{L}$. Integrating both sides of (7) gives

$$\int_a^b w\mathcal{M}zdx = -Z(w,z)|_a^b = \sum_{k=1}^{v} w_k f(c_k) + E(f), \quad (8)$$

and again in (8) the w_k are chosen such that (8) is made exact for a choice of trial functions $f_k(x) = \mathcal{M}z_k(x)$ to yield self-generated quadrature rules for a given weight function $w(x)$. An obvious choice is to fix the abscissae as equally spaced or Chebyshev points shifted to the approximate interval $[a, b]$, and $z_k(x) = x^{k-1}$ for $k = 1, 2, \ldots, v$.

For the trigonometric and Bessel functions, $\mathcal{L}$, $\mathcal{M}$ and Z can be determined as follows ([5,6]).

- For $\int_a^b f(x)J_m(rq(x))dx$ and $Q(f) = \sum_{k=1}^{v} w_k f(c_k)$:

$$\mathcal{L}w \equiv \left(\frac{qw'}{q'}\right)' + q'\left(r^2 q - \frac{m^2}{q}\right) w = 0, \mathcal{M}z = \mathcal{L}z, Z(w, z) = \frac{q(w'z - z'w)}{q'}. \tag{9}$$

- For $\int_a^b f(x)e^{ir_1 q_1(x)} J_m(rq_2(x))dx$ and $Q(f) = \sum_{k=1}^{v} w_k f(c_k)$:

$$\mathcal{L}w \equiv e^{-2ir_1 q_1} \frac{q_2}{q_2'} \Bigg\{ w'' + \left(\frac{q_2'}{q_2} - \frac{q_2''}{q_2'} - 2ir_1 q_1'\right) w' + \left[(rq_2')^2 \left(1 - \frac{m^2}{(rq_2)^2}\right) - ir_1 q_1' \left(\frac{q_2'}{q_2} - \frac{q_2''}{q_2'}\right) - (r_1 q_1')^2 - ir_1 q_1'' \right] w \Bigg\} = 0, \tag{10}$$

$$\mathcal{M}z = \mathcal{L}z, \quad Z(w, z) = e^{-2ir_1 q_1} \frac{q_2}{q_2'} (w'z - wz').$$

The weights w_j in the two quadrature rules are chosen so that the formulas are exact for a set of functions $f_k(x) = \mathcal{M}x^{k-1}$, $k = 1, 2, \ldots, v$ respectively.

Both these two methods are with error bounds [19,21] for $c_1 = a$ and $c_v = b$

- for $\int_a^b f(x)J_m(rx)dx (0 \notin [a, b])$, $E(f) = O\left(\frac{1}{r^{5/2}}\right)$
- for $\int_a^b f(x)\cos(r_1 x)J_m(rx)dx (0 \notin [a, b])$, $E(f) = O\left(\frac{1}{r^{1/2}(\max\{r_1, r\})^2}\right)$.

2 The Higher Order Generalized Quadrature Rule

The generalized quadrature rule can also be considered as a Filon method for the given irregular oscillatory weight functions $w(x)$. The generalized moment $I_k = \int_a^b w(x)\mathcal{M}z_k(x)dx = -[Z(w, z_k)]_a^b$ can be computed explicitly by Lagrange's identity. Let $f_k(x) = \mathcal{M}x^{k-1} (k = 1, \ldots, n)$. Then $\{f_k(x)\}$ is a sequence of linear independent basis functions.

Based on [7,8,18], a higher order generalized quadrature method for Bessel - trigonometric transformations [21]: Let s be some positive integer and let $\{m_k\}_1^v$ be a set of multiplicities associated with the node points $a = c_1 < \cdots < c_v = b$ such that $m_1, m_v \geq s$. Suppose that $v(x) = \sum_{k=0}^{n} b_k f_k(x)$ is the solution of the system of equations

$$v(c_j) = f(c_j), v'(c_j) = f'(c_j), \ldots, v^{(m_j - 1)}(c_j) = f^{(m_j - 1)}(c_j) \tag{11}$$

for every integer $1 \le j \le v$, , where $n = \sum_{k=1}^{v} m_k - 1$. The higher order generalized quadrature method is given by

$$Q_s^E(f) = \sum_{k=1}^{n} b_k I_k. \tag{12}$$

Theorem 1. *Suppose that $q_1'(x) \neq 0$, $q_2(x) \neq 0$ and $q_2'(x) \neq 0$ for all $x \in [a,b]$. If r_1 and r satisfy that $r \ge \alpha_1 \tau_1 r_1$ or $\alpha_1 r \le \tau_2 r_1$ for some constant $\alpha_1 > 1$. Then the error by the new generalized quadrature method approximating $\int_a^b f(x)\cos(r_1 q_1(x)) J_m(r q_2(x))dx$ is given by*

$$E(f) = |Q_s^E(f) - I(f)| = O\left(\frac{1}{r^{1/2}\max\{r_1, r\}^{s+1}}\right), \quad \max\{r_1, r\} \gg 1. \tag{13}$$

3 The Higher Order Asymptotic Method

Iserles and Nørsett [7,8] presented an asymptotic method for $\int_a^b f(x)e^{i\omega g(x)}dx$. The asymptotic method for m-vector function integral $I(F) = \int_a^b F(x) \cdot W(r,x)dx$ is based on the following asymptotic series [20]:

Assume that $W'(r,x) = A(r,x)W(r,x)$, where $A(r,x)$ is a nonsingular $m \times m$ matrix, and $B(r,x) = \left(\frac{1}{c(r)}A(r,x)\right)^{-1}$. Suppose that $W(r,x)$, $B(r,x) \in C^\infty[a,b]$ for $c(r) \ge \beta > 0$. Let

$$F_1(x) = F(x),$$

$$F_{k+1}(x) = (B^T(r,x)F_k(x))', k = 1, 2, \ldots.$$

The asymptotic method

$$Q_s^A(f) = \sum_{k=1}^{s} \frac{(-1)^{k+1}}{c(r)^k} \left(F_k(b) \cdot B(r,b)W(r,b) - F_k(a) \cdot B(r,a)W(r,a)\right) \tag{14}$$

is efficient for large $c(r)$.

Theorem 2. *Assume that $W'(r,x) = A(r,x)W(r,x)$, where $A(r,x)$ is a nonsingular $m \times m$ matrix, and $B(r,x) = \left(\frac{1}{c(r)}A(r,x)\right)^{-1}$. Suppose that $W(r,x)$, $B(r,x) \in C^\infty[a,b]$, $B(r,x)$ and its $s+1$ derivative are bounded uniformly for $r \gg 1$. Then*

$$E(f) = |I(f) - Q_s^A(f)| = O\left(\frac{\|W(r,x)\|_\infty}{c(r)^{s+1}}\right), \qquad r \gg 1. \tag{15}$$

Especially, for $\int_a^b f(x)\cos(r_1 x)J_\mu(rx)dx$, if there exists a positive constant α_1 with $1 < \alpha_1$ independent of r_1 and r such that $r_1 \ge \alpha_1 r$ or $r \ge \alpha_1 r_1$,

$$E(f) = |Q_s^A(f) - I(f)| = O\left(\frac{1}{r^{1/2}(\max\{r_1, r\})^{s+1}}\right), \quad \max\{r_1, r\} \gg 1. \tag{16}$$

4 The Levin-Type Method

The Levin-type method for

$$I(F)=\int_a^b \sum_{k=1}^{m} f_k(x)w_k(r,x)dx \equiv \int_a^b F(x)\cdot W(r,x)dx, \tag{17}$$

can be stated as:

Levin-type method for m-vector functions [20]: Let s be some positive integer and let $\{m_k\}_1^v$ be a set of multiplicities associated with the node points $a=c_1<\cdots<c_v=b$ such that $m_1,m_v\geq s$, and $\{\psi_k(x)\}_{k=0}^n$ be a set of linearly independent basis functions, where $n=\sum_{j=1}^v m_j-1$. Let $P(x)=(\sum_{k=0}^n a_k^{(1)}\psi_k(x),\ldots,\sum_{k=0}^n a_k^{(m)}\psi_k(x))^T$ satisfy

$$P'(c_j)+A^T(r,c_j)P(c_j)=F(c_j),\ \ j=1,2,\cdots,v, \tag{18}$$

and

$$\left[P'(x)+A^T(r,x)P(x)\right]_{x=c_j}^{(k)}=F^{(k)}(c_j),\ \ k=1,2,\ldots,m_j-1,\ \ j=1,2,\cdots,v. \tag{19}$$

The Levin-type method is achieved by

$$Q_s^L(F)=P(b)\cdot W(r,b)-P(a)\cdot W(r,a). \tag{20}$$

Remark 1. The simplest and most obvious choice for $\{\psi_k(x)\}$ is the standard basis of polynomials [14].

Theorem 3. *Let $W'(r,x)=A(r,x)W(r,x)$, where $A(r,x)$ is a nonsingular $m\times m$ matrix, and $B(r,x)=\left(\frac{1}{c(r)}A(r,x)\right)^{-1}$ for $r\gg 1$. Assume $W(r,x),\ B(r,x)\in C^\infty[a,b]$ and*

(i) $\frac{1}{c(r)}A(r,x)$ and $A^{(k)}(r,x)$ $(k=1,2,\ldots,s_0)$ in x are uniformly bounded for $r\gg 1$ and all $x\in[a,b]$, where $s_0=\max_{1\leq j\leq v} m_j-1$;

(ii) $B(r,x)$ and its $s+1$ derivatives in x are uniformly bounded for $r\gg 1$ and all $x\in[a,b]$.

Then

$$E(F)=|I(F)-Q_s^L(F)|=O\left(\frac{\|W(r,x)\|_\infty}{c(r)^{s+1}}\right),\qquad r\gg 1. \tag{21}$$

Theorem 4. *Let $I(f)=\int_a^b f(x)\cos(r_1x)J_v(rx)dx$ and s be some positive integer and let $\{m_k\}_1^v$ be a set of multiplicities associated with the node points $a=c_1<c_2<\cdots<c_v=b$ such that $m_1,m_v\geq s$. Suppose that $P(x)=(\sum_{k=0}^n a_k^{(1)}x^k,\sum_{k=0}^n a_k^{(2)}x^k)^T$, where $n=\sum_{k=1}^v m_k-1$, is the solution of the*

system of equations (18) and (19). If there exists a positive constant α_1 independent of r_1 and r with $1 < \alpha_1$ such that $r_1 \geq \alpha_1 r$ or $r \geq \alpha_1 r_1$, then the error for approximating $I(f) = \int_a^b f(x)\cos(r_1 x)J_\mu(rx)dx$ by the Levin-type quadrature $Q_s^L(f)$ is

$$I(f) - Q_s^F(f) = O\left(\frac{1}{r^{1/2}(\max\{r_1, r\})^{s+1}}\right), \qquad \max\{r_1, r\} \gg 1. \tag{22}$$

5 Filon-Type Method for $\int_a^b f(x)J_m(rx)dx$

Let s be some positive integer and let $\{m_k\}_1^v$ be a set of multiplicities associated with the shifted Chebyshev node points

$$c_k = a + (b-a)\frac{1+\cos(\frac{(k-1)\pi}{v-1})}{2}, \quad k = 1, ..., v$$

such that $m_1, m_v \geq s$. Suppose that $p(x) = \sum_{k=0}^n a_k x^k$, where $n = \sum_{k=1}^v m_k - 1$, is the solution of the system of equations

$$p(c_k) = f(c_k), p'(c_k) = f'(c_k), \ldots, p^{(m_k-1)}(c_k) = f^{(m_k-1)}(c_k), \quad k = 1, \ldots, v. \tag{23}$$

The Filon-type quadrature $Q_s^F(f)$ is defined by

$$Q_s^F(f) = \int_a^b p(x)J_m(rx)dx = \sum_{k=0}^n a_k I[k, m, r] \tag{24}$$

where

$$I[k, m, r] = \int_a^b x^k J_m(rx)dx, k = 0, 1, \ldots, n,$$

can be computed explicitly by special functions:

- **In the case that m is a nonnegative integer and $k - m$ is an odd negative integer:**

$$I[k, m, r] = \begin{cases} \dfrac{1}{r^{k+1}}\displaystyle\sum_{j=0}^n \frac{n!2^j}{(n-j)!} r^{k-j}\left[a^{k-j}J_{m-j-1}(ra) - b^{k-j}J_{m-j-1}(rb)\right], & b > a > 0 \\ \dfrac{1}{r^{k+1}}\left[\dfrac{n!2^j}{2^{m-n-1}(m-n-1)!} - \displaystyle\sum_{j=0}^n \frac{n!2^j}{(n-j)!}(rb)^{k-j}J_{m-j-1}(rb)\right], & b > a = 0, \end{cases} \tag{25}$$

where $n = \dfrac{k+m-1}{2}$. In the case $a = 0$, the moment is well-defined by its limit at 0 due to Abramowitz and Stegun ([1], p.360)

$$\lim_{x\to 0}\frac{J_\mu(x)}{x^\nu} = \lim_{x\to 0}\frac{\sum_{j=0}^\infty (-1)^j \frac{(\frac{x}{2})^{\mu+2j}}{j!\Gamma(\mu+k+1)}}{x^\nu} = \begin{cases} 0, & 0 \leq \nu < \mu \\ \dfrac{1}{2^\nu \Gamma(\nu+1)}, & \mu = \nu \geq 0, \end{cases}$$

where $\Gamma(\cdot)$ denotes the Gamma function.

– **In the other cases:** $I[k,m,r]$ can be calculated explicitly based on the Bessel functions $J_\nu(r)$ and first kind of Lommel functions denoted by *Lommel* $S1(\mu,\nu,r)$ ([Luke, p.22, p.85])

$$\begin{aligned}
&I[k,m,r]\\
&=\frac{k+m-1}{r^k}\left[bJ_m(rb)LommelS1(k-1,m-1,rb)\right.\\
&\left.-aJ_m(ra)LommelS1(k-1,m-1,ra)\right]\\
&-\frac{1}{r^k}\left[bJ_{m-1}(rb)LommelS1(k,m,rb)-aJ_{m-1}(ra)LommelS1(k,m,ra)\right].
\end{aligned} \tag{26}$$

Here $LommelS1(\mu,\nu,r)$ can be efficiently computed by MAPLE for small or moderate values of r based on its asymptotic ([Watson, p.345])

$$LommelS1(\mu,\nu,r)=r^{\mu-1}\sum_{k=0}^{\infty}\frac{(-1)^k(\frac{r}{2})^{2k+2}\Gamma(\frac{\mu-\nu+1}{2})\Gamma(\frac{\mu+\nu+1}{2})}{\Gamma(\frac{\mu-\nu+k+3}{2})\Gamma(\frac{\mu-\nu+k+3}{2})}.$$

For large values of r, $LommelS1(\mu,\nu,r)$ can be efficiently approximated by a truncated series of its asymptotic expansion ([Luke, p.74], [Watson, p.347, p.352])

$$\begin{aligned}
&LommelS1(\mu,\nu,z)\\
=\;& z^{\mu-1}\sum_{j=0}^{p-1}\frac{(-1)^j\Gamma(\frac{1}{2}-\frac{1}{2}\mu+\frac{1}{2}\nu+j)\Gamma(\frac{1}{2}-\frac{1}{2}\mu-\frac{1}{2}\nu+j)}{\left(\frac{1}{2}z\right)^{2j}\Gamma(\frac{1}{2}-\frac{1}{2}\mu+\frac{1}{2}\nu)\Gamma(\frac{1}{2}-\frac{1}{2}\mu-\frac{1}{2}\nu)}\\
&+2^{\mu-1}\Gamma(\tfrac{1}{2}+\tfrac{1}{2}\mu+\tfrac{1}{2}\nu)\Gamma(\tfrac{1}{2}+\tfrac{1}{2}\mu-\tfrac{1}{2}\nu)\cdot\left(-\sin(\tfrac{\mu-\nu}{2}\pi)J_\nu(z)+\cos(\tfrac{\mu-\nu}{2}\pi)Y_\nu(z)\right)\\
&+O(z^{\mu-2p}),
\end{aligned} \tag{27}$$

where $Y_\nu(z)$ is the second kind of Bessel function of order ν.

Theorem 5. *Let $f(x)$ be suitably smooth. Then the error for Filon-type method $Q_s^F(f)$ for $\int_a^b f(x)J_m(rx)dx$ is*

$$E[f]=I(f)-Q_s^F(f)=\begin{cases}O\left(\frac{1}{r^{s+3/2}}\right), & a>0,\\ O\left(\frac{1}{r^{s+1}}\right), & a=0\end{cases} \tag{28}$$

Acknowledgement. The project sponsored partly by NSF of China (No. 10771218) and partly by Program for New Century Excellent Talents in University, State Education Ministry, China.

References

1. Abramowitz, M., Stegun, I.A.: Handbook of Mathematical Functions. National Bureau of Standards, Washington, DC (1964)
2. Chung, K.C., Evans, G.A., Webster, J.R.: A method to generate generalized quadrature rules for oscillatory integrals. Appl. Numer. Math. 34, 85–93 (2000)
3. Clenshaw, C.W., Curtis, A.R.: A method for numerical integration on an automatic computer. Numer. Math. 2, 197–205 (1960)

4. Davis, P.I., Rabinowitz, P.: Methods of Numerical Integral Integration, 2nd edn. Academic Press, New York (1984)
5. Evans, G.A., Webster, J.R.: A high order, progressive method for the evaluation of irregular oscillatory integrals. Appl. Numer. Math. 23, 205–218 (1997)
6. Evans, G.A., Chung, K.C.: Some theoretical aspects of generalised quadrature methods. J. Complexity 19, 272–285 (2003)
7. Iserles, A., Nørsett, S.P.: On quadrature methods for highly oscillatory integrals and their implementation. BIT 44, 755–772 (2004)
8. Iserles, A., Nørsett, S.P.: Efficient quadrature of highly-oscillatory integrals using derivatives. Proc. Royal Soc. A 461, 1383–1399 (2005)
9. Levin, D.: Procedures for computing one-and-two dimensional integrals of functions with rapid irregular oscillations. Math. Comp. 38, 531–538 (1982)
10. Levin, D.: Fast integration of rapidly oscillatory functions. J. Comp. Appl. Math. 67, 95–101 (1996)
11. Levin, D.: Analysis of a collocation method for integrating rapidly oscillatory functions. J. Comp. Appl. Math. 78, 131–138 (1997)
12. Luke, Y.K.: On the computation of oscillatory integrals. Proc. Cambridge Philos. Soc. 50, 269–277 (1954)
13. Olver, F.W.J.: Asymptotics and Special Function. Academic Press, New York (1974)
14. Olver, S.: Moment-free numerical integration of highly oscillatory functions. IMA J. Numer. Anal. 26, 213–227 (2006)
15. Olver, S.: Numerical approximation of vector-valued highly oscillatory integrals. BIT 47, 637–655 (2007)
16. Piessens, R., Branders, M.: Modified Clenshaw-Curtis method for the computation of Bessel function integrals. BIT 23, 370–381 (1983)
17. Trefethen, L.N.: Is Gauss quadrature better than Clenshaw-Curtis. SIAM Review 50, 67–87 (2008)
18. Xiang, S.: Efficient Filon-type methods for $\int_a^b f(x)e^{i\omega g(x)}dx$. Numer. Math. 106, 633–658 (2007)
19. Xiang, S.: Numerical Analysis on Fast Integration of Highly Oscillatory Functions. BIT 47, 469–482 (2007)
20. Xiang, S., Gui, W., Mo, P.: Numerical quadrature for Bessel transformations. Appl. Numer. Math. (2007) doi: 10.1016/j.apnum.2007.07.002
21. Xiang, S., Gui, W.: On generalized quadrature rules for fast oscillatory integrals. Appl. Math. Comp. 197, 60–75 (2008)

Symbolic Computation of an Exact Solution of the Cauchy Problem for the System of Crystal Optics with Polynomial Data

Valery Yakhno[1] and Meltem Altunkaynak[2]

[1] Dokuz Eylul University, Electrical and Electronics Engineering Department, Buca, 35160, Izmir, Turkey
valery.yakhno@deu.edu.tr

[2] Dokuz Eylul University, Department of Mathematics, Buca, 35160, Izmir, Turkey
meltem.topcuoglu@deu.edu.tr

Abstract. A new method for computing explicit formulae of exact solutions of the Cauchy problem for the time-dependent hyperbolic system of crystal optics with polynomials entries is given. This method is based on symbolic computations of all Taylor coefficients for a solution of the Cauchy problem using the initial data and inhomogeneous term which have polynomial presentation with respect to space variables. Computing explicit formulae of solutions is implemented by Maple 10. A computational example is presented.

Keywords: Analytical method, symbolic computations, polynomial solutions, system of crystal optics, initial value problem.

1 Introduction

Modeling and simulation electric waves in anisotropic crystals have great recent and historical interest. The dynamics of electric wave propagations in anisotropic crystals is described by the partial differential equation system [1] called the system of crystal optics. This system is hyperbolic and it can be obtained from the Maxwell system (see, for example, [1], pages 603-612). Different methods have been used to study problems for the system of crystal optics. A solution of the initial value problem with smooth data has been obtained by Courant [1] using the reduction of the system to a scalar partial differential equation of the order four. Analytical methods of Green's function constructions have been studied for isotropic materials in [2,3]; for uniaxial anisotropic crystals in [4,5]; for biaxial anisotropic crystals in [6]. Burridge and Qian in [7] have obtained a formula for the fundamental solution of the system of crystal optics for biaxial crystals by the plane-wave approach. Yakhno in [8] has found an explicit formula for the Green's function by symbolic transformations in MATLAB. The obtained formula has been successfully used in [9] for the simulation of electric radiation in anisotropic crystals from a pulse dipole. Most of the methods for modelling

S. Margenov, L.G. Vulkov, and J. Waśniewski (Eds.): NAA 2008, LNCS 5434, pp. 596–603, 2009.

electromagnetic waves had been made by finite element and finite difference methods [10,11,12,13,14].

In the presented paper a new analytic method for computing explicit formulae of exact solutions of the Cauchy problem for the time-dependent system of crystal optics with polynomials entries is described. This method is based on symbolic computations of all Taylor coefficients for a solution of the Cauchy problem using the initial data and inhomogeneous term which have polynomial presentation with respect to space variables.

1.1 Problem Set Up

Let $x = (x_1, x_2, x_3) \in \mathbf{R}^3$, $t \in \mathbf{R}$; $\mathcal{E} = (\varepsilon_{ij})_{3\times 3}$ be a given symmetric, positive definite matrix with constant elements; $\mathbf{e}(x) = (e_1(x), e_2(x), e_3(x))$, $\mathbf{g}(x) = (g_1(x), g_2(x), g_3(x))$ be given smooth vector functions with components depending on $x \in \mathbf{R}^3$; $\mathbf{f}(x,t) = (f_1(x,t), f_2(x,t), f_3(x,t))$ be a given smooth vector function with components depending on $x \in \mathbf{R}^3$ and $t \geq 0$. The following initial value problem (IVP) is the main problem of the paper. Find a vector function $\mathbf{E}(x,t) = (E_1(x,t), E_2(x,t), E_3(x,t))$ satisfying

$$\mathcal{E}\frac{\partial^2 \mathbf{E}}{\partial t^2} + curl_x(curl_x \mathbf{E}) = \mathbf{f}(x,t), \quad x \in \mathbf{R}^3,\, t > 0, \tag{1}$$

$$\mathbf{E}(x,0) = \mathbf{e}(x), \quad \left.\frac{\partial \mathbf{E}(x,t)}{\partial t}\right|_{t=0} = \mathbf{g}(x), \quad x \in R^3. \tag{2}$$

1.2 Formula of a Solution When Data Are Zero Degree Polynomials

Let initial data and inhomogeneous term have the form of zero degree polynomials, i.e.

$$\mathbf{e}(x) = \mathbf{e}^0, \quad \mathbf{g}(x) = \mathbf{g}^0, \quad \mathbf{f}(x,t) = \mathbf{f}^0(t), \tag{3}$$

where $\mathbf{e}^0$, $\mathbf{g}^0$ are vectors with constants components, $\mathbf{f}^0(t)$ is a vector function which components depend on t only.

We find a solution of the problem (1), (2) in the form of the zero degree polynomial with respect to $(x_1, x_2, x_3) \in \mathbf{R}^3$:

$$\mathbf{E} = (E_1, E_2, E_3), \quad E_k = E_k(t),\, k = 1,2,3.$$

Remark 1. The fact that the solution (1), (2) under assumption (3) does not depend on space variable $x \in \mathbf{R}^3$ is natural. Really, taking $j = 1,2,3$ and differentiating (1), (2) with respect to x_j and denoting $\frac{\partial \mathbf{E}}{\partial x_j} = \mathbf{W}^j$ we find

$$\mathcal{E}\frac{\partial^2 \mathbf{W}^j}{\partial t^2} + curl_x(curl_x \mathbf{W}^j) = 0, \quad x \in R^3,\, t > 0, \tag{4}$$

$$\mathbf{W}^j(x,0) = 0, \quad \left.\frac{\partial \mathbf{W}^j(x,t)}{\partial t}\right|_{t=0} = 0, \quad x \in R^3. \tag{5}$$

The system (4) is hyperbolic [15]. Let us consider an arbitrary conoid of the dependence [1]. Applying energy inequalities in the conoid of the dependence (the uniqueness theorem) [1] we find $\mathbf{W}^j(x,t) \equiv 0$ inside of the conoid. Since the conoid is taking arbitrary then $\mathbf{W}^j(x,t) \equiv 0$ for $\mathbf{R}^3$, $t > 0$. This means that a differentiable solution of (1), (2) with data (3) does not depend on the space variables.

As a result of it the problem (1), (2) can be written as

$$\mathcal{E}\frac{\partial^2 \mathbf{E}}{\partial t^2} = \mathbf{f}^0(t), \quad t > 0, \tag{6}$$

$$\mathbf{E}\big|_{t=0} = \mathbf{e}^0, \quad \frac{\partial \mathbf{E}}{\partial t}\bigg|_{t=0} = \mathbf{g}^0. \tag{7}$$

Integrating (6), (7) we find the following formula for the solution of (1), (2)

$$\mathbf{E}(t) = \mathbf{e}^0 + t\mathbf{g}^0 + \mathcal{E}^{-1}\int_0^t (t-\tau)\mathbf{f}^0(\tau)d\tau, \tag{8}$$

where $\mathcal{E}^{-1}$ is the matrix inverse to $\mathcal{E}$.

2 Constructing an Explicit Formula of a Solution When Data Are p-Degree Polynomials

We assume that initial data $\mathbf{e} = (e_1, e_2, e_3)$, $\mathbf{g} = (g_1, g_2, g_3)$ and inhomogeneous term $\mathbf{f} = (f_1, f_2, f_3)$ have the following polynomial forms:

$$e_l(x) = \sum_{k=0}^{p}\sum_{m=0}^{p}\sum_{n=0}^{p} e_l^{k,m,n} x_1^k x_2^m x_3^n, \tag{9}$$

$$g_l(x) = \sum_{k=0}^{p}\sum_{m=0}^{p}\sum_{n=0}^{p} g_l^{k,m,n} x_1^k x_2^m x_3^n, \tag{10}$$

$$f_l(x,t) = \sum_{k=0}^{p}\sum_{m=0}^{p}\sum_{n=0}^{p} f_l^{k,m,n}(t) x_1^k x_2^m x_3^n, \tag{11}$$

where $l = 1,2,3$; p is a given nonnegative integer; $e_l^{k,m,n}$, $g_l^{k,m,n}$ are given real numbers; $f_l^{k,m,n}(t)$ are given continuously differentiable functions of t; $l = 1,2,3$.

We find the components of a solution $\mathbf{E}(x,t) = (E_1(x,t), E_2(x,t), E_3(x,t))$ of IVP (1), (2) in the form

$$E_l(x_1,x_2,x_3,t) = \sum_{k=0}^{p}\sum_{m=0}^{p}\sum_{n=0}^{p} E_l^{k,m,n}(t) x_1^k x_2^m x_3^n. \tag{12}$$

In this Section we obtain the recurrence relations for undetermined coefficients $E_l^{k,m,n}(t)$ and then, we describe a procedure of their successive recovery.

2.1 Recurrence Relations for $\mathbf{E}^{k,m,n}$

We note that for the symmetric, positive definite matrix $\mathcal{E}$ there exists an orthogonal matrix $\mathcal{S} = (S_{ij})_{3\times 3}$ such that $\mathcal{S}^T\mathcal{E}\mathcal{S} = \mathcal{D}$, where $\mathcal{D}$ is a diagonal matrix with nonnegative diagonal entries that are eigenvalues of $\mathcal{E}$; $\mathcal{S}^T$ is transpose to $\mathcal{S}$.

Letting $\mathbf{E} = \mathcal{S}\tilde{\mathbf{E}}$ and substituting this into (1), (2) and multiplying with $\mathcal{S}^T$ from left hand side we obtain

$$\mathcal{D}\frac{\partial^2 \tilde{\mathbf{E}}}{\partial t^2} + \mathcal{S}^T curl_x(curl_x \tilde{\mathbf{E}}) = \mathcal{S}^T \mathbf{f}(x,t), \quad x \in R^3,\, t > 0, \tag{13}$$

$$\tilde{\mathbf{E}}(x,0) = \mathcal{S}^T \mathbf{e}(x), \quad \left.\frac{\partial \tilde{\mathbf{E}}(x,t)}{\partial t}\right|_{t=0} = \mathcal{S}^T \mathbf{g}(x), \quad x \in R^3. \tag{14}$$

Using $\tilde{\mathbf{E}} = \mathcal{S}^T\mathbf{E}$ equation (12) may be written as follows

$$\tilde{\mathbf{E}}(x_1,x_2,x_3,t) = \sum_{k=0}^{p}\sum_{m=0}^{p}\sum_{n=0}^{p} \tilde{\mathbf{E}}^{k,m,n}(t) x_1^k x_2^m x_3^n, \tag{15}$$

where $\tilde{\mathbf{E}}^{k,m,n}(t) = \mathcal{S}^T\mathbf{E}^{k,m,n}(t)$, $\mathbf{E}^{k,m,n}(t) = (E_1^{k,m,n}(t), E_2^{k,m,n}(t), E_3^{k,m,n}(t))$.

Substituting (9), (10), (11) and (15) into (13), (14) we obtain

$$\mathcal{D}\frac{\partial^2 \tilde{\mathbf{E}}^{k,m,n}}{\partial t^2} = \mathcal{S}^T \mathbf{f}^{k,m,n}(t) - \mathbf{\Upsilon}^{k,m,n}\Big[\tilde{\mathbf{E}}^{k+2,m,n},$$
$$\tilde{\mathbf{E}}^{k,m+2,n}, \tilde{\mathbf{E}}^{k,m,n+2}, \tilde{\mathbf{E}}^{k+1,m+1,n}, \tilde{\mathbf{E}}^{k+1,m,n+1}, \tilde{\mathbf{E}}^{k,m+1,n+1}\Big](t), \tag{16}$$

$$\tilde{\mathbf{E}}^{k,m,n}(0) = \mathcal{S}^T \mathbf{e}^{k,m,n}, \tag{17}$$

$$\left.\frac{\partial \tilde{\mathbf{E}}^{k,m,n}}{\partial t}\right|_{t=0} = \mathcal{S}^T \mathbf{g}^{k,m,n}. \tag{18}$$

Here $\mathbf{e}^{k,m,n} = (e_1^{k,m,n}, e_2^{k,m,n}, e_3^{k,m,n})$, $\mathbf{g}^{k,m,n} = (g_1^{k,m,n}, g_2^{k,m,n}, g_3^{k,m,n})$, $\mathbf{f}^{k,m,n}(t) = (f_1^{k,m,n}(t), f_2^{k,m,n}(t), f_3^{k,m,n}(t))$ are known vector functions; the vector operators $\mathbf{\Upsilon}^{k,m,n}$ are defined by the following formulae

$$\mathbf{\Upsilon}^{k,m,n}\Big[\tilde{\mathbf{E}}^{k+2,m,n}, \tilde{\mathbf{E}}^{k,m+2,n}, \tilde{\mathbf{E}}^{k,m,n+2},$$
$$\tilde{\mathbf{E}}^{k+1,m+1,n}, \tilde{\mathbf{E}}^{k+1,m,n+1}, \tilde{\mathbf{E}}^{k,m+1,n+1}\Big](t) = \mathcal{S}^T \mathbf{B}^{k,m,n}(t), \tag{19}$$

where the components of the vector functions $\mathbf{B}^{k,m,n}(t)$ are defined by

$$B_1^{k,m,n} = T_2^{k,m,n} - T_4^{k,m,n}, \tag{20}$$

$$B_2^{k,m,n} = T_3^{k,m,n} - T_6^{k,m,n}, \tag{21}$$

$$B_3^{k,m,n} = T_5^{k,m,n} - T_1^{k,m,n}, \tag{22}$$

where

$$T_1^{k,m,n} = (m+2)(m+1)(S_{31}\tilde{E}_1^{k,m+2,n} + S_{32}\tilde{E}_2^{k,m+2,n} + S_{33}\tilde{E}_3^{k,m+2,n})$$
$$-(m+1)(n+1)(S_{21}\tilde{E}_1^{k,m+1,n+1} + S_{22}\tilde{E}_2^{k,m+1,n+1} + S_{23}\tilde{E}_3^{k,m+1,n+1});$$

$$T_2^{k,m,n} = (k+1)(m+1)(S_{21}\tilde{E}_1^{k+1,m+1,n} + S_{22}\tilde{E}_2^{k+1,m+1,n} + S_{23}\tilde{E}_3^{k+1,m+1,n})$$
$$-(m+2)(m+1)(S_{11}\tilde{E}_1^{k,m+2,n} + S_{12}\tilde{E}_2^{k,m+2,n} + S_{13}\tilde{E}_3^{k,m+2,n});$$

$$T_3^{k,m,n} = (m+1)(n+1)(S_{31}\tilde{E}_1^{k,m+1,n+1} + S_{32}\tilde{E}_2^{k,m+1,n+1} + S_{33}\tilde{E}_3^{k,m+1,n+1})$$
$$-(n+2)(n+1)(S_{21}\tilde{E}_1^{k,m,n+2} + S_{22}\tilde{E}_2^{k,m,n+2} + S_{23}\tilde{E}_3^{k,m,n+2});$$

$$T_4^{k,m,n} = (n+2)(n+1)(S_{11}\tilde{E}_1^{k,m,n+2} + S_{12}\tilde{E}_2^{k,m,n+2} + S_{13}\tilde{E}_3^{k,m,n+2})$$
$$-(k+1)(n+1)(S_{31}\tilde{E}_1^{k+1,m,n+1} + S_{32}\tilde{E}_2^{k+1,m,n+1} + S_{33}\tilde{E}_3^{k+1,m,n+1});$$

$$T_5^{k,m,n} = (k+1)(n+1)(S_{11}\tilde{E}_1^{k+1,m,n+1} + S_{12}\tilde{E}_2^{k+1,m,n+1} + S_{13}\tilde{E}_3^{k+1,m,n+1})$$
$$-(k+2)(k+1)(S_{31}\tilde{E}_1^{k+2,m,n} + S_{32}\tilde{E}_2^{k+2,m,n} + S_{33}\tilde{E}_3^{k+2,m,n});$$

$$T_6^{k,m,n} = (k+2)(k+1)(S_{21}\tilde{E}_1^{k+2,m,n} + S_{22}\tilde{E}_2^{k+2,m,n} + S_{23}\tilde{E}_3^{k+2,m,n})$$
$$-(k+1)(m+1)(S_{11}\tilde{E}_1^{k+1,m+1,n} + S_{12}\tilde{E}_2^{k+1,m+1,n} + S_{13}\tilde{E}_3^{k+1,m+1,n}).$$

In these expressions we assume that the components of the vector functions

$$\tilde{\mathbf{E}}^{p+2,m,n}, \quad \tilde{\mathbf{E}}^{k,p+2,n}, \quad \tilde{\mathbf{E}}^{k,m,p+2}, \quad \tilde{\mathbf{E}}^{p+1,p+1,n}, \quad \tilde{\mathbf{E}}^{p+1,m,p+1}, \quad \tilde{\mathbf{E}}^{k,p+1,p+1}$$

are equal to zero.

Equalities (16)-(18) are equivalent to the following relations:

$$\tilde{\mathbf{E}}^{k,m,n} = \tilde{\mathbf{F}}^{k,m,n}(t) - \int_0^t (t-\tau)\mathcal{D}^{-1}\boldsymbol{\Upsilon}^{k,m,n}\Big[$$
$$\tilde{\mathbf{E}}^{k+2,m,n}, \tilde{\mathbf{E}}^{k,m+2,n}, \tilde{\mathbf{E}}^{k,m,n+2}, \tilde{\mathbf{E}}^{k+1,m+1,n}, \tilde{\mathbf{E}}^{k+1,m,n+1}, \tilde{\mathbf{E}}^{k,m+1,n+1}\Big](\tau)d\tau, \tag{23}$$

where the components of the vector functions $\tilde{\mathbf{F}}^{k,m,n}(t)$ are defined by

$$F_l^{k,m,n}(t) = S_{l1}^T e_1^{k,m,n} + S_{l2}^T e_2^{k,m,n} + S_{l3}^T e_3^{k,m,n} + t\Big(S_{l1}^T g_1^{k,m,n} + S_{l2}^T g_2^{k,m,n} + S_{l3}^T g_3^{k,m,n}\Big)$$
$$+\frac{1}{d_l}\int_0^t (t-\tau)\Big(S_{l1}^T f_1^{k,m,n}(\tau) + S_{l2}^T f_2^{k,m,n}(\tau) + S_{l3}^T f_3^{k,m,n}(\tau)\Big)d\tau, \tag{24}$$

$l = 1,2,3;\ \ k = p, p-1, ..., 0;\ \ m = p, p-1, ..., 0;\ \ n = p, p-1, ..., 0.$

2.2 Procedure of Finding $\mathbf{E}^{k,m,n}$

We suppose that the components of the vector functions

$$\mathbf{e}^{k,m,n} = (e_1^{k,m,n}, e_2^{k,m,n}, e_3^{k,m,n}), \quad \mathbf{g}^{k,m,n} = (g_1^{k,m,n}, g_2^{k,m,n}, g_3^{k,m,n}),$$

$$\mathbf{f}^{k,m,n}(t) = (f_1^{k,m,n}(t), f_2^{k,m,n}(t), f_3^{k,m,n}(t))$$

are known for all $k = p, p-1, ..., 0;\ \ m = p, p-1, ..., 0;\ \ n = p, p-1, ..., 0$. The procedure of finding $\mathbf{E}^{k,m,n}$ consists of the sequence of the following iterative steps of computing some formulae from the others using the relation (23).

Step 0:

$$\tilde{\mathbf{E}}^{p+2,m,n} = \tilde{\mathbf{E}}^{k,p+2,n} = \tilde{\mathbf{E}}^{k,m,p+2} = \tilde{\mathbf{E}}^{p+1,m,n} = \tilde{\mathbf{E}}^{k,p+1,n} = \tilde{\mathbf{E}}^{k,m,p+1} = 0$$

when $k = p+2,\ p+1,\ ...,\ 0;\ m = p+2,\ p+1,\ ...,\ 0;\ n = p+2,\ p+1,\ ...,\ 0$. This fact follows from (15).

Step 1: using zero values from step 0 we compute formulae for

$$\tilde{\mathbf{E}}^{p,m,n},\ \tilde{\mathbf{E}}^{k,p,n},\ \tilde{\mathbf{E}}^{k,m,p},\ k = p,\ p-1,\ ...,\ 0;\ \ m = p,\ p-1,\ ...,\ 0;\ \ n = p,\ p-1,\ ...,\ 0.$$

Step 2: from the relations obtained on previous steps we compute

$$\tilde{\mathbf{E}}^{p-1,m,n},\ \tilde{\mathbf{E}}^{k,p-1,n},\ \tilde{\mathbf{E}}^{k,m,p-1},\quad k = p-1,\ ...,\ 0;\ \ m = p-1,\ ...,\ 0;\ \ n = p-1,\ ...,\ 0.$$

... ...

Step p: from the relations obtained on previous steps we compute

$$\tilde{\mathbf{E}}^{1,m,n},\ \tilde{\mathbf{E}}^{k,1,n},\ \tilde{\mathbf{E}}^{k,m,1},\quad \text{for}\quad k = 1,\ 0;\ \ m = 1,\ 0;\ \ n = 1,\ 0;$$

and $\tilde{\mathbf{E}}^{0,0,0}$.

Finally, components of $\mathbf{E}^{k,m,n}$ are found by $\mathbf{E}^{k,m,n} = \mathcal{S}\tilde{\mathbf{E}}^{k,m,n}$ for all $k = p, p-1, ..., 0;\ \ m = p, p-1, ..., 0;\ \ n = p, p-1, ..., 0$.

2.3 Constructing an Explicit Formula for a Solution of (1), (2) with Polynomial Data

Using the procedure described in previous Section and symbolic computations a solution $\mathbf{E} = (E_1, E_2, E_3)$ of the initial value problem (1), (2) is constructed. The implementation of this method has been made in Maple 10. The explicit formulae for the components of $\mathbf{E} = (E_1, E_2, E_3)$ have been constructed for arbitrary polynomial initial data and polynomial inhomogeneous terms. Using the direct substitution we have checked that constructed formulae give exact solutions of the initial value problem. We note that when a degree of polynomials is greater than 10 the formulae for components of $\mathbf{E}$ are cumbersome and take several printed pages. The robustness of the method is illustrated by the following example.

Example. Let $\mathcal{E}$ be arbitrary matrices defined as follows:

$$\mathcal{E} = \begin{pmatrix} 25 & 4 & 0 \\ 4 & 25 & 0 \\ 0 & 0 & 9 \end{pmatrix}$$

and let inhomogeneous term and initial data be given by $\mathbf{f} = 0$; $\mathbf{e} = 0$; $\mathbf{g} = (g_1, g_2, g_3)$, where

$$g_1(x) = {x_1}^2{x_2}^3x_3 + (x_1 + 2\,x_2)^4 + (x_2 + 3\,x_3)^3 + (2\,x_1 + 7\,x_2 + 18\,x_3)^3 \quad (25)$$
$$g_2(x) = 0,\; g_3(x) = 0.$$

Applying our method we compute the explicit formulae for the components of $\mathbf{E} = (E_1, E_2, E_3)$ and then we verify that $\mathbf{E}(x,t)$ is an exact solution of (1), (2). The formulae for the components of the solution are listed below:

$$E_1 = \frac{25}{1218}t^2{x_1}^4 + \frac{26797}{82418}t^4x_1 + \frac{50}{370881}t^4x_1\,{x_2}^2x_3 + \frac{25}{1218}t^2{x_1}^2{x_2}^3x_3 + \frac{900}{29}t^2x_1\,x_2\,x_3 + \frac{625}{1483524}t^4{x_1}^2x_2\,x_3 + \frac{100}{609}t^2{x_1}^3 + \frac{162}{41209}t^4{x_1}^2 + \frac{648}{41209}t^4{x_2}^2 + \frac{4300}{609}t^2{x_2}^3 + \frac{200}{609}t^2{x_2}^4 + \frac{6975}{58}t^2{x_3}^3 + \frac{100}{609}t^2{x_1}^3x_2 + \frac{50}{29}t^2{x_1}^2x_2 + \frac{648}{41209}t^4x_1\,x_2 + \frac{400}{609}t^2x_1\,{x_2}^3 + \frac{100}{203}t^2{x_1}^2{x_2}^2 + \frac{175}{29}t^2x_1\,{x_2}^2 + \frac{900}{203}t^2{x_1}^2x_3 + \frac{1175}{903466116}t^6x_2\,x_3 + \frac{22125}{406}t^2{x_2}^2x_3 + \frac{4}{1112643}t^4{x_2}^3x_3 + \frac{8100}{203}t^2x_1\,{x_3}^2 + \frac{56925}{406}t^2x_2\,{x_3}^2 + \frac{1694605}{1483524}t^4x_2 + \frac{2538}{41827135}t^6 + \frac{500}{677599587}t^6x_1\,x_3 + \frac{363358}{123627}t^4x_3$$

$$E_2 = -\frac{2}{609}t^2{x_1}^4 - \frac{4393}{41209}t^4x_1 - \frac{641}{1483524}t^4x_1\,{x_2}^2x_3 - \frac{2}{609}t^2{x_1}^2{x_2}^3x_3 - \frac{144}{29}t^2x_1\,x_2\,x_3 - \frac{25}{370881}t^4{x_1}^2x_2\,x_3 - \frac{16}{609}t^2{x_1}^3 - \frac{99}{41209}t^4{x_1}^2 - \frac{396}{41209}t^4{x_2}^2 - \frac{688}{609}t^2{x_2}^3 - \frac{32}{609}t^2{x_2}^4 - \frac{558}{29}t^2{x_3}^3 - \frac{1296}{203}t^2x_1\,{x_3}^2 - \frac{16}{609}t^2{x_1}^3x_2 - \frac{8}{29}t^2{x_1}^2x_2 - \frac{396}{41209}t^4x_1\,x_2 - \frac{64}{609}t^2x_1\,{x_2}^3 - \frac{16}{203}t^2{x_1}^2{x_2}^2 - \frac{28}{29}t^2x_1\,{x_2}^2 - \frac{144}{203}t^2{x_1}^2x_3 - \frac{1770}{203}t^2{x_2}^2x_3 - \frac{3365}{2710398348}t^6x_1\,x_3 - \frac{844}{1129332645}t^6x_2\,x_3 - \frac{25}{1112643}t^4{x_2}^3x_3 - \frac{4554}{203}t^2x_2\,{x_3}^2 - \frac{119086}{123627}t^4x_3 - \frac{277709}{741762}t^4x_2 - \frac{1551}{41827135}t^6$$

$$E_3 = -\frac{22}{609}t^4x_1 - \frac{230}{1827}t^4x_2 - \frac{197}{609}t^4x_3 + \frac{139}{150206805}t^6{x_1}^2 + \frac{64}{150206805}t^6{x_2}^2 - \frac{5027}{600827220}t^6x_1\,x_2 - \frac{25}{65772}t^4x_1\,{x_2}^3 + \frac{1}{10962}t^4{x_1}^2{x_2}^2 + \frac{41353}{7683979316580}t^8$$

3 Conclusion

In the present paper a new efficient method for symbolic computing explicit formulae for polynomial solutions of initial value problem with polynomial data and polynomial inhomogeneous term for the system (1) is described. The robustness of the method is confirmed by computational experiments. Our method can be used for verification of new numerical methods for the system (1).

As a further research we plan to extend our method for solving initial value problems for hyperbolic systems describing elastic, acoustic and electromagnetic wave propagations.

Acknowledgments. This work was supported by the Dokuz Eylul University of Turkey under the research grant 2006.KB.FEN.024.

References

1. Courant, R., Hilbert, D.: Methods of Mathematical Physics, vol. 2. International Science, New York (1962)
2. Haba, Z.: Green functions and propagation of waves in strongly inhomogeneous media. Journal of Physics A: Mathematical and General 37, 9295–9302 (2004)
3. Wijnands, F., et al.: Green's functions for Maxwell's equations: application to spontaneous emission. Optical and Quantum Electronics 29, 199–216 (1997)
4. Li, L.W., et al.: Circular cylindrical waveguide filled with uniaxial anisotropic media-electromagnetic fields and dyadic Green's functions. IEEE transactions on microwave and techniques 49(7), 1361–1364 (2001)
5. Gottis, P.G., Kondylis, G.D.: Properties of the dyadic Green's function for unbounded anisotropic medium. IEEE transactions on antennas and propagation 45, 154–161 (1995)
6. Ortner, N., Wagner, P.: Fundamental matrices of homogeneous hyperbolic system. Applications to crystal optics, elastodynamics, and piezoelectromagnetism. Z. Angew. Math. Mech. 84(5), 314–346 (2004)
7. Burridge, R., Qian, J.: The fundamental solution of the time-dependent system of crystal optics. European J. Appl. Math. 17(1), 63–94 (2006)
8. Yakhno, V.G.: Constructing Green's function for the time-dependent Maxwell system in anisotropic dielectrics. Journal of Physics A: Mathematical and General 38(10), 2277–2287 (2005)
9. Yakhno, V.G., Yakhno, T.M., Kasap, M.: A novel approach for modelling and simulation of electromagnetic waves in anisotropic dielectrics. International Journal of Solids and Structures 43, 6261–6276 (2006)
10. Monk, P.: Finite element methods for Maxwell's equations. Clarendon Press, Oxford (2003)
11. Cohen, G.C.: Higher-order numerical methods for transient wave equations. Springer, Berlin (2002)
12. Zienkiewicz, O.C., Taylor, R.L.: Finite elements method 1. Butterworth-Heinemann, Oxford (2000)
13. Cohen, G.C., Heikkola, E., Joly, P., Neittaan, M.P.: Mathematical and numerical aspects of waves propagation. Springer, Berlin (2003)
14. Werner, G.R., Cary, J.R.: A stable FDTD algorithm for non-diagonal, anisotropic dielectrics. Journal of Computational Physics 226, 1085–1101 (2007)
15. Ikawa, M.: Hyperbolic partial differential equations and wave phenomena. Translations of Mathematical Monographs, vol. 189. American Mathematical Society, Providence (2000)

On the Local Sensitivity of the Discrete-Time $\mathcal{H}_\infty$ Control Problem

A.S. Yonchev[1], P.Hr. Petkov[1], N.D. Christov[2], and M.M. Konstantinov[3]

[1] Technical University of Sofia, 1000 Sofia, Bulgaria
[2] Université des Sciences et Technologies de Lille, 59655 Villeneuve d'Ascq, France
[3] University of Architecture, Civil Engineering and Geodesy, 1046 Sofia, Bulgaria

Abstract. In this paper linear perturbation bounds are obtained for the linear matrix inequalities (LMI) arising in the discrete-time $\mathcal{H}_\infty$ control problem. The sensitivity analysis of the perturbed LMI is carried out in a similar way as for perturbed matrix equations, after introducing a suitable right hand side which is slightly perturbed.

1 Introduction

In many control problems, the design constraints have a simple reformulation in terms of linear matrix inequalities (LMI) [1]. The $\mathcal{H}_\infty$ control problems are a good illustration of this point since the $\mathcal{H}_\infty$ constraints can be expressed as a single matrix inequality via the bounded real lemma [6]. Although solutions of the $\mathcal{H}_\infty$ control problems can be found in terms of Riccati equations [2], the LMI-based approach for solving these problems has several advantages. First, it is applicable to all plants without restrictions on infinite or pure imaginary invariant zeros. Secondly, it offers a simple and insightful derivation of the problem solvability conditions [3]. In addition, the LMI solutions for the $\mathcal{H}_\infty$ control problems can be efficiently computed using reliable convex optimization algorithms [5,4].

In this paper we propose an approach to carry out linear sensitivity analysis of the LMI solutions for the discrete-time $\mathcal{H}_\infty$ control problem via introducing a suitable right hand side in the corresponding matrix inequalities.

We use the following notations: $\mathbb{R}^{m\times n}$ – the space of real $m \times n$ matrices; $\mathbb{R}^n = \mathbb{R}^{n\times 1}$; I_n – the identity $n \times n$ matrix; e_n – the unit $n \times 1$ vector; $M^\top$ – the transpose of M; $M^\dagger$ – the pseudo inverse of M; $\|M\|_2 = \sigma_{\max}(M)$ – the spectral norm of M, where $\sigma_{\max}(M)$ is the maximum singular value of M; $\|M\|_{\mathrm{F}} = \sqrt{\mathrm{tr}(M^\top M)}$ – the Frobenius norm of M; $\|M\|_\infty := \sup_{\mathrm{Re}\ s\geq 0} \|M(s)\|_2$; $\|.\|$ is any of the above norms; $\mathrm{vec}(M) \in \mathbb{R}^{mn}$ – the column-wise vector representation of $M \in \mathbb{R}^{m\times n}$; $\Pi_{m,n} \in \mathbb{R}^{mn\times mn}$ – the vec-permutation matrix, such that $\mathrm{vec}(M^\top) = \Pi_{m,n}\mathrm{vec}(M)$; $M \otimes P$ – the Kroneker product of the matrices M and P; $\mathrm{vec}(MXP) = (P^\top \otimes M)\mathrm{vec}(X)$ – column-wise vector representation of the multiplication MXP. The notation ":=" stands for "equal by definition".

S. Margenov, L.G. Vulkov, and J. Waśniewski (Eds.): NAA 2008, LNCS 5434, pp. 604–611, 2009.

2 Problem Statement

Consider the linear discrete-time system

$$\begin{aligned} x_{k+1} &= Ax_k + B_1 w_k + B_2 u_k \\ z_k &= C_1 x_k + D_{11} w_k + D_{12} u_k \\ y_k &= C_2 x_k + D_{21} w_k \end{aligned} \tag{1}$$

where $x_k \in \mathbb{R}^n$ is the state vector, $w_k \in \mathbb{R}^l$ is the exogenous input vector (the disturbance), $u_k \in \mathbb{R}^m$ is the control input vector, $z_k \in \mathbb{R}^p$ is the error vector, $y_k \in \mathbb{R}^r$ is the measurement vector and A, B_1, B_2, C_1, C_2, D_{11}, D_{12}, D_{21} are constant matrices of compatible size.

The suboptimal discrete-time $\mathcal{H}_\infty$ control problem consists in finding a control law u which leads to a bounded $\mathcal{H}_\infty$-norm of the transfer function matrix $T_{zw}(z)$ from w_k to z_k:

$$\|T_{zw}(z)\|_\infty < \gamma, \quad \gamma > 0. \tag{2}$$

In the optimal discrete-time $\mathcal{H}_\infty$ control problem one tries to find the infimum of γ (further denoted by γ_{opt}) which satisfies (2). The solution of the optimal $\mathcal{H}_\infty$ control problem corresponds to the best disturbance attenuation at the performance vector of the closed-loop system.

The $\mathcal{H}_\infty$ control problem (1), (2) is solvable if and only if there exist two symmetric matrices R^*, $S^* \in \mathbb{R}^{n\times n}$ satisfying the following system of LMI [3]:

$$\begin{bmatrix} \mathcal{N}_{12} & \vdots & 0 \\ \cdots & \cdots & \cdots \\ 0 & \vdots & I \end{bmatrix}^\top \begin{bmatrix} ARA^\top - R & ARC_1^\top & \vdots & B_1 \\ C_1RA^\top & -\gamma I + C_1RC_1^\top & \vdots & D_{11} \\ \cdots & \cdots & \cdots & \cdots \\ B_1^\top & D_{11}^\top & \vdots & -\gamma I \end{bmatrix} \begin{bmatrix} \mathcal{N}_{12} & \vdots & 0 \\ \cdots & \cdots & \cdots \\ 0 & \vdots & I \end{bmatrix} < 0 \tag{3}$$

$$\begin{bmatrix} \mathcal{N}_{21} & \vdots & 0 \\ \cdots & \cdots & \cdots \\ 0 & \vdots & I \end{bmatrix}^\top \begin{bmatrix} A^\top SA - S & A^\top SB_1 & \vdots & C_1^\top \\ B_1^\top SA & -\gamma I + B_1^\top SB_1 & \vdots & D_{11}^\top \\ \cdots & \cdots & \cdots & \cdots \\ C_1 & D_{11} & \vdots & -\gamma I \end{bmatrix} \begin{bmatrix} \mathcal{N}_{21} & \vdots & 0 \\ \cdots & \cdots & \cdots \\ 0 & \vdots & I \end{bmatrix} < 0 \tag{4}$$

$$\begin{bmatrix} R & I \\ I & S \end{bmatrix} > 0 \tag{5}$$

where $\mathcal{N}_{12}$ and $\mathcal{N}_{21}$ are the orthonormal bases of the null spaces of $\left[\, B_2^\top \; D_{12}^\top \,\right]$ and $\left[\, C_2 \; D_{21} \,\right]$, respectively. Here we assume that the optimal closed-loop performance $\gamma_{\rm opt}$ of system (1) is already obtained.

Suppose that the matrices A, B_1, B_2, C_1, C_2, D_{11}, D_{12}, D_{21} and the quantity γ in (3), (4) are subject to perturbations ΔA, ΔB_1, ΔB_2, ΔC_1, ΔC_2, ΔD_{11}, ΔD_{12}, ΔD_{21}, $\Delta\gamma_{\rm opt}$ and denote by $R^* + \Delta R$, $S^* + \Delta S$ the solution of the perturbed LMI system. The sensitivity analysis of the $\mathcal{H}_\infty$ control problem is aimed at determining bounds for ΔR and ΔS near the optimal value of γ, as functions of the perturbations in the data A, B_1, B_2, C_1, C_2, D_{11}, D_{12}, D_{21} and $\gamma_{\rm opt}$. In the next section we shall derive linear, condition number based bounds for ΔR and ΔS with respect to perturbations in A, B_2, C_2, D_{12}, D_{21} and $\gamma_{\rm opt}$.

3 Linear Sensitivity Analysis

The essence of our approach is to perform sensitivity analysis of LMI (3) and (4) in a similar way as for proper matrix equations after introducing suitable right hand sides which are slightly perturbed.

Consider first LMI (4). Its structure allows us to analyze only the perturbed inequality

$$\begin{aligned}&(\mathcal{N}_{21}+\Delta\mathcal{N}_{21})^\top\\ &\times\left\{\begin{bmatrix}(A+\Delta A)^\top(S^*+\Delta S)(A+\Delta A)-(S^*+\Delta S) & 0\\ B_1(S^*+\Delta S)(A+\Delta A) & 0\end{bmatrix}\right.\\ &\left.+\begin{bmatrix}0 & (A+\Delta A)^\top(S^*+\Delta S)B_1\\ 0 & -\gamma_{\rm opt}I-\Delta\gamma_{\rm opt}I+B_1(S+\Delta S)B_1\end{bmatrix}\right\}\\ &\times(\mathcal{N}_{21}+\Delta\mathcal{N}_{21}):=\bar{\mathcal{H}}^*+\Delta\bar{\mathcal{H}}_1<0\end{aligned}\tag{6}$$

where the matrix $\bar{\mathcal{H}}^*$ is obtained using the nominal LMI

$$\mathcal{N}_{21}{}^\top\begin{bmatrix}A^\top S^*A-S^* & A^\top S^*B_1\\ B_1{}^\top S^*A & -\gamma_{\rm opt}I+B_1{}^\top S^*B_1\end{bmatrix}\mathcal{N}_{21}:=\bar{\mathcal{H}}^*<0\tag{7}$$

and $\Delta\bar{\mathcal{H}}_1$ is due to the data and closed-loop performance perturbations, the rounding errors and the sensitivity of the interior point method used to solve the LMI.

Within first order terms the perturbed relation (6) may be written as

$$\bar{\mathcal{H}}^*+\Delta\bar{\mathcal{H}}_1=\mathcal{N}_{21}{}^\top\bar{\mathcal{V}}\mathcal{N}_{21}+\mathcal{N}_{21}{}^\top\bar{\mathcal{V}}\Delta\mathcal{N}_{21}+\Delta\mathcal{N}_{21}{}^\top\bar{\mathcal{V}}\mathcal{N}_{21}+\Delta\mathcal{N}_{21}{}^\top\bar{\mathcal{V}}\Delta\mathcal{N}_{21}\tag{8}$$

where

$$\begin{aligned}\bar{\mathcal{V}}=&\begin{bmatrix}A^\top S^*A-S^*+A^\top\Delta SA-\Delta S+A^\top S^*\Delta A+\Delta A^\top S^*A & 0\\ B_1{}^\top S^*A+B_1\Delta SA+B_1{}^\top S^*\Delta A & 0\end{bmatrix}\\ &+\begin{bmatrix}0 & A^\top S^*B_1+A^\top\Delta SB_1+\Delta A^\top S^*B_1\\ 0 & -\gamma_{\rm opt}I-\Delta\gamma_{\rm opt}I+B_1{}^\top S^*B_1+B_1{}^\top\Delta S^*B_1\end{bmatrix}.\end{aligned}$$

Using relation (7) one has

$$\begin{aligned}\Delta\bar{\mathcal{H}}_1=&\ \mathcal{N}_{21}{}^\top\Psi_S\mathcal{N}_{21}+\mathcal{N}_{21}{}^\top(\mathcal{H}^*+\Psi_S)\Delta\mathcal{N}_{21}+\Delta\mathcal{N}_{21}{}^\top(\mathcal{H}^*+\Psi_S)\mathcal{N}_{21}\\ &+\Delta\mathcal{N}_{21}{}^\top(\mathcal{H}^*+\Psi_S)\Delta\mathcal{N}_{21}\end{aligned}\tag{9}$$

where

$$\bar{\mathcal{H}}^* = \mathcal{N}_{21}{}^\top \mathcal{H}^* \mathcal{N}_{21},\ \Psi_S = \Theta_S + \Lambda_S,\ \Theta_S = \begin{bmatrix} A^\top \Delta S A - \Delta S & A^\top \Delta S B_1 \\ B_1{}^\top \Delta S A & B_1{}^\top \Delta S B_1 \end{bmatrix}$$

$$\Lambda_S = \begin{bmatrix} A^\top S^* \Delta A + \Delta A^\top S^* A & \Delta A^\top S^* B_1 \\ B_1{}^\top S^* \Delta A & -\Delta\gamma_{\text{opt}} I \end{bmatrix}.$$

Neglecting the second and higher order terms in (9) one obtains

$$\mathcal{N}_{21}{}^\top \Theta_S \mathcal{N}_{21} + \mathcal{N}_{21}{}^\top \Lambda_S \mathcal{N}_{21} + \Delta\mathcal{N}_{21}{}^\top \mathcal{H}^* \mathcal{N}_{21} + \mathcal{N}_{21}{}^\top \mathcal{H}^* \Delta\mathcal{N}_{21} = \Delta\bar{\mathcal{H}}_1. \quad (10)$$

Setting $\mathcal{H}^*\mathcal{N}_{21} = \tilde{\mathcal{N}}_{21}$ and $\mathcal{N}_{21}{}^\top \mathcal{H}^* = \tilde{\mathcal{N}}_{21}^*$ it follows that

$$\text{vec}(\Delta\mathcal{N}_{21}{}^\top \tilde{\mathcal{N}}_{21} + \tilde{\mathcal{N}}_{21}^* \Delta\mathcal{N}_{21}) = [(\tilde{\mathcal{N}}_{21}{}^\top \otimes I)\Pi_{(n+l),n^2} + (I \otimes \tilde{\mathcal{N}}_{21}^*)]\text{vec}(\Delta\mathcal{N}_{21}). \quad (11)$$

Denoting $\mathcal{H}^*\mathcal{N}_{21} = \tilde{\mathcal{N}}_{21}, \mathcal{N}_{21}{}^\top \mathcal{H}^* = \tilde{\mathcal{N}}_{21}^*$ $A^\top S^* = S^*_{A^\top}$, $S^* A = S^*_A$, $B_1{}^\top S^* = S^*_{B_1{}^\top}$ and $S^* B_1 = S^*_{B_1}$, relation (10) may be written in a vector form as

$$(\mathcal{N}_{21}{}^\top \otimes \mathcal{N}_{21}{}^\top)\text{vec}(\Theta_S) + (\mathcal{N}_{21}{}^\top \otimes \mathcal{N}_{21}{}^\top)\text{vec}(\Lambda_S) + \mathcal{N}_{S\Theta}\text{vec}(\Delta\mathcal{N}_{21}) = \text{vec}(\Delta\bar{\mathcal{H}}_1) \quad (12)$$

where

$$\text{vec}(\Theta_S) = \begin{bmatrix} A^\top \otimes A^\top - I_{n^2} \\ B_1{}^\top \otimes A^\top \\ A^\top \otimes B_1{}^\top \\ B_1{}^\top \otimes B_1{}^\top \end{bmatrix} \text{vec}(\Delta S) := N\text{vec}(\Delta S)$$

$$\text{vec}(\Lambda_S) = \begin{bmatrix} (I \otimes S^*_{A^\top}) + (S^*_A \otimes I)\Pi_{n^2} & 0 \\ (S^*_{B_1} \otimes I)\Pi_{n^2} & 0 \\ (I \otimes S^*_{B_1{}^\top}) & 0 \\ 0 & -e_{l^3} \end{bmatrix} \begin{bmatrix} \text{vec}(\Delta A) \\ \Delta\gamma_{\text{opt}} \end{bmatrix}$$

$$:= \begin{bmatrix} N_{\text{t1}}\ N_{\text{t2}} \end{bmatrix} \begin{bmatrix} \text{vec}(\Delta A) \\ \Delta\gamma_{\text{opt}} \end{bmatrix}$$

$$\mathcal{N}_{S\Theta} = (\tilde{\mathcal{N}}_{21}{}^\top \otimes I)\Pi_{(n+l),n^2} + (I \otimes \tilde{\mathcal{N}}_{21}^{*\top}).$$

Thus we have

$$N_{\text{s}}\text{vec}(\Delta S) + N_{\text{ts1}}\text{vec}(\Delta A) + N_{\text{ts2}}\Delta\gamma + \mathcal{N}_{S\Theta}\text{vec}(\Delta\mathcal{N}_{21}) = \text{vec}(\Delta\bar{\mathcal{H}}_1) \quad (13)$$

where

$$N_{\text{s}} = (\mathcal{N}_{21}{}^\top \otimes \mathcal{N}_{21}{}^\top)N,\ N_{\text{ts1}} = (\mathcal{N}_{21}{}^\top \otimes \mathcal{N}_{21}{}^\top)N_{\text{t1}},\ N_{\text{ts2}}(\mathcal{N}_{21}{}^\top \otimes \mathcal{N}_{21}{}^\top)N_{\text{t2}}.$$

It is well known [7] that the perturbation bound for the projector $\mathcal{N}_{21}$ may be written as

$$\|\Delta\mathcal{N}_{21}\|_2 \leq \|[C_2\ D_{21}]^\dagger\|_2\|[\Delta C_2\ \Delta D_{21}]\|_2. \tag{14}$$

Using the fact that $\|\mathrm{vec}(M)\|_2 = \|M\|_{\mathcal{F}}$, we finally obtain the relative perturbation bound for S^*

$$\frac{\|\Delta S\|_{\mathcal{F}}}{\|S^*\|_{\mathcal{F}}} \leq k_A^S\frac{\|\Delta A\|_{\mathcal{F}}}{\|A\|_{\mathcal{F}}} + k_\gamma^S\frac{|\Delta\gamma_{\mathrm{opt}}|}{|\gamma_{\mathrm{opt}}|} + k_{CD}^S\frac{\|[\Delta C_2\ \Delta D_{21}]\|_{\mathcal{F}}}{\|[C_2\ D_{21}]\|_{\mathcal{F}}} + k_{\mathcal{H}}^S\frac{\|\Delta\bar{\mathcal{H}}_1\|_{\mathcal{F}}}{\|\bar{\mathcal{H}}^*\|_{\mathcal{F}}} \tag{15}$$

where

$$k_A^S = \frac{\|N_{\mathrm{s}}^\dagger\|_2\|N_{\mathrm{ts1}}\|_2\|A\|_{\mathcal{F}}}{\|S^*\|_{\mathcal{F}}},\quad k_\gamma^S = \frac{\|N_{\mathrm{s}}^\dagger\|_2\|N_{\mathrm{ts3}}\|_2|\gamma_{\mathrm{opt}}|}{\|S^*\|_{\mathcal{F}}},\quad k_{\mathcal{H}}^S = \frac{\|N_{\mathrm{s}}^\dagger\|_2\|\bar{\mathcal{H}}^*\|_{\mathcal{F}}}{\|S^*\|_{\mathcal{F}}}$$

$$k_{CD}^S = \frac{\|N_{\mathrm{s}}^\dagger\|_2\|\mathcal{N}_{S\Theta}\|_2\|[C_2\ D_{21}]^\dagger\|_{\mathcal{F}}\|[C_2\ D_{21}]\|_{\mathcal{F}}}{\|S^*\|_{\mathcal{F}}}$$

can be considered as individual relative condition numbers of (4) with respect to the perturbations in the data.

In a similar way we can obtain a relative perturbation bound for the solution R^* of the LMI (3) In this case we consider the perturbed inequality

$$\begin{aligned}
&(\mathcal{N}_{12} + \Delta\mathcal{N}_{12})^\top \\
&\times\left\{\begin{bmatrix} (A+\Delta A)(R^*+\Delta R)(A+\Delta A)^\top - (R^*+\Delta R) & 0 \\ C_1(R^*+\Delta R)(A+\Delta A)^\top & 0 \end{bmatrix}\right. \\
&\left.+\begin{bmatrix} 0 & (A+\Delta A)(R^*+\Delta R)C_1^\top \\ 0 & -\gamma_{\mathrm{opt}}I - \Delta\gamma_{\mathrm{opt}}I + C_1(S+\Delta S)C_1^\top \end{bmatrix}\right\} \\
&\times(\mathcal{N}_{12} + \Delta\mathcal{N}_{12}) := \bar{\mathcal{E}}^* + \Delta\bar{\mathcal{E}}_1 < 0
\end{aligned} \tag{16}$$

where

$$\mathcal{N}_{12}^\top \begin{bmatrix} AR^*A^\top - R^* & AR^*C_1^\top \\ C_1R^*A^\top & -\gamma_{\mathrm{opt}}I + C_1R^*C_1^\top \end{bmatrix}\mathcal{N}_{12} := \bar{\mathcal{E}}^* < 0. \tag{17}$$

Here, instead of Θ_S and Λ_S we have

$$\Theta_R = \begin{bmatrix} A\Delta RA^\top - \Delta R & A\Delta RC_1^\top \\ C_1\Delta RA^\top & C_1\Delta RC_1^\top \end{bmatrix}$$

$$\Lambda_R = \begin{bmatrix} AR^*\Delta A^\top + \Delta AR^*A^\top & \Delta AR^*C_1^\top \\ C_1R^*\Delta A^\top & -\Delta\gamma_{\mathrm{opt}}I \end{bmatrix}$$

and thus

$$\mathrm{vec}(\Theta_R) = \begin{bmatrix} A\otimes A - I_{n^2} \\ C_1\otimes A \\ A\otimes C_1 \\ C_1\otimes C_1 \end{bmatrix}\mathrm{vec}(\Delta R) := M\mathrm{vec}(\Delta R)$$

$$\mathrm{vec}(\Lambda_R) = \begin{bmatrix} (R^*_{A^\top} \otimes I) + (I \otimes R^*_A)\Pi_{n^2} & 0 \\ (R^*_{C_1^\top} \otimes I) & 0 \\ (I \otimes R^*_{C_1}) & 0 \\ 0 & -e_{p^3} \end{bmatrix} \begin{bmatrix} \mathrm{vec}(\Delta A) \\ \Delta\gamma_{\mathrm{opt}} \end{bmatrix}$$

$$:= \begin{bmatrix} M_{\mathrm{t}1} \; M_{\mathrm{t}2} \end{bmatrix} \begin{bmatrix} \mathrm{vec}(\Delta A) \\ \Delta\gamma_{\mathrm{opt}} \end{bmatrix}.$$

Denote $\bar{\mathcal{E}}^* = {\mathcal{N}_{12}}^\top \mathcal{E}^* \mathcal{N}_{12}, \;\; \mathcal{E}^*\mathcal{N}_{12} = \tilde{\mathcal{N}}_{12}, \;\; {\mathcal{N}_{12}}^\top \mathcal{E}^* = \tilde{\mathcal{N}}^*_{12},$

$$\mathcal{N}_{R\Omega} = ({\tilde{\mathcal{N}}_{12}}^\top \otimes I)\Pi_{(n+p),n^2} + (I \otimes \tilde{\mathcal{N}}^*_{12})$$

$$M_{\mathrm{r}} = ({\mathcal{N}_{12}}^\top \otimes {\mathcal{N}_{12}}^\top)M, \;\; M_{\mathrm{tr}1} = ({\mathcal{N}_{12}}^\top \otimes {\mathcal{N}_{12}}^\top)M_{\mathrm{t}1}$$

$$M_{\mathrm{tr}2} = ({\mathcal{N}_{12}}^\top \otimes {\mathcal{N}_{12}}^\top)M_{\mathrm{t}2}.$$

Having in mind that

$$\|\Delta\mathcal{N}_{12}\|_{\mathcal{F}} \le \|[{B_2}^\top \; {D_{12}}^\top]^\dagger\|_{\mathcal{F}} \|[\Delta {B_2}^\top \; \Delta {D_{12}}^\top]\|_{\mathcal{F}}$$

we obtain the relative perturbation bound for R^*

$$\frac{\|\Delta R\|_{\mathcal{F}}}{\|R^*\|_{\mathcal{F}}} \le k^R_A \frac{\|\Delta A\|_{\mathcal{F}}}{\|A\|_{\mathcal{F}}} + k^R_\gamma \frac{|\Delta\gamma_{\mathrm{opt}}|}{|\gamma_{\mathrm{opt}}|} + k^S_{BD} \frac{\|[\Delta {B_2}^\top \; \Delta {D_{12}}^\top]\|_{\mathcal{F}}}{\|[{B_2}^\top \; {D_{12}}^\top]\|_{\mathcal{F}}}$$

$$+ k^R_{\mathcal{E}} \frac{\|\Delta\bar{\mathcal{E}}_1\|_{\mathcal{F}}}{\|\bar{\mathcal{E}}^*\|_{\mathcal{F}}} \tag{18}$$

where

$$k^R_A = \frac{\|M_{\mathrm{r}}^\dagger\|_2 \|M_{\mathrm{tr}1}\|_2 \|A\|_{\mathcal{F}}}{\|R^*\|_{\mathcal{F}}}, \;\; k^R_\gamma = \frac{\|M_{\mathrm{r}}^\dagger\|_2 \|M_{\mathrm{tr}2}\|_2 |\gamma_{\mathrm{opt}}|}{\|R^*\|_{\mathcal{F}}}, \;\; k^R_{\mathcal{E}} = \frac{\|M_{\mathrm{r}}^\dagger\|_2 \|\bar{\mathcal{E}}^*\|_{\mathcal{F}}}{\|R^*\|_{\mathcal{F}}}$$

$$k^R_{BD} = \frac{\|M_{\mathrm{r}}^\dagger\|_2 \|\mathcal{N}_{R\Theta}\|_2 \|[{B_2}^\top \; {D_{12}}^\top]^\dagger\|_{\mathcal{F}} \|[{B_2}^\top \; {D_{12}}^\top]\|_{\mathcal{F}}}{\|R^*\|_{\mathcal{F}}}$$

are the individual relative condition numbers of the LMI (3) with respect to the perturbations in the data.

4 Numerical Example

Consider the following system description

$$A_c = \begin{bmatrix} 0 & 1 \\ -k/m & -c/m \end{bmatrix}, \; B_{1c} = \begin{bmatrix} 0 & 0 & 0 \\ -pm & -pc/m & -pk/m \end{bmatrix}, \; B_{2c} = \begin{bmatrix} 0 \\ 1/m \end{bmatrix}$$

$$C_{1c} = \begin{bmatrix} -k/m & -c/m \\ 0 & c \\ k & 0 \end{bmatrix}, \; C_{2c} = \begin{bmatrix} 1 \, 0 \end{bmatrix}, \; D_{11c} = \begin{bmatrix} -pm & -pc/m & -pk/m \\ 0 & 0 & 0 \\ 0 & 0 & 0 \end{bmatrix}$$

$$D_{12c} = \begin{bmatrix} 1/m \\ 0 \\ 0 \end{bmatrix}, \; D_{21c} = \begin{bmatrix} 0 \, 0 \, 0 \end{bmatrix}$$

and $m = 3$, $c = 1$, $k = 2$, $pm = 0.4$, $pc = 0.2$, $pk = 0.3$. Here A_c, B_{1c}, B_{2c}, C_{1c}, C_{2c}, D_{11c}, D_{12c}, D_{21c} are the system matrices of a continuous-time system, which for the aim of the analysis is turned into a discrete one using sampling time of 0.01s.

The perturbations in the system matrices of the discrete-time system are chosen as

$$\Delta A = A \times 10^{-i}, \; \Delta B_1 = B_1 \times 10^{-i}, \; \Delta B_2 = B_2 \times 10^{-i}$$

$$\Delta C_1 = C_1 \times 10^{-i}, \; \Delta C_2 = C_2 \times 10^{-i}, \; \Delta D_{11} = D_{11} \times 10^{-i}, \; \Delta D_{12} = D_{12} \times 10^{-i}$$

$$\Delta\gamma_{\mathrm{opt}} = \gamma_{\mathrm{opt}} \times 10^{-i}, \; \Delta\bar{\mathcal{H}}_1 = \bar{\mathcal{H}}^* \times 10^{-i}, \; \Delta\bar{\mathcal{E}}_1 = \bar{\mathcal{E}}^* \times 10^{-i}.$$

The initial and perturbed LMI are solved using the LMI Control Toolbox of MATLAB [4]. The optimal closed-loop performance obtained is $\gamma_{\mathrm{opt}} = 0.4191$. Table 1 shows the relative perturbations and corresponding perturbation bounds of the LMI solutions R^* and S^* for different values of i.

Table 1. Relative perturbations and perturbation bounds of the LMI solutions

i	$\frac{\\|\Delta S\\|_{\mathcal{F}}}{\\|S^*\\|_{\mathcal{F}}}$	Bound (15)	$\frac{\\|\Delta R\\|_{\mathcal{F}}}{\\|R^*\\|_{\mathcal{F}}}$	Bound (18)
8	$0.1\ 10^{-6}$	$0.5\ 10^{-5}$	$0.7\ 10^{-6}$	$6.3\ 10^{-5}$
7	$0.1\ 10^{-5}$	$0.5\ 10^{-4}$	$0.7\ 10^{-5}$	$6.3\ 10^{-4}$
6	$0.1\ 10^{-4}$	$0.5\ 10^{-3}$	$0.7\ 10^{-4}$	$6.3\ 10^{-3}$
5	$0.1\ 10^{-3}$	$0.5\ 10^{-2}$	$0.7\ 10^{-3}$	$6.3\ 10^{-2}$
4	$0.1\ 10^{-2}$	$0.5\ 10^{-1}$	$0.7\ 10^{-2}$	$6.3\ 10^{-1}$

5 Conclusions

Linear sensitivity analysis of the LMI arising in the discrete-time $\mathcal{H}_\infty$ control problem is done. Condition number based perturbation bounds are obtained in a similar way as for matrix equations, introducing a slightly perturbed right hand side in LMI. A numerical example is presented illustrating the accuracy of the proposed LMI perturbation bounds.

References

1. Boyd, S., El Ghaoui, L., Feron, E.: Linear matrix inequalities in systems and control theory. SIAM Philladelphia 41(3), 358–367 (1996)
2. Doyle, J.C., Glover, K., Khargonekar, P.P., Francis, B.A.: State-Space Solutions to Standard H_2 and H_∞ Control Problems. IEEE Transactions on Automatic Control 34, 831–847 (1989)
3. Gahinet, P., Apkarian, P.: A linear matrix inequality approach to H_∞ control. Int. J. Robust Non. Contr. 4, 421–448 (1994)
4. Gahinet, P., Nemirovski, A., Laub, A., Chilali, M.: LMI Control Toolbox for Use with MATLAB. The MathWorks, Inc (2000)
5. Nesterov, Y., Nemirovski, A.: Interior–Point Polynomial Algorithms in Convex Programming. SIAM, Philadephia (1994)
6. Peterson, I.R., Anderson, B.D.O., Jonkheere, E.A.: A first principles solution to the non-singular H_∞ control problem. Int. J. Robust Non. Contr. 1, 171–185 (1991)
7. Steward, G., Sun, J.G.: Matrix Perturbation Theory. Academic Press, N.Y (1990)
8. Zhou, K., Doyle, J.C., Glover, K.: Robust and Optimal Control. Prentice-Hall, Upper Saddle River (1995)

Interpolation Method for a Function with a Singular Component

A.I. Zadorin

Omsk filial of Sobolev Mathematics Institute SB RAS,
Pevtsova 13, Omsk, 644099, Russia
zadorin@ofim.oscsbras.ru

Abstract. An interpolation formula for functions with boundary layer components is proposed. It is exact on the singular boundary layer component, that leads to uniform accuracy of the interpolation. It is shown, that the proposed formula can be used for interpolation of numerical solutions of boundary value problems with exponential and power layers.

1 Introduction

Methods of spline-interpolation for functions with bounded derivatives are well known, see for example [1]. But when we apply polynomial interpolation methods to functions with a boundary layer component, we see, that it leads to significant errors. In this article we consider an interpolation problem for a function with a boundary layer component. We offer an interpolation formula, fitted to boundary layer component and prove, that proposed formula has uniformly small error on arbitrary grids. The constructed interpolant may be used for an approximate derivative computation. It is shown, that the proposed formula can be used for an interpolation of functions with an exponential and power boundary layer components.

2 Construction of an Interpolant

The interpolation methods are well developed, see for example [1]. But we need to investigate such methods, if a function under interpolation has regions of large gradients. In this article we consider the question of interpolation for a function with boundary layer components.

Let the function $u(x)$ is smooth enough and has the form:

$$u(x) = p(x) + \gamma\Phi(x), \tag{1}$$

where the given function $\Phi(x)$ is bounded on interval $[0, 1]$, but has regions of large gradients. The function $p(x)$ is bounded together with it's first derivative and is unknown, and the constant γ is also unknown. The representation (1) holds for solutions of problems with boundary layers [3] - [7].

S. Margenov, L.G. Vulkov, and J. Waśniewski (Eds.): NAA 2008, LNCS 5434, pp. 612–619, 2009.

Let the function $u(x)$ is given in nodes of some mesh Ω_h :

$$\Omega = \{x_n : x_n = x_{n-1} + h_n,\ x_0 = 0,\ x_N = 1,\ n = 1, 2, \ldots, N\},$$

where $u_n = u(x_n),\ n = 0, 1, 2, \ldots, N$. The mesh Ω may be uniform or nonuniform, we require only, that function $\Phi(x)$ is monotone on every mesh interval $\Delta_n = [x_{n-1}, x_n]$.

The aim of the work is construction and investigation of an appropriate interpolation formula for a function $u(x)$ with a component $\Phi(x)$, having regions of large gradients.

Consider the linear interpolation formula:

$$u_L(x) = (u_n - u_{n-1})\frac{x - x_n}{h_n} + u_n,\ x \in \Delta_n. \tag{2}$$

According to [1] the estimate holds:

$$|u_L(x) - u(x)| \le \frac{h_n^2}{8} \max_{s\in\Delta_n} |u''(s)|,\ x \in \Delta_n. \tag{3}$$

It follows from (3), that if $|u''(x)|$ is uniformly bounded on $[0, 1]$, than the accuracy of the interpolation is of order $O(h_n^2)$, and for large values of $|u''(x)|$ the interpolation error can be significant.

It was shown in [9], that in the case $u(x) = \exp(-\varepsilon^{-1}x)$ and $\varepsilon = h_1$ we have $u_L(h_1/2) - u(h_1/2) \approx 0,0774$ for any small step h_1.

So, we need an interpolation formula with the property of uniform with respect to ε accuracy.

Consider the interpolation formula, exact on the singular component $\Phi(x)$:

$$u_\Phi(x) = \frac{u_n - u_{n-1}}{\Phi_n - \Phi_{n-1}}[\Phi(x) - \Phi_n] + u_n,\ x \in \Delta_n, \tag{4}$$

where $\Phi_n = \Phi(x_n)$. Let us estimate the accuracy of the nonlinear interpolation (4). Define

$$R_\Phi(u, x) = u_\Phi(x) - u(x).$$

Then

$$R_\Phi(u, x) = R_\Phi(p, x) = -\frac{\Phi_n - \Phi(x)}{\Phi_n - \Phi_{n-1}}(p_n - p_{n-1}) + p_n - p(x).$$

The function $\Phi(x)$ is monotone on the interval Δ_n, therefore:

$$|R_\Phi(u, x)| \le 2 \max_{x\in\Delta_n} |p'(x)|\, h_n. \tag{5}$$

So, if the derivative $p'(x)$ is bounded, than the formula (4) has interpolation error of order $O(h_n)$, uniformly with respect to the parameter ε.

We try to obtain more accurate error estimate.

We start with the presentation for $p(x)$:

$$p(x) = p_{n-1} + (x - x_{n-1})p'_{n-1} + Z(x), \quad Z(x) = \int\limits_{x_{n-1}}^{x} \int\limits_{x_{n-1}}^{s} p''(r)\,dr\,ds \tag{6}$$

to get:

$$R_\Phi(u,x) = \Big[x_n - x - \frac{\Phi_n - \Phi(x)}{\Phi_n - \Phi_{n-1}} h_n\Big] p'_{n-1} + \Big[1 - \frac{\Phi_n - \Phi(x)}{\Phi_n - \Phi_{n-1}}\Big] Z_n - Z(x),$$

where $Z_n = Z(x_n)$. We transform the expression in the first square brackets as follows: :

$$Q(x) = x_n - x - \frac{\Phi_n - \Phi(x)}{\Phi_n - \Phi_{n-1}} h_n = \frac{1}{\Phi_n - \Phi_{n-1}} \int\limits_{x}^{x_n} \Big[(\Phi_n - \Phi_{n-1}) - h_n \Phi'(s)\Big]\,ds,$$

which implies,

$$Q(x) = \frac{1}{\Phi_n - \Phi_{n-1}} \int\limits_{x}^{x_n} \int\limits_{x_{n-1}}^{x_n} \int\limits_{s}^{\tau} \Phi''(r)\,dr\,d\tau\,ds. \tag{7}$$

So, for the error of the interpolation formula (4) we have the presentation:

$$R_\Phi(u,x) = Q(x)p'_{n-1} + \Big[1 - \frac{\Phi_n - \Phi(x)}{\Phi_n - \Phi_{n-1}}\Big] Z_n - Z(x),$$

where $Z(x)$ and $Q(x)$ corresponds to (6) and (7) respectively.

If an estimate on $|\Phi''(x)|$ is known, then, using (7) one can get estimate for $Q(x)$:

$$|Q(x)| \le \frac{1}{|\Phi_n - \Phi_{n-1}|} \int\limits_{x}^{x_n} \Big[\int\limits_{x_{n-1}}^{s} \int\limits_{\tau}^{s} |\Phi''(r)|\,dr\,d\tau + \int\limits_{s}^{x_n} \int\limits_{s}^{\tau} |\Phi''(r)|\,dr\,d\tau \Big]\,ds. \tag{8}$$

Using (6), we can write an analogous estimate:

$$|Z(x)| \le \int\limits_{x_{n-1}}^{x} \int\limits_{x_{n-1}}^{s} |p''(r)|\,dr\,ds. \tag{9}$$

¿From the presentation for $R_\Phi(u,x)$, we get for the accuracy of formula (4):

$$|R_\Phi(u,x)| \le |p'_{n-1}| \max_x |Q(x)| + 2 \max_x |Z(x)|, \ x \in \Delta_n. \tag{10}$$

2.1 Calculation of Derivative

Using the such constructed interpolation formula (4), we find the formula for the derivative $u'(x)$:

$$u'_\Phi(x) = \frac{u_n - u_{n-1}}{\Phi_n - \Phi_{n-1}} \Phi'(x), \ x \in \Delta_n. \tag{11}$$

Formula (11) is exact for the function $\Phi(x)$. Let us estimate it's accuracy. Taking into account the representation (1), we get:

$$u'_\Phi(x) - u'(x) = \frac{p_n - p_{n-1}}{\Phi_n - \Phi_{n-1}} \Phi'(x) - p'(x).$$

Therefore,

$$u'_\Phi(x) - u'(x) = (p_n - p_{n-1}) \Big(\frac{\Phi'(x)}{\Phi_n - \Phi_{n-1}} - \frac{1}{h_n} \Big) + \frac{p_n - p_{n-1}}{h_n} - p'(x). \tag{12}$$

Let consider the well known formula, based on differentiation of the linear interpolant (2):

$$u'_L(x) = \frac{u_n - u_{n-1}}{h_n}, \ x \in \Delta_n. \tag{13}$$

Taking into account (1), we find:

$$u'_L(x) - u'(x) = (\Phi_n - \Phi_{n-1}) \Big(\frac{\Phi'(x)}{\Phi_n - \Phi_{n-1}} - \frac{1}{h_n} \Big) + \frac{p_n - p_{n-1}}{h_n} - p'(x). \tag{14}$$

Now we compare the errors (12),(14) if there is component $\Phi(x)$ with regions of rapid changes. We conclude, that formula (11) is more accurate for derivative calculation, than (13),

3 Exponential Boundary Layer

Let the function $u(x)$ is the solution of the boundary value problem:

$$\varepsilon u''(x) + a(x)u'(x) - b(x)u(x) = f(x), \ \ u(0) = A, \ u(1) = B, \tag{15}$$

where

$$a(x) \geq \alpha > 0, b(x) \geq 0, \ \varepsilon > 0,$$

and the functions $a(x), b(x), f(x)$ are smooth enough. According to [4] the solution of problem (15) has an exponential boundary layer near the point $x = 0$ and a representation (1) for $u(x)$ holds with

$$\Phi(x) = \exp(-a_0 \varepsilon^{-1} x),$$

and

$$|p^{(j)}(x)| \leq C_0 \Big[\varepsilon^{1-j} \exp(-\alpha \varepsilon^{-1} x) + 1\Big], \ 0 \leq j \leq 2, \tag{16}$$

$a_0 = a(0)$, $\gamma = -\varepsilon u'(0)/a_0$, $|\gamma| \le C_1$, where C and C_i, $i \ge 0$, are some positive constants, independent of the parameter ε and the mesh step.

The interpolation formula (4) in a case of such function $\Phi(x)$ was investigated in [9], where was proved, that

$$|u_\Phi(x) - u(x)| \le Ch_n, \quad x \in \Delta_n,$$

which corresponds to the estimate (5). In the present article we obtain more accurate estimate. Note, that $\Phi'(x) < 0$ for any x . It follows from here, that in the case of exponential boundary layer we don't need any restrictions on the mesh Ω.

Theorem 1. *Let*

$$u(x) = p(x) + \gamma \exp(-a_0\varepsilon^{-1}x)$$

and the inequalities (16) on $p(x)$ are fulfiled. Than for a some constant C and any interval Δ_n

$$|u_\Phi(x) - u(x)| \le C\frac{h_n^2}{h_n + \varepsilon}, \quad x \in \Delta_n. \tag{17}$$

Proof. Taking into account (16) in (9), we have:

$$|Z(x)| \le C_0 \int\limits_{x_{n-1}}^{x} \int\limits_{x_{n-1}}^{s} \Big[\frac{1}{\varepsilon}\exp(-\alpha\varepsilon^{-1}r) + 1\Big]\, dr\, ds$$

$$= C_0\Big[\frac{\varepsilon}{\alpha^2}\exp(-\alpha\varepsilon^{-1}x_{n-1})F(\tau) + \frac{(x - x_{n-1})^2}{2}\Big],$$

where

$$\tau = \frac{\alpha(x - x_{n-1})}{\varepsilon}, \quad 0 \le \tau < \infty, \quad F(\tau) = \tau - 1 + \exp(-\tau).$$

By the inequality

$$0 \le F(\tau) \le \frac{\tau^2}{1+\tau}, \quad \tau \ge 0,$$

we get:

$$|Z(x)| \le C_0\Big[\frac{h_n^2}{\varepsilon + \alpha h_n}\exp(-\alpha\varepsilon^{-1}x_{n-1}) + \frac{h_n^2}{2}\Big]. \tag{18}$$

Using (8)we can estimate $Q(x)$ as follows :

$$|Q(x)| \le \frac{1}{\exp(-\alpha\varepsilon^{-1}x_{n-1}) - \exp(-\alpha\varepsilon^{-1}x_n)} \times$$

$$\times \int\limits_{x}^{x_n}\Big[\int\limits_{x_{n-1}}^{s}\int\limits_{\tau}^{s}\frac{a_0^2}{\varepsilon^2}\exp(-a_0\varepsilon^{-1}r)\,dr\,d\tau + \int\limits_{s}^{x_n}\int\limits_{s}^{\tau}\frac{a_0^2}{\varepsilon^2}\exp(-a_0\varepsilon^{-1}r)\,dr\,d\tau\Big]\,ds.$$

Hence:

$$|Q(x)| \le \frac{a_0}{\varepsilon\Big[\exp(-\alpha\varepsilon^{-1}x_{n-1}) - \exp(-\alpha\varepsilon^{-1}x_n)\Big]} \times$$

$$\times \int_{x}^{x_n} \Big[\int_{x_{n-1}}^{s} \exp(-a_0\varepsilon^{-1}\tau)\Big(1-\exp(-a_0\varepsilon^{-1}(\tau-s))\Big)\,d\tau +$$

$$+\int_{s}^{x_n} \exp(-a_0\varepsilon^{-1}s)\Big(1-\exp(-a_0\varepsilon^{-1}(\tau-s))\Big)d\tau\Big]\,ds.$$

This implies

$$|Q(x)| \le \frac{a_0\Big[1-\exp(-a_0\varepsilon^{-1}h_n\Big]}{\varepsilon\Big[\exp(-\alpha\varepsilon^{-1}x_{n-1})-\exp(-\alpha\varepsilon^{-1}x_n)\Big]}\times$$

$$\times \int_{x}^{x_n} \Big[\int_{x_{n-1}}^{s} \exp(-a_0\varepsilon^{-1}\tau)\,d\tau + \int_{s}^{x_n} \exp(-a_0\varepsilon^{-1}s)d\tau\Big]\,ds.$$

Therefore,

$$|Q(x)| \le \exp(a_0\varepsilon^{-1}x_{n-1}) \int_{x}^{x_n} \Big[\exp(-a_0\varepsilon^{-1}x_{n-1}) - \exp(-a_0\varepsilon^{-1}s)+$$

$$+a_0\varepsilon^{-1}(x_n-s)\exp(-a_0\varepsilon^{-1}s)\Big]\,ds.$$

Increasing the difference of the exponents and using integrations by parts for the last term, we obtain:

$$|Q(x)| \le (x_n-x)\Big[1-2\exp(-a_0\varepsilon^{-1}h_n)+\exp(-a_0\varepsilon^{-1}(x-x_{n-1}))\Big].$$

So,

$$|Q(x)| \le 2(x_n-x)\Big[1-\exp(-a_0\varepsilon^{-1}h_n)\Big].$$

Note, that for $\tau > 0$ we have

$$1-\exp(-\tau) < \frac{2\tau}{\tau+1},$$

which implies

$$|Q(x)| \le \frac{4a_0h_n^2}{a_0h_n+\varepsilon}. \tag{19}$$

Using the estimates (18) and (19) in (10), we get (17). The lemma is proved . ◇

3.1 Estimate of the Derivative

The accuracy of formula (11) in the case $\Phi(x) = \exp(-a_0\varepsilon^{-1}x)$ was investigated in [9], where it was proved, that,

$$\varepsilon|u'_\Phi(x) - u'(x)| \le Ch_n, \quad x \in \Delta_n.$$

Note, that an application of a standard formula, based on polynomial interpolation, leads to significant error. For example, if $u(x) = \exp(-\varepsilon^{-1}x)$ then for $\varepsilon = h_1$

$$\varepsilon\left|\frac{u(h_1) - u(0)}{h_1} - u'(0)\right| = e^{-1}.$$

We see, that the relative error does not decrease, when $h_1 \to 0$.

4 Power Boundary Layer

Boundary value problems with a power boundary layer were investigated in some papers, for example, in [5]-[7]. Consider the problem from [6]:

$$-(\varepsilon + x)^2u'' + c(x)u = f(x),\ 0 < x < 1,$$
$$u(0) = A, \quad u(1) = B, \tag{20}$$

where

$$c, f \in C^1[0,1], \quad c(x) \ge 0,\ c(0) > 0.$$

According to [6] there exists unique solution of (20), and for the solution $u(x)$ the representation (1) is valid with

$$\Phi(x) = (1 + x/\varepsilon)^{-r}, \quad r = (\sqrt{1 + 4c(0)} - 1)/2, \tag{21}$$

and with next bounds on the derivatives:

$$|p^{(j)}(x)| \le C(\varepsilon + x)^{1-j},\ j = 1, 2, 3,\ x \in [0,1]. \tag{22}$$

In view of (21) the solution of the problem (20) contains power boundary layer near point $x = 0$.

Consider a particular solution of the problem (20)

$$u(x) = \left(1 + \frac{x}{\varepsilon}\right)^{-1}.$$

Then for $\varepsilon = h_1$

$$u_L\left(\frac{h_1}{2}\right) - u\left(\frac{h_1}{2}\right) = \frac{1}{4}.$$

So, the error of the linear interpolation does not decrease when $h_1 \to 0$.

The function $\Phi(x)$ monotone decreases on the interval $[0, 1]$, therefore, we don't need to do some restrictions on mesh steps, for example, it may be any uniform step, in order the interpolation formula to be applied. If we use interpolation formula (4), then according to (5), (22) the next estimate holds:

$$|u_\Phi(x) - u(x)| \le C_1h_n,\ x \in \Delta_n,\ \forall n.$$

On the same manner we can show expediency to apply formula (11) for calculation of derivative of a such function $u(x)$.

5 Conclusion

In summary,we proposed for functions with boundary layer components an interpolation formula. It is exact on the singular boundary layer components. The formula is applicable to numerical solution of singularly perturbed differential equation problems: see section 4 and [2,8]. Extension of the formula to functions of two variables and theoretical analysis are currently being studied.

Acknowledgement

Supported by National Fund of Bulgaria under project HS-MI-106/2005 and Russian Foundation for Basic Research under Grant 07-01-00729.

References

1. Ahlberg, J.H., Nilson, E.N., Walsh, J.L.: The Theory of Splines and their Applications. Academic Press, New York (1967)
2. Angelova, I.T., Vulkov, L.G.: Comparison of the two-grid method for singularly perturbed reaction-diffusion problems on different meshes. Amer. Inst. of Phys (in press)
3. Miller, J.J.H., O'Riordan, E., Shishkin, G.I.: Fitted Numerical Methods for Singular Perturbation Problems. World Scientific, Singapure (1996)
4. Kellogg, R.B., Tsan, A.: Analysis of some difference approximations for a singular perturbation problems without turning points. Math. Comput. 32(144), 1025–1039 (1978)
5. Kandilarov, J.D., Vulkov, L.G., Zadorin, A.I.: A method of lines approach to the numerical solution of singularly perturbed elliptic problems. In: Vulkov, L.G., Waśniewski, J., Yalamov, P. (eds.) NAA 2000. LNCS, vol. 1988, pp. 451–458. Springer, Heidelberg (2001)
6. Vulanovic, R.: On a numerical solution of a power layer problem. Numerical methods and approximation theory 3, 423–431 (1987)
7. Liseikin, V.D.: On the numerical solution of equations with power boundary layer. Comp. Math. and Math. Physics. 26(12), 1813–1820 (1986)
8. Vulkov, L.G., Zadorin, A.I.: Two-grid algorithms for difference schemes of exponential type for semilinear diffusion convection-dominated equations. Amer. Inst. of Phys. (in press) (in Russian)
9. Zadorin, A.I.: Method of interpolation for a boundary layer problem. Sib. J. of Numer. Math. 10(3), 267–275 (2007) (in Russian)

Characteristics of the Group Interest Network

Ning Zhang

Business School, University of Shanghai for Science and Technology, Shanghai, 200093, P.R. China

Abstract. By studying the behavior characteristics visiting world wide web of the specifically campus group, this paper constructed dynamic group interest network, which was a para-bipartite graph. The time features that the group user visited world-wide-web had been observed, the topological structure had been discussed. The degree exponents had been drafted for the one week data. Although the users' visiting time is random and their surfing pages were different but the interests of a majority of the campus group were accordant. The results indicate that the incoming degree distribution of the group interest network follows power law. And the group interest spectrum was basically steady. The visiting behavior of the campus group had their special disciplinarian.

1 Introduction

There are a considerable variety of complex systems in nature world and human society. The complex networks can be used to describe complex system and help us to find out the specialties of complex system further. Complex networks are available for studying a great deal of practical system, such as the World Wide Web, the Internet, the electrical power-grid networks, the biological nets and social networks [1-5]. An increasing amount of empirical evidence is indicating that the topological characteristics of practical networks are neither regular networks nor random networks[6], their topological structure belong to a kind of network with small world [7] and scale-free identity[8,9] that are totally different with the statistical characteristics of regular network and random network. The findings of complex network reflect the basic characteristics of many complex systems, bringing material breakthrough to these systems' research. For instance, scientist collaboration networks [10, 11, 12] can make us clear about the relationships among the scientists in different fields, which have short average path length but big clustering coefficient. That is to say, the scientist collaboration networks have the characters of good connectivity and strong clustering. The power law degree distribution of World Wide Web let it has dual-characteristics, robust and frangible [1, 3, 8]. These universal characteristics have significant theoretical meanings and engineering application values. The appearance of biological system and its diversification development are related with these characteristics. As for the engineering application, the values are obviously. On condition that we can identify all kinds of groups of 1370 million Internet users in China and the certain top tens web resources which was commonly concerned by users

S. Margenov, L.G. Vulkov, and J. Waśniewski (Eds.): NAA 2008, LNCS 5434, pp. 620–627, 2009.

among 0.843 million webs [14], then we can pick the top several main resources to attain abroad storage according to the orders. In this way, the power law distribution can ensure that the mainstream resources are able to meet the needs of the majority.

The effects of information in nowadays life become more and more important. How to conquer 'figure gulf' to let everyone share information resources fairly is always being the global issue and also a full concerned issue by our government. In the limited resource, we should to consider which kind of information resources are needed by different groups to let the individuals far from cities can share sunshine information[15,16] in the most proper way, which is different groups' interests problem. For interest model, we often use clustering analytical method, linearity regressive analytical method to constitute interest model and put up digging users interests according to word, text structure characters, paragraph and sorts expression ability [17]. For user's preference, we can use self-adaptive theory to study user's preference [18] and constitute user's preference model base one data cube [19]. For the data digging method, we can use Markov model to searching for user's behavior characters [20] and find out user interest profile according to the implicit feedback [21], and then combining web content and behavior analysis [22] or base on its searching history[23] to invest interesting model. There are plenty of searching engines and information filtration methods. But none of those researches deal with group users' interest structure characters, or explain group interest spectrum's structure and stability mechanism, or reveal the clustering phenomena and rules of group interests.

Our researches are aimed at studying group users' behavior characters of surfing on the net and topological characters of group interesting network by the method of complex network, finding out special characters of group interest network and it self's evolvement rules. This essay is only involved in introduction researching results. The further results will be published in other papers as the researches go deeper.

2 Data Analysis

The group users in this paper refer to the faculty and the students in our campus. They surf the net by local area network in their offices or dormitories, there are one fixed address for one office or dormitory, but more than one computer connected with the net and the user also more than one in one room, so we call these users as group users.

There are 3 sets of records were collected during the continuous period (see table 1). These data exist clearly periods. From dataset 1, we can see that the lower traffic volume appeared during 23:00Pm to 7:00Am in every day (see fig. 1). From dataset 2, we can see that the number of the users surfing the net reduced during legal holidays, and the web pages surfed by users in holidays was less than the working days, lower traffic volume appeared during the weekends in every week, the lowest traffic volume appeared at October golden week (see fig. 2, fig. 3). From dataset 3, we can see common fluctuating cycles (see fig. 4). By now our researches focus on dataset 1, there are 1023 users (IP address) and the total

Table 1. The data statistics

Data set	Begin time	End time
1	4:00 on Mar. 14,2006	3:59 on Mar. 20, 2006
2	4:00 on Aug. 30, 2006	3:59 on Aug. 08, 2006
3	4:00 on Nov. 19, 2006	3:59 on Nov. 22, 2006

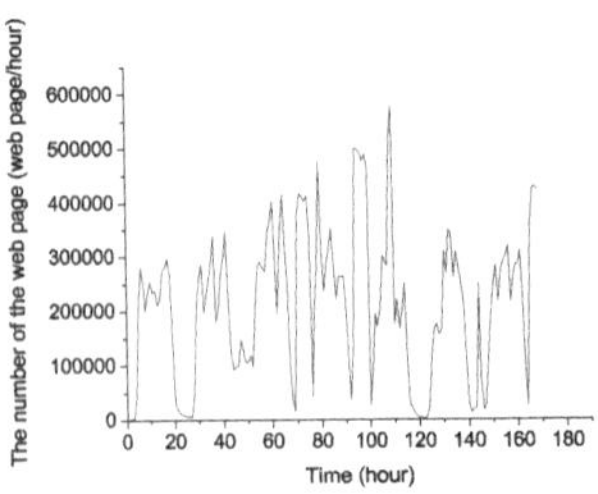

Fig. 1. The total traffic volume per hour of the group user in Data Set 1

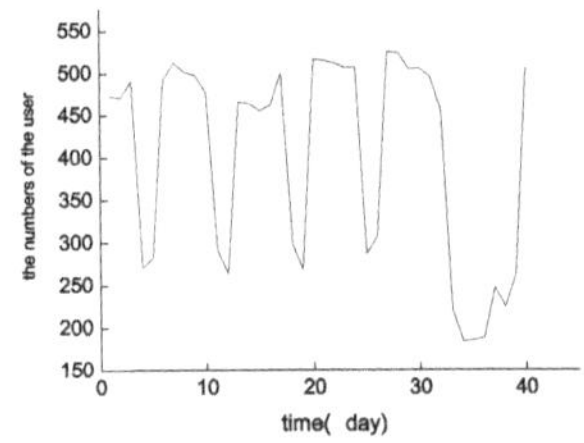

Fig. 2. The group user number per day in Data Set 2

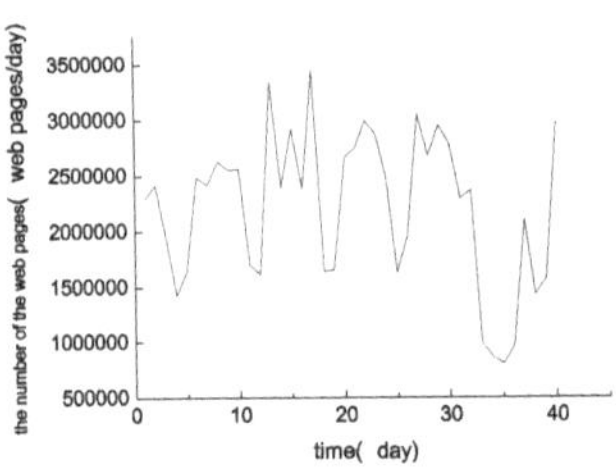

Fig. 3. The total traffic volume per day of the group user in Data Set 2

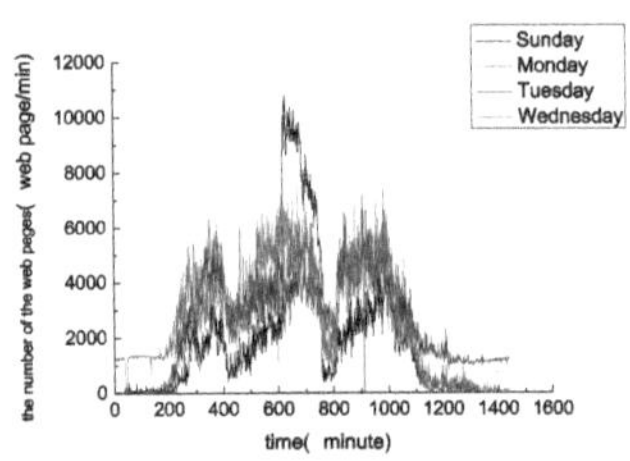

Fig. 4. The total traffic volume per minute of the group user in Data Set 3

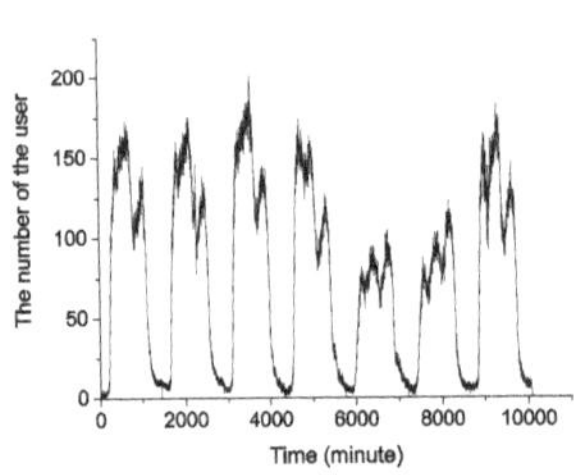

Fig. 5. The group user number per minute during one week in Data Set 1

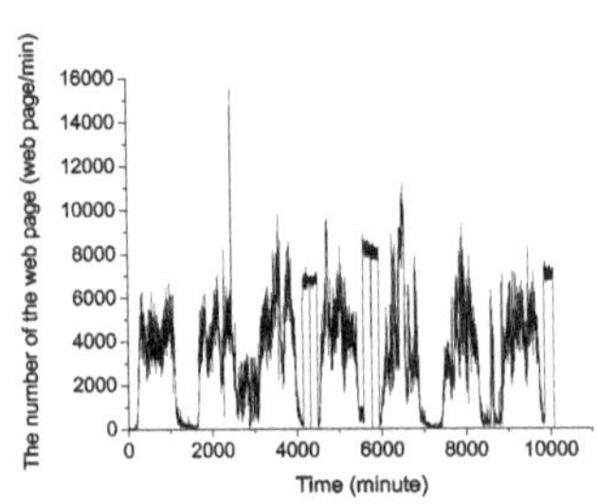

Fig. 6. The total traffic volume per minute of the group user during one week in Data Set 1

amount they visited web pages is 35935850. During a week, their average visiting web page number per hour is 213904, the highest volume of traffic per hour is 577911, appeared in Mar.18, 16:00 pm, and the lowest volume of traffic per hour is 1830, appeared in Mar.14, 4:00 am. In this week, the number of users visiting Internet per minute observed periodic and self-similar properties apparently, that the wave pattern are obviously similar during each workdays and different during weekend (see fig. 5). The web pages visited by group users per minute in this week are also regular (see fig. 6). We are still doing more meticulous researches in the universal character of the group behavior. The relevant results will be published in other papers.All these analysis results reflect the objective law that the campus group users visit Internet, that is to say, the lowest traffic volume of group user visiting Internet each day is corresponding to the dorms' black out time and whenever is the legal holiday that the teachers don't need to give a class and most students go home, the number of group user visiting Internet are lower and the visited web pages are lower. As a result of above, although the visiting time of each users surfing the Internet in the campus are random, we still can find some general rules of the group users to visiting Internet by data analyzing.

3 The Group Interest Network

Actually, group users visiting the Internet can be constructed to a dynamic complex network by itself, which is a virtual dynamic network. The Internet visiting by group users at each moment can be constructed to a group interest network. The above networks have the format of para-bipartite graph and contain two kinds of vertices in this essay, one is user vertex, which refers to group users, the other one is information resource vertex, which refers to the web site resource constructed by many web pages and need to be transferred to be seen. Since it's not like the bipartite graph with explicit two different kinds of vertices like the definition in the graph theory, we call it as para-bipartite graph. User visited the documents forms the relationship between users and information resources, which can be expressed by directed connection. And the complex relationship of many users corresponding to many information resources construct the group interest network (see fig. 7). According to the group interest network, we can get users relation network and resources relation network by the projection of users and resources respectively. We construct group users interest network by classifying the web pages group users visited to different web site according to group users accessing habit per day(see table 2), which is a directed dynamic complex network. Basing on the one week data, the network contained 1023 users and 60079 web sites in total during one week. Even though the number of user are different and the web sites they visited are different in every day, the in-degree distribution of group interest network have power law character (see fig. 8), which means that lots of web sites are with a few links (visiting amount), a few web sites are with a medium number of links and a very few noteworthy web sites are with a large number of links in this network. Researches indicate

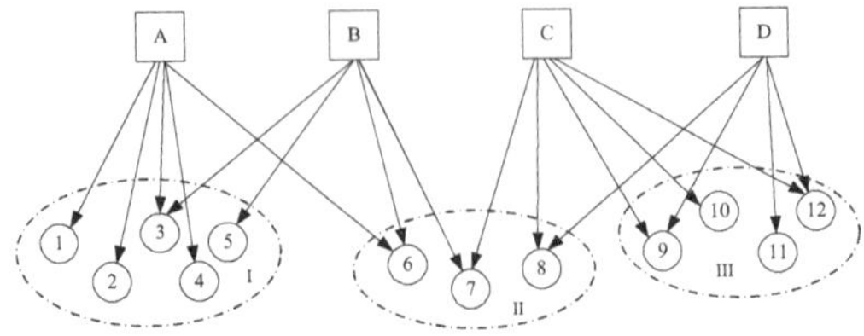

Fig. 7. Schematic illustration of the group interest network. The square nodes A-D represent group users, the circular nodes 1-12 represent web pages, Dash dot circles - represent resources (web sites).

Table 2. The statistics of the group user surfing the Internet

Week	User	Web page	Web site	Degree exponent
3.14(Tuesday)	658	3727905	16078	1.530477
3.15(Wednesday)	676	4361211	17491	1.533105
3.16(Thursday)	674	6068099	18233	1.520289
3.17(Friday)	647	6931486	18390	1.528237
3.18(Saturday)	381	5291921	12215	1.520854
3.19(Sunday)	393	3844392	14488	1.532267
3.20(Monday)	663	5710836	17584	1.521534

that the major users have the similar interests, and if we can conform the information resources interested by most individuals and use the abroad storage technique to maintain most peoples request for information, then we can let people to obtain sunshine information in the most economical way.

The in-degree distribution of group interest network follows power law, $p_{in}(k) \propto k^{-\gamma}$,$\gamma = 1.52$ (see fig. 8), so this network was scale free.

4 The Scale Exponent Fitting

The empirical researches always deal with data fitting. For the group interest network, we need to fit the scale exponent. Cumulate in-degree distribution of the group interest network has power law character (see fig. 9). Suppose

$$p(k) = Ck^{-\gamma} \tag{1}$$

then

$$\int_1^y p(k)dk = \frac{C}{\gamma - 1}(1 - k^{-\gamma+1}) \tag{2}$$

Based on the character of the distribution, to the enough larger N, we may obtain

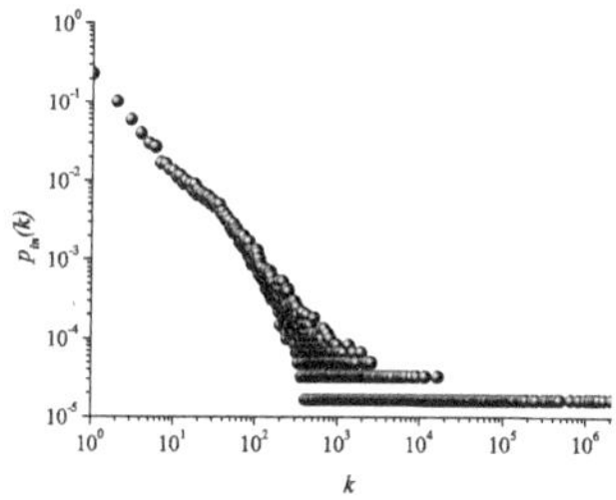

Fig. 8. The in-degree distribution of the group interest network

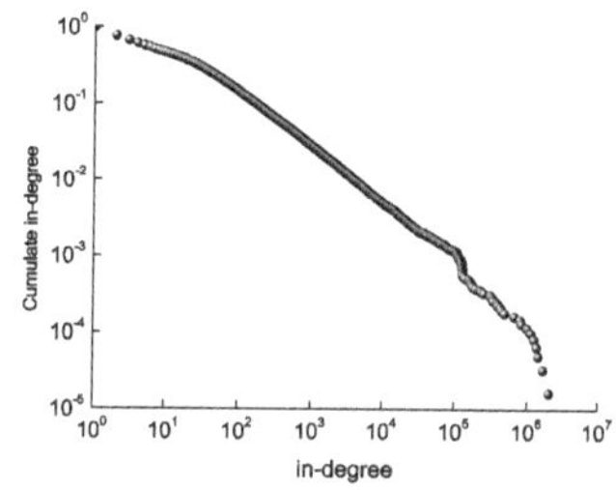

Fig. 9. Cumulate in-degree distribution of the group interest network (in one week)

$$\int_1^N p(k)dk = 1 \tag{3}$$

and

$$C = \frac{\gamma - 1}{1 - N^{-\gamma+1}} \tag{4}$$

hence

$$f(y) = \int_1^y p(k)dk = \frac{1 - y^{-\gamma+1}}{1 - N^{-\gamma+1}} \tag{5}$$

According to empirical data, we know that y and $f(y)$ are known quantity, if N is larger enough, $N^{-\gamma+1}(\approx 0)$, then

$$y^{-\gamma+1} = 1 - f(y) \tag{6}$$

$$-\gamma + 1 = \frac{ln[1 - f(y)]}{lny} \tag{7}$$

Therefore, $\gamma = \gamma(y)$ If $N^{-\gamma+1}$ can not be omitted, then we can use gradually approach method, hence

$$-\gamma^{k+1} + 1 = \frac{ln[1 - f(y)(1 - N^{-\gamma^{(k)}+1})]}{lny} \tag{8}$$

If $\gamma = \gamma(y)$ is a un-strict constant, there are several $\gamma(y_1), \gamma(y_2), ..., \gamma(y_n)$, then

$$\gamma = \frac{1}{n}\sum_{i=1}^{n}\gamma(y_i) \tag{9}$$

Compare the error between theoretic value $f(y)$ and the empirical value $f(y_i)$, $(1, 2, ..., n)$ in formula(5).

In practice, we can firstly calculate the in-degree of the every web site per day, secondly we calculate the in-degree frequencies and per centum of the different vertices, thirdly calculate the proportion of the accumulative frequency, and then we can get $f(y)$, and y is the in-degree. According to formula (7), calculate degree exponent γ . To different $f(y)$ and$y$, repeat these steps, we can get different $\gamma(y_1), \gamma(y_2), ..., \gamma(y_n)$. For $\gamma = \gamma(y)$ is an un-strict constant, we need calculate average value, in accordance with formula (9). The every day in-degree exponent of the group interest network (see table 2). To the integrate data of the week, we get $\gamma = 1.52197$.

5 Conclusions

According to many users to many information resources complex relationships, finding out the group users' general characters by studying group users' dynamic behavior characters is just at the beginning. This essay's study indicate that the group interest network's in-degree distribution belongs to power law distribution by using complex network's method to invest group users Internet visiting behavior. The given group's interest spectrum is basically stable and the visiting behavior of the campus group had their special disciplinarian.

Acknowledgment. This work was supported by Shanghai Leading Academic Discipline Project (No. S30501) and the Natural Science Foundation of Shanghai (06ZR14144).

References

1. Albert, R., Barabási, A.-L.: Statistical mechanics of complex networks. Rev. Mod. Phys. 74, 47–97 (2002)
2. Dorogovtsev, S.N., Mendes, J.F.F.: Evolution of Networks: From Biological Nets to the Internet and the WWW. Oxford Univ. Press, Oxford (2003)
3. Pastor-Satorras, R., Vespignani, A.: Evolution and Structure of the Internet: a Statistical Physics Approach. Cambridge Univ. Press, Cambridge (2004)
4. Newman, M.E.J.: The structure and function of complex networks. SIAM Rev. 45, 167–256 (2003)
5. Amaral, L.A.N., Ottino, J.M.: Complex networks-augmenting the framework for the study of complex systems. Eur. Phys. J. B 38, 147–162 (2004)
6. Erdös, P., Rényi, A.: On the evolution of random graphs. Publ. Math. Inst. Hung. Acad. Sci. 5, 17–61 (1960)
7. Watts, D.J., Strogatz, S.H.: Collective dynamics of 'small-world' networks. Nature 393, 440–442 (1998)
8. Albert, R., Jeong, H., Barabási, A.L.: Diameter of the World Wide Web. Nature 401, 130–131 (1999)
9. Faloutsos, M., Faloutsos, P., Faloutsos, C.: On power-law relationships of the Internet topology. Comput. Commun. Rev. 29, 251–262 (1999)
10. Newman, M.E.J.: Scientific collaboration networks: I. Network construction and fundamental results. Phys. Rev. E 64, 016131 (2001)
11. Newman, M.E.J.: Scientific collaboration networks: II. Shortest paths, weighted networks, and centrality. Phys. Rev. E 64, 016132 (2001)
12. Newman, M.E.J.: The structure of scientific collaboration networks. Proc. Natl. Acad. Sci. USA 98, 404–409 (2001)

13. Albert, R., Jeong, H., Barabási, A.-L.: Error and attack tolerance of complex networks. Nature 406, 378 (2000)
14. China Internet Network Information Center: Statistical Reports on the Internet Development in China (2007) (in Chinese), `http://www.cnnic.cn/html/Dir/2007/01/22/4395.htm`
15. Li, Y.P.: Sunshine information–conflict-free share structure. China engineering science 2, 24–27 (2000)
16. Li, Y.P.: Construct Broad-Storage grid. Computer world 37 (2005) (in Chinese)
17. Lin, H.F., Yang, Y.S.: The representation and update mechanism for user profile. Journal of computer research and development 39, 843–847 (2002) (in Chinese)
18. Lei, Y.S., Gan, R.C., Du, D.: A framework of adaptive vertical website based on user preferences. Computer Engineering 31, 18–20 (2005) (in Chinese)
19. Chen, J.L.: User's interest model based on data cube. Journal of Gulin university of technology 25, 84–88 (2005) (in Chinese)
20. Hu, Y.H., Zhao, H.J., Lu, H.R., Wang, H.J.: Research on extracting patterns from web user behavior. Computer Engineering and Design 27, 3416–3418 (2006) (in Chinese)
21. Sun, T.L., Yang, F.Q.: An approach of building and updating user interest profile according to the implicit feedback. Journal of northeast normal university 35, 99–104 (2003) (in Chinese)
22. Zhao, Y.C., Fu, G.Y., Zhu, Z.Y.: User interest mining of combining web content and behavior analysis. Computer Engineering 31, 93–94 (2005) (in Chinese)
23. Xu, K., Cui, Z.M.: User profile model based on user search histories. Computer technology and development 16, 18–20 (2006) (in Chinese)

Solving Ordinary Differential Equations by Simplex Integrals

Yongxiong Zhou[1,2] and Shuhuang Xiang[2]

[1] Department of Mathematics, Guangdong Ocean University, Zhanjiang, Guangdong 524088, China
[2] Department of Applied Mathematics, Central South University, Changsha, Hunan 410083, China
zhouyongxiong@126.com, xiangsh@mail.csu.edu.cn

Abstract. This paper is devoted to the proper discrete solution for ordinary differential equations, especially to oscillating solution. In contrast to Lipschitz condition, we define a new condition following that

$$\left|\int_{t_0}^{t_1} f(t)dt\right| \leq R \max_{\xi_1,\xi_2\in[t_0,t_1]} |f(\xi_1) - f(\xi_2)|$$

with small R for all t_0, t_1 in the correlative intervals. Under the assumption of this new condition, we obtain a new asymptotic formula

$$\phi_v(t) - Q_{v-1}(t) = O((Rh)^v),$$

where simplex integral $\phi_v(t)$ denotes

$$\int_{t_0}^{t} \cdots \int_{t_0}^{\xi_{v-2}} \int_{t_0}^{\xi_{v-1}} f(\xi_v)d\xi_v d\xi_{v-1} \cdots d\xi_1$$

and the $v-1$-th polynomials $Q_{v-1}(t)$ in which coefficient correspond to simplex integrals $\phi_{n_k}(t)$ with $n_k > v, k = 1, 2, \ldots, v$. In other words, the accuracy for approximation increasing rapidly as the integrable functions oscillate rapidly or for small step h while it's difficult for us to pursuit a polynomial to approximate a highly oscillatory function.

Applying this idea of approximation to ODE, this paper surveys the algorithmic issues. If ODE has the form

$$P_n y^{(n)} + P_{n-1} y^{(n-1)} + \cdots + P_1 y' + P_0 y = g(t),$$

where $P_n(t), P_{n-1}(t), \ldots, P_0(t)$ are arbitrary degree polynomials, then we can solve it by the recursive relation about simplex integrals altogether with approximate relation. Finally, numerical examples about Airy and Bessel equations illustrate the efficiency of this technique.

1 Introduction

The ratio of the change in a 'non-oscillatory' function to the corresponding change in its independent variable can be well approximated by various of rank

S. Margenov, L.G. Vulkov, and J. Waśniewski (Eds.): NAA 2008, LNCS 5434, pp. 628–633, 2009.

derivatives, representatively, Taylor expansion. In respect that, many numerical method had been developed to solve ordinary differential equations *(ODE)*, such as Runge-Kutta's. Unfortunately, polynomials are a very poor means to approximate rapidly-varying and highly-oscillating functions. We either require minute step size or a very high degree of a polynomial - and the former's local error which are caused by every sufficient small step can be contaminated into the typically large global error, the latter might lead to ill conditioning.

Instructively, integrator is reverse operator of differential. And quadrature of highly oscillating function *(solution of ODE)* is not our enemy but friends [1]. As many numerical analysts [1,2,3,4] showed that once numerical methods on quadrature are properly crafted, their accuracy increases with oscillating rapidly. We prefer to pursue a new way *(Definition 1)*, not Lipschiz condition, to sketch the mathematical foundations of a general approach in the case of oscillate solution.

2 Polynomial Expansion about Simplex Integrals

In order to solve the oscillatory ODE in new way, we need following condition and polynomial expansion about simplex integrals.

Definition 1. *Suppose that $f(t)$ is continuous in $a \leq t \leq b$, then we defined the condition as: there exits a positive constant R such that*

$$\left|\int_{t_0}^{t_1} f(t)dt\right| \leq Rh \max_{\xi_1,\xi_2\in[t_0,t_1]} |f(\xi_1) - f(\xi_2)| \tag{1}$$

for all $t_0, t_1 \in [a, b]$ with $h = t_1 - t_0$.

As the previous article ([2,4], etc) mentioned, oscillatory function $f(t)$ had been defined as there are plenty of extreme value points in the interval $[a, b]$. In general, the definition can be replace by Definition 1. For instance, $\sin(\omega t)$ is highly oscillating in $[0, 1]$ with $\omega \gg 1$. And Definition 1 reflects that for $R = 1/(h\omega)$,

$$\left|\int_{t_0}^{t_1} \sin(\omega t)dt\right| \leq Rh \max_{\xi_1,\xi_2\in[t_0,t_1]} |\sin(\omega\xi_1) - \sin(\omega\xi_2)|.$$

Let

$$\phi_0(t) = f(t),$$
$$\phi_{k+1}(t) = \int_{t_0}^{t} \phi_k(t)dt, k = 0, 1, 2, \ldots$$

It's well known that the multiple simplex integrals can also be written as the form of one-dimensional integrals by integrating by parts,

$$\phi_v(t) = \frac{1}{(v-1)!}\int_{t_0}^{t} f(\xi)(t-\xi)^{v-1}d\xi.$$

Theorem 1. *Under the assumptions in Definition 1, then there is a polynomial* $Q_{v-1}(t) = \alpha_0 + \alpha_1 t + \alpha_2 t^2 + \cdots + \alpha_{v-1} t^{v-1}$ *such that*

$$|\phi_v(t) - Q_{v-1}(t)| \leq R^v h^v \max_{\xi_1,\xi_2 \in [t_0,t_1]} |f(\xi_1) - f(\xi_2)|, \quad v = 1, 2, \ldots. \tag{2}$$

is true for all $t \in [t_0, t_1]$ *with* $h = t_1 - t_0$*. And the coefficients* $\alpha_0, \alpha_1, \cdots, \alpha_{v-1}$ *follow the system of equations*

$$\begin{cases} \sum\limits_{k=0}^{v-1} \alpha_k \int_{t_0}^{t_1} t^k dt & = \phi_{v+1}(t_1) \\ \frac{1}{1!} \sum\limits_{k=0}^{v-1} \alpha_k \int_{t_0}^{t_1} (t_1 - t) t^k dt & = \phi_{v+2}(t_1) \\ \quad \cdots\cdots\cdots & \cdots \quad \cdots \\ \frac{1}{(v-1)!} \sum\limits_{k=0}^{v-1} \alpha_k \int_{t_0}^{t_1} (t_1 - t)^{v-1} t^k dt & = \phi_{2v}(t_1) \end{cases}$$

With the above construction of the multiple integral's property, we get the main fact that the simplex integral $\phi_v(t)$ can be well approximated by a polynomial under the new condition!

Remark 1. Coefficients α_k can also correspond to more dimension simplex integrals $\phi_{n_k}(t)$ with different $n_k > v, k = 1, 2, \ldots, v$.

3 Solving Liner ODE with Polynomial Coefficients

This section corresponds to solve $P_n y^{(n)} + P_{n-1} y^{(n-1)} + \cdots + P_1 y' + P_0 y = g(t)$, where $P_n(t), P_{n-1}(t), \ldots, P_0$ are arbitrary order polynomials. According to this special form, we transform it into some recursive relations about simplex integrals about $y(t)$. Once numerical quadrature on simplex integrals $\frac{1}{(v-1)!} \int_{t_0}^{t} g(\xi)$ $(t - \xi)^{v-1} d\xi$ is sufficient accurately [1,2,3], we can obtain computing issue by virtue of these recursive relations and approximation formula (2).

Theorem 2. *For* v *stage method based on formula (2) and recursive relations with step size* h*, the error bounded by* $O(h^v)$*.*

Example 1: We commence from an example that has already featured in (Iserles and Nørsett) the solution of the *Airy equation*

$$y'' + ty = 0, \; t \geq 0, \quad y(0) = 0, y'(0) = 1.$$

And its exact solution can be represented as a linear combination of Airy functions

$$y(t) = \frac{1}{2} \Gamma \left(\frac{2}{3} \right) \left[3^{2/3} Ai(t) + 3^{1/6} Bi(-t) \right].$$

Let $t_0 = 0$, step $h = t_1 - t_0$, integrating the equation repeatedly,

$$y' + t\phi_1(t) - \phi_2(t) = 1,$$

$$y + t\phi_2(t) - 2\phi_3(t) = t,$$

$$\vdots$$

$$\phi_n(t) + t\phi_{n+2}(t) - (n+2)\phi_{n+3}(t) = t^{n+1}/(n+1)!$$

According to these recursive relations, we easily obtain discrete format for the approximate solution , e.g.

The four stage algorithm for example 1

$\phi_4(t_1)$ can be obtained by system of equations

$$\begin{cases} \phi_4(t_1) + t_1\phi_6(t_1) - 6\phi_7(t_1) = t_1^5/5! \\ \phi_5(t_1) + t_1\phi_7(t_1) - 7\phi_8(t_1) = t_1^6/6! \\ \phi_6(t_1) + t_1\phi_8(t_1) - 8\phi_9(t_1) = t_1^7/7! \\ \phi_7(t_1) + t_1\phi_9(t_1) - 9\phi_{10}(t_1) = t_1^8/8! \\ \phi_4(t_1) = \alpha_0 + \alpha_1 t_1 + \alpha_2 t_1^2 + \alpha_3 t_1^3 \\ \phi_4(t_1) = \beta_0 + \beta_1 t_1 + \beta_2 t_1^2 + \beta_3 t_1^3 \\ \phi_4(t_1) = \gamma_0 + \gamma_1 t_1 + \gamma_2 t_1^2 + \gamma_3 t_1^3 \end{cases}$$

where $\overrightarrow{\alpha} = \{\alpha_0, \alpha_1, \alpha_2, \alpha_3\}^T$, $\overrightarrow{\beta} = \{\beta_0, \beta_1, \beta_2, \beta_3\}^T$, $\overrightarrow{\gamma} = \{\gamma_0, \gamma_1, \gamma_2, \gamma_3\}^T$ are separately determined by

$$\begin{cases} \phi_5(t_1) = \int_{t_0}^{t_1}(\alpha_0 + \alpha_1 t + \alpha_2 t^2 + \alpha_3 t^3)dt \\ \phi_6(t_1) = \int_{t_0}^{t_1}(t_1 - t)(\alpha_0 + \alpha_1 t + \alpha_2 t^2 + \alpha_3 t^3)dt \\ \phi_7(t_1) = \frac{1}{2!}\int_{t_0}^{t_1}(t_1 - t)^2(\alpha_0 + \alpha_1 t + \alpha_2 t^2 + \alpha_3 t^3)dt \\ \phi_8(t_1) = \frac{1}{3!}\int_{t_0}^{t_1}(t_1 - t)^3(\alpha_0 + \alpha_1 t + \alpha_2 t^2 + \alpha_3 t^3)dt \end{cases}$$

$$\begin{cases} \phi_6(t_1) = \int_{t_0}^{t_1}(t_1 - t)(\beta_0 + \beta_1 t + \beta_2 t^2 + \beta_3 t^3)dt \\ \phi_7(t_1) = \frac{1}{2!}\int_{t_0}^{t_1}(t_1 - t)^2(\beta_0 + \beta_1 t + \beta_2 t^2 + \beta_3 t^3)dt \\ \phi_8(t_1) = \frac{1}{3!}\int_{t_0}^{t_1}(t_1 - t)^3(\beta_0 + \beta_1 t + \beta_2 t^2 + \beta_3 t^3)dt \\ \phi_9(t_1) = \frac{1}{4!}\int_{t_0}^{t_1}(t_1 - t)^4(\beta_0 + \beta_1 t + \beta_2 t^2 + \beta_3 t^3)dt \end{cases}$$

and

$$\begin{cases} \phi_7(t_1) = \frac{1}{2!}\int_{t_0}^{t_1}(t_1 - t)^2(\gamma_0 + \gamma_1 t + \gamma_2 t^2 + \gamma_3 t^3)dt \\ \phi_8(t_1) = \frac{1}{3!}\int_{t_0}^{t_1}(t_1 - t)^3(\gamma_0 + \gamma_1 t + \gamma_2 t^2 + \gamma_3 t^3)dt \\ \phi_9(t_1) = \frac{1}{4!}\int_{t_0}^{t_1}(t_1 - t)^4(\gamma_0 + \gamma_1 t + \gamma_2 t^2 + \gamma_3 t^3)dt \\ \phi_{10}(t_1) = \frac{1}{5!}\int_{t_0}^{t_1}(t_1 - t)^5(\gamma_0 + \gamma_1 t + \gamma_2 t^2 + \gamma_3 t^3)dt \end{cases}$$

And this leads to $y'(t_1), y(t_1), \phi_1(t_1), \phi_2(t_1), \phi_3(t_1)$ altogether with other recursive relations. More about its numerical accuracy compared with classical 4 stage Runge-kutta method (RK4 method, of order 4) [5,P138] in Fig.1.

Example 2: We consider to solve *Bessel like equation*

$$t^2y'' - ty' + (1 + t^2)y = 0, \quad y(10) = J_0(10), y'(10) = J_0(10) - 10J_1(10),$$

whose solution is $y(t) = tJ_0(t)$. Let $t_0 = 10$, step $h = t_1 - t_0$, integrating the equation repeatedly, we have

$$\begin{aligned} &t_1^2\phi_n(t_1) - (2n+5)t_1\phi_{n+1}(t_1) + (a_n + t_1^2)\phi_{n+2}(t_1) - (2n+4)t_1\phi_{n+3}(t_1) \\ &+b_n\phi_{n+4}(t_1) \\ =\ &(t_0^2 y'(10) - 3t_0 y(10))\frac{h^{n+1}}{(n+1)!} + t_0^2 y(10)\frac{h^n}{n!}, \end{aligned}$$

where $a_{-1} = 4,\ a_n = a_{n-1} + 2n + 5$ and $b_{-1} = 2,\ b_n = b_{n-1} + 2n + 4$. Fig. 2 show the accuracy for 4 stage method compared with RK4 method.

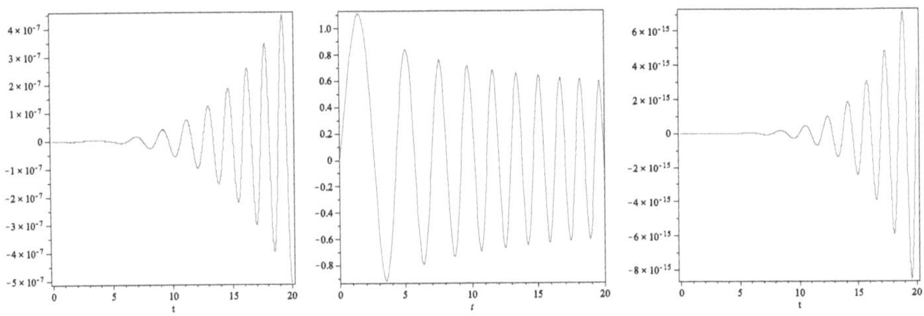

Fig. 1. The exact solution to Airy equation $y'' + ty = 0$ with initial condition $y(0) = 0, y'(0) = 1$ is in the middle. The left plot corresponds to the error of RK4 method with fixed step size $h = 1/100$. And with same step size, the right plot corresponds to order-four new method which based on the polynomials can well approximate to simplex integrals.

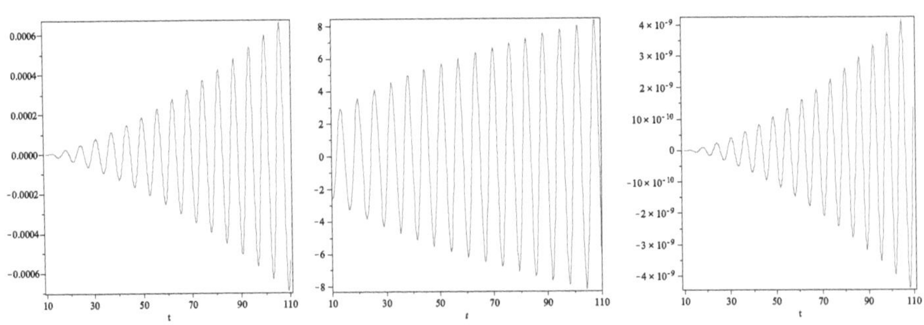

Fig. 2. The exact solution to Bessel like equation $t^2y'' - ty' + (1 + t^2)y = 0$ with initial condition $y(10) = J_0(10), y'(10) = J_0(10) - 10J_1(10)$ is in the middle. The left plot corresponds to the error of RK4 method with fixed step size $h = 1/10$. And with same step size, the right plot corresponds to order-four new method which based on the polynomials can well approximate to simplex integrals.

Remark 2. For example 2. Let $t_0 = 0$, there is a regular singularity at the origin. This problem can be settled by 'backward' simplex integrals instead of 'forward' ones.

References

1. Iserles, A., Munthe-Kaas, H.Z., Nørsett, S.P., Zanna, A.: Lie-group methods. Acta Nuerica, 1–148 (2005)
2. Iserles, A., Nørsett, S.P.: On quadrature methods for highly oscillatory integrals and their implementation. BIT 44, 755–772 (2004)
3. Levin, D.: Fast integration of rapidly oscillatory functions. J. Comput. Appl. Math. 67, 95–101 (1996)
4. Stein, E.: Harmonic Analysis: Real-variable methods, orthogonality, and oscillatory integrals. Princeton University Press, Princeton (1993)
5. Hairer, E., Nørsett, S.P., Wanner, G.: Solving Ordinary Differential Equations I. Nonstiff problems, 2nd edn. Springer, Berlin (1993)

Author Index

FSC
www.fsc.org
MIX
Papier | Fördert gute Waldnutzung
FSC® C083411

Kolibri 360 GmbH
Ferdinand-Jühlke-Straße 7
99095 Erfurt, Deutschland
produktsicherheit@kolibri360.de